Management of Horticultural Crops

Horticulture Science Series - 11

Part - I&II

Other Volumes of the Series

HORTICULTURE SCIENCE SERIES

Series Editor : Prof. K.V. Peter

Former Vice Chancellor and Presently Professor of Horticulture
Kerala Agricultural University, Thrissur - 680 656, Kerala

1. Aromatic Plants
   Baby P. Skaria, P.P. Joy, Samuel Mathew, Ancy Joseph & Ragina Joseph

2. Medicinal Plants
   Kurian Alice & M. Asha Shankar

3. Fruit Crops
   T. Radha & L. Mathew

4. Vegetable Crops
   Gopalakrishnan, T.R.

5. Spices
   E.V. Nybe and N. Mini Raj & K.V. Peter

6. Propagation of Horticultural Crops
   Rajan S. & B.L. Markose

7. Post Harvest Technology of Horticultural Crops
   Sudheer K.P. & V. Indira

8. Commercial Crops Technology
   Kurian Alice & K.V. Peter

9. Tuber and Root Crops
   Palaniswami P. & K.V. Peter

10. Flowers for Trade
    V.L. Sheela

12. Flowering Trees
    P.K. Valsalakumari, P.K. Rajeevan, P.K. Sudhadevi & C.K. Geetha

Horticulture Science Series Vol. 11

Series Editor:
Prof. K.V. Peter, Former VC and Presently Professor of Horticulture
Kerala Agricultural University, Thrissur - 680 656, Kerala

MANAGEMENT OF HORTICULTURAL CROPS

Part - I&II

Authors

T. PRADEEPKUMAR, B. SUMA
JYOTHIBHASKAR & K.N. SATHEESAN

Kerala Agricultural University, Thrissur - 680 656, Kerala

New India Publishing Agency
Pitam Pura, New Delhi- 110 088

ISBN 978-81-89422-49-3

Published by:
Sumit Pal Jain *for*

New India Publishing Agency

101, Vikas Surya Plaza, CU Block, L.S.C. Mkt.,
Pitam Pura, New Delhi- 110 088, (INDIA)
Phone: 011-27341717, Fax: 011-27341616
E-mail: newindiapublishingagency@gmail.com
Web: www.bookfactoryindia.com

Composed and Designed by NIPA

National Academy of Agricultural Sciences

Prof. M.S. Swaminathan
President

Foreword to the Series

A second green revolution aimed to meet the food requirements of about 1.8 billion people in India by 2010 is on the anvil. It is, however, imperative that the attainment of the food and nutritional security of this vast population should be combined with the preservation of clean environment. Horticulture encompassing fruits, vegetables, ornamentals, plantation crops, spices, tuber crops, medicinal and aromatic plants, besides harvest and post harvest technologies has a special role in providing food and nutritional security, rural poverty alleviation, employment generation as well as improving the quality of life in our country sides.

India, incidentally, is the world's leading producer of fruits and is followed by Brazil; in vegetables it occupies the second place after China. India is also the land of spices producing an array of such crops. Indeed, the varied eco-climatic conditions prevailing in different parts of this country make it possible to grow a disparate range of fruits, vegetables and ornamental plants at any time of the year.

NASC Complex
DPS Marg, P.O. Pusa
New Delhi - 110 012

Tel. 91-11-25846055
91-11-25846051
91-11-25846052
91-11-25846053
Fax 91-11-25846054
E-mail naas@vsnl.com
Website
//education.vsnl.com/naas
P.O. Box No. 11325

With the opening up of the world markets under the WTO regime, however, international competition for Indian horticultural products has reached unprecedented levels, necessitating strategies for quality control, branding and product development. In particular, production of organic vegetables, fruits, cut flowers, pepper, ginger, tea, coffee, rubber and cardamom has of late become demand-driven.

Among the inputs for higher productivity in horticultural crops, seeds and quality planting materials account for 20 to 30% of the potential productivity. Transgenic crops are a reality now; yet the technology for production of transgenics is a far cry in many situations. Furthermore, water is a limiting factor in several environments. Hence, water saving crops and those with high water use efficiency should be preferred over other crops and management practices. Likewise, the use of biodegradable plastics, biological control of pests and diseases, and new product development are receiving increasing attention, especially in the arena of horticultural crop management. Marketing, village knowledge centres, packaging industry, zero-energy cool chambers and a variety of economically viable technologies, however, remain to be streamlined.

The present series on **Horticulture Sciences,** edited by Professor K.V. Peter, Vice-Chancellor, Kerala Agricultural University, is a good compilation of the scientific knowledge and experience gained over the past 30 years in this important sector of Indian economy. Each volume in the series is authored by accomplished researchers such as Dr. T. Radha and Dr Lila Mathew (*Fruit Crops*), Dr. T.R. Gopalakrishnan (*Vegetable Crops*), Dr. Alice Kurian and Dr. K.V. Peter (*Commercial Crops Technology*), Dr. Alice Kurian and Dr. M. Asha Sankar (*Medicinal Plants*), Dr. Baby P. Skaria (*Aromatic Plants*), Dr. P. Palaniswami (*Tuber and Root Crops*), Dr. E.V. Nybe and Dr N. Mini Raj (*Spices*), Dr. S. Rajan and Dr Baby Lizy Markose (*Propagation of Horticultural Crops*), Dr. K.P. Sudheer and Dr. V. Indira (*Post Harvest Technology of Horticultural Crops*), Dr. V.L. Sheela (*Flowers for Trade*), Dr. T. Pradeepkumar, Dr. B. Suma, Dr. Jyothibhaskar, K.N. Satheesan & K.V. Peter (*Management of Horticultural Crops*). The New India Publishing Agency, New Delhi deserves special credit for publishing this useful series, especially in view of the fact that the Government of India is currently implementing the National Horticultural Mission.

Prof. M.S. Swaminathan, FNA, FRS
President, National Academy of Agricultural Sciences
and Chairman, National Commission on Farmers
New Delhi - 110012

डा0 एच.पी.सिंह
उप महानिदेशक
(बागवानी)
Dr. H.P. Singh
Deputy Director General
(Horticulture)

भाकृअनूप
ICAR

भारतीय कृषि अनुसंधान परिषद
कृषि अनुसंधान भवन–II पूसा, नई दिल्ली – 110 012
INDIAN COUNCIL OF AGRICULTURAL RESEARCH
KRISHI ANUSANDHAN BHAVAN-II
PUSA, NEW DELHI 110 012

FOREWORD

Horticulture sector, which includes. fruits, vegetables including potato, tubers and mushroom, spice, ornamental plants, medicinal and aromatic plants and plantation crops has played a crucial role for effective land use, nutritional security and employment opportunity. Efforts made for the horticulture research and development have been rewarding in terms of increased production, productivity, availability and export of horticultural produce. The country has emerged as a second largest producer of fruits and vegetables, and continue to hold a major share in export of cashew nut and spices. Currently the sector contributes 29.5 per cent to the GDP of agriculture.

In order to gain global acceptance and harness emerging potential, there is a need to capitalize on strength and enhance the competitiveness. Although we have succeeded in enhancing production and productivity, but quality and safety need much more upgradation. Several quality regulation standards are being adopted now which require a technological backup. In order to enhance the competition, new market rules have been framed and infrastructures are being developed. Contract farming, which enhances adoption of better technologies and investment, are being promoted. Thus, there is a need for a comprehensive strategies for production system management of horticultural crops. Dr. Peter and his colleagues have made efforts to provide their insight for management of horticultural crops. I am sure, their book **"Management of Horticultural Crops"** would be of immense value to all, those who are concerned with horticulture.

H.P. Singh

(H. P. Singh)

GENERAL INTRODUCTION

Horticulture provides different, interesting, and refreshing aspects in human life and environment. Over the time, definition of 'horticulture' has evolved as the crop science which deals with the production, utilization and improvement of fruits,vegetables, tuber crops, ornamental plants, spices and plantation crops including medicinal and aromatic plants. According to Janick(1979) horticulture can be defined as the branch of agriculture concerned with intensively cultivated plants directly used by people for food, for medicinal purposes, or for aesthetic gratification.

Main features of horticultural science are :

- Horticultural fresh produces are highly perishable.
- Horticultural crops generally require intensive cultivation warranting a large input, capital, labour and technology per unit area of land.
- Skilled operations are involved in propagation, fertilization, training, pruning, harvesting and marketing of horticultural crops.
- Horticultural crops are the rich sources of vitamins and minerals and rightly known as protective foods.
- Aesthetic sense or gratification is an exclusive phenomenon for horticultural science.

Horticultural crops are essentially high value crops ensuring maximum returns to the growers with multiple scope for value addition. Industry prefers these crops due to their consumer appeal. This sector has been providing the diversification of agriculture owing to better land use, creation of employment opportunities and overall nutritional security. World over horticultural crops received the back up from R&D and Industry and in post- globalization period, diversification of farm sector through newer opportunity to the horticultural science. Market driven agriculture prescribes quality in every stage of production and total quality management is a challenge to the horticulturist. The export market lays stringent parameters and even developed countries like USA and Australia some times fail in these aspects. Apart from phytosanitary concerns, production and post harvest treatments some times invite ban on export of fresh horticultural produces. In August 2003, India imposed a ban on the sale of fresh fruits and vegetables coated with wax. This action threatens U.S. horticultural exports to India, particularly apples and pears, which amounted to $9.5 million in 2004.

The role of technology is vital in production, harvest and post-harvest processes. Production technologies rendering possible different varieties, more diversified produces, improved yields, reduced quality losses from pathogens, using a fewer chemicals and improved nutritional value are essential for a more dynamic management of horticultural crops. Scientific horticultural crop management follows Good Agricultural Pactices (GAP) and hygienic Practices. These are the backbone for quality assurance and safety programs.

Good Horticultural Practices (GHP)

Include practices improving conventional production and produce handling methods - starting from selection of soil to be cultivated - and activities related to production and handling of fresh produce in the field. Practices also aim at reducing negative impacts of production systems on the environment, fauna, flora and workers' health. Applying GHP relies on the proper implementation of Integrated Crop Management (ICM) and Integrated Pest Management (IPM) programs with an aim of producing economically and viable products. To enforce ICM and IPM programs, practical researches on biological cycles of pests and diseases, infestation levels, propagation ways, economic damage thresholds, most appropriate and effective control practices, etc. are required. Good Manufacturing Practices and Good Hygienic Practices goes along with GHP for controlling post-harvest hazards affecting produce safety and having minimum effect on workers and the environment. With the development of hortibusiness, emphasis is shifted from the farm based production to value added chain and marketing of horticulture products.

Indian Scenario of Horticultural Production

World production of fresh fruits and vegetables was estimated in 2002 as 1244 million tonnes, with 772.7 tonnes of vegetables and 471.3 tonnes of fruits (FAOSTAT). In the last decade the overall rate of growth for vegetables has been more dynamic than for fruits, 5 % compared to 2.38 %. There are almost 180 families of fruits that are grown all over the world and citrus fruits constitute around 20% of world's total fruit production. In India, horticulture accounts for about 29.5 % of the country's agricultural GDP from 8.5 % area and 10 % of the total agricultural export earnings. India is the second largest producer of fruits in the world with an annual production of 43 million tonnes from an area of 4 million hectares and contributes to more than 9 % of global fruit production.

The diverse agroclimatic zones in the country make it possible to grow almost all varieties of fruits and vegetables in India. These crops form a significant part of total agricultural produce in the country comprising of fruits, vegetables, root and tuber crops, flowers, ornamental plants, medicinal and aromatic plants, spices, condiments, plantation crops and mushrooms. Major vegetables grown in India include potato, eggplants, tomatoes, cassava, cabbage, dry onions, cauliflower, pumpkin, okra and green peas, while fruits include mangoes, oranges, apples, grapes, pineapples and papaya. The contribution of horticultural crops to the national income is also perceptible by the annual foreign exchange earnings achieved through their export. Horticulture makes up 28.5 % of the country's agriculture GDP. Fresh fruit export during 2004 was valued $209.10 million against $159.31 million during the 2003. India is the largest producer of banana, mango and cabbage enjoying the highest productivity in grapes, sapota and banana. It also holds analogous status in vegetable production by being the world's second largest producer with over 90 million tonnes, next only to China and accounting for about 9.3 % of the world's vegetable production. Most grim area is the post harvest handling. Nation has been facing recurring post-harvest losses of horticultural crops of about Rs 51,500 crore/ annum for want of effective, easily accessible and economically viable means to prevent such losses. Now, wth the focussed attention given to horticulture, there has been spectacular change in terms of adoption of new technologies, production and availability of horticulture products.

Challenges in Horticulture Management

Although India is the largest producer of fruits in the world, the production/ capita is only about 100 g/day. However, it is estimated that more than 20-22% of the total production of fruits is lost due to spoilage at various post harvest stages. Thus the per capita availability of fruits is further reduced to around 80 g/day which is almost half the requirement for a balanced diet. It is estimated that around 20-25% of the total vegetables is lost due to poor post harvesting practices. Less than 2% of total vegetables produced in the country is commercially processed as compared to 70% in Brazil and 65% in USA. Value addition in fruits and vegetables sector is less than 2 % in India (Table 1) which is negligible compared to Brazil and Malaysia. With the development of technology and manpower, there is scope for improving value addition and income from horticultural sector.

Table 1: Extent of Value Addition in Fruits and Vegetables

Country	Extent of Value Addition (%)
Brazil	70
France	50
USA	60-70
Malaysia	83
Thailand	50
Israel	50
New Zealand	40
China	10
India	< 2

The working group for horticulture development, appointed by Planning Commission, Government of India in 2001 listed common constraints in respect of various sectors of horticulture in India and they are,

- Inadequate availability of disease free, high quality planting materials
- Underexploitation of micro-propagation techniques.
- Slow dissemination and adaptability of improved high yielding cultivars/hybrids.
- Inadequate facility for identification of nutrient deficiency and disorders
- Lack of diseases and pests outbreak forecast services.
- Unavailability of refined intensive integrated production systems
- Lack of quality standards.
- Lack of technologies in value addition.
- Lack of post harvest management technology and infrastructure
- Weak database and poor market intelligence.
- Poor marketing practices and infrastructure

- Instability of prices, with no support price mechanism. Inadequate technical manpower/human resource in farming system.
- Poor credit supply, high rate of interest coupled with inadequate crop insurance scheme.
- Ineffective transfer of technology and poor adoption of improved techniques.
- Poor linkage between Research and Development sectors, industries and farming communities.
- Late implementation of government policies and schemes.
- Absence of horticultural crop suitability map of India based on agroclimatic conditions depicting most suitable areas for optimum productivity of a particular crop.

It is estimated that India has 240 million acres of cultivable wasteland which can be brought under orchard crops without curtailing the area under food crops. The country has abundant sunshine throughout year, surplus labour and widely varied agro-climatic conditions, which offer high potential for successful and profitable commercial horticulture. Small land holdings and non-availability of good quality planting materials have been the major issues of concern. Poor farm management practices, necessitate strong need for extension education and training for the growers. New management strategy is required for horticultural crops in cooperative and contract farming for improved yield and quality in the long run. Maintaining global standards in post-harvest management practices such as grading, packaging, storage and transportation could help the country gain significantly in the world market. We need to create pockets of excellence since the country has a diverse agro climatic zone.

Potentials of Horticulture

Among horticultural commodities, India is exporting fresh fruits, vegetables, processed products of fruits and vegetables, cut & dried flowers, medicinal and aromatic plants, seeds, spices, cashew kernels and their products, tea and coffee. Export of tea leads all horticultural commodities followed by spices, coffee, cashew and processed fruits and vegetables. Among fruits, mango and vegetable onion has been the major crops exported from India. Agricultural and Processed Food Export Development Authority (APEDA) under the Ministry of Commerce promotes fruits and vegetables export and has identified fruits such as grape, mango, litchi, Kinnow in the selected regions for export promotion. Last decade witnessed tremendous boost in horticultural production. Mango production has improved substantially in Gujarat, Maharashtra, Andhra Pradesh and Uttar Pradesh. Maharashtra ranked first in citrus cultivation, covering 19.6 % of total citrus area. Three districts in that state - Nagpur, Amravati and Wardha of Vidarbha have over 14 million trees of citrus. Papaya, sapota, pomegranate, ber and aonla have registered growth both in area and production in recent years. In papaya, area expansion was high in the South, whereas ber and pomegranate were successful in Western Maharashtra. India tops the world in sapota production and high productivity was achieved in Karnataka (17.2 t/ha), followed by Maharashtra (16t/ha). In temperate fruits, apple alone accounts for more than 50 % of the total production, while the North-Western Himalayan region (J&K, H.P. and hills of Uttar Pradesh) covers 95 % of total area and 85 % of the total production of all temperate fruits.

Mango, grapes and walnut are exported in large quantities and pomegranate, sapota, banana, litchi, apple and strawberry in limited quantities. Mango pulp is important among the processed fruit products exported. India exports considerable quantity of onion and potato and limited quantity of vegetables like okra, brinjal, tomato and chillies. India occupies first position in the production of cauliflower, second in onion and third in cabbage in the world. Vegetable farming is highly remunerative for marginal and small farmers. In crops like radish and tomato, cropping has become possible throughout the year with the development of temperature tolerant varieties. Vegetable processing industry is picking up momentum in states like Punjab and Karnataka and at present, tomato, potato and green peas are mainly utilized in processing sector. Onion and potato stand first in export market and others including beans, peas, green chilli, broccoli, asparagus, mushroom, capsicum and lettuce enjoy good scope. Cucurbits are favorites with ethnic population. Less pungent white/yellow varieties of onion possess good demand in European markets. Although India exports considerable quantity of onions, level of production and prices prevailing within the country cause hindrances in promoting export on large scale. It is estimated that presently only 0.5% of the potato production of the country is processed. More than 80% of Indian potato is grown in the winter months when there is no potato crop in the Western temperate countries. Hence there is huge marketing advantage for Indian potatoes in export. The potential production zones of various horticultural crops are given in Table 2.

Table 2. Potential Production Zones of Various Horticultural Crops in India

Name of the crop	Key production zones
Fruits	
Mango	Maharashtra (Alphonsa), Gujarat(Kesari), A.P(Banganapalli), U.P, (Chausa and Dasheri), West Bangal(Himasagar)
Grapes	Maharashtra, A.P. and Karnataka
Litchi	W.B., Bihar, Uttrakhand, Parts of U.P, Parts of A.P
Cashew	Kerala, Maharashtra, Karnataka, NEH
Walnut	J &K , High hills
Mandarins including Kinnow	Maharashtra, NEH, Darjelling, Punjab and Rajasthan for kinnow
Pomegranate	Maharashtra, Karnataka, Rajasthan.
Vegetables	
Potato	W.B., U.P and Punjab
Onion	Kharif onion. Eastern states(Bihar, Orissa and W.B) Punjab, Maharashtra and Karnataka.
Chilli	Maharashtra, Rajasthan and A.P.
Bitter gourd and okra	Kerala, Maharashtra, W.B, Bihar
Spices	
Black pepper	Kerala, Maharashtra, and Karnataka
Ginger	Kerala, NEH including Sikkim and Darjeeling hills
Turmeric	Kerala, A.P., Meghalaya and Orissa
Cumin	Gujarat and Rajasthan
Large cardamom	Sikkim and NEH
Floriculture	
Rose	Maharashtra, and Karnataka
Orchid (Cymbidium)	Sikkim, Arunachal Pradesh and Darjeeling
Anthurium	Kerala, Karnataka and NEH
Cut greens	Kerala, Karnataka, A and N Islands and NEH
Dry flowers	Tamil Nadu, West Bengal

The floriculture industry in India covers trade of i) cutflowers ii) nursery and potted plants, seeds and bulbs iii) micropropagated plants and iv) essential oils from flowers for perfumes. It is estimated that annual growth rate of domestic trade of floriculture products is about 25-30 %. In domestic market, consumption of flowers in the southern states is much higher than in the northern region. The cut flower export value have shown a tremendous increase from Rs. 253.0 million in 1998-99 to Rs 2100 million in 2004-05 (*www.apeda.com*). This has come about with the establishment of a large number of export oriented cut flower units around Bangalore, Pune, Delhi and Hyderabad. The major product has been rose being grown by more than 90 % of commercial units followed by tropical orchids (dendrobium). The exports of medicinal and aromatic plants and their products have a tremendous potential particularly to advanced countries like U.S.A., Japan and Europe. Herbal medicine is explored as alternatives to antibiotics, steroids and hormonal drugs for avoiding side effects on body functions. The international market of medicinal plants related trade is estimated US $ 60 billion per year having a growth rate of 7 %. Another promising horticulture sector is spices and India commands 46% in global trade in terms of quantity and 28% in terms of value/ annum.

Agri Export Zones

Since 2002, the country is exporting fresh and processed fruits and vegetables through 39 agri export zones (AEZs) in 19 States. An investment of Rs 1,018.85 crore is expected from the private sector, the Central and the State Governments in these AEZs during the next three to five years. Moreover, the Government declared Vishesh Krishi Upaj Yojana (special agricultural produce scheme) in the foreign trade policy on August 31, 2004 for a five-year span, with a view to promoting export of fruits, vegetables, flowers, minor forest produce and their value-added products. Promotion of horticulture will lead to nutritional security, efficient input use, crop diversification, foreign exchange earnings and enhanced employment generation through agro-industries. There is a need for establishing quality control and assurance laboratories for horticultural produce in view of the strict sanitary and phyto-sanitary standards insisted upon by importing countries in the post-WTO regime. It is necessary, and in fact obligatory, on the part of Indian Government to create a level-playing field in the era of WTO for Indian agriculture to sustain itself and survive. When agri commodities were exported, they were subjected to strict SPS (Sanitary and Phytosanitary) checks, but when fresh fruits and vegetables were imported into India, they were not subjected to residue analysis which needs to be enforced.

Export of fresh fruits and vegetables declined from 1.29 million tonnes (mt) in 2003-04 to 1.24 mt in 2004-05. In terms of rupee value, the decline (for 2004-05) is said to be more pronounced – from Rs 1,457 crore to Rs 1,299 crore. Agri-Export Zones (AEZs) is projected to support the National

Horticulture Mission (NHM) with a view to increasing the share of horticulture exports from the extant level of 2 to 5 % in a five-year period, besides ensuring export receipts to go up from Rs 600 to 1,000 crores.

Governmental Initiatives in Horticulture

The Government of India, aware of this engine for economy growth has been making vigorous efforts to promote horticulture through various agencies under the control of ministry of agriculture and ministry of commerce. Horticulture development is accorded high priority from Eighth Five Years Plan onwards. The organisational support for horticulture programmes is available both in the Ministry of Agriculture and Ministry of Commerce at national level. At State level, Departments of Agriculture/Horticulture provide the required infrastructure. A number of Boards, Councils, Directorates, Foundation and Authorities are established under the Ministries at Central and State level to promote their products both in the domestic and export market.

Table 3. Five Year Plan Allocation in Horticulture

S. No.	Five year plan	Amount allotted Rs in crores
1	Fifth plan	7.61
2	Sixth plan	9.13
3	Seventh plan	24.19
4	Eighth plan	789.00
5	Ninth plan	1453.00
6	Tenth plan	5025.00

National Horticulture Mission (NHM)

National horticulture mission (NHM) with an outlay of Rs 2,300 crore is launched from the remaining period of Tenth Plan which will continue during Elevanth plan. NHM, which would be a Centrally-sponsored scheme under the Department of Agriculture and Cooperation, is expected to promote horticulture sector through regionally differentiated strategies. The objective of the NHM is to double horticulture production to 300 million tonnes (mt) by 2011-12 with technological interventions through forward and backward linkages from production to marketing and processing of horticultural produce. Under the mission about 33 lakh hectares of additional land would be brought under horticultural crops and 16 lakh hectares of senile plantation would be rejuvenated by end of Eleventh Plan. Alongside, about 21 lakh hectares would be covered under the integrated pest management and also about 2,040 nurseries would be established to make available good quality planting materials. NHM is cast to focus on horticultural research and horticultural development through generation of good quality seeds and planting materials, coverage of area with improved varieties and productivity improvement programmes. The effort is also to address post-harvest management and marketing by fostering infrastructure facilities for cold chains, market yards and market intelligence and value addition.

Various agencies involved in the promotion of horticulture in ministry of agriculture as well as ministry of commerce are given below.

Ministry of Agriculture

Horticulture Division

The Horticulture Division was created in September, 1980 by reorganization and redeployment of existing staff in the Crops Division. Policy planning for horticulture development is being done by the Horticulture Division in the Department of Agriculture & Cooperation (DAC) of Ministry of Agriculture (MOA) headed by a Horticulture Commissioner.

There are two Boards namely National Horticulture Board and Coconut Development Board and two Directorates namely Directorate of Arecanut and Spices Development and Directorate of Cashew and Cocoa Development supporting horticultural development activities in the Ministry of Agriculture.

National Horticulture Board (NHB)

Established in 1984 as an autonomous society under the Societies Registration Act, 1860 with a mandate to promote integrated development of horticulture, to help in coordinating, stimulating and sustaining the production and processing of fruits and vegetables and to establish a sound infrastructure in the field of production, processing and marketing with a focus on post-harvest management to reduce losses. National Horticulture Board has the mandate to take up programmes right from the stage of production to development of post-harvest infrastructure including marketing, processing and storage facilities for ensuring integrated development of horticulture. Major programmes of the Board include establishment of cold storage for horticulture produce and strengthening of post harvest inputs where back ended capital subsidy is provided. The board was also given the mandate of creating a database for horticultural crops. Other areas entrusted to the Board are providing market information service on fruits and vegetables, development of captive plantations for export and processing, market support operations for horticultural crops, development of export of horticultural products through equity participation, etc.

Coconut Development Board (CDB)

The Coconut Development Board was established in 1981 under the Coconut Development Board Act, 1979 with a mandate for integrated development of coconut industry in the country through promoting production, processing, marketing and product diversification of coconut.

Directorate of Cashewnut and Cocoa Development, Kochi

Established in,1966 with headquarters at Kochi with the mandate of development of cashewnut in the country. Development of cocoa was transferred to this Directorate during 1998. The Directorate is responsible for undertaking studies on various aspects of cashewnut and cocoa development including production, marketing, prices and other related problems.

Directorate of Arecanut and Spices Development, Calicut

Established at Calicut, Kerala in 1966 with the mandate to undertake development schemes on spices, medicinal and aromatic plants and arecanut at the national level. These schemes are implemented through State Governments, Agricultural Universities, ICAR Institutes as well as through Regional Research Laboratories under CSIR and monitored by the Directorate.

Ministry of Commerce

Ministry of commerce mainly supports the marketing and export of various horticultural commodities through various boards and authorities as detailed below.

Agricultural and Processed Food Products Export Development Authority (APEDA), New Delhi

APEDA was established in 1985 replacing the Processed Foods Export Promotion Council. It has the mandate of Export Promotion and Development of Fruits, vegetables and their products, Horticulture and Floriculture products, Herbal and Medicinal plants, Pickles, Papads and Chutneys, Cocoa products, Honey, Alcoholic and non-alcoholic beverages. APEDA is implementing a number of schemes for promotion of export of the above listed commodities.

Spices Board, Kochi

Established in 1987 with headquarters at Cochin, Kerala under the Spices Board Act, 1986. It is an apex body for export promotion of Indian spices and spice products. The Board is assigned responsibility of export development of 52 spices including pepper, chilli, ginger, turmeric, cardamom, coriander, cumin, fennel, fenugreek, celery, vanilla and saffron. Indian Cardamom Research Institute under the Spices Board carries out need based research programmes on crop improvement, production and also post-harvest techniques of cardamom.

Tea Board, Kolkatta

The Tea Act, 1953, established the Tea Board of India on April 1,1954. Tea Board is entrusted for increasing tea production and productivity, extension of area under tea cultivation, improvement in quality of tea, catalyzing co-operative efforts, backing up research and development, labour welfare, market development, export promotion and regulatory functions such as issuance of licenses and monitoring of various trade related activities.

Coffee Board, Bangalore

Established under the Coffee Act VII of 1942, with mandate of undertaking research, extension, development, quality up-gradation, market intelligence, external and internal promotion and labour welfare. During the pre-liberalised era, the Coffee Board was vested with powers for marketing the entire coffee produce under a unique pooling system. Consequent to depooling in the early 1990s, the Coffee Board withdrew from the direct marketing activities and focused in the areas of research, extension, development and market development programmes.

Rubber Board, Kottayam

Established under the Rubber Act, 1947. The functions and duties of the Board include promoting the development of rubber industry, advise the Government of India on all matters related to rubber industry including the import and export of rubber . Board promotes the development of rubber industry by undertaking, assisting or encouraging scientific, technological or economic research. Board also imparts training in improved methods of planting, cultivation, manuring and spraying and provides technical advice to rubber growers.

Non Governmental Organizations

Several non governmental organisations are also engaged in the development of horticulture crops. These include "Maha Grape", and 'Maha Mango' in Pune, Maharashtra. Similarly, Cadbury and CAMPCO, Mangalore (Karnataka) are dealing with the production and processing of cocoa and arecanut and also their marketing. HOPCOM, Bangalore (Karnataka) is a successful cooperative organisation for marketing of horticulture produce. The Society is under operation in the districts of Bangalore (Urban and Rural), Kolar, Tumkur, Mandya, Shimoga, Dakshina Kannada and Mysore. The main objective of the Society is to procure fruits and vegetables from member growers and supply to consumers through its own retail outlets. There are 405 retail outlets. The Society is handling around 44,000 metric tonnes of fruits and vegetables with a total turnover of Rs.400 million annually.

Vegetable and Fruit Promotion Council, Keralam (VFPCK)

This is a company registered under section 25 of Indian Companies Act established for the overall development of fruit and vegetable sector in Kerala. VFPCK is evolved from *Kerala Horticulture Development Programme (KHDP)* which was one of the most successful Agricultural Development Projects in India. VFPCK is a unique organisation having farmers, goverment and financial institutions as the major stakeholders. The organisation constituted for improvement of overall status of fruit and vegetable farmers, is professionally managed and has a result oriented approach. The Self Help Groups(SHGs) formed by the council form the base target groups for major interventions of the council like Transfer of technology, Rural credit, Group Marketing, Participatory technology development, Value addition and Export.

Conclusion

The average per capita availability of fruits and vegetables is inadequate to meet requirements for nutritional and protective diet of the population. With increasing income levels and demand the production of fruits and vegetables has to reach the level of 318 million tonnes by the year 2020. It requires to increase production of fruits and vegetables more than 2½ times by the year 2020. Of the total horticulture production, only 67 % is used for local consumption while 30 % is subjected to post harvest loss, 2 % for export and 0.8 % for processing. India has just one per cent share of world trade in fresh fruits and vegetables. Exports are plagued by issues like low investment in developing quality standards and targeting only ethnic/low discerning markets like middle east.

Market driven framing is yet to take off in horticulture sector. Prevalence of smaller holdings makes it more complex for the industry to support farm production side and industry very often ends up procuring from mandi where there is no control over the grades and quality which reduce processing efficiency. There is immense scope for developing horticultural produce market. Changing food habits, life styles and health consciousness and purchasing power has created an unprecedented opportunity for farmers and horticultural entrepreneurs. Production and promotion of regional brand names like MH grapes, Himachal apples, Kerala spices and Darjeling Tea protect the right of Indian horticultural farmers in world market.

CONTENTS

Part - II

PART - I

Chapter 1

FRUIT CROPS

Introduction

Fruits are undoubtedly God's gift to mankind. Fruits are protective foods as they contain vitamins, minerals and enzymes, necessary for better functioning of human body and for fighting diseases. History of fruits is as old as that of Adam and Eve who ate the forbidden apple. Date palm seems to be the earliest fruit cultivated by man as early as 7000 B.C. Pomegranate is also one of the oldest fruits which was under cultivation as early as 3500 B.C. During ancient times, monks survived solely on fruits and milk giving an indication about importance of fruits and antiquity of fruits in India. India is the place of origin of mango, banana, phalsa, jack fruit, bael, aonla and a few citrus fruits whereas all other fruits are introduced from foreign countries. Litchi, loquat, persimmon, peach and sweet orange are from China; papaya, guava, sapota, avocado, custard apple and pineapple are from tropical America; and apple, plum, pear, date palm, mangosteen, strawberry, cherry, almond, walnut, apricot, grape and pomegranate are from the Central Asiatic Regions.

Important Advantages of Fruit Culture are

1. From a unit of land comparatively more income is realized by growing fruits than growing any agronomic crops like wheat, rice, maize, etc.
2. Calorific value of food from unit area of fruit is more than cereals. Calorific value of banana is 16 times more than wheat.
3. Fruit farming is the bed rock of several industries like extraction of essential oil, pectin, fruit preservation, etc.

4. Fruit culture provides employment to rural people throughout the year.
5. Cultivation of fruit crops helps in maintaining ecological balance and checks pollution.
6. Being rich in minerals and vitamins, fruit crops assume great importance in nutritional security of masses.

Area and Production

India is the second largest producer of fruits after China, with a production of 47.68 million tonnes of fruits from an area of 4.18 million hectares. India's share in the world production of fruits was about 8% and India produced about 65% of the world mangoes and 11% of the world bananas, ranking first in both the crops. A large variety of fruits are grown in India, of which mango, banana, citrus, apple, guava, pineapple, papaya, sapota, grape and litchi are the major ones. Apart from these, fruits like mangosteen, annona, jackfruit, pomegranate, avocado, date palm in tropical and sub-tropical group and peach, pear, plum and strawberry in the temperate group are also grown over a sizeable area. Although fruits are grown throughout the country, major fruit growing states are Maharashtra, Uttar Pradesh, Andhra Pradesh, Karnataka, Bihar, Tamil Nadu, Orissa, Kerala, Gujarat, West Bengal and Madhya Pradesh.

Area and Production of Major Fruits in India

Fruits	Area (million ha.)				Production (million t)			
	1999-00	2000-01	2001-02	2002-03	1999-00	2000-01	2001-02	2002-03
Apple	0.23	0.24	0.24	0.25	1.05	1.23	1.16	1.47
Banana	0.49	0.47	0.47	0.68	16.81	14.14	14.21	16.82
Citrus	0.53	0.50	0.62	0.60	4.65	4.40	4.80	4.72
Grapes	0.04	0.05	0.05	0.06	1.13	1.06	1.21	1.15
Guava	0.15	0.15	0.15	0.22	1.71	1.63	1.72	1.78
Litchi	0.05	0.05	0.06	0.05	0.43	0.40	0.40	0.44
Mango	1.49	1.52	1.58	1.60	10.5	10.06	10.02	10.78
Papaya	0.06	0.07	0.07	0.08	1.67	1.79	2.59	1.85
Pineapple	0.07	0.08	0.08	0.09	1.02	1.22	1.18	1.31
Sapota	0.06	0.07	0.05	0.07	0.64	0.74	0.60	0.71
Others	0.63	0.23	0.63	0.48	5.89	6.49	5.11	6.65
Total	**3.80**	**3.89**	**4.00**	**4.18**	**45.50**	**43.14**	**43.00**	**47.68**

Source: Ministry of Agriculture, Govt. of India

Major Fruit Producing States in India

	Area (In 000' Ha)				Production (In 000' T)			
State / Ut's	**1996-97**	**1997-98**	**1998-99**	**1999-00**	**1996-97**	**1997-98**	**1998-99**	**1999-00**
Andhra Pradesh	406.7	414.5	378.6	449.2	5657.7	5899.1	4589.6	5175.4
Bihar	292.8	299.8	303.6	309.3	2752.2	3755.4	3797.2	3870.7
Gujarat	137.5	159.3	163.0	176.2	1820.0	2267.8	2293.5	2376.0
Karnataka	298.8	314.6	314.6	315.0	5133.6	5446.3	5446.3	5456.1
Kerala	195.8	195.8	233.1	187.8	1826.0	1826.0	1621.2	1184.5
Madhya Pradesh	59.5	62.4	63.1	67.4	1127.0	1184.0	1374.4	1536.1
Maharashtra	373.3	380.0	436.1	539.8	6333.1	6473.2	7521.7	8688.5
Orissa	206.6	227.4	249.4	204.9	1342.4	1511.8	1718.4	1202.9
Tamil Nadu	220.6	234.0	213.5	232.0	3862.7	3683.8	5447.6	5939.6
Uttaranchal	185.5	186.7	187.9	172.5	510.2	515.3	520.4	475.5
Uttar Pradesh	318.9	328.8	305.2	315.1	4045.1	4293.0	3097.8	3210.5
West Bengal	116.4	117.3	128.0	130.2	1035.1	1373.6	1536.0	1816.1
Others	767.0	781.3	750.7	697.4	5013.0	5033.9	5078.3	4564.1
TOTAL	**3579.4**	**3701.9**	**3726.8**	**3796.8**	**40458.1**	**43263.2**	**44042.4**	**45496.0**

Source: Database of National Horticulture Board, Ministry of Agriculture, Govt. of India

Area of Major Fruits in India (2002-03)

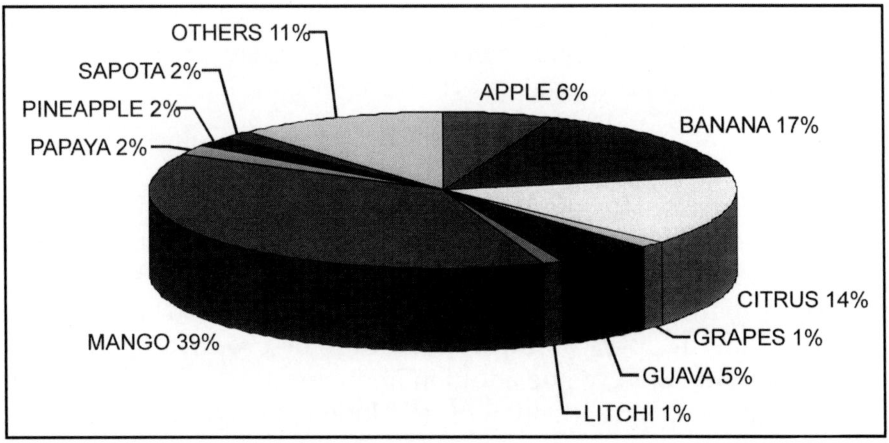

Production Share of Major Fruits in India (2002-03)

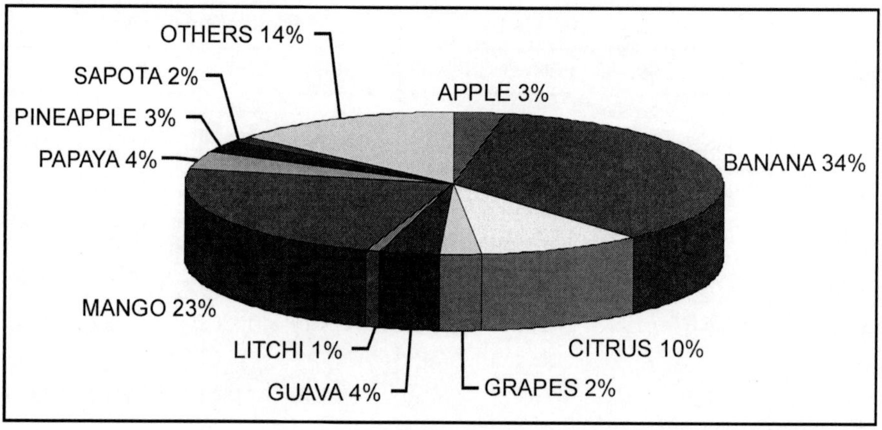

Potential

The development of fruit growing on a commercial scale came up only during the last 50 years with adoption of scientific cultivation practices. Area under fruit cultivation on scientific basis is increasing in spite of constraints observed due to incidence of diseases such as citrus decline, dieback in mango, panama wilt in banana and physiological disorders like mango malformation and erratic bearing in mango. During 1991-92, area under fruits was only 2.87 million hectare with a production of 28.6million tonnes which increased to 4.18 million hectare with a production of 47.68 million tonnes in 2002-03. Eventhough there was not much increase in area, production showed an increasing trend, mainly due to increase in productivity by means of scientific approach.

Five fruits viz., mango, banana, citrus, apple and guava together covered about 75% of total fruits produced in the country. However in lesser known fruits like pineapple, papaya, sapota and pomegranate, area expansion and production rise during the recent years were spectacular.

New vision and concerted efforts are required for change in variety mix and supply of quality planting materials from elite clones on indexed clonal rootstocks. High density planting, water management including micro-irrigation, integrated plant nutrient management and IPM strategy for plant protection are few of the areas which need greater R&D focus. Adoption of post-harvest management practices and infrastructure development for grading, packaging, pre-cooling and storage of produce needs focused developmental attention. Value addition and export promotion, particularly of apple are drawing due attention of developmental agencies in India.

It is estimated that India has 100 million ha of cultivable wasteland, lying idle, which can be brought under orchard crops without curtailing area under food crops. The country has abundant sunshine throughout the year, surplus labour and widely varied agro-climatic conditions, which offer high potential for successful and profitable commercial horticulture.

Scope of Fruit Growing

In India, there is ample scope for development of fruit industry as the demand for fruits is very high and farmers are able to get high economic returns from their land and above all, the agro climatic conditions prevailing in the country provide good scope for cultivation of different types of fruit crops in one part or other of the country making fruits available throughout the year. The main constraints observed are the high initial investment, high recurring expenditure and the long pre-bearing period of woody perennial fruit trees which mean that returns can be got only after 3 to 4 years. Apart from these constraints, these fruit trees require intensive cultivation practices and specialized methods of propagation involving root stocks and good quality scion, proper irrigation, timely application of fertilizers and manures and other cultural operations like weeding, mulching, training, pruning and growth regulator application. As the products are perishable in nature careful handling, transport and marketing are necessary. Lack of refrigerated transport has been a bottleneck in development of fruit industry. Majority of farmers in India are not in a position to cater to these requirements and this together with low purchasing power of consumers prevents development of fruit industry.

With growing urbanization, requirements for fruits in cities are increasing tremendously and there is need to boost fruit production. Presently, fruits are available at very high cost in the markets and are not within the reach of common man. Standard of living of people can be judged by consumption of fruits/ capita. Inspite of the fact that India is climatically favourable for production of a variety of tropical, sub-tropical and temperate fruits, per capita/ day consumption of fruits is the lowest (80 g) when compared to developed (250-450 g) countries like Switzerland, France, Sweden, Australia, Netherland, USA and Canada and even in a few developing countries (100-200 g) like Hungary, Germany, Philippines, Pakistan and Japan. Production has to be increased till the country becomes self-sufficient in fruits and price of fruit falls to a level at which all people, rich and poor can afford to buy it. This can be achieved only by increasing area under fruits and by rejuvenating existing orchards for which there is good scope in India.

Earlier, the fruit production was confined to pleasure gardens of kings and nobles. Commercial fruit production emerged in a real sense, after the end of second world war and it was adopted as a profession by many enterprising farmers. Fruit cultivation is labour intensive and is ideally suited to over populated countries of the world like India and China, where per capita cultivated land holding is less than one hectare. The multiple cropping system in the orchards also generates more avenues for employment apart from giving additional income. Like all other branches of horticulture,fruit production involves certain specialised skills in production, harvesting, marketing, canning, processing and many intricacies of the trade which require a touch of professionalism. Involvement and commitment of small and medium level entrepreneurs are highly desirable and are pre-requisites for increasing level of production, processing and profitable marketing of fruits.

More than 20-22% of total production of fruits is lost due to spoilage at various post harvest stages reducing the per capita availability of fruits. Fruit production in India recorded a growth rate of 3.9%, whereas the fruit processing sector grew at about 20%/ annum. Fruits are processed into various products such as fruit juices and concentrates, canned fruit, dehydrated fruit, jams and jellies. There exist over 4000 fruit processing units in India with an aggregate capacity of 12 lakh tonnes (<4% of total fruits produced). It is estimated that around 20% of production of processed fruits is meant for exports, the remaining caters to defence, institutional sectors and household consumption. Mango and mango-based products constitute 50% of exports.

With our rapid increasing population, pressure on land is likely to continue and size of holding remains small in foreseeable future. Area under cultivation does not bear any relationship to productivity where there is a considerable scope to raise it. Both quality as well as productivity need to be carefully upgraded through rigorous R&D programme. Moreover, fruit crops give much more production/unit of area and are certainly more paying than ordinary farm crops. Majority of fruit plants have very high and encouraging productivity like production of 50 tonnes of grapes, 45 tonnes of pineapple, 40 tonnes of papaya and 25 tonnes of banana from a hectare of land/ annum. It is reported that the horticultural crops with their meagre share of 2-3% of land area in most of the countries, contribute to 5-7% of gross food production and 10-35% of the national income from crop enterprise. The high productivity of fruit trees gives higher economic returns and also provides higher calorie/ unit area.

Keeping in view the above facts, there is enormous scope for increasing productivity of fruits in the country. The fruit production has great possibility in trade and in industrial development. In addition, the fruit crop fits well in crop diversification programmes in various scales, ranging from organised large orchards to a few trees in the waste land or backyards of home or elsewhere. Thus cultivation of fruits contributes to health, happiness and prosperity of people.

For further development of horticulture, special consideration should be given to following measures :

i. Fruit growers in principal fruit-growing regions should be assisted to organise themselves on cooperative lines for raising nursery plants, controlling pests and diseases, and for marketing fruits and fruit products.

ii. Suburban belts around large towns should be developed for raising fruits and the growers organised on a .cooperative basis, especially for purpose of marketing their produce. Steps should also be taken to popularise kitchen gardening in urban areas by supplying plants and technical advice.

iii. The preservation of fruit which was started on a small scale, should be expanded on modern lines. There are at present 467 factories with a total output of 10,000 tonnes valued at Rs. 1.63 crores. This is less than 2% of the total produce. Research on modern and indigenous methods of fruit preservation should receive increasing attention at hands of Central and State Governments.

The Governments of a few countries, besides providing facilities for research, give assistance and encouragement to fruit and vegetable preservation in following ways:

i. Sugar for fruit preservation is supplied at the world market rate,

ii. The import of machinery and other essentials is allowed either duty free or at concessional rates, and

iii. Concessions are granted in freight rates for transport of raw materials from the ports to the canning factories and of the finished products from canning factories to centres of consumption. With development of horticulture the adoption of similar measures in this country for encouraging preservation of fruit and vegetables will have to be considered. At a later stage, it is desirable to make available cold storage facilities at important markets and refrigerated railway wagons for transportation of horticultural produce.

The suggestion has also been made that a Fruit Development Board should be established in Food and Agriculture Ministry, responsible for developing the industry on a country-wide basis by enforcing quality standards, supervising research, maintaining statistics and giving technical advice to States which need it. This suggestion deserves consideration.

Classification of fruits

Based on the temperature requirements and response to different climatic conditions, fruit crops are classified into three broad groups;

1. *Tropical fruits :* are those which do not endure severe cold but can tolerate warm temperatures of about 40°C. These plants need strong sunshine, moderate warmth, fair degree of humidity and a very mild winter. They cannot stand frost or snow. Mango, banana, papaya, guava, pineapple, sapota, custard apple, dates and jack fruit belong to this group.
2. *Sub-tropical fruits :* like citrus fruits, litchi, pomegranate, grapes and avocado are intermediate in character. They need warmth and moderate humidity and can also tolerate mild winters. These plants are versatile in their ability to stand extremes of rainfall, humidity, sunshine and mildly tolerant of winter up to frost stage.
3. *Temperate zone fruits:* are commonly found in cold regions enjoying a mild and temperate climate. These fruit trees endure cold and go to rest or dormancy by shedding off all their leaves during winter. Apple, pear, peach, plum and strawberry are examples of such fruits.

The above classification indicates the broad differences in climatic needs of various fruit plants. This does not necessarily mean that a plant belonging to one climatic zone does not grow in other zones. For instance, grape vine is a fruit which goes to dormancy in temperate regions and remains evergreen in sub-tropical regions. It is grown in both these climatic zones as a result of acclimitization, cultural manipulations and breeding new varieties.

Annonaceous Fruits

The family Annonaceae comprises of 40 genera of which one important genus is *Annona.* It has 120 species of which six species produce edible fruits having economical importance and these are collectively called as annonaceous fruits. These species are *Annona squamosa, A. reticulata, A. muricata, A. cherimola, A atemoya* and *A. diversifolia.* All these are native to tropical America and were introduced from there to India. The fruits are syncarpia formed by fusion of pistil and receptacle into a large fleshy aggregrate fruit.

Among these species of *Annona* the most important and favourite one in India is custard apple (*Annona squamosa* L.). It is cultivated in Andhra Pradesh, Maharashtra, Karnataka, Bihar, Orissa, Assam, and Tamil Nadu. Besides India, it is common in China, Phillippines, Egypt and Central Africa. It grows unattended in several southern Indian States and is a promising tree of commercial importance. Several exclusive orchards with high yielding and choice varieties selected from Balanagar near Hyderabad have sprung up in Andhra Pradesh and Maharashtra in recent times.

Bullock's heart (*A. reticulata*) is common in the backyards of houses in South India but not seen on a commercial scale. Other annonas are cultivated only on a limited scale. Soursop (*A. muricata*) and atemoya (*A atemoya*) are grown in a few gardens as miscellaneous fruits. Cherimoya (*A. cherimola*) is seen on a limited scale in Assam and hills of south India.

Morphologically annonaceous fruits show some similarity but they differ in their fruit characters mainly taste, flavour, texture, pulp colour and shape. The fruits are generally used as fresh, while some products like custard powders, ice-creams are prepared from the fruits. Custard apple is high in calories and is a good source of iron. Besides high nutritive value, it also has high medicinal value. The bark and leaves contain an alkaloid, annonaine. A bark decoction is used to stop diarrohea, while the root is used in the treatment of dysentery. A decoction of the leaves is used as a cold remedy and to clarify urine. The fruits of Annona are haematinic, cooling, sedative, stimulant, expectorant, maturant, tonic. Oil extracted from the seeds is used as an eco-friendly insecticide.The seeds are abortifacient and are useful in destroying lice in the hair. The immature custard apple fruits are used as fish baits by the local people in Karnataka and in tribal regions. The leaf extract is used to repel honeybees by the tribals while extracting honey from the beehives.

Varieties

Being a cross-pollinated crop, custard apple has wide variations in forms and sizes of fruit as well as colour of the pulp. Depending on the external fruit colour, it is distinguished into green, red and yellow. But green ones are the most common and popular varieties that head to the market.

Climate and Soil

Annonaceous fruits in general require a tropical or near-tropical climate with summer temperatures from 25 ° to 41 °C, and mean winter temperatures above 15 °C. It is sensitive to cold and frost, being defoliated below 10 °C and killed by temperatures of a few degrees below freezing. It can be grown from sea level up to 1000m elevation. Custard apple requires hot dry climate during flowering and high humidity at fruit setting. Flowering occurs during May but fruit setting takes place on onset of monsoon. Low humidity is harmful for pollination and fertilization. It does the best in dry areas and it has high drought tolerance. In severe droughts, the tree sheds its leaves and the fruit rind hardens and will split with the advent of rain. Annual rainfall of 50-80 cm is optimum, though it can withstand higher rainfall. Rain's during flowering period is harmful.

The custard apple is not very particular about soil conditions and the tree thrives naturally in rocky terrain with shallow, gravelly, well-drained soils. However, they do extremely well in arable, red, sandy shallow soils slightly acidic in nature. Heavy soils and waterlogged fields are not suitable for its cultivation. It does well in calcareous soils with as high as 50% lime content. The tree is shallow-rooted and doesn't need deep soil.

The important features of the annonaceous fruits are given in a tabular form below.

Botanical name	Varieties	Climate	Propagation	Fruit character
A. squamosa	Red, Red-speckled, Crimson,Yellow, White-stemmed, Mammoth, British Guiana, Barbados Balangar,Washington, Mazar,Abd El Razik , Kakarlapahad	sea level up to 1000m elevation, Annual rainfall of 50-80 cm,	Seed and grafting on the same species or on *A.reticulata*	Globular medium sized fruits, sweet creamy white pulp and green skinned
A. reticulata	Tikal,Canul, Sartenaya,San Pablo,Benque, Caledonia,Chonox	between 0 and 1500 m amsl	Seeds	Large, heart shaped, sweet and creamy fruits with smooth thin skin and small seeds
A. muricata	-	between 0 and 1000 m amsl	Seed and grafting on *A.purpurea* or *A.montana*	Very large dark green fruits, heart shaped, fleshy spines, sweet and sour in taste
A. cherimola	Bay Ott, Chaffey, Dr White, Libby, Nata, Orton and Spain.	between 700 and 2400 m amsl	Seed and grafting on *A. squamosa* or *A. glahra;*	Medium sized fruits, sub globular, pale green, sub acidic
A atemoya	Red Sugar, M-2, Gefner	sea level up to 1000m elevation, Annual rainfall of 50-80 cm,	Seeds	Large sized green globular fruits, white smooth juicy pulp
A. diversifolia	Fairchild, Rosendo Pérez, Guillermo, Gramajo, Imery, Pajapita, Nilito, Roman, Genova white, Efrain.	between 0 and 1800 m amsl, an annual rainfall between 1000 and 1400 mm and very fertile volcanic soils.	Seeds after subjecting to gibberellic acid or hot water	Fruit colour and its texture range from slightly pasty to juicy, soft with concentrations of harder grain and sweet taste, with a typical aroma.

Propagation

Traditionally propagated by seeds, this fruit tree can be multiplied vegetatively by grafting and budding. . Seeds have a relatively long life, keeping well for 3 to 4 years. They germinate better a week after removal from the fruit than when perfectly fresh. Germination may take 30 days or more but can be hastened by soaking for 3 days or by scarifying.

Vegetative propagation is practiced where the crop is important and early fruiting is a distinct advantage. The seedlings of local custard apple have proved a good root stock for many improved varieties and hybrids. Seedlings may be budded or grafted when one-year old. In India, selected clones grafted on *A. reticulata* seedlings have flowered within 4 months and fruited in 8 months after planting out, compared with 2 to 4 years in seedlings. The grafted trees are vigorous, the fruits less seedy and more uniform in size. The custard apple itself ranks next to *A. reticulata* as a rootstock. In India, budding is the best when done in January, March and June. Results are poor if done in July, August, November or December unless the scions are defoliated and debudded in advance and cut only after the petioles have dehisced. Side-grafting can be done from December to May which requires good skill and the rate of success is less (60%). Shield-budding gives 75% success and is the only commercially feasible method. Cuttings, layers, air layers have a low rate of success, and trees grown by these techniques have shallow root systems and cannot endure drought as the seedlings do.

Planting

Planting is done during rainy season. Well-conditioned plants of about 30 cm height can be used for planting in the main fields. Pits of 60 x60 x 60 cm are ideal for planting the young seedlings or grafts. A spacing of about 5 x 5 m is ideal for commercial orchards. About 400 plants can be accommodated in a hectare. The pits are filled with ripe organic manure, vermi compost, powdered rock phosphate, powdered neem cake and a host of beneficial microbial bio-fertilizers to get the best establishment, growth and yield from the plants. With drip irrigation system, planting at 6 x 4 meters gives good growth and better fruit setting.

The gap filling should be done as early as possible. Stagnation of water during first monsoon should be taken care of, if the planting is done on impervious soils or ill drained soils.

Growth Regulator

For uniform flowering and for early flowering and also to check the flower and fruit drop and for improving fruit size, following growth regulators are used.

i. For defoliation and bringing the plants under uniform rest Ethryl (1000 ppm) is sprayed one month after harvest of the fruits.

ii. For better and early flowering, Biocil 1 ml per litre of water is sprayed just prior to flowering.

iii. NAA 10 to 20 ppm is sprayed just prior to flowering period to reduce the flower and fruit drop.

iv. During fruit development GA 50 ppm as foliar spray improves the fruit size and luster of the fruits.

Irrigation

In general, custard apple is grown as a rainfed crop, and no irrigation given. However, for early and bumper harvest of the crop irrigation during flowering i.e. from May should be given till regular monsoon starts. For better flowering and fruit setting, mist sprinkling is better over flood or drip system of irrigation as it helps to lower down the temperature and increase the relative humidity. Watering is done in the initial phase of establishment, and being a drought-tolerant crop, it can remain green and healthy even during long periods of dry spells. Irrigation during the dry season and once during ripening will increase fruit size.

Weeding

For good plant growth, weeding should be done to keep away the weeds. Intercropping with some legumes, peas, beans and marigold flowers are done by the growers. Normally, no crop is taken during winter as the plants go under rest.

Fertilizer Application

Generally, no manures or fertilizers are applied to rainfed crop. However, for early and bumper harvest with a good keeping quality, the following dose is recommended to a fully grown tree. Bonemeal 10 kg, 5:10:5 mixture of NPK 1 kg, micronutrient mixture 0.250 kg at the time of flowering, another dose of 10:26:26 or 19: 19: 19 mixture of NPK after fruit setting and a foliar spray with 8:12:24 10 g/l, twice during fruit development. Occassionally, zinc or iron or both deficiencies are noticed and can be taken care of by spraying chelated zinc or iron. On light soils, application of 60-80 kg organic manure and commercial fertilizer containing 3% N, 10 % P and 10% K/tree annually significantly increases flowering, fruit set and yield.

Pruning

Judicious pruning to improve shape and strength of tree must be done only in spring when the sap is rising, otherwise pruning may kill the tree.

Harvest and Yield

Custard apple starts bearing fruits from fourth year of planting, and yield declines gradually after the fifteenth year. It yields fruits during August to October in South India and during September to November in the Northern parts of the country. The custard apple is a climacteric fruit and harvested at the maturity stage when the fruit starts to change colour from green to its varietal colour shade. Immature fruits do not ripen. The inter-areolar space widens, and the fruits turn creamy white. The fruits are harvested manually when they are fully mature and still firm. About 5 to 6 pickings will be needed to harvest all fruits during a season. On an average, a well nurtured, seven-year-old tree will produce about 100 to 150 fruits, and a total yield of about 7 tonnes can be got from one hectare. With age, the fruits become smaller and it is considered the best to replace the trees after 10 to 20 years. The fruits will not ripen but just turn black and dry if picked before the white, yellowish or red tint appears between the rind segments, the first sign of separation. If allowed to ripen on tree, the fruit falls apart. The fruits ripen in about three days after plucking. The medium sized fruits of 250 g to 300 g with sweet pulp and pleasant aroma will have about 60 to 80 seeds each.

In bullock heart, the fruits are harvested after its colour changes from green to pink and still the fruits are firm. Within 2 to 3 days, the fruits will ripen and the skin becomes very soft which can be determined by feel. The skin is very thin and the fruit must therefore be handled carefully.

Storage

The fruits do not withstand cold storage as well as handling after ripening. Firm but mature fruits can be kept at 6°C temperature for about a week but such fruits loose aroma and attractiveness and also develop some bad taste. In India, mature fruits treated with 50-60 g carbide ripened in 2 days and thereafter remained in good condition only for 2 days at room temperature, while those packed in straw ripened in 5-6 days and kept well for 4 days.

Storage trials in Malaya indicated that ripening of custard apples can be delayed by storage at temperatures between 15°-20°C and 85-90% relative humidity, with low oxygen and ethylene levels. To speed ripening at the same temperature and relative humidity, levels of oxygen and ethylene should be high. Storing at 4°C for 5 days resulted in chilling injury.

Marketing

Most of the fruits of bullock heart is produced for family consumption and it is not commonly found in the markets for sale.

APPLE

Apple (*Malus pumila* Mill.) is the most important temperate fruit grown in India in terms of production and extent of area. It belongs to family Rosaceae. In India, Jammu & Kashmir and Himachal Pradesh are the two major apple producing states and apple based industries form the backbone of the economy of these states. At present, India has only a very minimal share in the total world apple production. Apple originated in South-West Asia and was introduced into the country by the British in the Kullu Valley of H.P. in 1865, while the colored 'Delicious' cultivars of apple were introduced to Shimla hills of the same State in 1917. The apple cultivar 'Ambri', is indigenous to Kashmir and was grown long before Western introductions.

About 99% of India's apple area falls under North Western Hill region, covering Jammu & Kashmir, Himachal Pradesh and Uttar Pradesh. In the North-Eastern Hill region, good quality apple is grown in a small area in Tawang belt of Kameng district in Arunachal Pradesh. The Tawang area is basically a rainshadow belt and therefore, permits a longer period of sunshine and freedom from heavy rains, making it ideal for apple. Apple is also grown in Sikkim and Nagaland but the production is not a major success. Although there has been 5-6 fold increase in apple production during the last 50 years, the productivity level is still very low (5.56 t/ha).Presently, a small quantity of apple produced in India is exported, mainly to Bangladesh and Sri Lanka.

Varieties

Systematic research on apple was started at various Research Institutes with assistance from Indian Council of Agricultural Research in late fifties, which now forms the nucleus of research in various states. Over 700 accessions of apple, introduced from USA, Russia, U.K., Canada, Germany, Israel, Netherlands, Australia, Switzerland, Italy and Denmark were tried and tested during the last 50 years. The Delicious group of cultivars predominated the apple market. In more recent times, improved spur types and standard color mutants with 20-50% higher yield potential are favoured.

The important selections are:

- Spur types - Red Spur, Starkrimson, Golden Spur, Red Chief and Oregon Spur.

- Color mutants - Vance Delicious, Top Red, Skyline Supreme.
- Low chilling cultivars - Michael, Schlomit.
- Early cultivars - Benoni, Irish Peach, Early Shanburry, Fenny
- Juice making cultivars - Lord Lambourne, Granny Smith, Allington Pippin.
- Scab resistant cultivars - Co-Op-12, Florina, Firdous, Shirean.

Apple varieties fall into two categories: diploids and tripolids. Diploids have plenty of good pollen and are self-fruitful. Triploids are self-unfruitful and become productive only when pollinated by using suitable pollinizer varieties. Even self-fruitful varieties have to be interplanted to get commercial crops through cross-pollination. Varieties selected for interplanting should sufficiently overlap in their blossoming periods. A few of the important varieties in various states are listed below:

J&K	H.P.	U.P.
Benoni, Irish Peach, Cox's Orange Pippin, American Apirouge, Golden Delicious Kashmir Valley Red Delicious (diploid), Baldwin (triploid), Ambri Kashmiri, White Dotted Red and Blood Red.	Tydeman's Early, Michael Molies Delicious, Schlomit Starkrimson, Granny-Smith, Red Spur, Top Red, Red Chief, Oregon Spur, Golden Spur,Red Delicious, Golden Delicious, Worester Pea rmain, Newton Wonder (all diploids), Cox's Orange Pippin (triploid), King of Pippins (No. 13), Starking (Royal) Delicious and Richared Simla Hills: \`Beauty of Bath' (triploid), Red Delicious', \`Jonathan', \`Rome Beauty' (all diploids), \`Early Shanburry', \`Red Astrachan', \`Red Sudeley', \`Stayman Winesap', \`Winter Banana' and \`Yellow Newton'  Kulu Valley: \`Ben Davis', \`Red Delicious', \`Golden Delicious', (all diploids), \`Cox's Orange Tippin', \`Blenheim Orange', \`Baldwin' (all triploids), \`Red Astrachan', \`King of Pippins' \`Yellow Newton' and \`Granny Smith'. Nilgiri: \`Rome Beauty' (diploid) and \`Irish Peach'.	Chaubattia Princess, Fenny Benoni, Red Delicious, Starking Delicious, Rymer, Buckingham Kumaon Hills: \`James Grieve', \`Jonathan', \`Rome Beauty' (all diploids), \`Blenheim Orange Pippin', \`Delicious', \`Early Shanburry', \`Golden Pippin', \`King of Pippins', \`Rymer' and \`Winter Banana'.

The hybdridization programmes in apple were initiated during sixties in Himachal Pradesh and Jammu and Kashmir to combine high dessert quality of delicious group and excellent keeping-quality and shape of Ambri, an indigenous cultivar of Kashmir. As a result of this programme, Lal Ambri (Red Delicious x Ambri) and Sunehari (Ambri x Golden Delicious) were released in 1973 in Jammu and Kashmir. In Himachal Pradesh, the hybridization work resulted in the release of 4 hybrids – Ambred (Red Delicious x Ambri), Ambrich (Richard x Ambri) and Ambstarking and Ambroyal (Starking Delicious x Ambri). Ambred, a late-maturing hybrid is the most promising. At Chaubattia in Uttar Pradesh, the hybridization programme for early maturity resulted in the release of Chaubattia Anupam and Chaubattia Princess (Early Shanburry x Red Delicious) as promising early maturity cultivars.

Climate and Soil

Apples are generally grown at elevations from 1600 to 2300m in India. In Kulu valley where the climate is cooler, it can be grown even at an elevation of 1200m. It requires chilling below 8°C for a period of two to three months to break the rest period. Temperature below -4°C is injurious. Areas which are frost free, receiving good sunshine during summer months and well distributed rainfall of 100-125cm throughout the growing season is most suitable for apple cultivation.

Fertile sandy-loam to clay loam soils with pH range from 5.5 to 7.5, free from water logging conditions and having good aeration are suitable for establishing deciduous fruit orchards.

Propagation

Apple is propagated mainly on seedling stocks of apple (*Malus pumila Mill*). Clonal rootstocks of M and MM series were introduced from East Malling Research Station, U.K. M-9, M-26, M-4, M-7, MM-106 and MM-111 were identified as promising. Merton 779 was recommended as a commercial rootstock for apple for Kumaon hills of Uttaranchal. However, clonal rootstocks have not found favour among apple growers. *In-vitro* protocols of M-9, MM-106 and MM-111 were developed. Clonal rootstocks are preferred only for high density orchards. Four rootstocks, namely, EMLA-111 for drought prone areas, EMLA-7 for slopy land and EMLA-106 for slopy and less clayey soil and EMLA-9 for high density planting on irrigated deep soils have been recommended. For raising rootstock seedlings, seeds of crab apple or commercial cultivars are stratified during December for 2-3 months at 2-5°C. One year old seedlings are used for budding/grafting.

Clonal rootstocks are raised through mound or stool layering. The mother plants are allowed to grow for one season and cut back to 3 cm from the ground level just before the growth begins. When new growth is about 10 cm, the shoots are covered with soil leaving the growing parts exposed. Rooted layers are cut off close to the ground level and planted in nursery beds for grafting/budding. In H.P. hills, different types of budding/grafting are recommended, chip budding in mid-June and mid-September, Whip and tongue and cleft grafting in February - March and T-budding in May-June.

Planting

Most of the apple cultivars especially of the Delicious group and its spur types and colour strains need adequate cross pollination for optimal yields. Research on the proportion and distribution of pollinizers indicated that 20-25% pollinizers are adequate for Delicious group of varieties. In marginal conditions where fruit set is adversely affected by adverse climatic conditions, the proportion of pollinizers should be increased to 33-50%. Studies showed that the apple cultivars grafted on seedling rootstocks by tongue grafting during February-March gave more than 90% grafting success. Mound layering was convenient for propagation of clonal rootstocks of apple, however, the propagation through hardwood cuttings was also quite successful. In apple, success of 70% was achieved in rooting of hardwood cuttings of crabs with IBA (2,500 ppm) and bottom heat treatment (21° ± 2° C) during dormant season in Chaubattia. Greater stress is now being given for propagation of healthy plant materials for improved productivity of new orchards. Virus infection is of great concern in apples as most viruses are graft transmissible, latent in expression, significantly reducing productivity and are incurable.

In flat lands/valleys, square, rectangular or triangular planting may be adopted. In hill slopes, planting on contours or terraces is recommended. In shallow slopy lands, small terraces (half-moon terrace), may be made to establish the plants and large scale disturbance of surface soil need to be avoided. Planting of deciduous fruit plants is done during winter from end of December to mid March. Planting is done in the centre of pits (1x1x1m cube or circle), prepared a month before planting. While refilling the pit, 50g of aldrin powder (15%) is mixed with soil. After planting, the young plants should be supported with stakes. Spacing varies from species to species and depends also on the type of rootstocks used. For apple raised on seedlings of crab or other commercial varieties of apple, a planting distance of 6x6m or 7x7m accommodating 277 to 205 plants/ ha is recommended for J&K hills. On clonal rootstocks like M4, M7, M26 and MM106, a spacing of 4.5 x 4.5m (555 plants/ha) is suggested. For spur type varieties and standard colored mutants of apple, high density planting on dwarfing rootstocks like M9, M4, M7 and MM107 is feasible. Fruit yield of 30-35 tonnes/ha was achieved in 12 year old orchards of color mutants of apple on MM106 under a planting density of 2222 plants/ha (3 x 1.5m) in cooler hills of H.P. Spacings recommended for apple are shown below:

Optimum Plant Spacing for Spur Types and Standard Color Mutants of Apple

Types	Rootstocks	Spacing (m × m)	No. of trees/ha
Spur Type	Seedling (Crab)	5×5	400
Spur Type	MM111, MM109	4×4	625
Standard Type	MM106, MM109	5×5	400
Spur Type	MM106, M7	3×3	1111
Standard Type	M9	2×2	2500

Source : Awasthi and Chauhan (1997)

Training and Pruning

Modified central leader system of training is recommended for apple on seedling rootstocks. The proportional heading back and thinning out system of pruning are adopted after the juvenile phase of plant growth. Spur pruning encourages vegetative growth and helps in new spur development in old plantations. For high density planting on semi-dwarfing and dwarfing rootstocks, spindle bush, dwarf pyramids and cordon system of training are suggested. Spindle bush on M7 and modified central leader on MM106 rootstocks are successful.

Fertilizer and Nutrient Management

Nutrient requirement varies from place to place. In Himachal Pradesh, widespread deficiencies of N, P, K, Ca, Mn, Zn and B are recorded. A fertilizer dose of 700:350:700 g of N, P and K for full grown bearing apple tree is recommended. A significant positive correlation was observed in available N, P, Ca and Mn in soil and fruit yield. Nutrient interaction studies showed that P is a strong antagonist of Zn. The studies indicated that leaf nutrient contents and available nutrients in soil are responsible for variability in fruit yields to the tune of 20.2 and 31.4% respectively. Recent studies in Himachal Pradesh showed that foliar application of urea (0.5%) in combination with Triton x - 100, given after 5 hr pre-treatment of foliage with ethyl oleate (2%) further increased N-absorption.

Water Management

Apple orchards in India are situated in rainfed sloppy lands where irrigation facilities are inadequate. However, success of apple culture depends mainly on uniform distribution of rain during the year or where supplemental irrigation during the critical periods is available. More critical periods of water requirement are April to August, and peak water requirement is after fruit set. In areas, where irrigation is not a limiting factor, 114 cm water during whole year through 19 irrigations are recommended. It includes one irrigation during December-January after manuring, one before flowering in April, followed by irrigation at 7-10 days interval during critical summer and later on at 3-4 weeks interval till onset of dormancy. In general, 8 irrigations are recommended for Starking Delicious apple during critical period.

Trickle irrigation at 75% of field capacity results in better tree growth and higher fruit yield in apple when raised on semi-dwarfing rootstocks. Experts also recommend drip irrigation in areas with scarce irrigation sources. Drip irrigation saves water considerably. Field trials indicated a total irrigation requirement of 3840 litres of water/ tree under conventional system of irrigation and under drip system 1695 litres of water was enough. The application efficiency under drip system is about 2.27.

Weed Management

In apple orchards, grass mulching (10cm thick) with one application of post-emergence herbicide Glyphosate (0.8 kg/ha) was effective. Gramaxone (1,000 ppm) or a mixture of 2, 4, 5 - T (100 ppm) + Gramaxone (500 ppm) is effective in controlling shrubby weeds at Mashobra. Diuron @ 2 kg/ha, Tok E-25 @ 4 litres/ha and Trafazine at 4 kg/ha checked weed growth in apple nurseries. Use of important herbicides in apple is given below:

Crop	Herbicide	Dose (kg/ha)	Time of application
Apple nursery	Diuron	0.8-4	Pre-emergence
	Nitrofen	0.5-1	Pre-emergence
	Simazine	0.2-4	Pre-emergence
	Terbacil	1-3	Pre-emergence
Apple orchard	Atrazine, Diuron, Simazine	2-6	Pre-and post-emergence
	Dalapon	4-10	Post-emergence
	Lasso	2.5	Pre-emergence
	Paraquat	0.75-1.0	Post-emergence
	Terbacil	1-3	Pre-and Post-emergence
	Tok E-25	1-2	Pre-emergence
	2, 4-D	1-2	Post-emergence
	2, 4-D (Na salt)	4	Post emergence
	Atrazine or Diuron + Glyphosate	4 + 0.8	Pre-and post-emergence

Mulching

Moisture conservation was maximum under grass mulching and, therefore, recommended for apple in H.P. Mulching with oak leaf was effective in U.P. Results of extensive studies carried out on mulching, herbicides and clean basin management in apple in Himachal Pradesh and Uttar Pradesh indicated that mulching followed by herbicidal application are the most effective for floor vegetation management and soil-moisture conservation. Oak leaves as mulch material are effective for growth and fruiting at Chaubattia whereas, hay mulch is beneficial in Himachal Pradesh. Black alkathene mulch in cooler climatic conditions is more effective in weed control, moisture conservation, reduction in fruit drop and improvement in fruit size, colour and quality. In warmer climatic conditions, the black alkathene mulch has adverse effect on root growth due to rise in soil temperature. A few of the effective floor management methods in apple are given below:

Effective Floor Management Methods In Apple	
Mulch materials	Methods of application
Grass	10 cm thick in tree basin after spring rains
Grass plus herbicide	10 cm thick mulch followed by glyphosate at 0.8 kg/ha
Composted conifer leaves	10 cm thick mulch
Composted conifer leaves plus herbicides	10 cm thick mulch followed by glyphosate at 0.8 kg/ha
Black alkathene	1.1 m wide 100 gauge thickness spread over basin

Nutrient Deficiencies

The deficiency symptoms of different elements are given below:

Nitrogen: The deficiency of nitrogen may cause restricted growth, narrowing of top-root ratio, less elongation of branches and reduction in size and colour of fruits. The bark of the tree may be reddish or yellowish green in early stages, but later developing tints of yellow.

Phosphorus: Shoots short, thin and upright. Leaves become small with dull purple bronze tints. Early defoliation of older leaves. Opening of buds in the spring is delayed. Lateral buds may remain dormant, therefore, a few lateral shoots appear.

Potassium: Deficiency of potassium is seen on spurs of old branches which progressed towards the younger leaves as the season advanced. The first evidence of scorch is the loss of normal green colour at the widest part of leaf margin, accompanied by a water-soaked or stained appearance, which was followed by necrosis. Severely affected trees were mostly stunted in growth with small-sized leaves. The fruits in affected trees do not develop properly, remain small in size with poor colour development.

Calcium: The young foliage of calcium-deficient plants showed an upward cupping of margins. In older leaves, the margins turn necrotic and they may fall. Death of growing points followed by die-back is also observed. In bitter pit, slight depressions appear on the skin usually towards the calyx end of fruit. These areas and the flesh below these areas turn brown and corky. Symptoms become more pronounced in the storage conditions. In cork spot, hard brown spots appear on surface skin. Flesh and these spots become hard due to abnormal tissue proliferation. Measles on bark result from a complex involving low calcium, high manganese and sometimes low boron. The deficiency can be corrected by giving one spray of calcium chloride (0.5%).

Magnesium: In severe cases, symptoms shown are marginal scorching of leaves. Severe defoliation of terminal shoots progresses from base to tip. Green colour starts fading on terminals of older leaves progressing interveinally towards the base and midrib, giving the very typical 'Herringbone' appearance to the leaves. In severe cases the fruit may fail to mature and drop early.

Zinc: Buds along shoots fail to develop. Leaves remain small and narrow (little leaf) tending to form rosette at tips. Foliage is sparse throughout the tree. Older leaves may fall in severe cases leaving tufts at terminals. Fruits become small, pointed and mis-shapen. The deficiency can be corrected by spraying $Zn\ SO_4$(0.5%) in two sprays at 15 days interval during May to June.

Boron: Symptoms normally appear on fruits before vegetative parts are affected. Fruits do not develop normally and become mishapened caused by depression usually underlaid by hard corky tissues which rapidly turn brown when exposed. Surface colour is dark green. Leaves become dark green, thick and brittle, starting at shoot tip. Boron deficiency is also known as corky pit, corky core, spot necrosis and drought spot. For correction of Boron deficiency, one or two sprays of Boric acid (0.1%) at 15 days interval in of June is sufficient.

Iron: The deficiency symptoms first appear on young leaves at tip of shoot. It is characterized by interveinal chlorosis of young leaves. In severe cases, the new leaves may unfold, but completely devoid of green colour. The veins, however, may turn green later.

Manganese: The characteristic chlorosis first appears at leaf margins and progresses interveinally toward the midrib. Early defoliation occurs particularly on the top of tree. Fruits remain poorly coloured. Controlled by giving one or two sprays of manganese sulphate (0.4%) at 15 days interval in June.

The following corrective measures for nutrient deficiencies are suggested for apple.

Corrective Measures for Nutrient Deficiencies in Apple

Elements	Chemical & Dose	Time of Spray
N	Urea, 5.0%	Pre-petal fall
Ca	Ca Cl_2, 0.5%	30-45 days before harvest
Zn	$ZnSO_4$, 0.5%	After petal fall
Mn	Mn SO_4, 0.4%	After petal fall
B	H_3BO_3, 0.1%	Before bloom and after bloom

A new dimension to nutritional research in apple is added recently by undertaking studies on role of biofertilizers, in particular mycorrhizal fungi. Symbiotic association of vesicular arbuscular mycorrhizae (VAM) in apple roots is well established which influences plant growth through improved availability of limiting soil nutrients. In general, VAM fungi increase phosphorus uptake by apple roots besides possibilities of increased uptake of micronutrients like Zn and Cu by certain selected strains of VAM. Four strains of VAM fungi *Glomus mosseae, G. fasciculatum, G. macrocarpum* and *G. versiforme* are tested on apple seedlings with and without addition of Zn in soil.The preliminary information indicated that *G fasciculatum* has the highest potential of mycorrhizal infection and colonization, seedling growth and biomass of apple seedling. The incorporation of Zn to soil up to 5ppm level was helpful in further improving colonization of VAM and seedling growth, indicating more availability of Zn from soil, which is otherwise given as foliar spray only in deficiency. The high levels of Zn in soil, ranging from 7.5 to 10 ppm, showed adverse effect on VAM fungi infection as well as subsequent seedling growth.

Intercropping

As apple is grown in hill slopes, seasonal intercropping accompanied by digging the soil is not encouraged to prevent soil erosion. Where slope of land is more than 10%, various grasses and legumes are grown as permanent covers. The tree basins are kept clean and sod culture in the orchards are practiced in the hills. The grasses are regularly cut when 7-10 cm high and the residue is left in the orchards for decomposition. Cultivation of legumes and grasses as intercrop are popular in orchards. The important legumes are red clover (*Trifolium pratense*), white clover (*T. repens*) and lucern (*Medicago sativa*). Among the grasses, orchard grass (*Dactylis glomerata*), fescues (*Festuca arundiceae*) and timothy (*Phleum pratense*) are common. Orchard grass is grown where moisture is high, while fescues are suitable for dry areas or southern aspects. In apple, sometimes filler trees of peaches or dwarf apple plants are grown.

At the young age of orchard plantations, vegetables like cole crops, potato and tomato are grown to supplement the income. However, this practice is abandoned when the orchard starts bearing after 5-7 years to avoid competition. In Shimla hills and Kullu Valley areas of H.P. cultivation of gladioli for flower and bulb production as intercrop are carried out in apple orchards since it tolerates shading.

Fruit Thinning

In apples, heavy bearing not only results in small-sized poor quality fruits but also sets in alternate bearing cycle. The judicious thinning suitably used at proper stage of fruit development can regulate cropping and improve fruit size and quality. In apples, chemical methods are employed using plant growth regulators, dinitro compounds and insecticides for fruit thinning, besides hand thinning which is cumbersome. Carbaryl (1,500 ppm) applied at 3 weeks after petal fall induced 61.5% thinning in Red Delicious apples whereas 2, 4, 5 -T could induce 37.6% thinning. In Golden Delicious, application of NAA (10 ppm) and Carbaryl (750 ppm) at petal fall was effective for optimal fruit thinning in Himachal Pradesh. At Chaubattia Carbaryl (1,000 ppm) induced optimal thinning and had appreciable effects on thinning.

Physiological Disorders

Bitter Pit: The symptoms appear as small, dry, brown pockets usually spherical in shape below skin and also in cortex. Cultivars like Northern Spy and Delicious are more prone to this disorder. Large fruits and immature ones are more susceptible than others. Heavy doses of nitrogenous fertilizers and heavy pruning favour bitter pit. Storage at 0-2°C under high humidity (85-95%) controls pitting.

Scald: It first appears as light mottling on greener surface of fruit. Darkening becomes more severe with elapsed time and ultimately extend to red surface also. Scald usually affects the skin only but in severe cases, it may extend to fruit flesh. The immature fruits are the most susceptible to scald which is aggravated by warmer temperature. Light scald fruit in storage may be severely affected during marketing.

Internal Browning: It is characterised by brownish streaks radiating into flesh from the core. It can be controlled by storage at 4°C. Controlled atmospheric storage with higher temperature can also be helpful in the control of this disorder. It is mainly associated with Yellow Newton apples.

Fruit Drop

Pre-harvest fruit drop in apple could be effectively checked with application of NAA (10 ppm) 20-25 days before harvest. Pre-harvest fruit drop is quite serious in early ripening cultivars like Tydeman's Early, Red Gold and Pippins, where loss of 40-60% crop load is often experienced. Mid-season Delicious group cultivars also experience 15-20% loss.

Most of the commercial varieties of apple are noticed to have the following three waves of fruit drop:

- Early drop
- June drop, and
- Pre-harvest drop

Application of NAA (10 ppm) before the expected fruit drop or 20-25 days before harvest can check fruit drop effectively.

Alternate Bearing

Apple cultivars Royal Delicious, Red Delicious and Golden Delicious have the tendency to bear heavy crop in a year followed by poor cropping season and setting a rhythm of irregular bearing.Once this pattern of irregular bearing sets in, the fluctuating rhythm of yield continues year after year. The alternate bearing can be checked to certain extent by adopting following practices :

i. Thinning of fruit in one year often increases the return bloom in following year. In a heavy cropping year, thinning is particularly essential for return bloom as trees become exhausted due to heavy crop.

ii. Nitrogen application early in the 'off' year has a slight tendency to promote annual bearing in trees having low-nitrogen level.

iii. Judicious pruning of spurs in winter preceding to 'on year' reduces crop loss.

iv. Application of growth retardant like SADH (2000 ppm) or paclobutrazol (150 ppm) in June-July before the flower bud differentiation takes place in an 'on year'.

Harvesting

Apple is harvested during July-Sept in Uttrakhand and during July-Nov in Himachal Pradesh and Jammu & Kashmir. Harvesting at the correct time is essential for production of quality apples. To ensure maximum storage life, apples should be harvested when mature but not yet fully ripe or overripe. If harvested before they have matured, apples will have poor eating quality and will be more susceptible to storage disorders such as scald, cork spot and bitter pit and may not ripen properly. Ripe fruit should be avoided because it continues to ripen in storage, rapidly becoming too soft and mealy for sale. Firmness and the level of soluble solids in the apple are good indicators of maturity to use in determining picking time. Apples are very susceptible to bruising and other forms of mechanical damage and therefore should be handled very carefully. Workers harvesting apples should be cautioned not to drop them or handle them roughly. Effects of bruising and scuffing cannot be reversed. Damage from rough handling accelerates deterioration, reducing the value of product. Apples are normally transported and stored in bulk boxes in the orchard. Each of these boxes holds about 600 pounds of fruit. Full boxes should not be allowed to sit for extended periods in direct sunlight nor for more than a few hours before cooling is started. They also should not be overfilled. When overfilled boxes are stacked, many apples throughout the box are bruised. Bulk boxes represent a substantial investment for most growers and processors. They should be kept clean and in good repair and should be stored indoors when not in use. Most bulk boxes made from hardwood have enough open space between the individual boards in the bottom and sides to ensure adequate air circulation and water drainage. For some of the newer boxes built from plywood, it may be necessary to drill holes to improve air circulation. Five to 8% of the lateral surfaces and 3 to 5% of the bottom should be open. Better circulation is provided by many small, well- distributed holes than a few large ones.

In apple, the number of days taken from full bloom to harvest are 132-134 days for Starking Delicious, 138-140 days for Red Delicious and 147-148 days for Golden Delicious varieties in the high hills (>2000 m) of H.P. In J&K State, the variety Maharaji requires 160 ± 3 days after full bloom for proper harvest maturity. Apple cultivar Granny Smith takes 180 ± 5 days for maturity. The maturity standards for different cultivars of apple are given below:

Maturity Indices for Commercial Cultivars of Apple

Cultivars	**Days from full bloom**	**Firmness (kg)**	**TSS (%)**
Royal Delicious	125 ± 5	8.2 ± 0.40	13.0 - 15.0
Red Gold	122 ± 3	8.3 ± 0.20	12.0 - 13.5
Red Delicious	134 ± 5	8.4 ± 0.40	10.0 - 14.0
McIntosh	135 ± 4	6.8 ± 0.25	11.5 - 13.5
Golden Delicious	148 ± 6	8.4 ± 0.40	12.0 - 14.5
Granny Smith	180 ± 5	8.7 ± 0.30	11.5 - 13.0

In apple, fruit colour development in warmer and lower (<1800 m) elevations is generally poor and fetch poor market price. Application of 2-chloroethyl phosphonic acid (Ethrel, CEPA or Ethephon) (1,000 ppm) about 10 days before harvest improves colour of fruit substantially in Himachal Pradesh but impairs shelf-life. In Jammu and Kashmir, ethephon (250 ppm) along with NAA (15 ppm) sprayed 20 days before expected harvest is beneficial in improving colouration of Red Delicious apples. An application of ethephon (500 ppm) can enhance maturity by 7-10 days when applied 3 weeks before harvesting along with NAA (10 ppm).

There is wide range of variation in fruit yield in different States and in different cultivars. Although average productivity of apple in H.P is 9.4t/ha, some well managed orchards of Delicious apple in Kotgarh area of the same State yield about 50 t/ha. Similarly, yield record of 35 t/ha is achieved in 12 years old plantations of coloured mutant apple cultivars on MM 106 rootstocks under high density planting system (2222 plants/ha) in a research station in Shimla hills of H.P.

Application of Dormex (hydrogen cynamide) (3%), 40 days before bud break, has enhanced flowering and improved fruit set in Starking Delicious apples. Excessive vegetative growth in warm apple growing areas can be checked by paclobutrazol (500-1,000 ppm), which can maintain the balance between vegetative and reproductive growth. Promalin containing gibberellins and cytokinins (30 to 60 ppm) sprayed at king-bloom stage is effective in improving oblong shape of Delicious apples which is a desirable characteristic for better marketing.

Storage

In H.P, two pre-harvest sprays of $CaCl_2$ (0.5%), followed by a surfactant like Tween 20 (0.03%) for 1-2 minutes are recommended for better shelf-life of apple. In apple, for obtaining better grade fruits of optimum size, a post-bloom spray of ethephon (100-200 ppm) 2-3 weeks after full bloom have been recommended both in H.P and J&K.

Apple grown in upper belts of Jammu and Kashmir can be stored for 90 days under ambient conditions whereas apples in lower belts can be stored only for 60 days. It is found that polythene wrapping of fruits significantly improves storage capacity of fast deteriorating Red Delicious apples. Influence of post-harvest infiltration of calcium chloride on keeping quality of Red Delicious apples was also investigated under various storage conditions. Fruits treated with calcium chloride (1%) for 5 minutes registered less pathological and physiological spoilage as compared to untreated fruits. In Himachal

Pradesh pre-harvest sprays of calcium chloride (0.5%) and post-harvest dip (4%) are recommended for improving shelf-life of apples. The pre-harvest sprays should be done 45 and 30 days before harvest, whereas, post-harvest dips should be given for 1-2 min using surfactant like Tween 20 (0.03%). The shelf-life of tray packed apple fruits in corrugated film-board cartons was extended up to 240 days in cold storage at 0° ± 1°C, when packed with a sachet of 'Purafill' or potassium permangnate crystals (5 g/kg fruit) or with small brick pieces impregnated with potassium permangnate (2%) solution.

Grading

In apple, 7 size grades are prescribed and dimensions of packing boxes, size of wrapping papers and number of layers of fruits for each grade are standardized. Presently CFB (Corrugated Fiber Board) telescopic tray pack cartons are used for apples and these are available with HPMC at their various packing stations. Plastic crates (both collapsible and non-collapsible) are also used for cold storage and for processing units.

Picking and packaging of fruits are done in different ways at different sites. In Uttaranchal hills, trees are generally strip picked into small wicker baskets or small gunny bags which are emptied into larger baskets for carrying to packing stations. Both Uttaranchal and J&K have developed packaging units for different temperate fruits.

Marketing

The biggest wholesale market for apple, Fruit and Vegetable is at Azadpur, in Delhi. About 70% of the total trade of apple is conducted through this market. The seasons of market arrivals from North Western States are as follows:

H.P.-July to October with a peak in August - September, J&K - August to November, with a peak in September - October, U.P. - June to October, with a peak in July - mid September.

There are a number of marketing channels, of which the predominant ones are:

a. Farmer - Pre-harvest contractor - Commission agent - Wholesaler - Retailer - Consumer.

b. Farmer - Forwarding agent - Commission Agent - Wholesaler - Retailer - Consumer.

c. Farmer - Commission agent - Wholesaler - Retailer - Consumer.

The sale through pre-harvest contractors is the most important system of marketing. Normally, the small orchardists sell their crop at flowering stage to contractors who organize plant protection practices, picking and packaging of fruits.

The medium and large orchardists prefer to market their produces through channels (b) and (c). In H.P., about 65% of total apple produced is marketed through these two channels. Some Grower's Co-operatives and Government Controlled Marketing Corporations (like HPMC in H.P.) are also involved in apple market. For example, in H.P, the volume of trading handled by Grower's Co-operatives is about 3.5 to 4.0%, while another 2-2.5% of the total produce is marketed by HPMC.

The National Horticulture Board (NHB) of Government of India regularly publishes and announces the wholesale price and market arrival figures of apple in different terminal markets for benefit of growers. The Himachal Pradesh Horticultural Produce Marketing and Processing Corporation Ltd. (HPMC) have set up modern fruit packing houses in different parts of the State. Each packing house has a capacity to grade and pack 55,000 to 2,75,000 cartons of fruits. Cold storage facilities at production centers and terminal markets at Delhi, Mumbai, Chennai and Calcutta are created. Similarly, the HPMC undertook large scale processing of temperate fruits like apples, peaches and others.

India presently exports a small quantity of apple (5.95% of total fresh fruit export), mostly to Bangladesh and Sri Lanka. Apple cultivars like Red and Royal Delicious, Ambri and other new coloured cultivars are suitable for export market. According to the Trade Year statistics of 1993, India exported only 7000 tonnes of apple during1993.

AVOCADO

Avocado (*Persea americana*) also known as butter fruit belongs to family Lauraceae. It is native of tropical America and is not a commercial fruit crop in India. It originated in Mexico and Central America, possibly from more than one wild species. It was introduced to India from Sri Lanka in early parts of 20th century. It is grown around Chennai and Bangalore but has never become very popular because of the preference for sweet fruits. and presently it is cultivated in a scattered way in tropical South Indian states like Tamil Nadu, Kerala, Karnataka and Maharashtra. The crop is also popular in the North Eastern Himalayan state (Sikkim), where it is grown successfully on hill slopes. The name 'Avocado' came from the Aztec name meaning 'testicle tree'.

Avocado is the most nutritive among fruits and is regarded as the most important contribution of the New World to human diet. The fruit is relished by many people, but not by all. The pulp is rich in proteins (up to 4%) and fat (up to 30%), but low in carbohydrates. The fat is similar to olive oil in composition and is widely used in the preparation of cosmetics. Avocados have the highest energy value (245 cal/100 g) of any fruit besides being a reservoir of several vitamins and minerals. Avocado is the fruit of promise in India.

Varieties

All three horticultural races, namely West Indian, Guatemalan and Mexican are tried in India and varieties Fuerte, Hass and Green performed well at different places. The cultivars of West Indian race are grown in localized pockets in Maharashtra, Tamil Nadu and Karnataka. In tropical and near-tropical areas, only West Indian race is well-adapted but its hybrids with Guatemalan (e.g. Booth selection) perform well and are valuable for extending the harvest season. In less tropical regions, hybrids of Guatemalan with Mexican race predominate since they combine the cold hardiness of the latter with the superior horticultural traits of both and also bridge the two seasons of maturity. Both the Mexican and Gautemalan races are grown successfully in Sikkim. The Mexican race is cultivated on mid-altitude hills. Ideal soil pH is 5-6.

The characteristics of major avocado races are:

a. Mexican race: Small (250 g) fruit type, ripening 6 to 8 months after flowering. Fruits have thin smooth skin with a large seed fitting loosely in the central cavity. Oil content is up to 30% in the fruits. Trees of this race are cold hardy. Gottfried, Duke and Pernod are the cultivars.

b. Guatemalan race: Fruits fairly large, weighing up to 600 g and borne on long stalks. The fruits ripen 9-12 months after flowering. Fruit skin is thick and often warty. Seeds, held tightly in the hollow of the fruit, are small. The oil content in fruits ranges between 8 and 15%. Linda, Taylor, Queen, Itsamma and Benik are the cultivars

c. West Indian race: Fruit medium in size and the fruit skin smooth, leathery and glossy. Fruits are borne on long stalks and require up to 9 months for ripening from date of flowering. Its seeds are large, fitting loosely in the cavity. The oil content of the fruit is low (3-10%). This race is suitable for warm tropical climates. Pollock, Waldin, Simmond, Fuchsia, Black Prince and Peterson are the cultivars.

Varieties, cultivated in India are Purple, Green, Fuerte, Hass, Pollock, Peradeniya Purple Hybrid, Trapp, Shambaganur, Bacon, Zutano, Booth 7, Sharwil, Ettinger, Round and Long. Among the several existing varieties, perhaps Fuerte is the most widely grown, but it is regarded as unsuitable for the tropics. The Purple and Green varieties were introduced into India from Srilanka in 1941. The varieties recommended for commercial cultivation are Pollock, Waldin, Nabal, Linda, Puebla, Gottfreid, Collins, Booth 8 and Fuerte. The avocado varieties grown in Sikkim are Fuerte, Pinkerton, Green and Hass.

Climate and Soil

Avocado is grown both in tropical areas experiencing summer rains and in humid sub-tropics with mild winter. Depending on race and varieties, avocados thrives and performs well in climatic conditions ranging from true tropical to warmer parts of temperate zone. Range of mean maximum and minimum temperature in the Niligiri Hills of Tamil Nadu where avocado is grown is 27.2 °C to 33.9 °C and 14.9 °C to 22.2 °C, respectively. In North Sikkim, the avocado growing area experiences temperature range from 12 °C to 30 °C with average rainfall of 2,000 mm. It cannot tolerate hot dry winds and frosts of Northern India. It can be grown up to an elevation range of 800-1,600 m amsl.

Avocados can be grown on a wide range of soils, but they are extremely sensitive to poor drainage and cannot withstand water-logging. They are intolerant to saline conditions. Optimum range of pH is from 5 to 7.

Propagation

In India, avocado is commonly propagated through seeds. Viability of seeds of avocado is quite short (2 to 3 weeks) but this can be improved by storing the seed in dry peat or sand at 5^0C. Removal of seed coat before sowing hastens germination. In Sikkim, all the trees grown are of seedling origin. Seeds taken from mature fruits are sown directly in the nursery or in polyethylene bags.When 6-8 months old, seedlings are ready for transplanting. Such seedling trees at 10-15 years produce 300 to 400 fruits.

Vegetative propagation is still not a common practice in avocados. However budding and grafting methods resulted in the establishment of selected varietal clones. Generally, vigorously growing seedlings irrespective of source are used as rootstocks. Side, veneer, cleft grafting or shield budding on 30-40 cm tall, robust, succulent seedlings are generally practiced.

The Gautemalan race is more sensitive to cold and has also proved more susceptible to high pH, chlorosis and to *Verticillium* wilt. West Indian stocks are preferred in warmer regions or where salinity is a problem. Need of *Phytophthora*-resistant rootstocks is recognized as the most important factor for the success of avocado cultivation. At the Fruit Research Station, Kallar, in the Nilgiri Hills of Tamil Nadu, layering and inarching gave up to 75% success, while in West Bengal, chip-budding is successful. In India, presently, there is no commercial nursery engaged in vegetative propagation of avocado, nor is there any initiative either at governmental or private level to undertake nursery production of avocado planting material.

Planting

If an avocado plantation is to be established in a relatively new area, the varieties to be selected for planting should belong to both A and B groups and their flowering must overlap. The proportion of A and B group varieties can be 1:1 or 2:1. Avocado is planted out to a distance of 6 to 12 metres depending on vigour of variety and its growth habit. For varieties, having a spreading type of growth, like Fuerte, a wider spacing should be given. In areas, prone to excess water, they should be planted on mounds as avocados cannot withstand waterlogging. In Sikkim, a planting distance of 10 x 10 metres on hills slopes (on half-moon terraces) is preferred and planting is done in June-July. Pits of 90 x 90 cm are dug during February-March, and filled with farmyard manure and top soil (1:1 ratio) before planting. In Coorg, a region of Karnataka state, mixed crop planting method is adopted in a coffee-based cropping system.

Training and Pruning

Pruning is rarely practised except with upright varieties such as Pollock. In spreading varieties like Fuerte, branches are thinned and shortened. Heavy pruning promotes excessive vegetative growth, consequently reducing the yield.

Fertilizer Application

Avocados need heavy manuring, and application of nitrogen has been the most essential. In general, young avocado trees should receive N, P_2O_5 and K_2O in a proportion of 1:1:1 and older trees in the proportion of 2:1:2. At a pH of above 7, iron deficiency symptoms may appear which may be corrected by applying iron chelate at the rate of 35 g/tree.

Various micronutrients (Fe, Zn, B) have profound influence on tree growth, nutrient uptake and yield of avocado. Integrated nutrient management with inorganic fertilizer, supplemented by organic manuring, is advocated for avocado.

In Sikkim, the soil is deficient in nitrogen, zinc and boron. Application of urea in two split doses, in March/April and September/October (just before and after onset of monsoon) is recommended. Foliar application of zinc sulphate (0.5%) may be undertaken in April-May, and other fertilizers applied in soil during March-April. A tentative fertilizer schedule recommended for Sikkim (Upadhyay and Srivastava, 1996) is given in a tabular form.

Fertilizer Schedule for Avocado in Sikkim

Manure/ Fertilizer (kg/tree)	Time of Application	Age of Plants (year)						
		1st	2nd	3rd	4th	5th	6th	7th & above
Farmyard manure	February/March	-	10	20	30	40	50	60
Urea	March/April	0.10	0.15	0.20	0.25	0.30	0.35	0.40-0.60
	September/ October	0.10	0.15	0.20	0.25	0.30	0.35	0.40-0.60
Muriate of potash	March/April	0.10	0.20	0.30	0.40	0.50	0.60	0.70-0.80
Bone-meal	March/April	0.30	0.40	0.50	0.60	0.70	0.80	1.00-1.50

Irrigation

Sprinkler irrigation improves the fruit size and oil percentage and also advances harvesting time. Irrigation at intervals of three to four weeks during dry months is beneficial to avocado. To avoid moisture stress during winter season, mulching with dry grass/dry leaves is desirable. Flooding is undesirable as it promotes root rot incidence.

Inter Cropping

In young orchards, suitable vegetable crops can be grown as intercrops to provide additional income to growers. This also helps to control weeds, lowers soil temperature, improves soil structure and fertility by way of adding organic matter and fixing nitrogen.

Harvesting and Yield

Avocado plants raised from seeds start bearing five to six years after planting. Mature fruits of purple varieties change their colour from purple to maroon, whereas fruits of green varieties become greenish-yellow. Fruits are ready for harvest when colour of seed coat within the fruit changes from yellowish white to dark brown. Mature fruits ripen six to ten days after harvesting. The fruits remain hard as long as they stay on the trees, softening takes place only after harvest.

The harvesting season of fruit is from July to September and a tree of 10-15 years old produces 300-350 fruits/ year. The yield ranges from 100 to 500 fruits/ tree.

In Sikkim, specifically, on an average of 300-400 fruits can be harvested from 10-15 years old trees. Fruits of Purple variety are harvested during July, and that of Green variety during September-October. In Tamil Nadu, July-August is the peak harvest time. The yield performance of avocado, both in tropical Southern India and humid sub-tropical North Eastern India is highly satisfactory.

Storage

Avocados do not ripen on the tree, and fruits soften only after they are picked. Fruits need to be picked carefully. They should be harvested at the correct stage of maturity, when they are still hard and have a minimum oil content of 12%. To fasten ripening process, put them in a closed plastic bag together with a banana, apple or pear. A ripened avocado can be judged by hearing the rattling sound of the seed when the fruit is shaken. Cut avocados turn brown when exposed and this can be prevented by sprinkling lemon or lime juice.

Marketing

In general, small to medium-sized fruits are the best for commercial production, especially for metropolitan markets. In India, fruits weighing 250 to 300 grams are preferred. Large fruits are suitable for local use especially by large families. Smooth, thin and green skin is preferred by consumers. The flesh should be virtually fibreless and of agreeable flavor and, for the dieter, of low oil content. The seed must be small and tight so as not to bruise the flesh during handling and shipping. The seed coats ought to adhere to the seed and not to the cavity. Hard and mature fruits are harvested and allowed to ripen during transport and distribution. Up to 14 days transport time is considered satisfactory, though unripe avocados can be stored for up to four weeks if the temperature is kept between 5.5 and 8°C.The fruit should ship well and stand cold storage. Presently, there is no organized marketing system in India for avocado as the production is small and production areas are scattered.

BANANA

Banana (*Musa* sp) is the fourth most important food crop in world after rice, wheat and milk products for human consumption. They are grown in

130 countries worldwide, more than any other fruit crop. Bananas are native to tropical south eastern Asia and are widely cultivated in tropical regions.Banana is known as common man's fruit in India, because there is large area under banana in our country, therefore its production is large and the prices are within the reach of common people. In popular culture and commerce, "banana" usually refers to the soft, sweet "dessert" bananas that are usually eaten raw. The bananas form another group of cultivars with firmer, starchier fruit, generally used after cooking known as plantains.Bananas and plantains belong to the banana family known as Musaceae. According to the taxonomist there are 25-80 species in the genus *Musa*. The genus is not only important for fruit production, but also has provided man with food, clothing, tools, and shelter prior to recorded history.

Banana is a nutritious, palatable and easily digestible fruit. It is available throughout the year. Technically banana fruit is a false berry. It is the oldest and commonest fruit known to the mankind and is known by the names 'Adam's fig' and 'Apple of Paradise'. The fruit is rich in carbohydrates, minerals such as calcium, potassium, Mg, Na and phosphorous. Other than fresh fruits, it can be consumed in various processed forms like chips, powder and flakes. Banana pseudostem is chopped and used as cattle feed and even the leaves are used as plates.

The naming of banana and plantain, and the distinction between the two types is quite confusing. All banana and plantain cultivars were derived from two main species: *M. acuminata* and *M. balbisiana* . Hybrids of *M. acuminata* and *M. balbisiana* are sometimes given the names *Musa paradisiaca* L., *Musa sapientum* L., or perhaps most accurately, *M. acuminata X M. balbisiana* Colla. However, a shorthand method of distinguishing hybrids and accurately representing their parentage was developed in 1950s, and is in common place today. Each type is given a 2 to 4 letter designation consisting of A's representing *acuminata*, and B's, representing *balbisiana*. For example, AA represents a diploid type derived only from *M. acuminata*, and AAB represents a triploid type with 2/3 *M. acuminata* and 1/3 *M. balbisiana* parentage. In general, the most important banana cultivars in the world are AAA, and plantains are mostly AAB, ABB or BBB.

In India, total area under banana is 3.70 lakh ha, next to Mango with a total annual production of 16.91 million tonnes (37% to total fruit production in India). Banana is a globally important fruit crop with 97.5 million tonnes of production. Among the States, Maharashtra contributes the maximum area of about 90,000 ha with a productivity of 60 t/ha. In Maharashtra, area under banana is concentrated in Jalgaon district with nearly 60% area and other districts growing bananas are Parbhani, Khandesh and Thane. Banana fruits from Jalgaon area are marketed in most parts of India, particularly in the north. Other important banana growing states are Kerala, Tamil Nadu, Assam and Andhra Pradesh.

Varieties

In India, banana is grown under diverse conditions and production systems.Large number of cultivars totaling about 70 are grown in different parts of the country. Selection of varieties is therefore based on a large number of characters catering to various kinds of needs and situations. Many cultivars have poor yield potential but are grown for its quality and farmers are compensated with higher price of the produce. The clonal situation prevailing in different regions suggests that varietal need cannot be unified and would need regional consideration. Investigations have also enabled to identify the cultivars for different situations to maximize the returns. However, around 20 cultivars viz. Dwarf Cavendish, Robusta, Grand Naine, Amritsagar, Grosmichel, Monsmarie, Dudhsagar, Monthan, Kanchikela, Batheesa, Poovan, Nendran, Red banana, Njali poovan, Rasthali, Virupakshi, Karpooravalli, Poomkalli, Chakia and Matti are widely cultivated.

Among these varieties, the types belonging to Cavendish group such as Dwarf Cavendish, Robusta and Grand Naine form the basis of commercial cultivation owing to high yield, wide market acceptability, short-crop duration and high economic returns/ unit area. But these cultivars are not common in coastal region due to their high susceptibility to sigatoka leaf spot disease.

Poovan : It is the most important commercial variety in Tamil Nadu, Andhra Pradesh and West Bengal. It is also known as Lal velchi in Maharashtra. It is resistant to Panama wilt and tolerant to drought having wider adaptability.

Njali Poovan : It is grown commercially in Karnataka. Fruits have good keeping quality and fetch double the price than other cultivars.

Rasthali It is a good variety significant in commercial production and its success largely depends on higher price it fetches but it is susceptible to Panama disease.

Virupakshi : Known as the hill banana, it is preferred for its fruits having unique aroma and taste. It is grown mainly under the perennial system of cultivation on hills of Tamil nadu and Karnataka.

Monthan : A culinary variety and a heavy yielder. Commercially cultivated in Tamil nadu, Kerala, Andhra Pradesh and Orissa.

Karpooravalli : Tolerant to drought, salt and wind. Very sweet fruits, ash coated in appearance and does not drop off when ripe and hence suitable for distant transportation.

Nendran : The most priced cooking and table variety grown in Kerala and Tamil Nadu. A large number of products are prepared from this variety. The fruit remains starchy even on ripening. About 14 ecotypes of Nendran are maintained in the banana research station at Kannara in Kerala.

Grand Naine: An introduction is gaining popularity and may soon be the most preferred variety due to its tolerance to biotic stresses and good quality bunches. Bunches have well spaced hands with straight orientation of fingers bigger in size. Fruit develops attractive uniform yellow colour with better shelf life and quality than other cultivars.

In banana, five hybrids are released; BRS 1, BRS 2, Co1, FHIA1 and FHIA3. Of these, BRS 1 and BRS 2 are medium statured in nature suitable for inter cropping in coconut gardens both under rainfed and irrigated conditons in Kerala. The variety Co1 retains the typical apple flavour of Virupakshi even when grown in plains. The hybrid FHIA1 is known for its resistance to Sigatoka and Panama wilt disease.

Climate and Soil

Banana is basically a tropical crop and is grown from sea level to 1000m above MSL. It grows well in temperature range of 13 - 38°C with RH regime of 75-90%. In India, this crop is being cultivated in climate ranging from humid tropical to dry mild subtropics through selection of appropriate varieties like Grand Naine. Chilling injury occurs at temperatures below 12°C. The normal growth of banana begins at 18°C, reaches optimum at 27°C, then declines and comes to a halt at 38°C. Higher temperature causes sun scorching. High wind velocity which exceeds 80 km per hr damages the crop.

Soil for banana should have good drainage, adequate fertility and moisture. Deep, rich loamy soils with pH 6-7.5 are the most preferred for banana cultivation. Ill drained, poorly aerated and nutritionally deficient soils are not suitable for banana. Saline soil and calcareous soils are also not suitable for banana cultivation. Avoid soil of low laying areas, very sandy and heavy black cotton with ill drainage. A soil that is not too acidic and not too alkaline, rich in organic material with high nitrogen content, adequate phosphorus level and plenty of potash are good for banana.

Propagation

Cultivated bananas are parthenocarpic, which make them sterile and unable to produce viable seeds. Due to lack of seeds, vegetative propagation method using suckers is resorted. Banana has a reduced underground stem known as corm, which bears several buds. Each of these buds develops into a new banana plant with an above ground pseudostem and a new corm known as sucker. Usually narrow leaved suckers (sword sucker) with some intact roots are removed and used for planting. The sword suckers weighing 700-1,000 g are optimum. Rhizomes with active lateral buds and dead central buds are preferred for distant transportation in western India. Double paring and shade drying followed by dipping in 0.5% monocrotophos and 0.2% bavistin are recommended to disinfect against nematode and soil-borne fungi. *In-vitro* propagation of banana has been achieved successfully and this method is becoming a commercial reality. However, the adoption of *in-vitro* propagated plants among growers has been slow mainly because of the high costs. However, adoption of drip irrigation and better management system, resulted in enhanced return from *in-vitro* plants. Uniformity in flowering, higher yield and early harvest of *in-vitro* plants have been demonstrated at different locations.

Planting

Prior to planting banana, growing green manure crops like daincha and cowpea and burying it in the soil is beneficial. The land can be ploughed 2-4 times and leveled. Use rotator or harrow to break the clod and bring the soil to a fine tilt. During soil preparation, basal dose of FYM is added and thoroughly mixed into the soil.

A pit size of 50cm x 50cm x 50cm is normally required. The pits are to be refilled with topsoil mixed with 10 kg of FYM (well decomposed), 250 g of neem cake and 20 g of carbofuran. Prepared pits if left open to solar radiation will help in killing the harmful insects, controlling soil borne diseases and also aids aeration. In saline alkali soil, where pH is above 8 potting mixture is to be modified to incorporate organic matter. Addition of organic matter helps in reducing salinity while addition of perlite improves porosity and aeration. Alternative to planting in pits is planting in furrows. Depending on soil strata, one can choose appropriate method as well as spacing and depth at which plant required is to be planted.

Planting of banana is done by two methods viz. pit method and furrow method. In low lying areas, planting is done on mounds. Adjust planting season depending upon local conditions avoiding periods of heavy monsoon and severe summer. Care should be taken to see that high temperature and drought do not coincide with bunch emergence. Usually planting is done during July-August in North India whereas in South-India, it can be done at any time of year except summer.

Planting 3 suckers at a spacing of 1.8 x 3.6m with 300:90:450g/pit of N: P_2O_5: K_2O respectively gave the highest cost: benefit ratio. Tissue-cultured plants performed better with higher yield with reduced crop duration and uniformity in fruits.

Tall varieties should be planted at 3 x 3 m whereas dwarf ones at 2 x 2 m apart. Plant suckers upright in the centre of the pit with 5 cm of the pseudostem seen above the soil level. Press soil around the sucker to avoid hollow air spaces.

Fertilizer Application

Banana is a heavy feeder and responds well to manuring. Banana is a quick growing and short-lived plant. Apply compost, FYM or green leaves at the rate of 10kg/plant at the time of planting. In Tamil Nadu, application of 100 g of N, 30 g of P_2O_5 and 300 g of K_2O/plant in three doses during the second, third and fifth month is recommended. In Maharashtra, 100 g N, 40 g P_2O_5 and 100 g K_2O/plant is recommended. Out of these, P_2O_5 and K_2O are applied at the time of planting and N is given in three split doses: third, fourth and fifth month after planting. In Kerala, the dose recommended for all banana varieties other than Nendran is 160-200 g of N, 160-200 g of P_2O_5 and 320-400 g of K_2O/ plant in two equal split doses; during the second and fourth months. For Nendran 190:115:300 NPK g/plant is to be applied in six split doses at monthly intervals starting from first month of planting to fifth month and the last dose after the complete emergence of the bunch.

Irrigation

During summer, irrigate once in three days. About 6-10 irrigations/ crop may be given depending upon soil conditions. Good drainage should be ensured to prevent water logging. Nendran variety requires 40 litres water/ plant once in two days.

Weed Management

Weeds cause approximately 60 - 70% yield loss in banana. Vegetative phase (1-6 months) is the most critical and weeds during this phase of growth cause considerable loss in productivity. Control of weeds during this period enhances fertilizer-use efficiency and yield. Suppression of weeds using intercrop, and control through integrated management result in increased production. Growing of cowpea [*Vigna unguiculata*] is the most effective in suppressing weed growth. Spray of glyphosate twice proved equally effective.

Mulching

Mulching basin of plant with paddy straw or dry leaves increased bunch yield.

Intercropping

Growing cowpea in the inter spaces not only helps in controlling weeds but also gives more returns by way of increasing fertility status of soil and getting higher yield. Cucumber and amaranth can be cultivated profitably with banana without affecting bunch weight. Elephant foot yam and greater yams can also be profitably inter cropped.

Growth Regulators

Under the present system of cultivation, yield of banana exhibits plateau in commercial plantation, and to increase yield, exogenous application of growth-regulators plays significant role. Among several bio-regulators tried at different locations, application of 2, 4-D (10-20 ppm) improved bunch weight.Bunch covering with polythene in November increased yield by 25-30%. Two sprays of KH_2PO_4 at fruit-development stage also increased bunch weight in Basrai banana. *Azospirillum* enhances height, pseudostem girth, and leaf production and leaf area of Poovan banana. In Elakki Bale, association of mycorrhiza increases yield.

Desuckering

Desuckering is an important operation in banana cultivation. In this, the unwanted suckers, which develop near base of plant, are removed. Cutting suckers from ground level and then pouring kerosene (2-3 drops) will kill the growing point of pseudostem. After bunch emergence, two suckers are to be retained.

Propping

It is an essential cultural practice to give proper support to the plants with bamboos. It is done to avoid falling down of plants due to high winds in tall cultivars.

Wrapping

For protecting fruits from sunburn, hot wind and dust, the bunch is to be covered. Wrapping is also done to improve colour of fruit which will help in fetching more price.Bunches can be covered using polythene sheets of different colour or using dried leaves.

Harvesting and Yield

Owing to the adoption of improved production technologies which include season of planting, selection of planting material, method of planting, high density planting, adoption of drip irrigation, nutrition and integrated management of pests and diseases, tremendous increase in production and productivity in banana have been observed in the past 50 years.

Fruits can be harvested when the bunches are about 75% mature for distant markets and this maturity is attained at 75-80 days after opening of the first hand. Harvest may be delayed upto 100-110 days for local markets. Usually dwarf varieties are ready for harvest in 12 to 15 months after planting, whereas tall varieties take 15 to 18 months. But this period may vary depending on the variety, age of the sucker, shade and the climate of that region. Signs of maturity of banana fruits are, fruit becomes plumpy and angles are filled in completely, when tapped gives metallic sound, drying off of top leaves and change in colour of fruits from deep green to light green.

Entire bunches are cut from pseudostems by hand, and carried on the shoulder or back to a nearby tram line for longer distance transport. The cutter leaves a portion of bare stalk as a handle for transporting to the packing house. Tall varieties like Poovan yield 15-25 tonnes/ha, while Dwarf Cavenshish yield 25-50 tonnes/ha. It can be stored at temperature slightly above 23^{0}C and relative humidity of about 85-95% for about three weeks.

Post Harvest Handling

Grading

Banana bunches are hung on ropes. Hands are cut into units of 4-10 fingers, graded for both length and width, and carefully placed in poly-lined boxes. Fruit are shipped by boat when green, and ripened by exposure to ethylene gas (1000 ppm for 24 hr) at their destination, in sealed "banana ripening rooms".

Storage

Fruit can be stored for a short time after ripening at temperatures not lower than 23^{0}C and relative humidity of about 85-95%, since fruits are susceptible to chilling injury.

Market

Most production for local sale is of green cooking bananas and plantains, as ripe dessert bananas are easily damaged while being transported to market. Even when only transported within their country of origin, ripe bananas suffer a high rate of damage and loss.

CITRUS

Citrus (*Citrus spp*) consists of a group of fruits belonging to Rutaceae or Rue family and cultivation of citrus fruits is termed as citriculture. The three economically important genera of this family are *Citrus, Fortunella* and *Poncirus,* of which *Citrus* is the most economically important one. Fruit of Citrus is known as hesperidium, a leathery rinded berry with juice sacks inside. The juice is a rich source of Vitamin C.

The various species of the genus *Citrus* are believed to be native to the subtropical and tropical regions stretching from eastern Asia southwards to Australia. The genus *Citrus* comprises of many species and hybrids of orange, lemons, limes, grapefruit, citrons, tangerines, citrange (a cross between trifoliate orange and sweet orange). Many types of citrus fruits found in the world breed freely in nature with each other resulting in production of new types. Some of these are indigenous to India and some are from China. However, citrus-types are separated into different botanical species. The hybrid process has also increased complexity of family. The range of citrus is enormous, taking in hundreds of varieties and subspecies to make it the most commonly propagated fruits cultivated in over 100 countries. Oranges account for about 65% of all citrus production, 15% goes to mandarins, 10% to lemons and limes, and 10% to grapefruit. Several other genera within Rutaceae family produce fruits classified as citrus, including the trifoliate orange (*Poncirus trifoliata*) which is a fairly cold-hardy deciduous relative that has inedible fruit and kumquat belonging to the genus *Fortunella* producing bitter fruits. Both the bitter kumquat and trifoliate orange are mainly used as ornamentals.

The important commercial citrus fruits grown in India are mandarin orange (*Citrus reticulata*), sweet orange (C. *sinensis)*], lime (*C. aurantifolia)* and lemon (*C.limon*).These are grown commercially in tropical, subtropical, arid, irrigated and mountaneous regions in varying soil and weather conditions. Although the citrus industry in India has faced many challenges, there is consistent increase in area and production owing to awareness for sustained production. Citrus is grown practically all over India. However, the states of Andhra Pradesh and Maharashtra have the largest share.

Varieties

Large number of species exist in the genus Citrus and they are classified by into different groups. For convenience sake, the important citrus species of commercial importance are grouped as follows.

A. Mandarin group (Loose skin orange)

1. *Citrus reticulata* (Santra)
2. *Citrus unshiu* (Satsuma mandarin)
3. *Citrus deliciosa* (Willow leaf mandarin)
4. *Citrus rashni* (Cleopatra mandarin)
5. *Citrus nobilis* (King)
6. *Citrus tangerina* (Tangerins)
7. *Citrus temple* (Temple)

B. Orange group (Tight skin orange)

1. *Citrus sinensis* (Sweet orange)
2. *Citrus aurantium* (Sour orange)
3. *Citrus myrtifolia* (Myrtle-leaf orange)
4. *Citrus bergamia* (Bergamot orange)
5. *Citrus natsudaidai* (Japanese summer grapefruit)

C. Acid group (Lime, lemon)

1. *Citrus aurantifolia* (Acid lime)
2. *Citrus limettioides* (Sweet lime)
3. *Citrus latifolia* (Tahiti or Persian lime)
4. *Citrus limonia* (Rangpur lime)
5. *Citrus limon* (Lemon)
6. *Citrus jambhiri* (Rough lemon)
7. *Citrus limetta* (Sweet lemon)
8. *Citrus volkameriana* (Volkamer lemon)
9. *Citrus kharna* (Kharna khatta)
10. *Citrus medica* (Citron)

D. Pummelo-grapefruit group

1. *Citrus grandis or maxima* (Pummelo or shaddock)
2. *Citrus paradisi* (Grapefruit)

E. Papeda group

1. *Citrus inchangensis* (Ichang papeda)
2. *Citrus junos* (Yuzu of Japan)
3. *Citrus latipes* (Khasi papeda)
4. *Citrus macroptera* (Melanesian papeda)
5. *Citrus histrix* (Mauritius papeda)
6. *Citrus macrophylla* (Alemow)

F. Other species group

1. *Citrus maderaspatana* (Kitchili or Vadukapuli)
2. *Citrus madurensis* (Calamondin)
3. *Citrus indica* (Indian wild orange)
4. *Citrus pennivesiculata* (Gajanimma)

Citrus groups	Important varieties
A. Mandarin group (Loose skin orange)	Mandarin-Coorg, Khasi, Nagpur, Desi Emperor, Kinnow, Sikkim, Mudkhed Seedless.
B. Orange group (Tight skin orange)	Sweet orange-Musambi, Sathgudi, Batavian, Blood Red, Malta, Pineapple, Hamlin, Jaffa, Valencia, Shamouti.
C. Acid group (Lime, lemon)	Acid lime-Kaghzi, Pramalini, Vikram, Chakradhar, PKM 1(Jai Devi),Selection 49,Seedless lime, Tahiti lime, Sai Sarbati, Tenali. Sweet lime- Mithachikna, Mithotra Lemon- Eureka, Villafranca, Lisbon, Lucknow seedless, Kagzi Kalan, Galgal, Pant lemon 1, Baramasi, PAU selection, Aurangabad Seedless
D. Pummelo- grapefruit group	Grapefruit- Marsh Seedless, Foster, Duncan. Pummelo- Chakaiya, Red fleshed

Among the large number of Citrus species, a good number are utilized for direct planting and many are used as rootstocks. In addition to these species, several hybrids among *Citrus* species, and between *Citrus* and *Poncirus* or *Fortunella,* are produced either naturally or through controlled breeding. A series of prefixes and suffixes are used to denote parents of such hybrids:

Intergeneric Hybrids

Citrange (trifoliate orange x sweet orange)
Citrumelo (trifoliate orange x grapefruit)
Citradias (trifoliate orange x sour orange)
Citrumuats (trifoliate orange x kumquat)
Limquats(kumquat x lime)

Interspecific Hybrids

Limonimes (lemon x lime)
Tangor (tangerine x sweet orange)
Tangelo (tangerine x grapefruit)

Climate and Soil

Citrus is grown in almost all parts of India clearly showing that it thrives well on a wide range of soil and climatic conditions. Although citrus trees do well in dry climate, with a rainfall between 75 and 125 cm, certain species, such as pummelo and certain mandarin oranges, thrive in heavy-rainfall areas of Konkan, Assam and Coorg. The optimum temperature for growth of citrus is between 25 and 30°C.However, some citrus species withstand high temperature even upto 50°C and low temperature below freezing point. High temperature during fruit maturation period causes poor pigment formation leading to poor colour development. High temperature also causes sunburn of fruits. Relative humidity plays an important role in fruit development. Low humidity favours better colour development while high humidity renders fruits to be more juicy with thin rind and more proned towards incidence of pests and diseases.

Citrus trees are grown in almost all kinds of soil, varying from heavy black soils to shallow open soils. The best soil for citrus is a medium textured soil of recent alluvial origin, uniform, reasonably deep and fertile, having good internal drainage and free from injurious salts. Some of the varieties of citrus seem to adapt them-selves to soil conditions better than others. They thrive in alluvial or medium black soils of loamy texture with good drainage. A hard substratum or a sticky impervious layer is very injurious. Soils having a high water-table should be avoided.

Propagation

Seeds of most of citrus species are polyembryonic in nature. Mandarins, acid lime and grape fruit have high polyembryony, whereas pummelo, citron and lemon are monoembryonic. Although citrus seedlings produce fruits identical to their parent tree due to polyembryonic nature, trees are generally budded onto various rootstocks to avoid long juvenile period of seedlings. Budding can be performed during most of the year, when pencil-sized, round bud sticks are available, and the bark readily slips on rootstocks. Bud unions on citrus are generally done little higher than many other tree fruits (3-4m above ground level) to avoid any contact of scion to ground. Commonly used rootstocks and their influence on resultant budded tree are given below.

Rootstocks	Yield/tree	Fruit size	Tree Vigour	Tolerance to frost damage to tree
Rough Lemon	high	large	high	poor
Sour Orange	mid	mid	mid	good
Cleopatra mandarin	low	small	high	good
Trifolate orange	low	small	low	good
Carrizo citrange	high	mid	high	fair
Troyer citrange	mid	mid	mid	good
Swingle citrumelo	high	mid	high	fair

The propagation of mandarin orange is largely through seed, except the Nagpur and Emperor varieties, which are propagated by budding. Like other citrus species, the seed is polyembryonic. Therefore, while propagating by seed, the sexual seedlings, which are usually stunted and poor are rogued out and the rest that are produced from the cells of the nucellus are allowed to grow. The seedlings, thus selected, are more or less uniform in growth and production. They are, however, late in bearing and remain tall and slender. Budded plants do not suffer from these defects. The santra orange is usually budded on rough lemon rootstock. The variety Emperor is budded on the kharna khatta rootstock. Studies at the Citrus Experiment Station, Coorg, of the Indian Institute of Horticultural Research, showed that trifoliate, rangpur lime, kodakithuli and troyer citrange are good rootstocks for mandarin.

Sweet oranges are usually propagated by budding and the most suitable rootstock is rough lemon. For the varieties Pineapple, Hamlin, Jaffa and Valencia, kharna khatta is the suitable root-stock. The trees are planted 6 to 7.5 metres apart each way in January or August- September in the North and in July-August in the South. The bud-union should be kept at least 15 cm above ground while planting. Sour lime is propagated mainly from seed. Budding on rough lemon rootstock, layering and marcottage are also practiced to some extent. The tree is susceptible to frost. It flowers twice a year in February-March and again in August. The main crop is obtained in August from first flowering. The second crop is ready in the following February. The propagation of sweet lime is done from mature wood cuttings, which root readily. It can also be propagated from seeds as seedlings are true to type.

Grape fruits are propagated by budding. However, owing to their polyembryonic nature, seedling-trees are also quite satisfactory. The rootstock most successfully employed in the Northern regions is kharna khatta and in the South and Bombay-Deccan is rough lemon.

Tolerance of rootstocks against salt is associated with differential uptake. The tolerant species has capacity to exclude chloride uptake from the system. Rangpur lime and Cleopatra mandarin were tolerant to salt. Rangpur lime exhibits high tolerance to drought.

Planting

In hills and humid regions, where plantings are generally done on steep slopes, land is properly terraced. In plains, where trees have to be irrigated, land should be leveled. Trees are usually transplanted during monsoon. In heavy-rainfall areas, the planting is generally done at end of heavy rains. In citrus, square and rectangular system of planting are generally practiced. In square system, mandarins are planted at a spacing of 5x5m (400 plants/ha), oranges 6x6m (278 plants/ha), limes and lemon 3x3m (1111 plants/ha) or 4x4m (625 plants/ha) and grape fruit and pummelo 7x7m (204 plants/ha). In rectangular system of planting, the spacing is modified as 4x6m (417 plants/ha), 5x7m (286 plants/ha), 3x5m (667 plants/ha) and 6x8m (208 plants/ha) respectively. High density planting is also practiced in citrus accommodating around 3000 plants/ hectare but life span of plants was considerably reduced. For planting, pits of size 50x50x50 cm are taken at recommended spacing. The pits are then filled with soil mixed with manure and seedlings planted. The best time for planting is beginning of rainy season.

Root Exposure

In Deccan region, root exposure is given to the tree. Water is withheld for about two months in advance of normal flowering season, and after about a month, roots are exposed by removing about 10 cm of soil in light soils and about 20 cm in heavy soils. After about 10 days, the soil is returned mixed with manure, and a light irrigation is given. After four or five days, a more copious watering is given, followed by 10 days later by full dose of water. In light soils, withholding of water without root exposure is sufficient to check vegetative growth and force blossoming. In southern India, no root exposure is given nor is it feasible. In the North, the root-exposure treatment is not necessary as the trees normally rest in winter and flower once a year. It should be mentioned that in most situations, root exposure of citrus trees is a devitalizing process and should be resorted to only under expert advice and direction.

Training and Pruning

Citrus trees in general require only very little training. Staking should be given to young citrus plants to help them grow, upright and straight. Young trees may be headed at 75cm height to induce branching and the trunk sprouts and suckers should be stripped off for the first 2 years. They may be defruited for a year or two to induce vegetative growth. At maturity, trees are mechanically hedged and topped to form hedges about 4m tall and wide.

Pruning young trees to build up a strong framework is recommended for mandarin. The bearing trees require little or no pruning. Undesirable growths, like water-shoots and crossing branches, should be removed once or twice a year. In Bombay-Deccan, root exposure or resting treatment is given to santra trees and sweet oranges to make them flower to order. Pruning of sweet orange trees begins in nursery. All branches that start within a few centimetres of the union are removed, leaving about half a metre of clean straight stem with a few well-placed branches. All unwanted branches are removed once a month during first year after planting, and once in two to three months in subsequent years. The bearing trees require little or no pruning. After the crop is picked, the branches touching the ground should be cut close to the laterals so that no stubs are formed. All diseased, injured and crossing branches, water-sprouts and dead wood should be removed periodically. Grapefruit trees require less pruning than orange-trees.

In sour lime, branches which are profusely growing on ground and water sprouts should be removed. Pruning operation should be attended yearly to provide a good framework to trees. Luxurient growth of branches in sweet lime cause overlapping and more height. It is, therefore, essential to prune the tree for maintaining even growth. In Haryana, moderate pruning in late November to early December results in higher fruit yield of sweet lime.

In lemon, plants should be trained to develop open centre. This can be achieved only by removal of twigs yearly. It is essential to prune lemons very tight annually to stimulate new growth and maintain production. Under humid conditions, drooping branches should be pruned to avoid infection by gummosis and foot rot fungi. Due to more dense foliage and heavier cropping, it gives out more dead wood than orange trees. Therefore, all such dead wood should be suitably pruned.

Off shoots or water sprouts are common phenomenon in Kinnow. These are characterised by long thorny growth with sparse foliage. They bear profusely but quality of fruit is rather poor. Trees under stress are more prone to produce sprouts. The water sprouts unnecessarily rob food material from plant. Judicious modification of these offshoots should be done from time to time by pruning and balancing them with rest of tree. Pruning is carried out early in spring after danger of frost has passed and before the tree starts a new growth cycle.

Fertilizer Application

Citrus needs adequate nutrients for better productivity, as the fruits are adapted to a wide range of soil types under varying climatic conditions. Lack of nutrients leads to decline. Leaf analysis of citrus orchards in different regions of the country was extensively done to understand the capability of soil to supply nutrients and also to use the leaf-nutrient status as an index for monitoring fertilizer use. The zinc is the most critical nutrient among the micronutrients, although nutrients like manganese and iron are found deficient and need attention. In heavy rainfall area, calcium and nitrogen nutrition need check. Thus, nitrogen, potassium, calcium and zinc nutrition are integral parts of nutritional management. Assessment of nutritional requirement for mandarin, sweet orange and acid lime under different agro-climatic conditions indicates that citrus requires 600-800 g N, 300-400 g P_2O_5 and 600-800 g K_2O plant/year for bearing trees. Experts advise that these nutrients should be applied in the form of cake and inorganic nutrients. Research findings revealed that application of 20 kg farmyard manure + 7.5 kg Neem cake and 50% dose of inorganic nutrients enhanced yield and quality of fruits. Generally, nitrogen is deficient in all the citrus growing areas of Punjab. The following fertilizer schedule is recommended for Punjab of which the whole quantity of farmyard manure is to be applied in December and nitrogen in two split doses (February and April-May).

Fertilizer schedule in Citrus

Age of the tree (years)	Farmyard manure (kg/tree)	Nitrogen (g/tree)
<3	5-20	50-150
4-6	25-50	200-250
7-9	60-90	300-400
>10	100	400-800

Zinc deficiency is widespread in citrus growing areas. To control zinc deficiency, spray the affected trees with zinc sulphate solution (0.3%) without addition of lime to spring flush in April-May, to summer flush in June and to late summer flush in August-September. Most appropriate time of foliar zinc application appears to be the stage when leaves are completing its expansion, i.e. 30-45 days old shoots. Since zinc deficiency becomes acute after first fruit bearing in fourth year, one maintenance spray of zinc sulphate solution (0.3%) should be given in April, right from start of third year. A gap of one week should be kept between foliar application of Bordeaux mixture and zinc sulphate solution.

Response to manganese, iron, copper and borax was also observed but results are not consistent which may depend upon status of these nutrients in soil. It is invariably observed that trifoliate orange as rootstock, is less efficient in nutrient absorption than Rangpur lime or rough lemon. Troyer citrange and carrizo citrange accumulate less manganese and appear to be useful in acid soils. The maximum root activity is within depth of 25 cm and at radial distance of 120 cm in mandarin. Similarly, in Kagzi lime also 80% roots are confined to radial distance of 120 cm. Application of fertilizer in this zone is beneficial for better fertilizer-use efficiency. Application of nutrients coinciding with time of flushing is better than the application when the plants are dormant. According to the Department of Agricultural Research and Education, ICAR, Government of India, in 1999-2000, application of 1,200 g N, 400 g each of P_2O_5 and K_2O/plant in mandarin recorded the highest yield at Akola (MH). A dose of 800 g N, 200 g P_2O_5 and 300 g K_2O/plant gave a yield of 23.9 kg/tree/year in acid lime. In sweet orange, a dose of 800 g N, 300 g P_2O_5 and 600 K_2O/plant/year gave a yield of 68.6 kg/tree/year at Tirupati (A.P).

Irrigation

Citrus plants respond well to irrigation. Water management constitutes an important input in citrus orchard, which determines productivity and longevity. Normally no irrigation is practiced in heavy rainfall areas of North-Eastern states. Plants which receive 1 or 2 irrigations during dry period have healthy growth and produce heavier fruits in these areas. At the same time, consecutive irrigation without definite period of soil-moisture stress is detrimental to crop yield, as such situation leads to more vegetative growth. Monitoring of soil moisture is effective for regulating flowering and fruiting in acid lime under tropical situation. Of late, drip irrigation is becoming very popular in water-scarce areas of Maharashtra, parts of Tamil Nadu and Andhra Pradesh. Many young citrus orchards were planned and planted with drip system of irrigation. The economic analysis favoured use of drip irrigation, which is viable. The system improved yield, uniformity in fruits and gave brilliant quality fruits in respect of shape, colour and edible quality. The growers apply the nitrogenous fertilizer through the drip which further economizes on fertilizer use.

Weed Management

Heavy weed infestation depletes soil moisture and competes for moisture and harbours pest and diseases during rainy season, which cause anoxia resulting in decline of tree. Mechanical weeding using hoe or spade or plough is common but to economize on use of manpower; experts advise integrated weed management.

Nutrient Deficiencies

One of the major causes of citrus decline in India is improper and inadequate nutrition of citrus plants. For proper growth and development of citrus, about 17 elements have important role to play. Apart from major nutrients like N, P, K, Ca, Mg, and S, citrus requires micro nutrients like Zn, Cu, Mn, Fe, B and Mo. Inadequate nutrition of citrus plants causes serious disorders and may eventually lead to decline of orchards. About 18 tonnes of citrus fruits remove about 21 kg nitrogen, 5 kg phosphorus, 41 kg potassium, 119 kg calcium, 3.6 kg magnesium, 2.3 kg sulphur, 40 g boron, 9 g copper, 50 g iron, 13 g manganese and 13 g zinc.

A brief description of the deficiency symptoms of some of the essential elements are given below:

Nitrogen: The deficiency symptoms of this element are easily recognized. The deficient trees show a general lightening of green of leaves. The leaves become distinctly yellowish green and some may even lose the green entirely when the deficiency persists for a longer time. The trees which remain severely deficient in nitrogen for long time, become stunted, have sparse foliage and produce dead-wood. Such trees bear very little fruit.

Phosphorus: The deficiency of phosphorous is rare in North India and is not easy to recognize in the field. The trees deficient in phosphorus have reduced growth and lack-lustre, smaller and narrower leaves than the normal trees. The fruit borne by such trees is rough in texture, having thick and coarse rind. The fruit has a hollow core and tends to drop prematuerly.

Potassium: The symptom of potassium appears as reduced growth and smaller leaves. The new growth is very weak and the twigs break very easily. The fruits remain small and have a very thin, smooth and well-coloured rind.

Magnesium: The deficiency symptoms usually occur in old citrus leaves when the fruit is mature. The symptoms first appear as chlorotic spots at the tips and between the veins. The leaves become abnormally thin, The yellow spots enlarge until an inverted V-shaped area remains green near base of leaf. In acute deficiency, the entire leaves become yellow and later drop prematurely, In North India, soils have adequate magnesium content.

Zinc: The symptoms of zinc deficiency appears on fully mature new leaves as irregular interveinal chlorosis commonly known as "mottled leaf". The terminal leaves become small and narrow referred to as "little leaf". Fruit bud formation is severely induced, twigs die back. The deficiency symptoms of zinc in citrus are widespread in Punjab.

Manganese: In young citrus leaves, it is hard to distinguish it from zinc deficiency, but when the leaves are full grown, the deficiency can be easily recognised. The deficiency manifests itself by green veins on a light green background. In mature leaves, bands or green develop along mature leaves, with light-green tissue. In severe cases of deficiency, dark-green leaf areas become dull green and the light-green areas turn grey or whitish. The leaves become thin and die-back of twigs starts.

Iron: With the iron deficiency, there is a general chlorotic condition of areas between veins and when deficiency persists, areas between veins become yellow. In such cases, therefore, the entire leaves become old. There is premature defoliation and in severe cases, twigs start dying back.

Copper. The copper status of the North Indian soil is usually low, however, its deficiency in citrus has not been noticed, Grapefruit seems to be more sensitive to copper deficiency than malta. The fruits develop symptoms of deficiency before the branches are affected. The fruits become misshapen, have dark-brown-gum-soaked eruptions, varying in size. There are gum pockets in the albedo. The leaves of deficient trees are deep green, over-sized and coarse. The shoots are long, vigorous and of S-shaped giving appearance of excessive nitrogen fertilization.

Intercropping

Intercropping should be done to utilize vacant land to generate income till the citrus plants become productive. Orchards start to decline if exhaustive intercropping is used indiscreetly. Choice of intercrop depends upon local conditions and climate. Based on benefit:cost analysis, potato, cabbage and peas were found beneficial in mandarin orchards in Darjeeling. Under Coorg conditions, mixed cropping with Robusta coffee was more remunerative than monoculture, though plant vigour of Coorg mandarin was comparatively suppressed.

Harvesting and Yield

Seedling trees of mandarins bear their first crop in eighth year and full crop from tenth year onwards. Budded trees start bearing from fourth year and full crop is from seventh year onwards. The harvesting periods differ in different parts of country. While picking fruit, stem-end should be cut close to fruit without damaging rind. Packing is done by putting fruits of different size/grades in separate wooden crates.

In sweet oranges, trees begin to bear fruits from fourth year onwards, but normal crops are borne from seventh year. Main harvesting season in North is December to February, whereas in the South, it is October to March. In Bombay-Deccan region, there are two main seasons, November to January for ambe bahar and March to May for mrig bahar. Picking may be done during the day, taking care that the stem is cut close to fruit without damaging rind. The fruits are washed, dried and graded for size and packed into wooden cases for disposal. The harvesting season for grapefruit is January to March in North and September to November in South. Picking, packing and other operations are same as that for orange. Quality and flavour of fruit are improved if stored before transporting.

All citrus are non-climacteric fruit, meaning that they ripen gradually over weeks or months and are slow to abscise from the tree. Maturity of citrus fruits depends upon climatic condition, scion rootstock and management practices. Extensive work was done on influence of climate, nutrition and rootstock on quality of fruit. Period of maturity is the shortest in acid lime (5-6 months), the longest in mandarin and sweet oranges (9 to 10 months) which is further influenced by heat unit and moisture. Attempts made to develop maturity standards based on TSS: acid ratio and external colour of fruit are indices for harvesting of fruits which go with experience of orchardist. External color changes during ripening, but is a function of climate more than ripeness and a poor indicator of maturity. Citrus is hand harvested, whether processed or marketed fresh. Mechanical harvesters are used for processed fruit in Florida and are increasingly in popular due to high labor costs and lack of labour availability.

Grading

For fresh fruit, standard packing line operations are used (in order): dumping, culling, washing, brushing, waxing, drying, grading (human), sizing, and boxing. For processed fruit, growers are paid for sugar content, based on juice analysis. Harvested fruits are culled for rot, and remaining fruit is washed prior to juicing. Juice is extracted by inserting a cylindrical strainer in center of fruit and compressing fruit hydraulically. Extracted juice contains some pulp and oils, which are separated from juice by centrifugation and screening. Juice is pooled into lots of various colors and sugar levels; some mixing is done to produce uniform product. Sales of frozen concentrate are outpaced by single-strength juice products in recent years, due to the superior flavor of the latter.

Storage

Citrus may be stored for periods of up to 1-2 months at low temperatures (0-8°C). Chilling injury is common in grapefruit, lemons, and limes when stored below 18°C, but rare in oranges and tangerines. A unique aspect of citrus is ability to store fruit on tree. Fruit reaches minimum maturity standards in early winter, but since they are non climacteric, they ripen slowly and will not soften or abscise for periods up to several months. Pre-harvest sprays of chemicals and growth-regulators increase shelf life of fruits. Calcium chloride or nitrate or spray of hormones (2, 4, 5-T) reduces weight loss and enhances TSS of fruits in storage, when applied 2 weeks before harvesting.Post-harvest diseases, caused by *Penicillium digitatum, P. italicum, Phomopsis citri, Diplodia natalensis* and *Alternaria citri* are major causes for deterioration of fruits during post-harvest handling and storage. Treatment of fruits with Bavistin (1,000 ppm) reduces post-harvest loss appreciably and extends shelf life to 25-26 days even at ambient temperature. Waxol along with Benomyl or growth-regulators was beneficial to extend shelf life of Kinnow.

Packing

Wrapping of fruits with grease-proof paper or perforated polyethylene bags, packing in wooden crate and storage at 0-3°C temperature and 86-90% relative humidity were optimum. Best storage temperature for Darjeeling mandarin was 1.5-3°C. Degreening at 26-28°C and 90-95% relative humidity with ethylene (500 ppm) was achieved in Nagpur Santra without affecting internal quality. Shelf-life of Nagpur Santra was extended by 3 weeks on wrapping fruits with heat-shrinkable polythene and Cryovate film (D9 55 and BDF 2001). These developments helped in export of mandarin.

DATE-PALM

The date-palm(*Phoenix dactlifera* L.) is the oldest among the cultivated tree fruits belonging to family Palmaceae. It is a monocotyledonous plant like banana. Date palm is indigenous to countries around the Persian gulf. Iraq, Saudi Arabia, Iran, UAE, Algeria, and Pakistan are the principal date-growing countries. In India, date is not grown commercially because it ripens in July when the monsoon rain starts and this results in the rotting of fruits. However, about 1.5 lakh seedling trees, are grown in parts of Kutch district of Gujarat State on boundaries of fields and water channels yielding fruits of variable quality. Almost whole of the fruit is sold as raw dates. However, a closely related date species (*Phoenix sylvestrs* (L).) is growing wild in almost all parts of the country. The fruit of this species is very inferior in quality and is not fit for eating. These palms have been 'tapped' for production of crude sugar.

At present India is importing 60,000 tonnes of dates every year. More than 40% of the amount is spent on import of Chhuharas (dry dates) and the rest on Khajoor (soft dates). To meet need of this fruit, the Indian Council of Agricultural Research sanctioned a Coordinated Scheme for the Improvement of Date palm in 1955 at Abohar (Punjab). This place provided congenial conditions for date cultivation due to scanty rainfall (23-30 cm) with sufficient canal irrigation facilities. Date palm is highly nutritious and is known to contains sugars, protein, fats and minerals. The flesh of fully ripe date (tamar) consists of 75-80% sugars (glucose and fructose).

Varieties

Date varieties are generally divided into 3 groups on the basis of flesh consistency i.e. soft, semi dry and dry. Soft dates are those which pass through rutab stage and remain soft when they reach tamar stage. These dates are mostly of the reducing sugar type. Semi-dry dates are those varieties which pass through a rutab stage but make a dry tamar. The sugar is generally of reducing type and most dates are of this type. Dry dates are those varieties which do not pass through a rutab stage. They are mostly of sucrose type.

Halawy, Barhee and Hayany are raw-eating dates. Medjool, Halawy, Zahidi, Khadrawy and Shamran are suitable for making Chhuhara. Khadrawi, Halawy and Shamran also yield good quality soft dates. A brief description of these varieties is given below.

Halawy: It is a soft date from Iraq. This variety is relished both in rutab and tamar stages. Total soluble solids range between 28 to 42%. It is an early variety yielding good crop and is somewhat more tolerant to rains. Dry dates (Chhuhara) of good quality and cured soft dates (Khajoor) can be prepared. Its fruit at tamar stage (fresh form) is very delicious. Average yield ranges from 50 to 80 kg/ palm.

Khadrawy: It also originated from Iraq. This variety proved successful under Abohar conditions. The trees of this variety are comparatively less tall and yield good crops. It can be used both for dry dates and soft dates. The yield ranges from 40 to 70 kg/ palm.

Shamran: It is a mid-season cultivar tolerant to high humidity. This variety yields prolific crops. The fruits can be cured into good quality. Fruit is medium to large, oblong to oval and yellow at rutab stage. Its yield/ palm is 40 to 70 kg.

Medjool: The variety originated in Morocco and has large sized fruits. It is late in ripening and has proved particularly good in preparing dry dates of attractive bold size and good quality. Its yield is 35 to 50 kg/ palm.

Barhee: It is a soft date from Iraq. The variety has proved extremely good for table use in fresh form (rutab stage). The rutab fruit contains about 32% TSS, has golden yellow colour and has a very pleasant taste because of low astringency and high pulp content. The trees of this variety are prolific yielders and the fruit is ready for eating late in the season i.e. first fortnight of August at Abohar. Its yield ranges from 60 to 110kg/ palm.

Hayany: It is a soft date originated from UAE. The berries have attractive deep colour. The fruits could be consumed as fresh in the rutab stage as ripening does not proceed further under Abohar conditions. The yield/ palm ranges from 30-40 kg.

Zahidi: It is a mid-season variety, slightly tolerant to rain or high humidity. The fruit is small to medium, obovate and yellow at rutab stage.

Climate and Soil

The date palm as a tree flourishes well under varied sets of climate. However, for proper maturity of fruit, date requires prolonged summer heat without rain or high humidity during the ripening period. A light shower with prolonged periods of cloudy weather and high humidity may cause more damage than a heavy rain followed by clear weather and drying winds. The mean temperature between the period of flowering and ripening of the fruit should be above 21°C rising to 27°C or higher for at least one month. For successful fruit maturation, nearly 3000 heat units are required. For successful cultivation of date palm, a long summer with high day and night temperature, absence of rain at the time of flowering and fruit setting, with low relative humidity and plenty of sunshine are essential. In India, no such location is available which could meet the requirement although fairly dry conditions are available in North-Western India. In different parts of India, fruit development period is cut short to 122-170 days and the fruit yet not full matured is harvested early before rains to avoid heavy fruit spoilage.

The date palm can be grown in soils containing more alkali or salts. It tolerates such condition better than many other fruit plants. In view of the large investment required to bring a date garden into bearing and maintain it on profitable production, sandy loam soil, 2 or 3 metres deep with good water holding capacity and drainage is the most desirable. Date palm tolerates high soil salinity.

Propagation

Dates can be grown either from seeds or from suckers (offshoots). When grown from seeds, approximately half of the palm will be male and produce only pollen. Its sex will be known only when it comes into flowering after four or five years. In this case, most of the female plants are inferior (not true-to-type) to the mother palm.

The date palm is always propagated through suckers (offshoots) for commercial plantations. The suckers usually arise near the ground around the trunk. These are separated from mother palms during March or August-September. The leaves are trimmed off 4-5 days before their detachment. The inner leaves should be cut back by one-half and the outer ones by two-thirds. The tender young unopened leaves near the central bud and parts of the bare stalks of the old leaves are necessary to protect the bud and it should be kept on the offshoot. The outer whirl of leaves may be fastened to afford protection from heat and cold to central bud and also to facilitate detachment and transport. Well rooted suckers weighing approximately 25 to 30 kg are removed using a chisel taking care that no injury is caused to mother palm. Cut surface should be coated with coal tar or some suitable material immediately and earthed up.

Planting

Date palm is perennial in nature and bears fruits for 40-50 years. Generally square system of planting is followed for facilitating easy intercultural operations and proper development of palms. The offshoots are planted 6 to 7 metres apart in the already prepared pits of 1 x 1 x 1 m size immediately after detachment. In this way 202 to 275 plants/hectare will be accommodated. If transported from long distance, the offshoot may be kept in the nursery for one season. Rainy season is the best suited for planting. Pits can be left open for weathering for two weeks and then filled with a mixture of top soil and FYM. While planting, care should be taken to see that the crown of the offshoot is 15cm high above the ground level. The soil around the offshoots should be pressed firmly. The field is irrigated immediately after planting. Thereafter, frequent light irrigations are given to keep the soil always moist. At Abohar, the offshoots removed in March survived better in direct field planting than those removed in August-September. However, the separation could be done both in February-March and August-September.

Fertilizer Application

Generally farmers do not apply manure to date palm. The trees certainly respond to manuring as indicated by increased vigour and growth. The results of studies conducted at Abohar show that 25 to 50 kg of farmyard manure and 1 to 2 kg ammonium sulphate/ palm should be added. The quantities can be reduced if some leguminous crops are grown as inter crops in the date palm plantation. Farmyard manure is applied during December-January. The nitrogen dose should be given a fortnight before flowering i.e. in first week of February.

Irrigation

Date palm is known as a drought resistant fruit tree and is able to survive for long periods without irrigation. However, continuous drought condition retards growth of plant. Date palm is highly tolerant to excessive irrigation and floods. Continuous stagnation of water or waterlogged conditions are injurious for its growth. To maintain maximum growth, the root zone up to 2-3 m should be kept moist and not allowed to dry. Light but frequent irrigation should be given after planting.

The full grown trees on lighter soils are usually irrigated 7 to 10 days interval during mid-summer and every 15 to 30 days interval during winter. On heavy soils, half the amount may be enough. Irrigation is withheld when ripening starts to facilitate harvesting, hastening fruit ripening and reducing fruit drop caused by high humidity.

Intercropping

Intercropping in date palm with suitable crops bring good income and also improves the fertility of the soil. During the first few years, intercropping can be practiced with no shortage of irrigation. Intercrops like green gram, peas, moong and lentil can be sown during summer. Inter cropping with vegetables in plantations located near the cities, can be practiced if sufficient irrigation and manuring facilities are available.

Fruit Thinning

Excess load of fruit may cause shriveling of berries, breaking of spathe stalks, more damage due to rain and humidity, delay in ripening and alternate bearing. It also result in the production of small fruits of poor quality. Therefore, it is necessary to thin out the fruits to retain only optimum quantity of fruits. This is usually accomplished either by reducing number of fruits on each bunch and or by removing some of the bunches.The number of fruit that a palm can safely carry depends on the cultivar, age, size and vigour of the palm and number of green leaves on it. Under normal conditions, 1-2 bunches in the 4th year and 3 - 4 bunches in the 5th year may be left. Normally 8-10 bunches/palm are retained in India. Small, defective and broken bunches should be removed.

Desirable number of fruits to be left is between 1300 and 1600/ palm depending on the variety. The thinning(%) generally done is 40-50 in Khadrawy, 50-55 in Halawy, 50-60 in Zahidi and Barhee. Ethephon (100-400 ppm) after 10 to 30 days from fruit set was effective in fruit thinning of cv. Hayany. The biennial bearing habit of treated palms was corrected by ethephon treatment (500-1000ppm) and it also hastened ripening of fruits.

Stages of Fruit Ripening

On the basis of fruit development and ripening, the date palm fruit can be divided into four stages (Pareek and Gupta, 1988). The Arabic names for these four stages are Chimri, Khalai, Rutab and Tamar. In Punjab, these stages are termed as Gandora, Doka, Dang and Pind, respectively. Kimri or Gandora is a stage when the fruit has grown a little and attains a clear apple green colour. The second stage Khalai or Doka is marked by turning of colour of fruit skin from green to yellow, finally spotted according to characteristic of the variety. In this stage, the fruit attained the full size. The third stage known as Rutab or Dang is characterized by fruits becoming more or less translucent and softening starts from distal end by darkening of skin from yellow or scarlet to brown or to black. The final stage in the ripening of date is called Tamar or Pind, when fruits are fully matured and has lost much of its water, enough to make sugar to water proportion sufficiently high to prevent fermentation. This stage is equivalent to that of raisin in grape.

Harvesting and Yield

Harvesting of dates generally depends upon weather conditions. The rain and high humidity in the atmosphere at ripening time is a limiting factor in its cultivation. When the climate is favourable, it is preferable to leave the fruit of most varieties on the palm until it reaches the stage of maturity at which it is to be consumed or stored. The changes associated with ripening and the period during which the fruit may be consumed extend from peak of khalal stage, when the fruit has its most intense red or yellow colour and maximum weight, to final tamar stage, when it has lost the greater part of its moisture content and will keep without special attention in storage. Most of the people like to eat the fruit in khalal stage. At least two varieties of dates, Halawy and Barhee, are liked most for eating in khalal stage. Dates are handpicked at the stage of maturity. All the dates in the same bunch do not ripen at same time and it is the practice to make several pickings to harvest the fruit during a season. Sometimes, when the season is favourable and more than 70% fruit is ripe, the entire bunch is harvested.

The researches conducted on curing of dates in Punjab (Abohar) revealed that rain and high humidity in the atmosphere at the time of ripening (July and August) do not allow ripening process on the tree to proceed satisfactorily beyond doka stage. If the fruit is retained on the tree, there is splitting and rotting. High humidity and rainfall also cause the fruit to drop. Therefore, the crop has to be harvested at doka stage. Though in comparatively drier seasons, partial crop on some trees may become dang (yellow and soft) but the quantity of such fruit is limited. The berries are removed and graded manually on the basis of size and colour.

Scientist working at Abohar (Punjab) has developed a satisfactory technique to transform date fruit at doka stage into dry dates (Chhuhara) of good quality. At least four varieties Halawy, Khadrawy, Shamran and Medjool yielded very good product. Chhuhara obtained from Medjool, which is a large-sized variety, is comparable with the high quality chhuhara imported from Middle East countries. The technique developed involved immersion of fruit at doka stage in boiling water for 6 to 8 minutes and then drying either in temperature-controlled oven (air-circulation type) for 80 to 120 hours at 48 to 50°C or in sun for 10-15 days if weather is dry. Thus, an average of 45% fruit product is obtained. Fruits at advanced doka stage or when they attain one-fourth, one-half or full dang (the berries become mellow and soft starting from distal end) can be converted into soft dates (khajoor) of good quality by drying either in the oven at 40°C or in the sun. Thus, for soft dates, only drying berries at partial or full dang stage is required with no other treatment and this way, a final product of soft khajoor ranging from 50-60% is obtained.

Artificial ripening of fruit at doka stage to transform it into dang stage was also attempted at Abohar. For these studies, berries were treated with 0.5% to 2.0% common salt (sodium chloride) and similar concentrations of acetic acid in combination with 1.0% salt. The fruits to be treated with salt were spread on polythene sheet and requisite quantity of salt applied by rubbing and smearing uniformly on the berries. The fruits which were given acetic acid plus salt treatments, were first dipped in solutions of desired acetic acid concentrations for two minutes followed by salt application by the method described above. Each treated lot was packed into wooden boxes lined with old newspapers and packed in laboratory at room temperature. The boxes were opened after 24 hours. From these studies, it was found that with 2%

salt, 60 to 70% of the doka fruit were transformed into dang. However, the dang obtained by this treatment was not as good in taste as that of naturally ripened on the tree, but still it was edible and generally acceptable on account of its having lost the astringency. But such product cannot be stored for more than 24 hours and, as such, should be consumed as early as possible.

GRAPE

Grape (*Vitis vinifera*) is the most important sub tropical fruit in the world and belongs to family Vitaceae. It originated from the Caspian sea area and is introduced into North India by the Persian invaders in 1300 AD, from where it was introduced into the South. Cultivation of grapes is one of the most remunerative farming enterprises in India.

In India, grape cultivation is concentrated mainly in Maharashtra, Karnataka, Andhra Pradesh, Tamil Nadu, Madhya Pradesh, Uttar Pradesh, Punjab, Haryana, Delhi and Rajasthan. Grape is cultivated over an area of 34,000 hectares with an annual production of 10 lakh tonnes. Although, the returns from grape cultivation, per unit area of land are very high, the area under grapes is not expanding fast owing to the high initial cost of establishing the vineyards and high recurring cost of production.

Varieties

In India, around 20 varieties are under cultivation. These are grouped into four based on the purpose for which they are mainly used.

Table grapes: Fruits are mainly eaten in fresh form. This includes white, red and black varieties. These are sweet in taste with a few seeds and the skin thin or thick. The important varieties are Anab e Shahi, Perlette, Bhokri, Thompson Seedless, Delight and Beauty Seedless.

Raisin grapes: Grapes that are dried for making raisins. Large, medium sized or small black or greenish white types are used. These types may or may not have seeds with high sugar content and thin or thick skin. The varieties are Thompson Seedless, Sultana, Red Corinth and White Corinth.

Juicy grapes: Types that are mainly used for preparing fresh juice and wines. Black or white varieties with distinct strong aroma and high sugar content are generally used. Muscat and Black Prince are the main varieties.

Based on colour of fruits and presence or absence of seeds, grapes are grouped into four.

(i) White coloured seeded fruits - Anab-e-Shahi, Dilkush

(ii) White coloured seedless fruits - Thompson Seedless, Pusa Seedless, Perlette, Tas-A-Ganesh, Sonaka and Manik Chaman

(iii) Coloured seeded fruits - Bangalore Blue, Gulabi

(iv) Coloured seedless fruits - Sharad Seedless and Beauty Seedless

Approximately 85% of the total production, irrespective of the variety, is consumed fresh. About 120,000 tonnes of Thompson Seedless and its mutants, namely, Tas-A-Ganesh, Sonaka and Manik Chaman are dried for raisins. Some 20,000 tonnes of Bangalore Blue are crushed to make juice, and 10,000 tonnes of Bangalore Blue, Cabernet Sauvignon, Chenin Blanc, Chardonnay, Merlot, Pinot Noir and Uni Blanc are crushed to process into wine.

Climate and Soil

Grape is a subtropical fruit that grows well in dry summer. The vines shed their leaves and rest in winter, put forth new growth in spring and mature in summer. Grape does not thrive in regions having humid summers. It tolerates frost during resting period, but succumbs to it readily during its growing period. Grape by nature is a deciduous fruit crop, grown extensively in temperate regions. Earlier, grape cultivation was a non-viable crop in the tropics. However, now because of the extensive research efforts, the world's highest productivity (96 tonnes/ha) is recorded from the tropical region of the country and more than 90% of the area under grapes in India is in this region.

In the North, where grape is cultivated mainly in Punjab, Uttar Pradesh and Himachal Pradesh, in Western and Southern India, where it is cultivated mainly in Maharashtra, Hyderabad, parts of Karnataka and Tamil Nadu, the vine grows throughout the year and bears two crops,.

Grape cultivation is unique in India and is grown under a variety of soil and climatic conditions in three distinct agro-climatic zones, namely, sub-tropical, hot tropical and mild tropical climatic regions.

Sub-tropical region: This region covers the North Western plains corresponding to 28° and 32° N latitude including Delhi; Meerut district of Uttar Pradesh; Hissar and Jind districts of Haryana; and Bhatinda, Ferozpur, Gurdaspur and Ludhiana districts of Punjab. Vines undergo dormancy and bud break starts in the first week of March while the rains arrive in the first week of June, and therefore, only 90-95 days are available from the initiation of growth to harvest. Consequently, 'Perlette' is the only early ripening variety grown in this region. Rain damage is a problem with Thompson Seedless in this region. Single pruning and a single harvest are the accepted practices here. It grows and fruits once a year in summer, and rests during winter.

Hot tropical region: This region covers Nasik, Sangli, Solapur, Pune, Satara, Latur and Osmanabad districts of Maharashtra; Hyderabad, Ranga Reddy, Mahbubnagar, Anantapur and Medak districts of Andhra Pradesh; and Bijapur, Bagalkot, Belgaum, Gulberga districts of Northern Karnataka lying between 15° and 20° N latitude. This is the major viticulture region accounting for 70% of the area under grapes in the country. Vines do not undergo dormancy and double pruning and a single harvest are the general practices in this region. Maximum and minimum temperature is 42°C and 8°C, respectively. The major problems in this region are soil and water salinity and drought. Berry growth is impaired and in certain locations pink blush sometimes develops on green berries due to temperatures that drop to 8°C. Thompson Seedless and its clones (Tas-A-Ganesh, Sonaka), Anab-e-Shahi, Sharad Seedless and Flame Seedless are the varieties grown in this region.

Mild tropical region: An area covered by 10° and 15° N latitude including Bangalore and Kolar districts of Karnataka; Chittoor district of Andhra Pradesh and Coimbatore, Madurai and Theni districts of Tamil Nadu fall in this region. Maximum temperature in a year seldom exceeds 36°C, while the minimum is about 12°C. Principal varieties are Bangalore Blue, Anab-e-Shahi, Gulabi, and Bhokri. Thompson Seedless is grown only with limited success. Except for Thompson Seedless, double pruning and double cropping system are followed and two crops (first in April and the second in August-September) are harvested in a year.

The grape grows the best on light, friable loamy soils with free drainage. Heavy soils are unsuitable. They can tolerate saline soils successfully.

Propagation

Grapes are multiplied exclusively by the rooting of hardwood cuttings. The cuttings for planting should be selected from one-year-old dormant wood from healthy bearing vines. No government agency is involved in the multiplication and supply of rooted cuttings. Growers themselves obtain the hardwood cuttings from elite vineyards and raise their own nurseries. Well matured canes obtained in September/October are selected. Cuttings of 4 nodes each with a thickness of 8 to 10 mm are made from the selected canes. The fresh cuttings are soaked in running water for 24 hours to leach out the water-soluble rooting inhibitors. The basal parts of cuttings are then dipped in a strong IBA (2,000 ppm) solution for five minutes before planting. It is also a practice to plant the cuttings *in situ* where three to four cuttings are planted at each spot in the main field. Soil drenching with chlorpyriphos (0.1%) is a practice to safeguard the cuttings against termite damage.

Of late, due to soil-borne problems, like nematodes, soil salinity and drought, use of rootstocks has become essential. Research carried out at IIHR, Bangalore indicated that Rootstock 1613 (nematode resistant) is well compatible with Anab-e-Shahi, without any reduction in fruit yield or quality. 'Dogridge' rootstock is being employed to combat soil and water salinity problems. Hardwood cuttings of the 'Dogridge' rootstock are subjected to rooting, preferably in polybags of 15 x 25 cm. Rooted cuttings of this rootstock are planted in the main field during February-March. The desired scion variety is then grafted /budded on rootstocks in the field by wedge grafting/chip budding. Wedge grafting is more common and the best time for operation is September-October, while June-July is suitable time for chip budding.

Planting

The land is tilled and levelled perfectly to have a gradient of less than 1% in any direction to ensure uniform discharge of water through the emitters of drip irrigation systems. Trenches of 75 cm width and 75 cm depth with a gap of 3 m between trenches are opened. They are closed with topsoil, up to a height of 45 cm after 15 days exposure to sun. The remaining gap is filled with a mixture of soil, cattle manure, single super phosphate, sulphate of potash and micro-nutrients. Usually, 50 kg of cattle manure, 2.5 kg of super phosphate, 0.5 kg of sulphate of potash and 50 g each of $ZnSO_4$ and $FeSO_4$ are added to the soil for every running meter length of trench. The best season for planting the rooted cuttings of cultivated varieties in the main field is September-October whereas for rootstocks, it is February-March.

Spacing generally varies with the varieties and soil fertility. For vigorous varieties it is 6 x 3 m or 4 x 3 m and 3 x 3 m or 3 x 2 m for less vigorous varieties. Wider spacing was followed for grapevines in the past and it was associated with long pre-bearing period, spent in development of permanent framework. The role of close planting in overcoming this problem was realized and as a result of which, spacing is reduced to 3 x 4.5 m from 3.5 x 7.0 m for Anab-e-Shahi and Bangalore Blue and to 1.2 x 1.8 m from 3.0 x 3.0 m for Thompson Seedless. However,too close planting of 1.2 x 1.8 m hampered fruit-bud fertility and fruit quality due to overcrowding of the shoots and poor light interception. A cane density of 60 was optimum for a bower area of 10 m^2. Therefore, the popular spacing in Maharashtra vineyards for Thompson Seedless and its derivatives is 1.8 x 2.4 m or 1.2 x 3.6 m. In Northern India, cuttings are made from one-year-old wood at the time of pruning in

February, when the vines are dormant. The cuttings are tied in bundles and stored in moist sand for about a month for callusing. The callused cuttings start well in the nursery. Elsewhere in India, they are obtained from pruning in October and planted in nursery for rooting. The cutting is ready for transplanting in January. It is also possible to raise a plantation by planting cuttings *in situ* in the field.

Training

For the system of training, bower was the most appropriate one and was associated with the highest productivity. This training system offers scope to develop canes up to the rate of 10 canes /m^2, which is required to obtain 50 tonnes/ha. With the technique to obtain as high as 3 bunches/cane, efforts have now shifted to less-expensive V and Y trellises to obtain more ventilated and more illuminated vine canopies. However, more than 80% of the area under grapes is still covered by bower grapevine training. Many training systems are in vogue in India, but the most popular are Bower, Telephone and Flat Roof Gable systems.

Bower system: Owing to the high production potential, bower was a very popular system of training in the past. It is highly suited for vigorous varieties like Anab-e-Shahi, Bangalore Blue and Gulabi. But in varieties like Thompson Seedless and Tas-A-Ganesh where vine vigour and excessive foliage density affect the productivity adversely, this system is not popular.

Telephone system: T-trellis is used in this system of training. With three top wires and 'T' shaped supports, the trellis looks like a telephone pole and wires and hence the name. This system is followed for moderately vigorous varieties like Thompson Seedless and other seedless cultivars in about 25-30% of the vineyard area in Maharashtra. Yields in this system are less than the bower. In very hot and dry places, sunburn of the berries and of the arms are experienced in summer.

Flat roof gable system: Combining the advantage of bower and the extended Y systems and eliminating their disadvantages, an inter-connected Y trellis forming a flat roof gable is being adopted. This system is particularly followed for vigorous vines (vines grafted on rootstocks). Bunches are protected from direct sunlight and well exposed to sprays of pesticides. The clusters hang within the reach of the worker of an average height. Owing to these advantages, this system is gaining popularity among the growers in Maharashtra, Andhra Pradesh and Karnataka.

Pruning

Pruning comprises the removal of living canes, shoots, leaves and other vegetative parts of the vine. In nature, grapevine shows acropetal growth and if left as such, its branches continue to elongate climbing with the help of tendrils. Acropetal growth is more pronounced under warmer conditions. When pruned keeping 8-10 buds, only the terminal 2 or 3 nodes usually put forth growth and those at proximal positions either do not sprout or produce only weak shoots bearing small bunches that do not ripen. Vine pruning regulates or encourages good yield and improves size and quality of fruit. The aim of pruning is to properly distribute the bearing wood over the vine and to regulate the crop maintaining its vitality for consistent productivity. Generally, pruning has a depressing effect on vine growth and reduces bearing. However, over-bearing also has similar effect and delays the ripening.

Pruning of the vines should be regulated to suit the bearing habit of each variety. Three distinct pruning practices are in vogue in relation to cropping in the three grape growing regions of the country. In the sub-tropical region, vines are pruned only once in December and the crop is harvested once. Half of the canes are pruned to renewal spurs and the rest to fruiting canes.

In hot tropical regions, vines are pruned twice but only one crop is harvested. All canes in a vine are pruned back to single node spurs in March-May to develop canes and the canes are forward pruned in October-November for fruiting. Number of nodes retained on a cane varies with variety and cane thickness. There is no scope to prune earlier than October and later than November due to unfavourable weather conditions.

In the mild tropical region, vines are pruned twice and the crop is harvested twice. In varieties like Gulabi and Bangalore Blue, which are fairly resistant to rain damage and in which fruit bud differentiation is not impaired by cloudy weather and rains, pruning is done at any time of the year. As a result, five crops are harvested every two years.

Methods of Pruning

(i) *Spur pruning:* Retention of 1-4 basal buds after pruning as fruiting canes is called spur pruning. For the most economical production of all varieties of grapes that produce medium sized clusters on shoots arising from buds at the base of the canes, spur pruning should be used.

(ii) *Cane pruning:* It is practiced by leaving more than four buds on a cane. It should be used with varieties of which the lower buds on the canes are usually sterile such as Thompson Seedless.

(iii) *Mechanical pruning:* Cane pruning with less number of canes/ vine of same length is being used for more varieties. It permits positioning of fruit for machine harvesting better than other system.

Pruning Types

(1) *Dormant pruning:* The principle pruning is done while the vine is dormant, between leaf fall in autumn and starting of buds in spring. In North Western plains, pruning is generally practiced from middle of January to first week of February. In tropical areas of Maharashtra, Andhra Pradesh, Karnataka, and Tamil Nadu, the vine continues growth all the year round and does not undergo dormancy. Pruning is, therefore, done twice a year, once in summer and again in winter. Summer pruning is done during March-April, except Tamil Nadu, where it is done in June when the canes are cut back to 1-2 buds to induce vegetative growth and is called back pruning or growth pruning. During winter, pruning is done between last week of September and first week of October in Northern districts of Karnataka and in December in Tamil Nadu when it is done to induce fruiting depending upon variety.

Amount of pruning depends on bearing behaviour of different cultivars. Number of canes/ vine and number of buds/cane vary from variety to variety. The intensity of pruning recommended for different varieties is as follows.

Variety	Number of canes to be left/vine	Number of buds to be left/cane
Perlette	40-50	4-6
Thompson seedless	40-50	8-10
Anab-e-Shahi	80-100	5 or 6

When a vine has reached the stage of full bearing, pruning consists of removing all the growth except (i) Bearing units: for production of fruit and new wood or fruit only, (ii) Renewal spurs: for renewal or production of wood for next year and (iii) Replacement spurs: In the older vines, for replacing or shortening of arms.

(2) *Summer pruning:* Summer pruning consists of removing buds, shoots or leaves, while they are green herbaceous and it is done while the vine is growing or active. Summer pruning has various uses, principally as follows:

a) Correction of shape: To direct growth into the parts that will form the permanent frame work of the vine such as the trunk, branches and arms and to keep these parts active and healthy. This is accomplished by such operations as disbudding, pinching and suckering.

b) Shading fruits: To increase shade on the fruit by topping, which promotes the upright position of shoots and growth of laterals.

c) Opening the vine to light: To open the vines and thus expose the fruit more favourably to light and air.

Fertilizer Application

As vineyard soils are located either in sandy loam or heavy clayey soil areas, the usage of organic manure has assumed high importance in India. Green-manuring may be done whenever feasible. In India, it is a common practice to apply massive doses of fertilizers as high as 2,000-3,000 kg each of N, P_2O_5 and K_2O/ha irrespective of the need of grapevines. However, long term experiments carried out at the IIHR, Hessarghatta revealed that a dose comprising 500 kg N, 500 kg P_2O_5 and 1,000 kg K_2O/ha/year for Anab-e-Shahi and of 300 kg N, 500 kg P_2O_5 and 1,000 kg K_2O/ha/year for Thompson Seedless is optimum.

The annual dose is fixed based on petiole analysis carried out at 45 days after spur pruning. When 40% of the annual dose is given through organic sources, the rest 60% is given in inorganic form. Sulphate of potash is the only source of potash used in place of muriate of potash, particularly in heavy clay soils. Application of soluble fertilizers through drip irrigation is practiced now. Under Maharashtra conditions where vines are pruned twice, application of 60% of annual dose of N, 50% of P_2O_5 at April pruning and rest of N and P_2O_5 and the entire K_2O at October pruning is recommended. Application of P after 30-40 days after pruning promoted fruiting in Anab-e-Shahi vines. Irrespective of the time of application, K promoted bud fertility. Application of K on 20th day after pruning improved the fertility status of proximal buds, while application on 50th day in the distal region. Early application of N *i.e.*, 20th day after pruning and later application of P and K 35-50th day after pruning, favoured fruit bud differentiation.

Application of nutrients through foliage is yet another method to increase efficiency of applied nutrients. It overcomes the soil and translocation factors in absorption of nutrients and envisages greatest economy of the nutrients. Foliar application of N as urea, P as calcium phosphate and K as potassium chloride resulted in the increased contents of N, P and K for improving yield and quality of fruit.

Irrigation

The grapevine needs regular irrigation in Southern India except at the time of pruning and monsoon. Irrigation is given at 6-8 days interval during fruiting period. In Madurai, the vines are irrigated four times a week during summer and two times a week during winter.

The irrigation timings should also be based on requirement of vines at the critical stages of bud burst, flowering and harvest. Other irrigations should be arranged to provide a reasonable moisture supply between these times. The water needs of vineyard are influenced by various factors.

The time to irrigate, number of irrigations and amount of water at each irrigation are determined by age and condition of vines and soil type. The climatic factors like temperature, humidity and wind influence greatly the water needs of vine yards. Young vines need more frequent irrigation than grown up vines. The vines trained on bower system need more water than those on kniffin or telephone or head system. The vines need not be irrigated during winter when they are remaining dormant. And so the soil moisture depletes to a very low level and for the vines to resume normal growth in spring, irrigation soon after pruning becomes necessary. Adequate water should be available to vines during the most critical period ie., between fruit set and maturity. The sugar accumulation in berries occurs during the hot period from mid May to mid June, during which there is great need for water. Stopping irrigation a few days before fruit maturity will improve the fruit quality.

According to estimates made in Uzbekistan, 1.5% of water is used from beginning of growth to flowering, 10% between beginning of flowering and fruit set, 43% during fruit growth and 45% during ripening, It is, therefore essential that the amount of irrigation water is regulated according to the stage through which the plant is passing.

Weed Management

Weeds between rows of vines are removed mechanically by tractor drawn implements. Within the rows, weeds are manually hoed and removed. Sometimes post-emergent weedicides, mainly glyphosate at about 2.0 kg/ha or paraquat at about 7.5 kg/ha is sprayed in fully grown vineyards.

Nutritional Deficiencies

Nitrogen: Nitrogen has a role in the flowering of grapes. Absence of Nitrogen in the fertilizer delayed the floral initiation in Anab-e-Shahi. It also influenced bunches/ vines and the mean bunch weight ultimately increasing yield/ vine. Need of this element is the greatest and response is the most rapid on sandy gravelly soils. On heavier soils, nitrogen availability and response occur more slowly. Crop yield may be reduced before deficiency symptoms of this element appear.

Phosphorus: Phosphorus has a clear cut role in floral bud initiation. Application of phosphorus promoted fruiting in Anab-e-Shahi vines. Increased levels of applied phosphorus were associated with increased fruit yield in Anab-e-Shahi and Thompson Seedless.

Potassium: Potassium was found to be favourable for inflorescence formation. Its application made those buds fertile which under normal conditions remained sterile through increased carbohydrate accumulation. Increased yields were obtained with increasing levels of K in Anab-e-Shahi and Thompson Seedless grapes. Potassium has an important role in the quality of grapes. Higher level of applied K was associated with higher total soluble solids content of grape juice. Deficiency of this element is associated with vineyard areas with very sandy soils. Potassium fertilizer must be placed near the active root zone. Measurable response to K fertilizer was obtained where visual deficiency symptoms appeared.

Magnesium: Magnesium requirement of vines is more during berry ripening. The deficiency is very common in vineyards of peninsular India, particularly in Anab-e-Shahi. The deficiency is mostly induced due to over manuring and is more prevalent in light sandy soils. Thus, Mg requirement is more during fruit development stage and in low Mg soil, it moves from leaf chlorophyll to the developing fruit causing chlorosis.

Calcium: Deficiency of calcium makes the berry skin susceptible to necrosis and rotting Higher levels of petiole calcium were associated with reduced yields in Anab-e-Shahi and Thompson Seedless. The exact role of Ca in vine production is, however, not clear. .

Micronutrients: Among micro nutrients, boron, zinc, iron, manganese and copper are important in viticulture. Boron is essential for regulation of carbohydrate metabolism. It is also important for pollen germination, growth of pollen tube and normal fruit set. Its deficiency is commonly known as 'hen and chicken disorder' of the clusters. Boron is delicate among the micro nutrients because of narrow range between its status of deficiency or excess. Shoot growth, bunch and berry size and sugar content of juice increased with boron application in Perlette grape. Boron spray, increased the protein and RNA contents of grape leaves. Thus, it seems to improve productivity and quality in grapes. Boron deficiency can be corrected by either soil application or foliar sprays of boron containing materials like boric acid or borax.

Zinc deficiency symptoms start appearing in mid summer when secondary or auxiliary growth begins. Secondary growth and the primary shoot tips are affected first. The main foliage symptoms of zinc deficiency are the chlorotic pattern, location and small size of affected leaves and particularly the widened angle formed between two basal lobes of leaf blade when the petiole is attached there, the primary veins tend to draw together, resulting in a fan like appearance. Foliar application of zinc improves vine growth, fruitfulness, fruit set, fruit yield and quality of berries.

Iron is needed for activation of several enzyme systems. Its deficiency impairs chlorophyll production resulting in characteristic chlorosis of young leaves. Deficiency of Fe is common in calcareous soils and in vines with weak root systems. Beneficial effects of iron on vegetative growth, fruit yield and quality of berries were reported. The concentration used ranged from 0.05 to 0.2% as iron sulphate.

Manganese serves as an activator for enzymes in growth processes and helps in chlorophyll formation. Its deficiency manifests in leaf chlorosis, common in acidic sandy soils. The deficiency can be corrected by spraying 0.2% neutralised manganese sulphate as soon as deficiency symptoms appear on veins.

Intercropping

Actually no intercrop should be sown in vineyards as it comes into bearing one year after planting and commercial crops start after 2-3 years of planting. Apart from this, the distance between vines is 3 x 3 metres and as such much space is not left between the plants for growing intercrops. Another factor is that almost entire plantations are trained on bower system of training. This structure with the spread of grapevines over it causes partial to complete shading in 3rd year. Also, the grapevine is a shallow rooted crop, therefore, root injury is quite possible in deep ploughing.

Keeping in view of above facts, intercrops should not be grown in vineyards. But grape growers have started planting of intercrops like radish, cucurbits, bhindi, brinjal, cabbage and potato for the first two to three years after planting the vineyards. Short duration crops like cucurbits are planted during summer when irrigation can be applied to these intercrops along with grapevine without any loss.

Growth Regulators

Plant growth regulators are well established as important cultural tools in grape production. They can control specific plant processes that cannot be conveniently or economically controlled by any other means. The growth regulators now can be applied either for promotion or reduction of growth, root initiation, bud or seed dormancy, fruit set, berry size, fruitfulness and total yields of seedless grapes, loosening of clusters, induction of seedless berries in seeded grapes and post harvest keeping quality of the berries.

The most commonly used auxins for rooting of grapevine cuttings are IBA and IAA. Soaking of basal ends of the cuttings in IBA (2500 ppm) solution for 15 seconds ensures better rooting of the cuttings. B-9 (1000 ppm) hastened termination of bud rest in grapes. Early bud break in Thompson Seedless grapevines can be achieved by dipping the canes in plastic cylindrical jar containing paclobutrazol solution (1500 ppm).When the flower bud clusters of the male vine were dipped in an aqueous solution of cytokinin (1000 ppm) before anthesis, they produced typical hermaphrodite flowers with normal fruit set. Spraying of SADH (2000 ppm) on Himrod just before anthesis resulted in 100% increase in fruit set. GA (75 ppm) 2-3 days after full bloom on Pusa Seedless variety increased bunch length, berry size and weight.

Maturity of grapes can be hastened by 6 days by dipping the clusters in ethephon (500 ppm) at the onset of ripening. Ethrel application (250 ppm) two months after set reduced the percentage of green berries to as low as 6.3 and 3.3 in Bangalore Blue and in Muscat respectively as against 30.9 and 25.9 in control. Ethephon (600 ppm) in Perlette at colour break stage advanced ripening.

For production of loose clusters pre-bloom GA sprays of 10 ppm and 15 ppm are given respectively on the 11th and 14th day after bud break for cluster elongation. Rachides of the clusters are trimmed to retain 8-10 berries, depending on number of leaves available/ cluster. Clusters are dipped in GA (30-40 ppm) when 10-20% of the flowers open in each cluster for berry thinning. GA (30-50 ppm) and Sevin (1000-2000 ppm) and ethephon (5-10 ppm) at the time of full bloom were very effective in flower thinning of Perlette grapevine.

CCC (2000 ppm) and IBA (100 ppm) were effective in checking flower and flower bud drop.Application of GA_3 (500 ppm) to Anab-e-Shahi variety 8 days before bloom decreased the bunch weight and number of berries while one-third of treated flowers developed seedless berries. Post harvest berry drop in Anab-e-Shahi grape is nearly 10% during transit and storage and NAA (100 ppm) was optimum for checking such type of losses. Shattering of berries during storage can be minimised with pre-harvest sprays of cycocel and kinetin at 4000 and 100 ppm concentrations, respectively.

Maleic hydrazide (500 ppm) sprayed at 5 leaf stage was the most effective in increasing the overall fruitfulness of the canes from 22.4 to 59.4%. CCC treatments significantly increased the number of bunches/ vine in Thompson Seedless. Maximum fruit set was recorded with CCC (600 ppm).

For increasing berry size, manual means are used to supplement chemical thinning to ensure adequate berry thinning and improving size of berries. Approximately 90-120 berries are retained per cluster depending upon number of leaves available to nourish it at 8-10 berries per every leaf depending on its size. Clusters are dipped in GA solution (40-50 ppm) once at 3-4 mm size of the berries and again at 7-8 mm size. When berry diameter is to be increased to more than 16 mm, clusters are dipped in a mixture of BA(10 ppm) + GA (25 ppm) or CPPU (2 ppm) + GA (25 ppm) or brassinosteroid (1 ppm) + GA (25 ppm) instead of GA alone at these two stages.

Canopy inter shading effects can only be manipulated by pruning alone yet the use of growth retardents have come to rescue. The excessive vegetative growth of vines can be curtailed by spraying the growth retardants like B-9 and phosphon-D at 0.12 and 0.05%, respectively at 10 leaf stage. Pac1obutrazol PP 333 (4-chlorophenyl-4, 4-dimethyl-2- (I, 2,3-trizol-I-YI) penton-3-01) inhibits synthesis of plant sterols and specially GA. Pre bloom sprays of paclobutrazol @ 1000, 1500, 2000 ppm in cultivars Roumi Red reduced shoot growth, internodal length but bud fertility was increased. The soil application @ 4 ml per vine was also equally effective. PP 333 was used as pre-bloom sprays and soil application. A single application of PP 333 was effective in reducing vegetative growth for a period of two years, if the rate of application was greater than 1.1 kg/ha.

In addition to the treatment with growth regulators, berry size and crispiness are increased by girdling. Width and depth of girdling are 1-1.5 mm. Girdling is done when the berries attain 4-5 mm diameter. Berry thinning and cluster thinning will help to maintain adequate leaf/fruit ratio, while girdling ensures a TSS content of 20°Brix.

Harvesting and Yield

Grape should not be harvested until they are fully ripe. The fruit is ready for picking after the berries near the tip have changed colour and have become sweet. There are many indicators to the proper stage of ripeness of grapes for picking but taste is the most valuable indication of ripeness of bunch. For this purpose, the greenest berries, which are always near the tips, should be tasted. If they are good to eat, the entire bunch is ripe for picking. Fully ripe bunches should be picked for local consumption while for distant markets, bunches should be picked when still hard ripe so that they do not spoil during transit. Ripe grape bunches should be clipped with scissors without injuring the berries. During handling, the natural bloom on berries must not be rubbed off. Harvesting should be done during the coolest period of the day. The picked fruits should not be exposed to the sun and, if it is to be sent to a distant market, it should be packed in sawdust. Trimming, grading and packaging should be done in shade. According to the Department of Agricultural Research and Education, ICAR, Government of India, in 1999-2000, grapes harvested from vines treated with hydrogen cynamide in October pruning did not have detectable residues and were safe for human consumption.

Well managed vineyard can be expected to start yielding after three years under North Indian conditions and even less than two years under tropical conditions of peninsular India. Vineyard can satisfactorily be maintained at least for more than two decades. The productivity/ hectare in India is much higher than in many countries. Yields as high as 100 t/ha in Anab-e-Shahi and 75 t/ha in Thompson Seedless were recorded in Hyderabad. However, quality of grapes is usually poor as a result of high yields. Approximately one million tonne of grapes is harvested annually in India. Grape is harvested almost all the year round. If not all the varieties, one or more varieties are always available at any given time of the year. Period of harvest and yield of different varieties is given below.

Variety	Yield (t/ha)		Period of Harvest
	Average	Potential	
Anab-e-Shahi	45	90	February-May, July, November-December
Bangalore Blue	40	60	January-March, June-December
Bhokri	30	50	November-December, June-July
Gulabi	30	50	January-March, June-December
Perlette	40	50	June
Thompson Seedless and other seedless varieties	25	50	January-April

However, major proportion of produce, mainly of Anab-e-Shahi, Thompson Seedless and its clones, is harvested during March-April from hot tropical region, which contributes more than 70% of total harvest.

Grading

Undesirable berries viz. unripe, over-ripe, small, mis-shapen, sun-burnt, decayed, soft and bruised should be trimmed. Bunches should be graded according to fruit quality, maturity and size and different grades should be packed in different containers. For nearby markets, mulberry or bamboo baskets can be used, whereas for distant market, wooden boxes of 4-5 kg capacity or Corrugated Fibre Board (CFB) cartons should be used.

Being a perishable fruit, it cannot withstand high temperature and low humidity prevailing at the time of harvesting which causes 25-30% losses through spoilage and cut down the already short storage life of grapes under ambient conditions. Thus after harvesting, it has limited possibility for transport to distant markets in a good condition which further restricts its marketability. It is, therefore, desirable to store sufficient quantity of grapes at harvesting time to spread its marketing over a longer period.

Packaging

For local markets, grapes are packed in bamboo strip baskets using newspaper and grape leaves as cushioning material. For distant markets, wood or corrugated cardboard boxes are used for packing. The size of packing is 6 or 8 kg in wood boxes and 2 or 4 kg in cardboard boxes. Wrapping of grapes in polyethylene bags is useful because it exchanges gases and allows permeability to water vapour with reduced moisture loss. It checks stem or berry shrivelling and provides good visibility of the packed produce. Sufficient perforations should be made in the polyethylene bags before packing. More recently in package SO_2 generators, both slow release and quick release type, have been introduced for commercial use. Popularly known as Grape Guard, the quick release type is a craft paper coated with a thin layer of a mixture of sodium bisulphite and a plastic polymer. The slow release (dual release) one is based on two poly plastic coated papers pasted together.

Storage

Strict cold chain is maintained right from harvesting by precooling and cold storage. Boxes are stored at 0-1°C temperature and 90-95% RH in cold storage. They are transported by sea in refrigerated containers.

Marketing

More than 80% of the total production is consumed as table grapes in India, and more than 70% of the total production is harvested in March-April, but the cold storage facilities are inadequate. Therefore, market gluts and fall of prices of grapes in March-April are common. Approximately, 2.5% (22,000 t) of fresh grapes are exported to the Middle East and European countries. The rest of the produce is marketed within the country. Grapes are exported through three different agencies viz., Grower Exporters, Growers Cooperatives and the Trader exporters. These agencies have established their own facilities for pre-cooling and cold storage in the vicinity of major production sites.

GUAVA

Guava (*Psidium guajava*) is an important fruit crop of India belonging to the family Myrtaceae. It has gained considerable importance because of its high nutritive value, availability at moderate price, a pleasant aroma and good flavour. It is the fourth most important fruit after mango, banana and citrus. Guava is native to tropical America where it occurs wild. It was introduced to India in the seventeenth century. Guava is one of the most common fruits liked alike by both the rich and the poor.

The fruit is relished when mature or ripe and freshly plucked from the tree. It is a good source of vitamin C, pectin, calcium and phosphorus. Fruit is used for the preparation of processed products like jams, jellies and nectar. Guava jelly puree is very popular for its attractive purplish red colour, pleasant taste and aroma. The puree can be used in juice, cakes, puddings, sauces, ice-cream, jam and jelly. Fruits can be preserved by canning. Good quality salad can be prepared from the shell (without seed core) of ripe fruits. Leaves of guava are used for curing diarrhoea and also for dyeing and tanning.

In India, presently, guava is grown throughout the country and is so much acclimatized that it appears to be a native of India.The area under guava cultivation in India increased by 64% (from 94 thousand ha in 1991-92 to 155 thousand ha in 2001-02) whereas the production increased only by 55%(from 11 lakh tonnes to 17 lakh tonnes). Allahabad in Uttar Pradesh has the reputation of growing the best quality of guava fruits in the world. Major guava producing states include Uttar Pradesh, Bihar, West Bengal, Maharashtra, Chhattisgarh, Tamil Nadu, Karnataka, Madhya Pradesh, Gujarat and Andhra Pradesh.

Varieties

More than 150 genotypes of *Psidium* sp are available in the Indian collections and are being maintained at several centers in the country in field gene banks. Some of the common varieties grown in India include 'Allahabad Safeda' (the most popular variety), 'Chittidar,' 'Sardar,' 'Seedless' and 'Red Fleshed.' The variety 'Lalit' has been released by the CISH, Lucknow, for commercial cultivation. Its fruits are medium sized (185 g) with attractive saffron-yellow colour and red blush. Its flesh is firm and pink with good blend of sugar and acid. It gives higher yield than popular variety 'Allahabad Safeda'. This is suitable for both table and processing purposes. Two hybrids, H 21 and H 136, have also been isolated for red pulp and soft seeds, and high TSS (12.5° Brix).

Important guava varieties cultivated in different states of India are given below:

States	Varieties grown
Andhra Pradesh	Allahabad Safeda, Lucknow 49, Anakapalli, Banarasi, Chittidar, Hafshi, Sardar, Smooth Green, Safed Jam, Arka Mridula
Madhya Pradesh	L-49, Allahabad Safeda, Gwalior-27, Hafshi, Seedless Chittidar
Jharkhad	L-49, Allahabad Safeda
Karnataka	Allahabad Safeda, L-49, Arka Mridula, Arka Amulya, Bangalore, Dharwar
Assam	Am Sophri, Madhuri Am, Safrior Payere
Bihar & Jharkhand	Allahabad Safeda, Apple Colour, Chittidar, Hafshi, Harijha, Sardar, Selection-8
Maharashtra & Gujarat	Nagpur seedless, Dharwar, Dholka, Kothrud, L-24, L-49, Nasik, Sindh
North-eastern States	Allahabad Safeda, Sardar, Red Fleshed.
Tamil Nadu	Anakapalli, Banarasi, Bangalore, Chittidar, Hafshi, Nagpur Seedless, Smooth Green
Uttar Pradesh	L-49, Allahabad Safeda, Lucknow Safeda, Apple Colour, Chittidar, Red Fleshed, Allahabad Surkha, Sardar, Mirzapuri Seedless, CISH-G-1, CISH-G-2, CISH-G-3
West Bengal	L-49, Allahabad Safeda, Dudhe Khaja, Gole Khaja, Kabli, Baruipur, Chittidar, Harijha, Sardar

In India, breeding work for guava improvement is done at several research institutions. At Fruit Research Station, Sangareddy, Andhra Pradesh, 2 hybrids, Safed Jam and Kohir Safeda, were selected from the crosses of Allahabad Safeda x Kohir and Kohir x Allahabad Safeda respectively and released. These hybrids are recommended for semi-arid tropical areas and have also been found suitable for juice purpose.

Climate and Soil

Guava is one of the hardiest fruit trees and it adapts itself to a variety of soils and climatic conditions. Even under most neglected conditions, it comes up well. Guava is grown in both tropical and sub-tropical regions from sea level up to 1,500 m. It tolerates high temperatures and drought conditions prevalent in North India during summer. However, it is susceptible to severe frost as it can kill the young plants. A cool winter induces heavy fruiting. An annual rainfall of about 100 cm. is sufficient but rainfall during the harvesting period deteriorates the quality of fruits.

Guava grows in all types of soils having pH ranging from 4.5 to 8.2. Good quality guavas are produced in river basins. The crop is sensitive to water-logging.

Propagation

Guava is mainly a self-pollinated crop though cross-pollination is also common. Propagation by seeds during early days gave rise to considerable variability which was utilized both by growers and horticulturists for selection of better types, and thus a large number of varieties came into existence and some of these types were later propagated vegetatively for commercial cultivation. Inarching, budding and air-layering are the vegetative methods of propagation commonly practiced. Propagation through root suckers, root cuttings and budding is sometimes successful.

Planting

Land is prepared during the summer season by ploughing, harrowing, levelling and removing weeds. Planting is done during the rainy season. June-July is the ideal time for planting the layers and seedlings. The plants are usually planted at a distance of 5-8 m. The exact planting distance is decided according to variety, soil fertility and availability of irrigation facilities. Standard spacing is 6 x 6 m. accommodating 228 plants /hectare. High density planting causes erect growth of branches making the plant tall, compact and also give higher yield/unit area in early years of fruiting. Square system of planting is generally adopted. Pits of 1x1x1m.size are dug before the monsoon and filled with a mixture of farmyard manure and soil.

The trials conducted under All-India Co-ordinated Fruit Improvement Project (ICAR) at different centres indicated that double-hedge-row system of planting with 800-900 plants/ha is beneficial for high yield/ unit area in Allahabad Safeda. But high density has adversely affected the quality of fruits, particularly total soluble solids and ascorbic acid. Spread of tree and number of branches/tree also decreased with increasing plant density.

Training and Pruning

Training of guava trees improves fruit yield and quality. Modified leader system of training is generally followed. The main objective of training guava plants is to provide a strong framework and scaffold branches suitable for bearing a heavy remunerative crop without damaging the branches. The flowers and fruits in guava are borne on current season's growth, hence, a light annual pruning is considered necessary to encourage new shoots after the harvest. Pruning is usually recommended after harvesting or in spring. Summer pruning is generally avoided as the plants get damaged due to sun burn. Dead, diseased, intercrossing branches and suckers coming up from the base and sides of the framework should be pruned back annually. Training of plants in young stage is essential to build a strong framework and to avoid weak crotches. The plants are trained as low headed trees to facilitate multiple hand pickings. Young trees require pruning several times a year to prevent the formation of long and slender branches. All flowers should be removed until the framework becomes strong enough.

Fertilizer Application

Even though guava is very hardy to soil and agro-climatic conditions, it shows good response to manuring in increasing fruit production. Research results have shown that guava is very responsive to the application of both organic and inorganic fertilizers. It was also observed that application of nitrogen stimulates growth and increases flower and fruit yields. Fertilization not only increased the yield but also improved the quality. Time of fertilizer application depends on the region and crop variety. In North India, fertilizer is applied in the first week of May for rainy season crop and in first week of July for winter season crop. The plants are manured twice a year, first during June-July and second by October.

The following are general doses of manures and fertilizers given to guava crop for obtaining higher fruit yield and better quality fruits.

Age of tree (years)	Farmyard manure (kg/plant)	Urea (g/plant)	Super phosphate (kg/plant)	Muriate of potash (g/plant)
1-3	10-20	150-200	0.5-1.5	100-400
4-6	25-40	300-600	1.5-2.5	600-1000
7-10	40-50	750-1000	2.0-2.5	1100-1500
10 and above	50	1000	2.5	1500

Guava generally suffers from zinc deficiency in India, but sometimes in certain areas, it suffers from boron deficiency also. In North India pre-flowering spray of boric acid (0.4%) and zinc sulphate (0.3%) corrected the deficiency. It also increased the fruit size as well as yield. Studies conducted at Gandhi Krishi Vigyan Kendra (GKVK) and IIHR, Bangalore, indicated common micronutrient deficiency of Zn. Soil application of 800 g $ZnSO_4$/ tree or foliar spray of $ZnSO_4$(0.5%) and boric acid (0.4%) 10-14 days before every flowering, effectively eliminated deficiencies in guava. At the University of Agricultural Sciences (UAS), Bangalore, bronzing was a common malady and the healthy leaves had narrow P: Zn, N : P and K:Zn ratios than the bronzed leaves. Soils in badly affected orchards showed very low pH and were poor in P, K, Zn and organic carbon. The poor nutrient status of N, P and Zn was the cause of bronzing. Soil applications of N, P, K, Zn at 200, 80, 150 and 50 g/plant/year respectively or foliar applications of these nutrients each at 2% for 4 months have resulted in reduction of bronzing.

Irrigation

Guava is mostly grown under rain fed condition. Irrigation has however, proved useful in cultivation. In North, bi-weekly irrigation during winter and also during April-June improved fruit size of the winter crop and increased fruit set and reduced fruit drop in both the seasons. In regions with 380-500 mm rainfall, additional 2,460 mm water is required through 8-10 irrigations, once in every 20 days in summer and every month in winter. Period during April-June was critical. One or two irrigations between the end of monsoon and harvesting (winter) are given in Northern India. In Southern India, irrigation throughout the year is necessary. Drip irrigation has proved to be very beneficial for guava as it resulted in substantial increase in size and number of fruits and also a saving of 60% irrigation water.

Weed Management

Weeds are usually removed by shallow cultivation. Green manuring is usually done during rainy season. Pre-emergence use of diuron (1.6 kg./ha.), oryzalin (1.67 litres/ha.), simazine (1.6 kg./ha.) or atrazine (1.6 kg./ha.) was effective in control of weeds in guava orchards.

Mulching

Dry leaves or straw are used as mulching material. Mulching can also be done either with black polycthylene sheet or with organic materials. Mulching the soil at least twice a year helps in conserving moisture and improving the fruit quality.

Intercropping

Leguminous crops or vegetables can be grown as intercrops during the first three years of planting provided irrigation facility is available. The growing of a green-manure crop during the rainy season, and clean cultivation during the rest of the year are recommended. Intercropping should be stopped when the trees are fully grown.

Growth Regulators

The winter crop is much superior in quality compared to the monsoon crop. Farmers often reduce monsoon crop by deblossoming to get a higher price. This is done by growth regulators like maleic hydrazide on spring flush of flowers. Growth regulators like NAA, and 2, 4-D are effective in thinning of flowers and also manipulating the cropping season.

Harvesting and Yield

The plants start bearing at an early age of 2-3 years but they attain full bearing capacity at the age of 8-10 years. The yield of a plant depends on its age, cropping pattern and the cultural practices. If both rainy and winter season crops are taken, more yields may be obtained in the rainy season. After 100-110 days of growth, guava fruits attain maximum growth, sucrose content and pectin. Guava fruits develop the best flavour and aroma only when they ripen on trees. Fruits must be plucked as they ripen. In most of the commercial varieties, the stage of fruit ripeness is indicated by colour development which is usually yellow. Plucking extends over several weeks.

Guavas are harvested throughout the year (except during May and June) in one or the other regions of the country. However, peak harvesting periods in North India are August for rainy season crop, November- December for winter season crop and March-April for spring season crop. In the mild climatic conditions of other parts of the country, the peak harvesting periods are not so distinct.

A 10 year old plant yields about 100 to 150 kg of fruits every year. Yields of 22,000 kg/ hectare have been reported.

Storage

Fruits are graded on the basis of their weight, size and colour. The fresh fruit has a short shelf life and distant marketing can be done only if it is properly stored. It can be stored for about 10 days at room temperature (18-23^{0}C) in polybags providing a ventilation of 0.25%. The shelf life can be extended upto 3 weeks by keeping them at low temperature of 5^{0}C and 75-85% relative humidity. Fruits of Safeda guava could be stored for 4 weeks in cold storage at 10°C. Irradiated fruits showed slower rate of ripening, sugar accumulation and weight loss during storage.

Packing

The shelf-life of fruits can be extended by waxing, pre-packaging and fungicide treatments. The fruits are either packed in boxes or in baskets of different sizes.The fruits are packed in baskets made from locally available plant material. For distant markets, wooden or corrugated fibre board boxes are used along with cushioning materials viz. paddy straw, dry grass, guava leaves or rough paper. Good ventilation is necessary to check build up of heat. Guava is a delicate fruit requiring careful handling during harvesting and transportation.

Transportation

Road transport by trucks/lorries is the most convenient mode of transport due to easy approach from orchards to the market.

Marketing

For long-distance marketing, it is necessary to harvest the fruit somewhat earlier. For local markets, fully yellow but firm fruits are harvested, whereas half yellow fruits are picked for distant markets. Fruits are harvested selectively by hand along with the stalk and leaves. Being perishable in nature, the fruits should be immediately sent to the local market after harvesting and only a small quantity need only to be sent to the distant markets. However, sometimes it is stored for a few days to meet the market demand. Ripening could be retarded, if mature green guava fruits are treated with MH and gibberellic acid or hastened by 2, 4-D, 2, 4, 5-T and ethephon. Majority of the cultivators sell their produce either through trade agents at village level or commission agents at the market.

JACKFRUIT

The jackfruit (*Artocarpus heterophyllus* Lam.) of the family Moraceae is native to south western India and Sri Lanka.The English name jackfruit is derived from Portuguese *jaca,* which is derived from Malayalam *chakka*. The jackfruit is a huge, spined and oval fruit that is believed to have been first cultivated in Indian rainforests. It is an excellent example of a food prized in some areas of the world and allowed to go to waste in others. It is cultivated at low elevations throughout India, Burma, Ceylon, Southern China, Malayasia and the East Indies.

In India, jackfruit is eaten raw or used in curries, soups, and stews. It also flavors various deserts and is a common ingredient in fruit salads. It is not a popular fruit among the well to do of India, and the fruit is very inexpensive there. Not much profit can be made from the cultivation of the jackfruit. However, Bangladesh claims it as its national fruit.The fruit itself is the largest fruit in the world. It can weigh up to 40 kg and be up to 1m long.

In South India, jackfruit is a popular food ranking next to mango and banana in total annual production. There are more than 1,00,000 trees in backyards and is grown for providing shade in betel nut, coffee, pepper and cardamom plantations. The total area under jackfruit in India is estimated as 26,000 ha. Government horticulturists promote the planting of jackfruit trees along highways, waterways and railroads to add to the country's food supply.

Varieties

In South India, jackfruits are classified into two general types: 1) Koozha chakka (soft fleshed) the fruits of which have small, fibrous, soft, mushy, but very sweet carpels; 2) Varikka chakka (firm fleshed) more important commercially, with crisp carpels of high quality. A variety known as honey jack, has sweet pulp and is claimed to be the best of all. Muttam Varikka is another important firm fleshed variety which is very much preferred in Kerala. Singapore or Ceylon jack is an early bearing type producing fruits in 1.5 to 2.5 years of transplanting. It was introduced to India from Srilanka and planted extensively in 1949. The fruit is of medium size with small, fibrous carpels which are very sweet. In addition to the summer crop (June and July), there is a second crop from October to December. Khaja Allahabad is a white fleshed heavy bearing variety grown in Uttar Pradesh.

In 1961, the Horticultural Research Institute at Saharanpur, India, collected the air-layered plants of varieties, 'Safeda', 'Khaja', 'Bhusila', 'Bhadaiyan' and 'Handia'. The Fruit Experimental Station at Burliar, established a collection of 54 jackfruit clones from all producing countries, and ultimately selected 'T Nagar Jack' as the best in quality and yield. The Fruit Experimental Station at Kallar, began breeding work in 1952 with a view of developing short, compact, many-branched trees, precocious and productive, bearing large, yellow, high quality fruits. In Assam, nurserymen have given names such as 'Mammoth', 'Everbearer', and 'Rose-scented' to preferred types. Horticulturists in Madras have found that hand-pollination produces fruits with more of the fully developed bulbs than does normal wind-pollination.

Climate and Soil

The jackfruit is adapted only to humid tropical and near-tropical climates. It is sensitive to frost in its early life and cannot tolerate drought. If rainfall is deficient, the tree must be irrigated. In India, it thrives in the Himalayan foothills and from sea-level to an altitude of 1,500 m amsl, in the south. It is stated that jackfruits grown above 1,200 m amsl,are of poor quality and can be used only for cooking.

The jackfruit tree flourishes in rich, deep soil of medium texture, sometimes on deep gravelly or laterite soil. It will grow in shallow limestone soils but the growth is slow. In India, the trees grow tall and thin on sand, short and thick on stony land. It cannot tolerate "wet feet". If the roots touch water, the tree will not bear fruit or may die.

Propagation

Propagation is usually by seeds which can be kept no longer than a month before planting. Germination requires 3 to 8 weeks but is hastened by soaking seeds in water for 24 hours. The seeds may be sown *in situ* or may be nursery-germinated and transplanted at 4 leaf stage. A more advanced seedling, with its long and delicate tap root, is very difficult to transplant successfully.

Air-layers produced with the aid of growth promoting hormones come to bear early and produce fruits of known quality. Propagation of selected clones is common by grafting (inarching, epicotyl and soft wood grafting) and budding (forkert, chip and patch) to seedling rootstocks of jack fruit itself. Epicotyl grafting performed during April- May at Konkan gave about 90% success and the grafts attained planting size within one year. At Calcutta University, cuttings have been successfully rooted only when etiolated shoots were treated with indole butyric acid (5,000ppm) and kept under mist. Tissue culture experiments have been conducted at the Indian Institute of Horticultural Research, Bangalore.

Planting

Soaking one-month-old seedlings in a gibberellic acid solution (200 ppm) enhances shoot growth. Gibberellic acid spraying and pasting increase root growth. In plantations, the trees are planted 9-12 m apart. Young plantings require protection from sunscald and from grazing animals. Seeds in the field may be eaten by rats. In Travancore, the whole fruit is buried, the many seedlings which spring up are bound together with straw and they gradually fuse into one tree which bears in 6 to 7 years. Jackfruit is a fairly rapid grower and lives as long as 100 years. However, productivity declines with age.

Training and Pruning

After harvesting, the fruiting twigs may be cut back to the trunk or branch to induce flowering in the next season. In the Cachar district of Assam, production of female flowers is said to be stimulated by slashing the tree with a hatchet, the shoots emerging from the wounds and branches are lopped every 3 to 4 years to maintain fruitfulness.

Fertilizer Application

Little attention was given to the tree's fertilizer requirements. Severe symptoms of manganese deficiency have been observed in India.

Irrigation

Protective irrigations are necessary for young plants at an interval of 12 to 15 days depending on soil and climatic conditions.

Harvesting and Yield

In Asia, jackfruits ripen principally from March to June, April to September, depending on the climatic region, with some off-season crops from September to December. Fruiting season lasts for about four months and fruits mature within 3 to 4 months after flowering. In Jamaica, an "X" is sometimes cut in the apex of the fruit to speed ripening and improve flavor. Seedling trees come to bearing after eight years with exception of Ceylon jack which bears in less than three years and also the grafted ones. In India, a good yield is 150 large fruits per tree annually. The tree attains its peak bearing at 15 to 16 years of planting and produces about 250 to 500 fruits. On an average 40 to 50 tonnes of fruit can be obtained from one hectare.

Storage

Jackfruits turn brown and deteriorate quickly after ripening. Cold storage trials indicate that ripe fruits can be kept for 3 to 6 weeks at 11.11-12.78°C and relative humidity of 85 to 95%.

Marketing

Jackfruits do not have a ready to buy market as seen in other fruits. Usually the market is confined to the use of the grower or it will be distributed among the neighbours and the major portion of the fruits produced during the rainy season mainly goes as a waste. Sometimes middlemen will approach the grower and make the deal at a very low rate for the fruits of that season and they sell it on the road sides as whole fruit or in pieces.

LITCHI

Litchi (*Litchi chinensis Sonn*) is believed to have originated in South China. It is an important sub-tropical evergreen fruit crop belonging to family Sapindaceae. As it is highly specific to climatic requirements, its cultivation is restricted to a few countries of the world.

Litchi is consumed in fresh form and is used in processing industry for making jams, jellies and squashes. Fruits are a rich source of Vitamin C, P, Ca and Iron. It also contains small amounts of Vitamins A and B. Processing of litchi is done in different forms. Canning of pulp, aseptic packing and ready to serve litchi juice are common. Dried litchi processing is not in practice. Pulp of litchi is aseptically packed and stored at 2-3°C for preparing litchi juice.

India and China account for 91% of the world litchi production but it is mainly marketed locally. India is the second largest producer of litchi in the world after China. In India, litchi ranks 7th in area and 9th in production among fruit crops but in value terms, it ranks sixth. There has been substantial increase in area and production of litchi in the last 50 years. In India, 4,28,900 tonnes of litchi is produced annually from 56,200 hectares. Evidently, production and productivity of litchi are constantly increasing in the country. Bihar is the largest litchi producing state in India where litchi is the livelihood for millions of people. The second largest litchi producing state is West Bengal followed by Tripura and Assam. Productivity is the highest in Bihar followed by West Bengal. It is also grown in the states of Uttar Pradesh, Punjab and Haryana.

Varieties

Although about 100 varieties of litchi are identified in the world over, only 15 of them are commercially cultivated in India. In Bihar, the recommended varieties include 'China, 'Purbi', 'Deshi', 'Bedana' and 'Dehra Rose'; in Uttar Pradesh, 'Rose-scented', 'Early Large Red', 'Kalkatia', 'Gulabi' and 'Late Seedless' and in West Bengal, 'China' and 'Muzaffarpur'.

Litchi varieties grown in India are highly variable under different climate and soil conditions. When distinguishing the cultivar, the shape of skin segments and protuberances are the reliable and stable genetic characteristics. Fruit size, shape and taste also vary and are influenced by factors other than genetic factors. Indian cultivars vary greatly in vegetative flushing pattern, flush colour and flowering ability. Based on these characteristics, cultivars were classified into five groups (Singh, 1998). Group A, which has 7 cultivars is the early group, B and C groups are mid-season, and group D is the late group. Only one cultivar, which is very late, is under group E and its cultivation is confined to Muzaffarpur.

Leaf colour along with shape and size of leaves is of importance in varietal identification. The fruit shape of litchi is very distinguishing. The round shape of Bedana is distinguished from the oblong shape of China or Shahi. The apex of fruit can be round, obtuse, blunt as in Shahi, or pointed as in China. The varieties can also be distinguished depending upon colour of the new flush and season of flushing. Shahi produces very light coloured flush while China has pinkish flush. Bedana has very dark pink flush. Emergence of panicle and its shape also differs. Shahi has long panicles while Bedana produces short and compact panicles. Colour of the litchi fruit is pinkish brown or dark red depending on cultivar. Colour of the skin varies and is also influenced by growing conditions.

Skin thickness depends on the cultivars. Bedana and China have very thick skin. Rose Scented and Shahi have very thin skin. Skin surface at maturity also varies being smooth, swelling, sharp and pointed. Protuberances of the skin (pericarp) can be smooth and sharply pointed. Presence and absence of seed as well as structure and size of seeds also vary from cultivar to cultivar, although it is influenced by environmental conditions. In Rose Scented and Bedana, a high proportion of chicken-tongued seeds (aborted seeds) are observed while China has bold seeds. In a recent selection cv. Swarna Rupa, a high proportion of fruits have small seeds (Singh and Yadav 1992).

Cultivars can be grouped into three categories based on period taken for fruit maturity namely, early, medium and late season. Fruit maturity also varies slightly and relatively from year to year depending upon prevailing weather conditions. The maturity period of fruit also varies depending upon agro-climatic regions. Some of the varieties in West Bengal come to maturity earlier, while the same varieties can be late in Uttar Pradesh. Due to production of litchi in varying agro-climatic conditions, maturity, fruit colour, shape and size are reported to be varying. Thus, there has been much confusion in the names of varieties. As a result, the same variety is called by different names at different locations (Ray et al. 1984).

Climate and Soil

Litchi cultivation is highly successful in areas having minimum temperature of 10°C from December to February and 38°C from April to June. However, temperature of 32° C during these months is optimum. It is highly specific to climatic requirement for its establishment, plant growth and fruiting, and consequently spread of area. A moist atmosphere, occasional rainfall, cool dry winter free from frost and hot winds are ideal for its cultivation.

Flower bud differentiation, flowering, fruit set, fruit quality and flavour development in litchi are influenced significantly by temperature and humidity and it has adapted well in the sub-tropics where summer months are hot and wet and winter months are dry and cool. Hot summers free from hot wind and winters free from frost are essential.

In litchi growing areas in India the temperature varies from 21 to 37.8°C during flowering and fruiting. It is observed that flower initiation in litchi requires comparatively low temperature. Seasonal variation in temperature is favourable for proper fruiting. A dry climate, free from rains for about 2 months before flowering induces flower bud differentiation, blossom and consequently give high production. In Chotanagpur, the fog free dry winter, mild sub-tropical summer and intermittent pre-monsoon showers during April-May have been observed to be highly favourable for blossoming, better aril development and improvement in fruit quality. The sub-tropical to mild temperate climate in the foothills and valleys of the Himalayas are also suitable for litchi cultivation. Depending upon the temperature rise after winter, the time of flowering and maturity is determined. No fruiting has been recorded when litchi has been grown in tropical conditions. However, on hills in southern states flowering is observed and harvesting commences in November-December.

In India, litchi is grown successfully on a wide range of soil types, which include sandy loams, laterite, alluvial sand, and calcareous soil, but the best litchi orchards are seen in alluvial sandy loam soils with good drainage and access to the water table. The performance of orchards is very poor on clay soil with poor drainage. Litchi grown in sandy soils has a root system network, while trees grown in clay soil have very poor root distribution. The pH of the soil in North Bihar range from 7.5 to 8, while in Jharkhand litchi grow well at a pH of 6 to 6.5. It grows well even in calcareous soil with 30% free lime content. However, in acidic soil of Jharkhand mycorrhizal activity is minimal which affects fruit yield and quality. The taxonomy, morphology and mycotrophic habit of mycorrhiza association with litchi was described by Pandey and Misra (1975) and their work also confirmed that litchi requires mycorrhiza to grow and produce better quality of fruits. Thus, it is often suggested that new orchards should be grown with the introduction of soil from old orchards.

Propagation

Litchi is generally multiplied by vegetative methods of propagation as plants raised through sexual method (by seed) grow slowly, have a long juvenile period and do not produce true to type fruits. However, earlier introduction in different parts of the country was perhaps through seeds, which enabled the selection of superior types and perpetuation of the cultivar through vegetative means. The most commonly practiced method of vegetative propagation is air-layering. Inarching, budding and grafting are also possible. One-year-old layers are planted nine metres apart in pits during the rainy season.

Planting

Orchard establishment is a highly specialized activity, which requires proper planning, selection of site, land preparation, layout, planting of saplings, as well as orchard protection and management.

Before layout, the land is cleared of bushes and other weedy vegetation and is leveled with a mild slope in the opposite direction of the water source. To improve the fertility of the soil organic matter is added. A green manure crop is grown and incorporated into the soil, which improves its fertility, moisture holding capacity and physical condition.

Pits of 90 x 90 x 90 cm in dimension are dug at the spacing decided for the orchard. Pit opening is normally recommended in April-May to have a sterilization effect for about 3 days. Before the onset of monsoon pits are filled with topsoil mixed with about 40 kg decomposed compost, 2 kg neem cake, 1 kg bone meal and 200-300 g muriate of potash. Incorporation of about 2 baskets of soil from the root zone of old litchi trees encourages the mycorrhiza growth. Then the soil is allowed to settle with the first few rains and leveled properly. Planting is done during June to July. At the time of planting, a hole is made in the centre of the pit at the marked point where the plant is fixed and the soil is pressed to remove air. Watering is done immediately after planting and the plant is regularly irrigated till it is properly established.

Litchi is an evergreen spreading tree, which attains the height of about 10-12 m at its full growth and development. Light penetration of its canopy is also desirable for proper fruiting, hence planting in square system at a distance of 9-10 m within and between the rows has been practiced. However, in an experiment conducted at the Central Horticultural Experiment Station, Ranchi, planting of litchi in a double hedgerow system at a distance of 4.5 x 4.5 x 9 m accommodating 329 plants/ha was found to be the best and gave higher yield of equally good quality fruits up to 16 years of plantation. High density planting adopting a double row system was also found superior at other locations in terms of yield and quantity of fruits.

Training and Pruning

Training of the plant in the initial stage is essential to provide the required framework. Unwanted branches should be pruned to provide definite shape and to promote growth of the trunk and crown of the tree. Three to four branches 60-75 cm from ground opposite to each other are allowed to form the proper frame of the tree. Further, crowded and crisscross branches are removed to facilitate better growth. The branches with narrow angles are also avoided as they are prone to breakage. Non-fruiting unproductive branches inside the canopy should also be pruned. Dried, diseased and scissors-shaped branches should also be periodically removed.

Light pruning after harvest is congenial for better growth, fruiting and yield. As the fruit is borne on growth of the previous year, the common practice is harvesting the fruit bunch along with 8-10 cm of twig to promote new flush and better bearing in the succeeding year.

Fertilizer Application

Among the several factors associated with production of litchi, balanced nutrition is the most important which determines productivity and quality. Litchi responds to exogenously applied manures and fertilizers and response varies depending upon cultivar (Kotur and Singh, 1993), climatic conditions and soil types. A survey conducted in the Doon valley indicated that 80% of orchards are low in N and P and need N and P application (Kunwar and Singh, 1993). In West Bengal, NPK was below the optimum level (Rao *et al*, 1985). However in Punjab N, K, Mg, Zn and Mn are in deficit range. Variation in nutrient content was also observed among varieties.

Field experiments conducted on different cultivars at different locations have clearly demonstrated the effect of a graded dose of NPK on growth, yield and quality of fruits. Application of 600-800 g N, 200-300 g P_2O_5 and 400-600 g K_2O/ plant is recommended for 12-15 year old trees. In general, application of FYM, potassic and phosphoric fertilizers in major litchi growing areas of the country is done during June-July, just after harvesting of crop. Nitrogen is applied in 2 split doses. The first dose is applied after fruit set, in March-April while the remaining half dose is applied immediately after harvesting of crop. Excessive application of nitrogenous fertilizer before flowering should be avoided. Phosphorus application at the time of flower bud differentiation improves flowering and fruiting. Application of cakes and manure are generally practiced to get better quality fruits. In general, litchi orchards maintained under higher doses of organic manure have better yield and quality as compared to orchards maintained with chemical fertilizers. In acidic soil, application of 10-15 kg lime/tree once in 3 years increased the yield.

Foliar application of zinc sulphate (0.1%) is done twice, 10-15 days before flowering for improving sex ratio and to reduce fruit drop. If a deficiency of zinc and magnesium is observed, application of 150-200 g $ZnSO_4$ and 150-200 g $MgSO_4$, respectively per plant during September has been beneficial. Foliar application of 0.1% borax, 2-3 times during period of fruit growth and development of the trees enhances fruit retention, minimizes cracking, improves fruit colour and sweetness, and advances maturity. Other micronutrients like Fe (Ferrous sulphate), Cu (Copper sulphate), Mg (Magnesium sulphate) are applied if deficiency symptoms are observed. Two to three sprays in a year are sufficient to maintain the trees in good health.

Irrigation

Litchi being an evergreen plant, maintenance of optimum soil moisture is critical for growth, development and fruit production. If rainfall is evenly distributed, litchi grows successfully and supplementary water requirement depends upon cultivar and evaporation demand. Water requirement ranges from 600-800 mm. Investigations carried out to determine the irrigation needs indicated clearly that irrigation is critical at fruit development stage to get better yield and quality of fruits. Interestingly, differential management of water in vegetative phase and reproductive phase is also suggested. To achieve faster growth of plant, no water stress should be permitted, while in the reproductive phase water stress is beneficial at the time of fruit bud differentiation. Light irrigation during summer and winter and cleaning of the basin are advocated. Irrigation at the intervals of 2-3 days during initial stage of plant establishment is essential. Further, the young plants should be irrigated during dry periods and winter months at intervals of 3-5 days. Mulching with dry leaves in the basin helps in better moisture conservation in young plants. Experiments conducted at Ranchi indicated that irrigation of plants at alternate day intervals, 6 weeks before harvesting improves fruit retention, encourages better fruit development, and minimizes the cracking, apart from quality of fruits. Certain physiological disorders like poor sex ratio, poor fruit set, heavy fruit drop and high fruit cracking, besides sunburn of fruits can be minimized with proper water management. The basin or flood method of irrigation is normally practiced. However, adoption of drip irrigation has been effective in the economic use of water and enhanced growth, especially in an area where water is scarce.

Weed Management

Weeds compete for water and nutrients during the initial year of growth. Depending upon location, more than 25 species of weeds are present in litchi orchards. Inter-culture operation and mulching suppress the weeds. Herbicides, although found effective for the control, are rarely applied. In mature orchards, the weed problem is not encountered due to heavy shade and natural mulching by falling leaves.

Mulching

Moisture conservation through mulching using dried weeds or black polythene sheet is useful. Trials were conducted to conserve moisture using farm residues and polythene sheets. Through adoption of mulching, frequency of irrigation is reduced. In a trial conducted at NRC on litchi, Ranchi, mulching with 3 irrigations was effective in reducing cracking and enhancing yield and quality of fruits (Singh, 1986). To check fruit cracking, mulching with 3-4 irrigations during fruit growth has been satisfactory.

Intercropping

Litchi is a slow growing plant and takes about 15-16 years to develop canopy and cover the area. During the initial period of establishment, the space between the plants can be utilized for planting intercrops. Planting of guava, custard apple, lime/lemon in the centre, between and within the rows of litchi gave additional income in the initial stage of planting without competing with the main crop. Papaya is also planted as filler plant at a spacing of 2.5 x 2.5 m. In between the plants in the initial stage, cowpea, french bean, okra, brinjal or other suitable crops of the regions are grown. In the mature litchi orchards, cultivation of partial shade loving plants (ginger, turmeric, elephant foot yam) is practiced successfully, which provides additional income.

Growth Regulators

Foliar application of the plant bio-regulator, NAA (20 ppm) at an interval of 10-15 days during period of fruit growth and development, twice or thrice, is useful for minimizing the fruit drop. IAA may be substituted for NAA, if the latter is not available. Spraying plain water four to six times in early morning hours of the day during the advanced stage of fruit growth and development is highly effective for achieving better growth with minimized fruit cracking.

Physiological Disorders

The splitting or cracking of fruits is quite common in almost all the litchi growing areas of India, particularly under dry conditions. In severe cases, the crop losses due to this malady may be as high as 20% or even more. Splitting of fruits is generally attributed to abrupt changes in atmospheric temperature, relative humidity and soil moisture conditions, following rain or heavy irrigations. This disorder occurs shortly before maturity when rains follow a prolonged period of drought during which the fruit gets checked in growth. Excessive soil moisture aided by fluctuations in temperature and humidity may aggravate fruit splitting.Not much work has so far been done on this problem However, the following measures arc generally considered appropriate to minimise fruit splitting.

1. Regular irrigation in orchards helps in maintaining growth and expression in fruit.
2. In the absence of rains during summer, water spray proves useful in keeping the ambient atmosphere of the fruit humid, as moisture has a good local effect on the fruit against splitting.
3. Some filler trees can be planted in quincunx system of layout in which the trouble is minimised.
4. The litchi plants should be trained to keep them low headed. Such trained plants with dense foliage, withstand more heat and desiccating winds as compared to tall trees and thus fruit splitting is the least in their case.
5. One or two rows of tall trees e.g. Safeda alternating with small sized trees having dense foliage (mulbery, Jamun, desi ber) may be planted around the orchard as windbreak. This wall of trees helps to create humid conditions in the orchard, which is favourable for developing fruits and thus splitting is avoided to a great extent.
6. Varieties which are less prone to splitting should be planted. Early varieties like Dehradun and Saharanpur split more than the mid-season or late varieties.
7. The litchi orchard should be planted near large bodies of water or in between tall and dense trees like mangoes.
8. The plants can be guarded immensely against high temperature by sowing Jantar or Arhar along the tree rows or in the periphery of the plant basin. These help to minimise splitting of fruits, beside protecting fruit trees from vagaries of hot weather.
9. NAA and 2,4,5-T (35-100 ppm) are effective in checking fruit splitting and increasing fruit size.

Harvesting and Yield

Litchi being a non-climacteric fruit requires to be harvested after attaining full maturity on the tree. Studies are conducted to determine maturity standard for different cultivars under different agro-climatic conditions. Fruits have a sigmoid pattern of growth. First the pericarp develops, then the seed and aril are formed and the seed turns green to brown. During maturity, acidity declines and TSS increases which is correlated with appearance and colour of fruit. Thus, colour of fruit is an important criteria to decide the harvesting stage (Singh and Yadav, 1988). The red pigmentation in litchi is associated with anthocyanin pigments, which develop better in the direction of good light penetration. Invariably pericarp colour and smoothness of pubicles are the best indicators. Depending upon the cultivar, 65-80 days are taken from fruit set to maturity.

In India, maturity of litchi commences first in Tripura, followed by West Bengal, then Bihar. The first and second weeks of May are times for harvest in the eastern region, while litchi of Bihar matures in the 3rd -4th week of May and continues up to first week of June. Litchi in Uttar Pradesh and Punjab is ready for harvest during 2nd - 3rd week of June. In Himachal Pradesh, litchi of the same cultivar is harvested in the last week of June. In most of the states, the best litchi orchards are seen along the rivers, big or small.

Fruits are harvested in bunches along with a portion of branch and a few leaves. At the time of harvesting, care is taken to harvest selected bunch, which has attained desirable maturity as determined by colour development and taste of pulp. For distant market, fruits are harvested when TSS attains 19° Brix and acidity 0.3 to 0.4%. The fruits are harvested early in the morning when temperature and humidity are congenial, to have longer shelf-life of the fruit. At the time of harvest, fruits are collected in a manner so that they do not fall on the ground. Use of mechanical tools for harvesting is practiced. The harvesting period is generally May-June, depending upon cultivar and location. However, in the hills of Southern India, litchi is harvested in November- December.

The yield of litchi varies according to the age of tree, agro-climatic condition and maintenance of the orchard. Usually about 80-150 kg fruits/ tree is obtained from 14-16 year old trees. However, from a fully grown tree, a yield of 160-200 kg/tree is recorded. Apart from management practices, bee keeping in litchi orchards increases the yield of quality fruits by 15-20%, since litchi needs cross-pollination. *Apis mellifera* is the commonly used bee in litchi orchards, which also provides additional income from honey.

Grading

Litchi fruits should be graded and then packed in mulberry baskets of suitable size. Soft dry grass or banana leaves should be placed in baskets before litchi clusters are piled up. The clusters should not be packed too deep or too high as the bottom fruit get crushed. The fruit clusters arc covered with some more grass or banana leaves and a piece of moist gunny bags should be wrapped around the basket and tied securely.

Storage

Litchi deteriorates very fast after harvest. Pericarp browning is a major post-harvest problem, which renders the fruit unmarketable. Browning is associated with desiccation. Peroxidase activity coupled with ascorbic acid oxidation enhances anthocyanin degradation. Techniques to reduce browning and maintain the red colour and prolonged storage life include sulphur treatment and packaging in perforated plastic bags and storage under cold conditions. Sulphur dioxide (SO_2) fumigation is used as a post-harvest treatment to reduce browning. SO_2 treated fruits have a bleached pericarp which turns uniformly pink in colour after 2-3 days. Fumigated fruits absorb 30-65% of applied SO_2. There is increasing concern about the residue of sulphur and the residual limit is only 10 ppm. For sea transportation 600-650g sulphur is recommended for a duration of 50-60 minutes, while for air transport 300-400 g sulphur for 30 minutes are advocated. Aril breakdown or softening of the aril involves a loss of turgidity and translucency where fruits become blunt in taste. The disorder starts near the pericarp and is prevalent at end of stem. Post-harvest decay also occurs due to bacteria, yeast and fungi. Irradiation of fruit reduces browning and post-harvest losses. Storage temperature of 2-5°C extends the shelf-life.Use of perforated polythene bags and storage at 3°C increase shelf-life. Controlled atmosphere storage is better for maintenance of freshness of fruits.

Marketing

A substantial amount of litchi produced in the country is consumed locally. Litchi, being highly perishable, is available for very short duration. Marketing of fruits is done in different forms. Growers rent their orchards to contractors, who in turn harvest and sell to local markets.

In chains of marketing, more number of middlemen reduces share of grower's income from the price of produce paid by consumers. Generally, for domestic market, litchi is packed in 10 kg boxes or baskets having a lining of litchi leaves. Considerable importance has been given to the packing of litchi for domestic markets. Now litchi is packed in 2-2.5 kg boxes and transported in cool-chain.

In the last few years, serious attempts are made to export litchi from India. Test consignments were initially sent by air and the technology for sea transport was perfected. The exportable litchi is packed in 2 to 2.5 kg or 5 to 6 kg boxes after sulphur treatment. Quality standards are managed as per standard developed by APEDA. The Agricultural Produce Export Development Authority (APEDA) has developed a procedure for production of quality litchi. The steps are: production > inspection of farm > harvesting > desolating and sorting > receipt at packhouse > acceptance of produce > sorting and grading > sulphur treatments > packing and cooling > palletization > storage > container loading and transportation. The strong cooperative marketing and infrastructure facilities developed are expected to promote marketing of litchi.

MANGO

Mango *(Mangifera indica* L.) is the most important fruit of India belonging to family Anacardiaceae. It is known as the king of fruits. No one knows the exact origin of mango but most believe that it is native to the Southern and Southeast Asian continent including India, Mynamar, Sri Lanka and Bangladesh. Reference to mangos as the "food of the gods" can be found in the Hindu Vedas. The name of the fruit came from the Tamil word *maangai* or the Malayalam word *maanga* and popularised by the Portuguese after their Indian exploration, hence the word 'manga' in Portuguese. Besides delicious taste, excellent flavour and attractive fragrance, it is rich in vitamins A&C. The tree is hardy in nature and requires comparatively low maintenance costs.

Mango fruit is utilized at all stages of its development both in its immature and mature state. Raw fruits are used for making chutney, pickles and juices. The ripe fruits besides being used as dessert are also utilized for preparing several products like squashes, syrups, nectars, jams and jellies. The mango kernel contains 8-10% good quality fat which is used for making soap and also as a substitute for cola in confectionery. The timber is used for making boats, furniture and flooring. Mango leaves are used to decorate the entrance of a household amongst Hindus.

Mango is grown over an area of 1.2 million hectares in the country producing 11 million tonnes. It accounts for 22% of total area (5.57 million ha) and 23% of total production of fruits (47.94 million tonnes) in the country. Though Uttar Pradesh has the largest area of 0.27 million hectares under mango, Andhra Pradesh has the highest productivity of 12 tonnes per hectare. Other mango growing states are Bihar, Karnataka, Kerala and Tamil Nadu. India ranks first among world's mango producing countries accounting for 52.63% of the total world's mango production of 19 million tonnes.

Varieties

There are more than thousand mango varieties in India. However, only about 30 varieties are grown on commercial scale in different states. Important mango varieties cultivated in different states of India.

States	Varieties
Andhra Pradesh	Banganpalli, Bangalora, Cherukurasam, Himayuddin, Suvarnarekha
Bihar	Bombay Green, Langra, Fazli, Himsagar, Kishen Bhog, Sukul, Bathua
Goa	Fernandin, Mankurad, Alphonso
Gujarat	Alphonso, Kesar, Rajapuri, Vanraj
Haryana	Dashehari, Langra, Bombay Green
Karnataka	Alphonso, Bangalora, Mulgoa, Neelum, Pairi
Kerala	Mundappa, Olour, Pairi, Neelum, Mulgoa, Muvandan, Alphonso,
Madhya Pradesh	Alphonso, Bombay Green, Langra
Maharashtra	Alphonso, Kesar, Mankurad, Mulgoa, Pairi
Orissa	Baneshan, Langra, Neelum, Suvarnarekha
Punjab	Dashehari, Langra, Chausa
Tamil Nadu	Banganapalli, Bangalora, Neelum, Rumani, Mulgoa
Uttar Pradesh	Bombay Green, Dashehari, Fazli, Langra, Safeda Lucknow, Chausa
West Bengal	Bombay Green, Himsagar, Kishen Bhog, Langra

Characteristics of Important Indian Varieties

1. *Alphonso :* This is the leading commercial variety of Maharashtra state and one of the choicest varieties of the country. This variety is known by different names in different regions, *viz.* Badami, Gundu, Khader, Appas, Happus and Kagdi Happus. The fruit of this variety is medium in size, ovate oblique in shape and orange yellow in colour. The fruit quality is excellent and keeping quality is good. It has been found good for canning purpose. It is a mid season variety
2. *Bangalora :* It is a commercial variety of South India. The fruit size is medium to large, its shape is oblong with necked base and colour is golden yellow. Fruit quality is poor. Keeping quality is very good. It is widely used for processing. It is a mid season variety.
3. *Banganapalli:* It is a commercial variety of Andhra Pradesh and Tamil Nadu and also known as Chapta, Safeda, Baneshan and Chaptai. Fruit

is large in size and obliquely oval in shape. The colour of the fruit is golden yellow. Fruit quality and keeping quality are good. It is a mid season variety and is good for canning.

4. *Bombai :* It is a commercial variety from Bihar state. It is also known as Malda in West Bengal and Bihar. Fruit size is medium, shape ovate-oblique and colour yellow. Fruit quality and keeping quality are medium. It is an early season variety.

5. *Bombay Green :* It is commonly grown in north India due to its early ripening habit. It is also called Malda in Northern India. Fruit size is medium, shape ovate oblong and fruit colour is spinach green. Fruit quality is good and keeping quality is medium. It is a very early variety.

6. *Dashehari :* This variety derives its name from the village Dashehari near Lucknow. It is a leading commercial variety of north India and one of the best varieties of our country. The fruit size is medium, shape is oblong to oblong oblique and fruit colour is yellow. Fruit quality is excellent and keeping quality is good. It is a mid season variety and is mainly used for table purpose.

7. *Fazli :* This variety is commonly grown in the states of Uttar Pradesh, Bihar and West Bengal. Fruit is very large, obliquely oval in shape. Fruit colour is light cream. Fruit quality and keeping quality are medium. This is a late season variety ripening by the middle of August.

8. *Fernandin :* This is one of the oldest varieties of Bombay. Some people think that this variety originated in Goa. Fruit size is medium to large, fruit shape is oval to obliquely oval and fruit colour is yellow with a red blush on shoulders. Fruit quality and keeping quality are medium. It is a late season variety mostly used for table purpose.

9. *Himsagar :* This variety is indigenous to Bengal. This is one of the choicest varieties of Bengal and has gained extensive popularity. Fruit is of medium size, ovate to ovate oblique. Fruit colour is yellow. Both fruit quality and keeping quality are good. It is an early variety.

10. *Kesar :* This is a leading variety of Gujarat with a red blush on the shoulders. Fruit size is medium, shape oblong and keeping quality is good. It is an early variety.

11. *Kishen Bhog :* This variety is indigenous to Murshidabad in West Bengal. Fruit size is medium, fruit shape is roundish oblique and fruit colour is yellow. Fruit quality and keeping quality are good. It is a mid season variety.

12. *Langra :* This variety is indigenous to Varanasi area of Uttar Pradesh. It is extensively grown in Northern India. Fruit is of medium size, ovate shape and lettuce green colour. Fruit quality is good. Keeping quality is medium. It is a mid season variety.

13. *Mankurad :* This variety is of commercial importance in Goa and in the neighbouring Ratnagiri district of Maharashtra. The variety develops black spots on the skin in rainy season. Fruit is medium in size, ovate in shape and yellow in colour. Fruit quality is very good. Keeping quality is poor. It is a mid season variety.

14. *Mulgoa :* This is a commercial variety of Southern India. It is quite popular among the lovers of mango owing to high quality of its fruit. Fruit is large in size, roundish oblique in shape and yellow in colour. Fruit quality is very good. Keeping quality is good. It is a late season variety.

15. *Neelum :* This is a commercial variety indigenous to Tamil Nadu. It is an ideal variety for transporting to distant places owing to its high keeping quality. Fruit is medium in size, ovate oblique in shape and saffron yellow in colour. Fruit quality is good and keeping quality is very good. It is a late season variety.

16. *Samarbehisht Chausa :* This variety originated as a chance seedling in the orchard of a Talukadar of Sandila district, U.P. It is commonly grown in northern parts of India due to its characteristic flavour and taste. Fruit is large in size, ovate to oval oblique in shape and light yellow in colour. Fruit quality is good, keeping quality is medium. It is a late variety.

17. *Suvarnarekha :* This is a commercial variety of Visakhapatnam district of Andhra Pradesh. Other synonyms of this variety are Sundari, Lal Sundari. Fruit is medium in size and ovate oblong in shape. Colour of the fruit is light cadmium with a blush of jasper red. Fruit quality is medium and keeping quality is good. It is an early variety.

18. *Totapuri :* This is a regular bearing and heavy bearing variety widely grown in South India. Fruit quality is medium but it has a typical flavour and fibreless cadmium yellow coloured flesh.

19. *Vanraj :* It is a highly prized variety of Vadodra district of Gujarat and fetches good returns. Fruit is medium in size, ovate oblong in shape and colour is deep chrome with a blush of jasper red on the shoulders. Fruit quality and keeping quality are good. It is a mid season variety.

20. *Zardalu :* This variety is indigenous to Murshidabad in West Bengal. Fruit size is medium, oblong to obliquely oblong and golden yellow in colour.

Fruit quality is very good. Keeping quality is medium. It is a mid season variety.

Hybrid Varieties

i) *Amrapali :* This hybrid is from a cross of Dashehari x Neelum. It is a dwarf, regular bearing and late maturing variety. The variety is suitable for high density planting as about 1600 plants may be planted in a hectare. It yields on an average 16 tonnes / hectare.

ii) *Mallika :* It is from a cross of Neelum x Dashehari. Its fruit is large in size, oblong elliptical in shape and cadmium yellow in colour. Fruit and keeping quality are good. It is a mid season variety.

iii) *Arka Aruna :* It is a hybrid between Banganapalli and Alphonso. It is dwarf, regular bearing and precocious. Fruits are large having attractive skin colour with red blush, free from spongy tissue.

iv) *Arka Puneet :* It is a hybrid between Alphonso and Banganpalli. It is a regular and prolific bearer. Fruits are medium sized having attractive skin colour with red blush and free from spongy tissue with excellent keeping quality.

v) *Arka Anmol :* This hybrid is from a cross of Alphonso and Janardhan Pasand. It is a regular bearer and good yielder. Fruits are medium sized having uniform yellow peel colour, excellent keeping quality and free from spongy tissue.

vi) *Arka Neelkiran :* It is a hybrid between Alphonso and Neelum. It is a regular bearer and late season variety with medium sized fruits having attractive, red blush free from spongy tissue.

vii) *Ratna :* This hybrid is from a cross of Neelum x Alphonso. Tree vigorous, precautious, fruits are medium sized, attractive in colour and free from spongy tissue.

viii) *Sindhu :* It is from a cross of Ratna x Alphonso. It is a regular bearer, fruits medium sized, free from spongy tissue with high pulp to stone ratio and very thin and small stone.

ix) *Au Rumani :* It is from a cross of Rumani x Mulgoa. It is precocious, heavy and regular bearing with large fruits having yellow cadmium skin colour.

x) *Manjeera :* This hybrid is from a cross of Rumani x Neelum. It is dwarf, regular and prolific bearer with firm and fibreless flesh.

Other hybrid varieties released are Alfazali, Sundar Langra, Sabri, Jawahar, Neelphonso, Neeleshan, Neeleshwari and PKM2.

Climate and Soil

Mango is very well adapted to tropical and subtropical climates. It thrives well in almost all the regions of the country from sea level to an altitude of 1500 m, i.e., from Cape Comerin to Himalayas. However, it cannot be grown commercially in areas above 600 m. Temperature, rainfall, wind velocity and altitude are the main climatic factors which influence its growth and fruiting. It cannot stand severe frost, especially when the tree is young. High temperature by itself is not so injurious to mango, but in combination with low humidity and high winds, affects the trees adversely.

Most of the mango varieties thrive in places with good rainfall (75 to 375 cm/annum) and dry season. The distribution of rainfall is more important than its amount. Dry weather before blossoming is conducive to profuse flowering. Rain during flowering is detrimental to the crop as it interferes with pollination. However, rain during fruit development is good but heavy rains cause damage to ripening fruits. Strong winds and cyclones during the fruiting season are not desirable as they cause excessive fruit drop.

Mango grows well on a wide variety of soils, such as lateritic, alluvial, sandy loam and sandy. Well drained, aerated and deep soils rich in organic matter with a pH range of 5.5 to 7.5 are the most suited for mango cultivation. The extremely sandy, shallow, rocky, water-logged, heavy textured and alkaline or calcareous soils are not suitable for mango cultivation. Although it grows very well in high to medium fertility soils, its cultivation can be made successful even in low fertility soils by appropriate management especially during early stages of growth.

Propagation

Mangoes are raised from seed or propagated vegetatively. Propagation from seed, though easy and cheap, is unable to perpetuate characters of the parent tree because most commercial varieties in India are cross-pollinated and monoembryonic. Plants also take more time to bear fruit. Accordingly, several methods of vegetative propagation have been tried with varying degree of success. However, it is essential to raise seedlings to be used as rootstocks. For this purpose, stones should be sown in June-July in beds mixed with well decomposed farm yard manure at the rate of 8-10 tonnes/hectare. Alternatively, 25 kg nitrogen (N) per ha may be applied in the form of urea, CAN or any other available inorganic source in two split doses at about two months interval after the leaves have become green. When the seedlings attain the age of 2-3 months, they should be transplanted well in prepared beds or pots. In this season, proper care should be exercised in irrigating the young transplanted seedlings. The seedlings should also be protected from frost by putting the pots under big trees or thatching the young seedlings in the field.

Plants are generally propagated using random seedling rootstocks. The polyembryonic rootstocks, however, have shown a promise in producing plants of uniform size and vigour. Moreover, these rootstocks have indicated possibility of inducing dwarfing and early bearing characters. Inarching or approach grafting, veneer grafting, stone or epicotyl grafting and soft-wood grafting are the popular vegetative propagation methods employed in mango.

Grafted trees grow more slowly than seedling trees and are often smaller. Grafted trees usually produce fruits in 3 to 5 years in dry areas, while seedling trees usually take at least five years to come into bearing. Mango trees can remain in production for 40 years or more and older trees may be top worked. Mangoes are not propagated from cuttings or by air layering because the resulting trees are weak rooted.

Planting

Land should be prepared by deep ploughing followed by harrowing and levelling with a gentle slope for good drainage. One year old healthy, straight growing grafts from reliable sources can be planted at the centre of pits along with a ball of earth intact during rainy season in such a way that the roots are not exposed and the graft union is above the ground level. Plants should be irrigated immediately after planting. In the initial one or two years, it is advisable to provide some shade to the young plants and also stake to make them grow straight.

Prior to planting, field should be deeply ploughed, harrowed and levelled. Pits of proper size should be dug at appropriate distances and filled by adding sufficient quantity of farm yard manure. The grafts to be planted should be procured from reliable nurseries a few days before actual transplanting.

The best time for planting all over India is during the monsoon when there is sufficient moisture in the atmosphere. In the area of heavy rainfall, the best time of planting mango is the end of the rainy season. In tracts, where rainfall is less, planting can be done in the early part of the monsoon for better establishment. The planting should be done in the evening to prevent the plants from withering due to excessive loss of water.

The planting distance varies according to variety, the fertility level of the soil and general growth conditions in the area. Spacing varies from 10 x 10 m, in the dry zones where growth is less, to 12 x 12 m, in heavy rainfall areas and rich soils where abundant vegetative growth occurs. New dwarf hybrids like Amrapali can be planted at closer spacing. For high density planting, the distance can be 5 x 3 or 5 x 2.5 or 3 x 2.5 or 2.5 x 2.5 m.

In locations where the soil is loamy and deep, pits of 0.5 x 0.5 x 0.5 m size may be dug at desired distances. However, in shallow and hill soils, the pits should be at least of 1 x 1 x 1 m size.The pits should be filled with the original soil mixed with 50 kg well rotten farm yard manure. In the top two-third portion, the proportion of the manure and soil may be kept as 1:3. To protect the soil from infestation of white ants, 200-250 g of aldrex dust may also be mixed. The pits should invariably be filled before the rainy season, so that there is maximum settling down before heavy rainfall.

The plant with its ball of earth intact should be taken out of the soil or pot. The plant can then be placed with the help of a planting board in the centre of the pit by excavating as much soil as necessary to accommodate the root-ball. The moist soil of the pit is then pressed all around the root ball to complete the planting process. A small basin is then made and the plant is properly watered. The planting should not be done so deep as to bury the graft-union in the soil or so high as to expose the upper roots. It is always better to adjust it at the same height/depth at which it was in the pot or the nursery bed.

Training and Pruning

Mango tree requires very less or no pruning. However, the training of the plants in the initial stages is very essential to give them proper shape. Specially when the graft has branched too low, the process of training becomes very important. At least 75 cm of the main stem should be kept free from branching and the first leader/main branch may be allowed after that. The main branches should be spaced in such a way that they grow in different directions and are at least 20-25 cm apart, otherwise there are chances of breakage due to smaller crotch angles and heavy top.

The branches which exhibit tendency of crossing and rubbing each other should be removed in the pencil thickness stage itself, otherwise they break by rubbing each other at a later stage and create complications. Secondly, if the centre are closed, the fruits produced are of poor quality having less colouration in the absence of sufficient sunlight.

By following the above practice and after giving proper shape to the trees, pruning can be limited to the removal of diseased, pest infested or dried shoots.

Fertilizer Application

Nutrient uptake in mango is from large volume of soils. Therefore, it is able to sustain growth even in low fertility soils. But, its efficient management involves the replenishment of the nutrients used-up by the tree for its growth and maintenance, harvested produce and natural losses from soils through leaching and run of. Even the under-nourished trees can be revived by suitable supplementation of nutrients through fertilizers. The idea of manuring bearing trees is also to secure regular fruiting. Recommendations based on very limited research and also on experience gained by the mango growers are given below. Manuring mango plant starts right from planting operation in the orchard. First application is made at the time of filling of the pits. Apply fertilizers and FYM at the rate indicated below.

Age of plant	FYM(kg/plant/year)	N: P_2O_5 :K_2O(g/plant/year)
1st year	10	20:18:50
2nd year	15	50:27:75
3-5 years	25	100:36:100
6-7 year	40	250:172:200
8-10 years	50	400:144:400
Over 10 years	75	500:360:750

Application of 75 kg well-decomposed organic manure should be given yearly once in May-June to create proper soil physical environment. Fertilizers may be applied in two split doses, one half immediately after the harvesting of fruits in June / July and the other half in October, in both young and old orchards, followed by irrigation if there are no rains. Foliar application of 3% urea in sandy soils is recommended before flowering. The application of micronutrients is not recommended as a routine. Need based applications are essential when these become limiting factor for production. It is advisable to apply micronutrients through foliar sprays. First of all, the weeds should be removed from the basins. The mixture of recommended dose of fertilizers should be applied in 25 cm wide and 25-30 cm deep trenches dug around the tree 2 m away from trunk.

Irrigation

Amount and frequency of irrigation depend upon the type of soil, prevailing climatic conditions, especially rainfall received, and its distribution and age of trees. No irrigation is required during the monsoon months unless there are long spells of drought. During the first year when the plants are very young with shallow root system, they should be watered every 2-3 days in the dry season. Trees in the age group of 2-5 years should be irrigated at 4-5 days interval. The irrigation interval could be increased to 10-15 days for 5-8 years old plants during dry season. In case of grown up trees, irrigation at 10 to 15 days interval from fruit set to maturity is beneficial for improving yield. However, irrigation is not recommended for 2-3 months prior to flowering as it is likely to promote vegetative growth at the expense of flowering.

Generally, intercrops are grown during the early years of plantation and hence frequency and method of irrigation have to be adjusted accordingly. It is advisable to irrigate the mango plants in basins around them which can be connected in series or to the irrigation channel in the center of rows. The intercrops need to be irrigated independently as per their specific requirements. In monocropping of mango also, basin irrigation is preferable with a view to economise water use.

Weed Management

Immediately after planting the mango, the weed problem may not exist, but it is advisable to break the crust with hand hoe each time after 10-15 irrigations. However, subsequent hoeing may be done depending on weed growth in the basin. If the intercrops are not being raised in the pre-bearing stage due to some reasons, the area between the basins should be ploughed at least three times a year, *i.e.*, pre-monsoon, post-monsoon and in the last week of November.

Intercropping

Mango orchard provides an opportunity for utilizing the land space to its maximum during initial years (up to 8-10 years) of establishment. Inter crops such as vegetables, legumes, short duration and dwarf fruit crops like papaya, guava, peach and plum, depending on the agro-climatic factors of the region, marketing facilities, levels of inputs and other local considerations can be grown. The water and nutrient requirements of the inter crops must be met separately. This enables the mango growers to raise extra income during the years when the main crop yields no / low returns. Some of the important crop rotations recommended are as follows:

1. Cowpea-potato, 2. Greengram-wheat, 3. Blackgram-wheat, 4. Greengram-gram, 5. Blackgram-gram, 6. Cowpea-wheat, 7. Cowpea-gram

Maximum monetary return can be obtained from cowpea-potato rotation.

Rejuvenation of Mango Orchard

In general, 40-45 years old mango trees exhibit decline in fruit yield because of dense and overcrowded canopy. The trees do not get proper sunlight resulting in decreased production of shoots. New emerging shoots are weak and are unsuitable for flowering and fruiting. The population of insects and pests and the incidence of diseases increase in such orchards. These unproductive trees can be converted into productive ones by pruning. Intermingling diseased and dead branches have to be removed first. At the end of December, the undesirable branches are marked and these branches are beheaded at 1.5 to 2.0 meter from distal end and the cut portions are pasted with copper oxychloride solution. During March-April, a number of new shoots emerge around the cut portions of the pruned branches. Only 8 to 10 healthy and outward growing shoots are retained at proper distance so that a good frame-work is developed in the following years. These rejuvenated trees are fertilized with 2.5 kg urea, 3.0 kg single superphosphate and 1.5 kg muriate of potash/ plant. The half dose of fertilizers is applied in February and the other half at the end of June. the plants are irrigated at an interval of 15 days especially in April, May and June for healthy growth of new shoots. In the first week of July, 150 kg of compost/ tree is also applied. Unwanted emerging new shoots are regularly removed to maintain the tree canopy and avoiding crowding of branches. It also helps in getting proper nourishment to retained shoots. Two years after pruning, new shoots come into bearing and the yield of fruit increases gradually.

Harvesting and Yield

Mangoes are generally harvested at physiological mature stage and ripened for optimum quality. Fruits are hand-picked or plucked with a harvester. During harvesting, the latex trickles down the fruit surface from the point of detachment imparting a shabby appearance to it upon storage. After harvest, the fruits are usually heaped under a tree on the ground. Bruised and injured fruits develop brown to black spots during storage making the fruits unattractive. Moreover, injuries to the peel or to the stalk end serve as avenues for invasion of microorganism and lead to rotting of the fruits. The post harvest losses in mangoes are estimated 25-40% from harvesting to consumption stage. If proper methods of harvesting, handling, transportation and storage are adopted, such losses could be minimized.

The mango fruits should be harvested at green mature stage. The harvest maturity in Dashehari and Langra cultivars reaches 12 weeks after fruit set, while in Chausa and Mallika, it takes about 15 weeks. The best way to observe maturity in mango is the colour of the pulp which turns cream to light yellow on maturity and hardening of the stone.

The harvesting in mango should be done in the morning hours and fruits should be collected in plastic trays and kept in shades. The fruits should not be allowed to fall on the ground as the injured fruits cause spoilage to other healthy fruits during packaging and storage. Fruits harvested with 8-10 mm long stalks appear better on ripening as undesired spots on skin caused by sap burn are prevented. Such fruits are less prone to stem-end rot and other storage diseases.

Fruits harvested by stick are injured / bruised due to impact resulting in decay, poor quality and attract low price. Traditional methods consume a lot of energy. To overcome these problems, a simple, low cost and portable mango harvesting device has been designed and developed at CISH, Lucknow. Mango fruits are taken into the pouch and held between the divider and knife and as the device is pulled, the blade cuts the pedicel. The fruits are then conveyed through a nylon chute to collect boxes without bringing down the device every time. This saves time and protects fruits from mechanical damage due to impact. It also protects operator's hand from the sap which oozes out from the point of detachment.

Grading

If the fruits are graded according to their size, weight, colour and maturity, both the producer and consumer are benefited. It was observed that bigger size fruits took 2-4 days more for ripening than smaller ones. Hence, packaging of smaller fruits with larger ones should be avoided to achieve uniform ripening. Immature, overripe, damaged and diseased fruits should be discarded. The fruits of mango can be graded according to their weight as follows.

1st grade : Fruit weight 320g or more

2nd grade : Fruit weight 270-320g

3rd grade : Fruit weight 200-270g

Packaging

Wooden boxes are commonly used for packaging and transportation of mango fruits. Under dynamic transport conditions nails come out due to vibration and puncture the fruits which result in bruising, decay and low price of fruits. Further, too much ventilation affects the quality of fruits due to shrinkage, loss in weight and colour. To overcome these problems, CFB Boxes of 5 kg and 10 kg capacity can be used for packing and shipping of mango fruits successfully and a better alternative to traditional nailed wooden boxes. The use of CFB boxes for packaging for the domestic market is also the need of hour due to scarcity of wood and environmental concerns of the country. Paper scraps, newspapers, etc., are commonly used as cushioning material for the packaging of fruits which prevent them from getting bruised and spoiled during storage and transportation. Polythene (LDPE) lining is beneficial as it maintains humidity which results in lesser shrinkage during storage. Wrapping of fruits individually with newspaper or tissue paper and packing in honey comb structure helps in getting optimum ripening with reduced spoilage.

Storage

Storage is essential for extending the consumption period of fruits, regulating their supply to the market and also for transportation to long distances. The mature green fruits can be kept at room temperature for about 4-10 days depending upon the variety. Shelf life of fruits could be extended by precooling, chemical treatments, low temperature, etc. The harvested fruits are precooled to 10-12°C and then stored at an appropriate temperature. The fruits of Dashehari, Mallika and Amrapali should be stored at 12°C, Langra at 14°C and Chausa at 8°C with 85-90 % RH. The fruits could be stored for 3-4 weeks in good condition at low temperature. The problem of chilling injury at low temperature can be overcome by keeping the fruits in 0.5% ventilated polythene bags.

Calcium infiltration is an improved technique of extending the storage life of Dashehari fruits. The fruits are kept in calcium chloride solution (4%) at sub- atmospheric pressure of 500 mm Hg for 5 minutes. The treated fruits can be stored at low temperature (12°C) for 27 days. It is a general practice to harvest fruits early in the season (premature stage) to capture early market. These fruits do not ripe uniformly without any ripening aid. Such fruits could be ripened uniformly by dipping in 750 ppm ethrel (1.8 ml/ litre) in hot water at 52+2°C for 5 minutes within 4-8 days under ambient conditions. Mature fruits can similarly be ripened with lower doses of ethrel for uniform colour development.

Fruits ripened artificially develop a better flavour and taste than tree ripened fruits. The fruit can be ripened quickly in about two days by keeping a small piece of calcium carbide inside a covered heap of mangoes. The acetylene gas emitted will ripen the fruits and gives a good colour.

Transportation

The truck is adopted as the most convenient mode of transport due to its easy approach from the orchards to the market. However, these trucks were not suitable for transporting this live material as they exert a lot of pressure on the fruits and do not possess temperature control devices. Therefore, it is imperative to design and develop suitable transport system. Reefer containers (Refrigerated vans) may be found useful for long distance transport and export purposes as they would help in reducing the post harvest losses.

MANGOSTEEN

Mangosteen (*Garcinia mangostana* L.) known as "Queen of Fruits" belongs to family Guttiferae. Its origin is in Malaya and can now be found in Northern Australia, Brazil, Burma, Central America, Hawaii, Southern India, Indonesia, Malaysia, Sri Lanka, Thailand, Vietnam, and other tropical countries. In India, it is generally cultivated in the lower elevations of Nilgiri hills and to a limited extent in Kerala and West Bengal. Mangosteen is one of the most widely recognized tropical fruits and has universal appeal because of its quality in color, shape and flavor. Demand often exceeds supply. The fruit is 2-3cm in diameter. A thick reddish-purple rind covers the aril or pulp, segmented like that of an orange. The white, moist, soft and juicy flesh is sweet and aromatic, has a high sugar content, but is low in vitamins and minerals. It is usually eaten fresh, but can be stored successfully for short period of time. It is also canned, frozen, or made into juice, preserves, and syrup. Mangosteen is also used as a pharmaceutical.

Varieties

Mangosteen is an obligate apomict in which reproduction is entirely asexual (parthenogenesis).It is found only as a cultivated female tree (male trees appear non-existent) and may be a fortuitous hybrid between two sister species (*G. malaccensis* and *G. hombroniana*) and is morphologically intermediate between these two species. Moreover, genetic variation is minimal and no cultivars are reported. It is suggested that all trees may have come from a single clone. There are no verified reports of pollen production, floral visitors which might be viewed as pollinators, nor of nectar production in mangosteen. The fruits from seedling trees are fairly uniform; only one distinct variation is known and that is in the Sulu Islands. The fruit is larger, the rind thicker than normal, flesh more acidic and the flavor more pronounced.

Climate and Soil

The mangosteen is ultra-tropical. It requires a wet lowland tropical climate and moist well drained soil high in organic matter. It is killed at temperatures below 5^0 C and sunburns occur at temperature above 38^0 C. It grows upto 1,500 m above sea-level. It ordinarily requires high atmospheric humidity and an annual rainfall of at least 127 cm and no long periods of drought.

The tree is not adapted to limestone and does the best in deep, rich organic soil, especially sandy loam or laterite. In India, the most productive specimens are on clay containing much coarse material and a little silt. Sandy alluvial soils are unsuitable. Sandy soil low in humus contributes to low yields. The tree needs good drainage and the water table ought to be about 1.8 m below ground level.

Propagation

Mangosteen is difficult to be propagated vegetatively and so most of the trees are produced from seeds, which remain viable only for a few days. The juvenile phase may last for 5 to 15 years depending on growing conditions (Richards, 1990).

Technically, the so-called "seeds" are not true seeds as there has been no sexual fertilization. When growth begins, a shoot emerges from one end of the seed and a root from the other end. But this root is short-lived and is replaced by roots which develop at base of shoot. The process of reproduction being vegetative, there is naturally little variation in resulting trees and their fruits. Some of the seeds are polyembryonic, producing more than one shoot. The individual nucellar embryos can be separated, if desired, before planting.

As percentage of germination is directly related to weight of seed, only plump, fully developed seeds should be chosen for planting. Seeds will lose viability in 5 days after removal from the fruit, though they are viable for 3 to 5 weeks in the fruit. Seeds packed in lightly dampened peat moss, sphagnum moss or coconut fibre in airtight containers remains viable for 3 months. Soaking in water for 24 hours increase speed and rate of germination. Generally, sprouting occurs within 3 weeks and completes in 6 weeks time.

Because of long, delicate taproot and poor lateral root development, transplanting is notoriously difficult and should be done before plants attain 60 cm height. Seedling survival is more if seeds are planted directly in the nursery than growing in containers and then transplanting to the nursery. The nursery soil should be at least 1 m deep. The young plants take 2 years or more to reach a height of 30 cm, when they can be taken up with a deep ball of earth and set out. Fruiting takes place in 7 to 9 years from planting.

Conventional vegetative propagation of mangosteen is difficult. Various methods of grafting have failed. Cuttings and air-layers, with or without growth-promoting chemicals, usually fail to root or result in deformed, short-lived plants. Inarching on different rootstocks is promising at first but later incompatibility is noticed.

Planting

Grown primarily in back yards and gardens, these attractive pyramidal evergreen trees grow to a height of 25 m. Effort was made to establish commercial orchards, however, obstacles yet to be overcome include difficulties in propagation, the slow growth period, problems in harvesting, and yield.

A spacing of 10-12 m is recommended. Planting is preferably done at the beginning of the rainy season. Pits 1.2 x l.2 x l.3 m are prepared at least 30 days in advance, enriched with organic matter and topsoil and left to weather. The young tree is put in place very carefully so as not to injure the root and given a heavy watering. Partial shading with palm fronds or by other means should be maintained for 3 to 5 years.

Training and Pruning

It is suggested that small inner branches be pruned from old, unproductive trees to stimulate bearing. In Thailand, the tree takes 12 to 20 years to fruit. In Panama and Puerto Rico, trees grown from large seed and given good culture come to bear in six years.

Fertilizer Application

Indian growers give each tree regular feeding with well-rotten cattle manure (45-90 kg) and groundnut cake or bone meal (5-7 kg) total, per year for getting good yield.

Irrigation

Some of the most fruitful mangosteen trees are grown on banks of streams, lakes, ponds or canals where roots are almost constantly wet. Dry weather just before blooming time and during flowering induces good fruit-set. Where a moist planting site is not available, irrigation ditches should be dug to maintain an adequate water supply and the trees are irrigated almost daily during the dry season.

Weed Management

Mulching preceeded by stirring of surface soil for keeping off weeds is an important cultural practice.

Mulching

In Malayasia and Srilanka, it is a common practice to spread a mulch of coconut husk or frond to retain moisture. A 40 cm mulch of grass restored trees that had begun dehydrating in Liberia.

Harvesting and Yield

Mangosteen usually flowers only once annually. However, flowering occurs twice annually following a dry period. Fruit set is highly variable. In India, there are two distinct fruiting seasons, one in the monsoon period (July-October) and another from April to June. The yield varies from tree to tree and from season to season. The first crop yields 200 to 300 fruits. Average yield of a full-grown tree is about 500 fruits. The yield steadily increases up to 30th year of bearing when crops of 1,000 to 2,000 fruits are obtained. In Madras, individual trees between the ages of 20 and 45 years, have borne 2,000 to 3,000 fruits. Productivity gradually declines thereafter, though the tree will fruit at 100 years of age.

Ripeness is judged by full development of color and slight softening. Picking may be done when fruits are slightly under ripe but they must be fully mature (developed) or they will not ripen after picking. The fruits must be harvested by hand from ladders or by means of a cutting pole and not be allowed to fall.

Storage

In dry, warm and closed storage, mangosteen can be stored for 20 to 25 days. Longer periods cause the outer skin to toughen and the rind to become rubbery; later, the rind hardens and becomes difficult to open and the flesh turns dry.

Ripened mangosteen keeps well for 3 to 4 weeks in storage at 4°-13° C. Trials in India have showed that optimum conditions for cold storage are temperatures of 4°-6° C and relative humidity of 85 to 90%, which maintain the quality for 49 days. It is recommended be wrap fruits in tissue paper and packed 25 to 30 fruits in light wooden crates with padding.

PAPAYA

Papaya (*Carica papaya* L) belonging to the Caricaceae family is one of the important delicious fruit crops grown in the tropical and subtropical parts of the world. Although a native of Mexico, it has spread to almost all the corners of the tropical world. It is a short duration crop and is easy to grow. In India, it is mostly consumed as a table fruit. It is consumed as a part of breakfast dessert and as fruit salad. It is also used in soft drinks, jams, ice-cream flavouring, crystallised fruit and also sold as canned cubes and juice in some countries. Ripe fruit is very delicious containing vitamin A and carbohydrates. As a raw fruit, it is used in cooking and a few preparations. Its latex is used as a papain in food and medicine industry.

It is grown all over India and is available round the year in the country. However, Karnataka, Orissa, Assam, West Bengal and Gujarat provide ideal climatic conditions for its growth. At one time, India was a leading producer of papaya in the world. Today, it is not because its production in other countries has gone up dramatically while in India the rise was at a much lower rate.

Varieties

Papaya research during the last 5 decades encompassed both basic and applied aspects. Research work concentrating mainly on crop-production technology and varieties suitable for different agro climatic regions of the country were carried out at different agricultural universities and ICAR institutions. Systematic papaya breeding work for development of improved varieties was initiated at TNAU and IARI in early sixties and subsequently at GBPUAT and IIHR in the seventies.

The first improved cultivar of papaya Coorg Honey Dew was identified in 1959 at the Citrus Experiment Station, Chethalli. Interdisciplinary research programme on crop improvement, production technology and plant protection is presently under implementation at the centres at Coimbatore, Pusa, Pune, Jalgaon, Rahuri and Bhubaneshwar with the support under the All-India Co-ordinated Research Project on Tropical Fruits of the ICAR, and through *ad-hoc* projects under ICAR AP Cess Fund Scheme.

Fifteen high-yielding varieties of papaya were developed in India, out of which nine are widely adopted. Some successful hybrids have also been developed and heterosis has also been noticed with a wide range of breeding materials. The Indian Institute of Horticultural Research, Bangalore developed a hybrid variety of papaya, 'Surya', which is a cross between 'Sunrise Solo' and 'Pink Flesh'. It produces medium size fruits of 600-800 gram with red pulp about 3.0-3.5 cm thick and a smaller fruit cavity. The hybrid is gynodioecious and does not require male plants for pollination. The seedlings, if planted at a spacing of 2 x 2m yield 125-150 tonnes of papaya per hectare. Some of the common varieties in India are

a. Pusa varieties: Pusa Delicious, Pusa Majesty, Pusa Giant, Pusa Dwarf, Pusa Nanha

b. Coimbatore varieties: Co1, Co2, Co3, Co4, Co5, Co6

c. Bangalore varieties: Coorg Honey, Pink Flesh
d. Pant nagar varieties: Pant 1
e. Other varieties: Ranchi, Washington, Honey Dew

Climate and Soil

Papaya grows well almost everywhere, except at altitudes higher than 1,500 metres and prefers a dry warm climate. Being a tropical crop, it favours high temperature and high humidity. It is very susceptible to frost and hail storm. Long days are favourable for good quality and flavour. Strong winds are highly detrimental to the trees as the hollow stems break easily. During flowering, high rains are injurious and cause heavy damage.

Even though papaya is adapted to a wide range of soils, it grows the best in loamy soil. Medium, fertile, well drained and lime free soils are preferred for papaya cultivation. Deep clayey soils that are prone to water-logging should be avoided. In heavy-rainfall areas, a prolonged stagnation of water near its stem is highly injurious. However, very shallow and very deep black soils are not suitable.

Propagation

Commercially papaya is propagated by seeds. The seeds loose viability in a short period and therefore the seeds should not be stored for more than a season. Germination of seeds takes place 2-3 weeks after sowing. Seeds treated with thiourea (100-200 ppm) and gibberellic acid (GA3 at 200 ppm) germinated better. Seed germination is the highest at an optimum temperature of 35°C. The seedlings in polybags are prepared. Due care is taken to avoid damping off of the newly germinated and young seedlings. The seedlings become ready for transplanting within 6-8 weeks. Storage of seeds at 10°C was the best for retention of seed viability. Viability of papaya seed can be maintained for 9 months at room temperature when stored in air-tight containers. Patch budding is quite successful (90%) in papaya but it is not practiced commercially.

Tissue-culture technique has been standardized for papaya but the technique is limited to research laboratories only. It is reported that plants raised through tissue-culture technique yield more fruit in Pusa Delicious when compared with the seedlings. Young papaya seedlings form callus profusely in Murashige and Skoog medium supplemented with auxins (NAA). Callusing was obtained from stem segments, roots and leaf tissue. Callus readily differentiated into leafy shoots in medium containing high ratio of kinetin and NAA (2.0: 0.2 mg/litre), while root formation was favoured by a low kinetin: NAA ratio (0.2: 0.5 ml/litre). Shoot tip culture from seedlings and lateral buds from female plants of Coorg Honey Dew variety was successful. The tissue-culture plantlets measuring 5.5 to 9.5 cm and hardened in plastic chamber survived better when planted in the field. The available information indicates that tissue-cultured papaya plantlets can be used with added advantage of uniformity in plant type.

Planting

Planting is done during the following seasons:

a. Spring season (February - March)
b. Monsoon season (June-July)
c. Autumn season (October-November)

While selecting the season for planting in a particular area, avoid periods of heavy rain, hot air, frost, etc. The pits of size 30 x 30 x 30cm are prepared in already selected and prepared field at a distance of 2.5 to 3 meters. The pits are filled with well decomposed FYM and NPK mixtures. Care is to be taken not to disturb the roots while transplanting the seedlings. A light or shallow ploughing or harrowing once or twice a year is necessary so also is the removal of weeds.

A spacing of 1.8 x 1.8 m is normally followed in most of the places. Field trials conducted indicate that a spacing of 1.8 x 1.8 m for varieties like CO. 1, CO.2 and Solo, and 2.1 x 1.2 m (3,968 plants/ha) for Coorg Honey Dew variety was optimum for Bangalore conditions. The spacing of 1.4 x 1.4 or 1.6 x 1.4 m is ideal for cultivar Pusa Delicious under sub-tropical conditions of Bihar. For the dwarf variety Pusa Nanha a closer spacing of 1.25 x 1.25 m was found satisfactory. For papain production, a spacing of 1.6 x 1.6 m was recommended based on field trials conducted in Tamil Nadu. In West Bengal under a spacing of 1.5 x 1.5 m (4, 444 plants/ha) fruit yield of 98.7 tonnes/ha could be recorded with papaya variety Ranchi.

Fertilizer Application

Nine kg of farmyard manure/ pit is applied at the time of planting followed by 35 to 45 kg of it every six months, once at the beginning of monsoon and again in winter. Inorganic fertilizers @ 200-250 g each of N, P_2O_5 and K_2O/plant are generally recommended for obtaining high yield in papaya. The N @ 200 g/plant was optimum for fruit yield at Pusa (Bihar), but papain yield increased with doses up to 300 g/plant at Coimbatore (Tamil Nadu). Inclusion of organic manures like farmyard manure (FYM), sheep manure, wood ash, neem cake and bone-meal, not only improved the yield but also produced better quality fruits.

Application of phosphatic fertilizers reduces acidity, while potash normally increases the TSS and total sugar. When K was adequate, TSS did not lower even with high N fertilization. The fruit quality in papaya as indicated by TSS was significantly influenced by K and N x K interactions.The lack of N and P causes reduction in growth of the papaya plant, while absence of K and Ca induces pronounced foliar symptoms.

Not many field trials were conducted to determine the micronutrient requirements of papaya. Among the micronutrients, B followed by Fe, Cu, Zn and Mo are important for proper growth and development. The B-deficient plants normally develop bunchy top and deformed fruits. Application of boron increased Ca in fruits which in turn increased the shelf-life of papaya. Boron application through soil or foliar spraying resulted in 20% increased fruit set. Deficiency of B and Zn in field conditions cause considerable yield loss. Foliar applications of Mn (0.25%) plus Cu (0.25%) at 60 and 90 days after transplanting helped in harvesting maximum quantity of fruits (94.7 tonnes/ ha) with high quality. Combined application of B (0.1%), Mn and Cu markedly improved fruit quality in papaya variety Ranchi under West Bengal conditions. Beneficial effects of mycorihiza were observed in papaya. Association of vesicular arbuscular mycorrhiza (*Glomus mosseae* + *Glomus fasciculatum* + *Gigaspore margarita)* helped in saving 25% of P recommended for papaya.

Some of the important fertilizer recommendations emerged from field trials conducted at different parts of the country are as follows:

i. A dose of 250 g N, 250 g $P_2O_5$and 500 g K_2O /plant/year applied in 6 bimonthly intervals was the best for Coorg Honey Dew variety and gave estimated fruit yield of 186.8 tonnes/ha in 28 months crop cycle at a plant spacing of 2.4 x 2.4 m (1, 736 plants/ha).

ii. A dose of 350 g N, 250 g P_2O_5 and 200 g K_2O /plant year applied in 6 split doses was the best for Solo variety spaced at 2 x 2 m (2,500 plants/ ha) under Bangalore (Karnataka) conditions.

iii. At Coimbatore, a dose of 200 g each of N, P_2O_5 and K_2O /plant in 4 split doses during 1, 3, 5 and 7 months after planting resulted in higher yield of papaya variety Co1. For papain production from Co2 variety, a dose of 250g N/plant/year in 6 split doses at bimonthly intervals commencing from the second month after planting was the best.

iv. A fertilizer dose of 200 g N, 300 g P_2O_5 and 600 g K_2O /plant gave the highest fruit (110 tonnes/ha) yield in papaya variety Ranchi under West

Bengal conditions. For maximum papain yield(94.45 g/fruit), fertilizer dose of 200 g N, 300 g P_2O_5 and 400 g K_2O was optimum.

Irrigation

For better growth, production and quality, the optimum soil moisture is to be maintained by irrigating the crop judiciously. Irrigation interval depends upon season, crop growth and soil type. In no case, water should be allowed to stagnate causing root and stem rot. Generally irrigation at 7-10 days interval during hot summer months is recommended for grown-up plants. Among various systems of irrigation, ring or furrow system of irrigation is preferred over the bed system. Bed system allows direct contact of irrigation water with the plant and thus predisposes the plant to collar rot disease. Irrigation in widely spaced furrows of 100 cm resulted in significantly higher fruit yield as compared to basin and flood methods of irrigation under Bangalore condition. Drip system of irrigation is beneficial as it helps in saving 50-60% water and not only that water use efficiency under drip is much higher.

Weed Management

Interculturing is mainly done to remove weeds during the early period of growth, weeding and hoeing in between rows also favour better aeration to root zone.Chemical weedicides like atrazine and simazine were phytotoxic for papaya. Gramoxone at 1.0 litre/ha controlled weeds and gave maximum fruit yield in papaya (91 tonnes/ha) under West Bengal condition. Paraquat (1.0 litre/ha) applied twice at 90 and 180 days after transplanting caused maximum reduction of weeds, improved plant growth, advanced flowering and resulted in higher fruit yield. In field trials conducted at IIHR, Bangalore, Diuron 2 kg ai / ha was effective in controlling weeds without any adverse effect on papaya. After ensuring one plant per pit, earthing up is done 30 cm in radius around the plants.

Intercropping

Low-growing vegetables of short duration may be taken as inter-crops. An occasional thinning of fruits is necessary to prevent overcrowding. Papaya may itself be grown as a 'filler' or in plantations of other crops where spacing is wide enough.

Harvesting and Yield

Papaya flowers in about four months after planting and fruits are ready for harvesting in another six moths. Except during winter, the trees continue to flower and fruit all the time. Papaya fruits should be harvested when colour of skin changes from dark green to light green and when one yellow streak begins development from the base upwards. Fruits in this condition will continue to ripen normally after harvest. Those fruits harvested before this stage will fail to show complete ripening, and those harvested after, are more susceptible to damage and bruising during handling. When the latex ceases to be milky and become watery the fruits are considered suitable for harvesting. Three to five pickings for one season will yield about 30-35 tonnes/ hectare.

Post Harvest Management

Papaya fruits are highly perishable and result in high post-harvest losses, if harvesting methods and handling techniques are inappropriate. From harvest, a shelf-life of four to six days under tropical conditions and up to three weeks at low temperature storage can be achieved with the correct harvest maturity, disease control measures, handling techniques and storage conditions.

Grading

Grading should be carried out as soon as possible after harvest and before packing. On arrival in the packing area, fruits should be washed in water to remove latex and debris. Washing, treatment and grading can be carried out mechanically or manually, depending on the volume of fruits. Grading in each carton is required in terms of size, sex (shape) and stage of ripeness. Female and hermaphrodite fruits cannot be mixed in the same carton because of the difference in fruit shape and size. Either all fruits must be of a similar size in each carton or the separation must be made for the degree of ripeness. Since the fruits are easily perishable, care should be taken to wrap the individual fruits in paper and finally packed in crates. Carton net weight is dependent on the importer, ranging from 3.5 to 5 kg and must not be overfilled during packing.

Storage

For the maximum marketing period of papaya, the fruit should be stored at 10° to 12°C. Temperatures below this range cause chilling injury and rapid deterioration in fruit quality. To develop ripening in papaya, fruits should be stored at 18° to 25°C and treated with ethylene gas at 100 ppm (0.01%) for 24 hours. Unripe papayas are sensitive to ethylene and will commence ripening if stored with ethylene producing commodities. Similarly, ripe papaya produces ethylene and will cause deterioration in ethylene sensitive crops.

Fruits harvested and placed to ripen at the recommended harvest stage (one yellow stripe) will ripen to 60 to 70% yellow colouration within four to six days under ambient tropical conditions (25° to 28°C). Fruits transferred to low temperature storage (10° to 12°C), when harvested at the one-stripe stage, will store successfully for 14 to 21 days, if post-harvest disease incidence can be controlled. When harvested at more advanced stages of ripening, the storage life will be significantly reduced.

Packing

Packing should be done in single layer, one-piece self-locking or two-piece full telescopic fibre board cartons. Internal packaging material should include shredded paper in the base of the carton and individual tissue wraps for each or alternate fruit. Individual labels can be attached to the fruit for appearance and recognition.

Transportation

Importers require fruits at specific stages of ripeness for optimum sales and this varies between 50 and 70% yellow colour depending on importer, market and time of year. For fruit to arrive in the importing country at correct colour stage, attention has to be paid to maximum and minimum colour stages on departure from the packhouse, the length of shipment period and the temperature in the importing country. Thus, colour stages on departure from the pack house for air shipment can vary from 20% to 50%, and the selections are generally more rigid during the summer months due to the rapid rates of ripening on arrival in importing countries. Fruits exported too green (less than 20% yellow colour) will fail to ripen adequately when temperatures in importing countries are low, particularly in winter months. Sea-shipment of papaya is possible when fruits are shipped at optimum harvest maturity, with one or two yellow streaks. Shipments should be made at 10° to 12°C and 85 to 95% relative humidity, in refrigerated holds or reefer containers. Post-harvest disease control is critical with sea-shipment, particularly anthracnose and Phytophthora. Colour development during sea-shipment usually increases from 10% to 40% during 10 days at 12°C, and will develop further during the customs, clearance and delivery period on arrival.

Marketing

All markets prefer small and medium hermaphrodite fruits with 50 to 70% yellow colouration. Small female fruits are acceptable to some importers. Selected markets and importers require medium and large fruits, particularly for ethnic or catering markets. Hermaphrodite fruits are preferred due to large cavity size and susceptibility to damage shown with large female fruits.

Papain Production

Crude papain is available as flakes or powder. Crude papain white, brown and spray dried are the three types of papain extracted from raw-mature papaya fruits. About 8-10 g of final produce can be obtained from a well developed fruit weighing about 2 kg. Tapping may be done 2/3 times preferably during cool hours. The fruits after latexing can be used as a table fruit or for preparing fruity or jams etc.

PEACH

Peach (*Prunus persica*) belonging to Rosaceae family requires the warmest climate among all other temperate fruits. In the plains of North India, cultivation of peach attained a significant position during the last few years and still there is scope for putting more area under this fruit. Peach is a native of Persia. The cultivated form of peach came from China.

Peach is a delicious, juicy and highly palatable fruit. It is a rich source of vitamin A, iron and protein. The fruit is generally consumed fresh but delicious squash can also be prepared. The fruit of some choice varieties can be canned with good success. The cultivars Shan-i-Punjab and Flordared are suitable for canning.

Varieties

A number of high quality sub-tropical peaches evolved in U.S.A., particularly, in Florida and California, were introduced at PAU, Ludhiana during 1968. Some of these cultivars like Flordasun, Shan-i-Punjab, Flordared, Nectarine and Sun Red have recorded excellent performance under these sub-tropical conditions. The cultivar FIordasun was the first among these cultivars to be recommended for commercial cultivation in the plains of Punjab.The main feature of this cultivar is that its fruits ripen very early, i.e., by middle of May at a time, when there is no other fresh fruit in the market and thus, fetches good price.

Pratap: It matures in 76 days, a week earlier than the Flordasun cultivar. The colour of its fruit is yellow with red blush and flesh colour is also yellow with red coloration. It yields 70 kg fruits/ plant and its average fruit weight and keeping quality are better than Flordasun. Also its trees remain smaller in size as compared with Flordasun.

Flordasun : The tree is vigorous and comes to bear within two years after planting in the field. It is a free-stone variety with yellow flesh. The fruit is medium in size (4-5 cm in diameter) and weighs about 80 g. It has excellent flavour, attractive skin colour and ripens evenly at tip and suture. The average yield is about 75 kg/tree/year.

Shan-i-Punjab: This is another early cultivar, maturing in the first week of May. Tree is vigorous in growth. It produces large fruits of 5-5.5 cm diameter, weighing about 90 g each. Colour of the fruit is yellow with red blush, juicy and sweet, with excellent taste, and free-stone. The fruit is quite firm in texture and can withstand transportation. In addition to its table use, this cultivar is suitable for canning. The average yield is about 70 kg/ tree.

Flordared: It is an excellent mid-season table peach maturing in beginning of June. Tree is vigorous in growth Fruit large, almost red at maturity, juicy with soft white flesh and free-stone. The average yield is about 100 kg/ tree.

Sun Red : A smooth skinned peach, which matures in middle of May and is, therefore, quite early. Fruits almost red in colour, very attractive but smaller in size. The flesh is firm, yellow and free stone. Like the cultivar Shan-i-Punjab, fruits of this cultivar can withstand transportation and are quite suitable for canning in addition to their suitability for table use.

Sharbati : The tree is spreading and vigorous. Fruits are large, greenish yellow with rosy patches, very juicy with excellent taste and flavour. Ripens in the first week of July. Average yield is 100 to 120 kg/ tree.

Khurmani: The tree is medium and upright in growth. It blooms in early February and fruit ripens slightly earlier than Sufeda. The fruit is large with white, soft and juicy flesh. Fruits weigh about 70g and is attractive with red colouration and is also slightly pointed at base.

Climate and Soil

In India, peach is mainly grown in mid hills at a height ranging from 1000-1600 metres above sea level and certain varieties that are grown in sub-tropical areas require only 250-300 hours of chilling during the year for proper flowering and fruiting. Peaches do well in wet and humid climate with cold winter and dry summer. The varieties of superb quality and good taste do not grow well in plains, because their chilling requirements are not met with. They also require protection from hot desiccating wind. On the other hand peaches do not grow well where winter temperature falls below the freezing point as it gets injured. The cultivar Flordasun has low chilling requirements of about 300 hours only at or below 7.2°C. The annual rainfall of 45 cm is sufficient for peach. Spring frost is more harmful for peach cultivation.

Peaches thrive well on light sandy soil. Deep fertile loam or sandy loam with good drainage is the best for successful peach growing. It does not tolerate imperfect drainage. Poorly drained soils resulted in spotty planting, short duration trees with low yield and high cost of production. Such soils should be rejected for peach cultivation.

Propagation

Peaches are commercially propagated by budding and grafting. Among budding methods, T -budding is commonly used. In case of grafting, the method employed depends upon thickness of stock. If scion and stock are of same size, tongue grafting is recommended. But when the stock is of more thickness, cleft or wedge grafting is done. Budding can be done from April to September and grafting is done during dormant season *i.e.* in November-January before scion and stock start sprouting.

Rootstocks for peaches are raised from seeds of Desi peach trees. The seeds of commercial cultivars like Sharbati and Khumlani are also used to grow the rootstock for peach propagation. Peach seeds require stratification for germination. Stratification of peach seeds can be done under natural conditions and under cold storage conditions at 10°C or below.

Peach seedlings from stratified seeds planted in the nursery beds during January become buddable in June, if they are grown in the fertile soil with full care. Budding of these seedlings continues upto September. The seedlings budded in June may attain saleable size upto next winter.But budding done in late season becomes saleable only in next winter. Peach seedlings budded during late September, do not sprout but remain green and these green buds put vigorous growth in coming spring. All the seedlings transplanted in January may not become buddable in the same year. In such case the seedlings left unbudded are grafted in following winter. The grafting can be done when seedlings are still in nursery beds or these can be uprooted and brought at one place for grafting.

The nursery plants should be saved from fungal diseases like Rhizoctonia. On seeing the attack of this fungus, the nursery beds should be applied with 25- 30 kg of Brassicol/ ha. For the rapid growth of nursery plants, regular irrigation at 7-10 days interval during summer months is required. A light dose of nitrogen (100 kg of CAN or Ammonium sulphate/ha) also helps in rapid growth of budded plants. Irrigation immediately after budding is very essential for better union of buds.

Planting

Peach should be planted by end of January, when the plants are dormant. Planting should be done after digging 1 x 1 x 1 metre pits refilling them with fertile top soil mixed with 40 kg of well rotten farmyard manure/ pit. Add 25 g of BHC 10% dust to each pit to prevent attack of white-ants. Peach plants should be planted at the distance of 6.5 metres apart, thus accommodating 225 plants in square system and 259 plants in hexagonal system/ hectare.

After planting, the young peach plants need staking to make them grow upright and straight. The supports should be strong, preferably of bamboo or other wooden logs and should be well fixed in the soil. The young plants should be watered at frequent intervals. Modified basin method is the best suited for young peach orchards. The stock sprouts should be removed carefully after every 10-15 days. The strings or tying materials tied at the bud or graft union may also be removed carefully to avoid constrictions. Training of young peach plants should be given utmost attention for developing proper framework and shape. If there is any indication of white ant attack, chlorphyriphos (1 ml/ litre of water) with irrigation should be given.

Training and Pruning

The peach is trained to the 'Modified Leader System'. For admitting more sunlight for better colouration of the inside fruits, 'Open Centre System' may be favoured. But under sub-tropical conditions, where sunshine is plenty and strong winds prevail during summer, this system is not suitable, in view of the weak frame-work and other obvious defects.

During first year plants should be headed back to a height of 90-100 cm at the time of planting in the field. All branches on plants are also cut back to two buds. After three to four weeks, plants develop many branches. Three to five laterals, which are well-oriented around the trunk, are selected as main scaffold branches and all other branches are removed. The lowest branch should not be below 40-50 cm from the ground level. In the second year very little pruning is done. The branches which are crowding in the centre or interfering with each other are removed. Dominance of central leader is maintained and no lateral is allowed to outgrow other. In the third year, the leader should be suppressed by cutting it back to a suitable lateral.

Peach fruits are borne on one-year growth. A small proportion of crop, however, is also borne on short-lived spurs. Pruning should be done so as to produce 50 to 100 cm of growth in young trees and 30 to 70 cm in old trees, annually. Some thinning-out of branches is also done to allow adequate sunlight for growth of fruits and development of good colour. Pruning of peaches should be carried out in early January.

Fertilizer Application

Age (years)	Farmyard manure (kg/tree)	*Nutrient (g/tree)			Fertilizer (g/tree)		
		N	P_2O_5	K_2O	Urea (46%)	Super phosphate	Muriate of Potash
1	10	90	30	90	180	190	150
2	15	180	60	180	360	380	300
3	20	270	90	270	540	570	480
4 & above	25	500	120	500	1000	760	830

Whole the quantity of farmyard manure along with phosphorus and potassium are given during December. Half of nitrogen should be given in January before flowering and the remaining half after fruit-set. The manure or fertilizer should always be applied by evenly broadcasting in the tree basins which should be sufficiently large and should encompass the entire canopy of tree. The manures should be thoroughly mixed with soil using a spade. Trees should be watered immediately after application of manures.

Irrigation

With onset of spring, trees put forth new growth flush. Consequently, they need much greater amounts of water than was being applied in winter. The young and old peach trees should be irrigated regularly and more frequently. The need for irrigation is going to build up further as the weather becomes warmer. During April as the season gets quite hot, both newly planted young plants and bearing peach trees require frequent irrigations. Watering of trees would depend upon-type of soil, age of plant, locality, etc. However, irrigation may be given at 8-10 days interval. The newly set plants would require watering at still shorter intervals.

For proper development of fruit size and quality, frequent irrigations should be given during April and May. The irrigation interval should be reduced to a week or so keeping in view the soil type, locality, prevailing temperature, etc. During November, the atmospheric temperature cools down considerably, irrigation to peach orchards should be given at longer intervals. The peach tree would become dormant and there will be less transpiration because of shedding of leaves. The irrigation at this time should be withheld, so that they enter into dormancy and become sufficiently hardened to withstand cool weather. By giving frequent irrigations at this time, the trees would continue to grow and become prone to winter injury.

Fruit Thinning

To improve prospectus of developing peach industry, it is essential to improve the marketable size and quality of fruits. Thinning is more desirable on mature trees making small annual growth than on young vigorous plants. Trees growing on light soils, deficient in moisture and nutrients, would also need fruit thinning. The heavy bearing cultivars like Pratap, Flordasun, Shan-i-Punjab, Florda Red, Sunred, Sharbati, and Sufeda need fruit thinning to minimise danger of limb breakage. Thinning may also be resorted when good-sized uniform fruits are needed for canning and fancy trade. The distance from fruit to fruit after thinning on shoots should be 15 cm and the average number of leaves/ fruit should be 20 to 25.

Harvesting and Yield

Peach comes into bearing within two years after planting in the field. The Flordasun and Pratap peach cultivars take 11 weeks to reach the stage of harvest maturity from the time of fruit-setting. The peak harvesting period of different peach cultivars in plains is from middle of May to middle of June. The pattern of increase in the length, diameter and fruit weight followed a 'Double Sigmoid Curve'. The fruits of Flordasun were green in earlier stage and changed to pinkish colour with red tinge at harvest stage.

The peach fruit being perishable in nature needs to be handled quickly when it is ready for harvest. Even a few hours delay results in considerable economic loss to the grower. Fruits picked at the right stage of maturity, on ripening, give excellent taste and flavour, characteristic of a particular cultivar. For successful orcharding, harvesting fruits at its proper stage of maturity is of utmost importance. For sending fruit to distant markets, they should be harvested when they attained good colour but are still hard-ripe.

The average yield of a full grown tree of different varieties varies form 70 to 120 kg. Sharbati is a high yielding cultivar and produces 100 to 120 kg fruit/ tree. The harvested fruits must be disposed off as early as possible. Peach fruits should be graded and then packed in mulberry baskets of suitable size. Soft dry grass should be placed in baskets beneath the fruits. Flordasun fruits harvested at optimum maturity can be kept in cold storage for about 2 weeks (3°C temperature, 85-90% R.H.) in perforated polythene bags. In apricot, plum and peach, three size grades have been prescribed. Smaller size universal CFB cartons are used for packing.

PEAR

Pear (*Pyrus communis*) is one of the superior temperate fruits belonging to the apple family and next only to apple in importance in area and production among temperate fruits in India. Pear is largely grown for consumption as a fresh fruit as it has good taste and flavour. It needs relatively less care due to its hardy nature which enables it to flourish well even on water-logged soils, where most other fruit trees fail to grow. It is also well known for its keeping quality and stands transportation well.

Because of wide range of its adaptability, the cultivation of this fruit has spread to the temperate and sub-tropical regions of Jammu and Kashmir, Himachal Pradesh, Punjab, Uttar Pradesh, Assam and South India. Pear is a very delicious and juicy fruit, rich in protein and minerals like phosphorus. It can be used in the preparation of several delicious products like pear preserve, pear jam, pear chutney, pear juice and canned pear. Pear preserve has attractive amber colour with characteristic flavour. Patharnakh and Kieffer varieties are good for canning.

Varieties

A number of good varieties exist in pears and the varieties belong to three groups-Asian, European and hybrids. More important and promising ones are as follows:

Patharnakh: It is an Asian pear (*Pyrus pyrifolia*) variety. This is the most popular variety because of heavy bearing and good keeping quality. The cultivation of Patharnakh pear received a boost as a commercial crop in Northern India and acreage under this fruit crop is steadily increasing because of its high yield potential and good economic returns. The fruit is tough and firm and can stand transportation very well for long distances without any spoilage. The fruit is medium round and green in colour with prominent dots. The flesh is gritty, crisp and juicy. It ripens in the last week of July. It is a heavy yielding variety with an average yield of 150 kg fruit/ tree.

Baggugosha : It is an European pear (*Pyrus communis*) variety. Trees are upright and vigorous and have irregular bearing habit. It has small, greenish, yellow fruits with tapering stem-end. The flesh is sweet and somewhat gritty. It ripens in August and has an average yield of 100 kg/ tree.

LeConte: It is a hybrid pear (*Pyrus communis* x *Pyrus pyrifolia*). LeConte is a variety having low chilling requirement. Tree medium in growth, upright and become spreading when fully mature. Trees bear good crop, but are blight susceptible. Its fruits are small to medium in size and pyriform in shape with alternate greenish yellow colour. It ripens by the end of July and yields about 60-80 kg fruit/ tree.

Kieffer: Kieffer is also a hybrid between *Pyrus communis* x *Pyrus pyrifolia*. It is a variety requiring longer chilling hours than LeConte and Patharnakh. Tree is very productive, widely adaptable, hardy and disease resistant. The fruit is medium to large, pointed at the ends, brownish red in colour and gritty, hard, well suited for canning. It ripens in August-September. Average yield 100 kg/ tree.

Climate and Soil

Pears can be grown on a wide range of climatic conditions as it tolerates temperature as low as -26°C during dormant period and as high as 45°C during growing period. Pear can be grown successfully in mid-hills at an elevation of 1,200 to 1,800 metres above sea level, where temperature ranges from 15-25°C in winter and 10-32°C in summer. An annual rainfall of 100-125 cm is adequate. The best flavour and keeping quality in Barlett pear develop in locations where summers are fairly hot and dry. Patharnakh pear is a very hardy tree and can be successfully grown in the sub tropical climate. Being deciduous in habit, it escapes damage from frost. Care must be taken to plant pear orchards in frost free locations. Absence of rains during flowering period is one of the pre-requisites for successful cultivation of pear. Rain at flowering not only washes away the pollen but also interferes with activity of pollinating insects, thus adversely affecting fruit-set. Areas exposed to excessive hot winds during summer also have an adverse affect both on fruit and foliage.

Pear should be grown on a land which is at a slightly higher level than the surrounding area to reduce chances of damage by frost. It grows well on a wide variety of soils provided these are well drained. The pear tree is hardy and can flourish even on rather inferior lands with relatively less care. It tolerates water logged conditions to some extent, however, the areas frequented by floods and where the water stagnates for many days during rainy season should never be selected for pear growing. The most desirable soil for pear is one which is of medium texture, deep, well drained, with a pH of 6.0 to 7.5 and has a water table below 200 cm around the year.

Propagation

The Patharnakh pear plants are raised both by budding and grafting on pear suckers and Kainth (*Pyrus pashia*) seedlings. Of the various methods of grafting, tongue and cleft grafting are the most common ones for pear propagation. When the scion and stock are of equal thickness, tongue grafting can be successfully employed, whereas when the stock is thicker than the scion, cleft grafting may be practiced. Tongue and cleft grafting are performed during January-February, when the plants are still dormant. T-budding is practiced from June to August. Pear plants can also be propagated through hard-wood cuttings.

Rootstocks are generally propagated from seeds or from cuttings. For raising rootstocks from Kainth seed, fully mature fruits are collected in September. These fruits are dumped for softening and easy extraction of seed. Seeds are subjected to stratification before they are sown in the field or on raised nursery bed. The stratification is completed in about a month. The sprouted seeds are then sown immediately in the nursery at a distance of 15 cm in rows 30 cm apart.

The seeds are also sown on raised beds during November-December. The stratification of seed on these beds is completed by natural process and the seedlings come out in about two months. These seedlings are allowed to grow on these raised beds for about a year and transplanted in the nursery in the following January. The distance from seedling to seedling and row-to-row is kept 15 cm and 30 cm, respectively. These seedlings become buddable during August-September and graftable in December-January.

The seedling rootstocks raised from Kainth seeds are not uniform in growth and vigour and take two years to become buddable or graftable. Therefore, clonal propagation becomes essential to get uniform and true-to-type rootstocks. For raising clonal rootstock, hard-wood cuttings are planted and they become graftable after one year, thus, one year is saved as compared to raising of seedlings.

Rootstocks can be raised from pear suckers also. This rootstock is widely used in Punjab for propagating pear. The root suckers (Batankan) are collected from underneath of old pear trees during October-December. These are headed back leaving two to three buds from ground level and planted 15-20 cm in rows 30 cm apart leaving 60 cm space after every two rows Out of the newly emerged shoots, single healthy shoot is retained which becomes buddable during August-September. Those suckers which do not become buddable during August-September are grafted in the following December-January with the scion cultivars. The pear suckers which have well developed root system are bench-grafted and directly planted in the nursery field.

Pear plants can also be raised through hard-wood cuttings. The cuttings are prepared from juvenile shoots during December. The cuttings are kept for callusing for about a month and are planted in the nursery field at a distance of 10 cm apart, in rows 30 cm apart. Treating the cuttings with IBA (100 ppm) for 24 hours prior to callusing gives better rooting success. The nursery fields are constantly kept moist with frequent and light irrigations. Flooding of the nursery should be avoided.

Clonal rootstock Quince A having semi-vigorous growth is now widely used. Newer clonal rootstocks like Old Home x Farmingdale series and BA 29C are under evaluation.

Planting

The best time for planting pear trees is from mid-January to first fortnight of February before they start new growth. One year old, healthy and vigorous plants should be used for planting. Older plants of 2-3 years age can also be planted with good success. Before the plants are put in the field, the orchard site should be properly laid out. The trees may be planted at a distance of about 7.5 metres apart by which 180 plants can be accommodated in one hectare by square system of planting. The pear may also be planted by the hexagonal system by which 207 plants are accommodated in one hectare.

Basins should be prepared immediately after planting. The level of soil near the trunk should be kept slightly higher than the level of basins so that water may not come in direct contact with the trunk. Irrigation should be given immediately after planting so that soil may settle down properly around the roots. After two or three days, second irrigation should be given. Subsequent irrigations should be given as and when required. The irrigation interval upto March may be 15-20 days, but later on as the season warms up, the interval may be reduced to 10-12 days upto June. The trunk of young plants should be white washed once in March, again in May and then in the rainy season to avoid injury from sun-bum and to prevent cracking of bark. Farmyard manure should be mixed in the basins at the rate of 20 kg/ tree in January-February. Irrigation should be given immediately after FYM is thoroughly mixed in the basins. To check the attack of white ants, trees should be given chlorpyriphos 1 ml/ litre of water once in April and then after rains. The surplus shoots not required for building the framework of trees should be removed. The shoots appearing on rootstock should also be removed, otherwise they will have an adverse effect on growth of plants. Trees should be protected from insect and pests by following proper spraying schedule.

Training and Pruning

Proper training and pruning of pear trees are essential for the development of proper framework and to control size of the tree for the convenience of horticultural operations, to maintain the vigour and growth of the tree, to spread the fruiting area uniformly on the tree, to secure fruit of good size and quality and to encourage regular bearing. Pear trees are usually trained according to. 'Modified Leader System'.

During the first year, low-headed trees are trained in such a way that the lowest branch is allowed to develop upto a height of 60 cm and then the top of plant is cut at about 90 cm. In case of high-headed trees, plants are headed back at a height of 1.2 to 1.5 metres. In the second year, the largest shoot in the centre is left as the leader but is cut back where immature portion starts. Three to five laterals (main scaffold branches), spaced at a distance of 15 to 22 cm from each other are kept in different directions. In three to four year old trees, the laterals left at the end of second season's growth is allowed to branch off to form new laterals, one or two of which is selected to form the secondary scaffold framework. In five year old trees, the leader is cut back to form a well-placed, outward growing lateral.

Pear trees bear fruit on shoots of limited growth which are called spurs. These spurs make only a few millimetres growth every year and bear fruit for 12-15 years. Therefore, the pruning of young bearing trees is only of corrective type for the first 10-12 years. Diseased, dried and criss crossing branches should only be removed. Severe pruning should not be done during this period which would lead to loss of fruiting spurs. Care should also be taken to avoid injury to the spurs during harvesting of fruit.

The older trees are to be rejuvenated for bringing back the bearing capacity. This is achieved by thinning out the entire bigger branches helping in opening up the trees for more light for better colour. The process of rejuvenation is completed in stages within three to four years. The trees again develop new fruiting spurs which will bear fruit for another 10 years. The pruning of pear should be done in early February.

Fertilizer Application

Pear trees should be adequately manured and fertilized so that they may make good vegetative growth and bear better crops of high quality fruit. Based on experimental evidence and experience, the following schedule of manuring and fertilization is recommended.

Dose/ tree				
Age of the tree (year)	Farmyard manure (kg)	CAN (25%N)	Superhosphate 16% P_2O_5	Muriate of potash (60% K_2O)
		*Fertilizer (g)		
1 - 3	10-20	200-600	200-600	150-450
4 - 6	25-35	800-1200	800-1200	600-900
7 - 9	40-50	1400-1800	1400-1800	1050-1300
>10	50	2000	2000	1500

The entire dose of farmyard manure, superphosphate and muriate of potash should be added in December. Half of CAN should be added in early February before flowering and the other half in April after the fruit-set.

Pear is susceptible to zinc and iron deficiency. On calcareous soils or where irrigation source is drainage water of cities, pear develops chlorosis due to lack of available iron, manganese or zinc. Boron deficiency may cause blossom blight in pears where the blossoms wilt and die but leaves remain normal. In severe cases, leaves also die and the limbs may or may not put on new buds for a few weeks. In copper deficiency, terminal leaves on current year's shoots are first affected. The leaf tips first become yellow and this is followed by necrosis. Later in the season, most of the new shoots develop die back. Due to zinc deficiency, the plants remain dwarf, slow growing and show little leaf symptoms. Shoot terminals exhibit rosette like appearance.

To correct zinc deficiency, zinc sulphate (1.5-2.0 kg/ 450 litres of water) is sufficient for spraying on non-bearing trees in April-May and bearing trees after fruit harvest. Iron chlorosis in young trees can be controlled by spraying chlorotic young trees with a solution of 0.5% ferrous sulphate. Boron deficiency can be corrected by broadcasting borax in the tree basins at the rate of 100 g/ tree in early fall. Such a treatment usually prevents boron deficiency for three years. Copper deficiency can be controlled by spraying copper sulphate at the rate of 0.5% after fruit harvest.

Irrigation

Water is needed during summer, which is the active growing period and when the young fruits are developing. Shortage of water during the month of April, May and June restrict tree growth and lead to formation of hard and small fruits of poor quality. During summer, pear should be irrigated at an interval of 7-10 days depending upon soil conditions. As regards the colour development, dessert and keeping quality of fruits, neither too little nor too much water is needed. If the plants suffer from extreme water deficiency, fruits may lack colour, become tough and somewhat meaty without crispness of fruits having higher water content. When water is excessive, the colour may again become less attractive and keeping quality deteriorates. Under relatively dry conditions, fruits develop higher acidity and better flavour. After harvesting of fruits in August, the trees should be irrigated at an interval of 20 days or so upto the end of October. After this, no irrigation may be given till January.

Weed Control

Weeds can be controlled mechanically by hoeing or by spraying the herbicides namely hexuron or glycel or gramoxone. Spray hexuron 80 WP (diuron) at 4.0 kg/hectare as pre emergence (first fortnight of March) after thorough cultivation so as to remove the established growing weeds. Gylcel 41 SL (glyphosate) (3 litres/hectare) or gramoxone 24WSC (paraquat) (3 litres/ hectare) should be sprayed as post-emergence when the weeds are growing actively up to a height of 15-20 cm. Dissolve the herbicides in 500 litres of water/ hectare so as to give a thorough coverage to the field. In case, there is emergence of weeds after rainy season, another spray of glycel or gramoxone at the same rate may be given. Spray glycel or gramoxone during calm day to avoid spray drift to foliage of fruit tree.

Intercropping

Leguminous crops such as moong, cowpeas, green gram and pea may be grown in young and non bearing orchards, which in addition to giving good income to the grower, improve the nitrogen status of soil by fixing in the atmospheric nitrogen with the help of nitrogen fixing bacteria. Some fruit plants such as peach, plum, papaya, and phalsa may also be interplanted as fillers. The fillers must be removed when the pear trees come into bearing. Additional doses of fertilizers should be added to filler trees according to their requirements, so that they do not compete for nutrients with permanent trees. Excessive watering to intercrops should be avoided as it is harmful to pear trees due to dormant condition in winter.

Fruit Thinning

Normally pear trees come into bearing at the age of six to seven years. Pear trees tend to overbear, resulting in smaller fruits. The number of fruits a tree will carry and develop to good marketable size varies with its vigour and growth conditions. One fruit/ cluster may be retained after thinning. In each of the commercial varieties, nearly 30 to 40 leaves/ fruit are essential as building materials for making a fruit. Many of the fruiting spurs are lost as a result of fruit harvesting with the help of hooks or downward pull of the fruit with hand. Hence a little or no fruit is set for the subsequent crop. To compensate this loss, the fruit set can be increased by spraying Gibberellic acid (20 ppm) at full-bloom stage. To check spur breakage, harvesting should be done with hand. Saving the spurs from breakage during fruit harvesting helps regular fruiting throughout the life of pear tree.

Harvesting and Yield

Harvesting of pear fruit at proper stage of maturity is very important. The criteria for maturity of pear fruits are changes in external fruit colour, decrease in firmness of fruit flesh and increase in total soluble solids (TSS) content. The fruit of Patharnakh pear reaches harvest maturity after 155 days of fruit-set. The size and weight of fruit continue to increase till harvest in August. Colour of fruit changes to light yellow at final stage of harvest. The optimum harvesting time of Patharnakh pear fruits in Punjab is from end of July to mid August.

The fruit should be picked carefully without any bruises and injuries. Care should be exercised during picking so that the fruit bearing spurs do not get damaged. Once the spurs are broken or damaged, that will take long time to become fruit-bearing spurs again and will ultimately affect fruit yield in the coming bearing seasons. For distant markets the fruit should be picked when it is still in a hard ripe stage and without injury and bruises. A small stalk (button) should be kept with the fruit to avoid fruit rot, due to fungus, in storage.

Grading

The fruits should be carefully sorted and graded before sending to markets. In this way, consumers get the quality of his choice, while the fruit grower earns a premium for his product.

Packing

When the fruit is to be despatched to distant places, it is generally packed in wooden boxes. For sending the fruit to short distances, baskets are largely employed. The base of wooden box is given thick layer of dried grass. A sheet of paper is then spread to cover the grass and the sides of the box. The fruit is arranged in the box in layers. The upper most layer is covered with paper, It is then followed by a layer of dried grass. Then, the lid is closed with the help of nails.

Marketing

The fruit growers normally do not conduct direct sale but generally auction the crop to contractors when the fruit is on the tree. The contractors pick the fruit according to their convenience and sell it in the market. Only a few orchardists market their crop directly in the main markets. To ensure due remuneration to orchardists, sales cooperatives need to be organised and such organisation should help in grading, packing and transportation of the fruit.

The facility of storage chambers can also play an important part in improving the marketing of pear fruit. Well managed pear orchards, may yield a net income of Rs.20000 to Rs. 25000/ hectare depending upon locality, soil, management practices, etc.

Patharnakh and other standard varieties keep well in cold storage, if properly handled. The fruits of Patharnakh pear can be stored for 90 days at 0 to 3°C temperature with 85-90% relative humidity after packing them in polythene bags. It is observed that the fruits harvested in the last week of July store better. The Patharnakh pear fruits can be successfully stored for 125 days by pre-harvest treatments with calcium nitrate (1.0 and 1.5%) one week before harvest. The post-harvest application of Benlate (500 ppm) also proves beneficial in increasing storage life of patharnakh pear fruits. The pear fruits particularly those of Patharnakh variety have a long post-harvest life and unique shipping quality and can be transported to distant markets like Bombay and Calcutta. This quality alone has encouraged several growers to take up pear cultivation, especially in far-flung areas.

PINEAPPLE

The pineapple (*Ananas comosus*) is one of the most popular tropical fruits belonging to family Bromeliaceae. This is a large, diverse family of about 2000 species. Like banana, it is one of the few important fruiting monocots.The origin of pineapple is the American continent, probably Brazil and Paraguay. It has spread throughout tropical and subtropical regions as a commercial fruit crop. The important pineapple growing countries of the world are the Hawaiian Islands, Philippines, Malaysia, Thailand, Brazil, Ghana, Kenya, Mexico, Taiwan, South Africa, Australia, Puerto Rico and India. Fruit of pineapple is botanically called sorosis.

Commercial cultivation of pineapple in India started only about four decades back. So although the conditions prevailing in large parts of India are ideal for pineapple cultivation, it does not hold any position of importance among the major fruits cultivated. It is being cultivated in high rainfall and humid coastal regions of peninsular India and hilly areas of North-Eastern region. Of late, it was shown that pineapple can also be grown commercially in the interior plains with medium rainfall and supplementary protective irrigation. It is grown in Assam, Meghalaya, Tripura, Mizoram, West Bengal, Kerala, Karnataka and Goa on a large scale, whereas in Gujarat, Maharashtra, Tamil Nadu, Andhra Pradesh, Orissa, Bihar and Uttar Pradesh on a small scale During 2001-02, India produced 1.26 million tonnes of pineapple from about 80,000 ha. India produces more than 8% of total world production of pineapple.

Pineapple is a tropical fruit. It is a good source of Vitamins A and B and rich in Vitamin C and minerals like calcium, magnesium, potassium and iron. The fruit contains a special enzyme called 'Bromelin' which digests protein. Fruit is used for table purpose, preparation of juices, slices, tit bits, squash, jam, mixed jam and candy. Dried waste after extraction (pomace) is a valuable cattle feed. Other byproducts are alcohol, calcium citrate, citric acid and vinegar. Leaves of pineapple are used for fibre preparation.

Varieties

Gaint Kew or Kew: It is the most important cultivar of Pineapple . It is ideal for canning industry. Flesh of the fruit is yellow, fibreless and juicy. The fruit is cylindrical in shape with average weight 1.9 kg. It is a late maturing variety and leaves lack spines.

Queen : It is the second important cultivar which is grown in India. It is an early maturing variety and an outstanding table variety. The fruits are small cylindrical, with average weight 1.2 kg. Fruit colour is golden yellow while the flesh is deep golden yellow, less juicy than Kew. Fruit has pleasant aroma and flavour.

Mauritius : It is a mid season variety from Queen group used exclusively for table purpose and grown in some parts of Kerala. Fruits of this variety are medium sized with yellow and red skin.

Jaldhup and Lakhat : Two indigenous small fruited varieties blended with acidity under Queen group grown in Assam.

Charlotte Rothchild : A small fruited spiny variety of the Spanish group with fruit characteristics and taste similar to Kew. Partly under cultivation in Kerala and Goa.

Climate and Soil

Pineapple is a tropical fruit. It requires moderate temperature and high humidity. The optimum temperate range is 21 to 23^{0}C. At low temperature, no fruit bud differentiation takes place.

Areas with heavy and well-distributed rainfall are the best for pineapple growth. Optimum rainfall is 1500mm per year although it grows in areas having 500mm to 5550mm of rainfall. The fruit grows well near the seacoast as well as in inland, so long as temperature ranges from 15.5 to 32.5^0 C. Low temperature, bright sunshine and total shade are harmful. It can grow successfully up to 1525m above sea level.

Slightly acidic soil with pH range of 5.5 to 6.0 is optimum for pineapple cultivation. The soil should be well drained and light in texture. Heavy clay soil is not suitable. It can grow in sandy, alluvial or laterite soil.

Fumigation of soil is practiced routinely, where nematode is a serious problem. Pineapples are naturally drought tolerant by virtue of being CAM (crassulacean acid metabolism) plants, opening their stomata at night for CO_2 uptake, and closing them during the day. They can be grown in seasonal wet/dry areas of the tropics that do not support less water efficient crops. However, supplemental irrigation is often provided to maximize yield in dry seasons. Black plastic mulch is used in intensive plantings for weed control and conservation of soil moisture.

Propagation

The choice of planting material is crucial as the performance of plants depends on the materials planted. It is always advisable to use uniform size material of monotype for getting uniform growth of plants, enabling uniform cultural operations and getting harvest at one time from such a field. Hence selection of right type and size of planting material is essential for commercial planting.

Pineapple is commonly propagated from crowns (on top of the fruit), or axillary shoots arising from either the base of fruit (slips) or base of plant (suckers). Growth is the best with suckers and slips. Plants grown from suckers produce fruits in about 18 months, whereas those from crowns take over two years. Therefore, crowns are not normally used. Suckers should be planted within 2 weeks after removing from mother plant. The planting material should be selected from healthy and disease free plants. Unrooted shoots can be placed into the soil and rooted in situ.

The size of the planting material affects production; as larger slips/suckers produced more vigorous plants, gave the highest yield and reduced the time from planting to harvest.Among the types and size of propagules tried, slips and suckers weighing around 350 and 450 g respectively were the best for yield and quality for Kew pineapple. Mass multiplication of propagule is vital to bring fresh area under cultivation. This is possible only when a number of plantlets can be obtained from a single mother plant, unlike a few suckers or slips. It is possible to use leaf cuttings from the crowns of Kew pineapple for multiplication of planting material. Total of 10-15 leaf cuttings are made from each crown. However, these cuttings will take even more time than crowns for flowering and thus are only recommended where planting material is not available.

Land Preparation

The lands should be thoroughly ploughed and pulverised to a good tilth before planting. It should be properly leveled to avoid water stagnation. Adequate drainage channels should be provided. In the hills, proper terracing is a necessity. A planting density of 43,500 plants/ per hectare can be followed, keeping a distance of 30 cm between plants ,60 cm between rows and 90 cm between beds. The system of planting varies depending on topography of land and rainfall. There are 4 planting systems in vogue, viz. flat-bed planting, furrow planting, contour planting and trench planting.

In high rainfall areas the density may be around 40000 to 44000 plants/ ha. In low rainfall areas with cool weather like that of Karnataka, even higher density of 63000 to 64000/ha is recommended. Pineapple is planted in double hedge system for convenient intercultural operations. For a density of 44000 plants/ ha , the spacing should be 90 x 30 x 60cm i.e. 90cm between two rows of adjacent beds, 30cm between plants in a row and 60cm between two single rows. High density planting is the latest advanced technology applied in pineapple cultivation. The other spacing followed are 120 x 60 x 40cm (41600 plants/ha.), 75 x 30 x 30cm (63000 plants/ha), 60 x 30 x 45cm (64000 plants/ha). To obtain a good yield from high density planting, application of fertilizers and treatment with flower-inducing chemicals are considered very important.

Planting

For planting, rainy season is the best. Planting is done at the beginning of monsoon in dry region and at the end of the monsoon in high rainfall areas. With irrigation, it may be at any time of the year. Suckers or slips are first cured by stripping off the lower leaves, followed by drying in the sun, or in partial shade for three to four days before planting. Planting material should be dipped in ceresan solution (4g in 1 lit. of water) or 0.2% Dithane M-45 before planting to protect plants against budrot. They are planted either in flat beds, where there is no danger of water stagnation, or in shallow trenches, which are filled as suckers grow and develop. Suckers should be planted at 10 - 15cm depth in 15-20cm deep hole. Care should be taken to see that the bud or 'heart' of suckers does not get buried. About 500g FYM or cow dung is to be added to soil of each hole. Pineapple may be planted using black polythene film as soil cover. This ensures total weed control and heavy crop.

Fertilizer Application

Pineapple is a shallow feeder with high nitrogen and potassium requirement. Since these nutrients are prone to heavy losses in the soil, practices relating to time of application and the form of fertilizer determine their efficient usage. Experts based on research trials conducted at a number of locations advise to give N and K_2O at 12 g each per plant. There is no need for P application. However, if the soils are poor in P, 4 g of P_2O_5 /plant can be applied. Nitrogen should be applied in six split doses. The first dose of N can be given 2 months after planting and the last dose 12 months after planting. Potash should be given in 2 split doses. Entire P and half the dose of K can be given at the time of planting and the remaining K, 6 months after planting. Application of fertilizer under rainfed conditions has to be done when moisture is available. Ammonium sulphate was a better choice for nitrogenous fertilizer than urea or calcium ammonium nitrate, as it resulted in the highest fruit weight and sucker production. There were no differences due to application of potassium as potassium sulphate or as muriate of potash.

Irrigation

Pineapple is mostly grown as a rainfed crop, but supplementary irrigation gives higher production. Irrigation in summer and winter keep the plants healthier. About 8 to 12 light irrigations during winter and summer months are beneficial. Irrigation is essential after planting or after manuring if there is deficiency in soil moisture. Fertilizers may be applied by spraying or drip method. Initial investment will be high but drip method is cost effective.

Earthing Up

Earthing up is an essential operation in pineapple cultivation aimed at good anchorage to plants. It involves pushing the soil into the trench from the ridge where trench planting is a common practice. As the pineapple roots are very shallow, the plants are eventually lodged especially under conditions of flat-bed planting in heavy rainfall areas. Lodging of plants when the fruits are developing would result in uneven development and ripening of fruits. This operation becomes more important in ratoon crops, as the base of the plant shifts-up, crop after crop. High-density planting would minimize necessity of this operation, as the plants prop each other preventing lodging

Weed Management

Controlling weed is the major interculture operation in pineapple field. Weeding should be done at least 3 to 4 times a year. Manual weeding can be partially eliminated by chemical weedicides like Bromacil @ 3kg and Diuron @ 2kg per ha. Application is done a few days after hand weeding. Before new weeds emerge, a second spraying @ 1.5kg per ha after 15 days is necessary.

To protect the fruit from sunburn, partial shade may be provided by planting arhar in between the beds. The spacing has to be adjusted accordingly. Covering the maturing fruits with rice straw or pineapple leaves will reduce both sunburn and bird damage.

Mulching

Mulching is one of the cultural practices aimed at weed control and soil-moisture conservation. Though mulching is not a common practice in India, use of dry leaves or straw is in practice in some areas of South India. Preliminary studies at IIHR Bangalore revealed that mulching with black polythene and saw-dust resulted in better growth of plants than that with white polythene and rice-straw mulching.

Desuckering

Suckers start growing with the emergence of inflorescence and slips grow with developing fruit. According to some of the studies, fruit weight increases with increasing number of suckers/plant, whereas the increased number of slips delayed fruit maturity. Crown size has no bearing on fruit weight or quality. Hence desuckering can be delayed as much as possible, whereas the slips are recommended to be removed as soon as they attain the required size for planting, Partial pinching of crown in Kew pineapple consisting of removal of the inner most whorl of leaflets along with the growing tips 18 months after fruit set gave better fruit size and shape.

Desuckering, keeping one or two suckers on the mother plant near the ground level and then removing all the others immediately after fruit harvest is important. Slips also should be removed. After desuckering, plants should be fertilized and earthed up.

Growth Regulators

Ripeness-to-flower stage is attained 11-12 months after planting and formation of at least 40 leaves. Normally 70 to 80% of plants flower in a year. Sometimes flower does not appear on time. Even after 15-18 months of growth under optimal nutritional and environmental conditions, as low as only 50 to 60% plants may come to flower. In such cases application of flower inducing chemicals is helpful. Yield may be increased by applying flower inducing chemicals. A pineapple plant produces only one fruit during its life time. Unsaturated gases like ethylene and acetylene in the form of smoke cause early flowering in pineapple. This can be achieved by applying acetylene saturated water or calcium carbide in the heart of the plant. About 50ml ethrel solution (10 ppm in combination with 2% urea and 0.04% sodium carbonate can bring forth flowering. NAA (25 ppm) also induces flowering. To delay harvest by a few days (10 - 15 days), Planofix (300 ppm) may be sprayed on the fruit just 60 to 70 days ahead of harvest. To ripen the fruit earlier by about 10 - 15 days, ethrel (500 ppm) may be sprayed on the fruit about one month before normal harvest.

Staggering of harvesting almost throughout the year is possible by the following means

i. using different planting materials.

ii. planting suckers and slips at regular intervals from July-December and

iii. applying flower inducing chemicals at desired time as stated above.

One main crop followed by two ratoons is the usual crop cycle in pineapple.Economic life of a pineapple plantation is around 4 years. After this, the plot should be uprooted and replanted.

Harvesting and Yield

The fruit is ready for harvest when the dark green colour becomes lighter and the deep seated eyes become shallow. Harvesting for local markets should be done at full maturity stage and for distance markets at 75-80% maturity stage. Harvesting should be done with a sharp knife severing the fruit stalk with a clean cut and retaining 5-7 cm of the stalk. Slips and a part of the crown are removed. Any mechanical injury on the fruit skin may cause the fruit to rot quickly. Great care is necessary in handling the fruit. Fruits are mainly harvested during July-August. However, a small crop is harvested during December to March also. By regulating the crop, harvesting is possible almost eight months a year. An average yield is 10 to15 tonnes/ hectare.

Pineapple fruits should be allowed to grow on the plant as late as possible for development of good flavour and colour on ripening. If harvested early, they may develop woody pulp on ripening.

For canning industry, only the cylindrical fruits of about 1.5kg size are considered good. Such fruits are obtained when peduncles are upright and plant density is around 40000/ ha. For canning, fruits are allowed to reach a more advanced stage prior to harvest, about ½ to 3/4th yellow. Fruits are de-crowned, cored and peeled, then sliced and canned. Slices damaged during processing are sold as chunks. About 60% of the fruit is recovered as slices and the remaining flesh on the core and skin are crushed for juice. The residue leftover, called pineapple bran, is used as livestock feed.

Post Harvest Management

Storage

Pineapples are sensitive to injuries due to compression or impact. Rotting starts and spread quickly from bruised or injured parts. The produce should be cooled as soon as possible after harvest. Cooling temperature is 7-10^{0}C for ripe and half-ripe ones. The relative humidity will be 85-95%. The same temperature and RH are kept in storage and transport. Cooling is done by forced air method. Storage period of pineapple is 2-4 weeks depending on the cultivar and ripening stage. Chilling injury is common at low temperatures and shows up only after removal from refrigeration as flesh browning and crown injury.

Packing

Fresh fruits are washed and waxed prior to boxing. Fungicides for post harvest disease prevention are in the wax. A short section of flower stalk is left to protect the base of the fruit during shipment.They are packed in full telescoping perforated CFB cartons.

Transportation

For transport, the fruits are wrapped up in straw and packed in bamboo baskets or crates in one or two layers.

Marketing

There is always a very good demand of pineapple in the internal markets. It is in high demand from the processing industry as well. Indian pineapple is exported to Nepal, U.K., Spain and U.A.E. The main products of export are canned slices, tit bits, juice and jam. Of the four main commercial cultivars grown in this country, Kew and Giant Kew are better suited for canning, while Queen and Mauritius cultivars are good for juice production.

PLUM

Plum (*Prunus salicina* Lindl) belonging to the family Rosaceae, is temperate in nature. It is a strong growing small tree. This fruit has found favour with orchardists because of their phenomenal yield potential and high economic returns. There are two types of plums- European (*P.domestica*) and Japanese (*P.salicina*). After evaluation, only Japanese plum is recommended for cultivation in India. The low chilling early ripening varieties of Japanese plum come into market when other fruits are scarce and thus had little competition in the market at that time. Being relatively a hardier tree, establishment of plum orchard is little easier than other fruit trees. In orchards of mango, litchi and pear, which come into bearing rather late, plum may be planted as filler tree.

The plum was first introduced in India by Mr. Alexander Coutts in his orchard at Mashobra (Simla Hills). The European settlers and missionaries started its cultivation in about 1870 in Kulu Valley and the Shimla hills (HP), Kumaon hills (UP) and Kashmir. The major plum producing countries are Russia, Romania, China, Yugoslavia, Germany, USA, France, Turkey, Bulgaria, Italy and Spain. In India, plums are grown on a commercial scale in Jammu and Kashmir, Himachal Pradesh and the hills of Uttaranchal. They are also grown to some extent in Nilgiri hills of South India. Besides being palatable and delicious, it has high nutritive value. It is one of the richest sources of vitamin B_1 (Thiamine). Plum fruit is also rich in vitamin A and riboflavin. It is a good source of sugars, proteins, carbohydrates and minerals like calcium, phosphorus and iron. The plum fruit can be utilized profitably for preparation of various kinds of products like jam, jelly and chutney. Plum fruit is used extensively for preparation of juice and squash. The firm and ripe fruit can be canned in sugar syrup.

Varieties

The varieties of Japanese plum are mainly cultivated in the temperate regions and there are about 280 varieties of which only a few are commercially cultivated.

Santa Rosa: Trees are upright, fruits are large and purplish crimson in colour. The flesh is amber with red near the skin. It is self-fruitful and prolific bearing cultivar. The fruit is very juicy and gives characteristic flavour.

Mariposa: An upright growing tree that requires cross-pollination for adequate fruit set. The fruit is heart shaped with greenish yellow fruit skin mottled with red. The flesh is red in colour of fair quality.

Satluj Purple: The fruit of this variety is significantly superior to Kala Amritsari in terms of fruit weight and size. Its average yield is 29.5 kg/ plant. It ripens a week earlier than other sub-tropical plum varieties and takes 81 days to mature. The colour of fruit at maturity is bright crimson and flesh is thick It is sweet in taste and covered with white bloom.

Kala-Amritsari: This is self-fruitful, the yield of which improves if pollinated with 'Titron'. The tree is low spreading but vigorous. The fruit is medium in size, rounded in shape, depressed on both ends and has brownish colour which turns dark-brown at full maturity. The flesh is yellowish, moderately juicy with sub-acid taste and is excellent for jam-making. Bears heavily with an average yield of 40-50 kg fruit/ tree.

Titron: The variety is self-fruitful but the yield improves if 'Alucha Early Round' is planted as a pollinizer. The tree is spreading in nature. Its fruit is small to medium, roundish with deep purple colour and thin skin. The flesh is yellow, moderately juicy and sub-acid in taste and is excellent for jam making.

Kataru Chak: This selection was made from the village Kataru Chak in Gurdaspur District (Punjab). The variety is partially self-fruitful but the yield improves if pollinated with 'Kala-Amritsari'. The fruit is large, heart shaped, purplish with a light white bloom. Pulp is juicy, creamish, moderately sweet, very good for jam-making and table purpose.

Alu Bokhara: The tree is upright, tall, vigorous and self unfruitful. 'Howe' is a good pollinizer for this variety. The fruit is large, having yellow skin tinted with red colour. Pulp is juicy and sweet.

Climate and Soil

The cultivars of Japanese plum require comparatively less of winter chilling i.e., about 700 and 1000 hours below 7.2°C. Most of the plum cultivars are resistant to winter cold and thus can be grown in more northerly regions of the Northern Hemisphere. The bloom period of plum tends to get delayed on the northern slopes thus helping them to escape spring frost injury. Plums do very well in the low and mid hill regions, ranging between 600 and 1600 metres above sea level. Some varieties of plum can be grown successfully even in the warmer region of the plains. This region is marked by an extreme summer temperature reaching 45°C and very low relative humidity that abruptly changes in first week of July by arrival of cool monsoon rains. In winter, the night temperature touches 0°C with occasional ground frosts. Plants get 250 to 300 chilling hours. Cool winter and warm summers are suitable for its successful cultivation.

The Japanese plum can even do well on inferior soils having shallow water table and high pH. For good growth and longer life of the plants, well-drained medium to deep loam soils are the most suitable. Plum propagated on peach rootstocks grows better on sandy loam soils whereas the heavy and wet soils are suitable for plants propagated on Kabul Green Guage rootstock of plum.

Propagation

Plums are generally propagated through hard-wood cuttings. In case, rootstocks are to be used, budding and grafting method are employed for its propagation. For producing seedling rootstocks, the plum stones are extracted from ripe fruits and stratified in field or in wooden boxes. The stratified seed is sown during end of January or first week of February in rows 30 cm apart with seed to seed distance of 10 cm. Light irrigation should be given immediately after sowing seed.

Plum rootstock can be propagated clonally from stem cuttings. For this purpose, cuttings of Kabul Green Guage plum are prepared during first week of December and are kept for callusing for about a month. These are planted in beds preferably during first fortnight of January. The cuttings are planted at a distance of 15 cm in rows which are 30 cm apart. The rooted cuttings are budded in following May-June or grafted in December-January. 'Kabul Green Guage' plums are used as rootstock in heavy and wet soils. The peach rootstock is recommended for sandy loam soils. Myrobalan B plum (*Prunus cerasifera*) is widely used as rootstock which is compatible with Japanese plum cultivars.

The rooted cuttings of Kala-Amritsari cultivar can be used directly without budding. The % of their successful establishment can be enhanced by dipping the basal portions (5-7.5 cm) of hard-wood cuttings for about 24 hours in IBA (100 ppm). The cuttings taken from sub-apical portions, planted after dipping in IBA and callusing for about a month give higher success (%). The highest rooting (%) of cuttings in Kataru Chak and Lalri cultivars was obtained when treated with IBA(100 ppm). The basal stem cuttings in these cultivars gave more rooting (%) as compared to sub-apical cuttings.

Budding by shield or T method is performed in April and July-August. The plants budded in April become saleable in January but those budded in July-August can only be sold after one and half year i.e. next January. The propagation by tongue grafting is performed when the stock and scion are still dormant (December-January).The plum can be propagated simultaneously by grafting on unrooted cuttings of Kabul Green Guage by dipping basal portion of cutting in solution of IBA (100 ppm) for 24 hours during first fortnight of January.

Planting

The plants are lifted from nursery with bare roots with help of a spade by digging deep trenches. Longer roots are trimmed to avoid injury during transportation. These plants are then tied in bundles of suitable sizes for easy handling. Plants of one year growth are preferred, as fewer roots are injured in the process of lifting and replanting and the desired framework can be attained by heading back the tree from required height. Trees older than two years should not be selected for planting. Plums should be planted in January when plants are in the dormant condition. The area is laid out and pits of 1 x 1 x 1 m are dug about a month before planting.

Pits are refilled with fertile top-soil mixed with 40 kg of well-rotten farmyard manure/pit. Add 25 g of BHC (10%) dust to each pit to prevent attack of white ants. Apply water soon after setting plants in pits. The plum plants may be planted 6 metres apart from row to row and plant to plant in square system, thus accommodating 275 trees/hectare.

The newly planted young plants, being very tender, need a lot of care and attention for their better survival and growth. They should be watered at frequent intervals but heavy irrigation should be avoided. The main stem should be cut back to desired height for scaffold branches for developing properly and to keep proper balance between root-system and growth of stem. The young plants need staking to make them grow upright and straight.

Training and Pruning

The trees should be trained to modified leader system. The plant is headed back to a height of 90 cm from ground level after planting in field. Two year old trees have a number of laterals of which one most favourably located around main stem is to be selected to make a proper framework. The training imparted to the three and four year old trees is designed to maintain proper balance between main scaffold limbs and secondary branches. The position of laterals maintained during previous growing season should not be disturbed during third growing season. During this period, water sprouts and undesirable twigs should also be removed. During fifth year, the leader should be headed back at a point where a suitable outgrowing lateral is present to check height of plant. Plum bears on one-year growth as well as on short spurs. To encourage fresh growth and to maintain a system of healthy spurs, light annual pruning should be done in January.

Fertilizer Application

Plum trees must be manured and fertilized judiciously for vigorous and healthy growth and optimum yield of good quality fruits. In Punjab, for one year old plum tree, 6 kg farmyard manure, 30 g N, 15 g P and 36 g K is recommended. Similar amount of farmyard manure and chemical fertilizers should be fed every year up to 6 years. Thereafter, the quantity of farmyard manure, N, P and K should be kept constant at 36 kg, 180 g, 90 g and 216 g, respectively. Farmyard manure and phosphatic and potassic fertilizer should be applied in December. The nitrogenous fertilizer should be split into two equal parts. Apply half dose of nitrogen in spring before flowering and remaining half one month later.

In Himachal Pradesh, newly planted plum is given 450 g calcium ammonium nitrate, 190 g super phosphate and 150 g of muriate of potash besides 10 kg of compost. Thereafter, the dosages are increased at the rate of 450 g of CAN, 190 g of SP and 150 g of MOP for each complete year of plum trees in orchard. The fertilizer doses are stabilized at the age of six years. Foliar application of urea @ 0.5-2.5% for correcting nitrogen deficiency is reported.

The deficiency of zinc is frequently observed in plums growing in sandy soils during summer. It can be corrected by foliar sprays containing 3 kg zinc sulphate in 500 litres of water.

Irrigation

The water requirement depends upon age and size of tree, type of soil, climatic conditions and cultural practices, followed. The irrigation should be avoided during flowering and fruit-setting stage. During April and May, frequent irrigations should be given for proper development of fruits. Irrigation should be stopped at the colour break stage to avoid heavy fruit drop. When there is rainless spell, light watering may be given.

Weed Management

Bromacil and atrazine at 2.5 to 7.5 kg/ha were effective to reduce weed population. Terbacil (3 kg/ha) and simazine (4 kg/ha) in plum orchards were effective in checking weeds. Both these herbicides also increased N, K and Ca content of leaf. In 12 years old Santa Rosa plum orchard, diuron and atrazine (4 kg/ha) proved effective in controlling weeds.

Intercropping

Intercropping can be successfully practiced during non-bearing period of trees and should not interfere with fruit trees. Need-based manures and fertilizers, irrigation and plant protection measures should be given independently to fruit trees and intercrops. The short-statured, shallow-rooted, leguminous crops like moong, mash, pear, gram and cluster bean can be sown as intercrops.

Thinning of Fruits

To improve prospects of plum cultivation, it is essential to improve marketable size and quality of fruits. There is generally excessive fruit set in plum. If all fruits are retained on tree, their size as well as quality may deteriorate and such small sized and poor quality fruits may not fetch good price in market. Thinning is more desirable on mature trees making small annual growth than on young vigorous plants. Trees growing on light soils, deficient in moisture and nutrients, would also need fruit thinning. The heavy bearing cultivar of plum like Kala Amritsari normally needs fruit thinning to minimize danger of limb breakage. Thinning may also be resorted when good sized uniform fruits are needed for canning industry and fancy trade. Thinning should be done after natural fruit drop in April. This operation should be completed before hardening of stones within fruits.

Hand thinning is commonly practiced. In this, first shake branches slightly to dislodge fruits which are likely to drop off naturally. If there is still surplus fruits, then start thinning from top to bottom of branches. Hold stem of fruit to be removed between thumb and second finger and pull it off gently. Main point to be kept in mind while thinning is to remove all the small, undersized, misshaped, insect or disease attacked fruits. For obtaining well coloured fruits of good size, the retained fruits should be spaced 5-8 cm apart depending upon variety.

Harvesting and Yield

Plum is a delicious fruit which ripens during summer when other kinds of fruits are not available in plenty. It is precocious and starts yielding fruits within 2-3 years of planting. The fruit reaches ripe stage in 12 weeks after fruit-set. The peak season for plum harvest is second fortnight of May. The varieties Kala-Amritsari, Titron and Kataru Chak ripen in the second week of May, whereas Alu Bokhara ripens in the first week of June.

On the basis of days from full bloom, picking dates of plum can be standardized. The best indices for judging the maturity of 'Titron' cultivar of plum are that specific gravity should be less than one with deep purple red colour of fruits. This cultivar should be picked in fourth week of May for sale in local markets and in third week for distant markets.

Kala-Amritsari bears heavily and is one of the most widely grown varieties of plum in plains of Punjab. The grown up tree of plum yields about 30-50 kg of fruits/ tree. Plum fruits develop the best dessert quality on the tree. When it is to be consumed locally, it should be allowed to ripen on trees. When required to be sent to distant markets, the fruits have to be picked a few days in advance of proper maturity. This keeps them in good condition during transportation and reach in the market in a firm and attractive condition. Change of colour for each variety would determine its stage of maturity e.g. dark brown colour in Kala-Amritsari at proper maturity.

Plum fruits are picked with hand in several pickings, as entire crop does not ripen uniformly at a time. While picking, care should be taken that spurs are not damaged. Care should also be exercised not to tear flesh of fruit or to destroy natural bloom. If the fruit is jerked in a direct pull from spur, fruit stalk may get torn out and flesh may become ruptured. While picking the fruit, give a gentle twist and let the stalk remains attached to it. Never permit threshing of branches with sticks for facilitating harvesting. Small baskets lined with paper or cloth or polythene should be used for placing fruit during harvest.

Grading

Grading should be regularly practiced for getting market value of fruit. Sorting of fruits to remove under-ripe, under-sized, over-ripe and damaged fruits must be done before packing. Fruits should be packed for sending to distant market immediately after picking as plum fruits soften rapidly after picking. Fruits should be packed in 5 kg wooden boxes instead of bamboo baskets to avoid losses in transportation.

Storage

Perishable nature of ripe plum fruit creates a serious problem for its storage, transport and marketing. The main reason for its limited cultivation in subtropical areas is poor keeping quality of its fruit. Storage in a fresh condition for extended period without loss in quality is, therefore, a very pressing problem.

The experiment conducted on storage behaviour of plum (Kala-Amritsari) showed that the ripe fruits of plum could be stored for 4 days in paper bags at room temperature after treating with 3-6% wax-emulsion. At the same time, fruits remained in good conditions up to 40 days in the commercial cold storage (temperature 3°C with relative humidity of 85-90%), when treated with 3-6% wax-emulsion for 30 seconds and stored in perforated polythene bags of 100 gauge thickness.

POMEGRANATE

Pomegranate (*Punica granatum*) is an ancient favourite table fruit of tropical and sub-tropical regions of the world. It belongs to family Punicaceae which includes only one genus and two species, the other being *P. protopunica* Balf (Socotra pomegranate). The name 'pomegranate' is derived from two Latin words *pomum* (apple) and *granatus* (seeded). Pomegranate is native of Iran and it is widely cultivated throughout India. The fruit is symbolic of plentifulness and it is very much liked for its cool, refreshing juice and valued for its medicinal properties.

Pomegranate cultivation is one of the most remunerative farming enterprises in India. It is an excellent choice under arid and semi-arid conditions as it tolerates drought. As a commercial crop pomegranate is grown to a limited extent in selected locations in many states. It is grown in Maharashtra, Karnataka, Andhra Pradesh, Uttar Pradesh, Gujarat, Rajasthan and Tamil nadu on a large scale. The total pomegranate production in the world is 10 lakh tonnes. India produces 5 lakh tonnes from an area of about 25,000ha but exports only 5000 tonnes, whereas Spain produces 1 lakh tonne and exports 75,000 tonnes and their export ends by December. India has the scope to export from January to June to European countries.

Varieties

In India there are several named cultivars. Preference is usually given to those with fleshy, juicy pulp around the seeds. Types with relatively soft seeds are often classed as "seedless". Among the best are 'Bedana' and 'Kandhari'. 'Bassein seedless and KVK - 1' are from Karnataka and 'Jalore Selection' are popular varieties in North India. In South India, 'Paper Shell', 'Spanish Ruby', 'Muscat Red' and 'Vellodu' have shown promise. In recent years, several new varieties are developed and new orchards have come with well-known improved varieties.

Ganesh : This is a selection from 'Alandi' which has revolutionized cultivation of pomegranate in Maharashtra state. This has soft seeds and pinkish flesh with juice of agreeable taste and it bears heavily.

Dholka : Popular variety of Gujarat with fruits of large size, rind greenish yellow, fleshy aril pinkish white, seeds soft, juice acidic. It is a medium cropper.

Jodhpur Local : Medium sized fruit with hard rind, fleshy aril light pink, sweet, juicy and seed moderately hard.

Bedana : Fruit medium to large in size, rind brownish green, fleshy aril light pink or whitish, very sweet, very juicy, soft seeded.

Kandhari : Fruit is large, deep-red, with deep-pink or blood-red, subacid pulp and hard seeds.

Alandi ('Vadki') : Medium-sized, with flesh red or pink, subacid pulp, very hard seeds.

Kabul : Large, with dark-red and pale-yellow rind; fleshy, dark-red, sweet, slightly bitter pulp.

Muscat Red : Small to medium, with thin or fairly thick rind, fleshy, juicy, medium-sweet pulp, soft or medium-hard seeds. The plant is a moderately prolific bearer.

Paper Shell : Round, medium to large, pale-yellow blushed with pink; with very thin rind, fleshy, reddish or pink, sweet, very juicy pulp and soft seeds. Bears heavily.

Poona : Large, with dark-red, gray or grayish-green rind, sometimes spotted, and orange-red or pink-and-red pulp.

Spanish Ruby : Round, small to medium or large; bright-red, with thin rind, fleshy, rose-colored, sweet, aromatic pulp, and small to medium, fairly soft seeds. Considered medium in quality.

Vellodu : Medium to large, with medium-thick rind, fleshy, juicy pulp and medium-hard seeds.

Muscat White : Large, creamy-white tinged with pink; thin rind; fleshy, cream-colored, sweet pulp; seeds medium-hard. Bears well. Desirable for commercial planting in Delhi.

Wonderful : Originated as a cutting in Florida and propagated in California in 1896. The fruit is obovate, very large, dark purple-red, with medium-thick rind; deep-red, juicy, winey pulp; medium-hard seeds. Plant is vigorous and productive.

The Japanese dwarf pomegranate, *P. granatum* var. *nana, is* especially hardy and widely grown as an ornamental in pots. The flowers are scarlet, the fruit only 5 cm in size but borne abundantly. Other ornamental cultivars are 'Multiplex' with double, creamy white blooms; 'Chico', double, orange-red; 'Pleniflora', double, red; 'Rubra Plena', double, red; 'Mme. Legrelle' and 'Variegata', double, scarlet bordered and streaked with yellowish-white.

Climate and Soil

Pomegranate prefers dry climate. During fruit development, prolonged hot and dry climate is required. A humid climate adversely affects formation of fruit. Optimum temperature congenial for fruit development is 38°C. It can be severely injured by temperatures below -11°C. In humid climate, the fruits are severely damaged by pomegranate butterfly and do not develop sweetness. It is winter hardy and very drought tolerant. It can grow up to an elevation of 1850 meters amsl. The tree adapts well to container culture and sometimes fruits in a greenhouse. Under subtropics, it is deciduous in nature while in tropics, it is evergreen. For the best results, pomegranate should be grown in full sun.

Pomegranate is a very hardy crop and thrives well in shallow rocky soils. It tolerates alkalinity and salinity and also thrives on calcareous soil. The best results are obtained in deep, heavy loam and well drained soils. It is sensitive to soil moisture fluctuations causing fruit cracking which is a serious problem of this crop.

Propagation

Pomegranate can be raised from seed but not be true to type and hence vegetative propagation is recommended. Cuttings root easily and plants from them bear fruits after about 3 years. Cuttings of 20 to 30cm length should be taken from suckers which spring from base of main stem of 1-2years old plant in winter. The leaves should be removed and cuttings treated with rooting hormone and inserted about two-thirds their length into the soil or into some other warm rooting medium. The plants will be ready for planting in 55-66 days. Plants can also be air-layered but grafting is seldom successful.

Planting

Rooted cuttings or seedlings are setout in pre-fertilized pits of size 60 x 60 x 60cm dug at 5m spacing in square system. Spacing of 5m between rows and 2m between trees gives maximum returns. Fill the pits with 20kg FYM. Planting should be done during monsoon. Watering must be done immediately after planting.

Planting distance depends on soil type and climate. Old orchards were planted at 3 to 6 m space. Nowadays with new dwarf varieties, a spacing of 4 x 4 m is recommended which accommodates 625 plants/ha.

Training and Pruning

Multi stem training is more prevalent and useful also because even if due to some reason, the plant losses one stem, it can survive. Usually the plant is allowed to retain 4 main stems from ground level. Extra stems should be removed continuously. The main stem should be topped at a height of about 70cm to reduce branching. The tree is given a balanced shape during the initial 2-3 years by proper selection of secondary and tertiary branches. Downward growing branches and crossing branches should be removed.

As the fruits are borne only at tip of new growths, for the first 3 years, it is better to shorten the branches annually to encourage production of maximum number of new shoots on all sides. This prevents straggerly development and helps to achieve a strong, well-framed plant. After the 3rd year, only limited pruning is to be done by removing the suckers and dead branches. The tree begins to bear in one year after planting out, but 2 1/2 to 3 years are more common. After 10 years, old main stems should be removed by cutting back to make it more productive.

Fertilizer Application

Higher yields can be obtained by application of 500-625g N, 250g P_20_5 and 250g K_20 to each tree/year. In tropics pomegranate flowers throughout the year in three distinct phases and time of fertilizer application varies with the phases. For ambe bahar (spring flowers which flowers in January - February) January / February, for mrig bahar (rainy flowers which flowers in June - July) June / July and for hasta bahar (which flowers in September - October) September / October are appropriate periods.

Irrigation

Irrigation requirements vary with type of crop. Water thoroughly at planting and again every few days for the first couple of weeks, after which the interval between watering can be lengthened gradually. To facilitate watering, construct a ring of soil several cms thick and high, a couple of meters in diameter, around the newly planted tree. Then just fill the ring with water as necessary. The ring will melt into the surrounding soil within a few month, by which time, the young plant will get established.

For mrig bahar crop, first irrigation in the middle of May followed by regular irrigation until monsoon sets in. Thereafter, in post monsoon period, copious and regular irrigation is essential for better development of fruits and for avoiding fruit cracking. For other crops weekly irrigation in summer and bi-weekly irrigation in winter should be followed. Once established, pomegranates can take considerable drought, but for good fruit production they must be irrigated.

Weed Management

To enable rapid establishment, all other vegetations should be removed within 30 to 60 cm of the young pomegranate. Turf grass can be killed initially with a systemic contact herbicide, followed by organic mulching to prevent weed and grass competition and to conserve soil moisture.

Intercropping

Since pomegranate plants take 4-5 years to come into good bearing low growing vegetables, pulses or green manure crops can be taken up as intercrops.

Flower Regulation (Bahar Treatment)

In subtropics, pomegranate flowers once in a year in spring while in tropics (Central and South India), it flowers throughout the year in three distinct phases with maximum intensity in rainy season. In spring, flowers are borne on mature wood of previous season growth while flowers are borne on current season's growth in June/July. Spring flowers (*Ambe-bahar*) give fruits in summer when the demand is maximum but of poor quality. Therefore, efforts should be directed to avail of rainy flowers (*Mrig-bahar*) so that fruiting period coincides with the time of maximum water availability in the soil and the crop is taken without irrigation. Trees usually shed leaves by March and remain dormant upto May. Then the field is ploughed, manure and fertilizers are applied and first light irrigation is given in the middle of May and thereafter one or two light irrigations are given until rains set in. The trees put up new growth, flower and fruit with crop availability in October-November. This is the most advisable approach in arid and semi-arid regions with limited resources.

Harvesting and Yield

The fruits ripen 6 to 7 months after flowering. The fruit cannot be ripened off the tree even with ethylene treatment. They should be harvested mature which can be judged by change in skin colour to slightly yellow and metallic sound when tapped. The fruit must be picked before over maturity when it tends to crack open if rained upon or under certain conditions of atmospheric humidity, dehydration by winds, or insufficient irrigation. Ultimate splitting is the natural means of seed release and dispersal.

A tree gives 20-25 fruits from third or fourth year and 100-150 after 10 years. A high density system with 'Ganesh' (1000 plant/ha) (5x2m) has been very profitable giving a crop of 50 fruits/tree of good quality. The fruits should not be pulled off but clipped close to the base so as to leave no stem to cause damage in handling and shipping. Too much sun exposure causes sunscald-brown, russeted blemishes and roughening of the rind.

Grading

After harvesting, fruits can be cured in shade for about a week so that the skin becomes hard and fruits can stand transportation better. Thereafter, the fruits can be graded according to weight

'A' grade - > 350 g

'B' grade - 200 to 350 g

'C' grade - < 200 g

Storage

Pomegranate is equal to apple in having a long storage life. It is the best maintained at a temperature of 0-5° C. The fruits improve in storage, become juicier and more flavorful and can be stored for a period of 7 months within this temperature range and at 80 to 85% relative humidity, without shrinking or spoiling. At 95% relative humidity, the fruit can be kept only for 2 months at 5° C, but can be kept for longer periods at 10° C as it avoids chilling injury and weight loss. After prolonged storage, internal breakdown is evidenced by faded, streaky pulp of flat flavor. The fruit ships well, cushioned with paper or straw, in wooden crates or, for nearby markets, in baskets.

SAPOTA

Sapota (*Manilkhara achras*) belonging to Sapotaceae family, commonly known as chiku, is mainly cultivated in India for its fruit value, while in South-East Mexico, Guatemala and other countries it is commercially grown for production of chickle. The fruit is a native of Mexico and other tropical countries of South America. The area under sapota cultivation has increased from 27 thousand hectare in 1991-92 to 52 thousand hectare in 2001-02 and production from 3.96 lakh tonnes to 5.94 lakh tonnes. Average productivity is 11.4t/ha.Sapota is mostly grown in Gujarat, Maharashtra, Karnataka, Tamil Nadu, Andhra Pradesh and Kerala.

The fruit is a fleshy berry ,variable in shape, size and weight and is a good source of digestible sugar (15-20%) and an appreciable source of protein, fat, fibre and minerals (Ca, P and Fe). Sapota pulp is used for making sweets and halwas. It is also an ingredient of fruit salads and milk shakes. Chickle, the gum like substance extracted from its stem latex, forms the base for making chicklet and chewing gum.

Varieties

Important varieties cultivated in different states of India are listed below :

State	Varieties grown
Andhra Pradesh	Cricket Ball, Kalipatti, Calcutta Round, Pala, Kirthibharathi, Dwarapudi, PKM-1, Jonnavalasa I & II, Bangalore, Vavi Valsa
Bihar	Baramasi
Gujarat	Kalipatti, Pilipatti, Cricket Ball, PKM-1
Karnataka	Cricket Ball, Kalipatti, Calcutta Round, DHS-1, DHS-2
Maharashtra	Kalipatti, Dhola Diwani, Cricket Ball, Murabba
Orissa	Cricket Ball, Kalipatti
Tamil Nadu	Pala, Cricket Ball, Guthi, Co 1, Co 2, PKM-1
Uttar Pradesh	Baramasi
West Bengal	Cricket ball, Calcutta round, Baramasi, Baharu, Gandhevi Barada

Climate and Soil

Sapota, being a tropical crop can be grown from sea level up to 1200 m. It needs warm (10-38^0 C) and humid climate (70% relative humidity) for growth and flowering. It can be cultivated throughout the year. Coastal climate is the best suited for its cultivation. It does not tolerate low temperatures, even if of short duration.

Alluvial, sandy loam, red laterite and medium black soils with good drainage are ideal for cultivation of sapota. At higher elevations in places in Punjab and Haryana, it gives only one crop during summer.

The state-wise growing belts are mentioned below:

State	Growing belts
Andhra Pradesh	Prakasam, Guntur, West Godavari, Nellore, East Godavari
Gujarat	Valsad, Navsari, Surat, Junagadh, Bhavnagar
Karnataka	Kolar, Bangalore, Chikkamangalur, Belgam, Raichur
Orissa	Baleswar, Cuttack, Kendrapara, Jagatsinghpur, Puri, Khurda, Ganjam
West Bengal	South 24-Parganas, Kalimpong I & II, blocks of Darjeeling district.

Propagation

Sapota was mainly propagated by seed during earlier times and it took seven to eight years to come into bearing. This created a wide variation in populations, greatly reducing the possibility of harvesting genotypes that were uniform in size, weight and quality.

Vegetative methods of propagation adopted are air layering, grafting and budding. At present, softwood grafting method using khirni (*Manilkara hexandra*) is commercially adopted as it gives 93% success. It is an economically viable, faster, efficient and best technique. This technique improves productivity characteristics and halves the period taken from planting to harvesting. In this way, incompatibility problems between stock and scion are corrected and the desirable characteristics of scion are combined with the special qualities of the stock.

Planting

The land is ploughed two to three times and in the case of undulating land, it is divided into terraces and leveling is done. Tall and thick growing trees viz. mango, jamun, tamarind, silver oak and casuarinas are established on the wind-ward side or on all sides of the orchard. The plants for windbreak may be planted at a distance of 1.5 to 1.8 m in a row.

Planting can be done in any season provided irrigation facilities are available. Grafts are usually planted in the beginning of the rainy season. In areas which receive heavy rainfall the crop can be planted as late as September. On an average, 130 plants are planted/ha at a spacing of 8.5m. High density planting with a spacing of 5 x 5 m up to the age of 13 years has been adopted successfully.

In light soils, pits of 60x60x60 cm size are taken, whereas in heavy and gravely soils pits of 1x1x1 m. size are made in April-May and exposed to sun for a period of fifteen days. The pits are later on filled with top soil mixed with equal quantity of well-rotten compost or farmyard manure and 3 kg super phosphate and 1.5 kg muriate of potash. The pits are then left to monsoon rains for settling and planting is done at appropriate time.

Grafts, budded plants or layers are planted one in each pit and care is taken so that the bud joint or graft is at least 15 cm. above the ground level. After planting, stakes are provided to avoid wind damage. Young plants are protected from the sun by making dry grass thatch on the top and the three sides excepting the south-east for sunlight. Square system of planting is recommended. Contour planting is recommended in case of sloping land.

Training and Pruning

No definite training system has been developed for the plant. Plants raised through inarching require training for appropriate shape and framework development. Most of the trees are trained in central leader system. Pruning is done to expose the trees to sunlight and to remove the dead and diseased branches. It is mainly done to regulate the vegetative growth to improve the productivity and quality of fruits.

Fertilizer Application

The crop responds well to manuring. A tree requires per year about 50 kg of farmyard manure,1 kg N(1.5 kg. under rainfed conditions), 0.5 kg P_2O_5 and 0.5 kg K_2O. The dose is regulated on the basis of age of tree and status of nutrients in soil especially that of P and K. Under rainfed conditions, fertilizers are applied before the onset of monsoon. Under irrigated conditions, it is applied in two splits, one half at the beginning of monsoon and the remaining half in the post-monsoon period (September-October).The whole quantity of organic manure should be given along with the first dose of fertilizers. In case of Zn and Fe deficiency, application of organic manures and also spraying of $ZnSO_4$ and $FeSO_4$ at the rate of 0.5% are beneficial.

Irrigation

Irrigation is provided at an interval of 30 days in winter and 15 days in summer. Drip irrigation is beneficial in saving 40% water with 70-75 % higher income. This system is laid out with 2 drippers spaced 50 cm. from tree at an initial stage during the first two years and then 4 drippers about 1 m. away from the tree till it attains five years of age.

Weed Management

The problem of weeds is common in young orchards. Application of 2 kg bromacil and 2 kg diuron/ha as pre-emergence spray is effective in controlling weeds for a period of 10-12 months.

Intercropping

Inter cropping with vegetables and fruits may be taken up in the first four years of planting. Banana, papaya, pineapple, cocoa, french bean, peas, tomato, brinjal, cabbage, cauliflower and cucurbits are recommended for inter cropping depending upon the climate and irrigation facilities available.

Growth Regulators

Fruit drop is a very serious problem in sapota. Spraying with GA_3 (50-100 ppm) at the time of flowering is quite effective for getting better fruit set and also preventing fruit drop.

Harvesting and Yield

Sapota starts bearing from third year of planting but economic yield is obtained from 5th year onwards. The two main seasons of flowering are October- November and February-March and the two corresponding harvesting seasons are January-February and May-June. Sapota takes four months from flowering to maturity of fruits.

The fruits are hand picked or harvested with special harvester which has a round ring with a net bag fixed onto a long bamboo. In high density planting, the production increases from 10.0 tonnes /ha to 15.0 tonnes/ha from the 5th to the 7th year. Thereafter, the yield stabilizes from 8th to 15th year at 20.0 tonnes/ha.

Post Harvest Management

Grading

Grading is mainly based on size and shape of fruits. The fruits are graded into three categories depending on their size viz. large, medium and small.

Storage

The fruits are highly perishable and can be stored under ordinary condition for a period of 7-8 days after harvesting. At a storage temperature of 20^0 C, the storage life can be increased up to a period of 21-25 days by removing ethylene and adding 5-10% CO_2 to the storage atmosphere. The fruits are dipped in GA (300 ppm.) + Bavistin (1000 ppm.) solution at a pre-packing stage to extend the storage life of the fruits.

Sapota being a climacteric fruit has to be ripened artificially. Unripe fruits can be ripened by applying ethephon (1000 ppm.) at 20-25 ^{0} C. Ripened ones can be stored at 2-3^0 C and 90-95% RH for a period of six weeks.

Packing

Fruits are packed in cardboard boxes of 10 kg capacity with rice straw as padding material and ethylene absorbents. Bamboo baskets with banana leaves as lining materials and covered with net are used for carrying the produce from farm to local market. In some cases, the produce after harvesting is packed in wooden containers and transported to urban areas for sale. Wooden boxes prevent damage of ripened fruits and fetch better price in urban areas.

Transportation

Road transport by trucks/lorries is the most popular mode of transport due to easy approach from orchards to market.

Marketing

Several intermediaries like wholesalers and commission agents are involved in marketing of fruit. The farmer realizes around 35% of wholesale price in the secondary market. The fruit is mostly consumed indigenously. Export constituted only a very minor fraction of production (0.2% in 2001-02). Low volume export of sapota is due to non-ideal post harvest practices, transport procedures, lack of proper storage facilities, outdated handling practices.

STRAWBERRY

Strawberry (*Fragaria* sp.) is a native of temperate region, but varieties are available which can be cultivated in sub tropical climate. It belongs to family Rosaceae. The strawberry is the most widely adapted of the small fruits. It is grown on a limited scale because of its highly perishable nature. It has a tremendous scope for cultivation near towns and canning units where the produce can be utilized immediately after harvest. Strawberry is the most profitable fruit crop in the shortest possible time as compared to other fruits. The fruit is delicious, attractive, has a pleasant aroma and a delicate flavour. It is also nutritious and beneficial for anaemic persons. The wider adaptation of the strawberry plant in regions with wide variation in climate permit harvesting and marketing of the fruit during greater part of the year.

Strawberry is a soft, delicious and a highly perishable fruit taken afresh in several ways. It is used for making excellent ice creams and jams on account of its rich aroma, and is also a good source of vitamin C. The fruit is an enlarged receptacle bearing numerous seed like achenes.

In India, it is generally cultivated in the hills. Its main centre of cultivation is Nainital and Dehradun in Uttaranchal, Mahabaleshwar (Maharashtra), Kashmir Valley, Bangalore and Kalimpong (West Bengal). In recent years, strawberry is cultivated successfully in plains of Maharashtra around Pune, Nasik and Sangali towns.

Varieties

A large number of varieties are available for cultivation and they are octaploids. For the hilly areas, varieties Royal Sovereign, Srinagar and Dilpasand are suitable. Some of the introductions from California, such as Torrey, Tioga and Solana are also suitable. The variety found successful in Bangalore is named Bangalore and it performed well at Mahabaleshwar also. For the North Indian plains, Pusa Early Dwarf which has dwarf plants, large firm wedge-shaped fruits, is recommended. Another variety with rich aroma but softer fruits is Katrain Sweet. Some of the varieties found successful in warmer parts of the U.S.A. are: Premier, Florida-90, Missionary, Blakemore, Klonmore and Klondike. Some of these are successfully cultivated in Indian plains.

Chandler: Fruit very large, flesh and skin firm, desert and processing quality good, TSS/acid ratio 9.97, tolerant to viruses and excellent runner producer.

Seascape: Fruit large, flesh and skin firmness medium, desert and quality excellent, processing quality good, TSS 7.4 %, acidity 0.7 % and TSS/acid ratio 10.68. Ripening starts by the first week of April.

Tigra: Fruit very large, flesh and skin firm, desert and processing quality good and tolerant to virus. TSS 7.4 %, acidity 0.6 % and TSS/acid ratio 9.25.

Selva: This variety has the capacity to produce off-season fruit. Fruit shape is conic to blocky; large and firm. TSS 7 %, acidity 7 %.

Pajaro: A very good variety for summer season. The fruit is large, symmetrical, attractive and firm. Good for processing and runner-producer.

Climate and Soil

It can be grown in plains and hills in tropical, subtropical and temperate climates. Strawberry thrives the best in temperate climate. It grows upto an elevation of 3000m from mean sea level. It is a short day plant which requires exposure to about 10 days of less than 8 hours sunshine for initiation of flowering. In winter, the plants do not make any growth and remain dormant. The exposure to low temperature during this period helps in breaking dormancy of plant. In spring, when days become longer and temperature rises, the plants resume growth and begin flowering. From the standpoint of response to length of light period, strawberries are placed in two groups: (1) Everbearing varieties - varieties that develop flower buds during both long and short light periods and (2) June bearing varieties - varieties that develop flower buds during the short light periods only and most of the commercial varieties come under this group.

Strawberry requires a well-drained medium loam soil, rich in organic matter. The soil should be slightly acidic with pH from 5.7 to 6.5. At higher pH, root formation is poor. Presence of excessive calcium in the soil causes yellowing of the leaves. In light soils and in those rich in organic matter, runner formation is better. Strawberry should not be cultivated in the same land for a number of years. It is preferable to plant it in green manured field. Alkaline soils and soils infected with nematodes should be avoided.

Propagation

Strawberries are propagated by two methods, planting of runners or crowns. Early formed runners (runners that are formed after the blooming season) are vigorous, have a good root system and are more productive and are preferred for planting. The plants may be allowed to set as many runners as possible but not allowed to set any fruits. All the plants with good root system should be utilised to set a new plantation. Given the best attention and care, a single plant usually produces 12 to 18 runners.

For large scale multiplication of virus free plants, tissue culture technique is used. One million plants can be produced from a single meristem within a period of one year. Immature embryos, anthers and meristamatic growing tips can be used as explants.

Planting

The land for strawberry planting should be thoroughly prepared by deep ploughing followed by harrowing. Liberal quantities of organic manure should be incorporated in the soil before planting. Strawberry can be planted on flat beds, in the form of hill rows or matted rows, or it can be planted on raised beds. In irrigated areas, planting on ridges is advised. The usual practice is to plant on raised beds of size 4 x 3 meters or 4 x 4 meters. The planting distance should be 45 cm from plant to plant and 60 to 75 cm from row to row. In the hills, transplanting is done in March-April, September-October, but in the plains, the months of January-February are the best. At Mahabaleshwar, normally strawberry is planted during November-December. Late planting results in reduction in yield during the first year.

The plants should be set in the soil with their roots going straight-down. The soil around the plant should be firmly packed to exclude air. The growing point of the plant should be just above the soil surface. The roots of strawberry plants spread out close to the surface. During planting, the plants should not be allowed to dry out and should be irrigated immediately after planting. Hoeing should be done lightly and young plantation should be kept weed free.

Planting time is considered as one of the most important factors for successful strawberry cultivation. Traditionally, under north-Indian plains, strawberry is planted after the second week of October, which results in availability of fruits for a month or so. In Northern plains of India, Chandler strawberry when planted in mid-September favoured vigorous growth, enhanced flowering and fruiting and also resulted in production of the best quality, largest fruits and the highest yield. With the use of plasticulture techniques, the planting time can be enhanced, which facilitates early and higher production, and enhances availability of fruit for a longer period.

Training

For strawberry, four training systems are followed- hill system, spaced row system, matted row system and plastic mulch system. In India, generally matted row system is followed.

Hill system - Practiced for cultivars producing a few runners. All the runners are removed and so the mother plants get all the available nutrients. Though cost of production is high and yield is low, they bear large sized fruits.

Spaced row system - Cultivars producing moderate runners are trained in this system. Runners are not removed and the number and spacing of new plant are regulated. Three to four plants are allowed to develop from each runner. The plant size will be more and the plant produces more yield.

Matted row system – An economical and the simplest method. The mother plants are planted at a spacing of 45 x 90 cm apart in rows and runners are allowed to grow in such a way that it covers all the inter space around the mother plant in the form of a mat. In this method more number of plants can be accommodated per unit area resulting in higher yield. Overcrowding of plants should be avoided to prevent fruit rot.

Plastic mulch system – A modified hill system where black or green plastic film is used as a mulch for controlling the weeds and to conserve the soil moisture. The plants come to fruiting earlier and the fruits are clean and rotting of fruits will be less.

Fertilizer Application

For good yield and quality, manures and fertilizers are to be applied. Addition of organic matter to the soil, in the form of 50 tonnes of farmyard manure/ hectare is highly desirable. It improves water holding capacity of soil and also gives better runner formation. Farm yard manure may be supplemented by chemical fertilizers to make up total quantity of nitrogen from 84 to 112 kg/ hectare, phosphorus 56 to 84 kg/ hectare, and potash 56 to 112 kg/ hectare. The Phosphatic fertilizer should be incorporated into the soil before planting. The nitrogenous fertilizer should be applied in two doses (three weeks after planting and again at the time of flowering) and potash at the time of flowering only. Application of adequate amounts of nitrogen gives higher yield of early berries.

Irrigation

Since strawberry is relatively shallow-rooted, it is susceptible to conditions of drought. Planting early in autumn allows plants to make good vegetative growth before onset of winter. However, it is necessary to ensure that newly planted runners are irrigated frequently after planting; otherwise mortality of plants becomes high. During September and October, irrigation should be given twice a week if there is no rain. It may be reduced to weekly intervals during November. In December and January, irrigation may be given once every fortnight. When fruiting starts, the irrigation frequency ought to be increased. At this stage, frequent irrigation gives larger fruits.

Trickle and sprinkler system of irrigation is popular now. Water and energy are saved by the trickle method. Not only that it helps in better water supply during winter, less disruption of picking schedule and less rotting of fruits. Sprinkler irrigation is beneficial in areas where there is heat stress and also in areas requiring frost control.

Weed Management

Weeds are very harmful for strawberry crops, because they compete with the fruit plants for nutrients and water. It is important to keep the field clean during cultivation, either by using herbicides or by hoeing to kill the weeds. Light earthing is profitable for strawberry plants to conserve more water close to the crown. Hoeing is required at 15-days interval during flowering and runner formation. Strawberry is a dicot plant so a pre-emergence weedicide can also be used to control weeds like STOMP (6 l/ha) or basalin (2.5l/ha).These weedicides should be applied at least 25 days before transplanting.

Mulching

Mulching is the most important cultural practice in strawberry, which influences plant growth, fruit yield and quality. In cold climate, the soil is covered with mulch in winter to protect roots from cold injury. The mulch keeps the fruits free from soil, reduces decay of fruits, conserves soil moisture, lowers soil temperature in hot weather, protects flowers from frost in mild climates and protects plants from freezing injury in cold climates. Several kinds of mulches are used, but the commonest one is straw mulch. The name strawberry is derived from this fact. The straw of wheat, dry grass, black and transparent polyethylene can be used for mulching in fields. Plants mulched with black polyethylene had the best growth, fruit weight, yield and quality compared with those mulched with other mulching materials. It saves irrigation water, prevents growth of weeds and keeps the soil temperature high. Before mulching, the soil should be dusted with an insecticide to protect the strawberry from insects.

Growth Regulators

In strawberry, application of IBA (100ppm) or Morphectin (50ppm) 10days before flowering induced more number of runners. When GA_3 (50ppm) was sprayed four days after flowering, the yield increased by 31 to 41% and also with spraying of maleic hydrazide (0.1-0.3%). Morphectin (50ppm) improved the fruit size.

Harvesting and Yield

Plants planted during September-October start bearing from April onwards. Successive crops can be taken; however, the yield and quality deteriorate after two seasons. The fruit is harvested along with the stalk when 50% of the surface has attained red colour. Generally the fruits ripen within a month from flowering. The fruit should be picked daily in warm weather. Farmers can get 15 to 20 pieces of 250 g each from one plant. The fruit ripens during late February to April in the plains and during May and June at high elevations like Mahabaleshwar, Nainital and Kashmir. Harvesting should be done early in the morning in dry conditions. Washing the fruit, bruises it and spoils its lustre. The yield varies according to season and locality. A yield of 20 to 25 tonnes/ hectare is excellent, though yields upto 50 tonnes per hectare are reported under ideal conditions. From one hectare farmers can get a net profit of Rs 4.0 lakhs by selling fruits and plants.

Storage

Strawberries are harvested in small trays or baskets. They should be kept in a shady place to avoid damage due to excessive heat in the open field. Strawberries can be stored in cold storage at 32°C upto 10days. Afterwards, they loose their fresh bright colour, showing shriveling and deterioration in flavour. Strawberries can be frozen for their processed products or as dessert.

Packing

Since fruit is highly perishable, it is packed in flat shallow containers of various types (card board, bamboo, paper trays etc.) with one or two layers of fruits.

Transportation

For distant marketing, strawberries should be pre cooled at 4°C within two hours of harvesting and kept at same temperature. After pre cooling, they are shipped in refrigerated vans.

Marketing

For local market, the fruit should be harvested when fully ripe, but for transport to distant markets, it should be harvested when still firm and before colour has developed fully all over the fruit.

Strawberries are highly perishable and hence a great deal of care in harvesting, handling and marketing require to be organized. Usually the fruit is picked in early morning and sent to market in afternoon of same day or is picked in late afternoon, stored overnight in a cool place, and sent to market, the following morning.

Export Potential of Indian Fruits

Although India is one of the world's largest fruit producers, current export volume of fresh fruits remains small owing to a number of constraints. These constraints are related to production practices, post-harvest methods, issues related to supply chain, market access and government policies. Most of the fruits produced in India are consumed domestically. Export volumes represent less that 1% of total domestic output, as well as less than 1% of world exports. More than 80% of fruit exports from India are tropical fruits, including mangoes, guava, pineapples, and papayas. India was the second largest exporter of mangoes in the world in 2004, after Mexico, shipping 1, 80,000 tonnes or 21% of the world's total. Exports of papaya and pineapple amounted to 3,550 tonnes and 1,624 tonnes, respectively. Indian fruit production, yields and exports grew faster than average for the world; production at 5.33% compared to 2.2%; yields at 2.14% compared to 0.95%and exports at 8.21% compared to 2.0%. India also imports a sizable quantity of fresh fruits, but these are mainly temperate fruits like apples and pears, and a large volume of dried fruits like dates and raisin. Total imports of both fresh and dried fruits reached 2, 16,400 tonnes in 2003 and were valued at US$ 47.1 million.

Foreign exchange earnings from fresh and processed fruits (including dried and canned fruit as well as pulps and juices) amounted to US$ 140 million in 2003, an almost two fold increase (185%) over the 1996 - 1998 average. Mango is the main fruit, exported from India. Export earnings from fresh mangoes totalled US$ 80 million in 2003, which was 61% of India's total fruit exports. About 50% of processed fruit exports were mango-based products, such as mango pulp for juice mixes, and condiments including pickles and chutneys. However, on a comparative basis globally, only 2% of fruit production in India is processed compared to 70% in Brazil, and 83% in Malaysia. It is estimated that about 25 to 30% of fruit production is wasted due to lack of post-harvest infrastructure.

The Near East, the Far East and Europe are the main export destinations for fresh tropical fruits and processed products from India. Processed products like fruit juice, fruit pulp and pickles are mostly exported to Russian Federation and the Near East. Other markets for processed fruits are the United Kingdom, United Arab Emirates, Saudi Arabia, Kuwait, Germany, United States, Netherlands and Switzerland. India also exports dried mango slices, puree, paste and juice to many of the same markets.

Trade in fruits has become more important over the last decades. Knowledge of export to different countries is essential to plan the export, considering prices and other related matters. The composition, volume and direction of this trade have changed as income and insistence on quality have grown on demand side while technology and trade agreements have influenced the income side. Lower prices and greater availability of produce year round, in tune with increasing income have enhanced the array of fruits in the global market. In addition, other factors like greater concern for healthy diet, improved methods of handling and transportation have further favoured fruit trade globally.

Developed countries continue to dominate global trade in fruits and new varieties will find their way into diets of relatively rich people everywhere. Global imports are forecasted to reach 4.3 million tonnes by 2010 with 87% or 3.8 million tonnes to be coming from developed countries. The European Union remains the worlds largest import market followed by the U.S, together accounting for 70% of import demand. Europe remains the main market for tropical fruit, with France a major importer and Netherlands, a major point for shipment of imported fruits in Europe.

The export market for fresh fruits is highly competitive among the top exporters. Gaining access to foreign markets is critical to countries that are large exporters. Free trade agreements are means to provide increased market access and encourage more of exports. Top exporters use various export promotion and marketing techniques to increase market share in foreign markets in addition to the negotiated trade agreements.

Problems for Export

India plays only a marginal role in the export of fresh fruits in most of the leading markets as the Indian exporters have not succeeded in establishing direct linkages with buyers in importing countries, as a result of which a large portion of exports are further processed and re-exported by other countries. Constraints, faced by Indian exporters are lack of exportable varieties, lack of proper post-harvest treatment facilities, lack of packing houses at the farm and ports, lack of refrigerated trucks or containers and high cost of obtaining export certification.

The problems can be classified into two, one inherent to producing country regarding production and supply and the other relating to importing country regarding their requirements. In India, the largest fresh fruit produce exporter records annual sale of about 500 million rupees only. This low volume export reduces the economics and makes export uncompetitive. These exporters are not able to maintain their position in a consistent manner and mainly rely upon opportunities they get then and there. And this has led the exporters to concentrate on upgrading skills, product innovation, quality improvement, brand building and low investment.

In developed countries private food safety and quality standards are now well established. These standards which are operating along with normal regulatory systems have become mandatory in the so called rich countries. Countries in developing world are now put into much more trouble as they are finding it more difficult in selling their produce to developed countries as these standards have now become a global phenomenon.

The main technological constraints that India is facing today is high cost of production resulting from low productivity of crops due to small fragmented holdings, un-irrigated area, old and genetically inferior plantations, lack of availability of good quality planting materials and high incidence of pest and diseases. Further, the quality of produce in many a times is far from satisfactory level. The post harvest losses still continue to be high. The future growth of fruit industry largely depends on novel and competitive technology. Advantage of several frontier areas like biotechnology, protected cultivation, computer aided management of inputs, integrated nutrient management, leaf nutrient analysis, biofertilizers and integrated pest management are to be exploited. A change in both the content and approach of research is the need of the hour and this can be taken up with cooperation of private sector on aspects such as production of quality planting materials, production of hybrids, green house production of quality fruits, off season production, biotechnology, value added products, product diversification and export.

The above mentioned constraints can be overcome by careful consideration, formulation and implementation of relevant actions step by step. The line of action should be such that first a few products or a few markets should be concentrated. The products should be selected based on demand and which can be produced making the best use of the resources available in India and can be presented in a differential and appealing way with a low cost benefit ratio. To solve the problem regarding market in targeted countries, knowledge regarding market demand of Indian fruits in overseas market is essential. For this purpose institutionalize the market intelligence network for exporters, harmonize with international standards and develop required infrastructure for certification and testing.

Market Intelligence

A strong data base of current and potential exporters has to be prepared first for taking part in national decisions. These exporters should be provided with information regarding major importing markets, main competitors, existing tariff structures, non-tariff barriers, likely changes in content of WTO requirements and current status of quality standards and food regulation in targeted markets.

Harmonization with the International Standards, Certification & Testing

India to become globally competitive, the level of quality has to rise. Variations do exist in quality standards and regulation adopted in different countries which may lead to trade conflicts and disputes. So to avoid this, the solutions suggested are to carry out post arrival testing of Indian products in importing country by a recognized international agency, motivate importing countries to set up export certification offices in India, compel food testing laboratories in India to obtain accreditation from international agencies by part funding from the Government, to introduce certified zones like pesticide free area, disease free area, pollution free area to facilitate production of quality products thereby adding more value to products and also promoting production of purely organic fruit crops.

Future Thrust Areas

The production and supply of fruits should be in line with requirement of importing countries which require control and monitoring of quality standards of both the raw material and processed products. To achieve this, steps are to be taken one by one. First enables direct farmer- importer linkages by amendment of APMC Act, setting up of independent world class food testing and inspection infrastructure with presence of exporters, devise systems of processing based on internationally accepted norms apart from quality standards of table grade products, encourage investments in infrastructure to improve quality and support private sector initiatives for investing in specialized transport infrastructure like refrigerated vans through specific financing schemes for this purpose.

Integration of Government Schemes

The Government through various ministries and allied agencies, offers support to exporters through various schemes to part-finance specific investment requirements. The allied agencies include Ministry of Agriculture, Ministry of Food Processing, APEDA, MPEDA, Coffee Board, Tea Board, Export Inspection Council, etc. To avoid duplication of efforts, it is necessary to integrate the support extended by governmental agencies. All the issues related to supply, market access and Government policies are inter-connected and as a comprehensive solution to these issues, the concept of modern terminal markets emerged. Government of India and several State Governments have already decided upon the establishment of such terminal markets.

Future Line of Action

- Integrate all schemes offered for export promotion through various ministries and allied agencies.
- Strengthen food processing infrastructure in identified Agri-Export Zones(AEZ).
- Encourage food testing laboratories to get accreditation from international agencies.
- Setting up of independent world class food testing and inspection infrastructure with the involvement of exporters.
- Devise an alternate system of processing of quality products based on internationally accepted norms.
- Promote aggregation of exports to meet minimum order requirement of importers.
- Develop a strong market intelligence system to aid exporters to take right decisions.
- Introduce certified zones like pesticide free area, disease free area, pollution free area to facilitate high value exports.
- Promote certification of organic farming for different crops.
- Build global brands in the name of India.
- Support research and development wing for upgrading technology for sustenance and growth of export market share.

Fruit Exports from India

Mangoes, grapes and banana are the three important fruits exported from India for which specifications for export are formulated. The major destinations are Bangladesh, Saudi Arabia, the United Arab Emirates, Yemen, Netherlands, Kuwait, Nepal and the United Kingdom.

Export Specifications for Mango

Total export from India	60,551.32 million tonnes (2003-2004)
Varieties grown in Maharashtra	Alphonso, Kesar, Payari, Vanraj, Rajapuri, Gulabi
Area under crop in Maharashtra	4,19,836 ha.
Maharashtra's Production	6,15,910 million tonne
Available in	February to June
Export	Alphonso export around 13,000 million tonne per year. Major Export to Middle East, UK, Netherlands
Varieties in demand Internationally	Kent, Tomy Atkin, Alphonso, Kesar

Export Specifications

		Middle East	**Netherlands /Germany**	**U.K.**
Variety	Alphonso	Wt: 200-250 g	Wt: 250-300 g	Wt: 250-300 g
	Kesar	Wt: 200-250 g	Wt: 225-250 g	Wt: 225-250 g
Packing		1 Doz/2.5 kg	1 Doz/2.5 kg	1 Doz/2.5 kg
Storage Temperature		13°c	13°c	13°c
Export		By Sea	By Air	By Air

Export Specifications for Grapes

Total export from India	26783.83 million tonnes (2003-04)
Share of Maharashtra	70 – 80% (Approx.)
Varieties grown in Maharashtra	Thompson Seedless, Sharad Seedless, Flame Seedless, Tas-A-Ganesh, Manik Chaman, Sonaka
Area under crop in Maharashtra	35236 ha
Maharashtra's Production	9,88,722 million tonne (2002-03)
Availability	December to April
Major exports to	Middle East, UK, Holland, Germany
Varieties in demand internationally	Thompson Seedless, Emperor, Ruby Seedless, Red Globe, Christmas Rose, Crimson Seedless, Calmeria, Rouge Ribier, Exotic, Fantasy Seedless

Export Specifications

	Countries		
Variety	Middle East	Holland/Germany	U.K.
Thompson Seedless	Berry Size: 15mm Colour: amber	Berry Size: 16mm Colour: white/amber	Berry Size: 18mm white
Sharad Seedless	Berry Size: 15mm black	Berry Size: 16mm black	Berry Size: 18mm black
Flame Seedless	-----	Berry Size: 16mm pink	Berry Size: 18mm pink
Packing	1 Kg	4.5 Kg /9 kg	4.5 Kg /9kg
Storage Temp.	0-1 °C	0-1 °C	0-1 °C

Export Specifications for Banana

Varieties grown in Maharashtra	Grand Naine, Shreemanti, Dwarf Cavendish, Basrai, Robusta, Lal Velchi, Safed Velchi, Rajeli Nendran, Red Banana
Availability	Round the year
Major export potential	UAE, Gulf countries
Varieties in demand Internationally	Grand Naine, Cavendish
Area under Cultivation (Maharashtra.)	57,000 ha.
Production	36,07,590 million tonne

Export Specifications

		Middle East
Variety	Grand Naine Cavendish	Colour : Green, Weight of Bunch : 2.5 Kg Fruits preferably straight
Packing		13 Kg
Storage		13-14 °C
Transport		By Sea

Agri Export Zone

Promotion of agricultural exports is important for creating conditions for providing remunerative prices to farm products. Though India ranks as the world's second largest producer of fruit crops, the country cares less impressively on processing fronts. Lack of adequate storage facilities is one of the main problems faced by food processing industry leading to almost

30-40% wastage of production. Simultaneously, most of the varieties grown in India are simply not suitable for processing. Therefore the concept of Agri Export Zones developed by APEDA aims to improve levels of food processing to reduce waste, increase marketability and help farmers to enjoy higher value of realization. It takes a comprehensive look at a particular produce/product located in a contiguous area for purpose of developing and sourcing raw materials, their processing/packaging, leading to final exports. Thus, the entire effort is centered on cluster approach of identifying potential products, the geographical region in which these products are grown and adopting an end-to-end approach of integrating entire process right from stage of production till it reaches market. It will also identify/enlist difficulties/ problems encountered at each stage. These difficulties could be procedural in nature or may relate to a particular quality standard. A package needs to be developed to suggest solutions to these problems. Corrective time bound measures will then be delegated to specific government agencies. The APEDA ensures that the measures are completed on schedule. Export quality products will have specific niche markets identified and will be promoted extensively. Under the programme, international Agricultural experts may also be invited to demonstrate modern harvesting technologies.

Operation of the Concept

The entire approach of promoting the Agri Export Zone would have to be taken on a project mode. This would mean that the State Governments would need to identify potential export products which could be selected for development with a cluster approach. State Governments will have to evolve projects which are feasible and are possible to be implemented immediately. They have also to conform to indicative guidelines given below. The States will forward such project proposals to APEDA which will conduct the initial scrutiny of proposals .If found feasible, APEDA provides necessary guidance in preparing the detailed project report. This report, after preliminary scrutiny, will be placed before the Steering Committee, constituted under the chairmanship of Commerce Secretary, Government of India with following members:

1. Director General of Foreign Trade, Member
2. Joint Secretary (EP Agri Division, DOC) Member
3. Joint Secretary (Deptt. of F.P.I.,MOA) Member
4. Joint Secretary [Infrastructure Division, DOC] Member
5. Executive Director, NHB Member

6. Representative of DG, ICAR Member
7. Director (Finance, Deptt. of Commerce) Member
8. Chairman, APEDA Convenor

Once the project proposal of a State is approved by the Committee, an MOU would be signed between APEDA (on behalf of the Central Government) and the State Government for providing possible assistance at each stage of project. The responsibilities of State government would also be defined in the MOU, a draft of which is under preparation.

Measures Envisaged to Promote Exports

Financial Assistance

Both Central and State Governments and their agencies are providing a variety of financial assistance to various agri export related activities. These extend from providing financial assistance for Training and Extension, R&D, Quality Upgradation and Infrastructure and Marketing. Thus, whereas Central government Agencies like APEDA, NHB, Dept. of Food Processing Industries, Ministry of Agriculture provide assistance, a number of State Governments have also extended similar facilities. All these facilities would have to be devoted and extended to promote agri exports from proposed Zones in a coordinated manner. Additional features like providing grants from Market Access Initiative fund could also be considered by state governments.

Fiscal Incentives

Benefits under Export Promotion Capital Goods Scheme, which were hitherto available only to direct exporters, have now been extended to service exporters in the Agri Export zones. Thus, even service provided to ultimate exporters will be eligible for import of capital goods at a concessional duty for setting up of common facilities. They shall fulfil their export obligation through receipt of foreign exchange from ultimate exporters who shall make payments from their EEFC account.Exporters of value added agri products will be eligible for sourcing duty free fuel for generation of power, provided cost component of power in the ultimate product is 10% or more and input-output norms are fixed by advance licencing committee of the DGFT. In view of the power intensive nature of most of the value addition, almost all exporters of value added agriculture produce will become eligible for such facility. Similarly, input-output norms can also be fixed for sourcing other inputs, like fertilizer, pesticides etc. duty free for cultivation purpose.

Anticipated Benefits

- Strengthening of backward linkages with a market oriented approach.
- Product acceptability and its competitiveness abroad and in the domestic market.
- Value addition to basic agricultural produce.
- Bring down cost of production through economy of scale.
- Better price for agricultural produce.
- Improvement in product quality and packaging.
- Promote trade related research and development.
- Increase employment opportunities.

Fruit Processing Industry in India

At present only 1% of fresh fruits and vegetables are processed in India as compared to 70% in Brazil and USA and upto 80% in ASEAN countries like Malaysia. While many units are operating in small scale sector in the country, most of them are not capable of catering to international standards in an organized fashion. One such organisation operating in the North East is North Eastern Regional Agricultural Marketing Corporation (NERAMAC) Ltd., Assam. Besides marketing agricultural products, it is also engaged in processing of pineapple into concentrate.

The market data revealed a positive indication for aseptically packed banana puree in local and international market. In Europe, USA and Middle East, there is good demand for banana puree, provided it is of good quality supplied in bulk aseptic packaging and the rates are competitive.

Food processing industry started receiving a great deal of attention from the policy makers. In India, very little effort is made in streamlining production, procurement, processing and marketing of perishable horticultural products specially fruits and vegetables. There is a change in the trend of consumption of fruit pulps during the last few years in India as demand for bulk packaging of fruit pulp concentrate, banana puree has increased as compared to consumer packs. India as well as many countries in Europe, Australia and Middle East are interested in bulk packaging for import of fruit juice/pulp concentrates and puree as they are using concentrates and puree in the following :

Blending for Fruit Juices and Drinks

- Ice cream
- Yoghurt
- Baby foods
- Ketchup, Sauces and Soups
- Dressing of pastries

The current industrial policy environment is congenial for development of food processing. The central government is making efforts to accelerate pace of development in food processing industries.

There is a big international market for fruit concentrates, tomato paste and banana puree. Trade centres around USA and Italy, Chile and Mexico cater to the large USA market while Italy, Greece, Spain and Portugal cater mostly to European market (France, Germany etc.). With total world trade in fruit and vegetable juices amounting to almost USS 5000 in 1990, the fruit juice industry has become one of the world's major agri businesses. Its importance for developing countries is emphasized by the fact that these countries account for roughly 40% of total world exports. Japanese market is growing rapidly as a result of liberalisation. Other non-traditional markets are also expected to expand, partly because of growing health awareness and higher standards of living. Consumers in most of the countries are showing a growing interest in tropical fruit juices and pulps. After pineapple, the best selling tropical fruit juices and pulp/puree are passion fruit, mango and banana. Besides, banana puree is used in baby foods and ice creams.

There are good reasons to believe that the fruit juice business will remain a growth industry for a long time. First, per capita consumption of fruit juices and nectars is still fairly low in most markets, including several of the world's largest markets. One of the main reasons for expectations of growth is that consumers are becoming more health conscious, a trend that started some years ago. Product development, including the introduction of new flavours and blends, and innovations in packaging by bottlers/packers and retail organisations, together with advertising and sales promotion and more aggressive marketing, are all factors that will help push up per capita consumption. At the same time, increasing amount of fruit juice raw material are used in other food and beverage products, including dairy products like yoghurt and yoghurt drinks, as well as health drinks, desserts and baby foods.

USA is the world's largest import market for fruit concentrates. There is a marked increase in consumption of juices and pulps from guava, passion fruit, mango, papaya and tamarind. Banana, guava and mango fruit concentrates are popular with semi-solid baby foods and are used in baby food mixtures. USA is also the largest importer of banana puree.

Present market for banana puree on an average is around 1,50,000 tonnes/annum in the world. The average growth every year is around 3-5%. However, if we take also the future demand for banana puree in India alone it could be around 50,000 tonnes/ annum given proper promotion and better quality.

While pre-cooling (cold chain) and surface coating dominate in the short run, CAIMA packaging and irradiation technologies emerge in the long run for preservation and extension of shelf life.

While marketing of fruits and vegetables is expected to be dominated by cooperatives and middle men in short term, organised direct sourcing supermarkets are likely to emerge in the long term. Frozen and dehydrated products, fruit juices, pickles and other forms of preserves are likely to emerge as popular processed products. Change in consumer taste, food habits and life style, convenience, nutritional value and purchasing power are the likely reasons for preference of processed products. While the level of processing would hover around 5-10% in the next 10 years, 15-20% of fruits and vegetables may he processed in the long term. The share of sectoral consumption for processed fruits and vegetables in the long term would be as follows:

- Domestic - 30%
- Institutions - 40% (including defence)
- Exports - 30%

While the small scale processing units would dominate in the short term, an advent of large/medium scale units is likely in the long term.

The summary of future technology and business scenario for the fruits and vegetables sector is presented elsewhere.

Aspect	Short Term	Medium Term	Long Term
Mechanization in Harvesting	Essentially manual as land holdings are likely to remain small	Will continue to be manual	
	Semi-mechanized means for bulk crops like potato and onion	Increased use of farm tools, implements & tractors	Limited use of mechanized harvesters for potato & onion to be cultivated on large farms
	Use of hand tools for fruits		
Post Harvest Treatments	Selective use of pre-cooling for high value crops like Strawberry/Grapes	Increased use of pre-cooling techniques like • Mobile Pre-cooling • Ventilated Pre-cooling • Evaporative Cooling	Extensive use of mobile pre-cooling
	Use of irradiation not likely	Use of broad spectrum chemicals to continue	Limited use of irradiation for potato, onion & spices
			Use of broad spectrum chemicals to be curtailed
Storage	No significant change from currently followed practices in storage at field level	5-10% of fresh produce to be channelized through cold chain	20-25% of fresh produce to be channelized through cold chain
	Less than 5% of produce would be channelized through cold chain		
	Use of CA/MA storage only for high value perishable produce	Very Limited use of CA/MA Storage	Extensive use of CA/MA Storage
Transportation	Mainly by trucks and by cart load	Increased use of Reefer trucks	Reefer trucks and CA/MA containers for transport
Packaging	Traditional packaging methods – (Jute bags, Rattan baskets, modern boxes) to continue	Traditional methods to continue	Traditional methods to continue but use of plastic packaging to grow at the cost of wood based packaging
	Use of CFB Cartons, shrink wrapping for exports		Use of CA/MA packaging to increase significantly
			Emergence of packing stations

Marketing and Distribution	No significant change from current practices	Increased role of cooperatives and private institutions	Institutionalisation of marketing & distribution network
	Middlemen would continue to play dominant role		
Processing	Level of processing (as % of total production of fresh product) would about 5-8%	Level of processing about 8-12%	Level of processing would be about 15-20%. Focus to shift from exports to domestic consumption
	Consumption to be dominated by institutional/defence and export segments		Increase in share of consumption by household sector
Industry Structure	Small scale sector to dominate		Small scale units to emerge as subsidiaries/ancillaries to large scale units
Processing Technology			
Dehydration	Sun drying and forced air drying		Freeze drying
Concentration	Predominantly batch pan concentration in the small scale sector	Use of low temperature vacuum evaporation	High temperature short time evaporation will also be used
	Low temperature vacuum evaporation for large processing units		
Freezing	IQF	IQF	IQF
	Plate Freezing		Plate Freezing to decrease
Thermal Processing	Canning to dominate	Canning as well as use of aseptic filling	Use of aseptic filling
	Introduction of aseptic filling		
Packaging of Processed Products	No significant change from current practices	Increase in use of food grade plastics	Use of food grade plastic and aseptic laminates
		Aseptic packaging to gain popularity	
Exports	Will account for about 1-2% of world trade	To account for about 2-3% of world trade	To account for about 3-5% of world trade
R&D	Thrust on extension services	Thrust on developing new high yielding varieties	Increasing role of private sector in R&D
	Thrust on development of varieties with extended shelf life	Fruits & Vegetables with specific characteristics suited for processing to be developed	Increasing role of private sector in R&D.

Marketing of Fruits

Marketing plays a key role in the post harvest handling operations of fruits. A perfect and efficient marketing system covers all aspects of handling from the stage of harvesting till the commodity reaches the consumer. Marketing of perishable fruits presents more problems as compared to other agricultural commodities. Generally, the interest of the producer and consumer are poorly served. The growers get less return for their produce and consumers have to pay more than what is necessary. Due to the presence of middlemen, price of fruits are 50-100% higher in terminal markets (mandis) than in growing areas. The middlemen manipulate the situation by artificially creating gluts and offering lower price to growers. They also falsely reject the fruits as substandard and indulge in other malpractices. The co-operative systems play a very important role in the marketing of fresh produce.

India is one of the most important fruit producing nations in the world, accounting for 10% of all fruits and nearly 40% of tropical fruits produced globally and the largest producer of mango in the world. Fruit production in India increased at an average annual growth rate of 3% between 1995 and 2004, while harvested area and yields increased by 2.4 and 1.5%, respectively. Fruit production increased steadily from 36 to 47 million tonnes, between 1995 and 2004, largely as a result of supply side policy incentives. It contributes 10% on average to the gross value of total agricultural output in India, and about 13% of the total export earnings from major agricultural products. The production of important fruits in India over a period of 10 years from 1994 to 2004 shows an increasing trend.

Production of Selected Fruits in India

	Average (1992/1994)	2002	2003	2004
	('000 tonnes)			
Bananas	9718	16820	16820	16820
Mangoes	10108	10640	10780	10800
Oranges	1743	3120	3070	3070
Apples	1205	1160	1470	1470
Lemons and Limes	863	1440	1420	1420
Pineapples	956	1180	1310	1300
Grapes	684	1210	1150	1200
Papayas	470	700	700	700
Pears	127	200	200	200
Peaches and Nectarines	83	150	150	150
Grapefruit and Pummelos	83	140	142	142
Plums	55	80	80	80

Source: FAOSTAT

Favourable growing conditions permit the production and supply of a large variety of tropical fruits throughout the year. These are consumed not only in the regions where they are grown but throughout India. Terminal markets in major Indian cities receive large quantities of tropical fruits, which are then sold through retail dealers in the other cities and towns. Total apparent consumption of fruits in India was 39 million tonnes in 2002, about 8 million tonnes more than in 1995, with consumption of fresh tropical fruits estimated at 16.3 million tonnes. Annual per capita consumption of fruit was estimated to be 37 kg in 2002, a 12% increase over 1995, with major tropical fruits at 13.3 kg (bananas alone at 7.7 kg), temperate fruits at 4.2 kg and other fruits at 5.9 kg. Research studies into the consumption of tropical fruits in India showed that among major states, Kerala, Tamil Nadu and Goa have the highest level of consumption of fruits both for rural and urban areas. These states are followed by Punjab, Haryana, Delhi and Union Territories of Chandigarh in the North. Bihar and Orissa in the east and Rajasthan and Madhya Pradesh in central India have much lower levels of consumption.

Population growth in India was at 1.67%/year between 1995 and 2004. According to official statistics, population growth will decline at 1.3%/year by 2015 when the total population of India reaches 1.6 billion. Research studies revealed a sizable potential to expand fruit consumption in rural areas in states with currently low per capita consumption rates and growing urban areas. Indian economic development is strong recently with GDP growth averaging 5.6% between 1990 and 2003, largely as a result of robust domestic demand. Inflation is contained and exports are strong. These facts contributed to a 3.75% rise in per capita income between 1990 and 2003, with household expenditure for fruits and vegetables increased about 5%/year over the same period. Annual per capita fruit consumption also increased from 30 kg in 1992 to 37.7 kg in 2002.

Research conducted by the Indian Agriculture Research Institute, New Delhi indicated that proportion of population that makes up the highest income group consumes about six times more fruits than the lowest income group, in both rural and urban areas. Urban households spend 2.25% of their income on fresh and dried fruits, while rural households spend 1.8% of their income on same items. The study also predicts that, lower income groups are more likely to account for most of future growth of fruit market in India.

The interest in healthier diets led to an expansion in consumption of all fruits, of which a large share was that of tropical fruits. Temperate fruits are consumed mostly by high income group, while tropical fruits consumption is more wide-spread among all income groups. Large quantities of fresh unripe mangoes are used in household preparation of pickles and chutneys. Consumption of litchi and sapota is mainly concentrated among middle and high income group. The tourism industry is also contributing significantly to rise in tropical fruit consumption. Growth in air travel and tourism in India has enabled foreigners to experience taste of a large variety of tropical fruits. This contributes positively to a rising market demand for tropical fruits in Asia, Europe and North America. Another factor contributing to rising consumption is domestic promotional activities. The Government of India has carried out various nutritional education programmes to raise awareness on importance of a varied diet, and benefit of fruit consumption. In addition, promotional programmes carried out by fruit juice processing industry through television, newspapers and in-store promotional activities encouraged overall demand for fruits.

Prices of fruit have increased more than the General Price Index between 1999/2000 and 2003/2004: the All Commodities Wholesale Price rose by 21%, while the Fruits wholesale price increased by 39%. A substantial price differential exists between wholesale and retail prices. This might be due to rising costs in processing, distribution, and marketing, or due to a non-competitive market structure. High retail prices constitute a major constraint for increasing consumption, particularly among middle to low income household groups.

Prior to 1997, imports of fresh and dried fruits, with exception of dates and figs, were prohibited. Following liberalization of trade in agricultural commodities, imports of all fresh fruits (except citrus, grapes and litchis) are permitted generally on Open General License (OGL) basis. In case of South Asian Association for Regional Cooperation (SAARC) countries, an import tax of 44% is levied, while for other countries it is set at 45.6%. There are no reported export duties on fruits from India.

Fruit consumption in India is anticipated to increase by about 4%/year according to projected growth rates for income, population, and trends in food preferences. In addition, growth in demand for Indian fresh and processed tropical fruits is strong in recent years and this trend is likely to continue in medium term. This holds considerable potential for domestic producers and suppliers of tropical fruits in India, and in Indian market, for the overseas suppliers that are able to compete effectively with domestic supplies. Growth in fruit production and processing sub-sectors provides employment opportunities and income generation, and contributes positively to food security. Furthermore, tropical fruit production provides an alternative option for crop diversification and support to rural entrepreneurs engaged in supply of inputs for production and purchasing of fruit for marketing. However, to fully realise this potential, requires changes in value chain, particularly to compete in quality sensitive international markets. Development of efficient post-harvest handling infrastructure, transportation facilities and quality of produce must be priorities, especially for fruits that are grown for lucrative export market. Improvement to existing system is challenging and requires considerable support from Government, research institutions, and technology suppliers to develop appropriate facilities at all stages of the value chain.

Significant development has taken place in international trade regime for fair play and non-discriminatory trade. This has bearing for Indian economy. Many countries are interested in Indian market, because of market access provision in the WTO regime. The increasing consumption of fruits coupled with increase in income will certainly help such nations to export until India takes remedial measures to increase productivity and improve quality of products. Many efforts are needed to exploit the world market potential especially for temperate fresh fruits when India have tough competition in fruits from USA, France, Belgium, Australia, Italy, Spain etc.

Disease and Pest Management in Fruit Crops

Important Disease and Control Measures of Major Fruit Crops

Apple and Pear		
	Name of Disease	**Control Measures**
1	Scab	At silver tip stage spray Dodine (100g/l)/ Mancozeb (400g/l)/ Captan (300g/l) At pink bud stage spray Contaf (30ml/l)/ Systhane (40 g/l)/ Baycor (50g/l)/ Rubigan (40ml/l)/ Mancozeb + sulphur (300 + 200g/l) At petal fall stage spray Bavistin (50g/l)/ Contaf (20 ml/l)/ Topsin M (50g/l) At pea stage spray Contaf (30ml/l)/ Baycor (50g/l)/ Rubigan (40ml/l)/ Synthane (40g/l)/ Mancozeb (300 g/l)/ Captan (300g/l) At fruit development stage spray Bavistin + Mancozeb (25 + 250g/l)/ Mancozeb + Topsin M (250 + 50g/l)/ Captan (300g/l) At fruit development stage (second spray) spray Mancozeb (300g/l)/ Captan (300g/l)/ Baycor (50g/l)/ Rubigan (40ml/l)/ Systhane (40g/l)

		About 15-20 days before harvest spray Mancozeb (300g/l)/ Captan (300g/l) Before leaf fall spray Urea (5 kg/l)
2.	Shoot-fruit blight	Uproot affected seedlings and destroy, Drench the nursery with Thiram (3g/l), Change nursery site regularly and fumigate
3.	White root-rot	Provide proper drainage of orchard, Remove infected roots and apply Chaubattia paste to cut ends of roots, Drench soil with Carbendazim (1g/l) at interval of 15 to 20 days during monsoon
4.	Powdery mildew	Prune and destroy affected terminals, Spray wettable sulphur (3g/l)/ Carbendazim (500mg/l)/ Karathane (500mg/l) during late dormancy, bud swell, petal fall stages and 2 weeks later.
5.	Collar rot	Remove soil around collar portion during November-December and expose to sun, Remove affected bark and apply Chaubattia paste, Drench soil with Mancozeb (3g/l)/ Copper oxychloride (500mg/l)/ Ridomil MZ(3g/l)
6.	Cankers	Spray of Copper oxychloride (0.3%) or Captan (0.2%) after harvest and at bud-swell stage
7.	Crown gall	Affected plants to be destroyed
Avocado		
1.	Fruit spot	Spray Carbendazim/ Thiophanate methyl (1g/litre)/ Mancozeb (2g/litre) during rainy season
2.	Leaf spot	- do -
3.	Fruit rot	- do -
4.	Root rot	Avoid water logged condition, Mix soil with Metalaxyl before planting/ drench soil with Metalaxyl
5.	Anthracnose	Spray Carbendazim/ Thiophanate methyl (1g/litre)/ Kavach (2g/litre) during rainy season
6.	Stem end rot	- do -
Banana		
1.	Panama wilt	Eradication of the infected plants, Growing resistant varieties like Dwarf Cavendish, Drench soil with Vapam (0.85%) and Mercuric chloride (3000 ppm) in nursery, Apply Carbendazim (1g/litre)
2.	Bunchy Top	Uproot and burn infected plants, Spray 2,4- D to control weeds.

3.	Sigatoka leaf-spot	Spray Carbendazim/ Thiophanate methyl (1g/l)/ Calixin (0.7ml/l)/ Kavach (2g/litre) during rainy season
4.	Anthracnose	Spray Carbendazim/ Thiophanate methyl (1g/l)/ Calixin (0.7ml/litre)/ Kavach (2g/litre) in humid weather
5.	Leaf blight	Spray Carbendazim/ Thiophanate methyl (1g/l)/ Calixin (0.7ml/l)/ Kavach (2g/l) during February/March
6.	Cigar end rot	Spray Thiophanate methyl (1g/l) as soon as fruits are formed
7.	Tip over	Spray Sreptocyclin 300 ppm + Copper oxychloride/ Carbendazim (4g/1) during rainy season
8.	Stem end rot	Spray Carbendazim/ Thiophanate methyl (1g/l)/ Kavach (2g/1) as pre harvest spray and post harvest dipping
9.	Mosaic	Use disease free planting material
10.	Kokkan	Rogue infested plants
Citrus		
1	Gummosis	Soil drenching with Bordeaux mixture (1g/l) and Mancozeb (2g/l)
2.	Twig blight	Prune blighted twigs, Spray Carbendazim (1g/l) and Kavach (2g/l)
3.	Die-back	Prune dead twigs, Spray Carbendazim (1g/l) and Kavach (2g/l)
4.	Bacterial canker	Spray Copper oxychloride (2g/l) during rainy season
5.	Powdery mildew	Apply Calixin (0.5ml/l) and wettable sulphur (2g/l) in September-March
6.	Stem rot, foot rot, leaf fall and fruit rot	Soil drenching with Bordeaux mixture (1g/l), Spray Mancozeb (2g/l)
7.	Pink disease	Prune affected trees and paste cut ends with copper fungicides. Apply Mancozeb/ Copper oxychloride (3 g/l)
Custard apple		
1.	Leaf spot	Spray Carbendazim/ Thiophanate methyl (1g/litre)/ Mancozeb (2g/litre) during rainy season
2.	Fruit rot	- do -

Date palm		
1	Graphiola leaf spot	Remove and burn the affected leaves, Spray the trees with 2:2:250 Bordeaux mixture (2 kg copper sulphate + 2 kg unslaked lime mixed in 250 litres of water).
2.	Fruit rot	Spray the bunches with a mixture of 5% Fahana (ferric dimethyl dithiocarbamate) in sulphur.
Grapes		
1	Downy mildew	Spray Bordeaux mixture (1g/l)/Copper oxychloride (2g/l)/ Mancozeb (2g/l)
2.	Powdery mildew	Apply Wettable sulphur (2g/l) during cloudy nights
3.	Anthrac-nose	Spray Bordeaux mixture (0.8g/l)/Copper oxychloride (0.25g/l)/ Carbendazim (1g/l)
4.	Red rust	Apply Wettable sulphur (2g/l) in September-October and February-March
5.	Bacterial leaf spot	Spray Streptocyclin (500 ppm)
Guava		
1.	Wilt	Liming of the pits, Drench soil with Carbendazim (1 g/l), Spray Difoltan (0.3%), followed by Dithane Z-78 (0.2%) at one-month interval
2.	Anthracnose	Apply Carbendazim (1 g/l)
3.	Fruit canker	Three or 4 sprayings of 1% Bordeaux mixture.
4.	Fruit rot	Apply Mancozeb (2 g/l) and Carbendazim (1 g/l) during rainy season
5.	Leaf spot	Apply Mancozeb (2 g/l) and Carbendazim (1 g/l) during November/December
Jack fruit		
1.	Pink disease	Prune affected trees and paste cut ends with copper fungicides. Apply Mancozeb/ Copper oxychloride (3 g/l)
2.	Stem rot	Prune affected trees and paste cut ends with Bordeaux paste. Apply Dithane M-45/ Copper oxychloride (3 g/l)
3.	Fruit rot	- do -
4.	Male inflorescence rot	- do -
5.	Leafspot	Apply Mancozeb/ Copper oxychloride (3 g/l) during rainy season

Litchi		
1.	Powdery mildew	Apply Wettable sulphur (2g/l)/ Carbendazim (0.5 g/l)
2.	Anthrac-nose	Apply Carbendazim (1 g/l)/ Mancozeb (2 g/l) during rainy season
3.	Leaf spot	- do -
4.	Red rust	Apply Wettable sulphur (2g/l) in September-October and February-March
Mango		
1.	Powdery mildew	Two to three sprays of Wettable sulphur (0.2%)/ Karathane (0.1%) at 10-15 days interval
2.	Anthracno-se	Two sprays of Wettable sulphur (0.2%)/ Bavistin (0.1%) at fortnight interval
3.	Malforma-tion	One spray of 200 ppm NAA in October followed by deblossoming at bud burst stage in December - January.
4.	Fruit drop	Regular irrigation during fruit development Timely and effective control of pests and diseases and spraying 20 ppm NAA at pea size of fruits.
Mangosteen		
1	Thread blight	Spray Carbendazim /Thiophanate methyl (1g/l)/ Kavach (2g/l)
2.	Canker	Spray of Copper oxychloride (0.3%)/ Captan (0.2%) after harvest and at bud-swell stage
Papaya		
1.	Stem rot or collar rot	Good drainage, Spraying of Bordeaux mixture (1%)
2.	Powdery mildew	Apply Bavistin (Carbendazim) (0.1%) twice at 15-day interval
3.	Viral Diseases (leaf curl, mosaic)	The spread of the disease could be minimized by controlling aphid vector through application of Malathion, Pre-immunize seedlings with milder strain of the virus, Grow disease resistant varieties
4.	Nematode	Apply neem-cake (1 kg)/ Carbofuran (2 kg)
Peach and plum		
1	Leaf-curl	Prune affected shoots, Spray Copper oxychloride (2g/l)/ Carbendazim (1g/l) during late dormancy, Spray should be repeated at bud swelling stage

2.	Shot hole	Spray Captan (2g/l) at leaf fall or bud swelling stage.
3.	Powdery mildew	Prune diseased shoots, Spray wettable sulphur (3g/l) before the blossoms open, at petal fall and again two weeks later.
4.	Bacterial gummosis	Clean the wound and apply Mashobra paste, Spray Streptocyclin (500 ppm) before the onset of rains, Spray Copper oxychloride (3g/l) after leaf fall
5.	Rust	Spray Mancozeb (3g/l) at 20 days interval from petal fall
Pineapple		
1.	Heart rot	Provide good drainage, Proper selection and prophylactic treatment of planting material with Dithane Z-78 (0.3%)/ Foltaf (0.04%), Spray affected plantations with Dithane or Foltaf.
Pomegranate		
1.	Leaf spot	Apply Mancozeb (2 g/l) during rainy season
2.	Fruit rot	Apply Carbendazim (1 g/l) and Mancozeb (2 g/l) during September-October
Sapota		
1.	Leaf spot	Apply Dithane M-45/ Copper oxychloride (3 g./l.)
2.	Base rot	Drench soil with Kavach (2g/l) and Copper oxychloride (3 g/l.)
3.	Heart rot	Dip suckers in Copper oxychloride (3 g/l), Drench soil with Dithane M-45 (2g/l)/ Copper oxychloride (3 g/l)
4.	Anthrac-nose	Apply Dithane M-45/ Copper oxychloride (3 g/l.)
Strawberry		
1.	Root rot	Spray Captan and Thiophanate methyl (2g/l)
2.	Powdery mildew	- do -
3.	Anthrac-nose	- do -
4.	Viral diseases-	Raise strawberry nursery in the hills to control the vectors

Important Pests and Control Measures of Major Fruit Crops

Apple		
	Name of pest	**Control measures**
1.	San Jose scale	Biologically control using natural predators like *Aphytis procila, Chilocorus bijugua, Pharoscymnus flexibilus* and *Coccinella septempunctata* and endoparasitoid *Encarsis perniciosi*

2.	Woolly apple aphid	Soil application of Phorate/ Carbofuran granules during May and October/November checks incidence and spread of pest, Foliar spray of Chlorpyriphos (0.02%)/ Fenitrothion (0.05%)/ Dimethoate (0.03%)/ Phosphamidon (0.03%), Malling Merton rootstocks like MM 106, MM 109, MM 111 and Merton 793 were found resistant to this pest.
3.	Root Borer	Drench basins with Chlorpyriphos (0.04%)/ dust with Folidol dust @ 25 kg/ha in September, Trap and kill adults during June-September.
4.	Blossom Thrips	Spray Chlorpyriphos (0.04%)/ Fenitrothion (0.05%) at pink-bud stage if population exceeds 15/flower bud.
5.	Codling Moth	Mass pheromone trapping (25 traps/ha), collection and destruction of over wintered pupae in cocoons during April to June, deep burying of fallen fruits during August, Two sprays of DDVP or Phosphamidon (0.04%) in June- July at an interval of 2-3 weeks.
6.	European Red Mite	Spraying of Dicofol (0.05%) followed by Malathion (0.05%), Biological control using predators like coccinellids, predatory mite and anthocorid bugs.
Avocado		
1.	Mealy bug	Prune and burn the infested shoots, Spray Dimethoate (0.05%)/ Chlorpyriphos(0.05%) at early stage of infeststion.
2.	Mite	Spray Phosalone (0.05%) or Dimethoate (0.05%) once or twice depending on the severity.
3.	Scale	Prune and burn the infested shoots, Spray Chlorpyriphos (0.05%) during severe infestation.
Banana		
1.	Banana rhizome weevil	Cut and remove the outer layer of the rhizome and sundry for 3-4 days after smearing with cowdung and ash.
2.	Pseudostem-borer	Uproot and burn the infested plants, At initial stage insert 3 tablets of aluminium phosphide (0.5g/plant) in the pseudostem, Apply Quinalphos (0.05%)/ Chlorpyriphos (0.03%)/ Carbaryl (0.2%).
3.	Banana aphid	Uproot and burn the infested plants, Spray Phosphamidon (0.05%)/ Dimethoate (0.05%).
4.	Spindle leaf miner	Spray Dimethoate (0.05%) on the spindle.
Citrus		
1.	Psylla	Prune affected shoots and destroy, Spray Monocrotophos (0.05%)/ Dimethoate (0.05%), First spray should coincide

		with emergence of new flush. Repeat once or twice at 10-12 days interval.
2.	Black fly and White fly	Avoid close planting and water logging, Affected plant parts should be pruned and destroyed, Avoid excessive use of nitrogenous fertilizer and irrigation.
3.	Red scale Brown scale Green scale	Good orchard sanitation, removal of weeds and destruction of ants, Field release of predatory ladybird beetle, *Cryptoleamus montrouzieri* @ 10 beetles/plant and release of exotic parasite, *Leptomastrix dactylopii.*
4.	Citrus nematode	Apply neem-cake, Under glasshouse conditions a fungal parasite, *Paecilomyces lilacinus* reduces the nematode population.
5.	Mealybug	Prune affected shoots and destroy, Destroy ant colonies by dusting Quinalphos/ Phosalone dust around tree trunk, Spray Neem oil (2%)/ Pongamia oil (2%) once or twice at 21 days interval.
Custard apple		
1.	Mealy bug	Prune and burn the infested shoots, Spray Dimethoate (0.05%)/ Chlorpyriphos(0.05%) at early stage of infestation.
2.	Fruit-borer	Collect and destroy all fallen and infested fruits
3.	Soft scale	Prune and burn the infested shoots, Spray Chlorpyriphos (0.05%) during severe infestation
Date palm		
1.	Date Stone bettle	Spraying the plants with Malathion (0.075%), Regularly remove dropped fruits and destroy them.
2.	Nitiduled beetle	- do -
3.	Cigar hoeing beetle	Fumigate the dates with Methyl bromide at a dose of ½ kg/ 27 cubic meter for 24 hours.
4.	Termites	Apply 10% BHC@ 30g before planting in each pit.
Grape		
1.	Flea beetle	Spray Carbaryl (0.15%)/ Quinalphos (0.05%)/ Dichlorvas (0.1%)/ Phosalone (0.05%) from the fourth day until the emergence of leaves.
2.	Thrips	Spray Phosphamidon (0.05%)/ Carbaryl (0.125%)/ Phosalone (0.05%)/Malathion(0.05%), Prophylactic spray to be given once in five days from the initiation of bloom to berry set.
3.	Mealy bug	Avoid spraying broad-spectrum insecticides particularly synthetic pyrethroids, Spray Dichlorvas (0.1%) + Neem oil

		(0.2%) / Tridemorph (0.1%), Release *Cryptolaemus montrozieri* beetles (mixed population of grubs and adults) at 8,000-10,000/ hectare when the berries start softening.
4.	Leaf hoppers	Spray Carbaryl (0.15%)/ Fenitrothion (0.04%)/ Phosalone (0.05%) /Quinalphos (0.05%), A mixture of Quinalphos (0.05%) + Phosalone (0.05%) is more effective on the nymphs while Tridemorph (0.1%) is effective on the adults.
Jack fruit		
1.	Shoot and fruit borer	Prune affected shoots and destroy, Bag young fruits with perforated alkathene bags, Spray Carbaryl (0.1%)/ Phosalone (0.05%)
2.	Mealy bug	Plough inter spaces in November and dust 2% methyl parathion@200 g/ tree near the trunk, Fix 20 cm wide 400 gauge polythene strips around the trunk with grease applied on the lower edge in January as prophylactic measure, Two sprays of monocrotophos (0.04%) at 15 days interval
3.	Brown bud-weevil	Destroy all infested shoots, buds and fruits, Spray Carbaryl (0.1%)/ Quinalphos (0.05%) during flower bud emergence and fruit formation stage
4.	Aphid	Spray Mahua oil (2%) or Dimethoate (0.03%)
Litchi		
1.	Litchi mite	Prune affected twigs/branches and burn, Two sprays of Karathane (0.05%) at 7-10 days interval, Apply neem cake to reduce the incidence of pest.
2.	Shoot borer	Prune and burn affected twigs, Apply paste of crude carbolic acid (130ml), soft soap (1kg) and hot water (3.7 litres) to holes in the bark and plug the holes.
3.	Bark eating caterpillar	Clean the infested area and plug holes with Petrol/ Nuvacron/ Formalin soaked material, Spray Carbaryl (0.1%)/ Chlorpyriphos (0.04%) on tree trunk, Inject Dichlorvos (0.1%) into bored holes
4.	Trunk borer	- do -
5.	Fruit Borer	Collect and destroy all fallen and infested fruits, Spray Phosalone (0.05%)/Fenthion (0.05%) alternatively at 10 days interval at early stage of fruiting.
Mango		
1.	Mango hopper	Two sprays (at panicles emergence and at pea size of fruits) of Carbaryl (0.15%)/ Phosphamidan (0.05%), Biological control using *Beauveria bassiana*.

2.	Mealy bug	Plough inter spaces in November and dust 2% methyl parathion@200g/ tree near the trunk, Fix 20 cm wide 400 gauge polythene strips around the trunk with grease applied on the lower edge in January as prophylactic measure , Two sprays of Monocrotophos (0.04%) at 15 days interval, Biological control using *Beauveria bassiana*.
3.	Fruit fly	Spray Malathion (0.1%) emulsion/suspension containing 2% sugar at monthly intervals from initial fruit set to harvest
4.	Mango stem borer	Apply paste of crude carbolic acid (130ml), soft soap (1kg) and hot water (3.7 litres) to holes in the bark and plug the holes.
5.	Leaf feeders	Apply Carbaryl (0.1%) .
6.	Shoot midge	Apply Carbaryl (0.1%)/ Dimethoate (0.05%).
Mangosteen		
1.	Leaf-eating caterpillar	Spray 1kg of Carbaryl (Sevin 50% WP) in 500 litres of water.
2.	Small ant	Spray Chlorpyriphos(0.05%) during severe infestation.
3.	Mite	Spray Phosalone (0.05%)/ Dimethoate (0.05%) once or twice depending on the severity.
Papaya		
1.	Fruit fly	Collect and destroy all fallen and infested fruits, Spray Deltamethrin (0.003%)/ Dimethoate(0.05%) when fruits are fully mature.
2.	Grass hopper	Dust foliage with Carbaryl/ Phosalone dust during morning hours.
3.	Aphid	Destroy virus affected plants, Prophylatic spray of Dimethoate (0.3%)/ Methyl demeton (0.05%).
4.	Red spider mite	Infested leaves should be destroyed, Spray Dicofol (0.04%)/ Dimethoate(0.05%).
5.	Stem borer	Destroy the infested shoots, Inject Dimethoate(0.05%) into the stem using syringe .
6.	Grey weevil	Dust foliage with Carbaryl (5%) dust, Drench soil around the tree trunk with Chlorpyriphos (0.1%).
Peach		
1.	Peach leaf curl aphid	Spray 500 ml of Rogor 30 EC (Dimethoate) in 500 litres of water immediately after the fruit-set, Repeat spray after 15 days.

2.	Peach black aphid	Spray 500 ml of Malathion 50 EC in 500 litres of water on the colonies on the stem and limbs as soon as the pest congregates on these parts. Spray should be done when the sun is high.
3.	Mite	- do -
4.	Defoliating beetles	Spray in the evening with 1kg of Carbaryl (Sevin 50% WP) in 500 litres of water, Repeat spray after 5 or 6 days.
5.	Peach fruit fly	Burry the infested fruits at 60 cm depth, Hoeing of orchard should be done in May and June to expose the pupae which are present mostly at 4-6 cm depth, In April-May, spray the bait consisting of 0.1% Malathion+ 10g sugar/jaggery (in 1 litre of water) on the foliage of trees.
Pear		
1.	Leaf-hoppers	Spray 500 g of Sevin/Hexavin 50 WP (Carbaryl) or 250 ml Rogor 30 EC (Dimethoate) in 250 litres of water.
2.	Aphids	- do -
3.	Hairy caterpillar	Collect and destroy the feeding young caterpillars. Phosalone 35 EC / Chlorpyriphos EC 2 ml/l
4.	Bark eating caterpillar	Remove the webs and apply 100g BHC suspension (BHC 50% WP) in 10 litres of water into the holes with wash bottle during September-October and again in January-February.
5.	Mites	Spray 250 ml Rogor 30 EC (Dimethoate) in 250 litres of water.
Pineapple		
1.	Fruit eating beetle	Do not allow fruit to ripen in the field, Damaged fruits should be plucked and destroyed
2.	Mealy bug	Dip healthy suckers in Fenitrothion (0.05%)/ Phenthoate (0.05%) emulsion for 15 minutes before planting, Apply Phorate granules@ 1.75 kg/ha, 3 months after planting, Destroy weeds and ant colonies
3.	Termite	Do not retain dried suckers in the field, Drench the field with Chlorpyriphos (0.1%)
Pomegranate		
1.	Pomegranate butterfly	Collect and destroy all fallen and infested fruits, Bag fruits after single spray of Dimethoate (0.05%) with muslin cloth or butter paper
2.	Bark eating caterpillar	Prune dead shoots, Spray Carbaryl (0.1%)/ Chlorpyriphos (0.04%) on tree trunk, Inject Dichlorvas (0.1%) into bored holes
3.	Fruit fly	Spray Carbaryl (0.2%) (200g/100 litre)

4.	Whitefly	Spray Dimethoate (0.05%) at the initiation of infestation, Repeat spraying after fortnight
5.	Mealybug	Prune affected shoots and destroy, Destroy ant colonies by dusting Quinalphos or Phosalone dust around tree trunk, Spray Neem oil (2%) or Pongamia oil (2%) once or twice at 21 days interval
6.	Aphid	Single spray of neem seed kernel extract(2%) at the initiation of new flush
Sapota		
1.	Leaf webber	Spray Phosalone 35 EC/ Chlorpyriphos 20 EC (2 ml/litre)
2.	Hairy caterpillar	- do -
3.	Bud worm	- do -
Strawberry		
1.	Red spider mite	Spray Monocrotophos (0.05%) + Wettable sulphur (0.25%)/ Dicofol (0.01%)
2.	White grubs	Deeply plough the field, Treat soil before planting with Chloradane or Heptachlor (5%) dust at the rate of 50 kg per hectare
3.	Cut worms	- do -
4.	Hairy	Spraying with Malathion (0.5ml/ litre) on the appearance of caterpillars.

REFERENCES

Awasthi, R,P and Chauhan, P,S. 1997.Apple and Pear (In)*50 years of Crop Science Research*, ICAR Publication, New Delhi

Bal, J, S. 1992. *Fruit Growing*, Kalyani Publishers, New Delhi. 318p

Chadha, K.L.2006. Fruit Crops. *Handbook of Horticulture*.ICAR, New Delhi.107-338

FAO: *Report of the Second Session of Sub-Group on Tropical Fruits*: CCP:BA/TF 99/17.

Garg, Anand: (Ed.): ITC-HS Classification of Import Items with Indian Import Tariff, Vol.3: Published by Business Database Publishing Company: Rohini, Delhi.

Ghosh, Saurindra, P. 1999. *Avocado Production in India* Food and Agriculture Organization of the United Nations. 1999.

Government of India. 1995. Department of Statistics: National Sample Survey Organization: Level and Pattern of Consumer Expenditure: 5th Quinquennial Survey: 1993-94: NSS Fiftieth Round: Report No. 402.

Government of India. 2003. Planning Commission: *Ninth Five Year Plan (1997-2002): Development Goals, Strategy and Policies*, Vol. L.

Kaul, G.L.1986. *Horticultural Crops in India*. Anmol Publications, New Delhi, 1986.

Kunwar, R. and Singh, R. 1993. Note on nutritional survey of litchi orchards in Doon Valley of Garhwal Hills of Uttar Pradesh. *Progressive Horticulture* 25:164-5.

Pandey, S. and Misra, A.P.. 1975. Mycorrhiza in relation to growth and fruiting of *Litchi chinesis Sonn. Journal of Indian Botanical Society*. 54:280-293.

Pareek, O, P and Gupta, O, P. 1988. Packaging of ber,date palm and phalsa. (In) *Packaging of Fruits and Vegetables in India* (Venkataraman,L.et.al., Eds). Agric. Hort.Soc, Hyderabad.pp.91-103.

Rao D.P., Mukherjee, S.K. and Ray, R.N. 1985. Nutritional Studies in litchi orchards in different districts of West Bengal. *Indian Horticulture* 42:1-7.

Ray, P.K., Sharma, S.B. and Mishra, K.A.. 1984. Important litchi cultivars of Bihar. *Indian Horticulture*. 30:9-13.

Singh H.P. and Yadav, I.S.. 1988. Physico-chemical changes during fruit development in litchi cultivar. *India Journal of Horticulture* 45:212-218.

Singh H.P. and Yadav, I.S.. 1992. Studies on seed germination in litchi cultivars affected by physiological maturity.(*In*) *National Seminar on Recent Developments in Litchi Production* 30-31 May, RAU, Pusa; 16p.

Singh, A.1986. *Fruit Physiology and Production*, Kalyani Publishers, New Delhi.513p

Singh, H.P. 1998. *Genetic Diversity, Breeding and Utilisation of the Genepool of Litchi. Tropical Fruits in Asia* (Arora and Rao, Eds) IPGRI. 171-184.

Singh, R.1992. *Fruits*. National Book Trust, India.231p

Subrahmanyam, K.V.1981. Economics of production and marketing of important fruits and vegetables in Karnataka, Andhra Pradesh and Tamil Nadu. Final Report, UNDP Project, IIHR, Bangalore.

Vir, S. 1986. Horticulture intervention: towards prevention of micronutrient deficiency in India. (*In*) *Strategies for Horticulture Development in India*, the Ministry of Agriculture, Government of India, New Delhi.

Chapter 2

VEGETABLE CROPS

Introduction

India has a long history in the cultivation of vegetable crops and it sustains the livelihood of a large section of population directly or indirectly. Vegetable crops offer high returns, have a potential to alleviate the poverty of people, provide nutritional security and essential raw materials for agro industries. These crops also offer greater rural employment opportunity and have high potential for foreign exchange earning. India bestowed with varied soil and high climatic conditions has excellent potential for growing both tropical and temperate vegetables. India is the second largest producer of vegetables in the world after China. The total area under vegetables is estimated 5.9 million ha which contributes more than 2.5% of total cropped area. The production is estimated 90.83 milion tonnes. China having a production of about 278.59 million tonnes is the largest producer of vegetable crops in the world. The annual growth rate in vegetables is of the order of 2.6%. Potato, brinjal, tomato, chilli, cabbage, onion, cauliflower,okra and peas are grown on commercial scale. West Bengal, Uttar Predesh, Bihar, Orissa and Karnataka are the leading vegetable producing states. Maharashtra, Gujarat, Karnataka, Uttar Pradesh and Andra Pradesh are the leading states in onion production where as West Bengal, Uttar Predesh and Bihar are the leading potato producing states. Bengal, Orissa, Bihar and Assam are the leading brinjal producing states. Area under cabbage has increased rapidly as a result of increased use of F_1 hybrids. The impact of liberalized seed import policy is seen in the production of vegetables, which has increased from 58.5 million tonnes during 1991-92 to 90.83 million tonnes during 1999-2000. However, productivity of vegetables is still lower in India than the world average productivity of about 15.7 t/ha. The average productivity of tomato in India is merely 157.10 q/ha. Productivity in USA is 587.62 q/ha, in Greece 498.6 q/ha, in Italy 465.55q/ha and in Spain 466.29 q/ha. Experimental productivity in vegetables is, however, much higher which means that there is scope for increasing production of vegetables further (Table 1).

Table 1 : Average Yield of Important Vegetable Crops

Crop	Developed countries	Developing countries	World	India (t/ha)	
	(t/ha)	(t/ha)	(t/ha)	On-farm Trials	Coordinated Trial
Beans(green)	15.9	6.7	8.2	2.0	7.5
Cabbage	25.2	17.6	21.6	14.3	36.0
Cauliflower	16.8	10.6	13.6	14.2	26.6
Chilli	15.1	13.9	14.5	6.4	13.6
Cucumber	33.7	18.9	25.0	15.8	30.0
Onion	7.2	3.7	6	10.3	16.5
Peas(green)	18.3	11.6	13.7	11.3	34.0
Tomato	12.9	15.8	14.7	12.7	17.6
Watermelon	7.6	6.2	6.9	2.5	11.5

It is estimated that per capita availability of vegetables is about 175g/day after deducting the quantity exported (about 0.5 million tonnes) and post harvest losses of about 30%. This is very low, compared with recommended dietary allowance of about 280 g/capita/day for a balanced diet. The vegetable requirement of India is estimated 185 million tonnes by 2009-2010 and 220 million tonnes by 2020. There is increasing trend in vegetable production in view of increased urbanization, thus more demand and also more awareness about balanced diet. Main reason for low production is low productivity of vegetables.

Constraints in Production and Marketing

Despite the progress shown in many areas of vegetable production, further development needs to be taken up to tackle many problems which still continue to be there. The productivity and quality of many vegetable crops are much below the potential demonstrated. Yields are much below the world average. Poor management technique and post harvest handling system contribute to the low quality of final produce and high losses.

1. Narrow Adoptability of Improved Varieties

Improved high yielding varieties with wider adaptability, tolerance/ resistance to biotic and abiotic stresses are still a few. Institutes/ Universities have developed promising varieties of tropical vegetables but their effective adoption and further improvement to fit local agro-ecological conditions are often weak.

2. Non-Availability of Quality Seed

Good quality seed in adequate quantities is still lacking. In tomato, cabbage and many other vegetables, hybrid seeds are being supplied by seed companies in private sector, the cost of which is high for small farmers. Seed production of public sector bred hybrids is not yet organized which hampers diffusion of this vital technology.

3. Biotic Stresses

Biotic stresses including pests, diseases and weeds cause havoc in tropical environment and crop failure or devastation from pests and diseases is now a regular phenomenon in India. Integrated Pest Management (IPM) measures need to be adopted for economically and ecologically viable vegetable cultivation. Effective IPM methods are also not developed for all the crops

4. Non-Availability of Cost Effective Technology

Vegetable cultivation is highly labour intensive. Cost of cultivation is thus very high. Further, many times, labourers are not available in time for sowing of seed, transplanting or weeding/harvesting. Mechanization has not started in vegetables except in potato. Appropriate machines for sowing or transplanting are not available.

5. High Post Harvest Losses and Unorganized Marketing

Due to perishable nature of most vegetables, losses are inevitable when produce remains physiologically active even after harvest. Loses vary from 15-30% depending upon the period and type of vegetables. Marketing system does not cater to special needs of handling perishables. Losses start at farm level when adequate precautions in pre-harvest practices are not taken by farmers. There has been no adequate support for market infrastructure development, causing enormous post harvest losses in various stages of handling. Proper facilities for storage, grading, packing and disposal are not available.

6. Lack of Market Support and Market Intelligence

Many times there is price fluctuation due to over production, under production and lack of storage facilities. Farmers thus are not assured of remunerative return for their produce. There are market distortions like high and injust market charges, delayed payment and lack of open auction system. Authenticated data on vegetable area, production, arrivals and prices of vegetables are not available which do not allow scientists, policy makers and farmers to prioritize areas of action.

7. Lack of Credit Support

Banks are, though available in the markets, but their mode of operation is not of use to small and marginal farmers as there is no concessional credit facility to them.

8. Non-Availability of Quality Standards and Short Term Policy

Quality standards are not available in many vegetables. Wherever these are available , they are obsolete. This affects domestic marketing and also export.Since export policy in some of the major export oriented crops is short term, buyers do not have confidence on India. Short term export policy affects farmers and the country as a whole as neither farmers are assured of remunerative price nor the country gets foreign exchange.

9. Low Thrust on Natural Resource Management

The levels of production, efficiency and productivity are dependent mainly on natural resource base and the interaction of farming practices with environment. Many times adverse weather conditions like high temperature, short days, high humidity and heavy rains and water logging conditions adversely affect the crop. Locally adoptable technologies need to be developed with farmers participation for overcoming these problems. Conservation of water and soil should be a top priority.

Tips for Realization of Maximum Productivity

Maximum productivity in vegetable crops can be realized by adopting following techniques.

Obtain seeds of high yielding varieties/ hybrids from reliable sources. Treat the seeds with effective fungicides, if they are not treated. Raise seedlings in poly house/ green house/ shade/ net house. Sow the seeds in treated / sterilized growing medium @500, 600 seeds per sq.meter. Apply plant protection measures as per need of the crop to control insect pests and diseases of nursery. Irrigate the nursery by wateringcan attached with nozzle. Prepare the field by chiseler for easy penetration of roots to a great depth. Transplant seedlings at right age in rows either in flat beds or raised beds as per situation. Always apply balanced doses of organic and inorganic manures and fertilizers. Apply light irrigation after transplanting by mean of drip or wateringcan. Irrigate the crops as per need and avoid heavy flooding. Train the crops as per requirement through suitable devices. Pinch the early flowers and fruit for proper growth of the plant. Spray insecticides and fungicides as per need at right time. Use micro nutrients if there is deficiency. Use plant growth regulators as per need of the crop to get early and high yield. Harvest the fruits at ideal horticultural maturity and at frequent intervals to boost the yield.

BHINDI

Bhindi,(Syn name. okra, lady's finger and gumbo.), *Abhelmoschus esculentus* (L) Moench, is a warm-season fruit vegetable grown in the tropical and subtropical countries. The major bhindi (okra) producing states are Uttar Pradesh, Bihar, West Bengal, Orissa, Assam, Andhra Pradesh and Karnataka. It is an annual erect herb, hairy, with 3 to 5 lobed palmately cordate leaves. The plants bear pyramidal pods. The dehydrated bhindi is a processed product for preservation and export. Bhindi seeds form a nutritious ingredient of cattle feed and are sources of vegetable oil.

Soil and Climate

Crop is adapted to warm and humid climate. It is sensitive to frost and extremely low temperatures. 24°C -30°C is the ideal temperature and higher temperature results in flower drop. It can be grown year round in moderate climate. Usually raised as summer and *kharif* crop and better yield is obtained in *kharif* crop. Can be grown on range of soil from sandy to clay and light, well-drained, rich soils are ideal. Favourable pH is 6.0–6.8 and exhibits some tolerance to salt.

Varieties

Important commercially cultivated varieties in different parts of the country are:

Varieties	Source	Features
Arka Anamika	IIHR, Bangalore	Upright plants, Fruits are dark green with 5 prominent ridges and comparatively less smooth surface. Suitable to South India. Resistant to yellow-vein mosaic, Yield- 12-15 t/ha.
Parbhani Kranti	Marathwada Agricultural University (MAU), Parbhani	Tall, single-stemmed plants with dark green foliage. The fruits are smooth, dark green,slender, 5-ridged with long beak. Yield 9-12 t/ha.
Punjab 7	PAU, Ludhiana	Tall plants with pigmented stem. Fruts are medium-long, green, 5-ridged, slightly furrowed with less pointed beak. Yield 6-10t/ha.
Pusa A4	IARI, New Delhi	Semi tall plants having short internodes (2-4cm). The fruits are 5-ridged,dark green, 12–15cm long having excellent shelf-life. It is resistant to yellow-vein mosaic virus and tolerant to aphids and jassids. Yield 10-18 t/ha.
Varsha Upahar	HAU, Hissar	Dark green fruits. Resistant to YVMV. Yield 12-13 t/ha
CO-3	TNAU, Coimbatore	F_1 Hybrid, Light green fruits, Very long, 7-8-ridged. It is moderately resistant to yellow-vein mosaic virus. Yield16-18 t/ha.
Kashi Vibhuti	IIVR, Varanasi	Attains 60-70 cm height in rainy and 40-50 cm in summer season, bears 2-3 branches with short internodal length. Resistant to YVMV and OLCV. Yield 17-18 t/ha. Released for cultivation in Rajasthan, Gujarat, Haryana and Delhi
Kashi Pragati	IIVR, Varanasi	Tall plants(130-175 cm) with 1-2 branches. Fruits 8-10 cm long. Resistant to YVMV and OLCV. Yield 13-19 t/ha. Released for cultivation in Chhattisgarh, Orisa and Andhra Pradesh.
Kashi Satdhari	IIVR, Varanasi	Tall plants(130-150 cm) with 1-2 branches. Fruits 13-15 cm long with 7 ridges. Resistant to YVMV under field condition. Yield 11-14 t/ha. Suitable for UP, Bihar and Jharkhand
Kashi Bhairav	IIVR, Varanasi	Hybrid, Medium tall with 2-3 branches, Dark green fruits with 10-12 cm . Resistant to YVMV and OLCV under field condition. Yield 20-22 t/ha. Released for entire bhindi growing region of India.
Kashi Mahima	IIVR, Varanasi	Tall plants(130-170 cm). Green fruits, 12-14 cm long . Resistant to YVMV and OLCV. Yield 20-22 t/ha. Suitable for Punjab, UP, Bihar, Jharkhand, Chhattisgarh, Orissa, Andhra Pradesh and Maharashtra.

The F_1 hybrids viz. Vijay, Varsha, Panchali, AROH-221, SOH-77, Karishma, Mahabeej, Anokhi, Ever Green, Biokeerti and NOH-24 developed by private seed companies are also popular among farmers.

Field Preparation and Planting

Bhindi is a direct sown crop. The seed rate of 18–22kg/ha for spring-summer and 8–10kg for *kharif* season crop is optimum. Plant population of 40 000 plants/ha produces near maximum yields, if the plants are well spaced. Soaking the seeds in water for 6-12 hrs before planting or fungicides such as bavistin improves the germination. Seed treatment with Imidocloprid @2.5g/kg seed or trichoderma @ 5g/kg of seed are also beneficial. Seeds are sown in ridges taken at 45 to 60 cm apart maintaining 30-45 cm within row spacing. The soil should be well-prepared and organic matter, P, K and half of N should be thoroughly incorporated.

Nutrient Management

FYM should be applied at the time of field preparation @ 20t/ha. Average nutrient removal from one ha. is around 79 kg N, 32 kg P_2O_5, 89 kg K_2O 15 kg MgO and 29 kg CaO. Fertlizer recommendation depends on soil factors and variety used for cultivation. A fertilizer dose of 150 kg each of N,P and K is recommended for Pusa Sawani as a finding in the AICVIP trials. The P is applied at planting, together with 25 % of the N and 60 % of the K, followed by two applications of N and K_2O approximately 20 and 40 days after seedling emergence. Split application of N at every third picking is beneficial for increasing yield by increasing frequency of harvest and size of pod. Foliar spray of zinc @ 2 mg/l and Molybdenum @ 20 mg/l also increases the fruit yield and appearance.

Aftercare

Irrigation at 4-5 days interval is recommended for summer crop. Furrow and drip system of irrigation are beneficial. Proper weed management in okra could save up to 90% crop losses due to weeds. Application of Fluchloralin (Basalin 48 EC) @ 1.2kg/ha and Pendimethalin (Stomp 30 EC) @ 0.75kg/ha as post sowing and pre-emergence soil surface spray gives initial control of dicot weeds. One hand weeding may be needed in *kharif* crop in addition to application of weedicides.

Harvesting

Pod should be harvested at tender stage. They are ready for harvest in about 45-60 days after seed sowing depending upon variety and season. For processing industry and fresh fruit export, 6–8cm long fruits are sorted out. Export market requires tender, dark green, straight, short (6–8cm) fruits which readily snap at tip. Longer fruits are used for fresh market. Harvesting on every alternate day is advisable. Harvesting in the morning is convenient. For distant markets, harvesting late during evening and transporting during night is also advised. Pods should be handled with care; rubber gloves should be used during harvesting and handling, Care should be taken during harvesting so as to have approximately 1 cm of stem should remain attached to the pod.

Grading, Packaging and Storage

After harvesting fruits, they are stored under shade. Pods showing discolouration, bruising, chemical residue or insect damage are removed during grading procedures. Fruits may be graded by hand according to size. All pods meeting the size specifications can be loose packed into cartons. Harvested fruits are packed in baskets and covered with gunny bags before transporting to local markets. For exports, the fruits are loose packed in well ventilated 2.5-5 kg fiber board cartons.

Shelf life of Okra can be extended upto 8-10 days by storing them at 7-10°C with 95% relative humidity. It is chilling sensitive and exposure to temperatures below 7°C for a short time results in discolouration, pitting and decay. Similarly temperatures above 10°C cause yellowing, toughening and rapid decay. Storage with other fruits and vegetables which produces ethylene should be avoided. Controlled Atmosphere (CA) storage is also recommended for maintaining green color and to reduce decay of the damaged pods. The CO_2 concentration of 4-10% is recommended along with optimum storage temperatures.

Bulb Crops

Onion

Onion (*Allium cepa*) is a popular vegetable grown for its pungent bulbs and flavorful leaves. It is widely grown throughout the world. The area, production, productivity and per capita availability of onion in the country have improved during the past decade by about 50, 11, 239 and 66%, respectively. India now ranks first in area under onion (4.21 lakh hectares) and second in production (5.97 million tonnes) after China and third after the Netherlands and Spain in exports. The development of varieties suited to *kharif* season has brought about the revolution in onion production in the country. In India, it is grown in states of Maharashtra, Gujarat, Uttar Pradesh, Orissa, Karnataka, Tamil Nadu, Madhya Pradesh, Andhra Pradesh and Bihar.

Soil and Climate

Onion grows well under mild climate. The best performance can be obtained in a mild weather without extremes of cold and heat and excessive rainfall. The optimum temperature is 13 to 24°C before bulbing and 16 to 21°C during bulbing. Low temperature in the early stages favours bolting, whereas a sudden rise in temperature favours early maturity in *rabi,* resulting in small-sized bulbs. The bulb onion cultivars are grouped into short, intermediate, and long-day types based on requirement of photoperiod. Short-day onions (12 to 13 hour threshold) are generally mild, soft fleshed, and suitable for storage. Long-day onions (over 14.5-hours threshold), if grown in the lower latitudes, will not form bulbs, and only green onions would be produced. In contrast, short-day types grown in the higher latitudes will bulb very quickly and will be little more than sets in size. Most of the improved cultivars in India are short-day types; long-day cultivars include yellow, white and red globes. The intermediate-day cultivars (13.5 to 14.5-hour threshold), are relatively soft-fleshed and used primarily for fresh trade. They are grown in areas of mild temperatures lying between 32 and 38° latitudes.

Onions can be grown successfully on any fertile, well-drained and non-crusting soil. The optimum pH range, regardless of soil type, is 6.0 to 6.8. Onions do not thrive in soils below pH 5.8 because of trace element deficiencies, or occasionally, aluminum or manganese toxicity. Sandy soil needs more and frequent irrigation and favours early maturity, whereas heavy soils give rise to mis-shapen bulbs causing problem in digging of bulbs. Salt concentration above 4 mmhos/cm^2 inhibits vegetative growth of most of its cultivars.

Varieties

A number of varieties have been developed in India. Onion varieties can be broadly classified into 2 groups—common onion and multiplier onion. Common onion can be further grouped into 3 sub-groups based on colour of skin—red, yellow and white.

Onion Varieties Developed in India

Name	Source	Characteristic Features
Pusa Red	IARI	Red coloured bulbs, flattish round shape, yield 25-30 t/ha. Very good in storage, wider adaptability.
Pusa Ratnar	IARI	Deep red, round, large, fair in storage, yields 30-40t/ha
Pusa Madhvi	IARI	Light red colour, good keeping quality, yields 30-35 t/ha
Punjab Selection	PAU Ludhiana	Light red, yield 20t/ha
N-53	MPKV, Rahuri	Deep red coloured, yield 15-20 t/ha, suitable under kharif season
N-2-4-1	MPKV, Rahuri	Light red, yield 25-30t/ha, good in storage
Baswant 780	MPKV Rahuri	Bright red, yield 20-25 t/ha suitable for *kharif* season
Arka Niketan	IIHR	Light red, yield 33t/ha, good for storage
Arka Kalyan	IIHR	Dark red, yield 33t/ha suitable for *kharif* season
Arka Bindu	IIHR	Bright dark red, ready in 100 days, yield 25t/ha, suitable for export, particularly to Malaysia and Singapore.
Agrifoud light red	NHRDF	Light red, yield 30 t/ha, good in storage
Agrifound Dark red	NHRDF	Dark red, yield 20-25 t/ha suitable for growing under *kharif* season
Agrifound Rose	NHRDF	Dark red, yield 19-20t/ha, suitable for export
Udaipur 101	Udaipur University	Dark red, yield 20-30t/ha
Udaipur 103	Udaipur University	Dark red, yield 25-30 t/ha
Punjab Red Round	PAU Ludhiana	Red coloured, yield 30t/ha
Hisar-II	HAU Hissar	Light red, yield 20 t/ha
Kalyanpur Red Round	CSAUA&T Kanpur	Light red, yield 20 t/ha
White varieties		
Pusa White Flat	IARI	Flattish round shape, TSS 11-12%, drying ratio 9:1, yield 33-35t/ha. Suitable for dehydration purposes-Very good for green onion purposes and good in storage.
Pusa White Round	IARI	Roundish flat shape, TSS 12-13%, drying ratio 8:1, yield 30-33 t/ha, very good in storage. Suitable for dehydration purposes and green onion.

Contd...

Punjab-48(S-48)	PAU Ludhiana	Flattish round, yield 30t/ha, suitable for dehydration.
Udaipur 102	Udaipur University	Roundish flat, TSS 12%, yield 30-35t/ha suitable for dehydration.
N-257-9-1	Niphad	Roundish flat shape, yield 25-30 t/ha, good in storage. Suitable for dehydration.
Yellow varieties		
Early Grano	IARI	Large globular bulbs, good for salad purposes, yield 50-60 t/ha, good for green onions, poor in storage.
Brown Spanish	IARI Regional Stn Katrain	Long day type, can only be grown in hills, maturity 160-180 days, yield 28t/ha, keeping quality 6 months
Multiplier onion		
CO-1	TNAU Coimbatore	Light red, medium sized, 7-9 lateral bulbs, maturity 65 days, yield 12 t/ha
CO-2	TNAU Coimbatore	Light red, 7-9 lateral bulbs,60-65gram clump, maturity 65 days, yield 12 t/ha, showed resistance to thrips and purple blotch
CO-3	TNAU Coimbatore	Pink coloured, 8-10 lateral bulbs, 75 gram clump, bigger sized bulbs, yield 16 t/ha, maturity 65 days, liked by consumers,has medium resistance to trips.
CO-4	TNAU Coimbatore	8-10 lateral bulbs,90 gram clump, maturity 60-65 days, yield 18 t/ha
MDU-1	TNAU	10-11 lateral bulbs, 75 gram clump, yield 15 tonnes/ha, maturity 60-65 days, good keeping quality and resistant to lodging.

The F1 hybrids released from IIHR Bangalore, viz Arka Kirthiman and Arka Lalima are tolerant to purple blotch, basal rot disease and thrips. 'Punjab Selection' and 'N 2-4-1' were good for dehydration. 'Pusa Red' and 'Punjab Red Round' were the best suited for storage at room temperature (up to 5 months). The exotic varieties are White Portugal, Silver Skin, Australian Brown Sweet Spanish, Red Italian, California Early Red, Yellow Bermuda, Ebenezer and Mountain Danvers. The white skinned varieties are mild and good flavoured as compared to the red varieties that are relatively more pungent but keep better owing to presence of catechol and protocatechuic acid in the skin.

Planting

Three systems of planting are employed:

1. Direct seedling is preferred and gives excellent results where the season is sufficiently long to provide early prebulbing growth.

2. Transplanting is commonly followed during *rabi crop* and transplants normally have three to five well-formed leaves at transplant time. Transplant leaves are pruned during growth prior to field setting, facilitating handling and increasing plant hardiness.
3. Sets are used in some areas to ensure large bulb size and uniform maturity. Sets are small dry bulbs, approximately 12 mm in diameter, produced during the previous season by seeding thickly or growing under conditions favouring rapid bulbing. Planting sets are preferred for production of *kharif* onions

Nursery Management

About 10–12kg seed in *rabi* and 12–15kg seed in *kharif* are required to raise seedlings for planting in one hectare. Best time for sowing early-*kharif* crop is February-April in South, May-June in Maharashtra and other parts, August-September for late-*kharif* and October-November for *rabi* in plains all over the country. In hills (higher altitudes), March-April is an ideal sowing time. Nursery bed of 3 m × 0.6 m size may be raised up to 15–25 cm with a distance of 70 cm between the beds to facilitate intercultural operations. The nursery beds should be irrigated 15-20 days before sowing and covered with 250 gauge transparent polythene for soil solarization. Application of *Trichoderma viride* @ 1,250 g / ha is also recommended to manage damping off and raise healthy seedlings. The seeds can be pelleted with Azospirillum @ 100 g/kg. of seeds, sown in lines at a depth of 1-2 cm and covered with top soil. These beds are then covered with straw or dried grass and watered by rose can. The seeds germinate in 7-8 days. Then the straw mulch is removed and the seedlings are irrigated. The seedlings are ready for transplanting 6–7 weeks and 8–9 weeks after sowing during *kharif* and *rabi* respectively. Seedlings 20–25cm in height are ready for transplanting. In multiplier onion, bulbs are first separated and then bigger-sized bulbs are used for planting.

Direct Sowing

Onions can also be grown by direct seeding. Seeds are sown 6 to 18 mm deep in heavy mineral soils, deeper in light mineral soils and mucks. Using coated seed and precision seeding, the seeding rate can be adjusted easily for projected bulb size. For normal storage onions, seeds are spaced 7.5 cm apart. When small boiling, pickling or pearl onions are desired, spacing would be reduced to 2.5 cm in the row. Large bulb size is promoted by spacings of 10 cm or more.

Field Preparation, Transplanting and Fertilizer Application

The field is ploughed 4-5 times and 25 tonnes of FYM are applied at the last ploughing. Then beds and channels of convenient sizes or ridges and furrows are formed at a spacing of 30 cm. A row to row and plant to plant spacing of 15 cm x 10 cm are recommended for Pusa Red variety when grown in Jabalpur. Sets can be planted on ridges spaced at 30 cm for *Kharif onion.* For an yield of 30 tonnes of onion/ha, crop removes 85 kg of N, 42 kg of P_2O5 and 130 Kg of K_2O. Dosage of fertilizer varries with region, soil type, season and variety. AICVIP has recommended application of 150 kg N, 60 kg P_2O5 and 60 Kg K_2O for Pusa Red during *rabi* season. Just before transplanting, half dose of N, full doze of P and half dose of K are applied on both sides of ridges. The unbranched root system is very inefficient in P uptake unless the root is mycorrhizal. Acid soils should be limed, since the crop does not tolerate acidity well. Mn deficiency may occur in alkaline soils. All fertilizers should be applied close to surface to be within reach of the shallow root system. On 30th day after transplanting, remaining quantity of N and K are applied as top dressing. For *Kharif* onion, NHRDF recommended 150 kg N, 40 kg P_2O_5 and 50 kg K_2O/ ha for bulb production of N 53 onion at Karnal.

Weed Management

Onions are not good competitors with weeds. Manual weeding is costly and an integrated approach with a combination of cultural and chemical method is more beneficial. One hand weeding in the initial stage followed by application of Baseline @ 2 kg a.i/ha or Stomp @ 3.5 litres/ha immediately after transplanting are quite effective to control weeds.

Irrigation

Water requirement of onion is less in initial stage and further irrigation frequency depends on growth stage, soil and climate factors. Fields that suffer growth retardation produce excessive numbers of doubles or splits, reducing the number of Grade 1 bulbs. Furrow irrigation is generally used. Light sandy soils can also be irrigated with overhead systems. Onions at bulbing stage utilizing substantial amounts of water there should not have any moisture stress during this period. Frequent light irrigation is preferred due to shallow root spread. Watering should be stopped prior to harvesting.

Harvesting, Curing and Storage

The harvesting stage depends on purpose for which it is grown. Green onions are harvested 45-80 days after planting, while bulb onions are ready in 125-150 days after planting. Bulb onions are ready for harvest when the leaves collapse. Bulb maturity can be accelerated by withholding irrigation or by pruning the root system. Field-dry maturity is indicated when bulb neck is completely dry to the touch and tops are fallen over. Best time to harvest rabi onion is one week after 50% tops have fallen over. In *kharif* season, since tops do not fall, bulbs are harvested soon after the colour of leaves change to light yellow and when bulbs develop red pigmentation on them.

Since onion bulbs are normally formed at the soil surface, it is sometimes possible to pull the mature bulbs by hand. Usually it is harvested by loosening the bulbs with a fork or hoe before lifting them. The harvested crop is left in the field for a few days until the tops are dry. The leaves are cut leaving about 2-2.5 cm tops above the bulb after complete drying After harvesting, crop should be left in the field for a few days for drying of leaves. Common onion varieties yield up to 25.0–30.0 t/ha and multiplier onions 15.0–18.0 t/ha.

Harvested bulb is subjected to curing for the development of natural dormancy and to dry the onion sufficiently. Curing is a drying process intended to dry off the necks and outer scale leaves of the bulbs to prevent the loss of moisture and the attack by decay during storage. The essentials for curing are heat and good ventilation, preferably with low humidity. Onions are considered cured when neck is tight and the outer scales are dried enough to make rustling sound. If onions cannot be dried in the field, they can be collected in trays, which are then stacked in a warm, covered area with good ventilation. In cool, damp climates, onions in bulk-ventilated stores are dried with artificial heat blown through the bulk at a temperature of 30°C. A properly cured onion will have a dry shrunken neck and dry outer scales. The respiration rate of a cured bulb is lower than that of an uncured bulb. Fully mature bulbs are harvested and artificially cured by exposure to temperature up to 35°C in low (less than 50%) relative humidity. Air movement must be provided at the rate of 1 cubic foot/minute/cubic foot of onions (60 cubic meters/hour/cubic meters of onions). Immature onions require twice the rate of air exchange. Following curing, the temperature of stored onions is lowered gradually to 0°C, or slightly higher, with the relative humidity at 60 to 70 percent. Air exchange in the storage facility is important to prevent any condensation on the bulbs. Also, when the bulbs are removed from the storage, they should be conditioned for several days at 20 and 50% relative humidity.

Grading

Onions after curing are graded manually before they go in to storage or for marketing.. It has been observed that if onions are stored after proper sorting and grading, losses in storage are reduced. For local market, the onions are graded based on their size. Extra large onion (>6 cm dia.), Medium (4-6 cm dia.) and Small (2-4 cm dia.).

Packaging

Onions are packed in jute (hessian) bags for transporting to yard . For safe handling, 40 kg open mesh jute bags should be used in domestic market. For export, onions are packed in 5-25 kg size open mesh jute bags. Bangalore Rose and multiplier onions are packed for export in 14-15 kg wooden baskets. Nylon net bags, when used for packing resulted in less storage loss because of good ventilation.

Storage

Onions should not be stored unless adequately dried either in the field or by artificial means. It is necessary to dry the neck tissue and outer scales until they rustle when handled, otherwise the bulbs will rot in storage. Sprouting in onion is controlled by temperature. The temperature between 10-25°C increases sprouting. Rooting is influenced by relative humidity .Under ambient conditions, the onions are stored at a temperature of 30-35°C with Relative Humidity of 65-70%. In cold storage, temperature is maintained at 0-2°C while the Relative Humidity is kept at 60-75% with adequate air circulation. CA storage is not recommended. Onions are damaged by < 1% O_2 and 10% CO_2.

Freshly harvested onions are dormant and will not sprout for a variable period of time depending on cultivar. Storage will prolong this dormancy. Sprouting will increase in storage temperatures above 4.4°C. To reduce the frequency of sprouting after the rest period, onions may be field treated with maleic hydrazide (MH-30) at 2.2 to 3.4 kg/ha when the tops are still green but beginning to senescence. Cultivars intended for long-term storage should be firm with a thick dry neck; free from greening, root growth, sunburn, or freeze damage; and well covered with dry scales. Bulbs with fleshy and soft necks are susceptible to persistent rot, especially, if storage humidity exceeds 70%.

Use of irradiation improves shelf-life of onions. Exposure of onion bulbs after harvesting when bulbs are in dormant stage with 60-90 Gy inhibit their sprouting regardless of crop season, environmental condition and type of storage. However, to reduce the microbial and other losses, combined use of irradiation with improved storage and providing the irradiation facilities at production level are to be considered.

In India, the farmers practice different storage methods. The onions are bulk stored in special houses with thatched roof and side walls are made up with bamboo sticks or wire mesh for good air circulation. In North India, the sides are also covered with gunny cloth. Onions are stored in these sheds by spreading them on dry and damp proof floor or racks. Periodical turning of bulbs or removal of rotten, damaged and sprouted bulbs should be done. Well-ventilated improved storage structures with racks or tiers having two or three layers of bulbs would be desirable for proper storage.

Marketing and Export

Nasik is the biggest onion market of the country which normally controls and regulates the onion price and supply. In private trade for domestic consumption, producers sell onions to commission agents in regulated markets which are now in operation extensively throughout the country. The commission agents sell the onion to sub-wholesalers or directly to retailers, who in turn, sell to the consumers. A second channel is the cooperative channel in which producers market their onion through cooperative marketing societies at the village level. NAFED and the other canalizing agencies procure their requirements from cooperative societies and rarely buy from commission agents or wholesalers in markets. This benefits the small producers as they collectively receive a better price for their product

Grading and classification are necessary for both local market and export. Delhi market prefers big-sized bulbs; Calcutta, Patna and Lucknow prefer medium-sized bulbs and Bangalore, Bhopal, Jabalpur and Hyderabad markets have demand for medium to big-sized bulbs. Markets of Bhubaneswar, Guwahati and other North-Eastern states prefer small-sized onions. Jute bags are used for onion packaging for sending to distant markets by trucks, trains or even by air. Generally 40kg capacity jute bags are used for transport within the country, whereas for export, jute bags of 8–25kg capacity are used. Onions should be packed in 14–15kg capacity cane baskets for export purpose.

Exports of onions from India are not free but are permitted only through certain designated canalizing agencies. NAFED is responsible for fixing the minimum export price (MEP) of onions, which is done on a monthly basis. The Price Fixation Committee of NAFED decides this price. Factors such as market trends, world prices and domestic prices, and margins are considered for arriving at the MEP of onion. Exports are concentrated to neighboring countries in South and South East Asia and the Middle East. The quality standards of onion are fixed by Agmark, although now it is not mandatory to obtain Agmark certificate for onion export. It is necessary to maintain the quality by observing the standards. The grades, big, medium, small and mixed are followed for different types of onions which depend on requirements from importing country. In general, Middle East countries demand light red to dark red colour, European countries and Japan demand yellowish /brown colour onions having mild pungency, 3-4 cm sized onions are preferred in Bangladesh, 4-6 cm sized are preferred in Middle East and Far East, while European and Japan prefer 6-7 cm sized onions. Rose onions of 2.5 - 3.5 cm size and multiplier onions of bigger size with attractive red colour bulblets are preferred for export. European countries and Japan prefer yellow big-sized onions. The demand in these countries is from February and up to May, when it is very easily possible to cultivate onion in Nasik area of Maharashtra.

Economic Analysis

Cost analysis of Onion cultivation in Karnataka is given below.

Costs and returns structure of Onion production/ha

S No	Particulars	Belgaum	Dharwad	Bijapur	Raichur	Gulbarga	Overall
A	*Variable costs*						
	1. Seeds	1332.65 (7.95)	16858.65 (9.35)	2352.00 (7.99)	2567.45 (8.23)	1991.99 (7.50)	1985.95 (8.14)
	2. Farm yard Manure	2277.79 (13.59)	2735.81 (15.18)	3366.90 (11.43)	3276.33 (10.50)	1319.50 (4.97)	2595.27 (10.64)
	3. Fertilizers	2807.55 (16.75)	2526.03 (14.01)	4003.15 (13.59)	4486.75 (14.38)	1908.07 (7.18)	3146.31 (12.89)
	4. P.P Chemicals	251.22 (1.50)	210.00 (1.16)	418.71 (1.42)	690.93 (2.21)	130.12 (0.49)	340.20 (1.39)
	5. Human labour	5103.35 (30.44)	5828.40 (32.33)	10970.40 (37.26)	12736.00 (40.81)	11619 (43.75)	9251.47 (37.91)
	6. Bullock labour	1960.00 (11.69)	1960.00 (1087)	636.00 (5.56)	1084.50 (3.47)	1680.50 (6.33)	1664.20 (6.82)
	7. Irrigation	-	-	2525.47 (8.58)	2116.27 (6.7)	2546.13 (9.59)	2395.96 (9.82)
	8. Repairs and maintenance	65.77 (0.39)	60.99 (0.34)	203.04 (0.69)	179.95 (0.58)	187.22 (0.70)	139.39 (0.57)
	9. Interest on variable cost	965.58 (5.76)	1783.3 (6.06)	1783.30 (6.06)	1899.67 (6.09)	2993.64 (11.27)	1738.59 (7.12)
	Sub-total	14764.21 (88.06)	16057.36 (89.08)	27259.01 (92.58)	29037.85 (93.05)	24376.77 (91.79)	22299.04 (91.39)

Contd....

B	*Fixed costs*						
	1. Depreciation	55.83 (0.33)	76.05 (0.42)	223.87 (0.76)	211.39 (0.68)	219.33 (0.82)	157.29 (0.64)
	2. Rental value of land	1750.00 (10.44)	1700.00 (9.43)	1750.00 (5.94)	1750.00 (5.61)	1750.00 (6.59)	1740.00 (7.13)
	3. Land revenue	13.02 (0.08)	13.02 (0.07)	13.02 (0.04)	13.02 (0.04)	13.02 (0.05)	13.02 (0.05)
	4. Interest on fixed cost	181.88 (1.08)	178.91 (0.99)	198.69 (0.67)	197.44 (0.63)	198.23 (0.75)	191.03 (0.78)
	Sub-total	2000.73 (11.94)	967.98 (10.92)	2185.58 (7.42)	2171.85 (6.95)	2180.58 (8.21)	2101.34 (8.61)
C	*Total cost of cultivation*	16764.94 (100.00)	18025.34 (100.00)	29444.59 (100.00)	31207.70 (100.00)	26557.35 (100.00)	24400.38 (100.00)
D	*Returns*						
	Yield (q)	99.04	116.72	211.80	196.62	205.76	165.99
	Rate(Rs/q)	336.67	350.33	458.73	486.67	471.00	420.68
	Gross returns	33343.80	40890.58	97159.00	95689.05	96912.96	69828.67
	Net returns	**16578.86**	**22365.18**	**67714.41**	**64421.35**	**70355.61**	**45429.29**

Figures in parenthesis indicate percentage to total cost of cultivation
Source : Balappa and Hugas, 2003

Garlic

Garlic (*Allium sativum* L) is an important bulb crop grown and used throughout India primarily as a spice or condiment. It is mainly grown in India in Gujarat, Madhya Pradesh, Orissa, Maharashtra, Uttar Pradesh and Rajasthan. Small quantities of very good quality garlic are also produced in Ooty and Kodaikanal in Tamil Nadu but almost the entire produce is consumed locally. Recently a few farmers in Punjab and Haryana have also started cultivating garlic and the yield recorded in these states have been far higher than the yields in the traditional growing areas.

Climate and Soil

Garlic is a frost-hardy plant requiring cool and moist period during growth and relatively dry period during bulb maturity. Bulbing takes place during longer days. Exposure to low temperature subsequent to bulb formation favours the process. There are 2 types of varieties, one suitable for long-day conditions (around 30° N latitude) and the other for short-day conditions. In India, mostly short-day types are grown. Garlic can be grown on a variety of soils, but thrives better on fertile, well-drained loamy soils. The soils having pH of 6–7 are suitable for good crop. Highly alkaline and saline soils are not suitable for garlic cultivation.

Varieties

Name	Source	Features
Agrifound White	NHRDF	The bulbs are compact, silvery white with creamy flesh. Bigger elongated cloves with 20-25 in numbers. Diameter 3.5 to 4.5 cm. Yield 13.0 t/ha
Yamuna Safed	NHRDF	The bulbs are compact, silvery white skin with creamy flesh. Diameter 4.0 to 4.5 cm. Sickle shaped cloves, 25-30 in number. Yield 15.0-17.5 t/ha
Yamuna Safed-2	NHRDF	The bulbs are compact attractive white creamy flesh, bulb diameter 3.5-4.0 cm, Yield 15.0-20.0 t/ha. The variety is recommended for Northern India
Yamuna Safed-3	NHRDF	Bulbs are creamy white and bigger sized (5-6cm dia). Average yield 175-200 q/ha. The variety is suitable for export.

G 282 develops good bulbs in Southern hills and Northern plains. Agrifound Parvati is a long day type and can be grown only in northern hills.

Planting

Garlic is propagated by single clove. Cloves of 8-10 mm diameter give increased yield of better quality. Care should be taken to select bigger cloves from outer side of bulbs. About 500 kg cloves of 8-10 mm diameter are required to plant one hectare. Cloves can be planted through dibbling in small beds, or furrows or can be broadcasted. It is planted from August to October in Madhya Pradesh, Maharashtra, Karnataka and Andhra Pradesh, and from September to November in Northern plains of India. The planting distance varies from 10-15 cm x 8-10 cm and wider spacing results in production of bigger cloves.

Nutrient Management

Garlic responds very well to organic manures. For a normal soil, 15 tonnes of farmyard manure, 100 kg N, 50 kg P and 50 kg K/ha through chemical fertilizer is recommended. Entire P and K along with half of N should be applied as basal dose. Remaining N should be top dressed at one month after planting. Micronutrients also increase its yield potential.

Irrigation

In general, garlic needs irrigation at 8 days intervals during vegetative growth and 10-15 days during maturation. As the crop matures (when the tops first begin to break over or become dry), stop irrigation to allow field to dry out. Continued irrigation during crop maturation phase causes roots and bulb scales to rot. This discolours bulbs and exposes outer cloves and decreases market value of bulbs. Irrigation after long spell of drought results in splitting of bulbs. Excessive irrigation results in sprouting.

Weed Management

Manual weeding is tedious and expensive and often damages the plants. First weeding is done one month after planting and second after one month. Hoeing the crop just before formation of bulbs (about two-and-a-half months after sowing) helps in setting of bigger sized well filled bulbs. Pendimethalin @ 3.5 litres or Goal @ 0.25 kg/ha and one hand weeding gives good control of broad leave weeds.

Harvesting, Curing, Storage and Marketting

Crop should be harvested when its tops turn yellowish or brownish and shows signs of falling. Bulbs mature 130–180 days after planting depending on cultivar, soil and season. In India, harvesting is done manually by hand digger and bulbs are taken out along with tops and wind rowed for curing. The yields of bulbs vary from 10.0 to 20.0 t/ha depending upon cultivar and region and high yield is obtained in regions enjoying mild climate and long day length.

Garlic needs to be dried and properly cured before storage. Drying is done to remove excess moisture from outer skin and neck to reduce storage rot, while curing is an additional process of drying to remove excess moisture and to allow the colour development and help the bulbs to become compact and go into dormant stage. It is done for about a week in the field for drying. Bulbs are covered along with their tops to avoid damage to bulbs from sun. These are also cured for 7-10 days in shade either with tops or after curing the tops by leaving 2.5 cm above the bulbs and removing the roots. Artificial curing can be done by passing hot air at 27-35°C through curing room. It takes about 48 hours for complete curing process, if humidity is between 60 - 75%.

After curing, the bulbs are run over a grader or graded manually. The thick-necked, splitted, injured, diseased or bulbs with hollow cloves are sorted out. Size grading is done after sorting depending upon demand of market. For local markets, bulbs are graded based on size as follows-

1. Big (20-30 mm dia.)
2. Medium (15-25 mm dia.)
3. Small (10-120 mm dia.)

Government of India prescribed certain grade designations for different qualities of garlic for export. In India, garlic bulbs are packed in open mesh jute bags for domestic use. It is packed in bags of 90 and 40 kg capacity each in Andhra Pradesh, Karnataka and other garlic growing states. As per the garlic grading and packing rules, 18 and 25 kg packing are done in perforated 10 ply corrugated cardboard boxes for export.

Garlic can be kept in good condition for 1-2 months at ambient temperatures (20°-30°C) under low relative humidity (60 to 70 %). However under these conditions, bulbs will eventually become soft, spongy and shriveled due to water loss. For long-term storage, garlic is the best maintained at temperatures of -1°C to 0°C with low relative humidity (60-70%). Good airflow is also necessary to prevent any moisture accumulation. Under these conditions, garlic can be stored for more than 9 months. Garlic odour is easily transferred to other products and should be stored separately. High humidity in the storages and improper curing will favour mould growth and rooting. CA storage with high CO_2 (5-15%) is beneficial in retarding sprout development and decay during storage at 0-5°C. UV light treatment for 30 minutes further reduces loss to 8% in cold stores for 150 days storage. Irradiation with 2-6 krad of cobalt 60 gamma rays is also recommended for controlling sprouting in storage. Storage life of garlic is prolonged and loss in weight is also reduced by spraying MH (3000 ppm) 3 weeks before harvesting. Pre-harvest spraying of carbendazim (0.1%) and disinfection of premises for handling and storage of garlic also reduce post-harvest losses.

Cole Crops

Cole crops are essentially Mediterranean vegetables and highly popular across the globe. Among the cole crops cauliflower (*Brassica oleracea* var. *botrytis*) and cabbage (*B. oleracea* var. *capitata*) are extensively grown in India while Brussels sprouts (*B. oleracea* var. *gemmifera*), Sprouting broccoli (*B. oleracea* var.*italica*) and Knolkhol (*B. oleracea* var. *gongylodes*) are grown in limited pockets. Though grown as cool season vegetables, with the advent of tropical types which are more tolerant to temperature and humidity, cultivation has spread to nontraditional areas. The major cole crops producing states are Uttar Pradesh, Orissa, Bihar, Assam, West Bengal, Maharashtra and Karnataka. Cabbage is a small, leafy biennial producing a compact globular mass of smooth or crinkled leaves wrapped over each other known as head. The outer leaves are generally larger than the inner. The stem is short and stout. Plants flower generally after winter. Cauliflower is a low growing herbaceous plant, with loosely arranged leaves and bearing terminal heads or curds yellow to creamy white in colour which are the prefloral fleshy apical meristem. The stem is short and stout. Cole crops in general have similar cultural requirement.

Soil and Climate

Among the cole crops, cauliflower is the most temperature sensitive vegetable. Vegetative, curding and reproductive phases of cauliflower are influenced by temperature. There are tropical and temperate cauliflower which widely differ in temperature requirement for curd formation and development. Temperature requirement varies from 5°- 28°C. Successful curding in cauliflower requires a temperature of 12- 22 °C. Cabbage withstands extreme cold and frost better than cauliflower. The optimum temperature for heading in cabbage is between 15-20 °C. Heading broccoli prefers 15°-20°C for heading and as in cabbage, the heads become loose with rise in temperature.

Cole crops can be grown on a variety of soils from sandy loam to clayey loam soils which are well-drained and rich in organic matter and nutrients. The optimum soil pH is between 6 to 7. Cole crops particularly cauliflower are highly sensitive to deficiency of boron and molybdenum. Acidic soil reduces availability of Mo while at higher pH reduces the availability of boron.

Varieties

Cauliflower

Cauliflower cultivars grown in India can be classified in to two broad groups, *viz.,* Indian cauliflower or tropical cauliflower and early temperate type known as snow ball groups or late or European types.

Variety	Source	Features
Kashi Kunwari	IIVR, Varanasi	Early season variety, tolerates high ranfall, Semi dome white compact curd weighing 300-450 g. Total yield 30-35 t/ha.
Pant Gobhi 3	GBPUA &T	Early season variety (September maturity), tolerant to a mean temperature of 27 °C for curd inititation and development, can be sown during June in North India

Contd...

Pusa Deepali	IARI, New Delhi	Early season variety (October maturity), tolerant to a mean temperature of 25 °C for curd inititation and development, can be sown during June in North India erect tall plants, medium size compact white curds which are almost free from riceyness.
Pusa Hybrid 2	IARI, New Delhi	F_1 hybrid, Mid-early, tolerant to a mean temperature of 20 °C for curd inititation and development, good compact curd, can be sown during July- Aug. in North India
Pusa Shubhra	IARI, New Delhi	Mid- late variety, tolerant to a mean temperature of 16 °C for curd inititation and development, compact creamy white curd. can be sown during Aug- Sept. in North India
Pusa Snowball-1	IARI, Katrain	Late variety, tolerant to a mean temperature of 10-16 °C for curd inititation and development, Big white curd. can be sown during Sept-Oct in North India
Pusa Snowball-K1	IARI, Katrain	Late variety, very popular due to its high yield, curd quality maintains even if harvesting delayed. It is resistant to black rot.

Cabbage

Variety	Source	Features
Pusa Muktha	IARI, Katrain	Heads are medium sized, 1.5-2.0 kg, slightly flattish round, compact, resistant to black rot, Yield 20 t/ha
Pusa Drumhead	IARI, Katrain	Flat head, solid, weighing 3.0–4kg each. It is tolerant to black-leg or dry-rot.Yield 30 t/ha
Pusa Agethi	IARI, New Delhi	Early tropical variety, Forms head at 15-30 °C, medium sized head, tolerant to high temperature, flat round compact head 600-1200g, Yield 33 t/ha
Sri Ganesh Gole	IAHS, Bangalore	F_1 hybrid , round and compact head weighing 2-2.5 kg. It stands up to 10-15 days in the field, Yield- 75-87.5 t/ha.

Contd...

Quisto	Sandoz	F_1 hybrid , tolerant to high temperature. Compact round, dark green colour head.. Head can stand in the field conditions up to 70 days after maturity.

Brussels Sprout

Variety	Source	Features
Jade cross	NSC	An early, short-stemmed and high-yielding Japanese F1 hybrid, suitable for a single harvest.
Hilds Ideal	IARI, Katrain	Medium-tall plants (55–60cm) and bear 45–50 sprouts weighing 7–8g each. Sprouts are green and compact with good flavour. About 250–400 sprouts are obtained from a single plant. Yield - 10.0–15.0 t/ha

Knol-khol

Variety	Source	Features
White Vienna	NSC	Early variety with globular, light green, medium-sized knobs with creamy-white tender flesh having delicate flavour. Its plants are dwarf, leaves and stems are medium green. Matures 55–65 days after transplanting Yield - 15.0–20.0 t/ha.
Purple Vienna	NSC	Bears purplish blue knobs with greenish-white flesh. It has purple leaves. It requires 55–65 days for knob formation. Yield-17-20 t/ha.
Large Green	NSC	Bears green, round, large-sized knobs with delicately flavoured with white flesh. It is ready for harvesting in 76 days . Yield - 22.5–25.0 t/ha.

Turnip

Variety	Source	Features
Pusa Kanchan	IARI	Round, red colored root with creamy-yellow flesh. Short top. Roots can be kept

Contd...

		for a longer time in field without becoming spongy.
Pusa Swarnima	IARI	Flattish round roots with creamy-yellow skin and pale-amber coloured flesh. Medium top. It is suitable for growing from June to October in hills and October to December in plains. It matures in 65–70 days.
Pusa Sweti	IARI	Early maturing variety, Attractive, white roots, mature 45–50 days after sowing. Suited for October sowing in plains.

Nursery

Cole crops are generally raised as transplanted crop and nursery management is important for raising good quality seedling. The nursery beds should be well prepared by addition of well rotten FYM/vermicompost @ 10 kg/m^2. Prepare bed of 3 m long, one-meter width and 15 cm height and 20-25 beds will be required for transplanting one ha main field. Around 400-500 g seed is required/ha, seeds should be treated with Captan or Thiram (3g/kg) or Bavistin (2g/kg) to get rid of fungal diseases. Sow the seeds in shallow furrows at 10 cm apart and 1.0-1.5 cm deep. It is advisable to cover the bed with mulching materials like dried grasses or polyethylene. The mulch should be removed as soon as the seedlings start emerging. The drenching of nursery with Captan or Thiram (0.2%) prevents post-emergence damping off. The seedlings are ready for transplanting in 3–6 weeks after sowing. In case of cauliflower for early maturity group, 5–6 weeks old seedlings and in mid season and late types 3–4 weeks old seedlings should be transplanted.

Field Preparation, Transplanting and Nutrient Management

Field is ploughed 2-3 times to a fine tilth and ridges are taken at a spacing of 60-75 cm. Recommended plant density ranges from 28700 to 40000 plants/ha depending on season and variety. Seedlings are transplanted in ridges at a spacing of 30-60 cm.

FYM is incorporated at the time of field preparation @ 20 t/ha. A fertilizer dose of 180 : 50: 50 kg NPK/ha is recommended for cabbage variety Pride of India. For cauiliflower NPK @ 120 : 60 : 60 kg/ha is recommended for mid season varieties. F_1 hybrids respond to higher dosage of nutrients. Half of N and whole of P and K should be applied as a basal dose and the rest of N in 2–3 split doses. Weekly spraying of 1–2% urea 20 days after transplanting is advised for good growth. Special attention must be paid to B and Mo. The most common effect of B deficiency is hollowing of the stem due to cracking of the pith, and browning , commonly known as brown rot. Mo deficiency occurs in acid soils and distribution of Mo salts to the soil or in foliar sprays will be useful. K deficiency can also occur, causing shortening of internodes, thickening and curling of lamina and purple pigmentation along the leaf veins. Some of the cole crops like cauliflower require a high magnesium and deficiency leads to chlorosis when grown on high acidic soil. Liming of soil should be carried out at least in 2 years for correcting soil reaction.

Aftercare

Cole crops are highly sensitive to stress and any check during vegetative stage results in poor quality product. Earthing up should be carried out after one month of transplanting. Optimum moisture supply is essential during growth and curd development. Furrow irrigation can be advocated when planting is done on ridges or raised beds . The intercultural operations should be done regularly to keep the crop free from weeds and for aeration of the root system. Very shallow hoeing should be done to remove weeds and avoid injury to the roots. Preplanting application of Stomp (Pendimethalin) or Basalin (Fluchloralin) @ 2.0-3.0 kg a.i/ha is recommended to control weeds.

Harvesting and Storage

Cauliflower is harvested when curds are compact, attains proper size and retains original colour. As the curds develop and approach the right stage for harvesting, bend one or two inner leaves inward at the midribs (some cultivars do this automatically) to protect the curd from sunlight and keep the curd white. Curds exposed to sunlight turn yellow. Curds should be harvested at the right stage of maturity, when the curd is compact and smooth in appearance. Storage of cauliflower is generally not recommended for more than 3 weeks. Wilting, browning, yellowing of leaves and decay are likely to increase, if stored above 5° C. Slightly immature and compact heads keep better than more mature ones. Cauliflower is very sensitive to temperature fluctuations during storage and such fluctuation causes darkening of curds. Exposure to ethylene during storage causes more rapid leaf yellowing and abscission. Controlled atmosphere (CA) is not recommended as it induces off odours and flavours

Cabbage heads should be harvested when they attain full size. After that, they burst or loose. Cabbage should be handled carefully from field to storage and only solid heads with no yellowing, decay or mechanical injuries should be stored. Before storage, all the loose leaves should be trimmed away leaving heads with 3-6 tight wrapper leaves. Heads can be stored for 3-6 weeks, at 0° C with a relative humidity of 98% for early variety, while the late harvested crop is stored at 0° C with 95% relative humidity. Late cabbage can be stored for 7-8 months at controlled atmosphere at 0° C with 2.5-3% O_2 and 5% CO_2.

In Brussels sprouts, 3-4 harvestings are done depending on compactness and size of sprouts. Sprouts can be stored at 0–0.5°C for 3–5 weeks. The knolkhol knobs should be harvested at right tender stage (5-8 cm diameter) as overmature knobs turn tough and woody within a few days. Turnip roots are harvested when they are 5–10 cm in diameter depending upon variety. The roots become tough and fibrous, if harvesting is delayed. The roots can be stored in fresh form up to 16 weeks at 0 °C with 90–95% relative humidity.

Economic Analysis

Cost analyses of major cole crops viz, cauliflower and cabbage are given below

Cost of Cultivation of Cauliflower

S. No	Particulars	Category of vegetable growers			Overall average
		Small	Medium	Large	
1	Operational cost				
	(i) Family labour	6946.89	1543.75	771.88	5398.42
	(ii) Hired labour	132.03	5403.15	6946.90	1771.82
	(iii) Machinery	385.94	385.94	385.94	385.94
	Total operational cost	7464.86 (37.79)	7332.84 (32.92)	8104.72 (33.96)	7551.18 (36.40)
2	Material cost				
	(i) Seed	1659.53	1559.53	1459.53	1559.53
	(ii) Manure & fertilizers	3187.37	4723.88	5495.76	3716.72
	(iii)Plant protection chemicals	1507.19	1543.75	1389.38	1480.79
	(iv) Irrigation	1421.05	1235.20	1930.00	1438.83
	Total Material cost	7775.14 (39.36)	9162.36 (41.13)	10274.67 (43.05)	8195.87 (39.48)
3	Total working cost	15240.00 (77.15)	16495.20 (74.14)	18579.39 (77.85)	15752.05 (75.88)

Contd...

4	Other costs				
	(i) Depreciation	716.66	1763.16	1265.42	992.56
	(ii)Interest on working cost	883.92	956.72	1077.60	919.42
	(iii)Land revenue	17.33	17.33	17.33	17.33
	(iv) Rental value of land	2666.67	2666.67	2666.67	2666.67
	(v) Interest on fixed cost	247.10	367.39	277.54	328.62
	Total cost	4531.68 (22.94)	5771.27 (25.94)	5304.56 (22.22)	4924.74 (23.72)
5	Total cost of cultivation				
	(i) Cost $A_1 = A_2$	9911.02	17688.66	20167.86	12382.94
	(ii) Cost B_1	10158	18056.05	20445.40	12711.56
	(iii) Cost B_2	1280.46	20705.39	13094.74	15360.90
	(iv) Cost C_1	17105.01	19599.80	21217.28	18109.98
	(v) Cost C_2	19754.35	22249.14	23866.62	20759.32
	(vi) Cost C_3	21729.79	24474.05	26253.28	22835.25
	Net return over				
	$A_1 = A_2$	27770.79	22449.93	17401.20	24690.75
	B_1	27524.11	22081.73	17124.62	24360.03
	B_2	24875.43	19433.54	14474.38	21711.22
	C_1	20577.46	17888.53	13703.07	16313.35
	C_2	17927.20	20538.11	16352.01	18963.65
	C_3	15952.15	15664.10	11316.43	14237.70

Source : Wadhwani and Bhogal (2003)

Cost of cultivation of Cabbage (Rs/ha)

S No	Particulars	Category of vegetable growers		Overall average
		Small	Medium	
1	Operational cost			
	(i) Family labour	6632.96		6803.19
	(ii) Hired labour	1195.16		1363.31
	(iii) Machinery	383.06		384.61
	Total operational cost	78211.18 (41.08)		8554.11 (41.29)
2	(i) Seedling	1647.18	1659.53	1666.72
	(ii) Manures & fertilizers	2938.19	2671.31	2933.35

Contd...

	(iii) Plant protection chemicals	1379.03	1389.38	1395.39
	(iv) Irrigation	1281.52	1235.20	1260.47
	Total Materials cost	7245.92 (36.25)	6955.60 (32.30)	7255.93 (35.03)
3	Total working cost	15457.10 (77.34)	15819.89 (73.46)	15810.04 (76.32)
4	Other costs			
	(i) Depreciation	716.66	1763.16	992.56
	(ii) Interest on working cost	896.51	917.55	916.98
	(iii) Land revenue	17.33	17.33	17.33
	(iv) Rental value of land	2666.67	2666.67	2666.67
	(v) Interest on fixed cost	247.10	367.39	328.62
	Total cost	4544.27	5732.10	4005.18
5	Total cost of cultivation			
	(i) Cost $A_1 = A_2$	1057.64	11570.86	10933.72
	(ii)Cost B_1	10704.74	11938.25	11262.34
	(iii)Cost B_2	13354.08	14587.59	13911.68
	(iv)Cost C_1	17337.70	18885.14	18065.53
	(v)Cost C_2	19987.64	21534.48	20714.87
	(vi)Cost C_3	21985.74	23687.93	22786.36
	Net return over			
	$A_1 = A_2$	49312.09	50673.84	48217.92
	B_1	49063.65	50306.42	47890.91
	B_2	46414.43	47658.83	45241.72
	C_1	42430.44	43360.00	41087.86
	C_2	39781.22	40710.25	38438.67
	C_3	37782.45	38557.59	36366.16

Source : Wadhwani and Bhogal (2003)

Cucurbits

Cucurbits are the most important group of warm season vegetables grown in India. Fruits are used as salad (Cucumber), cooked vegetable (gourds), dessert fruit (water melon, musk melon) or in confectionary(ash gourd). Ribosomal inhibitor proteins such as trichosanthin, luffaculin, mommordicin etc isolated from cucurbits show the theraputic use of these crops in future. Seeds are rich in oil and protein and comparable to legumes. Most of the cucurbits are seed propagated and have long taproot system with viny stem.

Ash Gourd

Ash gourd (syn. hairy melon, winter melon, ash pumpkin, white pumpkin, Chinese preserving melon, wax gourd, white gourd, petha) *Benincasa hispida* Cong, is an important warm season cucurbit vegetable grown for its fruit used in confectionery and ayurvedic medicinal preparations. Ash gourd is grown throughout the old world tropics and less commonly in the new world. An enzyme extracted from the ash gourd juice was successfully used in place of calf rennet in the production of Cheddar cheese.

Soil and Climate

Crop prefers warm, humid tropical climate with a mean temperature requirement of 22°–35°C. Though a deep loamy soil is best suited, it can be grown on heavy clay as a rainfed crop. The optimum pH is 6.5–7.5. Ash gourd can be successfully grown during January-March and September-December. For rainfed crop, sowing can also be started after receipt of first showers during May-June.

Preparation of Land and Sowing

Spacing depends on variety, soil and region. Pits of 60 cm diameter and 30-45 cm depth are taken at 4.5 x 2.0 m spacing. Pits of size 45cm × 45cm × 45 cm can also be dug at 2.0–2.5m × 1.0–1.5m apart. Well rotten FYM and fertilizers are mixed with topsoil in the pit. Seed rate of 0.75-1.0 kg/ha is optimum.

Seeds are sown at the rate of four or five/pit. Remove unhealthy plants after two weeks and retain two or three plants/pit. Raising seedlings in polybags and then planting in pits prepared in main field gives good yield. Two healthy seedlings can be planted in each pit. The seed rate can be reduced by this method.

Manuring

A basal dose of farmyard manure (20 tonnes/ha) should be given. Then N (40–60kg), P (50–60kg) and K (60–80kg) should be topdressed. The N should be applied in 2-3 split doses at the time of vining and at the time of full blooming.

Aftercultivation

During initial stages of growth, irrigate at an interval of three or four days. Irrigate on alternate days during flowering and fruiting. For trailing, spread dried twigs on the ground. Conduct weeding, and raking of the soil at the time of fertilizer application. Earthing up may be done during rainy season.

Varieties

Varieties of Ash gourd

Varieties	Source	Features
Kashi Dhawal	IVRI	Oblong fruits, Av. fruit wt-12 kg, ready for picking in 10-105 days. Preferred variety for sweet preparation, Av yield 55-60 tonnes/ha
APAUShakthi	APAU	Fruits long, cylindrical, yield 30.0-35.0 t/ha in 140-150 days duration
CO1	TNAU	Fruits large 8–10kg in weight, suited for cultivation in Tamil Nadu and Kerala. Crop duration is 140-150 days, yield 20.0–25.0 t/ha.
CO2	TNAU	Fruits small, oblong, weight 2-3kg each, crop duration 120-130 days and yield 30.0-34.0 t/ha

'Okinawa No.1' and 'Kurokawa Early' are popular varieties in Japan. 'S 1' is recommended for Punjab and 'Thudiyalur' is popular in Maharashtra.

Harvesting

At full maturity, the ashy bloom covering the fruit slowly drops off slowly and fruits become ready for harvesting 90–100 days after sowing. Fruits can be stored in a well-ventilated room for 4–5 months. The average yield is 25.0-30.0 tonnes/ha.

Bitter Gourd

Bitter gourd (syn. bitter cucumber, balsam pear), *Momordica charantia* L is an important cucurbit fruit vegetable grown mainly in tropics. The alkaloid momordicosoides impart the bitter taste to the fruit. Fruits rich in iron (1.8 mg/100 g), calcium (20 mg/100 g), phosphorus (55 mg/100 g), vitamin A (210 IU/100g) and vitamin C (88 mg/100g) form an ideal diet to the diabetic patients. Maharashtra, Kerala, Karnataka and Tamil Nadu are major bitter gourd-growing states. It is a luscious,climbing vine with slightly fuzzy stems with dark green, deeply lobed leaves and yellow, monecious flowers

Climate and Soil

Optimum temperature for growth and flowering is 25°-30°C. The minimum temperature should not go below 18°C. More tolerant to low temperature compared to other gourds. High temperature (above 40°C) induces more male flowers combined with high incidence of viral diseases.

June–July and December–January are sowing time in South India, whereas January–February in North India. For the rainfed crop, sowing can be started after receipt of first few showers during May-June. A well-drained, loamy soil with a pH of 6.5–7.0 is ideal for its cultivation.

Varieties

Numerous hybrids and open-pollinated varieties are available. Hybrids usually produce higher yields, but their seeds are relatively expensive and must be purchased for every planting. Generally, there are three types of varieties:

1) small, 10–20 cm long, 100–300 g, usually dark green eg : Arka Harit 2) long, 30–60 cm long, 200–600 g, light green in color with medium size protuberances eg : Priyanka 3) triangular fruit type, cone-shaped, 15–30cm long, 300-600 g, light to dark green with prominent tubercles.

Varieties in Bittergourd

Varieties	Source	Features
Arka Harit	IIHR, Bangalore	Spindle shaped green fruits, smooth regular ribs, crop duration 120 days, Yield 15.0 tonnes/ha
Pusa Vishesh	IARI	Long fruits, medium size, Av. Fruit wt. 120 g, first harvest 55-60 days after sowing, yield-15 tonnes/ha
Coimbatore Long Green	NSC	Long white fruits, suitable for rainy season, Crop duration 115-120 days, Yield 20 tonnes/ha
Preethi	KAU	Medium size white fruits with spines, Av. Fruit length 30 cm, Av. fruit wt-310 g, Pre –bearing period 5 days, duration 4-5 months, Yield 15 tonnes/ha
Priyanka	KAU	Spindle shaped white fruits with smooth spines, Av. Fruit length-25 cm, Fruit wt. 0.30 Kg, pre-bearing period-55 days, duration-4-5 months. Yield 28 tonnes/ha
Pusa Hybrid-1	IARI	Long , glossy green fruits, pre-bearing period-55-60 days, suitable for spring-summer season, Yield-20 tonnes/ha

A good number of F_1 hybrids which are early- and high-yielding, have been identified. COBgoH.1(MC.84 x MDU.1) is a popular F_1 hybrid developed from TNAU with an yield potential of 51.29t/ha. Reshma, Shreya, INDAM-49, BSS 405 are the popular hybrids from private sector.

Seed Rate and Planting

Direct seeding is the most common method of planting .Seed rate for direct sowing is 4–5.5 kg/ha. Seed rate can be reduced to 2.5-3.0 kg by transplanting as polybag seedlings. Here optimum plant density differs with variety and usually ranges from 2,500 to 11,000 plants/ ha. Seeds can be sown in pits, beds or in channels. Pits are taken at a spacing of 2 x 2 m or 1.5 x 0.75 cm or 2.5 x 1.5 cm depending upon the variety. For bed planting, on raised beds, sow two or three seeds/ hole at a depth of 2 cm. Space holes at 40–60 cm apart in rows spaced at 1.2–1.5 m apart. Seeds can also planted in channels. Long channels 60 cm wide are formed. Along these channels, pits of 45 cm × 45 cm × 45 cm size are dug. A spacing of 1.5–2.0 m from row-to-row and 1.0 m from hill-to-hill is maintained. For transplanting, sow seeds in small plastic bag using a potting mix that has good water-holding capacity and good drainage. Plant two or three seeds/ container and thin to a single seedling when they have four to six true leaves. Seedlings are ready for transplanting, 15–20 days after sowing. Transplant seedlings into the field at spacings similar to those used for direct seeding method.

Manuring

Bitter gourd requires a balance of nutrients from organic and chemical fertilizers. Fertilizer application rates depend on soil type, fertility level, and soil organic matter.

Recommended fertilizer for OP varieties under average management is given below,

Andhra Pradesh	:	100 : 100 : 50 kg N , P_2O_5 and K_2O/ha
Assam	:	40 : 30 :30 kg N , P_2O_5 and K_2O/ha
Kerala	:	70 : 25 : 25 kg N , P_2O_5 and K_2O/ha

Apply FYM @ 20-25 t/ha as basal dose. In clay or heavy texture soils, the entire amount of P, and one-third of N and K is applied before planting, either by broadcasting and tilling or by banding a few cm deep and to the side of the plant row in the bed. The balance of N and K is applied in two or more sidedressings. In sandy soils, fertilizer application consists of a basal application followed by four sidedressings at fortnightly intervals.

Aftercultivation

Conduct weeding and raking of soil at the time of fertilizer application. During the initial stages of growth, irrigate at 3-4 days interval, and alternate days during flowering/fruiting. Irrigation at 15 mm CPE (approximately at 3 days interval for sandy loam soils) is more economical than irrigating once in two days especially during summer months. The growing vines are trained on stakes to reach the erected pandal/ trellies. Earthing up is done during rainy season.

Spraying of MH (150–200ppm) or Ethrel (250ppm) at 2-leaf stage (twice) at weekly intervals increase production of female flowers. Foliar spray of micronutrient mixture viz, B 100 ppm, Zn 100 ppm, Mo 50 ppm, Cu 100 ppm, Fe 100 ppm, Mn 100 ppm improves the sex ratio and fruit production significantly.

Harvesting

Harvesting commences from 70–75 days after sowing. Tender, immature fruits should be harvested. Yield 10-15t/ha.

Bottle Gourd

Bottlegourd, *Lagenaria siceraria* (Mol.) Standl. is grown throughout India and its fruits are available in the market round the year. It is diuretic and has cardiotonic properties. Popular as summer crop in North Indian plains, it is cultivated extensively in Uttar Pradesh, Punjab, Gujarat, Assam, Meghalaya and Rajasthan.

Climate and Soil

Crop prefers humid and semi-dry climate. Night and day temperature of 18°–22°C and 30°–35°C respectively are optimum for proper growth and high fruit set. January-March and Sept-December are the ideal sowing seasons. For the rainfed crop, sowing can also be started after receipt of first few showers during May-June. Loam or sandy-loam soil is most suitable, though crop is adoptable to wide range of soils. As other cucurbits, soil should be rich in organic matter and with good drainage. Salinity and alkalinity adversely affect the crop.

Varieties

Varieties in Bottle Gourd

Varieties	Source	Features
Kashi Bahar	IVRI	F_1 Hybrid, early maturing
Arka Bahar	IIHR	Straight fruits, medium-sized, weighing about 1kg each at edible stage, skin light green. Suitable for mild climate- Yield 25 tonnes/ha
Punjab Round	PAU	Its plants are vigorous and prolific-bearer. The fruits are spherical, tender and shining. Yield 20 tonnes/ha.
Pusa Manjari	IARI	Hybrid, fruits are round, green, tender and attractive. Recommended for cultivation for spring sowing
Pusa Meghdut	IARI	Hybrid, yields early and is suitable for both spring and summer sowing. Fruits are long, light green, tender and attractive.
Pusa Naveen	IARI	Suitable for spring-summer and rainy seasons in northern plains. The fruits are straight, cylindrical and free from crook-neck. Yield - 30 tonnes/ha.
Pusa Hybrid-3	IARI	Slightly clubbed green fruits, without neck, pre-bearing period-55-60 days, Av.yield 42.5 tonnes/ha in summer and 47.0 tonnes/ha in rainy season.
Narendra Rashmi	NDUAT	Bottle shaped fruits, shallow neck, Av. fruit wt 1 kg, Yield 30 tonnes/ha
Rajendara Chamatkar	RAUBAC	Fruits are long (50 – 60 cm), uniform, green and bottle-shaped, the yield is 20–25tonnes/ha

Sujata, Shambhu, Swathi, Soumya,Vikrant, Divya, NDBG 4, NDBG 1, Phule BTG 1, and NBBH-48 are popular F_1 hybrids.

Land Preparation and Sowing

Pits of 60 cm diameter and 30-45 cm depth are taken at 3 m x 3 m spacing for growing on pandals. For trailing on the ground, make pits at a spacing of 2 m in rows 3-4 m apart. Well rotten FYM and fertilizers are mixed with topsoil in the pit. Seed rate, 3-4 kg/ha. 60-75 kg N, 25-50 kg P and 25-50 kg K is required along with 15-20 t of FYM/ha.

Aftercultivation

Spring-summer crops are allowed to spread between channel/rows while rainy season crop are spread on pandal. The marketable fruit yield increases in this system of training. During the initial stages of growth, irrigate at an interval of three or four days. Irrigate on alternate days during flowering and fruiting periods.

New Trends in Cultivation

Application of Ethrel (200 ppm) induced more female flowers, early ripening, high yield and good fruit size.Spraying of MH (200 ppm) was the most effective to induce female flowers at lower nodes followed by NAA (100 ppm). Hand pollination increased yields from 5.4 to 8.9 kg/plant during summer and 3.4 to 6.6 kg/plant during monsoon seasons. The cost: benefit ratios were 1:3.7 during summer and 1 :2.6 during monsoon.

Harvesting and Yield

Ready for harvesting in 55–75 days after sowing. The fruits should be harvested within 3 days of the shedding of the small hairs present on the skin. Yield of open-pollinated varieties ranged from 20–25 tonnes/ha, while the yield of hybrids is more than 30 tonnes/ha.

Cucumber

Cucumber (*Cucumis sativus* L) is a warm-season fruit vegetable used as salad and for pickling purpose. In developed countries, it is grown as a glasshouse vegetable and in developing countries as an open-field vegetable. Cucumber forms an essential item of dietary in the West.

Climate and Soil

Cucumber is a a warm-season crop grown mainly in tropical and subtropical regions. Crop prefers slightly low temperature compared to melons and hence it is grown extensively in subtropical regions. Growth and yield are promoted at a temperature of 18°–24°C. A well-drained loamy soil is preferred by cucumber. Lighter soils which warm quickly in spring are usually utilized for early crop. Cucumber prefers a pH of 5.5–6.8. The long tap root system of cucumber is adapted for growth of plants in river beds.

Varieties

Earlier varieties were developed through introduction. Now many HYVs and F_1 hybrids are developed in public and private sector.

Varieties in Cucumber

Varieties	Source	Features
Japanese Long Green	IARI Regional station, Katrain	A temperate cultivar, it is suited for cultivation in hills and lower hills. Its extra early fruits mature in 45 days. Fruits are 30–40cm long, with light green and crisp flesh
Pusa Sanyog	IARI, New Delhi	It is an early and high-yielding hybrid. Fruits are 28–30cm long, cylindrical and dark green with yellow stripes and crisp flesh. Fruits mature in 50 days. It performs well in temperate regions only.
Poinsett	NSC	Fruits are dark green and 20–25cm long. It is resistant to downy mildew, powdery mildew, anthracnose and angular leaf-spot. Av. yield 18-22 t/ha
Pant Shankar Khira 1	GBPUAT, Pant Nagar	F_1 Hybrid. Cylindrical long fruits (20 cm), green with light strips. First picking 50 days after sowing, yield 20 tonnes/ha
Swarna Poorna	HARP, Ranchi	Long cylindrical fruits, medium size, light green, Tolerant to powdery mildew, Yield 30-35 t/ha

Varieties bearing female flower in early nodes, bushy plant type, self staked habit, higher female: male sex ratio, more flesh thickness, desirable flesh colour, fruit shape as per local preference, less or negligible cucurbitacin content, uniformity of fruit, less seediness, medium fruit size, better transportability, resistance to parasitic and non-parasitic diseases and adaptation to once over harvest are preferred for commercial cultivation.

Land Preparation and Sowing

Sown in October–November. In plains of north-eastern India, sowing time is from November to March when weather is comparatively dry and in hills, April–May. Seed rate, 2.5–3.5kg/ha.

Sown in furrow, bed or pit . In furrow sowing, furrows are made at 1.0–1.5 m apart. Usually sowing is done on the top of the sides of furrows and vines are allowed to trail on organic/plastic mulches spread on the ground.

The hills should be made at a distance of 0.5–0.75m. About 5–6 seeds are sown in a pit and finally two vines are retained.

Manuring and Fertilization

Cucumber responds well to manuring and fertilization. Crop yielding 20 tonnes/ha removes 39: 27: 70 kg NPK and 10 kg MgO from one ha. Generally well-rotten farmyard manure @ 15–20 tonnes/ha is applied at the time of land preparation. Cucumbers are sensitive to Mg deficiency and respond to Mn and Cu applications. In sandy loam soils with pH 6.5, apply 80 kg/ha N, 45 kg/ha P_2O_5 and 85 kg/ha K_2O. It is advisable to give half the recommended dose of N and K through top dressing. Hybrids require a higher dose of fertilizer and NPK @ 200 : 125 : 125 kg/ha is recommended from MPKV, Rahuri. Spray the 0.25 % boron solution at the six leaf stage and at the flower bud initiation stage.

Aftercare

During early stage, the crop should be kept weed-free. At the time of topdressing with nitrogenous fertilizers, weeding and earthing-up are done. Pruning improves yield and fruit quality. Pruning all secondary shoots up to 5 nodes and afterwards pruning to a single stem plus 100 cm × 60 cm spacing gives highest yield. Staking of plants particularly in rainy season, is helpful in checking the rotting of fruits.Crop should be irrigated during dry weather at an interval of 3-5 days depending on soil type. Two sprays of ethrel (150-200 ppm) at 2-3 leaf stage and a fortnight later increase the female flowers and yield.

Harvesting and Yield

Harvesting stage is attained within a week from anthesis of female flowers. For salad or slicing cucumber, dark green skin colour should not turn brownish-yellow or russeting. Harvesting can be done at 2 days intervals. Yield varies with system of cultivation, cultivar and season. Yield ranges from 8.0–20 tonnes/ha. Harvested fruits are, cleaned, graded and packed in gunny bags or bamboo baskets. Refrigerated storage at 10 ° C and 85 % RH can prolong the storage life of cucumber to 10-14 days.

Melons

Melons are annuals with climbing, creeping or trailing vines of length up to 3 m. The dessert types quench thirst and add to the nutrient content of main diet. The non-dessert types are used as vegetables. The leaves of melons are shallow lobed and more or less rounded. Flowers are large, yellow, fragrant and edible. Fruits are extremely variable in size and shape. They may be ellipsoid to globose, with or without longitudinal grooves. Skin may be soft or hard yellow, green, cream or orange in colour and plain, netted or echinate (prickly) in texture. Flesh inside varies from white to cream-yellow, orange or green. At present most of the commercial types have thin skin and thick orange pulp. Fruits are many seeded pepo.

Musk Melon

Climate and Soil

Musk melon(*Cucumis melo*) prefers dry cool climate and cannot tolerate frost. High humidity increases the incidence of foliar diseases. Sowing starts from November to January as forcing methods in riverbeds. In northern plains, sowing is done in February while in December–January in southern plains. A well-drained, loamy soil is ideal. Long tap root system permits its cultivation in riverbeds. Crop is sensitive to acid soils. A soil pH between 6.0 and 7.0 is ideal.

Varieties

Varieties in musk melon

Varieties	Source	Features
Arka Jeet	IIHR	The fruits are small, round with flat ends, weighing 300–500g each, skin smooth and orange to orange-brown; flesh white with medium soft-textured, excellent flavour, juicy, very sweet (15–17% TSS); rich in vitamin C. An early-maturing variety, its average yield, 15 tonnes/ha
Arka Rajhans	IIHR	An early-maturing variety, fruits are round to slightly oval, weighing about 1kg each. Rind is creamy-white and netted; flesh white, thick, firm and sweet (11–14% TSS); shelf-life and transport qualities are good. The average yield, 28.5 tonnes/ha
Hara Madhu	PAU	late-maturing variety, fruits are round, tapering towards stalk-end, weighing about 1kg each; flesh light green, juicy and very sweet (13% TSS); keeping quality poor; average yield 12.5 tonnes/ha.
Hisar Madhu	HAU	Its vines are long, providing round fruits; rind brownish-yellow with green sutures; flesh orange, juicy and sweet. The average yield is 10.0 tonnes/ha

Hisar Saras	HAU	With short vines, Hisar Saras matures early with round fruits; skin netted; flesh greenish-white and sweet. It is tolerant to down mildew. The average yield is 8–10 tonnes/ha
Punjab Hybrid-1	PAU	Its vines are vigorous and long, providing fruits early. Fruits are round, weighing 800g; rind netted, light yellow and quite thick, flesh thick, juicy, orange flavoured and sweet (12% TSS). Its average yield is 16.0 tonnes/ha.
Pusa Madhuras	IARI	A mid-season variety, its fruits are roundish-flat, weighing a kg or more; skin pale-green, sparsely netted with dark green stripes; flesh salmon-orange, juicy and sweet (12–14% TSS); keeping quality poor. The average yield is 15 tonnes/ha.
Pusa Rasraj	IARI	An early-maturing hybrid, fruits oval; rind greenish-yellow, smooth and sutureless; flesh green. The average yield is 17.3 tonnes/ha
Pusa Sharbati	IARI	An early-maturing variety, its fruits are round; skin netted; flesh salmon-orange, firm, thick and moderately sweet (11–12% TSS); seed cavity small, keeping quality good, suitable for river-bed cultivation and northern India. The average yield is 15.0 tonnes/ha

Land Preparation and Sowing

Seeds should be sown on raised beds or furrows. Bed of 3-4 m width is prepared and planting is done on both sides of beds. Hill-to-hill distance of 50–60cm should be kept. The planting can also be successfully done in long furrows. The seed rate of 1kg/ha is optimum for sowing in field by dibbling method, or for transplanting in polythene bags. Transplanted crop matures 15–20 days earlier than the direct-seeded one.

Manuring and Fertilization

Crop removes about 50-120 kg N, 15-25 kg P_2O_5, 50-200 kg K_2O, 70-100 kg CaO, and up to 20 kg MgO/ha. Highest rate of nutrient uptake occurs 4-6 weeks after planting, i.e. a few weeks before maximum rate of dry matter accumulation. FYM, 25–40 t is required/ha which should be incorporated in soil 10–15 days before sowing. In soils containing at least 3 % organic matter, an irrigated crop yielding 30-35 t/ha should receive a basal fertilizer application, placed along the rows before planting, of about 100 kg N, 100 kg P_2O_5, 150 kg K_2O/ha, followed by a topdressing of 50-100 kg/ha each of N and K_2O during fruit swelling. Topdressing is generally not recommended for unirrigated crops. Protected crops usually require three topdressings. Melons are very sensitive to Mg and deficiency (< 0.3 % in dry matter) results in shortage of chlorophyll, browning of tissues and chlorosis of leaves. In acid soils, excess of Mn can occur; a concentration of this micronutrient of >1000 ppm in the leaves causes problems of phytotoxicity - yellowing of leaves, brown spots on stems, and necrosis. The crop is also sensitive to low availability of micronutrients Fe, Mo and B.

Harvesting and Yield

Rind colour is an indication of maturity. Harvesting at the proper stage is of major importance in marketing good quality produce. The fruits for distant markets should be harvested as they reach the half-slip maturity, to avoid losses from the over-ripeness and decay. For local markets, harvesting should be done at 'full-slip' stage. Yield 13-15 tonnes/ha

Oriental Pickling Melon

Oriental pickling melon, *Cucumis melo* var. *Conomon* is an important vegetable in South India. It is grown for immature fruit and has an important role in religious functions. The ideal growing seasons are January-March and September-December.

Varieties

Mudicode, Arunima and Saubhagya are high yielding varieties of oriental pickling melon released from KAU.

Land Preparation and Sowing

Pits of 60 cm diameter and 30-45 cm depth are taken at a spacing of 2 x 1.5 m. Well rotten FYM and fertilizers are mixed with topsoil in the pit and four or five seeds are sown in a pit. Remove unhealthy plants after two weeks and retain three plants/ pit. Seed @ 0.5 – 0.75 kg is required/ ha.

Manuring

Apply FYM @ 20-25 t/ha as basal dose along with half dose of N (35 kg) and full dose of P_2O_5 (25 kg) and K_2O (25 kg/ha). The remaining dose of N (35 kg) can be applied in two equal split doses at the time of vining and at the time of full blooming.

After Cultivation

During initial stages of growth, irrigate at an interval of 3-4 days. Irrigate in alternate days, during flowering and fruiting. For trailing, spread dried twigs on the ground. Conduct weeding and raking of the soil at the time of fertilizer application. Earthing up may be done during rainy season. Ethrel application at 3-4 leaf stage improves emergence of female flowers and yield.

Harvesting and Yield

Fruits are ready for harvesting in 70-75 days after sowing. Rind colour is an indication of maturity. Yield varies from 15-30 tonnes/ha.

Water Melon

Watermelon, *Citrullus lanatus* (Thunb.), is of ancient cultivation in the Mediterranean region and was grown in Egypt in remote times. It reached India in prehistoric times. It was introduced to New World in post-Columbian times and is now widely cultivated in the tropics. In India, water melons are mainly cultivated in Maharashtra, Karnataka, Tamil Nadu, Punjab, Rajasthan, Madhya Pradesh, and Uttar Pradesh. It is a trailing annual vine with stems 15 m or more in length and branched tendrils having yellow flowers bearing globular to oblong fruits.

Climate and Soil

It tolerates warm humid climate better than musk melon. Ideal temperature is 30-35°C. Prefers warmer days and cooler nights. Can not with stand frost. Higher temperatures are beneficial during ripening. The ideal season is Dec-April. Sandy loams are the best for early crop, while loams have high-yielding potential. Ideal pH - 6.5–7.0.

Varieties

Varieties in Water Melon

Varieties	Source	Features
Sugarbaby	IARI NSC	Introduction from the USA. Fruits weighing 4–6 kg each. Fruits are round in shape, having bluish bloom on dark green skin and black-green stripes. The fruits ripen 85–90 days after sowing.
Ashahi Yamato	IARI	Introduction from Japan. Fruits weighing 6–8 kg each. Fruits have striped light green skin and deep pink, crisp flesh. Ripen in 90–95 days after sowing.
Arka Jyothi	IIHR	F1 hybrid. The plants bear round fruits weighing 6–8 kg each, light green skin with regular dark green stripes, flesh crimson-red colour.
Arka Manik	IIHR	Fruits are somewhat round (oval) and weigh about 6 kg each. Skin colour is light green with dull green stripes. The flesh is deep red, very sweet. It is resistant to

Contd...

		powdery mildew, downy mildew, tolerant to anthracnose and blossom-end rot. It stands transport.
Fuken	HAU	Heavy yielding, adoptable to humid hot climate, resistant to anthraconose

Besides, there are a number of extensively cultivated hybrids from private sector which include, Madhu, Milan, Mohini, MHW 4, MHW 5, MHW 6, MHW 11, MHW 12, MHW 15, NS 246, NS 295, Suruchi, Samtrupti, Amruth and Century 2.

Sowing

Seeds are sown in hills on raised sides of furrows in upland and in trenches or in pits in the river-beds. Two plants/ hill in trenches or furrows and 4 plants/ pit are retained for growing. A long vining type requires row and plant spacing up to 3.5 and 1.2m in river-beds while medium vining type like Sugar Baby requires 2.0 m × 1.0 m spacing in upland. Seed rate: 1 to 1.5 kg/ha.

Manuring

Crop yielding 15 t/ha removes an average of 56 kg N, 16 kg P_2O_5, 100 kg K_2O, 25 kg MgO and 98 kg CaO. N or Mg deficiency reduces fruit-set. 90-100 kg N, 60 kg P_2O_5 and 60 kg K_2O are recommended per ha. Dolomitic limestone should be applied to offset Mg or Ca deficiencies. Pre-plant broadcast applications are recommended in preference to band-application to prevent seedling salt injury.

After Cultivation

Apical shoot is pinched and 2–4 side shoots are retained. Fruit thinning is useful and retaining of 2 fruits/ vine improves fruit size and fruit yield. Spray with NAA (100 ppm)/MH (200 ppm) results in heavier and larger fruits. Application of ethrel (200 ppm) improves yield in watermelon.

Harvesting and Storage

Crop is ready for harvesting 90–120 days after sowing depending upon cultivar and season. Change in colour of the portion of fruit which rests on ground from white to creamy in case of light skin colour, or to yellow in case of dark skin colour, is a useful guide of maturity. Watermelons are graded according to their size for local market. Distinction among grades is based predominantly on external appearances. However, watermelons should be symmetrical and uniform in appearance. The surface should be waxy and bright in appearance devoid of scars, sunburn, transit abrasions or other surface defects.

The fruits are transported by road in bulk by stacking them on dried grass in trucks.

Watermelons can be stored for 14 days at 15°C. For short-term storage or transit to distant markets (>7 days), watermelons can be stored at 7.2°C with 85-90% relative humidity. Extended holding at this temperature induces chilling injury. Many watermelons are still shipped without pre-cooling or refrigeration during transit. These fruits must be utilized for prompt market sales as quality declines rapidly under these conditions. Watermelons should not be stored with apples and bananas as the ethylene produced during storage from these fruits hastens softening and development of off flavour to watermelons.

Pointed Gourd

Pointed gourd (syn. parwal), *Trichosanthes dioica* is one of the most nutritive cucurbit vegetables. Fruit contains 2 % protein, 0.3 % fat, 2.2 % carbohydrate, 153 IU Vitamin A and 29 mg Vitamin C per 100 g of fruit. It is a perennial trailing vegetable, grown for its tender immature fruits. The fruits are small, pointed at both the ends with white stripes. It is easily digestible, diuretic and laxative.

Climate and Soil

Prefers tropical climate. Grows well in high rainfall areas. It remains dormant in winter and grows vigorously during spring. It grows best in well-drained sandy loams Heavy soils are not suitable for its cultivation.

Varieties

Varieties in Pointed Gourd

Varieties	Source	Features
Faizabad Parwal 4	NDUAT	Recommended for reclaimed sodic soils. The fruits are light green, spindle-shaped with tapering ends. Suitable for pandal system of cultivation.
Rajendra Parwal 1	RAUBAC, Sabour	Suitable for diara land cultivation. Its fruits are big-sized striped and green in colour. The average yield is 14.0–15.0 tonnes/ha
Rajendra Parwal 2	RAUBAC, Sabour	Suitable for diara cultivation. Fruits are big, dark green with whitish stripes. Average yield - 15.0–17.0 tonnes/ha.

Contd...

Swarna Alaukik	HARP, Ranchi	Produces light green fruits with blunt ends. The fruits are thin skinned and good for vegetable as well as for preparation of sweets. Yield-23.0–28.0 tonnes/ha
Swarna Rekha	HARP, Ranchi	Fruits are greenish-white, striped, long and tapering on both sides. Average yield is 20.0–23.0 tonnes/ha on vertical bower system.

Planting

Propagated through cuttings and root suckers. Vine cuttings are collected in October and planted for rooting in polythene bags filled with potting mixture in such a way that 2–3 nodes are outside the bag. Keep it in shade for 3 months. Sproutings can be seen after 10 days onwards. These can be transplanted in the mainfield after 3 months. Root cuttings collected from 3–4 year old plants can also be used for planting. Three node root cuttings can be used for direct planting or for polytube planting. In bower system, 8,000 cuttings are required to raise one hectare crop with a spacing of 1.0 m × 1.5 m, whereas in flat system, 4,500–5,000 cuttings are required spaced at 1.0 m × 2.0 m. Ensure a minimum of 10% male plants in the field for fertilization and fruit setting.

Manuring and Fertilization

Apply farmyard manure @ 20 tonnes/ha as basal dressing. NPK @ 90: 60:40 kg/ha is followed in UP while 150:60:80 is applied in Orissa.Half the quantity of N and full of P and K should be applied at the time of planting in small basins. Remaining half of N should be applied 60 days after planting.

Training and Pruning

Training of vine on vertical staking or on bower system facilitates effective pollination, increased fruit set and yield compared to trailing over the ground. Vines should be pruned before winter as the crop passes through dormant phase in winters. Vines are pinched at 30-45 cm length from basin.

Harvesting and Yield

Vine starts bearing fruits 5 months after planting and continues to yield for 4-5 years. Fruits should be harvested from two weeks after anthesis.

Pumpkins and Squashes

Pumpkins and squashes, *Cucurbita* spp. (syn. vegetable marrow, musky gourd) have been associated with man in the Americas for at least 10,000 years. The 5 cultivated species viz, *Cucurbita pepo* L. (summer squash), C. *moschata* Duch. ex Poir. (pumpkin), C. *maxima* Duch. (winter squash), C. *mixta* Pang. (winter squash) and C. *ficifolia* Boucha. (fig leaf gourd) were selected by American Indians long before 1492 and they were staple items of their diet. Introduced to India from the South American centre of crop origin by foreign navigators and emissaries, pumpkin is grown throughout India.

Pumpkin *Cucurbita moschata* Poir.

Climate and Soil

Prefers tropical climate. A temperature of 24°–30°C is optimum. It can not stand frost. Pumpkin can be successfully grown during January-March and September-December. Grown in various kinds of soils but sandy loam to loam soils are ideal. Slightly tolerant to acidic soils.Optimum pH, 5.5–6.8.

Varieties

Varieties in Pumpkin

Varieties	Source	Features
Kashi Harit	IVRI	Green, spherical fruits, Av. fruit wt.2.5-3.0 kg, pre-bearing period70-75 days-Av. yield 36-40 tonnes/ha.
Pusa Vishwas	IARI	Light brown, spherical fruits with thick, golden-yellow flesh, weighing up to 5.0kg. Yield-40.0 tonnes/ha.
Pusa Vikas	IARI	Small fruits, weighing 2.0- 2.5kg each, round, flat with yellow flesh. Early, matures in 115 days. Yield 30.0 tonnes/ha.
Pusa Hybrid-1	IARI	F_1 hybrid. Fruits are round, flat, weighing on an average 5.0kg each. Mature in 120 days, yield- 45.0 tonnes/ha.
Co 2	TNAU	Early variety bearing green fruits.Yield-25.0 tonnes/ha.
Co 1	TNAU	Late-maturing variety. Globular fruits weighing 7–8kg and flattened at the base. Yield-25.0 tonnes/ha.
Arka Chandan	IIHR	Round fruits with flat blossom end and medium in size (2–3kg). It has good cooking and keeping qualities. Yield-32.5 tonnes/ha.
Ambili	KAU	The plant has a spreading habit. Fruits are flat, round and green, weighing about 6 kg each. Yield- 30 tonnes/ha.

Land Preparation and Sowing

Seeds are sown directly in the field on raised beds or in furrows or trenches or pits. Pits of 60 cm diameter and 30-45 cm depth are taken at a spacing of 4.5 x 2.0 m. Well rotten FYM and fertilizers are mixed with topsoil in the pit. Seed rate is 1.0 to 1.5 kg/ha. Four of five seeds are sown/ pit. Remove unhealthy plants after 2 weeks and retain three plants/ pit.

Manuring

FYM should be applied @ 20-25 t/ha as basal dressing. In North India, NPK @ 100 :50 :50 is recommended/ha. Half the dose of N and K should be applied through top dressing.

Aftercultivation

Pre-emergence application of butachlor (2 kg a.i/ha) effectively controls weeds. Application of 2-chloroethane phosphonic acid (200 ppm) at 4-5 leaf-stage and 7-10 days later are observed to increase fruit set.

Harvesting and Yield

Fruits are ready for harvesting from 75–180 days after sowing depending on variety and season. Mature fruits have long storage life and can be easily transported to distant markets. Yield is 25.0 to 40.0 tonnes/ha.

Ridge Gourd and Smooth Gourd

Ridged-gourd, *Luffa acutangula* roxb. (four-angled gourd, angled loofah) and smooth gourd, *Luffa cylindrica Roem.* (dish rag gourd) are important warm-season fruit vegetables. Both are climbers, stem acutely 5-angled, hairy, tendrils 3-fid or more, leaves 5-7 angled or shallowly lobed. The fruits are used only in the immature stages. On maturity, the fruits become bitter and inedible.

Differences Between Ridge Gourd and Smooth Gourd

Ridge gourd	**Smooth gourd**
Leaves are shallowly lobbed and larger	Leaves are deeply lobbed and smaller
No white patches on leaf	white patches on leaf
Male flowers smaller	Male flowers larger
Disc shaped and not tomentose ovary	Cylindrical and tomentose ovary
Angled and 10-ribbed fruits	Smooth fruits
Seeds black and pitted	Seeds black, white flat and not pitted

Climate and Soil

Prefers warm hot climate, the optimum temperature being 25°–30°C. Can not stand frost. Sandy loam soil, rich in organic matter is most-suited for higher yield. Grown as *kharif* and summer crop.

Varieties

Varieties in Ridge Gourd and Smooth Gourd

Varieties	Source	Features
Ridge gourd		
Arka Sumeet	IIHR	Cylindrical long fruits, 55 cm long and 2.5 cm girth, fruit wt. 380 g, yield 20 t/ha
Co1	TNAU	Green, long (45-50 cm) fruits with broader stylar-end. Average fruit weight- 700g. Yield 14.0 tonnes/ha in 125 days.
Co2	TNAU	Very long fruits (90–100cm), each weighing 800g. Yield 25.0 tonnes/ha in 120 days
Pusa Nasdar	IARI	An early variety with light green, medium-sized club-shaped fruits. Yield 15-20 tonnes/ ha.
Swarna Manjari	HARP, Ranchi	Long medium size fruits, deeply ridged, tolerant to powdery mildew, Yield18-20 t/ha
Smooth gourd		
Pusa Chikni	IARI	Early variety, smooth dark green fruits, yield 12-15 t/ha
Pusa Sneha	IARI	Suitable for long distance transport, dark green fruits with hard skin and soft flesh, Yield 12 t/ha

Land Preparation and Sowing

Seeds are sown in pits, beds or channels. Seed rate, 3-4 kg/ha. Long, raised beds are formed with furrows 50–60 cm wide. The seeds should be sown on both the edges of beds. In each hill, 3–4 seeds are sown. Retain only 2 healthy seedlings 2.5 m × 1.0 m apart. In South India, pits of 45 cm × 45 cm × 45 cm are dug at a spacing of 2 m x 2 m and filled with farmyard manure and top soil.

Manuring : Similar to that of bittergourd.

Aftercultivation

Application of morphactin-fluresol at 100 ppm to 6 weeks old plants increased proportion of pistillate to staminate flowers. The fruits/plant and yield/plant were also greater than in untreated plants. A single spray of 2-chloro ethyl phosphonic acid (240) ppm at 2-3-leaf stage lowered the number of both staminate and pistillate flowers in monoecious variety, but increased the number of perfect flowers in the hermaphrodite variety. Seed treatment with naphthalene acetic acid (NAA) slightly promoted the formation of pistillate flowers.

Harvesting and Yield

Ready for harvesting in 15-17 days after anthesis. Yield-10-15 tonnes/ha.

Snake Gourd

Snakegourd (*Trichosanthes anguina* L) is an annual creeping cucurbit, remarkable for its fringed corolla and slender, angled and pubescent stems. It occupies a pride of place among vegetables grown particularly in South India. The fruits are harvested and cooked green and have a fairly good nutritive value.

Season

Crop prefers tropical, warm and humid climate. It cannot tolerate frost and grows very well under 25°–30°C. A loamy soil with a pH of 6.0–7.0 is ideal for its cultivation.

Varieties

Varieties in Snake Gourd

Varieties	Source	Features
APAU Shwetha	APAU	Fruits long with whitish background and green stripes. Yields 28.0-30.0 tonnes/ha in 125-130 days.
CO-2	TNAU	Fruits short, 30cm long, stout, pale-greenish-white. Vines less-spreading, it can be planted closely (1.5mx 1.0m), the yield potential being 25.0 tonnes/ha.
MDU-1	MDU	F1 hybrid, fruits medium in length (50–60cm) with white stripes over a green background, can yield up to 35.0 tonnes/ha
PKM-1	TNAU	Fruits extra long (200cm), the yields being 25.0 tonnes/ ha in 135-140 days

Seed rate: 3.0-4.0 kg/ha

Planting, manuring and aftercultivation : Same as that of bittergourd

Harvesting and Yield

Fruits are ready for picking 70–80 days after sowing. Harvesting can be done at an interval of 5-7 days. Average yield 18.0–20.0 t/ha.

Cost Analysis

Cucurbits are the crops with assured market demand and cultivation of water melon, bitter gourd, cucmber and musk melon very often fetches premium price for the growers. Certain agencies such as Vegetable and Fruit Promotion Council of Keralam (VFPCK) promotes the cultivation of these vegetables facilitating bank loan and marketing.

Cost of Cultivation for Cucurbits Requiring Pandal (per ha)

S.No.	Item	Men (Rs.150)	Women (Rs.100)	Total Cost (Rs.)
A	Land preparation			
1	Digging and levelling	50		7500
2	Taking pits	20		3000
3	Opening drainage channels	25		6250
B	Planting materials and planting			
1	Seeds @ 2.5kg/ha			1100
2	Sowing		20	2000
C	Manures and manuaring			
1	Organic manures@ 25t/ha (Basal)			25000
2	Urea			7575
3	Phosphate			4125
4	MOP			2040
5	Application	20		3000
D	After cultivation			
1	Interculturing		20	2000
2	Irrigation	30		4500
3	Pandal materials			12000

Contd...

4	Construction of Pandal	10		1500
E	Plant protection			
1	Chemicals/other materials			7500
2	Application	10		1500
H	Harvest			
1	Harvesting	10		1500
I	Rent for lease land/year @ Rs.50/cent			12500
	Total cost of cultivation			**104590**

Source : VFPCK, Keralam

Cost of Cultivation for Vegetable Non Pandal (per ha)

S No	**Item**	**Men (Rs.150)**	**Women (Rs.100)**	**Total Cost (Rs.)**
A	Land preparation			
1	Digging and levelling	50		7500
2	Taking pits	20		3000
3	Opening drainage channels	25		6250
B	Planting material and planting			
1	Seeds @ 2.5kg/ha			1000
2	Sowing		20	2000
C	Manures and manuring			
1	Organic manures@ 25t/ha (Basal)			25000
2	Urea			7575
3	Phosphate			4125
4	MOP			2040
5	Application	20		3000
D	After cultivation			
1	Interculturing	20		2000
2	Irrigation	30		4500
E	Plant protection			
1	Chemicals/other materials			7500

Contd...

2	Application	1 0		1500
H	Harvest			
1	Harvesting	10		1500
I	Rent for lease land/year @ Rs.50/cent			12500
	Total cost of cultivation			**90990**

Source : VFPCK, Keralam

Legume Vegetables

These are highly nutritious vegetables. They enrich soil by transforming atmospheric nitrogen. Cow pea, French beans and peas are the major legumenaceous vegetables cultivated in India.

Cowpea

Cow pea is grown as a vegetable and as a pulse crop. Pulse type (*Vigna unguiculata* var. *radiata*) and dual purpose type (*Vigna unguiculata* var. *cylindrica*) are grown on a large scale compared to vegetable type *Vigna unguiculata* var. *sesquipedalis.* The typical vegetable cowpea is characterized by long pods (>30 cm), stringless pods, fleshy pod pericarp, thin and long seeds, and higher monosaccharide: polysaccharide ratio. Pole and bush types are also available in vegetable cowpea.

Soil and Climate Factors

Cowpea can be grown on all types of soils, but it does not thrive well in highly acidic or alkaline soil. Crop grows profusely in loamy or slightly heavy soils. Early harvest is possible in light soils and higher yields in slightly heavy soils. It is a warm season crop and can be grown both in spring and rainy season in the plains. Cowpea can be grown at a temperature range of 20-35°C. Higher temperature (> 35°C) causes flower drop while low temperature (<15°C) results in poor flowering and fruitset.

Varieties

Name	**Source**	**Features**
Sel 263	PAU, Ludhiana	Bush type, early maturing, can be grown in both spring and rainy seasons. Pods are green, thick, fleshy, tender and medium (20 cm long); resistant to mosaic and golden mosaic viruses; Yield - 22 t/ha.

Contd...

Pusa Komal	IARI, New Delhi	Bush type, early. Pods are light green, tender, 25-30 cm long; suitable for both spring, summer and rainy season cultivation; resistant to bacterial blight. Yield- 10 t/ha.
Pusa Rituraj	IARI, New Delhi	Bush type, Thin pods, 22–24cm long. Photo and thermo insensitive and can be grown during rainy as well as summer season. Dual purpose variety.
Arka Garima	IIHR, Bangalore	Pole type variety, growing to a height of 2.5–3m. It is more suitable for rainy season and poor seed formation is observed during summer. Pods are light green, long thick, round, fleshy and string less; yield- 18 t/ha.
Arka Samrudhi	IIHR, Bangalore	
Lola	KAU Vellanikkara	Pole type. Light green colour pod with purple tip. Pod length-53 cm-Yield-20 t/ha
Anaswara	KAU Vellanikkara	Semi trailing, Light green colour pod. Pod length-28 cm-Yield-12.5 t/ha

Field Preparation, Sowing and Manuring

Cow pea is grown as a direct seeded crop and it can be grown on ridges or pits depending on growth habit of each varieties. Pole type varieties require wider spacing and ridges may be taken at a spacing of 1-2 m. Bush types may be dibbled at 10-15 cm distance on the ridges taken at 50-60 cm apart. A seed rate of 20-25 kg/ha is required for bush types while 5-6 kg seeds/ha is sufficient for pole types. Cow pea can be grown during rainy and summer seasons.

Manuring and Fertilization

FYM should be applied as basal dosage @ 15–20 tonnes/ha at the time of field preparation. Cowpea crop removes approximately 40-50 kg of nitrogen, 20-30 kg of phosphorus and 40-60 kg of potash/ha. Being a leguminous crop it has low nitrogen requirement. Depending on soil fertility and variety, 20–25 kg/ha N along with 50–70 kg/ha of P_2O_5 and K_2O may be applied as basal dose. The doses of P and K may be based on soil test values to economize on fertilization. Seed inoculation with *Rhizobium* culture is beneficial. In acidic soils application of lime @ 250 kg/ha is beneficial.

Aftercare

There should be sufficient moisture at the time of sowing. While irrigation can be avoided during rainy season, irrigation at 7-10 days interval is required in summer season. Cow pea is sensitive to waterlogging. In lighter soils, irrigation at 75% available moisture is optimum. Weeding, hoeing and earthing up are necessary to keep weeds under control. It is done manually one month after sowing. For chemical control of weeds Alachlor at @ 1 kg ai/ha in 1000 liters water should be sprayed before sowing or Pendimethaline @ 1.0 kg ai/ha in 1000 liters of water may be sprayed within 2 days after sowing. The pole types have to be raised on trellies and local available material should be used for minimizing cost of cultivation. Training of vines helps in spraying operations and harvesting. Spraying of maleic hydrazide (50–200ppm) before flowering and NAA (15 ppm) , 15 and 30 days after sowing improve pod set and yield.

Harvesting

Crop is ready for harvesting in 45-50 days after sowing. Harvesting should be done at 3-4 days interval for getting good quality tender green pots with high yield. Bushy cultivars give 4-5 pickings while from climbing type cultivars, 8-9 pickings can be harvested. For vegetable purpose, pods are harvested 15 days after pod set. Depending on season and variety, grain crop matures in 75–125 days. For seed purpose, the harvested pods are dried for a few days before threshing. The threshed seeds are also dried sufficiently before storage in a cool and dry place.

French Bean

The *Phaseolus* beans are valued grain legumes or pulse crops in many tropical countries. In temperate countries, they are mainly grown for fresh pod consumption and for processing as a frozen vegetable. The vegetable type french bean (*Phaseolus vulgaris* L.) is known by many names such as snap bean, salad bean and green bean. Haricot bean, dry bean and navy bean are applicable only to pulse types while string bean, dwarf bean and pole bean are applicable only to particular growth forms. It is largely grown in hilly areas and cultivated in northern plains as autumn or spring crop.

Soil and Climate Factors

French bean is a tender, warm season vegetable that cannot tolerate frost. The favorable temperature for crop growth, flowering and fruiting is 18-23°C and if temperature drop below 15°C, growth and development are adversely affected. Seeds do not germinate below 15°C and plants drop blossoms in hot or rainy weather. Grow best on well-drained soils in a pH of 5.5 to 7.5. Excessively wet soils encourage root diseases and nutrient problems. Beans do not tolerate excess acidity or salinity.

Varieties

Name	Source	Features
Kashi Param	IIVR, Varanasi	Bush type, Fleshy pods, 14.7 cm length, Yield 12-14 t/ha.
Pusa Parvati	IARI, New Delhi	Bush type, pods ready in 45-50 days. Pods are 15-18 cm long, round, tender, stringless. Yield- about 8.0-8.5 t/ha.
Arka Komal	IIHR, Bangalore	Plants are erect, bushy with bright green stem and foliage, mature in 50 days after sowing
Contender	IARI, Katrain	Bush type, Takes about 50-55 days for first picking. Pods are round, green, 13-14 cm long, stringless, meaty and slightly curved. Yield- 8.0-9.0 t/ha.
Pant Anupama	GBPUA &T	Bush type, with concentrated fruting, Moderately resistant to bean mosaic and angular leaf spot. Yield- 9 t/ha.

Field Preparation and Sowing

Field is ploughed to good tilth and seeds are sown in ridges taken at a spacing of 30-45 cm. A plant density of 100000 to 330000 plants/ha (bush beans) can be accommodated in case of bush types and 20000 - 35000 plants/ha in pole beans. About 50–75kg seed is enough for one hectare crop of bush bean, while 25–30kg is sufficient for pole type varieties.

Manuring and Fertilization

FYM should be applied as basal dosage @ 15–20 tonnes/ha at the time of field preparation. Organic manure is supplemented with chemical fertilizers to obtain good yield. Application of N early in growth is important to promote growth before effective atmospheric N fixation by nodule bacteria. The crop is sensitive to Mg deficiency. The crop is sensitive to excess of B and to deficiencies of Cu, Mo and particularly Zn. For realizing high yield N:P:K @ 50:75:75 kg/ha is recommended. Full dose of P,K and half dose of N as basal and remaining half dose of N is applied at flowering stage as top dressing. Seed inoculation with strains of *Rhizobium phaseoli* favors quick nodulation and high yield.

Aftercare

Pole beans need to be trailed on support. Intercrop such as maize or okra can be used as support for growing trailing types. Earthing up should be done along with top dressing. Crop prefers light irrigation and flowering and pod setting time are critical. Avoid wetting the plant while applying water for minimizing leaf diseases. If possible, irrigate the soil using furrows, drip lines, or soaker hoses. One hand weeding and one shallow hoeing should be done for weed control. For chemical weed control, use Alachlor or pendimethalin @ 1 kg ai/ha in 1000 liters of water within 48 hours of seed sowing.

Harvesting

Beans are harvested at tender stage (about 8–12 days after pod setting). Generally 2-3 harvests are possible in bush type and 5-7 in pole type. Pole beans continue to produce pods as long as the plant remains healthy. Delayed harvesting results in fibrous pods. Yield of bush varieties range 5-8 t/ha whereas, Pole type cultivars yield 8-12 t/ha. Fresh pods can be stored in a good condition for 7-10 days at 4-7°C with 95% relative humidity. In USA, more than 80% crop is processed. Processors have strict requirements for specific varieties and for seed purpose, white colour is preferred. The varieties with fleshy pods are widely used for commercial freezing.

Pea

Green peas (*Pisum sativum* L) cultivated as vegetable, and are marketed fresh, canned, or frozen while ripe dried peas are used whole, split, or made into flour. Green peas are the number one processed vegetable specifically in UK and USA. Green foliage of garden pea is also used as vegetable in parts of Asia and Africa. It is a tender annual and vigorous, knee-high vines. Leaves are modified into tendrils used for attachment to support and climbing. The plants bear white flowers and pods having a number of seeds. In India pea is extensively grown in the temperate zone; but restricted to cooler altitudes in the tropics and winter season in the sub-tropics. The major pea growing states are Uttar Pradesh, Bihar, Haryana, Punjab, Himachal Pradesh, Orissa and Karnataka.

Soil and Climate Factors

Cool temperature is favored for normal growth. Peas grow the best at mean temperature of 13-18°C. The optimum temperature for seed germination is about 22°C. However, it germinates even at 5°C. The crop requires well-drained soils which allow a good symbiotic activity on roots. Preferred pH is 6.0-7.5. Calcareous soils may cause plant chlorosis.

Varieties

Name	**Source**	**Features**
Kashi Nandini	IVRI, Varanasi	Early variety, bears 8-9 cm pods, takes 55 days to first picking. Green pod yield 11-12 t/ha. Tolerant to leaf miner and pod borer.
Kashi Udai	IVRI, Varanasi	Early variety, Pods are 9-10 cm long, Takes 60-65 days to first picking. . Green pod yield -10-11 t/ha.
Kashi Shakti	IVRI, Varanasi	Medium maturing variety, Pods are 10-10.5 cm long. Green pod yield -14-16 t/ha.
Arkel	IARI, New Delhi	Dwarf plant. Early variety, Takes 60-65 days to first picking. Pods are well filled and attractive. Green pod yield -5-6 t/ha.
Harbhajan	IARI, New Delhi	Early variety, Takes 50 days to first picking. Green pod yield 3.0-4.0 t/ha. Seed yield 0.6–0.7t/ha. It is highly susceptible to powdery mildew disease.
Bonneville	IARI, New Delhi	Medium maturing variety, Green pod yield -10-12 t/ha. . The seed yield is 1 t/ha. The variety suffers from the attack of powdery mildew.
Jawahar Mattar 5	JNKV, Jabalpur	Medium maturing tall variety, Green pod yield -11 t/ha. Matures in 130–140 days. The seeds are wrinkled and yellow, yield is 1-2 t/ha. Resistant to powdery mildew.
JP 19	JNKV, Jabalpur	Mid-season garden pea. Dwarf plant. Green pods are large and flat and can be eaten fresh like french beans. The green pod yield 10.0–11.0 t/ha.

Field Preparation and Planting

Field is ploughed to a fine tilth and seeds are sown in ridges or on flat beds. Seed drill is commercially used for placing seeds. The seeds should be sown 2–3cm deep in the soil. The drilling or dibbling can be done in rows placed 30cm apart. The spacing of 10cm from plant-to-plant is ideal. Seed requirement of early dwarf varieties is 125-150 kg/ha and that of medium tall variety is 100-120 kg/ha. Seed inoculation with pea nodule bacteria culture is found beneficial. The seeds may be treated with fungicides like thiram or captan (3 g/kg seed) or Bavistin (2 g/kg seed) to save the crop against wilt. If both the treatments are to be given, first the seeds are treated with fungicide followed by inoculation with rhizobium culture.

Manuring and Fertilization

Pea fixes atmospheric nitrogen through root nodules. Crop requires sufficient organic matter in soil and FYM should be incorporated during field preparation @ 10-15 t/ha. Pea crop approximately removed 50-70 kg of nitrogen, 20-30 kg of phosphorus and 40-60 kg of potash/ha. For average management, 40-50 kg N, 50-60 kg P_2O_5 and 40-50 kg/ha K_2O are recommended. Full does of phosphorus and potassium and half nitrogen are applied as basal dose and remaining N is applied 30-40 days after sowing as top dressing. Adequate P benefits the Rhizobium symbiosis. Low K may cause lack of colour in pods. Foliar application of 0.1% ammonium molybdate is beneficial.

Aftercare

Peas prefer cool dry climate and water requirement is less. It can be grown under rain fed condition provided sufficient moisture is available in the field. Light irrigation at the time of pod setting and pod filling stage improves the yield. Being a direct seeded crop, weed competition at initial stage drastically reduces the yield. Herbicides have been found beneficial in controlling weeds. Usually 1–2 hand-weedings are sufficient. Weeds can also be controlled by early mechanical means but deep harrowing damages the roots. Pre-emergence application of 1 kg alachlor or 1 kg methabenzthiazuron or 1 kg pendimethalin may take care of weeds.

Harvesting

Peas should be harvested at tender stage for ensuring good quality with high sugar content. During maturity, sugar content decreases rapidly and there is an increase in starch and other polysaccharides. Picking should be done as soon as green ovules are fully developed and pods still not over mature. In early variety, 2-3 pickings and in mid season varieties 3-4 pickings are done at 7-10 days interval during the season. Dry peas are harvested when the pods are completely dry and can be threshed directly in the field by a combine. For dry peas timely harvest is important for maintaining quality and is usually done when the seed moisture content is less than 13%.

Grading, Packing and Storage

Generally the edible-pods should be uniformly bright green (light to deep green but not yellow-green), fully turgid, clean, and free from damage. The Bureau of Indian Standards has fixed four grades for peas used for processing depending on the seed size. The pods are generally packed in gunny bags for sending the produce to short distance markets. For long distance markets, the bamboo baskets are generally preferred. Fresh unshelled peas may be kept for two weeks at 0°C with 90-95 % relative humidity. Peas are highly perishable and will not maintain good quality for more than 2 weeks. Wilting, yellowing of pods, loss of tenderness, development of starchiness and decay are likely to increase following storage beyond 14 days. Frozen peas can be stored at 23 to 18°C for one year when packed in gas tight packaging.

Pea is among the four important cultivated legumes next to soybean, groundnut, and beans. Pea is the predominant export crop in world trade and represents about 40% of the total trade in pulses. The major exporting countries, excluding the European Economic Commission (EEC), are Australia, Canada and the USA. In India frozen peas marketed by Mother Dairy India Ltd (MDIL) under the brand name 'safal' pea is highly popular.

Root Vegetables

Carrot (*Daucus carota* L.), radish (*Raphanus sativus* L), beet root (*Beta vulgaris* L) and turnip (*Brassica rapa*) are the important root vegetables grown in India. With the evolution of tropical types in carrot and radish, its cultivation is now spread to non traditional areas. Root vegetables are nutrient rich and comparatively free from pests and diseases. Some of the root vegetables like carrot produced in India has good export potential.

Soil and Climate

Root vegetables in general are better adopted to cool or moderate climate. Asiatic types in carrot and radish tolerate higher temperature. In general, root which is the economically important plant part attains the best flavour, texture and size at 10 to 15°C. Growth is maximum at a day temperature of 20-25°C, but at 25-30°C, the water requirement of plant increases considerably and root deformations occur. Tropical type of radish forms roots even at a temperature of 35-40°C. Crop is very sensitive to low temperature. If they are exposed to relatively low temperature of 4.5-10°C for 15 days or more, bolting is likely to occur before the roots reach marketable size. Being root vegetables, these crops prefer loose friable soil. Crop grows luxuriously in well-textured soils, rich in decomposed organic matter and do not tolerate acid, alkaline or saline soils

Varieties

Important commercially cultivated varieties in different parts of the country are:

Carrot

Commercially Important Carrot Varieties Grown in India

Name	Source	Features
Pusa Kesar	IARI	Red coloured root, self coloured core, suitable for early sowing, matures in 90-110 days. Yield 30 t/ha.
Pusa Meghali	IARI	It has long orange coloured tapering roots with self-coloured core. It is suitable for both, early cultivation (August-September) sowing and late sowing (October-November). Yield 25-30 t/ha
Sel 233	PAU Ludhiana	Orange root, light orange flesh. Yield-20-25 t/ha
Early Nantes	NSC	Temperate type, cylindrical roots, with blunt end, maturity-110-120 days
Pusa Yamdagini	IARI, Katrain	Temperate type, Orange flesh, self coloured core

Radish

Commercially Important Radish Varieties Grown in India

Name	Source	Features
Pusa Reshmi	IARI	Asiatic type, Suitable for mid September to early October sowing. It has white tapering roots of medium to long size (30-45 cm).
Pusa Chetki	IARI	Asiatic type, adoptable to warmer growing condition, Roots medium long stumpy, pure white mildly pungent with soft texture, ready for harvesting in 40-45 days after sowing. Yield-20.0-25.0 t/ha.
Japanese White	IARI	Asiatic variety, matures in 45-50 days. The root is cylindrical and about 20 to 30 cm long, 5 cm in diameter, skin white, less pungent and with a blunt end. Suitable for October to December sowing.
Arka Nishant	IIHR	Asiatic variety. Roots are medium sized (25 cm x 3.4 cm), marble white, crisp texture, pleasant aroma, free from early bolting, pithiness, splitting and forking. Matures in 45-55 days, Yields 20.0-30.0 t/ha within 45-55 days.

Contd...

Punjab Safed	PAU, Ludhiana	Asiatic type, Roots are white, smooth and tapered mild in taste, 30-40 cm long and 3-5 cm thick. It is a quick growing type with roots remain edible for 10 days after attaining full size. It takes 40-50 days for root formation.
Pusa Himani	IARI, Katrain	European type, roots are white, 30-35 cm long, sweet flavoured, Matures in 60-65 days. Suitable for hills and mild climate

Varieties of Radish Suitable for Whole Year Growing in North India

Variety	Period of sowing	Period of harvesting
Pusa Desi	Mid August to mid October	Late September to early December
Pusa Reshmi	Mid September to mid November	Late October to early January
Japanese White	Mid October to mid December	Mid December to early March
Pusa Himani	Mid December to late February	Mid February to mid to late March
White Icicle and RRWT	Late October to February	Mid November to March
Pusa Chetki	Early April to mid August	Early May to mid or late September

Beet root

Commercially Important Beet Root Varieties Grown in India

Name	Source	Features
Detroit Dark Red	NSC	Roots are perfectly round with deep red skin,. Flesh is deep red, fine-grained, tender and corrosive when eaten raw, but exceptionally sweet when boiled. Yield 15.0–20.0 t/ha. It is suitable for salad making and storage.
Crimson Globe	NSC	Roots are globular to flattened globe, medium red with little shoulders. Suitable for canning and pickling. Flesh is dark red, with hardly any zones and non-corrosive in taste when eaten raw. This is a heavy yielder (>200q/ha) and suitable for salad-making and storage.

Turnip

Commercially Important Turnip Varieties Grown in India

Variety	Source	Features
Pusa Kanchan	IARI, Katrain	Round, red colored root with creamy-yellow flesh. Short top. Roots can be kept for a longer time in field without becoming spongy.
Pusa Swarnima	IARI, Katrain	Flattish round roots with creamy-yellow skin and pale-amber coloured flesh. Medium top. It is suitable for growing from June to October in hills and October to December in plains. It matures in 65–70 days.
Pusa Sweti	IARI, Katrain	Early maturing variety. Attractive, white roots, mature 45–50 days after sowing. Suited for October sowing in plains.

Field Preparation and Planting

Field is ploughed to good tilth and seeds are sown on ridges or raised beds. The soil should be free from clods and trashes as it may result in forking and misshaping of roots.

Seed rate

Beet root : 7.5-8 kg/ha

Carrot : 5-6 kg/ha

Radish : 10-12 kg/ha

Turnip : 4-5 kg/ha

Seeds are sown on ridges/raised beds taken at spacing of 30-45 cm. Seeds are dibbled on lines in ridges and a plant spacing of 10–15 cm within the line is maintained by thinning. Seed treatment with bavistin or captan resulted in better seedling emergence and controls the damping off in initial stage. Seed pelleting with biofertilizers like azospirillum and VAM improve the plant stand and early growth. Shallow sowing about 1.5 cm deep gives good germination. Radish seeds germinate within 5–10 days of sowing in properly moist soils while carrot seeds take 15-20 days for germination. In the case of beet root, most of the varieties produce multigerm seeds and 3-4 seedlings emerge from the seed ball. Hence thinning is important for maintaining optimum plant population and quality of roots.

Nutrient Management

Apply well decomposed organic matter to prevent forked root development. Excessive N may also be partly responsible for splitting or forked roots and heavy N treatments also promote foliage growth at expense of root growth. A basic dose of 25 t/ha of well-rotton farmyard manure or compost at the time of soil preparation is beneficial. Quantity of NPK can be reduced depending on the organic matter content of soil. All the root crops need potassic fertilizers in plenty as these advance tuber and root development. The application of 60-100 kg N:40-50 P:50-70 kg K/ha is sufficient for beet root production. In the case of carrot, 25–60kg N, 25–50kg P and 90–110kg K/ha is recommended depending upon fertility of soil. Radish requires a dosage of 50–80kg of N, 50kg of P and 50kg of K/ha. Application of 70-100 kg N and 50 kg each of P and K is required for turnip. A full dose of P and K and one-third of N may be applied as basal dose. Second and third doses of N may be applied as topdressing with equal interval depending on crop growth, 30 days after sowing.

Aftercare

Being direct sown crop, weeding is an important operation and application of stomp at the rate of 3.5 litres/ha immediately after sowing helps to control weeds. Pre-emergence application of TOK E-25 @ 2 kg/ha is also effective in controlling weeds. Care should be taken to ensure proper moisture at the time of application of Stomp. One or two earthing up ensures proper root growth and helps to check weed growth.

In the light soils, the first irrigation may be given soon after sowing which may be followed by next at 4 to 6 days interval depending on soil moisture available. Water deficiency leads to reduction in root size.

Harvesting

Tropical carrot is harvested on attaining a diameter of 2.5–4 cm at the upper end. A light irrigation may be given a day before harvesting to facilitate lifting of roots. Delay in harvesting makes roots fluffy and unfit for consumption. The roots are trimmed and washed before sending them to market. Carrots for processing purposes are left in the ground for a longer period.In general, yield of tropical type ranges from 25–30 t/ha, while European types produce around 10–15 t/ha. Beet root is topped in the field when they attain a diameter of 6.5–8 cm and packed in polybags under cool, and moist storage conditions. Yield varies from 25-30 t/ha. Unlike carrot and beet root, radish should be harvested at tender stage and edible maturity period varies from 25 to 60 days depending upon the cultivar. Top is also used for cooking. Delay in harvesting results in pithiness and bitterness of roots. Yield of tropical type ranges from 15.0–20.0 t/ha while European cultivars produce 5–7 t/ha. Turnip roots are harvested when they are 5–10 cm in diameter depending upon variety. The roots become tough and fibrous, if harvesting is delayed.

Grading, Packing and Storage

After harvesting, topped carrots can be stored at temperatures of 0 - 4.4°C with 93-98% relative humidity for 6 months. Untopped carrots can be stored only for 10-14 days at 0°C. The air in the store house or cellar should be kept humid to prevent wilting and withering of roots. Beetoot is washed to remove soil adhered to roots and then graded according to their size. Removal of tops increases shelf-life of beets by reducing water loss during transit and storage. Topped beets, ideally stored at a temperature of 0°C with 90-98% relative humidity can be kept for 4-6 months. Bunched beets can be stored for 10-14 days at 0°C. Beet roots can also be stored in pits 90–150 cm deep. Before storing, pit should be dry. After filling the pits with roots, a layer of dry soil (30cm) is laid on roots. Then this is covered with dry grass and finally by polythene to protect from moisture, rain, snowfall or low temperature. Like this beet roots can be stored for 4–6 months.

The radish should be uniform and of similar shape for the variety, well formed, smooth, firm but of tender texture. It should also be free from harvest damage, decay, disease or insects. Bunched radish tops should be fresh in appearance and turgid. Radish cannot be stored at room temperature for more than 2-3 days. However in cold storage, topped radish, if packed in polybag with a few holes for aeration, can be stored at 0° C with 90-95 % relative humidity for about 2 months . Atmospheres of 1 to 2% O_2 and 2 to 3% CO_2 are slightly beneficial in maintaining quality of topped radish when storage temperatures are 5 to 7°C. Turnip does not store well. It can be stored for 2-3 days under room conditions. In cold storage, the topped turnips are stored at 0° C with 95% relative humidity for several weeks. Turnips are usually stored in slatted crates or bin that allow good circulation

Solanaceous Vegetables

Most important group of vegetables comprising tomato, egg plant, chilli, sweet pepper and potato, belongs to family Solanaceae having similar cultural requirement except potato.

Tomato

Popular vegetable in India, grown for ripe fruits. India has an area of 4.58 lakh ha under tomato cultivation producing 75 lakh tonnes. Processing industry heavily depends on tomato and leading tomato growing states are Karnataka, UP, Maharashtra, Haryana, Punjab and Bihar.

Varieties

Varieties are grouped into:

(1) fresh market type - fruits are usually red but vary in color, shape, and size;

(2) processing type- fruits have intense red color with 2-3 locules and high pericarp thickness and high solids content.

Tomato varieties are also classified according to plant habit: determinate, semi-determinate, or indeterminate. Determinate and semi-determinate varieties produce stems that end with a flower cluster. Determinates are short and bushy while semi-determinate types grow slightly taller. Indeterminate varieties produce new leaves and flowers continuously and attain a viny growth requiring staking. Harvesting period extends over 3 months.

Variety	Source	Features
Kashi Amrit	IIVR, Varanasi	Determinate variety, round fruit, 108 g, tolerant to TLCV, yield 62 t/ha
Pusa Gaurav	IARI, New Delhi	Determinate, processing type, oblong fruits, Yield-35 t/ha
Pusa Hybrid-2	IARI, New Delhi	Round fruits, medium size, Good keeping quality, field resistance to nematodes , Yield 55.0 t/ha
Hisar Arun	HAU, Hisar	Determinate type, early maturing, round fruits, 70 g, Crop duration120-130 days. Yield -25 t/ha
Arka Vikas	IIHR, Bangalore	Determinate, Oblate fruits, 85 g, tolerant to mosaic, yield 35-40 t/ha
Sakthi	KAU, Trichur	Semi determinate type, flat round fruits, medium size, 48 g, resistant to bacterial wilt, yield 32 t/ha
Arka Vishal	IIHR, Bangalore	Indeterminate, Fruits are large (140 g) with green shoulder. Suitable for fresh market and salad. Crop matures in 165 days. Average yield is 75 t/ha.
Arka Vardan	IIHR, Bangalore	Indeterminate, large fruit (140g) with green shoulder. Suitable for fresh market and salad. Nematode resistance. Crop matures in 160 days. Average yield is 75 t/ha.
Arka Shreshta	IIHR, Bangalore	Semi - determinate with light green foliage. Fruits medium large (70-75 g), round with light green shoulder. Fruits firm with good keeping quality (17 days) and long transportability. Suitable for both fresh market and processing. Resistance to bacterial wilt. Crop matures in 140 days. Yield is 76 t /ha.
Arka Abhijit	IIHR, Bangalore	Determinate with dark green foliage. Fruits round, medium large (65-70 g) with green shoulder. Thick fleshed fruits have good keeping quality (17 days) and long transportability. Resistance to bacterial wilt. Crop matures in 140 days. Yield - 65 t /ha.

F_1 hybrids are highly popular in tomato. A few of the hybrids released from public sector are , COTH-1(TNAU), Arka Vardhan, Arka Vishal, Arka Abhijit (IIHR), Pusa Hybrid-1, Pusa Hybrid2 (IARI) and Swarna Baibhav (HARP, Ranchi).

Private companies are making steady fortune from sale of F_1 hybrid seeds and some of the popular hybrids from private sector are, Rasmi, Vaishali (IAHS, Bangalore), Avinash2 (Syngentia), Sonali, Gulmohar (Mahyco), Swaraksha, Uttam (Namdhari Seeds), Meenaksi, Tolstoi(Beejo sheetal), Sun 496 (Sungrow seeds, Delhi).

Soil and Climate Factors

Tomatoes grow the best under temperatures of 20–27°C. Fruit setting is poor when average temperatures exceed 30°C or fall below 10°C. Most tomato varieties are adapted to cool dry season. Hot set varieties can be used for growing in summer and wet season for exploiting offseason market. However, successful offseason production requires a combination of a suitable variety and special management practices, such as use of fruit-set growth regulator, raised beds or shelters to manage excessive rainfall. Prefer a well-drained soil and the crop is sensitive to water logging. Optimum soil pH is 6.0-7.0.

Nursery

Commercially raised as transplanted crop. About 300-350 g of seeds of OP varieties and 125-175 g of F_1 hybrid will be required to plant one hectare. Sow seeds 0.5 cm deep. Seedlings emerge within 8 days at the optimal soil temperature of 20-30°C. Seedlings grown in trays with individual containers are healthier and more vigorous compared to those grown in beds.

Seedbed method : Choose a well-drained area not recently cropped with a solanaceous crop. Burning a 3-4 cm layer of rice straw on the seedbed before sowing and forming a raised seedbed of 15 cm or higher to improve drainage will reduce incidence of soilborne diseases. Solarization is also useful to reduce incidence of damping off. The seeds should be treated with Thiram or Bavistin @ 2 g/kg. Sow the seeds in rows 6 cm apart at a rate of 750-900 seeds/sq.m. Cover the bed surface with a thin layer of straw mulch. Water the seedbed regularly so that it is moist but not waterlogged. Thin the seedlings 2-3 days after the first true (non-cotyledon) leaves appear. At the 2-leaf stage, irrigate seedlings once with a 0.5% ammonium sulfate solution or 0.25% urea solution. Application of neem cake / castor cake / neem leaf / castor leaf / pongamia leaf / calotropis leaf @ 400 g/m2 of nursery bed followed by application of spore suspension of a nematophagus fungus *Paecilomyces lilacinus* or

Verticillium arbuscular mycorrhizae fungi or *Glomus fasciculatum* or *G. mossae* will reduce the nematode infestation and increase the multiplication of endomycorrhizae and its colonization of tomato root, which in turn, could protect the crop from nematodes in main field also. Seed treatment with azospirillum also effective in promoting seedling growth. Application of copper oxychloride and furadan is recommended for preventing incidence of damping off and pest attack.

Mainfield Planting

Transplant 4 weeks old seedlings in late afternoon or on a cloudy day to minimize transplant shock. Spray of cycocel (200 ppm) and zinc sulphate (0.25%) + proline at the time of transplanting improves hardening. Tall, spindly seedlings should be placed deeper. Press soil firmly around the root, and water around the base of the plant to settle the soil. Irrigate the field as soon as possible after transplanting. Plant spacing depends upon cropping system, soil type, and plant habit. Various spacings followed are 60 x 30-45 cm, 75 x 60 cm and 75 x 75 cm. The field is ploughed three or four times. After last ploughing, 20-25 tonnes of farmyard manure have to be applied. Ridges and furrows are formed at a required spacing.

Nutrient Management

Tomato plants should be fertilized with organic manure and chemical fertilizers to produce high yields. N requirements are moderate during foliage growth until fruit set. P is very important for vigorous growth and fruit production. Root inoculation with *Glomus intraradices* (vesicular-arbuscular mycorrhizal fungi) improves growth through improved P uptake and decreased *Fusarium oxysporum* disease incidence. K is required for fruit set and enlargement. Important growth stages for nutrient uptake include seedling establishment when maximum nutrient efficiency uptake occurs, and early fruiting when the maximum rates of K accumulation take place. Tomato should be supplied with Mg in soils deficient in this nutrient. Ca deficiencies may cause "blossom end rot" or apical necrosis.

Fertilizer uptake efficiency by a crop is highly variable and depends upon many factors including fertilizer form and placement (surface versus incorporated versus banded), as well as irrigation and other management practices. The hybrids require more amount of nutrients to express their full yield potential and the recommendation is 250:250:250 kg NPK/ha.. In the tropics, for OP varieties, common fertilizer application rates are 60-120 kg N/ha, 60-140 kg P_2O_5/ha, and 60-120 kg K_2O/ha. It is recommended that half of the fertilizer be applied as basal dose and the remaining fertilizer be added as top dressing during first fruit-set and subsequent harvest. Fertilizer applications through a drip irrigation system may improve the efficiency of N uptake.

Water Management

Crop is the most sensitive to water deficit during flowering, after transplanting and during fruit development, and the least sensitive during vegetative growth. The root zone of young transplants is shallow so irrigation should be frequent and just enough to recharge the root zone. The root zone enlarges during crop development which permits less frequent but heavier irrigation Tomato plants are sensitive to water logging and flooded fields should be drained within 1-3 days.

Training and Pruning

Indeterminate varieties should be staked to facilitate pruning, pinching, harvesting, and other cultural practices. Determinate varieties should be staked in wet season to prevent fruit contact with soil. Staking or trellising tomato plants with bamboo poles, wood stakes, or other sturdy material provides support and keeps the fruit and foliage off the ground. Staking increases fruit yield and size, reduces fruit rot, and eases spraying and harvesting. Pruning is the selective removal of side shoots to limit plant growth and this will cause fruit to mature earlier and grows to greater size and uniformity. Pruning improves air circulation within the canopy, which reduces foliar diseases, and facilitates spraying and harvesting. Indeterminate varieties should always be pruned so they do not produce too much vegetative growth. The degree of pruning varies according to the season and variety. Indeterminate varieties can be pruned to one stem during the cool-dry season and maintain two branches/ plant (main branch and second branch below the first fruit cluster) during the hot-wet or hot-dry seasons. Two branches provide more shade to the fruit, thereby reducing sunburn and blotchy ripening.

Weed Management

Weeds should be controlled in tomato crops because they compete for light, water, and nutrients. Sometimes, they host tomato diseases, such as tomato yellow leaf curl virus. Mulches suppress weed growth on the beds. Remove weeds from furrows by pulling or hoeing. In the main field, a pre planting application of 1 litre of fluchloralin (as basalin 2lit/ha) dissolved in 500 litre of water is done just before transplanting. The field is to be irrigated immediately and planting of seedlings has to be taken up. Post emergence spray of Sencor (0.5 kg a.i/ha) is also effective.

Aftercultivation

Earthing up and top dressing of N and K fertilizer is done 30-35 days after transplanting. Immediately after earthing up, the field is copiously irrigated. Foliar application of 0.1% $CaCl_2$ during fruitset and maturity help to prevent fruit cracking.

Harvest

Depending on final usage, tomatoes have to be harvested at specific stages of maturity. Green stage (when tomato is fully developed but green) is suitable for transport to distant markets; pink stage (when the tomato is partially red or pink but not fully ripe) is the best suited for local markets; ripe stage (the tomato mostly is red and begins to soften) is better used immediately and full ripe stage (fully red and soft) is most suited for processing purposes. Post-harvest technologies such as use of cold stores, delay of ripening and spray drying can go a long way in preserving tomatoes, particularly when either the prices shoot up or hit rock bottom.

Grading

During grading of fruits, damaged, rotten and cracked fruits should be removed. Only healthy, attractive, clean and bright fruits should be selected. On the basis of size of fruits, three grades are formed: small (<100 g), medium (100-255 g) and large (> 255 g). Retailers normally do size grading for local market. Internal urban markets have differential prices for size grades as against ungraded fruit. Bureau of Indian Standards specified 4 grades viz., Super A, Super, Fancy and Commercial for tomato crop.

Packaging

Using good and suitable packaging material is important to preserve harvested tomatoes for a reasonable time. Technologies such as minimal processing, hurdle technology and use of ethylene-absorbent packaging material will help in shelf life extension of tomato. For local markets, fruits are packed in bamboo baskets or plastic crates. Plastic crates can be conveniently stacked one on the other and a contoured rim keeps the product safe and natural and allows sufficient air circulation. The packing should ensure careful handling i.e. rigid enough to protect fruits from being crushed. For exports, fruits are packed in cardboard telescopic boxes with capacities of not more than 15 kg, should be used.

Storage

The main objective in storage after harvest is to control rate of ripening to extend marketing period. As tomatoes are chilling sensitive, the recommended storage temperatures differ depending on fruit maturity. A storage temperature of13° C with 90-95% relative humidity is recommended for slow ripening. At this temperature, most varieties keep in good condition for 2-3 weeks and change colour very slowly. In cold storage, unripe tomatoes can be stored for 4 weeks at a temperature of 8-10° C with 85-90 % relative humidity. Fully ripe fruits are stored at 7° C with 90% relative humidity for 1 week.

Cost of Cultivation

Tomato is a remunerative crop and cost analysis in tomato cultivation with determinate and indeterminate F_1 hybid is given below

Estimated costs and returns of tomato per acre in Kolar district during 2000-01.

S No	Items	Long duration variety		Short duration variety	
		Qty	Value(Rs)	Qty	Value(Rs)
1	Land preparation		1650 (6)		1650 (7)
2	Planting material(No.of seedlings)	7000	2800 (11)	7000	3500 (14)
3	FYM (Tractor Load)	1.5	900 (4)	1.5	900 (4)
4	Fertilizers A. Urea (Kg) B. DAP (Kg) C. Potash (Kg) D. 19:19:19 (Kg)	 50 50 50 100	 230 (1) 475 (2) 225 (1) 400 (2)	 50 50 50 50	 230 (1) 475 (2) 225 (1) 200 (1)
5	Man days@ Rs. 60/- day	35	2100 (8)	25	1500 (6)
6	Woman days@ Rs. 35/- day	75	2625 (10)	60	2100 (9)
7	Plant Protection Chemicals	-	2500 (12)	-	2200 (8)
8	Irrigation cost@ Rs. 100/- acre inch	20	2000 (8)	20	2000 (8)
9	Interest on working capital@ 12.5%	-	1976 (8)	-	1810 (7)
10	Rental value of Land(for one year)	-	7500 (29)	-	7500 (31)
11	Cost of cultivation	-	25881 (100)	-	24090 (100)

Contd...

12	Transportation cost @ Rs 500/ tones	-	17500	-	10000
13	Cost of production(1 to 12)	-	43381	-	34090
14	Yield (in tonnes)	50	-	35	-
15	Gross returns (Price @ Rs 1.5/ Kg)	-	75000	-	52500
16	Net returns (Rs)	-	31619	-	18410

Figures in paranthesis indicate percentage
Source : Lokesh *et al.*, 2005.

Brinjal

This is a typical Indian vegetable grown in all seasons and in more areas compared to tomato and chilli. The major Brinjal producing states are Uttar Pradesh, Orissa, Bihar, Assam, West Bengal, Maharashtra and Karnataka. The white brinjals have a few medicinal properties. Brinjal (syn. egg plant, garden egg, guinea squash, cubergine) is a warm-season fruit vegetable highly susceptible to frost. It is an erect bush with large, fuzzy leaves and stiff, hard stems. The plants usualy usually grow 45 to 60 cm high and bear oval to long shaped fruits purple or greenish in colour.

Soil and Climate

A deep fertile silt loam or clay loam soil with a pH of 5.5 to 6.0 is the best for growth and development of brinjal. A sandy loam soil is ideal when an early yield is desired. Heavy clay and saturated soils should be avoided due to build-up of root-rotting diseases. Grows very well in a temperature range of 25°-30°C. Very low temperature especially below 15°C will affect growth of plant as well as fruit quality. Crop tolerates drought and excessive rainfall, but struggles to grow when temperatures exceed 30°C. At high temperature and humidity, brinjal becomes more vegetative. Heavy rains during flowering and fruit set hinder crop growth and retard yield.

Varieties

Selection of variety is an important consideration in the cultivation of brinjal as local preference varies considerably from region to region.

Popular Varieties of Brinjal

Cultivars	Bearing maturity	Fruit		Average yield (t/ha)	Recommended for the region; special features
		Shape	Colour		
Pusa Kranti	Medium	Long	Dark-purple	35	All over India
Pusa Bhairav	Early	Long	Dark-purple	30	Resistant to Phomopsis fruit rot
Pusa Purple Cluster	Medium	Long cluster	Violet purple	30	Hilly regions and mild climates, tolerant to bacterial wilt
Pusa Anupam	Medium	Long, Cluster	Purple	30	Resistant to Phomopsis fruit rot
Pusa Uttam	Early	Oval	Dark-purple	40	Northern plains, Gujarat and Maharashtra
Pusa Upkar	Early	Round	Dark-purple	40	-do-
Pusa Bindu	Early	Small, oval-round	Violet purple	30	-do-
PH-4	Medium	Long	Purple	30	All over India
Pb. Neelam	Medium	Round	Purple	30	-do-
Pb. Barsati	Medium	Long, cluster	Purple	45	-do-
Pb. Sadabahar	Early	Long	Black purple	30	Punjab
Hisar Shyamal	Medium	Round	Purple	35	Haryana
Hisar Jamuni	Medium	Oblong	Dark-purple	35	-do-
Pant Samrat	Early	Long	Purple	35	All over India, tolerant to bacterial wilt
Pant Rituraj	Medium	Oval-round	Violet purple	35	All over India
NDB-25	Medium	Long	Purple	35	Uttar Pradesh
Vaishali	Medium	Oval	Bicolour	25	Maharashtra
Pragati	Medium	Oval	Purple	35	-do-
Arka Keshav	Medium	Long	Purple	45	Resistant to bacterial wilt
Arka Nidhi	Medium	Long	Purple	48.5	Resistant to bacterial wilt
Arka Neelkanth	Medium	Long	Purple	43	Resistant to bacterial wilt, No bitter principle

Contd...

Surya	Late	Oval	Purple	30	Resistant to bacterial wilt
Swetha	Medium	Long	White	30	Resistant to bacterial wilt
Haritha	Late	Long	Light green	62	Resistant to bacterial wilt

F_1 Hybrids of Brinjal.

Hybrids	Source	Bearing maturity	Fruit		Average yield (t/ha)	Recommended for the region; special features
			Shape	Colour		
Arka Navneet	IIHR	Early	Oval	Dark purple	60	Mild climates of southern India
Pusa Hybrid-5	IARI	Early	Long	Dark purple	50	Northern central and southern states
Pusa Hybrid-6	IARI	Early	Round	Dark purple	45	Northern and central India
Pusa Hybrid-9	IARI	Early	Round	Dark purple	56	Maharashtra and Gujarat states
NDBH -1	Faizabad	Early	Long	Dark purple	52	Northern and central India
NDBH -6.	Faizabad	Medium	Round	Dark purple	53	Maharashtra and Gujarat states
ABH -1	GAU	Early	Oval	Purple	37	Northern and central India
ABH -2	GAU	Early	Oval	Purple	39	Maharashtra and Gujarat states
Neelima	KAU	Early	Oval	Purple	65	Bacterial wilt prone area

Popular F_1 hybrids marketed by private seed companies include Supriya, Sourabha, Suphal (IAHS), Kalpatharu, Ravaiya (Mahyco), Apsara, Nisha (Namdhari), Kanhaiya and Novkiran (Sungro seeds).

Nursery

About 250-300 g of seeds (treated with 2g. of Thiram/bavistin) are required to raise seedlings to plant one hectare. Preparation of nursery bed and nursery care are same as that described for tomato.

Field Preparation and Transplanting

The field is prepared by ploughing four or five times. At the last ploughing, 25-30 tonnes of farm yard manure are applied. Ridges and furrows are formed at a spacing of 60- 75 cm. Healthy and stocky seedlings which are free from diseases and shoot-and fruit-borer infestation having 3–4 leaves are transplanted. The spacings generally recommended are 60 cm × 45 cm, 60 cm × 60 cm, 75 cm × 60 cm and 75 cm × 75 cm depending upon maturity and spread of plant. Highly vigorous and late-bearing varieties are given wider spacing compared with dwarf, upright and early types. Two kg in each of Azospirillum and Phosphobacteria can be mixed with 40 kg of sieved farm yard manure and applied evenly for one hectare.

Nutrient Management

Brinjal should be fertilized with organic manure and chemical fertilizers to produce high yields. Nutrition of brinjal resembles that of tomato and depends on soil nutrient status, season and variety used for cultivation. Generally, 100–120kg N, 75–85kg P and 45–50kg K are applied. Foliar application of 1% urea increases growth and fruit set. The fertilizer is applied along one side of ridges as band and mixed with soil. Part of N and K fertilizer should be topdressed at the time of earthing up one month after transplanting.

Water Management

Irrigation is most critical immediately after transplanting, flowering and fruit set. A lack of water during this period could lead to development of blossom end rot and malformed fruit. Reduction of fruit size and yield are also caused by moisture stress. Brinjal is a medium-rooted crop with a root zone depth of 90 cm and irrigation up to 45 cm depth is recommended for good growth.

Training and Pruning

Staking is recommended one month after transplanting with a bamboo or coir/nylon rope to support the plant from fruit load. Pruning is recommended to produce bright-colored, high quality fruit. Maintain three branches/ plant: two branches from primary division of main stalk and one branch below this division. All the other lateral branches should be removed periodically. Application 2, 4-D (5 ppm), GA (50ppm), and 4 CPA (20 ppm) increase the fruit set and ultimately the yield. The main limitations to high yield in brinjal are incidence of phomopsis blight and attack of fruit-and shoot-borer.

Harvesting, Grading and Storage

Eggplant fruits are harvested once they have reached sufficient size for marketing (usually three to four weeks after flowering). Harvesting is done by hand using a sharp knife or clippers, leaving the calyx attached to the fruit. Yield is in the range of 30 to 40 tonnes/ha. Eggplant does not have a long storage life and should be marketed immediately after harvest. The fruits are graded according to their size and colour. The Bureau of Indian Standards recommended three grades for brinjal, viz., Super, Fancy and Commercial.

Fruits are generally sorted by size and color, and packed into either baskets or cartons. Eggplant can be stored safely for 7 to 10 days at 7-10 °C and 90-95% relative humidity. It is subject to chilling injury when stored at temperatures below 7 °C.

Chilli

Chilli (syn. pepper) is an important cash crop in India where it occupies 0.915 million hectares producing 1.018 tonnes of dry fruits. In addition to hot chilli, bell pepper (capsicum), and paprika are also grown on a large scale. Andhra Pradesh is the leading producer of chilli in India, contributing 25 % of area and 40 % of production.

Soil and Climate

Chilli is better adapted to hot weather than is sweet pepper, but fruitset is affected when night temperatures are greater than 24°C. Optimum day temperature ranges from 20 to 30°C. When the temperature falls below 15°C or exceeds 32°C for extended periods, growth and yield are usually reduced. It is insensitive to photoperiod. Grows the best in a loam or silt loam soil with good water-holding capacity, but can grow on many soil types, as long as the soil is well drained. Soil pH should be between 5.5 and 6.8.

Varieties

Important varieties and hybrids of chilli

Variety	Source	Features
Aparna	RARS, APAU, Lam	Pods 8.3 cm long, cylindrical with semi-cup shaped calyx and blunt tip. Retain colour in storage. High pungency (0.7%). Average yield (dry) 11.7q/ha.
Bhagyalakshmi	RARS, APAU, Lam	Immature pods olive-green turning dark red on ripening,Yield (dry) 11.7q/ha under rainfed and 50q/ha under irrigation. Pods 7-8cm long and 3cm in girth

Contd...

G3	RARS, APAU, Lam	Pods, 6-7cm long and 3cm in girth. Immature green and mature red fruits remain smooth after drying, suitable for rainfed and irrigated conditions, pungency high, average yield (dry) 10.6q/ha
Arka Lohit	IIHR	Dual purpose, highly pungent chilli with an average yield of 3.5 tonnes/ha (dry)
Pusa Sadabahar	IARI	Pods erect, 8cm long and 6-14/cluster. Ripe pods bright red, pod yield (dry) 15-20 q/ha, Resistant to CMV, TMV, leaf-curl viruses
Co2	TNAU	Pods small, 2.4cm long with dark red colour having 7-8 seeds/pod, Crop duration of 200 days, Average yield (dry) 22.50q/ha
Co3	TNAU	Suitable for high-density planting, Dual purpose chilli, with a yield potential of 15-18 tonnes/ha (green chilli) or 35q/ha (dry chilli)
K2	TNAU	Tall plants, semi spreading, long fruits with blunt tip, tolerant to thrips, yield 1.8 t/ha (dry chilli)
Pant C1	GBPUA&T	Fruits erect, small-sized, highly pungent. Yield (dry) 12q/ha. Tolerant to mosaic and leaf-curl virus
Pant C2	GBPUA&T	Fruits pendant, average yield (dry) 14q/ha. Tolerant to mosaic and leaf-curl virus
Jawahar 218	JNKV, Jabalpur	Pods erect, early and high-yielding, average pod yield (dry) 18-22q/ha. Suited for green and red chilli production. Tolerant to leaf-curl and fruit-rot
Musalwadi	MPKV, Rahuri	Fruits small, dark green with black patches. Dual purpose chilli. It is tolerant to die-back and powdery mildew.

Contd...

Punjab Lal	PAU, Ludhiana	Pods erect, medium-sized, dark green, turn dark red at maturity. Rich in capsaicin (0.7%), Resistant to TMV, CMV and leaf-curl virus. Moderately resistant to fruit-rot and die-back. Avarage fruit yield (fresh) 100-120q/ha
CH-1	PAU, Ludhiana	F_1 hybrid, Medium tall plant (92.6 cm) bearing medium size fruits(6.62 cm). Light green fruits turning deep red on ripening, highly pungent, tolerant to fruit rot . Yield 25 t green chilli /ha.
Ujjwala	KAU, Vellanilkkara	Chilli with erect fruits borne in clusters. Fruit length 6.17 cm, fruit wt. 2.54 g, fruits/ cluster-9-10, Yield-18 t green chilli/ha. Resistant to bacterial wilt.
Anugraha	KAU, Vellanilkkara	Solitary, linear, pendulous fruits which are light green at immature satge. Av. Fruit length 12 cm, Av fruit wt 3.6 g, Yield 27 t green chilli/ha.
Kashi Anmol	IIVR, Varanasi	Detereminate plant (60-75 cm height), pendant dark green fruits, Yield 25 t/ha in 150-160 days
Kashi Vishwanath	IIVR, Varanasi	F 1 hybrid, tall plants (110-120 cm), pendant long fruits (10-12 cm), yellowish green in immature stage and turns bright red at maturity, suitable for dry chilli production, yield 20 t/ha in 220 days
Kashi Early	IIVR, Varanasi	Hybrid suitable for green as well as dry fruit production, pendant long (8-9 cm) fruits , dark green at immature stage turning bright red on maturity. Yield 30 t/ha in 220 days.

VNR-277, NCH-168, BSS 450, ARCH-32, Arya Rajlexmi and Sitara are promising Chilli hybrids grown at National level.

California Wonder, Arka Gaurav, Bharat Hybrid and Pusa Deepthi are a few the popular bell pepper varieties grown in India.

Nursery

Around 1 kg of seed is required for transplanting to one ha. To minimize seed transmission of fungal pathogen, soak seeds in warm water (50°C) for 30 minutes, rinse them in cold water, and dry before sowing. Apply a fungicide seed coating, with 0.1 % Benomyl or Thiram . To minimize seed transmission of virus, soak 2 g of seeds in 10 ml of 10% (w/v) trisodium phosphate for 30 min. The potting mixture should have good water-holding capacity and good drainage. In raised soil bed, the soil should be sanitized by burning a 5-cm thick layer of rice straw or other dry organic matter on the bed. This also adds small amounts of P and K to the soil for the seedlings. Sow seeds in shallow raws taken at 5 cm apart or broadcast the seeds lightly in a seedbed. Other treatments are similar to that described for tomato.

Applications of biofertilizer as seed treatment and seedling treatment are beneficial. For seed treatment, 200 g of azospirillum should be mixed with 200 ml of boiled rice water and shade dried for 30 minutes before sowing. For seedling treatment 400 g of azospirillum should be mixed in 20 litres of water and used for seedling dip treatment for 15 minutes.

Field Preparation and Transplanting

The field is prepared by ploughing four or five times. At the last ploughing, 25-30 tonnes of farm yard manure are applied. Ridges and furrows are formed at a spacing of 60- 75 cm depending on season and variety. Seedlings are ready for transplanting in five-six weeks after sowing. The ideal seedling should have 4–5 true leaves, disease-free, stocky, and has no flowers.

Nutrient Management

Chilli yielding 21 t/ha of fresh fruit removes 70 kg N, 16 kg P and 92 kg K from one ha. As a basal dressing, 60 kg of phosphorus and 30 kg potash are applied along with 40 kg nitrogen along one side of ridges and mixed with soil. At least 50 % of total fertilizer N should be NO3-N. Maximum absorption of nutrients occurs in the first 8 to 14 weeks of growth and again after the first fruit removal. N should be top dressed @ 40-50 kg/ha in split application at the time of flowering and in the fruit initiation stage. Improved N use efficiency and greater yields are achieved when N is applied under drip irrigation system. Blossom-end rot results from Ca deficiency which may be corrected with foliar sprays of calcium chloride or calcium nitrate. The crop is also sensitive to Mg deficiency and has a low salt tolerance, but root inoculation with vesicular-arbuscular mycorrhizal fungi can improve growth under salt stress conditions. Bellpepper responds better than chilli to higher dosage of nutrients and top dressing with N on each harvest will improve the yield.

Aftercare

Chilli is shallow-rooted and havs low tolerance to drought or flooding. Fields should be irrigated if there are signs of wilting at midday. Furrow or drip irrigation are better and overhead irrigation should be avoided as wet leaves and fruits promote disease development. Plants cannot tolerate flooding and fields should be drained quickly after heavy rain. Chilli plants wilt and die if they stand in water for more than 48 hours. Phytophthora blight and bacterial wilt cause total crop loss following prolonged flooding. Earthing up and weeding are done 30-40 days after transplanting. Foliar application of antitranspirent such as triacontanol 1.25 ppm increased the photosynthetic efficiency and thereby the yield. Foliar spray of NAA (50 ppm) (equal to 1 ml of Planofix mixed with 4.5 lit. of water) improved the fruitset and reduced flower and fruit drop.

Harvesting and Processing

For fresh use, chilli can be harvested either at green immature or red mature stage. It takes about 55–60 days after flowering for fruits to fully ripen, depending on temperature, soil fertility, and cultivar. Fruits can be harvested at weekly intervals. Fresh chilli fruits should not be washed unless they will be kept cool (10°C) until sold. At typical tropical ambient temperature and humidity (28°C and 60% RH), fruits will last unspoiled for 1–2 weeks. Anthracnose is the major cause of fresh fruit spoilage. For dry chilli, it's important to preserve red color of mature fruits. Drying them in the sun is a common practice, but this will bleach the fruits. Whole chillies can be dried by spreading on perforated rectangular Aluminium trays @ 5 kg/m ^{2}. It took about 15 days to dry at room temperature of 20-25 °C and relative humidity of 34-50%. Prickling the chillies longitudinally reduced the drying time to 12 days and blanching in 2 % K_2CO_3 solution reduced it to 7 days.

Marketing

Major marketing centers in chilli are located in Maharashtra, AndhraPradesh, Karnataka and Tamil Nadu.

Major Chilli Marketing Centres in India

State	Centres
Maharashtra	Nasik, Ahmed Nagar, Sholapur, Nanded, Amaravathi, Lasalgon
Andhra Pradesh	Guntur, Warangal, Hyderabad, Cuddapath, Vijayawada, Rajmundri, Nellore
Karnataka	Dharwad, Mysore, Hassan, Bangalore, Bellary, Ranibennur, Hubli, Byadgi
Tamil Nadu	Pollachi, Ramanand, Madurai, Trichi, Thani, Dindigul, Virudunagar

All the predominant marketing channels operating for other vegetables are also operating for chilli. Marketing Federation (Markfed), at the instance of State Governments is entering the market to prevent falling prices. The commission agents still take a major share of consumers price. Cost of cultivation in chilli is going up due to high cost of labour and plant protection chemicals. The hired human labour consumes 18-22% of total cost of cultivation in Andhra Pradesh

Breakup of Total Production Costs in Chilli for one ha (Rs)

Items	Small farm	Large farm	Average
Hired human labour	2505 (18.85)	3502 (22.18)	2779 (20.54)
Bullock labour	685 (5.15)	507 (3.48)	596 (4.41)
Machine labour	82 (0.61)	307 (2.69)	195 (1.44)
Seeds	363 (2.73)	370 (2.69)	367 (0.271)
Manures and fertilizers	1825 (13.74)	2258 (16.41)	2042 (15.09)
P.P chemicals	2470 (18.58)	2815 (20.15)	2643 (19.54)
Irrigation	125 (0.94)	125 (0.91)	125 (0.92)
Depreciation	216 (1.62)	182 (1.32)	199 (1.47)
Interest on working capital	431 (3.24)	499 (3.63)	465 (3.44)
Cost A1	8702	11115	0411
Interest on fixed capital	151(1.13)	126 (0.91)	139 (1.03)
Rental value of owned land	2770(20.84)	2747 (19.96)	2759 (20.39)
Cost B	11623	12988	12309
Family labour	1665(12.53)	774 (5.63)	1219 (9.02)
Cost C	13287	13762	13528

Source: Prasad *et al*, 1980

Export

India has immense potential to grow chillies. India is the major exporter of chillies and around 8 % of total production is exported. WANA region (West Asia and North Africa), Common wealth of Russia and other E.E. countries and Western Europe, USA, Canada, Japan and Australia are major destinations of Indian hot chillies. The share of EU market in the total export of Indian chillies is around 10%. Chilli oleoresin and curry powder have also good export potential. Following detection of Sudan 1 Red and aflatoxin in consignments of chillies exported by certain exporters in Mumbai, the

European Union had mandated that each consignment of hot chilli and chilli products imported into EU must accompany an analytical report demonstrating that the chilli products do not contain Sudan-1. Exporters intending to export chilli powder or any other spice product containing chilli, excluding chilli oil and oleoresin, to any destination, should get the certificate from Spices Board. The tolerance limit for chillies fixed by the European Union is 10 ppb aflatoxin while it is 15 ppb in the US.

Potato

India is the world's fourth largest producer of potatoes in terms of both production and area (1.50 million ha). However and despite significant increase in productivity /hectare from 6.59 tonnes (1949) to 16.71 tonnes, productivity/ hectare is low in comparison to potato production in Europe and the USA. It is extensively grown in North Indian plains contributing 85% of the total acreage and remaining 15% lies in the hills & plateau. With the inception of new processing firms, potato cultivation is spreading making bounties to farmers. Potato is a short duration crop that fits well in different multiple and inter-cropping systems. Potato protein has a well balanced amino acid content and biological value of its protein is comparable to that of eggs and milk.

Soil and Climate

Light fine friable soils near neutral pH is ideal for potato cultivation. Maximum area under potato is in alluvial soils, followed by hill, black and red soils. Well drained coarse or sandy loam to loamy soils, rich organic matter are ideal for growth of potato tubers.It tolerates pH in the range of 5.5 to 8.0. Potato is a short day crop and requires cool climate for tuberisation. Generally potato crop is raised in India when maximum day temperatures are below 35^{0}C and night temperatures are not above 20^{0}C. Night temperature is critically important to tuber development and should not exceed 20°C (68°F), a factor which limits the season in many areas in the south, as frost limits the season in higher altitude regions of the North. Crop is grown during autumn/winter/ spring seasons in plains and during summer/autumn in the hills. In areas enjoying mild climate, potato can be raised throughout the year.

Indian Potato Varieties

Central Potato Research Institute, Simla has bred 34 high yielding varieties adapted to the short-day conditions and short growing periods (60-110 days) of India. Among the many characteristics of these varieties,a few show resistance to late blight, potato cyst nematodes and wart disease, others show tolerance to virus and others are suitable for processing. CPRI currently produces about 2,600 tonnes of basic seed/annum, which is then distributed to various State Department's of Agriculture for multiplication before distribution to certified seed growers.

Salient features and zones of adaptability of different varieties released over last 20 years.

Variety	Salient Features and Adaptability
Kufri Lauvkar	Early maturing and rapid bulking under warmer conditions. Plateau region of peninsular India.
Kufri Dewa	Medium-maturing, good keeping quality and resistant to frost. Tarai area of Western Uttar Pradesh (Uttrakhand).
Kufri Badshah	Medium-maturing, resistant to both late and early blight and PVX. North Indian Plains and plateau region of peninsular India.
Kufri Bahar	Medium-maturing and heavy yielder. North Indian plains.
Kufri Lalima	Medium-maturing, with red tuber. North Indian plains.
Kufri Sherpa	Medium-maturing, resistant to late blight and immune to wart. Hills of West Bengal.
Kufri Swarna	Medium maturing, resistant to late blight and cyst nematode. Southern hills.
Kufri Megha	Medium maturing and late blight resistant. Hilly region of Meghalaya.
Kufri Ashoka	Shortduration (75 days). Plains of central and eastern Uttar Pradesh, Bihar and West Bengal.
Kufri Jawahar	Medium-maturing, resistant to late blight and ideal for inter cropping. Punjab, Haryana, and the plateau regions of Madhya Pradesh, Gujarat and Karnataka.
Kufri Sutlej	Medium-maturing and resistant to late blight. Western and central Indo-Gangetic plains.
Kufri Pukhraj	Medium-maturing, and resistant to late blight. Northern plains and plateau region.
Kufri Chipsona 1	Medium-maturing, resistant to late blight and excellent for processing North Indian plains.
Kufri Chipsona 2	Medium-maturing, resistant to late blight and excellent for processing. Uttar Pradesh and Bihar.

Contd...

Kufri Giriraj	Medium to late maturing and resistant to late blight. North Western hills.
HT/92-61	Hybrid, Heat tolerant, resistant to leaf hopper and mites, high drymatter content, suitable for making French fries.
JW-160	White Hybrid, having field resistance to late blight, excellent keeping quality
MS/92-2105	Red skinned high yielding hybrid, oval attractive tubers, field resistance to late blight
SM/87-185	Late blight resistant, white tuber hybrid with high dry matter content and better keeping quality
Kufri Surya	Early maturing, heat tolerant and hopper-burn resistant potato variety with oblong tubers, white smooth skin and pale yellow flesh.
Kufri Arun	Medium maturing, main season, high yielding table potato variety suitable for cultivation in north Indian plains. Tubers are red, oval with shallow to medium eyes and creamy-light yellow flesh, and having good keeping quality.

Field Preparation and Planting

Field should be brought to a fine tilth by deep ploughing and braking clods. Soil borne pathogens and nematodes are major limiting factors in cultivation of potato. Soil solarization adopting deep ploughing and turning of soils during summer months help in checking this menace. After burying of green manure, the fields are properly leveled and proper provision for good drainage is provided in the fields. Before planting, fields are ploughed with mould board plough or disc harrows followed by one or two tillings with tiller.

Time of sowing differs in plains and hills of India. In plains, potato is classified as monsoon crop (July-August), early winter crop (mid-September-mid-October), main crop (mid-October-mid-November) and seed crop (December-January). In hills, potato is classified as early crop (January-February), main crop (March-April) and late crop (July-August).

Propagation

Potato is traditionally propagated through tubers. Seeds account for nearly 40% of cost of potato cultivation. Seed quality is important for maintaining high productivity. The eyes on surface of tubers contain axillary buds. Potato tubers have a dormancy of nearly 8-10 weeks. In order to use freshly harvested tubers for planting, dormancy needs to be broken. The whole tuber is subjected to 2-3 shallow cut and soak for 1 hour in 10ppm GA3 solution ((15-25^{0}C) with 10kg thiourea/ litre. The tubers are then dried in shade and planted the next day. Pre-sprouting of seed stock is desirable as it helps in increasing number of main stems and consequently yield. It also ensures quick, uniform and full generation. Seed stocks are taken out 10-15 days before planting from cold stores and are kept for 24 hrs in pre-cooling chambers. Thereafter, they are spread in cool and shady place for encouraging sprouting.

Tubers can be planted by cutting into pieces for reducing seed cost. It should be ensured that, each piece is having one eye and weighs approximately 40 g. Dip treatment in mancozeb (0.2%) for 10min and subsequent drying in shade for 24-48 hrs prevent cut end decay in soil.

Planting

Tubers can be planted on ridges. At first, furrows are made and filled with FYM and fertilizing mixture and cover it with top soil for making of ridges. The seed tubers are dibbled 5-7cm deep on north of each ridge. Row spacing is usually from 45 to 60 cm. Optimum depth of planting depends on temperature and moisture of soil, probable weather following planting, and mode of conducting field operations later. Fertilizer drill and automatic planter are used for planting potato seeds. Nandekar (2005) reported the highest net return of Rs. 68,482/ ha and cost benefit ratio 1:3.16 with small size tuberlets (<10g) planted on high fertility level (180 :150:150 kg NPK/ha.) and the second best treatment was 10-20g tuberlets at 120:100:100 kg NPK/ha which recorded net return Rs.63,496/ ha with cost benefit ratio 1: 3.13.

True Potato Seed (TPS) Technique

In potato, seed alone costs 40% of production cost and True Potato Seed (TPS) Technique was developed as an alternative technology for reducing cost of cultivation and ensuring disease free planting material. It is easy to store and transport. However TPS has higher vulnerability to environmental stress when plants are young. More farming operations are required and the crop has high heterogeneity in plant type. Two high yielding TPS varieties viz. TPS C3 and DPS/1/13 are bred. TPS is produced by hand pollination of desired female lines with pollen collected from designated male parent. Berries are harvested 6-7 weeks after pollination and allowed to ripen at room temperature. Hybrid seeds are extracted from berries, washed clean and dried to 5-6% moisture. Potato crop can be raised either through transplanting seedlings in the field or by seedling tubers produced in the preceeding crop season. Approximately150g TPS is required for raising 1 ha crop. Seed should be treated with 0.01% benlate/bavistin or any systemic fungicide to avoid attack of soil borne fungus. Seeds are sown in rows 5 cm apart at a depth of 0.5 cm. Foliar application of 0.1% urea after two leaf stage is beneficial for quick growth and vigour of seedlings. Seedlings are ready for transplanting after 25-30 days of sowing.

Nutrient Management

Potato responds well to the external application of manures and fertilizers. The fertilizer requirement varies with soil and previous crop taken in field. The requirement of major nutrients in different soil types are as follows:

Soil type	Recommended doses (kg/ha)		
	N	P	K
Alluvial	180	80	110
Black	115	45	50
Red	120	115	135
Hill soils	115	135	95

Farm yard manure should be applied at the time of field preparation @ 30 t/ha. Nitrogen is the first limiting nutrient in potato production directly affecting tuber yield in all soil groups. It increases roots, foliage and tuber growth. Nitrogen is applied in two split doses i.e. half at time of planting and remaining half at time of earthing up. Both calcium ammonium nitrate (CAN) and ammonium sulphate are good sources of nitrogen to potato crop, yet urea, though less efficient, still being cheaper, can also be applied either alone or in combination with CAN in 1:1 ratio on N basis. Phosphorous and potassium are the other two essential elements in potato production. Phosphorous increases the tuber yield by increasing yield and number of medium size tubers whereas potassium increases the number of large size tubers. The application of P and K in furrows in full dose at the time of planting gives the best result. Water soluble phosphate fertilizers like superphosphate and DAP are the most suitable for potato. Similarly potassium sulphate is a better source of K than muriate of potash. Micronutrient requirement can be met by applying zinc sulphate , manganese sulphate, ammonium molybdate and borax as per soil test data before last ploughing. Combined application of neem seed powder @1.2 t ha^{-1}, karanj cake @ 1.0 t ha^{-1}, farm yard manure @ 6 t ha^{-1}, commercial formulation Biomax @ 0.6 t ha^{-1} and Biomas @1.5 t ha^{-1} along with 75 % of recommended dose of NPK fertilizers increased tuber yield by 10 %.

Water Management

Potato is sensitive to soil moisture. Total water requirement of potato crop is 450-550 mm and varies with factors like soil types, climate and variety. The most critical growth stages for water management are stolon formation and tuber initiation. Drought during this critical development stage reduces size of tubers, hence reduces yield. Too much irrigation leads to crust formation in heavy soils which will hinder the sprout to come out. In case of channel and furrow irrigation, watering at an interval of 10 days is sufficient and frequency needs to be increased in summer months depending on crop stage. Other irrigation methods like Sprinkler and drip irrigation methods increase water use efficiency by 40-50%.In hills, frost damage can be decreased by irrigation during night. Soil moisture can be conserved by mulching which will also induce quick germination and suppress weed growth. Locally available material such as paddy straw, wheat husk, maize or jowar stalks or farm refuse act as good mulch and is applied on ridges.

Weed Management

Weeds can be effectively managed by a combination of cultural and chemical methods. Cultural methods include hoeing and weeding when the crop is about a month old followed by earthing up. Pre emergence application of alachlor, linuron, mettribuzin, nitrofen, oxyflorofen and simazine are effective in controlling weeds. Post emergence application of paraquat at about 5% emergence and propanil at about 2-3 leaf stage are also effective in checking weed growth. Application of glyphosate (0.694 kg ha^{-1}) as a single pre-harvest foliar spray at 60 days after planting was effective in suppressing sprout growth during storage at 18°C. Glyphosate application generally did not affect tuber yield but increased incidence of tuber cracking in Kufri Jyoti from 15% (control) to 20–39%.

Harvesting, Post Harvest Handling and Storage

As the potato plants become mature and the tubers are fully formed, the leaves become gradually yellowish and then brownish, and finally the plants die. It is always better to harvest the crop after these signs are evident in the field. Haulms are cut and tubers are dug out after 10 days for favoring hardening and increasing shelf life. After harvesting potatoes are kept in heaps in cool places for another 10-15 days for drying and curing of skin; Before grading, all the cut damaged and rotten tubers are removed. The tubers are graded and packed in gunny bags according to size. Seed tubers are treated with 3%boric acid solution for 30 minutes for protecting against soil borne pathogens.

The tubers for seed purpose should be stored under refrigeration where temperature is maintained at 2-4^{0}C and the relative humidity is maintained as high as possible without making the potatoes wet. Before planting, they may be spread out in the shade to dry and to be kept in diffused light. The potato tuber for consumption also is ideally stored at 4^{0}C as higher temperature increases the risk of sprouting. Application of Malic hydrazide (MH), liquid formulation containing 30 % MH equivalent(3 kg of MH equivalent/ ha) given as foliar spray to the standing crop at 4 weeks prior to crop maturity can reduce sprout growth significantly.

Storage requirements of potato vary with purpose for which potatoes are stored. The methods generally used for storage of potatoes are-

Room Storage

The room should have a high plinth and good ventilation. The doors and ventilators should be fitted with insect-proof wire-nets. Potatoes can be stored in single layers on sand. Frequent examination is necessary to discard tubers showing rotting symptoms.

Pit Storage

In this system, the tubers are stored in pits 60-75 cm deep and 2.5 metre long and 1 metre wide. These pits are made in a cool shady place. Water is sprinkled inside the pit to achieve the cooling effect. After two days, neem leaves, dry grass or sugarcane trash is lined all round the pit from inside. Bamboo chimneys of 1.5 metre length are placed inside the pit 1 metre apart for facilitating evaporation of the moisture deposited due to transpiration of stored tubers. Pits are then filled with tubers leaving 15 cm on top followed by a 30 cm layer of dry grass. A thatch is provided over the pit as protection from rain and sun.

Cold Storage

This is the best method of storing potatoes. At optimum conditions, potatoes should have good quality after storage of 3 to 5 weeks. The best temperature and humidity conditions for potatoes are as follows-

Intended Use	Temperature (° C)	RH (%)
Seed	2.4	95
Table	7	98
Processing	8-12	95

Potatoes in cold storage are spread over the shelf to a thickness of not more than 30 cm. Individual shelves are about 45 cm high and 15 cm space is allowed between two layers. Sprouting is often a serious problem in storage of potatoes. Temperature, humidity, variety and maturity affect sprouting. Isopropyl N-Chlorophenyl Carbamate (CIPC) is effective as sprout suppressant. Its effectiveness is greater when CIPC treated potatoes were stored under refrigeration under 8-12°C as against non-refrigerated storage.

Marketing

Potatoes are semi perishable and bulky and, therefore their transport to long distances is a problem. Often potatoes rot during transit because of high temperatures. The problem is compounded further due to shortage of transporting wagons. The total cold storage capacity in the country is adequate only to store nearly 35-40% total potato production. Their functioning is many times not up to the mark and are marred by various malpractices. The markets in potato producing belts in northern India are not properly integrated with major consuming markets of metropolitan cities in other regions. Because of these factors marketing of potato is a complicated process and potato prices are prone to high fluctuations and artificial glut situations.

Value Added Products

Potatoes can be easily processed into dehydrated and canned products like chips, flares, French fries, finger chips, granules, disc, cubes, flour etc. on industrial scale. Processing is largely confined to developed countries. However in recent years, the processing industry is also picking up in India. It is desirable to avoid gluts and consequent difficulty of storing large quantities of potatoes during periods of high temperatures after harvest in the plains. India is on its way to have a 'crispy revolution' due to emerging growth and rapid progress in potato processing sector. Growing urbanization and changing food habit have made the potatoes leave the tables and take the refuge in packets or pouches at everybody's pleasure across the generation gap. Value-added processed products are opening up newer market avenues in international agricultural trade, and as a result, the farmers are finding it highly remunerative to grow potatoes even in areas traditionally indifferent to potato cultivation. The multinationals operating in India are now increasingly opting for indigenous processing varieties, which are fast replacing the exotic American and European varieties. Presently, the global potato export stands at 13.27 million tonnes per year amounting to about US$ 4.75 billion, and in this international export market processed potatoes comprise 4.12 million tonnes per year with an annual turn over of US$ 2.90 billion. To be competitive in global potato trade, which still continues to be a European affair, the country needs to accelerate the development and diffusion of suitable quality-enhancing technologies in the processing sector. However, institutional innovations must accompany the technological progress for attaining a leading role in international (potato) processing trade. Potato processing proves to be an important option for India's becoming self-reliant in sustaining the food and nutrition.

Export

It is estimated that 61.47% potato production are used for table purpose, 21% as seed, 0.5% processed and only 0.03% are exported while about 17% are lost in post harvest handling, marketing and storage. India although holds 5th place in production after China, Russia, Poland and USA its share in export is very low. In world potato export markets India does not stand mainly due to high price and poor quality of produce. India exported maximum of 61012 tonnes potato during 2004-05. India exports mainly to Malaysia, Nepal, Sri Lanka, Turkey and UAE. India also exports a little quantity of processed potatoes.

Production Technology of High Value and Exotic Vegetables

Vegetable consumption and preference of Indian urban population changed dramatically during last decade and with the globalisation, exotic vegetables started receiving a major place in dining table. Capsicum, gherkin and baby corn are of special importance in this context and in India periurban farming of these vegetables has tremendous potential.

Baby Corn

This is a distinct form of corn (maize) where young, unfertilized, tender cobs (ears) are harvested, hence called baby corn. Harvesting is done at silk emergence stage and dehusked cobs are consumed. Dehusked cobs are very nutritious and its nutritional value is comparable with other important vegetables consumed in India

Nutritional value of baby corn (dehusked cob) compared to other vegetables (100 g fresh vegetable)

Component	Baby corn	Cauliflower	Brinjal	Tomato	Cucumber
Moisture (%)	89.1	90.3	98.5	94.1	96.4
Fat (g)	0.20	0.04	0.20	0.20	0.20
Protein (g)	1.90	2.40	1.00	1.00	0.60
Carbohydrate (mg)	8.20	6.10	5.70	4.10	2.40
Calcium (mg)	28.0	34.0	30.0	18.0	19.0
Phosphorus (mg)	86.0	50.0	27.0	18.0	12.0
Vitamin (IU)	64.0	95.0	130.0	735.0	0.00

Source: Baby Corn Production Technology in Thailand- A Success Story, APAARI publication

Baby corn production in maize belt of India stands better promise for export (fresh/canned) trade, high income generation and for creation of employment in agriculture sector through canning and dairy industries.

Variety : Most sweet corn varieties have been developed for long days in northern latitudes. However, sweet corn hybrids suitable for the tropics are now available . Baby corn production is picking up in India and important varieties recommended for cultivation are VL 42,(open-pollinated variety), MEH 114 (hybrid) and CoBC 1 (synthetic variety). Ideal variety should be early and produce at least 3 cobs/plant. Kernel at immature stage should be yellow with uniform row arrangement.

Soil and Climate Requirement

Optimum mean day temperatures are around 25° C. Temperatures in excess of 35° C can severely damage pollination, especially if accompanied with moisture stress from just before silk development extending through pollination. Baby corn can be grown round the year in South India and in North India, good quality cobs can be produced in all seasons except peak summer. It is adoptable to most soil types from- sandy to heavy clay. Nevertheless, cultivation on saline and alkaline soils should be avoided and soils with god drainage preferred.

Cultural Practices

Baby corn can be cultivated like that of dent corn. In areas, where demand of baby corn is uncertain, cultivars of corn should be grown following the package of practice recommendations of dent corn and only lower cobs can be harvested as baby corn. However, commercial production requires specific growing strategies. Plant population can be increased with 1,10,000- 1, 50,000 plants/ha than that of dent corn. Seeds should be sown at the distance of 40-60 cm x 10-15 cm, depending upon season and variety. During *kharif* season, crop should be raised on ridges and summer on furrows. Detasseling should be done before flowering for getting more number of well developed cobs.

Harvesting

Harvesting is critical as a delay of even a single day in picking of young cobs degrades the quality and taste of cobs, thus decreasing its market acceptance. Cobs should be harvested every day in the morning from the date of silk emergence for fresh market and 2 or 3 days after silk emergence for processing purpose. Silk emergence and cob size vary with varieties and one has to determine size of young cob initially by tasting it during first picking and later by feeling the cob by hand to asses the size of the cob inside. Generally, 6-7 days pickings are required to harvest complete crop in 45-60 days. Under good management practises, hybrid variety gives 6-8 tonnes/ha husked cob yield with 15-20 % recovery of dehusked tender cobs. Apart from cob, 25-30 tonnes/ha green plant yield can be obtained which can be fed to cattle or used as green manure.

Post Harvest Management

Cobs should be precooled immediately after harvest for saving water loss and maintaining quality of tender cobs. De husked cobs are classified on the basis of their length as large (10-13 cm), medium (7-10 cm) small (4-7 cm) and mixed broken cobs. Dehusked cobs are packed in perforated plastic punnets and these are further packed in to well-ventilated cartons. It is highly perishable and requires proper storage at 5-7 ^{0}C with 90% relative humidity. Low temperature also reduces the conversion rate of simple sugar to starch, sugar loss can be 4 times lesser at 0 ^{0}C than 12 ^{0}C. At room temperature (30 ^{0}C), sugar loss can be up to 80% after 24 hours of harvesting, 50% after 3 days at 10 ^{0}C and 2% after 4 days at 0 ^{0}C.

Prospects

Indian market price of dehusked cobs varieties from Rs 50-90/kg. Estimated total cost of cultivation comes around Rs 30,000-50,000/ha. In a short period of approximately 2 months, a net income of Rs 45,000-50,000/ha can be obtained. Baby corn production can be taken up in Agri Export Zones (AEZ) and as periurban farming in other areas. Apart from the profitability, consumption of this vegetable should be promoted based on its merit as the nutrient rich vegetable.

Gherkin

Gherkin' literally means a small cucumber and belongs to the family cucurbitaceae. The gherkin plant is a creeper which can grow up to a height of 3m. Gherkins are now an important commercial and garden vegetable. They are grown for their edible fruit but are preferred in the pickled form. Although, many vegetables, including beetroot, cabbage and peppers are sold in the pickled form, gherkins are the leading pickled vegetable in world. Gherkin industry in India is primarily concentrated in the three southern states of Karnataka, Andhra Pradesh (AP) and Tamil Nadu. Karnataka accounts for almost 60% of the gherkin production and the whole production is based on contract farming. The entire volume of gherkins produced is exported, with little or no domestic demand, except for the five star hotels which serve pickled gherkins as an accompaniment.

Soil and Climate Requirement

Gherkins ideally grow in areas where the temperature ranges between 15 and 35^{0}C. Gherkins are a perennial, short duration crop (75-90 days) enabling farmers to harvest two to three crops per year. Crop prefers deep well fertile soil rich in organic matter.

Cultural Practises

Land should be ploughed deeply and leveled the field to fine tilth. Open straight furrows at an interval of 1. 2m. Add well decomposed farm yard manure 15-20t/ha & mixture of Neem cake, DAP, MOP and Magnecium Sulphate into furrows and then cover with soil by opening another furrow next to it. Plant four rows of fodder maize around the field which will prevent pest and disease incidence and also act as wind break. Sow single seed per hill at an interval of 30 cm and not deeper than 2 cm. Irrigate the field immediately after sowing. Germination will occur in 6-7 days after sowing. Prunning has to be carried out between 15 to 20 days after sowing. Remove all the buds, flowers and fruits in the first 3 to 4 bottom nodes without damaging to the plant. Pruning helps main shoot to grow vigorously and helps in good aeration and in turn avoids downy mildew disease.

Harvesting

Harvest the fruits every day as per the prescribed grades. The smaller gherkin (5-7 cm) fetches premium price in market. The stalk of the fruit should remain intact with the plant. After harvest, spread the fruits as a thin layer on a plastic sheet.

The defective fruits should be segregated and disposed properly in the waste pit. Good fruits should be hand graded as per the predetermined grades. Fruits will be graded using sieves.

Prospects

To promote exports, the Government has abolished the APMC (Agricultural Produce Marketing Committee) cess levied on gherkins. Financial institutions provide Rs. 18,000 as loan per acre of the gherkin crop. The Centre has removed cess on the purchase of raw materials for processing, and Agri Export Zones (AEZ) for gherkins have been set up in Karnataka. Gherkin is the only crop successfully grown under the contract farming system. Under the partnership between farmers and firms, the latter supply inputs and provide production services, credit, technology, and assured market and prices to farmers,

Companies pay for gherkins depending upon the quality. The price of gherkins of first-grade quality is Rs. 11/ kg. The smaller the gherkin, the higher the price. On average, each firm provides employment to about 2,000 farmers, who are paid Rs. 8,000 to Rs. 9,000 per tonne of gherkins.

Asparagus

The Asparagus is one of the most delicate and lucious perennial plants of the garden. Its erect branching stems, 15-180 cms in length, which are known as spears are used in soups and other products. Asparagus beds once established will continue to produce for many years. It is a good diuretic and is also used in Cardiac dropsy and gout. It is indigenous to Europe and Siberia from where it was introduced to other parts of the world.

Varieties

It belongs to Liliaceae family. The varieties are divided into two groups one with purple spears and the other free from purple colour. The cultivated varieties are

1. Argentevil Early – a selection from the Dutch purple type.
2. Argentevil Late – Dutch type.
3. Connivers Colossal – an American light green type.
4. Marther Washington – an American variety with roundish tipped bud.
5. Palmetto – Argentevil type pea green in colour.
6. Sutton's Giant Fresh – Argentevil type with stout buds.

The recommended variety of I.A.R.I is Perfection.

Soil and Climate Requirement

Crop loves cool climate and in India, it can be successfully grown in temperate hills. The best soil is light and medium loams or sandy loam soils, well supplied with organic matter. It can even be grown in heavy clay soil with deep cultivation and well drainage. The optimum pH required for growing asparagus is between 6.0 and 6.7. Asparagus can be grown on soil, which has a salt content, too high for many other crops.

Cultural Practices

8-10 kg of seeds are necessary for one hectare. The seed is sown in early spring or during winter and takes 3 to 4 weeks to germinate and 8-10 weeks to produce 'Crowns'. The rows are spaced at about 1.5 to 2 meters and 45 to 60 cm distance is kept within rows. The cultivated asparagus is a dioecious plant, e.g. male and female flowers are produced on separate plants. In addition to dioecious plants, a small percentage of seedlings develops hermaphrodite plants. Low growing vegetables may be grown as intercrops between the asparagus rows to compensate the expenses in the 1st and 2nd year. Cole crops, Beans, green crops etc. can be grown, provided enough nutrients are applied for both crops.

In light soil, along with 15 tonnes of FYM, application of 150 kg each of nitrogen and phosphorus and 200 kg of potash is recommended. After the harvest of spears, 25 to 35 cart-load manures together with 100 kg each of Murate of potash and Superphosphate should be applied to the soil.

In the summer season, the crop needs irrigation once a week, while irrigation is not necessary from August to January. Weeds between the rows can be controlled by proper cultivation. Frequent shallow hoeing keeps down the weed population. After the harvest of spears a thorough cultivation should be done destroying all the weeds. The broad-leaved weeds can be controlled successfully by means of 2,4 –D before and after the harvest, avoiding spray to asparagus tops.

Harvesting

Harvesting is done from the 3rd year onwards. However, if the plantation is vigorous, the 1st cutting may be done from the 2nd year after the crowns are set out. The 1st harvesting should be done within 2 to 3 weeks. Yield of plantation in full harvest is on an average about 3500 to 4000 kg/ hectare each year.

Prospects

Asparagus is a delicate vegetable now served in star hotels. Indian elites like this vegetable and cultivation in high ranges has lot of scope once entered a tie up for marketing.

Production Technology of Minor Exotic Vegetables

Production technology of some of the promising exotic vegetables are given below.

Vegetable	Planting time	Seed rate(g/ha)	Planting distance (cm)	Nutritional requirement		Yield (q/ha)
				FYM (tones/ ha)	NPK(kg/h a)	
Broccoli	Oct.(plains)	450-500	60 x 60 60 x 45	20	150:80:60	150-200
Romanesco cauliflower	Oct.-Nov.(plains) Sep.-Apr.(hills)	450	60 x 45	10-12	120:80:60	150-200
Chinese cabbage	Oct.-Nov.(plains) Sep.-Apr.(hills)	400	45 x 30	10-12	120:50:60	450-500
Red cabbage	Oct.-Nov.(plains) Sep.-Apr.(hills)	450	45 x 45	20	100:100:100	600-700
Savoy cabbage	Oct.-Nov.(plains) Sep.-Apr.(hills)	300-400	45 x 45	10-12	150:80:60	400-500
Brussels sprout	Oct.-Nov.(plains) Aug.-Apr.(hills)	450	45 x 45	10-12	120:60:50	100-150
Kale	Oct.-Nov.(plains) Sep.-Apr.(hills)	300-400	60 x 40	20	150:75:00	150-250
Collards	Oct.-Nov.(plains) Aug.-Apr.(hills)	350	60 x 45	10-12	150:60:00	250-300
Endive	Oct.-Nov.(plains) Sep.-Apr.(hills)	150-200	30 x 20	10	80:60:40	150-200
Sugar leaf chicory	Oct.-Nov.(plains) Sep.-Apr.(hills)	150-200	30 x 20	10	80:60:40	150-200
Leek	Oct.-Nov.(plains) Aug.-Apr.(hills)	1.5kg	60 x 45	25	200:120: 100	300-400

Source: Pandey and Rai, 2005

Organic Vegetable Production

Vegetable production in India is spreading across large, small and marginal farmers adopting different systems of cultivation such as market garden, peri-urban garden, truck garden, kitchen garden, seed production and poly house cultivation. While kitchen garden more or less follows the principles of organic vegetable production, commercial vegetable cultivation in other systems rely on heavy use of chemical inputs. Organic farming ensures safe vegetables and good agricultural practices that are environment friendly and promotes ecological stability. Developing organic agriculture, as high value export-business option is yet another emerging opportunity in India. Because of the historical context of de facto organic farming practices, India is globally seen as a potential source of future organic exports, along side China, from Asian region. Major concern in commercial vegetable production and need to switch over to organic production are,

i) Chemical fertilizer consumption in India will rise to 30 m. tonnes by 2025 AD from the current level of 14 m. tonnes.

ii) The demand for pesticides would double from the current 0.8 lakh tonne

iii) There is heavy draining on exchequer on account of import subsidies given to fertilizers and pesticidal raw materials.

iv) Leached nitrates, phosphates and agrochemicals reach waterbodies making them unfit for drinking. While 45 mg NO2 per litre is the upper limit for potability, water bodies in India contain levels beyond this. In addition , manufacturing units of chemical inputs inflict severe environmental damage.

v) Pesticide residues in fresh vegetables for human consumption have many times crossed the safe limit

vi) Food chain is affected leading to complicated diseases in mammals.

vii) Bio-diversity of flora and fauna are being lost continuously

Organic vegetable production differs from "conventional" production in the areas of input management manily fertilization and pest control. It is a holistic production system, in harmony with nature. Emphasis is placed on the use of a balanced fertilising programme, use of crops and varieties well-adapted to the environment, fertile soils of high biological activity, adapted rotations, companion planting and green manures. The organic production system prefers to use natural and organic materials and methods, and avoids using practices and synthetic chemicals that may be detrimental to environment. Although the cost of certification and the time and labor involved in managing the system are high, returns can also be, on average, 20% higher than conventionally produced products, provided that a market exists. Main issues in organic vegetable production India are related to the Standards, Inspection, Certfication , Accreditation, Input and Market (Export/Domestic).

Standards for Organic Agriculture

The organic agriculture (cultivation as well as processing) is regulated by a complex set of standards. Globally 61 standards are available and important Standard includes - IFOAM, CODEX, EU, Demeter, JAS, Naturland etc. At international level the basic standards for organic agriculture defined by IFOAM (International Federation of Organic Agriculture Movements) are most relevant. These standards define the principle requirements for the production of organic producers. The standards set by IFOAM are the obligatory basis for detailed production standards set by private certification bodies, which are evaluated on behalf of IFOAM with in the frame of the IFOAM Accreditation Programme. About 16 private certification bodies are accredited by IFOAM world wide. Out of the IFOAM Basic Standards also official regulations for the organic industry were developed in different countries. National Standard in India is under NPOP (2002) (General Principle, Recommendation, Standard, list (permitted/non-permitted). Standards are mainly prescribed for processes which cover - cultivation, fertilisation and pest/disease management. These standards are under review and are in an effort to harmonise with EU regulation. There are 11 certfication agencies in India , which include Ecocert (G) (Aurangabad), Skal (N) (Bangalore), IMO (S) (Bangalore), SGS (S) (Gurgaon), Naturland (G) (Gurgaon), Lacon (G) (Cochin), Indocert (S) (Cochin), APOF (Bangalore), ISCOP (Coimbatore), IRFT (Mumbai), Bioinspectra (S) (Cochin). Their certification is mandatory for export market. These are accredited by National Accreditation Boards (APEDA, other Commodity Board and Ministry of Agriculture).

Inspection and Certification

Every producer, processor and exporter who intend to produce and sell organic products shall fullfill all requirements laid down in the internationally valid standards for organic agriculture as well as in existing regulations in the country of importation. At least once per year, all production units have to be inspected by the certification.

Conversion Plan

Prior to the conversion of a farm or a processing plant, it is of utmost importance to acquire sufficient information about all certification body with the objective to develop a conversion plan for the operation. Furthermore, prior to the first inspection, it has to be clarified which information is needed by the certification body and/or which records have to be taken constantly (eg. Bought-in of raw materials and production means, sales documentation etc.) . The producer must be aware of the conversion requirements (besides all detailed standards for production) at first. For example, the EEC-regulation for organic agriculture (EEC) 2092/91 requires a conversion period of 24 months and products can be sold as organic in conversion after 12 months of organic production; however, for most of the products, it is very difficult or nearly impossible to sell 'products in conversion'. This conversion requirement is therefore very important for the short-and midterm planning of any farm operation.

Crop Production Standards

Organic vegetable production is based on a holistic approach. Maintenance of soil health is given utmost priority in production. Soil should be covered by mulch round the year and incorporate annual and perennial green fallow periods into the crop rotation along with planting of legumes, cowpea, daincha, sunhemp to regenerate the soil. Develop mixed cultivation system (e.g. okra and cow pea, beans and potato, methi, beans and carrot). Clearing of land through means of burning organic matter, e.g. slash-and burn, straw burning shall be restricted to the minimum. Relevant measures should be taken to prevent erosion, salt accumulation of soil, excessive and improper use of water and the pollution of ground and surface water. Hedge rows (one or more rows of trees/ shrubs) with a wide range of functions should be grown which will serve wide functions like wind breaker, prevention of erosion, habitat for natural enemies of pests, timber for construction and fuel, fodder and pruning materials for mulching and / or composting, protection against side contamination with pesticides from conventional farms nearby etc.

Choice of Crops and Varieties

All seeds and plant materials should be certified organic. Species and varieties cultivated should be adapted to the soil and climatic conditions and be resistant to pests and diseases. When organic seed and plant materials are available, they shall be used. The certification programme shall set time limits for the requirement of certified organic seed and other plant materials. When certified organic seed and plant materials are not available, chemically untreated conventional materials shall be used. Inter cropping and mixed cropping are preferred over pure cropping. Leguminous vegetables should be mixed with cole crops, cucurbits, bhindi and bulb crops. Local adoptable cultivar should be given preference.

Nutrient Management

Biodegradable material of microbial, plant or animal origin produced on organic farms should form the basis of fertilisation programme. Sufficient quantities of biodegradable material of microbial, plant or animal origin should be returned to the soil to increase or at least maintain its fertility and the biological activity within it. Usually, it is in the form of animal manures, plant manures, cover crops, compost, or mixed organic fertilizer. Crop residues should be used for making compost and applied along with FYM and bio-gas slurry. FYM, poultry manure and compost from outside sources are not encouraged in organic certification. The supply of nitrogen(N) in organic farming is usually provided by legumes through a symbiosis with nodule bacteria, these plants are capable of fixing atmospheric nitrogen and making it available to plants. Chilean nitrate and all synthetic nitrogenous fertilisers, including urea, are prohibited in organic vegetable system.

On organic farms emphasis, is placed upon increasing the availability of phosphate content of the soil for the plants. This is achieved by a biological conversion of insoluble to soluble P- compounds in the soil. In fertile soils with sufficient organic matter, the conversion to soluble P- compounds is the most efficient. Mineral fertilisers shall be applied in their natural composition and shall not be rendered more soluble by chemical treatment. Rock phosphates are natural deposits of phosphate in combination with calcium and yield phosphorus very slowly. Broadcast the material over the soil surface and work into the topsoil at least three weeks before planting. Potassium is widely distributed in nature, occurring in rocks, soils, tissues of plants and animals, and water of seas and lakes. For organic vegetable production, materials such as wood ashes, tobacco stems, seaweed, potash salts, greensand, and ground rock potash are used alone or in combinations.

The following strategies are important in organic farming to ensure a sufficient supply of K for the plants:

1. Regular applications of organic matter will improve absorption of potassium in the upper soil layers, where it can be reached by the plants' roots.
2. Use of deep- rooting plants to mobilize K in lower soil layers.
3. Integration of plants with a high K- uptake in mixed cultivation systems (e.g. bananas on coffee plantations).

4. A permanent mulching layer, especially in the wet tropics in order to reduce leaching of K.
5. In arid regions with soils poor in K, it can be useful to mix pulverized rock containing mineral clay into the compost. In case of potassium deficiencies showed by soil analysis, it is permitted to use certain potassium salts with lower chlorine content (Muriate of potash/ potassium chloride is not allowed). Wood ash from untreated wood is also allowed.

Lime can be applied for correcting soil acidity and sources include limestone, dolomite, and shell. All these forms must be finely ground to provide maximum benefit to the soil and plants. Dolomite is preferred due to its content of both calcium and magnesium. Apply lime well in advance of planting date, preferably 2 to 3 months before the garden is planted.

Animal husbandry enables the recycling of organic matter to be further optimized in agro eco systems. In practice, the necessary C:N ratio for each of the material used can only be reached when a sufficient quantity of manure is added. Rock phosphate can be used by mixing it directly into the compost. Furthermore, VAM fungi can also dissolve phosphates in a compost. The following points should be cleared up before organic material is used from sources other than the farm itself:

1. The origin of all organic material must be ascertained.
2. The importation of organic material into the farm system is limited, with the main proportion being produced on site.
3. Every instance of organic material purchase (both type and amount must be approved before hand by the certification body).

Pests, Diseases and Weed Management

Weeds, pests and diseases should be controlled by a number of preventive cultural techniques which limit their development, e.g. suitable rotations, green manures, a balanced fertilising programme, early and predrilling seedbed preparations, mulching, mechanical control and disturbance of pest development cycles. The natural enemies of pests and diseases should be protected and encouraged through proper habitat management of hedges, nesting sites etc. An ecological equilibrium should be created to bring about a balance in the pest predator cycle.

Products used for pests diseases and weed management, prepared at the farm from local plants, animals and micro-organisms, are allowed. The use of synthetic herbicides, fungicides, insecticides and other pesticides is prohibited. Thermic weed control and physical methods for pests diseases and weed management are permitted. Soil solarization is recommended for controlling nursery diseases, nematodes and other soil borne pathogens. The bio-agent *Trichoderma* can be used for seed treatment, potting mixtures and main field. The use of synthetic growth regulators and synthetic dyes are prohibited. The use of genetically engineered organisms or products are prohibited.

Following practices will help to reduce losses without use of chemical pesticides.

1. Plant resistant varieties
2. Use seed from disease-free plants.
3. Select pest-free transplants.
4. For cutworms, place a cardboard of tinfoil collar around plant stems at ground level.
5. Spade garden early so vegetation has time to rot before planting.
6. Use a mulch to protect vegetables touching the soil.
7. Clean up crop refuse early.
8. Use polytube seedlings in cucurbits for ensuring better crop stand.
9. Keep out weeds which harbor insects and diseases.
10. Summer fallowing (clean cultivation) and flooding helps control nematodes.
11. Hand-pick insects.
12. Water in morning so plants are not wet at night.
13. Dispose of severely diseased plants before they contaminate others.
14. Some insects, like cabbage worms, may be killed by spraying with natural preparations such as *Bacillus thuringiensis*. Spraying commercial formulations of *Bacillus thuringiensis* (Bt) controls fruit borers and other caterpillar pests of crop.
15. Follow crop rotation
16. Crotolaria spectabalis and marigolds, when planted as cover crops and intercrops, tend to reduce many kinds of nematodes.

17. A good garden mulch tends to reduce *damage* caused by nematodes.
18. Use sprays and other preparations containing naturally occurring materials. Eg: Neem, Pyrethrin, rotenone, and ryania . These give control to insects under certain conditions.
19. Natural predators should be encouraged wherever possible; however, predators raised in captivity, then released into the garden area are usually ineffective.
20. Insecticidal soaps, made from fatty acids tend to work well for some insects under average conditions.
21. Insect traps, baited with pheramone lures, work well in some instances. Many of these have sticky adhesives to catch insects.
22. Solar fumigation is effective in reducing soil-borne problems such as nematodes.
23. Application of *Bacillus macerans* or *B. circulans* (1.2 x 10.6 cells/pit) before sowing controls root- knot nematode in bhindi and brinjal. Inoculation of VA mycorrhizal fungi and *Pseudomonas fluorescence* in soil reduces infestation of root-knot nematode in crops.

Some of the Plant Based Insecticides are given below

Preparation	Dose	Pest
Neem Leaf extract	2-5%	Leaf feeders of amaraths like *Spodoptera litura* & leaf webber, aphids of vegetables
NSKE (Neem seed kernel extract)	3-5%	Aphid of brinjal Fruit & shoot borer of okra, brinjal Pod fly of cowpea American serpentine leaf miner (ASLM) of vegetables (cucurbits, brinjal, cowpea)
Neem oil	2.5-10%	Pea aphid, ASLM, Epilachna beetle, mites of vegetable
Neem cake	250kg/ha	Shoot and fruit borer of bhindi, brinjal
Custard apple- *Annona squamosa* Leaf extract + cow's urine	10% + 10%	Leaf feeders of crops
Clerodendron- *Clerodendron infortunatum* -Plant extract	4-8%	Tobacco caterpillar, Epilachna beetle ofbrinjal, cucurbits, leaf roller of okra
Vilaitulsi- *Hyptis suaveolens*- Plant extract	10%	Aphid and soft bodied pests of vegetables
King of bitters (Kiriyanth)- *Andrographis paniculata*- plant extract	10%	Jassids of cucurbits, leaf caterpillars

Contd....

Contd....

Chromolaena (Eupatorium) odorata - Leaf	Incorporation in the soil @ 15 t/ha two weeks before planting	Nematodes of vegetables
Chinese chaste tree (Notchi)- *Vitex negundo* -Leaf extract Dried powdered leaf	5-10% 1:100	Fruit & shoot borer of okra, Leaf caterpillars of cucurbits, Stored product pests
Lantana - *Lantana camara* -Plant extract + cow's urine	10% + 10%	Soft bodied pests

For protected structure coverings, plastic mulches, fleeces, insect netting and silage rapping, only products based on polyethylene and polypropylene or other polycarbonates are allowed. These shall be removed from the soil after use and shall not be burnt on the farmland. The use of polychloride based products is prohibited.

Post Harvest Handling

Handling and processing of organic products should be optimised to maintain the quality and integrity of the product and directed towards minimising development of pests and diseases. Organic and non-organic products shall not be stored and transported together except when labelled or physically separated.

Besides storage at ambient temperature, the following special conditions of storage are permitted

- Controlled atmosphere
- Cooling
- Freezing
- Drying
- Humidity regulation

Post harvest pests and diseases should be controlled through use of physical barriers, sound, ultra- sound, light, and UV-light, traps (including pheromone traps and static bait traps), temperature control, controlled atmosphere and diatomaceous earth. Processing methods should be based on mechanised, physical and biological processes. The use of material for packaging shall be eco-friendly. Recycling and reusable systems shall be used wherever possible. Biodegradable packaging materials shall be used.

Labelling

When the full standards requirements are fulfilled, products can be sold as "produce of organic agriculture" or a similar description. The name and address of the person or company legally responsible for the production or processing of the product shall be mentioned on the label. Product labels should list processing procedures which influence the product properties in a way not immediately obvious. Additional product information shall be made available on request. All components of additives and processing aids shall be declared. To get processed foodstuff certified as organic, it is not sufficient to use only organic ingredients and a couple of other aspects have to be taken into account. In this context the use of ingredients of non-agricultural origin (eg. Food additives, carriers, flavourings, water and salt, microorganisms preparations, minerals and vitamins), the use of processing aids (eg. tannic acid, silicon dioxide gel, gelatine), the use of disinfectants as well as the application of specific processing methods (eg.ionising radiation is prohibited) is regulated in a very restrictive manner. Hence the processors interested in the processing of organic food stuff are recommended to contact an accredited certification body in time, to develop a conversion plan.

Protected System of Vegetable Cultivation

Polyhouse/Greenhouse structures were developed for crop production during periods when growing outdoors was not feasible. Polyhouses are framed structures covered with transparent or translucent material to grow crops under partial or fully controlled environmental conditions to get maximum productivity and quality produce. Polyethylene/plastic film covered greenhouses are being widely used in recent years throughout the world, as they are lighter in weight, cheap and easy for installation. Greenhouses can be designed to optimize the penetration of incoming solar radiation, temperature, relative humidity, carbon dioxide and air movement. In highly specialized crop production, all conditions must be controlled uniformly in every part of the greenhouse for maximum net returns. Hi-tech polyhouse technology controlling all the above factors, requires huge initial investment and production cost is also very high. The small and medium farmers cannot afford the technology due to high capital investment. Hence, low cost poly houses built with locally available materials like bamboo and UV stabilized (UVS) transparent polyethylene have more scope for vegetable production. In India green house technology is used to reap the benefit of offseason vegetable market. The production of vegetables like tomato, sweet pepper, broccoli, okra etc has resulted into higher yield and offseason production which fetched premium prices in the market.

Scope of Polyhouse Cultivation of Vegetables in India

Protected cultivation of vegetables has increased the yields by many folds. Capsicum production has increased from 40 tonnes in an open field to 100 tonnes in a greenhouse in a crop of six months. Similarly tomato yield has been increased to about 180 tonnes per hectare. The major fraction of cost, however, is that of the structure, which amounts to about 70% of the cost of production of vegetables. Therefore, to make it profitable and a sustainable farm activity, cost effective structures are to be developed, which will trigger the growth of protected cultivation in larger areas. Concerted efforts are required to bring down cost of structures utilizing locally available material like wooden poles or granite pillars in place of steel tubes wherever feasible and reducing the cost of production by making use of locally available less expensive organic substrates as growing media and optimizing the use of costly inputs like water, nutrients and need based plant protection chemicals. Greenhouse technology has made growing of vegetables possible even in Leh-(Ladakh) area, where sub-zero temperature makes any kind of plantation almost impossible.

The distinct advantages of growing crops under low cost polyhouses are:

- High productivity per unit area can be achieved through high quality crop management
- Off-season vegetables can be grown which fetch high prices in the market
- Off-season healthy nursery can be raised which in turn will yield early crop
- Good quality produce, free from any blemishes
- Easy to protect the crops against pests and diseases and extreme climatic conditions.

Off Season Vegetable Production

During monsoon in South and East India and winters in north India, rain, temperature and solar radiations are sub-optimal for growing vegetables like tomato, capsicum and cucurbits. The festive season after monsoon and winter offers premium price to these vegetables. In tomato, low temperature and low radiation cause puffiness and blotchy ripening. Heavy rain causes flower fall, leaf and fruit rotting. Low cost polyhouse claded with UV stabilized (UVS) polyfilm without any cooling facility can be used for growing high value vegetables like hybrid tomatoes, capsicum and cucumber profitably.

The protected environment could be well adopted in home gardens at slightly high altitude where winter is prolonged. A small-sized polyhouse can be made over uncultivable landscape, even on roof/terrace receiving sunlight for growing chilli, tomato, brinjal, summer squash and okra in pots and other containers.

Off-Season Nursery Production

Poly houses can be used for raising polyhouse seedlings of cucurbits, tomato and capsicum which in turn can be transplanted to outdoors during favourable weather conditions. The environmental conditions particularly increase temperature inside polyhouse, hasten the germination and early growth of warm season vegetable seedlings for raising early crops in spring-summer. Thus phases of heavy monsoon and cold winter can be overcome. This technology fetches the bonus price due to marketing of produce in the off-season. Asparagus, sweet potato, pointed-gourd and ivy gourd are sensitive to low temperature. The propagating materials of these vegetables can be well-maintained under polyhouse in winter season before planting their cuttings in early spring-summer season for higher profit.

Management of Polyhouse Microclimate

Shading the polyhouse with shade film/gunny cloth reduces the reception of solar radiation inside polyhouse during summer months, which helps in reducing the temperature. Providing mist irrigation reduces the temperature and increases humidity to a considerable extent during summer months. During winter months by providing limited ventilation and irrigation, temperature and humidity can be regulated considerably. Providing irrigation through drip system is desirable, as it reduces the humidity build up inside polyhouse after irrigation. Application of fertilizers through irrigation (fertigation) helps in saving the quantity of fertilizers and labour besides improving the yield. The size of a polyhouse depends on expertise of grower, investment capacity and market. It is estimated that area ranging from 200–500 sq.m under greenhouse is sufficient for retail business by single person.

Site Selection

Polyhouse is ideally suitable for peri-urban vegetable farming. Place should be well connected with market, both for its input supplies and sale of its produce. There should be sufficient availability of good quality water and the place should be well drained. The proposed site/location should be fairly shadow free. At the same time, the area should have sufficient wind break on all sides to minimize the wind damage to the proposed structure.

Structure Materials

A greenhouse structure has three distinct segments i.e. frame, glazing material and control/monitoring equipment. All the three components have different designed life periods. Whereas a greenhouse frame is designed for a service period of 15 to 25 years and glazing materials have life span of 5 years. Galvanized mild steel pipe as a structural member in association with wide width u.v.stabilised polyethylene film is a common option selected by greenhouse designers. A 600 to 800 gauge thick polyethylene film can safely withstand normal wind loads prevailing in many parts of the country. A single piece polyethylene film to cover greenhouse is preferred due to material economy, easy handling and improved environmental control. Glass has excellent light transmittance and weather resistance but has the disadvantages of brittleness, high cost and heavy weight. Polyethylene and polyvinyl chloride (pvc) films are in use as greenhouse covers. Polyethylene has high light transmittance but lacks durability. U.V. stabilized polyethylene film available in 100 to 250 micron thickness and width of 7m, will last two to three years, depending on quality and local climate. PVC with an ultraviolet inhibitor has a life span of 4 years. It is available in 200 to 800 micron thickness and width up to 120 cm. Fiber glass reinforced plastic (FRP) panels are easy to install.

The Characteristics of the Main Materials used to Cover Green House

	Flexible			Rigid		
Characteristics	**Polyethylene (0.08 mm)**	**PVC (0.1mm)**	**Corrugated PVC (1-2mm)**	**Methyl polymethylacrylate (4mm)**	**Stratified polyester (1-2 mm)**	**Glass (1.7mm)**
Density	0.92	1.3	1.4	1.18	1.5	2.40
Refraction index	1.512	1.538	-	1.489	1.549	1.516
Percentage stretching to destruction	400-500	200-500	50-100	low	low	none
Durability	2 years	2-3 years	High	High	High	Very high
Transparency (%) (380-760 mm)	70-75	80-87	77	85-93	70-80	87-90

Depending upon the environment control facilities and the cost involved in the construction, the polyhouses are categorised as low cost, medium cost and high cost. The low cost polyhouse does not have any control system. It has only the supporting structure, which may be made of GI Pipe, angle iron or even bamboo. UV stabilized 200 micron thick plastic film is normally used for coverage. The medium cost polyhouse made of iron structure may have double layer of UV stabilized plastic film with cooling pad/fan or heating arrangements depending on this requirement. It may also have misting and shading facilities including UV stabilized cladding film/poly carbonate sheet, heating/cooling and humidification system, drip irrigation system, auto control mechanism etc.

Shade Nets

Greenhouse curtain systems are called shades, screens and even blankets. These are movable panels of fabric or plastic films used to cover or uncover the space above the plants. They are used for heat retention, shade (and the cooling effect of shade), and day length control or blackouts. Shade clothes are typically green, black, white or aluminized. The aluminized strips reflect light out through the roof of greenhouse. This reduces the cooling load under the shade considerably. Shading is necessary for many greenhouse crops because direct radiation at noon increases the plant temperature by 5-6° C.

Ventilation

Proper ventilation is the key factor in controlling temperatures and relative humidity, while providing CO_2 enrichment for good crop development. There are two basic systems of ventilation, natural and mechanical. Natural ventilation is the method of lowering the air temperature by wind or the natural exchange of hotter inside air with cooler outside air. This is done through roof and side-wall vents aided by naturally occurring breezes. Greenhouse design and location are very important to ensure that this method provides effective cooling. Ventilation, the mechanical method of lowering the air temperature is accomplished by drawing cooler outside air into the greenhouse. Naturally ventilated greenhouses have to be protected through insect screens to restrict the entry of insects and reduce the pesticide use. Insect screens are designed to prevent the entry of three major small insects that damage many crops are thrips, aphids and whiteflies. Insect sizes are: 192 microns- flower thrips, 340 microns- aphids, whiteflies-462 microns and leaf miners-640 microns. A 40 mesh nylon net means 40 strands in a linear inch, which means the strands are placed at a distance of 635 microns.

Cooling Systems

While ventilation may be used for cooling during autumn and spring seasons, other methods have to be employed for cooling during summers.

Roof shading, Water film on the greenhouse cover, Evaporative Cooling (EC), Fan and Pad System and low and high pressure mist systems are usually employed for cooling polyhouses.

Fan and Pad System

A forced cooling system uses large capacity low speed exhaust fans and inlet vents. The recommended air change capacity is 1.5 times the greenhouse air volume exchange per minute. For example, if the greenhouse volume is 30,000 cubic feet, then 45,000 cubic feet of air per minute has to be removed. It is adaptable to both large and small greenhouses. Low velocity and large volume fans draw air through wet fibrous pads mounted on the opposite side or end wall of the greenhouse. The outside air is cooled by evaporation to 20°C of the wettable temperature. Greenhouse air temperature reduction ranging from 3-12°C can be achieved by F & P system over uncooled greenhouse sections and depends on radiant flux density, relative humidity, and amount of shading compound applies. Various materials, viz; gravel, pine bark, straw, burlap, aspen wood fiber (shredded *Populous tremuloides* mats) and volcanic rock(1-4 cm in dia) are used as pads. Most greenhouse will require about one meter of pad height for every 20 m of pad to fan distance. Placement of exhaust fans also affects the distribution of cooled air. They should not be located more than 2 m apart or warm areas will develop

Selection of Cultivar

Early maturing and high yielding hybrids of high value vegetables like tomato, capsicum and cucumber are ideal for growing under polyhouse. In case of insect pollinated vegetables like cucumber, parthenocarpic hybrids are preferred. The selection of cultivars and hybrids depend on plant type and their growth behaviour. Tomato cultivars and hybrids should be of indeterminate type. The plants are grown upright as a single stem rather than bush. The cucumber cultivars should be gynoecious (only female flowering habit), with dark green parthenocarpic (seedless) fruits free of bitterness. Training and pruning are an important operation in polyhouse vegetable production.

Tomato

Tomatoes can be raised on beds, prepared parallel to the length of green house and trained on two over head wires at the height of 3 m fixed on each bed parallel to the length of beds. In case of tomatoes, 30 days old seedlings can be transplanted at a spacing of 60 cmx 60 cm. Regular training, pruning and trellising should be done by allowing only one main branch on each plant. Fertigation should be given through low pressure drip pipes. Plants can be harvested for 7-8 months continuously and an average yield of 4-5 tonnes can be expected from 500 sq.m during this period. The economics of tomato production under indigenously-designed naturally ventilated green house was given else where. The cost of production of tomato was calculated by taking into account total operational and fixed cost of greenhouse. The net profit (Rs 75,520/ 500 m^2 area) and cost: benefit ratio was worked out as 1: 2.0.

Economics of Tomato Cultivation under Naturally Ventilated Green House

Operations	**Units**	**Cost (Rs 500 m²)**
Fixed cost/ infrastructure		
Depreciation on fixed cost + interest on invested money	Indian green house @ Rs 500 m²	40,000.00
Fixed cost of water source	Rs 5,700/year for	5,700.00
Irrigation water cost	Low pressure drip Rs 1,500/year	1,000.00
Operational cost		
Tractor	Rs 250/ hour,4 hours	1,000.00
Para plough	Rs 20/ hour,4 hours	80.00
Rotavator	Rs 60/ hour,2 hours	120.00
Labour(for all operations)	Rs 100/person/day	16,000.00
Fertilizers/chemicals	For 500 m² area	4,000.00
Soil treatment	Formaldehyde @ 15/litre (100 litre/500 m²)	1,500.00
Seed / seedling cost	Seed	3,000.00
Plastic ropes	For 500 m² area	1,000.00
Electricity charges	Electricity is not used in the structure	Nil
Total cost of production	For 500 m² area	74,900.00
Total production(quintal)		75.21q
Break even cost		9.95/kg
Economics		
Gross income	From 500 m² area	1,50,420.00
Net profit	From 500 m² area	75,520.00
Cost: benefit ratio		1:2.01

Source: Singh *et al.*, 2005

Capsicum

Four lobbed peppers with square shaped fruits are in demand for growing under polyhouse. Colour of fruits vary from, green, violet, red and orange. Some of the cultivars suitable for green house production are, Mazaruka(red), Fiesta (yellow), Torkel(Red), Bharat (red), Golden Summer (yellow) and Indira (red). Capsicum is grown on beds under drip fertigation system. Drip laterals of 16 mm diameter, 2 litre water discharge/dripper/hour and 30 cm distance between two drippers are laid on beds. Seedlings (35 days old) are transplanted in beds at a spacing of 30 cm x 60 cm. Plant is initially trained on two stem and these branches are trained upon strings to the mainwire running on length of rows on 2-2.5 m height. Every internode produces two shoot, of which one is pruned by leaving one flower and two leaves. Fertilizer is applied through drip and N:P:K solution (5:3:6) is applied. Calcium is applied through calcium nitrate and micronutrients through magnesium sulfate, zinc sulfate and chelated iron. Capsicum can be grown for a period of 7-8 months under low cost polyhouse with natural ventilation and is highly profitable for peri-urban farming.

Cucumber

Parthenocarpic slicing cucumber varieties are preferred for growing under polyhouse. Variety Hasan and Sarig are ideal for summer and rainy season while Satis, Dinar, Nun 9729 and Nun 3019 are suitable for winter season. Here also, seedlings are used for cutting short the time for harvesting. Polybag seedlings are transplanted at a spacing of 60 x 30 cm. Fertigation is recommended as in capsicum. Harvesting starts from 70 days after transplanting and 3-4 harvests are possible. Cultivation of cucumber was found to be more profitable than tomato.

Economics of Cucumber Cultivation Under Naturally Ventilated Green House

Operations	Units	Cost (Rs 500 m^2)
Fixed cost/ infrastructure		
Depreciation on fixed cost + interest on invested money	Indian green house @ Rs 500 m^2	40,000.00
Irrigation water cost	Low pressure drip Rs 1,500/year	1,500.00
Operational cost		
Tractor	Rs 250/ hour,4 hours	1,000.00
Para plough	Rs 20/ hour,4 hours	80.00

Contd...

Rotavator	Rs 60/ hour,2 hours	120.00
Labour(for all operations)	Rs 100/person/day	9,000.00
Fertilizers/chemicals	For 500 m^2 area	3,000.00
Soil treatment	Formaldehyde @ 15/litre (100 litre/500 m^2)	1,500.00
Seed / seedling cost	Seedling @ Rs 4/seedling x 1,000 seedling/ 500 m^2/ crop	12,000.00
Plastic ropes	For 500 m^2 area	500.00
Electricity charges	Not required for green house	Nil
Total cost of production	For 500 m^2 area	69,700.00
Total production(quintal)		80q/500 m^2/ (three crops)
Break even cost	Per kilogram	8.71/kg
Economics		
Gross income	From 500 m^2 area	1,60,000.00
Net profit	From 500 m^2 area	90,300.00
Cost: benefit ratio		1:2.29

Source: Singh *et al.*, 2005

Vegetable Seed Production

Vegetable seed production is a remunerative enterprise compared to fresh vegetable production. However, production of genetically pure and quality pedigree seed requires high technical skill and specialization. The producer should be familiar with genetic and agronomic principles of seed production. For most purposes, the important consideration to seed producers is the extent of cross-pollination. Vegetables are classified into 3 groups-self pollinated, often cross-pollinated and cross-pollinated-based on the extent of cross-pollination. The extent of cross pollination ranges from 0 to 5% in self-pollinated, 5-12% in often cross-pollinated and more than 12% in cross-pollinated vegetables. The information on extent of cross-pollination is important to decide appropriate breeding methods to develop and improve the crop, to maintain genetic purity of seed, and to decide safe isolation distances from close relatives of the crop.

The important factors for maintaining genetic purity and maximizing seed yield are:

1. *Control of Seed Source :* The use of seed of an appropriate class and from an approved source are necessary for raising the seed crop. Four classes of seeds, namely, breeder's, foundation, registered and certified seeds

have been defined by the Association of Official Seed Certification Agencies (AOSCA) :

Breeder's Seed : Breeder's seed is the genetically pure seed produced by the concerned breeder or by the institution which is used for the production of foundation seed.

Foundation Seed : Foundation seed is also genetically pure seed produced from breeder's seed under strict supervision. Foundation seed is the source of registered and/or certified seed.

Registered Seed : Registered seed is the progeny of foundation seed that is so handled as to maintain satisfactory genetic identity and purity, and that has been approved and certified by a certifying agency. This class of seed should be of a quality suitable for production of certified seed.

Certified Seed : Certified seed is the progeny of foundation or registered seed. Certified seed is so handled as to maintain satisfactory genetic identity and purity and that has been approved and certified by the certifying agency.

2. *Isolation :* One major factor during the course of seed production is to ensure that the possibility of cross-pollination between different cross-compatible plots or fields is minimized. In addition to the question of cross-pollination, adequate isolation also assists in avoiding admixture during harvesting and the transmission of pests and pathogens from alternative host crops. Vegetable seed crops can be isolated by time and by distance and minimum isolation distance required for production of each class of seed is given.

3. *Selection of crop and variety:* Seed multiplication of popular local varieties should be given priority. Knowledge about the photo and thermosensitivity of the variety is a prerequisite for successful vegetable production.

4. *Cultural practices:* the cultural practices like adjusting sowing time for seed crops such that their maturity may fall in the drier part of the season, taking necessary steps for timely control of insect and disease onslaughts, etc. also play a key role in obtaining good quality seed.

5. *Roguing of seed crop :* Removal of off-type plants in the seed crop is termed as roguing. Not only the off-types but the diseased and abnormal plants are also to be removed for esuring genetic purity of seed. The number of roguing required for the seed crop will vary with the kind of vegetables, purity of the seeds sown, nature of the previous crop etc.

6. *Harvesting* : The quality and yield of seed depend on harvesting at correct mature stage. In cucurbits, okra and brinjal, fruits can be retained for extraction of seed after 2-3 harvests for vegetable purpose. The seed processing is necessary to dry the seeds to safe moisture level, remove/ reduce to extent possible the various undesirable materials, weed seeds, other crop seeds, deteriorated/ damaged seeds; uniform size grading and seed treatment to upgrade the overall seed quality. In its common usage, seed processing refers to all steps necessary for preparation of harvested seed for marketing, viz. handling drying, shelling, preconditioning, cleaning, size grading, treating, packaging, etc. The prescribed limit is given below.

Day Length and seed Yield of Important Vegetables Grown in India

Vegetable	**Day length**	**1,000-seeds weight (g)**	**Seed rate (kg/ha)**	**Duration (months)**	**Crop yield (t/ha)**	**Seed yield (kg/ha)**
Bell pepper	N*	5.5	0.4	4	7.5-10.0	60-100
Chilli	N	3.3	0.5	9	12.5-15.0	200-600
Tomato	N	3.3	0.4	5	20.30.0	80-150
Brinjal	Sh/N	4.0	0.8	7	20.0-30.0	200-300
Ash gourd	N	70.0	2.0	5	27.5-32.5	250-400
Muskmelon	N	25.0	2.0	4	20.0-25.0	300-600
Cucumber	N	25.0	3.0	4	20.0-25.0	300-800
Pumpkin	N	170(70)	2.0		425.0-32.5	500-800
Bottlegourd	N	150	3.0	6	25.0-30.0	400-600
Ridged-gourd	N	90	3.0	6	20.0-25.0	250-400
Bittergourd	N	60	5.0	4	20.0-30.0	250-400
Cowpea	Sh/N	220	20	5	125-175	700-2500
Sword bean	Sh	4000	60	9	150-175	1300-2500
Dolichos Bean	Sh/N/L	330	140	9	150-175	900-2000
Pea	N	170	120	3	100-150	1500-2500
Turnip	L	3	1.5	6'	250-300	600-1200
Cabbage	N	4	0.6	6'	250-300	600-1200
Cauliflower	N	3	0.3	6'	250-300	150-300
Radish	n/L	10	10	6	200-250	800-1200

Contd...

Amaranth	Sh/N	0.3	1.0	4	150-200	500-1500
Bhindi	Sh/N	50	6	6	100-125	400-1000
Onion	N	3.0	4	9*	200-300	400-800
Leek	L	2.5	3	12	100-125	200-600
Carrot	L	1.2	1	9	125-150	300-1000
Asparagus	N	40	4	3 years	10-15	400-1000
Beet	L	16	10	9*	100-150	800-1200
Lettuce	L	0-8	0.3	6	100-150	125-200
Spinach	L	10	20	6	100-150	600-1200

*N, day neutral; L, long day; Sh, short day; * -period after optimum vegetative growth

Prescribed Field Standards (Isolation Requirements, Minimum Field Inspections and Maximum Permitted off-types) for Vegetable seed Production

Crops	**Minimum isolation distance(m)**			**Minimum field inspections**		**Maximum permitted off-types (%)**	
	F	**C**	**To be isolated from**	**No.**	**Stages of crop growth**	**F**	**C**
Potato	5	5	Other varieties	3	45 days after planting, just before haulm cutting	0.2	0.5
Tomato	50	25	Other varieties	3	Before flowering, flowering to immature fruit stage, mature fruit stage	0.1	0.2
Brinjal	200	100	Other varieties	3	Before flowering, flowering to immature fruiting, mature fruit stage	0.1	0.2
Chilli	400	200	Other varieties and chillies from sweet pepper	3	Before flowering, flowering to fruiting, mature fruit stage	0.1	0.2
Sweet pepper	400	200	Other varieties and sweet pepper from chillies	3	Before flowering, flowering to fruiting, mature fruit stage	0.1	0.2
Okra	400	200	Other varieties and wild *Abelmosclus* spp	3	Before flowering, flowering to fruiting, mature fruit stage	0.1	0.2
Pea	20	10	Other varieties	3	Before flowering, flowering, edible pod stage	0.1	0.2
Other beans like cowpea, French bean, cluster bean, hyacinth bean etc	50	25	Other varieties of the crop	2	Before flowering, flowering to fruiting	0.1	0.2

Contd...

Cabbage	1600	1000	Other varieties of the crop and other crops like cauliflower, Brussels sprouts, Knolkhol, kale, collards	3	Head formation stage, fully developed head stage, flowering stage	0.1	0.2
Cauliflower	1600	1000	Other varieties of the crop and other crops like, cabbage, sprouting broccoli, Knolkhol, kale, collards	4	Before initiation of curds, curd initiation on stage, flowering stage	0.1	0.2
Onion	1000	500	Other varieties	3	20-30 days after sowing, fully developed head stage, flowering	0.1	0.2
Radish	1600	1000	Other varieties of the crop	3	20-30 days after sowing, at lifting and replanting, flowering(root-to- seed method)	0.1	0.2
Turnip	1600	1000	Other varieties of the crops and other crops, like Chinese cabbage, rape seed,mustard and rutabaga	3	20-30 days after sowing, at lifting and replanting, flowering(root-to- seed method)	0.1	0.2
Carrot	1000	800	Other varieties	4	20-30 days after sowing, at lifting and replanting, flowering, fully devel oped umbel stage (root-to- seed method)	0.1	0.2
Beet	1600	1000	Other varieties of the crops and other crops, like palak, sugar beet, swiss chard	3	20-30 days after sowing, at lifting and replanting, flowering	0.1	0.2
Amaranth	400	200	Other varieties and wild *Amaranthus* spp	2	Fully developed stage before flowering, flowering stage	0.1	0.2
Palak	1600	1000	Other varieties of the crops and other crops, like beet, sugar beet, swiss chard	2	Fully developed stage before flowering, flowering stage	0.1	0.2
Musk melon	800	400	Other varieties and other crops like long melon, snap melon	3	Before flowering, flowering to fruiting, mature fruit stage	0.1	0.2
Pumpkin	800	400	Other varieties and other crops like summer squash and winter squash	3	Before flowering, flower ing to fruiting, mature fruit stage	0.1	0.2
Bitter gourd, Bottle gourd, Sponge gourd, Snake gourd, Water melon, Cucumber, Tinda, etc	800	400	Other varieties of the crop	3	Before flowering, flowering to fruiting, mature fruit stage	0.1	0.2

Specified Seed Standards for different Vegetable Crops

Crop	Minimum physical purity (%)		Minimum germination (%)		Maximum moisture(%) in ordinary container		Minimum moisture(%) in vapour-proof container	
	F	C	F	C	F	C	F	C
Tomato	98	98	70	70	8	8	6	6
Brinjal	98	98	70	70	8	8	6	6
Chilli	98	98	60	60	8	8	6	6
Sweet pepper	98	98	60	60	8	8	6	6
Okra	98	99	65	65	10	10	8	8
Pea	98	98	75	75	9	9	8	8
Cowpea	98	98	75	75	9	3	-	-
Hyacinth bean	98	98	75	75	9	9	7	7
French bean	98	98	75	75	9	9	7	7
Cluster bean	98	98	70	70	7	9	-	-
Cabbage	98	98	70	70	7	7	5	5
Cauliflower	98	98	65	65	7	7	5	5
Knolkhol	98	98	70	70	7	7	5	5
Chinese cabbage	98	98	70	70	7	7	5	5
Onion	98	98	70	70	8	8	5	5
Radish	98	98	70	70	6	6	5	5
Turnip	98	98	70	70	6	6	5	5
Carrot	95	95	60	60	8	8	7	7
Beet	96	96	60	60	9	9	8	8
Amaranth	95	95	70	70	8	8	6	6

Contd...

Palak	96	96	60	60	9	9	8	8
Fenugreek	98	98	70	70	8	8	-	-
Lettuce	98	98	70	70	8	8	6	6
Cucumber	99	99	60	60	7	7	6	6
Muskmelon	99	99	-60	60	7	7	6	6
Longmelon	99	99	60	60	7	7	6	6
Watermelon	99	99	60	60	7	7	6	6
Pumpkin	99	99	60	60	7	7	6	6
Summer squash	99	99	60	60	7	7	6	6
Winter squash	99	99	60	60	7	7	6	6
Bitter gourd	99	99	60	60	7	7	6	6
Bottle gourd	99	99	60	60	7	7	6	6
Ridge gourd	99	99	60	60	7	7	6	6
Sponge gourd	99	99	60	60	7	7	6	6
Snake gourd	99	99	60	60	7	7	6	6
Round melon	99	99	60	60	7	7	6	6

F=Foundation C= Certified Seed.

Potential of Vegetable Seed Industry

Seed production of open pollinated vegetable varieties is under taken by farmers since ages. Last decade witnessed tremendous progress in the development of high-yielding open-pollinated varieties and hybrids. Seed industry in private sector is flourishing and there are some pockets like ranibannur in Karnataka where local growers master the expertise in seed production. Profit from seed production is given below.

The development of Indian seed industry began only after independence. In India, seed production programme was started during 1942-43 in Katrain and Kashmir valley. Organised seed production in the country became possible with the formation of National Seed Corporation (NSC) during 1963, which paved the way for a sound seed industry to come. NSC was the first agency to certify seeds of all the crops. Subsequently Seed Act 1966 was enacted in the Parliament as a means to protect quality of seed and Seed Rules 1968 came into force throughout the country from October, 1968 to regulate the seed related issues.

Returns from Seed Production of Major Vegetable Crops

Crop	Estimated seed yield(kg/ha)	Seed Sale rate (Rs/ha)	Income (Rs/ha)	Estimated cost of production of seed(Rs/ha)	Returns (Rs/ha)
Tomato	125	1,000	125,000	35,000	90,000
Brinjal	200	400	80,000	35,000	45,000
Chilli	200	400	80,000	35,000	45,000
Okra	1,200	70	84,000	28,000	56,000
Vegetable pea	2,500	50	125,000	25,000	100,000
Radish	800	150	120,000	38,000	82,000
Cowpea	1,200	70	84,000	25,000	59,000
Frenchbean	1,200	80	96,000	25,000	71,000

Source : Singh *et al*, 2006

India ranks 10th position in the world regarding production of seeds. In the case of vegetables, major crops in which seed production is being carried out are tomato, cabbage, gourds, melons, brinjal, cauliflower and chilli. In gourds, India is the largest producer of seeds. Twenty-four national and multinational companies are involved in the production and distribution of gourds seeds.

Estimated vegetable seed requirement in India is given elsewhere. Thirty % of OP varieties requirement is met by private companies, 15 % by public sector and rest from farmer's own seeds. F_1 hybrid seed production is more or less became a monopoly of private seed industry in India. Vegetable seeds, which are low volume but high value, have attracted private investment. In fact, vegetable seeds perform a catalytic function in the transition of seed industries from public sector involvement and commercialization. The introduction of hybrid seed in vegetable crops like tomato and cabbage has more than doubled the productivity of these crops from hardly 20 tonnes/ha to around 60 tonnes/ha and resulted in increasing the production. The production of okra, peas, cabbage and tomato have increased by 111.44, 77.95, 40.96 and 23.96 % respectively after the introduction of hybrids and improved varieties. In October 1988, Government announced a new policy on seed development, under which import of seeds and planting materials of some items are allowed under Open General License (OGL) at a much reduced customs duty. Since 1988, about 50,000 tonnes of vegetable seeds, particularly of tomato and cabbage, were imported under the OGL scheme. From 1992 onwards, seeds and planting materials are allowed to be imported free of duty. Many new national and multinational private sector companies entered the seed business in a very big way. Some of them are Novartis, Cargill, Hoechst, Mahyco, Indo-American Hybrid seeds, Nath, Hindustan Lever, Nijjer, Suttons, Harrisons Malayalam, Southern Petrochemical Industries Corporation Limited and Indian Tobacco Company.

F_1 hybrids

Even though first hybrid of vegetable crop in India, bottle gourd- Pusa Meghdoot- was developed by IARI in 1971, it could not reach the farmers because no concerted effort was made to multiply these seeds in large quantities by public sector. In 1973, Indo- American hybrid Seed Co., a private sector institution, released their first tomato hybrid 'Karnataka' and first capsicum 'Bharat'. An aggressive field level campaign was launched by the company. Now the technology is finally spreading at a faster pace and large number are being added every year. Hybrids of tomato and cabbage are most visible (31%) followed by egg plant and okra. Though price is 20 to 30 times high, farmers preferred these hybrids for their superiority in yield, quality, uniformity and earlieness. Some of the popular F 1 hybrids released for cultivation are given else where.

Hybrids Recommended for Release and Cultivation

Crops	Name of identified hybrids
Tomato	ARTH-4, MTH-6, ARTH-3, Pusa hybrid-2, Arka Vardhan, NA-501, DTH-4, KT-4, NA-601, FMH-2, BSS-20, AVinash-2, HOE-303, Sun-496, DTH-8, Swarm Baibhav (CHTH-1), ARTH-128, KTH-2
Brinjal	Arka Navneet, Pusa, Hybrid-6, Pusa hybrid-5, ARBH-201, NDBH-1, ABH-1, MHB-10, MHB-39, NDBH-6, ABH-2, Phule Hybrid-2, Pusa Hybrid-9, ARBH-541,, PBH-6, JBH-1, BH-1, BH-2, Kashi Sandesh
Chilli	HOE-888, Arch-236, Sungro-86-235, Arch-228
Capsicum	KT-1, SEL-11, Lario, DARL-202.
Okra	Pusa Hybrid-1, Dch-541; Synthetics: Synthetic-1, Early Synthetic. Shri Ganesh Gol, Nath-401, BSS-2, Nath-501. Quisto; Pusa Synthetic
Carrot	Hybrid-1
Muskmelon	Hybrid- M-3, MHY-5
Water melon	Arks Jyoti
Bitter gourd	Pusa Hybrid-2
Cucumber	PCUCH-1
Bottle gourd	PBOG-2, PBOG-1, NDBH-4.

Hybrid Seed Production Technique

To exploit the phenomenon of heterosis the basic requirements are presence of heterotic combination, flower size and colour, pollen production capacity of male plant and longer duration of stigma receiptivity, easiness in emasculation and pollination, attraction to insect as a means of pollen transfer, seed setting and their economic feasibility in production and adoption.

Three Steps in F_1 Hybrid Seed Production

Three steps involved in F_1 hybrid seed production are

1. *Development of inbred line and their production:* The inbred lines are developed through exploiting inbreeding depression and fixation of the desired traits in them. The seeds of the developed lines are produced in isolation or by hand pollination.
2. *Testing of combining ability:* The combining ability (gca/sca) is tested by line x tester or diallel cross method.
3. *Production of F_1 hybrid seed:* Techniques for production of F_1 hybrid seeds are as follows
   i. *Hand emasculation and hand pollination:* This technique is applicable for limited scale produciton, since a lot of trained labour are required in pinching, pollen collection and hand pollination, as followed in tomato, capsicum, brinjal and okra.
   ii. *Hand emasculation and pollination by insect:* The male flowers from female lines are pinched off day before of anthesis regularly, which honeybees and other insects (voluntary) use as a pollinizer. The male and female lines are grown in alternate rows. The fruit set on female lines are of hybrid and harvested for seed extraction. The planting ratio varies within the crops e.g. summer squash 3:1 and 4:1 in muskmelon and cucumber but depend upon the population of bees in plot. This technique is used in bottlegourd, pumpkin, muskmelon, cucumber, summer squash and bittergourd for hybrid seed production.
   iii. Exploitation of male sterility system: Male sterlity system has been utilized for commercial hybrid production in onion and muskmelon (Punjab Hyb.-1).

   Male sterility is the best exploited for commercial F_1 hybrid seed production in onion. Three parental lines, namely A, B and C are used for hybrid seed production. The female parent (A line) has a constitution of 'Smsms', while line B has a genetic constitution of 'Nmsms'. Since the fertile cytoplasm factor is not transferred to the female, the A line remains male sterile when crossed with the B line, which differs only by its sterile cytoplasmic factor. The B line is used to maintain seed supply of male-sterile A line. Both factors, genetic sterility and cytoplasmic sterility, must be present for the plant to be male sterile. Line C is the fertile male parent, an imbred, ideally diverse from line A or B for ensuring heterotic cross.

The genetic male sterility in muskmelon is controlled by single recessive gene (msms). For hybrid seen production, the male sterile line is used as female parent. Since genetic male sterile line is maintained in heterozygous forms, 50% fertile plants are to be removed at flowering. The other 50% having non-dehiscent empty anther are retained in female rows. The female and male are grown in 4:1 ratio. However, to maintain the good plant population in female rows, it is suggested that seed parent should be sown with double seed rate. It is also advised that female line seedling should be raised in polythene bags and transplanted at flower appearance to avoid the fertile plants in female rows. The pollination is done by honey bees and 1 to 2 medium sized hives are good enough to ensure the good pollination and fruit set at female row. The male sterile line is maintained in heterozygous form by crossing with maintainer line under adequate isolation distance or under cover.

iv. *Use of gynoecious sex form* : The gynoecious sex forms have been commercially exploited in hybrid seed production of cucumber (Pusa Sanyog) at IARI R.S. Katrain and in muskmelon (MH-10) at PAU, Ludhiana. For hybrid seed production, female and male rows are planted in 4:1 ratio. The female (seed parent) bears only female flowers and pollination in done by insect (honeybee). To ensure the good fruit and seed recovery, the sufficient population of honeybee 1 to 1½ colony of medium size has to be kept at the boundary of seed production plot to boost the amount of crossing. The parental lines i.e. male parent maintained by selfing (mixed pollination) and rouge out undesirable plants before contamination take place. The female lines i.e. gynoecious lines maintained by inducing the staminate flower through sprays of silver nitrate (200 ppm) at two to four true leaf stage and then selfing is carried out. It was observed that 10-11 male flowers appear per 100 nodes. The performances of gynoecious lines are unstable under high temperatue and long photo period conditions because of their thermo-specific responses for gynoecious stability. Hence, gynoecious cucumber did not receive much attention in the tropical countries. However, a few true breeding tropical gynoecious lines in cucumber and muskmelon are developed at IARI. As a result of development of true breeding lines, muskmelon hybrid 'Pusa Rasraj' was developed. These homozygous gynoecious lines are maintained by using GA3, (1500ppm) or silver nitrate (200-300 ppm) or silver thio sulphate (400 ppm) to induce staminate flowers at two and four true leaf stage. Homozygous lines are planted in strict field isolation. The gynoecious lines are crossed with monocious male parent to produce F1 hybrid.

v. *Hybrid seed production through chemical induced sex expression:* The hybrid seed can also be produced in cucurbits by application of chemicals for attaining sex of cucurbits. Specific chemicals are known to induce femaleness and maleness as desired. The spraying of ethrel (2-chloro-ethyl-phosphonic acid) (200-300 ppm) at two and four true leaf stage and another at flowering is useful for inducing pistilate flower successively in the first few nodes on female in bottlegourd, pumpkin and squash for F1 seed production. The row of male parent is grown side by side of female and natural cross pollination is allowed. In the absence of insect, hand pollination is possible when two sexes are separate. Four to five fruit set at initial nodes are sufficient for hybrid seed. The complete suppression of male flowers in squash can be achieved at higher concentration of ethrel(400-500 ppm) applied twice and has made hybrid seed production comparatively easier and nearly 56% of total squash seed produced in USA are of F_1 hybrid. The other chemicals like GA3 (10-25 ppm) in cucumber, MH-(100 ppm), ethephone (600 ppm) in squash induce female flowers.

Popular Hybrids in Private Sector

Crops	Hybrids
Tomato	Vasihali, Lerica, NS-815, Avinash 2, Rashmi, Ramya
Brinjal	Suphal, HOE-808, Kalpatharu, Ravaiyya
Chilli	Agni, NS.1101, HOE-808
Capsicum	Indira, Bharat, Lerio, NS.33, Hira
Cabbage	Cabbage No.8, Questo, NS.22, Hari Rani Gol, Krishna, Vishesh

Overseas Seed Market

The major vegetable seeds exported during 2002-03 include radish seed (2.62%), peas (15.59%), cabbage (20.92%), and tomato (2.68%). In value terms, the share of tomato was 20.12 % and that of cabbage seeds was 8.38%. India exports vegetable seeds to over 40 countries. In terms of volume, leading importing countries are Pakistan (45.96 per cent), Bangaladesh (11.48%), USA (5.62%), Italy (2.25%), Jordan (0.9%), Sri Lanka (1.71%), UAE (2.78%) Japan

(3.36%), Saudi Arabia (3.51%) and Netherlands (2.86%). Among the seeds imported, radish occupied first in quantity (19.91%) followed by cabbage (11.63%).

Institutional Support

National Agricultural Research System, Seed division GOI and Directorate of Plant Protection Quarantine and Storage work closely with the private sector in working logistics and assist in providing it a wide range of varieties/ hybrids. The public institutions meet demand of indented quantities of breeder seed at reasonable price. This has considerably helped the upcoming seed organizations to benefit from the value additions made by these institutions. The varietal testing programme run by the AICRP (All India Coordinated Resaech Project) as a standard practice, include private sector for trial of new materials. The project has identified a number of F_1 developed even by private breeders. Other institutions viz. National Horticulture Board, APEDA, State Seed Certification Agencies, State Regulatory Agencies and State Department of Horticulture also play their assigned roles with seed industry.

Issues in Vegetable Seed Sector

1. Seed Sales, Quantities and Prices

Private seed industry works under a highly competitive environment. For acquiring greater brand equity, most companies now depend on their proprietary hybrids which are priced high to recover the investment. Bulk items like peas, okra, beans etc. are priced low. When compared, a bulk of 1000 tonnes of peas may equal I ton of hybrid tomato, in value. Prices move on a wide scale (1-100) between OP varieties and premium hybrids.

2. Seed Marketing Systems and Bottle Necks

Inorder to push the products, the seed companies now conduct trials and simultaneously test market their products direct to users sometimes involving village dealers. All modern marketing tools like road shows, news papers/ periodicals, cinema slides, short films, radio and TV ads are pressed in to action.

3. Low Seed Yield

As the seed production is concentrated in interior country side, seed growers do not have an easy access to advance technology and very often financial resources do not warrant its optimal use. Controlled conditions to combat vagaries of nature are not affordable which all result in low seed yield. Clearly the breeders during the course of final evaluation need to involve the agronomists, the pathologists and entomologists so as to recommend appropriate crop/seed production technology.

4. Low Seed Replacement Rates

Seed replacement depends on: (a) reproducibility- European vegetables need cold stimulus and do not reproduce in lower altitudes and latitude making 100% replacement inevitable: (b) hybridity -progenies of hybrids must segregate to warrant 100% replacement; (c) varietal obsolescence- new varieties consistently replace the old ones (the pace of replacement varies according to costs and benefits); (d) quality-regional influence on the physical, physiological, as well as genetic quality (d) economics- certain regions have high multiplication ratios making retention of seeds an un-viable proposition, especially when fresh vegetables fetch a good price; (e) vagaries of nature-occasional crop failures necessitate out- sourcing and there by replacements; (f) institutional interventions- government and /or non-government institutions selectively pursue farmers to adopt new varieties through various market intervention schemes.

India still has a low seed replacement rate owing to all reasons mentioned above.

5. Intellectual Property Protection (PVP, Patents)

Being a signatory to the Agreement on Trade Related Intellectual Property Rights (TRIPS), India is committed to enact a legislation for plant variety protection. In Indian Plant Variety Protection and Farmers Right Act 2001, farmers role is honoured not just as a cultivator but also as a conserver of the agricultural gene pool and as a breeder who has bred several successful varieties. This act will safe guard the interest of vegetable breeder and growers at the same time protects the growers from the clutches of MNCs. As to patents, India is not yet ready for serious discussion on plant patents. However, increasing investment in biotechnology research, particularly by the private sector, is perceived to provoke a fresh debate.

Export of Vegetables

The Agricultural and Processed Food Products Export Development Authority(APEDA) has identified traditional vegetables like onion, potato, okra, bitter gourd, chilli and nontraditional vegetables like asparagus, celery, sweet pepper, sweet corn, babycorn, green peas, frenchbeans, cucumber and gherkin and cherry tomato having good export potential. Among other vegetables, potato, okra, tomato, cucumber and gherkins, chillies frenchbeans, capsicum, bittergourd and bottle gourd are being exported. Vegetables are also being exported in processed form. The total export of vegetables in processed form is over Rs.231 crores annually.

Export Demand of Vegetables

India having varied agroclimatic conditions can produce and supply the vegetables to different markets from one to the other parts comfortably round the year if necessary infrastructural facilities are created for precooling and transport. Indian vegetables can compete for middle east where the demand is round the year. In European markets, the demand is from November-December to April- May when there is no local production or even stored produce is not available.

Potential Area of Production

Export vegetable production is now limited in certain pockets for meeting the demands of the foreign markets. The details of some of the crops and area of production are mentioned as under. AEZ operating in states of Maharashtra, Karnataka, Punjab and Andhrapradesh give emphasis to the export of vegetables.

Vegetable Production Centres for Export Market in India

Crops	Area of production
Okra	Nasik, Ozar, Saikheda, Dindhori, Kolhar, Naraingaon and Sholapur in Maharashtra.
Chillies	Pen, Alibaugh and Choul in Raigad distt. Dindhori, and Niphad and Igatpuri taluka in Nasik district of Maharashtra
Watermelon	Panvel near Mumbai, Maharashtra state.
Bottle gourd and Bittergourd	Nasik and Pune distts. in Maharashtra
Gherkin	Nasik in Maharashtra and Bangalore in Karnataka
Capsicum	Nasik, Pune and Satara distt in M.S
Babycorn	Bangalore in Karnataka , Nasik and Pune in M.S
French bean and cluster bean	Dindhori in Nasik distt Wai in Satara distt and Naraingaon in Pune, Dhule and Ahmednagar.
Onions(big)	Nasik, Pune and Satara distt in M.S, Periyar and Coimbatore distt in Tamil Nadu, Budaun in U.P

Contd...

	and Patna. Biharsharif in Bihar, Bhavnagar and Rajkot in Gujarat.
Onion(small)	Kolar and Bangalore in Karnataka and Cuddapah in A.P
Onion(multiplier)	Anna, Madurai, Salem and Coimbatore distts of Tamil Nadu
Garlic	Indore and Mandsaur in M.P, Ooty in Tamil Nadu, Jamnagar and Rajkot in Gujarat and Kullu in H.P
Tomato	Nasik and Pune distts. in Maharashtra and Bangalore in Karnataka
Potato	Jalandhar and Ludhiana in Punjab, Kurukshetra and Karnal in Haryana, Ooty in Tamil Nadu and Indore in M.P

Source : Pandey, 1998.

Factors Affecting Export of Indian Vegetables

1. Quality of the Cultivar

The requirement of foreign markets in terms of varieties and qualities differ from domestic requirements. The requirements also differ between different foreign markets. European markets, Japan Australia, USA etc require different qualities and varieties than Gulf countries and South East Asian countries. For example, in case of onions, European and Japanese markets require yellow/ brown colored mild pungent onions of big size where as Gulf countries and South East Asian countries require light red to dark red colored strong pungent onions of varying size. South East Asian countries and Srilanka also demand small rose and multiplier onions. Some of the specific requirements of size, color, shape etc. are given here under:

Specific Features of Vegetable Crops in Demand for Export Market

Crops	Specific requirement
Okra	Green tender; 6-9cm long
Chillies	Green 6-7 cm long
Cluster bean	Green tender; 7-10 cm long
Bitter gourd	Green, 20-25 cm long having short neck

Contd....

Bottle gourd	Light green, straight cylindrical in shape, 25-30cm long
Gherkin	Green small sized having 160-300 fruits/ kg in premium grade
Tomato	Round, medium size, red color in Middle East, Cherry tomatoes in European Countries
French bean	Straight 10-12 cm long, round green pods in bush beans, Flat beans having 12-13 cm length and pods are also in demand in European markets
Small onion	Dark red 2-3 cm and round shape
Multiplier onion	2.5-3.5 cm sized bulblets of bright red colour
Garlic	White, round, 5 cm and above, bigger cloves of 10-12 mm and above with 10-15 number. For Bangladesh and Sri Lanka 4-5cm size bulbs are also acceptable.
Potato	White, oval, 4.5 to 6.0 cm size. Bangladesh demands red types and that Iran & Iraq demand potatoes with yellow flesh.

Source : Pandey, 1998.

2. Storage, Packaging and Transport

Inorder to maintain the quality of produce, fresh vegetables need to be stored under ambient temperature after proper post harvest treatment. Major export of vegetables is from Mumbai. Fresh vegetables are packed in basket or open mesh jute bags in the evening and brought by trucks to Mumbai where, sorting, grading and packing are done in cartons before air lifting in the morning. Garlic and potato are also brought in open mesh jute bags to Mumbai where, final sorting, grading and packing are done according to requirement. *Kharif* and late *kharif* onions are cured, sorted, graded and packed in open mesh jute bags of required size at Lasalgon, Pimpalgaon, Saikheda, Sinnar, Chandward and Manmad in Nasik district or Chakan in Pune district and Lonand in Satara district of Maharashtra. Rabi onions are also cured, sorted, graded and packed in open mesh jute bags at Dindhori, Pimpalgaon, Saikheda, Niphad, Lasalgon, in Nasik and Chakan in Pune or stored in loose after curing and then sorted, graded and packed in open mesh jute bags of different sizes depending upon the requirement of different markets. Some exporters in Nasik and Pune districts are precooling the vegetables for 5-6 hours after packing and then passing to cold storage. Depending upon the demands, the produce is sent in refrigerated containers. The open mesh jute bags should have 9 x10 mesh per sq. inch and cartons should have perforations for air circulation. Tissue paper or polythene sheets are used for wrapping to reduce evaporation and thereby prolonging shelf life of the fresh vegetables like asparagus, capsicum, chillies, brinjal and okra. Specific requirements in respect of storage, packing material size and handling temperature for some of selected vegetables are else where.

Storage, Packing Material size and Handling Temperature for Selected Vegetables

Crops	**Temp. and RH during storage and transportation**	**Packing material**	**Size (kg)**	**Remark**
Asparagus	Cooled immediately after harvest and kept at 2° C throughout transporatation and storage, RH 95%	Packed in bundles or wrapped in paper or plastic or packed in cartons of 500g size and then kept in corrugated fiber board boxes	5-6	Export by air
Brinjal	Cooled as soon as possible after harvesting. 7-10° C temp. RH 90%	One or more layers, tissue paper between layers. Each brinjal may be wrapped with polythene film and then kept in corrugate fiber-board boxes.	5-6	Requires minimum ventilation, export by air
Capsicum	Cooled with 24 hrs after harvesting. Optimum temp. for green 7-10° C and red 4-6 C RH 94-95%	Wrapping in polythene 5-6 preferred, waxing is also done then packed in corrugated fiber board boxes.	5-6	Needs minimum ventilation when cooled, export by air.
Chillies	Cooled as soon as possible after harvesting. temp 7-10° C. RH 90 -95%	Tissue paper is used as filling material then packed in corrugated fiber board boxes	2-5	Needs minimum ventilation when cooled, export by air.
Melons	Cooled as soon as possible after harvesting cantaloupe at 20° C whereas honey dew at 7-10° C, RH 85-90%	Necessary to use filling 5-10 material to avoid mutual bruising, tissue papers are used. Use trays for separation and then corrugated fiber board boxes	5-10	Export by air
Okra	8-10° C temp., RH 90 95%	Jumble packed, normally inserts are not used, packed in corrugated fiber board boxes	2-5	Life 1-2 weeks, export by air, ther should be sufficient ventilation.

Contd....

Onion	0-2° C temp. and 65-70% RH under cold storage conditions, can be kept for 5-6 months	Packed in plastic netted or open mesh jute bags	8-25	Good ventilation is required when transported/ stored under ambient condition, export by ship and road.
Garlic	0-2° C temp and 65-70% RH under cold storage conditions, can be kept for 5-6 months	Packed in plastic netted bags/ open mesh jute bags or corrugated fiber board boxes having opening for ventilation.	2-5	A good ventilation under ambient transportation/ storage conditions, export by ship, road and air
Potatoes	Should not be cooled immediately after curing at 13-18° C for 10-14 days. RH 90-95%	Packed in open mesh jute or plastic. Occasionally packed in corrugated fiber board boxes.	2.5-25	Life 5-8 months under cold storage conditions, export by ship.
French bean	Cooled as quickly as possible after harvesting at 7-10° C, RH 90-95%	Jumble packed or parallel in corrugated fiber board boxes.	2.7-6	No free moisture, llife 1-2 weeks, exports by air, free ventilation.

Source : Pandey, 1998

3. Competing Countries

Thailand, Jordan, Lebanon, Syria, Kenya, Zimbabwe, Guatemala, China, Argentina, Indonesia, Egypt, Turkey, Iran, Cyprus, Australia, New Zealand and Holland are the other countries which export vegetables to different countries.

4. Export Policy of the Government

Fresh vegetables export other than onion is allowed freely. Under this category any person may export vegetables to any country except to a country where export is prohibited by law enforced from time to time. Government has also started the policy of giving cash incentives of 10-20% depending upon the kind of vegetable being exported. Onion export is canalized through National Agricultural Co- operative Marketing Federation of India (NAFED) under the present policy of export. In this system the exporters are required to register with NAFED and get NOC for export. Other vegetable export is on consignment basis under OGL (Open General License).

The export of vegetables requires proper care from the stage of harvesting to the stage of export. India has good scope to increase the export manifold particularly from Maharashtra and Karnataka which have good climate. Export markets need to be surveyed for ascertaining varietal and quality requirement. The research and development programmes need to back these efforts. Promotion of integrated processing complexes with well developed infra structure close to producing pockets will have to be develop. Infrastructure will also have to be developed at the ports and airports for proper handling and storage. Organic vegetables will get an edge in future.

Economics of Vegetable Production

Vegetable production in India is unevenly distributed across the country, with 57% of the total vegetable area being concentrated in just four states-Uttar Pradesh, Tamil Nadu, Orissa, and Bihar. At the national level, about 3.7% of the country's total area under all crops is devoted to vegetables (including potato, tapioca, and sweet potato). Tubers and bulbs are the largest group, accounting for 45% of the country's total vegetable area; the major vegetables in this group are tapioca, potato, and onion. All leaf and flower crops have a total share of only 12% in the total area under vegetables. Cabbage and cauliflower are the main vegetables in this group. The fruit group contributes about 28%, with eggplant at 8%, and lady's finger, chilli, tomato, peas, and beans each having a 5% share in the total vegetable area. Erratic rains contribute to high variation in vegetable yields as 50% of the vegetable is under rainfed cultivation. Scarcity of water and rise in temperature cause a major threat to vegetable cultivation. Critical analysis of factors of production is required to frame out a national policy in this regard. Substantial improvements in production can be achieved even with the existing technology by bridging the gap between farm-level and experimental yields through extension, training, and educational programs. Researchers should focus on generating stabilizing technologies (e.g., stress-tolerant and pest-resistant varieties), and management practices (e.g., protected cultivation, optimum use of water) to counter environmental effects

Economics of vegetable production are important to all the stakeholders in vegetable cultivation right from farmers, marketing men and consumers. Consumer end price of vegetable is more influenced by the supply forces rather than farm production. Generally, vegetable prices rose faster at the retail level than at the farm gate or wholesale level. At the national level, the weighted average monthly market price index of all vegetables (excluding onion) was the highest during June-August, rainy months, and the lowest during January-March, which are cooler months. The increase in prices during the low supply period was about 120% of the lowest price during the highest supply period, and the corresponding difference in market arrival in the lean supply period was about 55% of the highest supply period (Subramanian *et al.*, 2000).

Market Integration and Seasonality

Indian markets across states seemed relatively less integrated, as suggested by wide variability in seasonality of a vegetable across markets. For example, prices of eggplant in Calcutta were the highest in October, while in the low range in the Delhi and Madras markets. A similar situation was seen to prevail for other vegetables and markets. Therefore, integrating markets by providing information on market arrival and prices can help to reduce seasonality in Indian markets. Seasonal advantage of different growing pockets in India can thus be exploited for the benefit of farmers as well as consumers.

Factors in Cost of Vegetable Production

Vegetable cultivation concentrates more on irrigated area than does cultivation of cereal crops (Singh and Sikka 1992). On average, cultivation of one hectare of vegetables requires about 10% more labor than cultivation of one hectare of rice; however, labor needs for vegetable cultivation are about twice those for cultivation of wheat. The amounts of fertilizer applied to vegetables (300-500 kg/ha) are two to three times those used on rice or wheat. Except for onion, fertilizer use is generally higher than the recommended level. For example, the recommended level of fertilizer for tomato, eggplant, and cauliflower for the Karnataka state is 275 kg/ha, whereas it is even lower for Tamil Nadu. Pesticides, too, are used in far greater quantities on vegetables than on cereals. However, there is wide variation in fertilizer and pesticide use among vegetables grown in different states.

Costs and Returns

Costs and returns of major vegetables in the major growing states are reported in elsewhere. Production costs for rice and wheat are included for comparison.

Wide variation was observed in costs and returns of different vegetables grown in different states, depending upon the social and physical infrastructure, and policy environment. Despite these variations, however, the benefit-cost (B-C) ratio (net benefit/total cost) was positive and generally high for all vegetables in all states.

Economics of Vegetable Production in India

Crop	State	Operational cost (Rs/ha)	Fixed cost (Rs /ha)	Total cost (Rs /ha)	Yield (t)	Cross Returns (Rs /ha)	Net benefit (Rs /ha)	B-C ratio	Cost/kg (Rs)
Potato	Bihar	4286	3224	7510	11.9	9195	1685	0.22	0.63
	Himachal Pradesh	4020	2313	6333	7.8	8255	1922	0.30	0.81
	Tamil Nadu	7812	2476	10288	18.3	13750	3462	0.34	0.56
	Uttar Pradesh	6787	2147	8934	18.7	12627	3693	0.41	0.48
	West Bengal	7684	3295	10979	19.2	12000	1021	0.09	0.57
Onion	Karnataka	2169	2334	4503	8.3	6709	2206	0.49	0.54
	Maharashtra	5205	2560	7765	14.5	11729	3964	0.51	0.54
	Tamil Nadu	3368	2200	5568	8.2	8353	2785	0.50	0.68
Cabbage	Himachal Pradesh	9230	3250	12480	24.2	19844	7364	0.59	0.52
	Karnataka	7968	1782	9750	18.2	16380	6630	0.68	0.53
	Tamil Nadu	12882	2684	15566	27.1	24379	8813	0.57	0.57
Cauliflower	Haryana	13540	2180	15720	11.2	36960	21240	1.35	1.40
	Uttar Pradesh	13011	1973	14984	14.8	48100	33116	2.21	1.01
	Tamil Nadu	17236	1763	18999	8.9	28587	9588	0.50	2.13
Egg plant	Andhra Pradesh	3254	1914	5168	10.0	12300	7132	1.38	0.52
	Tamil Nadu	6837	3965	10802	24.1	27844	17042	1.58	0.45
Tomato	Karnataka	5583	2527	8110	19.3	30880	22770	2.81	0.42
	Madhya Pradesh	16481	5039	21520	15.7	32467	10947	0.51	0.37
	Tamil Nadu	8436	2316	10752	14.9	31290	20538	1.91	0.72
Chilli	Andhra Pradesh	9966	3000	13662	3.0	18453	4791	0.35	4.53
Garlic	Assam	22535	17059	39594	6.0	78783	39189	0.99	6.59
Ginger	Kerala	225540	636	26323	20.0	60000	33677	1.28	1.31
Aveage of all vegetables		9583	3211	12969	14.7	24949	11981	0.92	1.22
Rice		7440	2745	10185	3.9	13830	3045	0.30	2.61
Wheat		5658	2706	8364	4.0	14300	5936	0.71	2.09

Source : http://www.avrdc.org

Vegetables generate higher net benefit than rice, except for potato production in Bihar, Himachal Pradesh and West Bengal, and onion in Karnataka and Tamil Nadu. On average, vegetable cultivation generates about four times higher net benefit than does rice cultivation, and about a 100% higher benefit than does wheat. The B-C ratio for vegetables is also generally higher than that of rice and wheat . The relative profitability of some vegetables would be even higher if the crop duration were taken into consideration, because some vegetables occupy land for a shorter time. The cost of producing 1 kg of any vegetable is lower than producing the same quantity of rice or wheat.

Marketing

In India 90-98% of vegetables are sold (the exceptions being root and tuber crops, where a large proportion is saved for seed). The Indian Institute of Horticulture Research (IIHR 1989) conducted a market study in Karnataka, Andhra Pradesh, and Tamil Nadu. About 92% of the vegetable farmers in Karnataka and 62% in Andhra Pradesh and Tamil Nadu sell their vegetables through commission agents. However, in Andhra Pradesh 22.6% of the farmers sell their output to wholesalers in markets and 15.4% sell their output to retailers in the fields. In Tamil Nadu, 21% of farmers have pre-harvest contracts (all farmers growing cabbage and carrot), one-tenth sell vegetables to retailers (all farmers growing cauliflower) who take the output from their fields, and 5% (farmers growing eggplant) choose wholesalers at the market. In Karnataka, only 3% of farmers make pre-harvest contracts, mostly for cabbage and cauliflower, and 5% (mostly lady's finger farmers) sell their output to retailers in their fields.

Marketing Channels

Seven market channels for vegetables can be identified

I. Producer → commission agent → shipper → secondary whole sale trader → retailer → consumer

II. Producer → commission agent → primary whole sale trader → retailer → consumer

III. Producer → commission agent → retailer → consumer

IV. Producer → primary whole sale trader → retailer → consumer

V. Producer → cooperative (or regulated market) → primary whole sale trader → retailer → consumer

VI. Producer → cooperative (or regulated market) → retailer → consumer

VII. Producer → cooperative (or regulated market) → consumer

Channel I, involving the shipper, is important for long- distance markets such as onion in Maharashtra, which serves the northern states and the export market, and potato in Haryana, which reaches the south. Commission agents usually buy on behalf of wholesaler traders. Whenever a co-operative society or regulated market is successful, Channel V becomes important. Channels VI and VII are the least important because they serve only the immediate neighbourhood of production areas.

Market Efficiency

For perishable commodities, quick and efficient disposal of produce is very important to avoid loss of quality and value. In India, the marketing infrastructure is not developed sufficiently to avoid such losses, so vegetable production is mostly concentrated near consumption centers (peri-urban areas), and is mostly for immediate sale. If any grading and processing done, it is done by traders and packers, not by farmers (Singh and Sikka 1992). Cooperatives might do some grading on behalf of growers, but their role in the sale of vegetables is limited to a few isolated cases, such as the Nilgiris Vegetable Growers Cooperative Marketing Society, a cooperative in Maharashtra and Punjab. Marketing of vegetables is generally in the hands of private traders.

Producers' Share in Retail Prices

Producers' share of the consumers' Rupee was relatively high, ranging from 40 to 67% for tapioca, cabbage, cauliflower, pea, chilli, garlic, and ginger. A review of the literature on marketing margin also suggests a decline in producers' share of the consumer's price in all vegetables reviewed except peas (Singh and Kharwal,1995 and Dahiya,1995). Farmers who sold their vegetables directly to consumers received higher shares of the consumers' price than did those who sold their produce through agents. The farmers' share is higher in the off-season than in the harvesting season (Vasisht *et al.*, 1995). The share only marginally differs for the places connected and not connected by a road (Yadav and Pant, 1995). The farmers' share can be as high as 87% in onion, 68% in cabbage, and 62% in cauliflower, when they sell their output direct to consumers. But the share can fall to below 50% as more agents are involved in marketing (Mishra *et al.* 1995).

Marketing Cost

Marketing cost varies from place to place and among commodities. For example, marketing cost to bring vegetables (based on the average for all vegetables) from field to market place varies from 90.3 Rs/t in Tamil Nadu to

212.5 Rs/t in Karnataka. Similar variation is observed for individual commodities.

Vegetable Marketing Cost for Cultivators (Rs/t) in Different States in India

Items	Crops							Average of all vegetables	
	Tomato	Brinjal	Bhindi	Cabbage	Cauli-flower	Carrot	Beans[1] Gourd	Rs	percent
Karnataka									
Transport cost	68.7	75.4	56.6	86.3	92.2	92.1	85.8	68.6	38.3
Loading/unloading	15.2	9.2	6.4	14.2	29.9	15.5	13.5	14.2	6.7
Commission charges	117.1	102.3	108.2	107.1	106.0	88.2	118.4	106.1	49.9
Packing charges	13.3	6.6	607	5.9	10.6	6.1	5.9	7.7	3.6
Tax and other costs	3.2	4.4	3.7	1.1	-	1.7	3.3	3.2	1.5
Total	217.5	197.9	181.6	214.6	238.7	203.6	226.9	212.5	
Andhra Pradesh									
Transport cost	36.2	45.8	26.9	-	-	-	57.2	46.1	34.2
Loading/unloading	10.0	22.7	4.5	-	-	-	10.7	14.9	11.1
Commission charges	55.6	39.6	107.1	-	-	-	84.9	62.3	46.2
Packing charges	7.3	11.4	6.4	-	-	-	16.4	11.6	8.6
Tax and other costs	-	-	-	-	-	-	-	-	-
Total	109.1	119.6	144.9	-	-	-	171.2	134.9	
Tamil Nadu									
Transport cost	28.0	25.4	24.8	45.7	70.0	76.0	NS [3]	30.7	34.0
Loading/unloading	6.2	5.1	8.2	8.7	5.0	3.2	NS	6.2	6.9
Commission charges	47.7	18.8	30.7	42.0	120.8	101.4	NS	35.8	39.7
Packing charges	1.6	1.0	25.6	48.3	1.3	2.4	NS	10.8	12.0
Tax and other costs	-	6.8	-	-	-	-	NS	6.8	7.5
Total	83.5	57.1	89.3	144.7	197.1[2]	183.0	NS	90.3	
Uttar Pradesh									
Transport cost	32.7	-	-	44.5	27.6	70.7	-	43.8	30.0
Packaging	82.8	-	-	33.0	45.6	59.9	-	55.3	37.9
Commission charges	24.0	-	-	22.7	36.4	25.4	-	27.1	18.6
Grading	11.3	-	-	7.4	18.1	16.9	-	13.5	9.2
Tax and other costs	8.1	-	-	4.7	4.4	7.3	-	6.1	4.2
Total	158.9	-	-	112.3	132.1	180.3	-	145.9	

Source : http://www.avrdc.org

1. French beans in Karnataka and Ridge gourd in Andhra Pradesh
2. Based on only one sample
3. NS = Not studied

Commission charges and transportation are the major marketing costs for farmers. On average these activities consume 39.7% and 34.0%, respectively, of the total marketing cost in Tamil Nadu, and 49.9% and 38.3% in Karnataka. In Uttar Pradesh, transportation, packaging, and commission charges are major costs.

Disease and Pest Management in Vegetable Crops

Phytophagous insects and plant diseases are major constraints in vegetable production throughout the world. These are especially important in the tropics and subtropics because environmental conditions are often conducive year-round for growth and development of hosts and pests. Integrated pest management (IPM) approach is required for effectively combating these biotic stresses. Besides being sustainable, the IPM approach reduces production costs and makes safe vegetables available to consumers at affordable prices. At the same time, it reduces the risk that chemical pesticides poison humans and the environment. Most celebrated strategy of this approach is the use of resistant varieties. Cultural methods like delayed planting, early planting, crop rotation with non- host plants and summer fallowing is found effective Biological agents such as fungi, bacteria and actinomycetes are promising against soil-borne diseases viz, root rot, damping off, collar rot and wilt of vegetable crops. Antagonistic fungi also kill some parasitic nematodes infecting vegetables. The *Trichoderma* is effective against several pathogens like *Pythium, Fusarium, Sclerotium, Macrophomina, Rhizoctonia*. Damping off caused by *Fusarium oxysporum* f.sp.*lycopersici, Phytophthora parasitica* and *Pythium debaryanum* in tomato has been effectively controlled by *T. viride*. *Trichoderma* can easily be isolated from any soil, and the strain should be resident of particular soil-ecosystem for effective use in that area *Aspergillus niger* has been antagonistic to *Fusarium oxysporum* causing wilt of muskmelon and watermelon. *Ampelomyces quisqualis* is effective against powdery mildew fungi like *Erysiphe* and *Uncinula*. *Sporidesmium sclerotivorum* is effective against *Sclerotinia* spp. infecting vegetables. Combination of *T. viride* and *P. fluorescence* is very effective to control *Phytophthora* blight in chilli. Among the bacteria, *Pseudomonas fluorescence* has shown efficacy in checking growth of *R. solani, Sclerotinia sclerotiorum* and *Phytophthora megasperma*. Many of the antagonistic bacteria also act as plant growth-promoting rhizobacteria(PGPR) in plants against vegetable diseases. Three antagonistic mechanisms are involved in biological control of soil borne plant pathogens, viz. competition, antibiosis and mycoparasitism. Bioagents are usually applied through soil and seed treatment. Seed treatment is preferred in direct sown crops, where as soil application and seed treatment are feasible in nursery crops. Biological seed treatment can be done by priming on seed, soaking seeds and seedling dipping depending upon the nature of biocontrol agents. The *Trichoderma* @ 6-10g/kg of seed should be used for seed treatments but spore concentration should be in between 10^6 and 10^9/ml. Dosage depends on inoculum of pathogen present in soil and nature of seed. Green manuring along with soil application of *Trichoderma* @ 5kg/ ha give good control of most of the soil-borne diseases in *Kharif* season grown vegetables. Root dipping of vegetable seedlings in *Trichoderma* formulation @ 1% for 10-25 minutes is very effective in checking root roting after transplanting in *kharif* season. Drenching of *Trichoderma* @ 1% near collar region should be carried out 20 days after transplanting to check root rot, collar rot and wilt incidence. The fungicides which are completely insensitive to antagonists, but sensitive to pathogen, can be used together to increase efficiency of disease management strategies. Copper hydroxide is absolutely insensitive for bioagents up to 400 ppm. Captan, chlorothalonil and propineb show tolerable growth up to 200 ppm after 7 days of incubation. Soil application of carbofuran followed by seed treatment by antagonists can be integrated for insect pest and disease management with out adverse effect on seed germination. Cypermethrin, fenvalerate and glyphosate are safe pesticides for integration with *Trichoderma.* Important diseases and control measures of major vegetables grown in India is given else where.

Management of Insect Attack on Vegetable Crops

Insect pests are responsible for reducing about 40% yield in vegetable. There is change in vegetable pest scenario. Many pests have adapted on to new hosts, developed resistance to pesticides and often there are secondary outbreaks. As the vegetables are harvested at shorter intervals, chemical method of plant protection has become risky and hazardous. Integrated Pest Management (IPM) would provide an acceptable and affordable basis for pest control in vegetables. Integrated Pest Management strategies generally rely on biological defences against pests before chemically altering the environment. Here crop protection is considered as one aspect of agro-ecosystem management and addresses the economic, ecological and social issues of pest control. The input intensive high yielding varieties/hybrids with good agronomic traits and yield attributes have gradually changed the pest situation in vegetables. Many pests have adapted new hosts, developed resistance to pesticides and often there are secondary outbreaks. New pests like American serpentine leaf miner *Liriomyza trifolii* (burges) showed severity in tomato hybrid growing pockets. Earlier gall midge of brinjal a minor pest is gradually becoming a regular problem in brinjal, whereas *Hellula undalis* and red spider mite have intensified the attack. *Plutella xylostella,* a lepidopteran pest of cruciferous vegetables have already developed resistance to most of the insecticides. Among the various methods of pest control, chemical method still enjoys first choice because of its quick and certain action. Consequently, vegetables consume about 13-14% of the total pesticide consumption in India, although only 2.6% of cropped area falls under vegetables.

Important Diseases and Control Measures of Major Vegetables Grown in India

Crop/ Name of Disease	**Symptom**	**Control Measures**
Tomato		
Damping Off (*Pythium aphanidermatum*)	Diseases in the nursery.occurs in two stages, i.e. the pre -emergence and the post - emergence phase. The infected tissues become soft and water soaked. The seedlings topple over or collapse.	Seed treatment with fungal culture *Trichoderma viride* (4 g/kg of seed) or Thiram (3 g/kg of seed), Soil drenching with Dithane M45 (3 g/litre of water) helps to reduce the disease incidence.
Early Blight (*Alternaria solani*)	Occurrs on the foliage at any stage of growth. First observed on plants as small, black lesions mostly on older foliage. Spots enlarge, concentric rings in a bull's eye pattern can be seen in the center of the diseased area..Infects the fruit, concentric rings are also present on the fruit.	Removal and destruction of the affected plant parts. Practicing crop rotation helps to minimize the disease incidence. Spray withDifolatan (0.2%), Dithane M-45 (0.2%) or Bavistin (0.1%).
Buck Eye Rot (*Phytophthora parasitica*)	Appears as a greyish green or brown water soaked spot on fruit touching the soil. As the spot enlarges, the surface of lesion assumes a pattern of concentric rings of narrow, dark brown and wide, light brown bands.	Ensure good drainage conditions. Stake plants and remove foliage and fruits touching ground. Spraying with Difolaton (0.3%) 4 times at an interval of 10 days effectively controls the disease.
Late Blight (*Phytophthora infestans*)	Lesions produced on the leaves are at first irregular, geenish-black and water-soaked. These areas enlarge rapidly, becoming brown, and under humid conditions, develop a white moldy growth near the margins of the diseased area on the lower surface of the leaves or on stems. Fruit lesions occur as large, green to dark brown lesions, mostly on the upper half of the fruit	Follow crop rotation with non solanaceous crops. Adopt seed treatment with Thiram (2-3 g/kg of seed). Spray with metalaxyl+mancozeb combination at 0.25 % concentration at an interval of 15 days starting with first visible symptom, This should be followed by weekly spray of 0.2% mancozeb.
Fusarium Wilt (*Fusarium oxysporum* f. sp. *lycopersici*)	Symptom appears as clearing of the veinlets and chlorosis of the leaves. The younger leaves may die in succession and the entire plant may wilt and die in a course of a few days. On grown up plants, petiole and the leaves droop and wilt. At later stage, browning of vascular system occurs.	Remove and destroy effected plant. Soil drenching with Carbendazim plant (0.1%) and seed treatement with the Thiram (2.5 kg/ha) is effective. Follow crop rotation with a non-host crop such as cereals.
Powdery Mildew (*Leveillula taurica*)	Severe during dry seasons. A white powdery coating of the fungal growth appears on the leaf surface. Infected leaves become stiff, and narrow. Affected fruits are smaller in size.	Spraying with Karathane (0.1%) or Wettable Sulphur (3 g/ litre of water) twice at an interval of 10 days.

Contd....

Bacterial Wilt (*Ralstonia solanacearum*)	Relatively high soil moisture and soil temperature favour disease development. Characteristic symptoms of bacterial wilt are the rapid and complete wilting of normal grown up plants. Lower leaves may drop before wilting. Pathogen is mostly confined to vascular region,.	Grow resistant varieties, Ensure good drainage, Follow crop rotations with non host crop viz cereals and legumes. Seedling treatment with Streptocycline (1 g/40 litres of water) for 30 min effective in the initial stages of growth.
Tomato Mosaic Virus (TMV)	Characterized by light and dark green mottling on the leaves often accompanied by wilting of young leaves in sunny days. The leaflets of affected leaves are usually distorted, puckered and smaller than normal. The affected plant appears stunted, pale green and spindly. The virus is spread by contact with labour, touching of infected plants with healthy ones, plant debris and implements.	Seeds from disease free healthy plants should be selected for sowing. Soaking of the seeds in a solution of Trisodium Phosphate (90 g/litre of water) a day before sowing helps to reduce the disease incidence. In the nursery all the infected plants should be removed carefully and destroyed. Crop rotation with non solanaceous crops should be undertaken.
Tomato Leaf Curl Virus (TLCV)	Disease is transmitted by whitefly (*Bemisia tabaci*). It is one of the most devastating diseases of tomato, characterized by severe stunting of the plants with downward rolling and crinkling of the leaves. The newly emerging leaves exhibit slight yellow colouration and later they also show curling symptoms. Older leaves become leathery and brittle. The nodes and internodes are significantly reduced in size. The infected plants look pale and produce more lateral branches giving a bushy appearance.	The affected plants should be removed and destroyed. Soil application of granular insecticide like Furadan (1 kg a.i./ha) at the time of sowing checks whitefly population. Another dose of Furadan (1.5 kg a.i./ha) is given one week after transplanting. 2-3 foliar sprays with Dimethoate (0.05%) or Monocrotophos (0.05%) at 10 days intervals control the white fly population. Barrier crops like maize, jowar, bajra are good to protect the crop from TLCV infection by checking incoming viruliferous whiteflies. Mulching with polythene sheets of white, blue, grey and black colours are effective.
Tomato Spotted Wilt Virus (TSWV)	The spotted wilt virus is transmitted through thrips (*Thrips tabaci, Frankliniella schultzi* and *F. occidentalis*). This disease causes streaking of the leaves, stems and fruits. Numerous small, dark, circular spots appear on younger leaves. Leaves may have a bronzed appearance and later turn dark brown and wither. Fruits show numerous spots about 4.5 cm in diameter with concentric, circular markings. On ripe fruit, these markings are alternate bands of red and yellow.	The affected plants should be removed and destroyed. Alternate or collateral hosts harboring the virus causing this disease are removed at the time of weeding or earthing up operations to minimize the spread of the disease. Checking the population of thrips can reduce the disease incidence. 2-3 foliar sprays with Dimethoate (0.05%) or Monocrotophos (0.05%) at 10 days intervals control the thrips population.

Contd....

Brinjal		
Damping Off (*Pythium* spp., *Phytophthora* spp., *Rhizoctonia* spp.)	Same as tomato	Same as tomato
Phomopsis Blight (*Phomopsis vexans*)	Causes Damping off of seedlings. When the leaves are infected, small circular spots appear which become grey to brown with irregular blackish margins. Lesions may also develop on petiole and stem, causing blighting of affected portion of plant. Symptoms on the infected fruits appear as minute, sunken dull and dusky spots whtch later merge to form rotten areas. The flesh of severely infected fruits rots.	Grow resistant varieties, Adopt field sanitation, destruct infected plant material and follow crop rotation. Seed treatment with Thiram (2 g/kg seed) protects the seedling inthe nursery stage. Spraying with Dithane Z-78 (0.2%) or Bordeaux mixture (1%) effectively controls the disease in the field.
Leaf spot (*Cercospora melongenae*)	Symptoms are characterized by chlorotic lesion, angular to irregular in shape, later turning greyish-brown. Severely infected leaves drop off prematurely, resulting in reduced fruit yield.	Remove and destroy affected plant parts. Foliar spray with Bavistin (0.1%) or Chlorothalonil (2 g/litre of water) is useful for disease control.
Alternaria Leaf Spots (*Alternaria melongenae*)	The disease causes characteristic leaf spots with concentric rings. Severely affected leaves drop off. Large deep-seated spots appear on fruits. The infected fruits turn yellow and drop off prematurely.	Removal and destruction of affected plant parts and spraying the affected plants with Bavistin (0.1%) is useful for disease control.
Verticillium Wilt (*Verticilium dahliae*)	The disease attacks the young plants as well as mature plants. Young plants show dwarfing and stunting . Infection after the flowering stage results in development of distorted floral buds and fruits Dark brown discoloration of the xylem vessels is observed in roots.	Crop rotation with bhindi, tomato and potato should be avoided. Soil application and foliar application with Benlate (0.1%) are effective .
Bacterial Wilt (*Ralstonia solancearum*)	Bacterial wilt disease causes severe problem in brinjal cultivation. The characteristic symptoms of the disease are same as that reported in case of tomato.	Same as tomato.
Little Leaf of Brinjal	Disease is transmitted by leaf hopper (*Cestius* (*Hishimonus*) *phycitis* and *Amrasca biguttula biguttula*). The leaves show a reduction in size and are malformed with yellow colour. Disease affected plants are generally shorter in stature bearing a large number of branches, roots and leaves than healthy plants. The petioles get shorter considerably, many buds appear in the axil of leaves and internodes get shortened thus giving the plants a bushy appearance.	Adopt sanitary measures including the eradication of susceptible volunteer crop plants from a previous crop. Use of barriers of trap crops and early removal and destruction of infected plants are also recommended. The sowing time can be adjusted to avoid the main flights of the beet leafhopper. Spraying Malathion (2ml/litre of water) starting with the appearance of the leaf hoppers controls their population.

Contd....

Mosaic	This is a viral disease caused by Potato Virus Y and transmitted by aphids (*Aphis gossypi* and *Myzus persicae*). Symptoms are mosaic mottling of the leaves and stunting of plants. The leaves of infected plants are deformed, small and leathery. Plants show a stunted growth when infected in the early stages.	Destroy the infected plants and susceptible weed hosts. In the nursery, aphids can be controlled by application of Carbofuran (1 kg a.i./ha) in the nursery bed at the time of sowing seeds followed by 2-3 foliar sprays of Phosphamidon (0.05%) at an interval of 10 days.
Chilli		
Damping off *(Pythium aphanidermatum*	Same as tomato	Same as tomato
Anthracnose / Die back and Fruit rot (*Colletotrichum capsici*)	Symptoms appear mostly on ripened fruits. Circular, water-soaked and sunken spots with black margins develop on fruits. The sunken spots are covered witha pinkish mass of fungal spores. As the disease advances, the spots spread, forming concentric markings with dark fructifications representing the fungal acervuli. The fruits with many spots drop off prematurely resulting in heavy loss of yield.	Adopt seed treatment with 3g Captan or Mancozeb per kg seed. The secondary spread can be checked by spraying the plants with copper fungicides at fortnightly interval beginning from the time, the first bunch of fruits is almost mature.
Bacterial leaf spot *(Xanthomonas vesicatoria)*	Yellowish green spots appear on young leaves, on older leaves they are dark and water soaked. Lesions on leaf begin as circular, water soaked spots may develop straw coloured centres and dark margins. Severely spotted leaves turn yellow and drop, defoliation is common. On the fruits, small blister - like spots occur becoming warty in appearance later.	Follow seed treatment and crop rotation. The nurseries and plants in the mainfield may be sprayed with a copper fungicidesuch as Bordeaux mixture. Spray 2 or 3g of Plantomycin or Paushamycin mixed with 30gs of Copper Oxychloride (blitox or phytolon) per 10 litres of water twice or thrice at 15 days interval depending on severity of the disease.
Bacterial Wilt (*Ralstonia solanacearum*)	Same as tomato	Same as tomato
Leaf curl	Leaves curl towards midrib and become deformed. Stunted plant growth due to shortened internodes and leaves greatly reduced in size. Flower buds abcise before attaining full size and anthers do not contain pollen grains.	The virus is generally transmitted by whitefly. So control measures of whitefly in this regard would be helpful
Potato		
Wart (*Synchytrium endobioticum*)	Periodic flooding and lack of proper drainage and aeration favours the disease. The disease is characterised by 'cauliflower-like' warty growths on tubers, stolons and stem bases but not roots. Under wet conditions, it may be seen in the form of	Control is possible only by cultivation of immune varieties. No effective treatments are available and cannot be applied on large field scale. However, preventive measures like practicing long crop rotation (5 years

Contd....

	greenish-yellow crust on the stems and leaves at or near the soil level. Diseased tubers may show formation of either one or more tumours. Such tubers sometimes are completely transformed into warty mass.	or more), using disease free potatoes as seed material and burning of wart affected lumps and potato peelings are effective in checking spread of disease.
Soft Rot (*Erwinia. carotovora* subsp *carotovora*)	Causes huge losses in storage. Excessive moisture, high temperature, excess nitrogen, tuber injuries and poor ventilation during storage are the important factors helping disease development. Initially a small area of tuber tissue around lenticel or stolon attachment point becomes water soaked and soft. Under low humidity, the initial soft rot lesions become dark and sunken. Under high humidity, the lesion may enlarge and spread to larger area. Tubers in advanced stages of decay are usually invaded by other organisms and the decaying tissue becomes slimy with foul smell and brown liquid ooze. The tuber skin remains intact and sometimes the rotted tubers are swollen due to gas formation.	Avoid excess irrigation and nitrogen, provide proper drainage Cultural methods such as adjust planting time to avoid hot weather during plant emergence and harvesting the crop before soil temperature rises above 28° C is recommended. The crop should be harvested only when the tuber skin is fully cured. Care should be taken to avoid injury to tubers. Treating the tubers with 3% boric acid for 30 min and drying them under shade minimize infection in the storage. The treated tubers should be stored either in well-ventilated cool stores or cold stores.
Charcoal Rot (*Macrophomina phaseolina*)	Fungus infects the tubers in the soil through proliferated lenticels and injuries, Black spots appear around the lenticels and eyes which enlarge into patches extending deep into the tuber flesh.	Harvesting the crop early before the soil temperature reaches 28° C can check the disease.
Bacterial Wilt or Brown Rot (*Ralstonia solanacearum*)	The earliest symptom is slight wilting of the top, which is soon followed, by total wilting. In advanced stage, if the base of the stem of the affected plants are cut transversely and squeezed, the bacterial mass is seen to ooze out as a dull white slimy mass on the cut surface.	Disease free seed tubers obtained from disease free areas should be used for planting. Splitting of the tubers at the time of the planting should be avoided. Application of bleaching powder (12kg/ha) mixed with fertilizer in furrows while planting reduces wilt incidence. Practice crop rotation.
Black Scurf and Stem Canker (*Rhizoctonia solani*)	Symptom comprises of dark brown to black irregular lumps sticking on the surface of tubers. These irregular lumps are closely adhered to the tuber surface and do not wash off easily. Other symptoms are skin	Adopt tuber disinfection and improved cultural practices. In the hills, tuber treatment with an organomercurial compound & soil application of PCNB @ 30kg/ha is the

Contd....

	cracks on tuber, crater like depressions, pitting, stem-end necrosis and shape deformity. The disease often causes sprout injury both in storage and in fields after planting. The emerging sprouts when infected later develop cankers causing girdling of stem bases. Such affected plants show upward rolling of leaves with pinkish or purplish margin.	most effective. In the North India plains, treatment of the diseased seed with thiabendazole, acetic acid +zinc Sulphate, Carbendazim and Boric Acid effectively controls the disease. Crop rotation with maize or 'dhaincha' also checks the disease build up.
Common Scab (*Streptomyces* spp.)	Causes disfiguring of tubers, thereby reducing the market value and increasing peeling losses. Cuases abrasion of skin (russetting) and corky lesions on tuber and concentric corky rings around lenticels. In case of severe attack, dark brown lesions may develop on roots and stolons.	Use disease free seed tubers. Treat seed tubers with organomercurial compounds (0.015%for 20 minutes) or Boric Acid (3% for 30 minutes).The same treatment is repeated before the storage of tubers. Crop rotations with wheat, pea, oats, barley, lupin, soybean, sorghum and bajra advisable.
Late Blight (*Phytophthora infestans*)	The disease affects all plant parts, viz., leaves, stems and tubers. It appears on leaves as small pale green spots, which enlarge into large water soaked lesions. A white mildew (cottony growth) ring forms around the dead areas on the lower side of leaves. In dry weather, water soaked areas turn necrotic brown. On stems, light brown elongated lesions are formed which may encircle the stem. Tubers develop reddish brown, shallow to deep, dry rot lesions..	Use disease free seed tuber. If the weather conditions (temperature 10-20° C, RH>80%) are favourable for the disease development, irrigation should be stopped immediately. When the disease affects 75% crop foliage, the haulms should be destroyed. Protective sprays with a contact fungicide, viz., Mancozeb (0.2%) before appearance of the disease is effective. Spray may be repeated at 8 to 10 days interval. In case of severe blight attack, give one or two sprays of Metalaxyl (0.25 %).
Early blight(*Alternaria solani*)	The disease mainly infects leaves and tubers. Initially the symptoms occur on lower and older leaves in the form of small (1-2 mm) circular to oval brown spots. Later lesions become large and angular. Mature lesions on foliage look dry and papery, and often have the concentric rings, looking like bulls eye. The symptoms on the tuber comprise of brown, circular to irregular and depressed lesions with underneathflesh turning dry, brown and corky. Lesions enlarge during storage and affected tubers later become shriveled.	Use disease free seed tubers for raising the crop. The crop must be given balanced doses of fertilizers, especially nitrogen. Spraying the crop with urea (1.0%) at 45 days after sowing and giving subsequent sprays 8-10 days after the first spray is helpful. In the hilly regions, spraying of Copper Oxychloride (0.30%) and Bordeaux mixture (1.0%) is recommended.

Contd....

Potato Leaf roll Virus (PLRV)	The primary symptoms are confined to top young leaves, which roll and turn slightly pale. Infected plants have characteristic pale, stunted and upright appearance with rolling of lower leaves that turn yellow, brittle and are leathery in texture.	The disease is managed by using virus-free seed potatoes. Multiply virus-free seed in aphid free areas. Population of aphid vectors is controlled by application of suitable contact/systemic insecticides.
Bitter gourd, water melon and other cucurbits		
Powdery Mildew (*Sphaerotheca fuliginea*)	White powdery residue appear on the upper leaf surface. On the lower surface of the leaves, circular patches or spots appear. In severe cases, these spread, coalesce and cover both the surfaces of the leaves and spread also to the petioles, stem, etc. Severely attacked leaves become brown and shrivelled and defoliation occurs. This disease is favored by high humidity and tends to occur on older leaves first.	Bavistin (1 g/litre of water) or Karathane (0.5 ml/litre of water) should be sprayed immediately after the appearance of the disease. 2-3 sprays are to be given at an interval of 15 days.
Fusarium Wilt (*Fusarium oxysporum* f. sp. *niveum*)	The leaves show yellowing, loose turgidity and droops. Eventually, the plantdies. In older plants, leaves wilt suddenly and vascular bundles in the collar region become yellow or brown.	Three sprays of Karathane (6 g in 10 litres of water) or Bavistin (1 g /litre of water) immediately on appearance of initial symptoms at 5-6 days interval controls the disease.
Downey Mildew *Pseudoperonospora cubensis*	Prevalent in areas of high humidity, especially when summer rains occur regularly. The disease is first seen as yellow angular spots on the upper surface of the leaves. Under conditions of high humidity, whitish powdery growth appears on the lower surface of the leaves.	Disease can be controlled with Ridomil (1.5 g/litre of water) which must always be used simultaneously with a protectant fungicide such as Mancozeb (0.2%) to prevent the development of resistant strains.
Bitter Gourd Mosaic	Small irregular yellowish patches are seen on the leaves. Some leaves show vein clearing in one or two lobes of the leaf and severely infected plants show reduction in leaf size and elongation and/or suppression of one or two lobes. Young developing leaves are completely distorted and malformed. The virus is transmitted by five species of aphids.	Spraying the crop just after germination with Monocrotophos (0.05%) or Phosphamidon (0.05%) at 10-days intervals prevents aphid vectors.
Bitter gourd Witches' Broom	Malformation and proliferation of axillary buds are the characteristic symptoms. Diseased plants show many abnormally little leaves, which fail to attain full size. The plants bear many flowers and blossom earlier than healthy plants. Flowers on infected plants show characteristic green, phyllody symptoms.	Application of Carbofuran (1.5 kg a.i./ha) at the time of sowing seeds followed by 5-6 foliar sprays of either Phosphamidon (0.05%) or Monocrotophos (0.05%) at 10 days interval, to control the vector population. Spraying of Oxytetracycline hydrochloride solution (500 ppm) at weekly intervals suppresses the disease symptoms

Contd....

Anthracnose (*Colletotrichum* sp.)	Symptoms appear on the leaf as black spots, which later turn pink in colour. In case of severe infection, the disease spreads to the developing fruit.	Repeated sprayings at 5-7 days interval with Dithane M 45 (0.2%) or Foltaf (1 g/litre of water) effectively controls the disease.
Okra		
Cercospora Leaf Spot (*Cercospora malayensis, C. abelmoschii*)	C. *malayensis* causes brown, irregular spots and C. *abelmoschi* causes sooty black, angular spots. Both the leaf spots cause severe defoliation and are common during humid seasons.	Spraying Mancozeb or Zineb 2 g or Carbendazim 1 g/l can control the disease
Fusarium Wilt (*Fusarium vasinfectum)*	The conspicuous symptom is a typical wilt, beginning with a yellowing and stunting of the plant, followed by wilting and rolling of the leaves. If a diseased stem is split lengthwise, the vascular bundles appear as dark streaks. When severely infected, nearly the whole stem is blackened.	Follow crop rotation, Drench the field with Copper oxychloride @3g/litre of water. Treat the seeds with Carbandazim or Mancozeb @ 3g/kg seed
Powdery Mildew *Erysiphe cichoracearum*	Greyish powdery growth occurs on the under as well as on the upper surface of the leaf causing severe reduction in fruit yield.	Effectively controlled by dusting finely ground sulphur at 30 kg/ha or spraying wettable sulphur 3 g/litre of water or Dinocarp 1 ml/litre of water three to four times at 15 days intervals.
Vein - Clearing/Yellow Vein Mosaic Bhendi yellow vein mosaic virus	Yellowing of the entire network of veins in the leaf blade is the characteristic symptom. In severe infections, the younger leaves turn yellow, become reduced in size and the plant is highly stunted. Infection may start at any stage of plant growth. Infection restricts flowering and fruits, if formed, may be smaller and harder.	The virus is transmitted by the whitely (*Bemisia tabaci),*So control of this vector will give better results. Parbhani Kranti, Arka Anamika and Arka Abhay can tolerate yellow vein mosaic. Spraying monocrotophos (1.5 ml/litre of water) can restrict the disease spread.
Onion		
Damping-off (*Fusarium oxysporum* f.sp.*cepae; Pythium* sp.; *Sclerotium rolfsii* and *S. cepivorum* and *Colletotrichum* sp.)	The disease is more prevalent during *kharif* season and causes about 60-75% damage. Two types of symptoms are observed- Pre-emergence damping-off: Results in seed and seedling rot before these emerge out of the soil. Post-emergence damping-off: The pathogen attacks the collar region of seedlings on the surface of soil. The collar portion rots and ultimately the seedlings collapse and die.	Adopt seed treatment with Thiram @ 2g/kg of seed. The topsoil of nursery should be treated with Thiram @ 5g/m^2 area of the soil and nursery should be drenched with the same chemical @ 2g/litre of water at fortnightly interval. Soil solarization by spreading 250 gauge polythene sheet over the bed for 30 days before sowing and application of bio-control agent *Trichoderma viride* in soil @ 1.2kg/ha is also found effective.
Purple Blotch (*Alternaria porri*)	Hot and humid climate with temperature ranging from 21-30°C and relative humidity (80-90%) favour the development of the disease. It is more common in *kharif* season.	Use of healthy seeds for planting and crop rotation of 2-3 years with non-related crops checks the disease. Spraying Mancozeb (0.25%) or

Contd....

	purple coloured centres. The lesions may girdle leaves/stalk and cause their drooping. The infected plants fail to develop bulbs .	transplanting at fortnightly interval reduces the disease incidence.
Stemphylium Blight (*Stemphylium vesicatorium*)	Serious problem in Northern parts of the country especially in the seed crop Fungus attacks leaves and flower stalks. The symptoms appear as small yellowish to orange flecks or streaks in the middle of the leaves, which soon develop into elongated spindle shaped spots surrounded by pinkish margin. The disease appearing on the inflorescence stalk causes severe damage to the seed crop.	Field sanitation and collecting and burning of crop residues minimize the spread of infection. Spraying Mancozeb (0.25%) with sticker triton on appearance of disease at fortnightly interval controls the disease.
Downy Mildew (*Perenospora destructor*)	The disease is worst in damp conditions and seen on late planting of the crop. Application of higher doses of fertilizers and more watering increased disease severity. Symptoms appear on the surface of leaves or flower stalk as violet growth of fungus, which later becomes pale greenish yellow and finally the leaves or seed stalks collapse.	Bulbs meant for seed crop should be exposed to sun for 12 days to destroy the fungus. Spraying with Zineb (0.2%), Karathane (0.1%) or Tridemorph (0.1%) also gives good control of the disease.
Onion Smut (*Urocystis cepulae*)	Disease appears as elongated dark, slightly thickened areas near the base of seedlings. The affected leaves bend downwards abnormally. On older plants, numerous raised blisters occur near the base of the leaves. The lesions on plant at all stages often expose a black powdery mass of spores.	Treating the seeds with Captan or Thiram (2.5g/kg of seed) before sowing controls the disease. Seed bed treatment with Methyl Bromide (1 kg/25 m .) is also effective.
Onion Smudge (*Colletotrichum circinans*)	It occurs on white onion varieties and reduces the market value of the bulbs. The disease is characterized by small dark green to black spots, which appear on the outer scales.	Thorough curing of the bulb after harvesting and storing the bulbs in well ventilated rooms can control the disease.
Black Mould (*Aspergillus niger*)	Common in onions stored in hot climates where the temperature ranges between30-45° C. It is characterized by the black powdery mass of spores that appear on the exterior of the scales. The black spore masses are also seen on inner scales. It reduces the market value of the bulbs.	Cure bulb for two days. These bulbs should be further dried in shade for 10-15 days before storage. Care should be taken to avoid injury to the bulbs during post harvest handling. The crops should be sprayed with Carbendazim (0.2%), 10-15 days before harvesting.

Contd....

Onion Yellow Dwarf	Caused by onion yellow dwarf virus. Transmitted mechanically as well as by insect vectors. The symptoms are severe stunting of the plants, dwarfing and twisting of the flower stalk. The affected leaves and stems change their normal green colour to various shades of yellow and leaves tend to flatten and crinkle and as a result bend over.	Removal and destruction of the diseased plants check the spread of the disease. Healthy bulbs should be used for seed production. Spraying of Malathion (0.1%) or Metasystox (0.1%) to control the vectors checks further spread of the disease.
Anthracnose (*Colletotrichum gloeosporiodes* or *Colletotrichum circinans*)	The symptoms appear initially on the leaves as water soaked pale yellow spots, which spread lengthwise covering entire leaf blade. The affected leaves shrivel and droop down.	Field sanitation and destruction of infected crop debris help in reducing the disease. Mancozeb (0.25%), Carbendazim (0.1%) or Thiophanate Methyl (0.1%) as foliar spray is effective.
White rot (Sclerotium rolfsii):	Leaf tips show yellowing and dieback. Scales, stem plate and roots get destroyed. The bulbs become soft and water soaked. White fluffy or cottony growths of mycelium with abundant black sclerotia resembling mustard grain are seen on the infected bulbs.	Seed treatment with Thiram (4 g/kg of seed) and drenching of soil with Mancozeb (0.25%) are effective in controlling the disease. Application of bio-control agents like *Trichoderma viride* to the soil reduces the disease inoculum. Follow crop rotation.
Cabbage & Cauliflower		
Black Rot (*Xanthomonas campestris* pv. *campestris*):	Common in a warm and wet climate. Infected seedlings become yellow. Leaves may be affected on only one side of a seedling. The classic symptom of black rot is caused by local infection that results when bacteria enter leaves through natural openings of leaf margins. The infected tissue turns pale green-yellow and then turns brown and dies. Affected areas are usually wedge- or V-shaped. These areas enlarge as the disease progresses, and severely affected leaves may drop off.	Use tolerant varieties. Considerable reduction in disease has been observed when seeds are treated with Agrimycin-100 (100ppm) or Streptocycline (100 ppm). Planting should be done on raised beds to facilitate drainage. Cultivation in the fields where crucifers have been continuously grown during last 2 years should be avoided.
Downy Mildew (*Perenospora parasitica*):	Serious in nursery and it can also appear in field planting. High humidity, fog, drizzling rains, and heavy dew favour the disease development and spread.The first symptoms are small, light green-yellow lesions on the upper leaf surface, later showing on the undersurface. The spots turn yellow as they enlarge. During periods of high humidity, a grayish white moldy growth is developed on the undersurface of the leaf. Later the leaf may become papery and die. Cabbage heads develop sunken black spots.	Spraying the seedlings in the nursery beds with Copper Oxychloride (0.3%) is effective. The first spray should be given as soon as the seedlings appear. Subsequent sprayings are given at weekly intervals until the plants are transplanted in the field. For controlling the disease in the field, the crop is sprayed with Copper Oxychloride (0.5%).

Contd....

Leaf Spot and Blight (*Alternaria brassicae* and *A. brassiciola*)	It is a destructive disease on seed crop. Older leaves are more susceptible. The initial symptoms are in the form of small dark yellow spots on the leaf surface. Later on, the spots enlarge to circular areas with concentric rings and possibly surrounded by yellow halos. In severe cases, the entire plant defoliates. Violets to tan spots develop on infected cabbage seed pods which intensifies in wet weather.	Use of disease free seeds, practicing proper crop rotation and seed treatment with hot water (50° C for 30 minutes) helps to minimize the disease incidence. Crops grown for seed purpose should be sprayed at full bloom, pod set and pre- harvest stage with Captan (0.2%) or Copper Oxychloride (0.5%) for the control of disease.
Yellows or Fusarium Wilt (*Fusarium oxysporum* f. sp *conglutinans*)	The disease affects the seedlings in nursery stage, however plants exhibit symptoms 2 to 4 weeks after transplanting. Disease development is promoted by warm weather conditions. Initial symptom appears as the development of yellowish green colour on one side of the plant. A lateral warping or curling of the stem and leaves occurs. The lower part of the leaf blade adjoining the petiole or midrib wilts and dies. The lower leaves turn yellow and later the upper leaves are affected.	Use resistant varieties. However, as a preventive measure, the vulnerable stage of the young seedlings to the infection can be avoided by very early sowing of cabbage.
Black Leg (*Phoma lingum*)	Pale, irregular spots develop on leaves, which later become ashy gray with scattered black dots on the surface. Stem lesions are elongated with purple borders near the ground level and extend below the soil surface, causing a black rot of lower stem and roots. Severely affected plants remain stunted and finally wilt. As plants mature, they fall sideways. Seed crop symptoms include occasional cankers on stem bases and spots may appear on overwintered leaves.	Disease free seeds should be used for planting. As the main infection is through seeds, hot water treatment of seeds is recommended. Cultivation in the fields where crucifers have been continuously grown during last 2 years should be avoided. Seedbeds and seed plots should be regularly inspected for obvious foliar infections. Seedlings before transplanting should not be dipped in water.
Clubroot (*Plasmodiophora brassicae*)	Roots develop clubs (swellings) that can be 12-15cm wide. Affected seedlings do not show any root swellings until about 3 weeks after infection. Infection in the nursery stage results in death of seedlings. When plants are attacked at a later stage, they wilt in hot weather but partly recover at night. Finally leaves become stunted, yellowish and prematurely bolt in hot weather.	Use only uninfected seedbeds and clean equipment. Long rotations (6 years or longer) help prevent a pathogen buildup and reduce disease incidence. Finely ground limestone is thoroughly mixed into the soil six weeks before planting to raise the soil pH above 7.0.
Sclerotinia rot/ White Mould (*Sclerotinia sclerotiorum*)	Infections may occur on the stem at the ground level and on the leaves at their bases. The infections begin as water-soaked, circular areas, which soon become covered by white, cottony fungal growth. The affected tissue becomes soft and watery as the disease progresses. The fungus eventually colonizes the entire cabbage head and produces large, black, seedlike structures called sclerotia on the diseased tissue.	Follow crop rotation with non host. Mechanical injuries to cabbage heads during harvesting operations should be avoided.

Contd....

Pea		
Fusarium Wilt (*Fusarium oxysporum* f.sp. *pisi*)	Symptom is expressed as drooping of the plants followed by sudden death. Pod formation is severely affected. In collar regions of the wilted plants, necrosis and discolouration can be seen. Vascular browning can be seen in advanced stage.	Follow long crop rotation. Treating the seeds with Carbendazim (2 g/kg of seed) protects the seedlings during the initial stages of growth. Spraying the crop with Bavistin (1%) helps to control the disease.
Bacterial Blight (*Pseudomonas syrinagae* pv. *pisi*)	Water-soaked lesions appear on the stems, leaves and pods. Lesions on the leaves appear as round-oval or irregular, reddish brown spots with translucent centre and a dark brown margin. Several such lesions may coalesce on the leaves to produce a blighted appearance. On the stem brown linear streaks are observed.	Use of healthy and disease free seeds for planting helps to minimize the disease spread.
Powdery Mildew (*Erysiphe polygoni*)	Infection is first apparent on the leaves as small slightly darkened areas, which later become white powdery spots. These spots enlarge and cover the entire leaf area. Severely infected leaves may become chlorotic and distorted. Affected pods are small in size and malformed.	As soon as the disease symptoms are observed, dusting Sulphur (2.5kg/ha) or spraying Calixin (0.5 l a.i/ha) at 15 days interval helps to control the disease.
Radish		
Alternaria Blight (*Alternaria raphani*)	Small, yellowish, slightly raised lesions first appear on leaves. Infection spreads rapidly during rainy weather, and the entire pod may be so infected that the stylar end becomes black and shrivelled. The infected seed fails to germinate.	Hot water treatment of the seeds, Regular spraying with Difolatan (0.3%) or Dithane M 45 (0.2%) or Ridomil (0.1%) controls the disease effectively.
Radish Mosaic Virus (RMV)	The symptoms first appear as small, circular to irregular, chlorotic lesion in between and adjacent to the veins.. It is transmitted through aphids.	Controlling aphids with 2-3 foliar sprays of either Dimecron (0.05%) or Monocrotophos (0.05%) at 10 days interval check the spread of disease.
Radish Phyllody	The disease is transmitted by jassid *Orosius albicinctus*. The diseased plant assumes a dull grey to light violet colouration. The symptoms of the disease appear at the time of flowering when all the floral parts become green violet and leafy.	One or two sprays of Monocrotophos (0.05%) or Phosphamidon (0.05%) or Oxydemetan Methyl (0.02%) are done to eradicate the jassids. Thimet 10-G (1.5 kg a.i./ha) is also recommended.
Carrot		
Cercospora Leaf Blight (*Cercospora carotae*)	The symptoms first appear along the margins of the leaves, often causing the leaves to curl. Spots inside the leaf edges are small, roughly circular, and tan or gray to brown with a dead center. As the lesions increase in number and size, the entire leaflet withers and dies. The fungus attacks younger leaves and plants in preference to older ones.	Use disease-free seed. Early applications of Foltaf (0.2%), Copper Oxychloride (0.3%), starting at the first sign of infection, effectively control leaf blights.

Contd....

Alternaria Blight (*Alternaria dauci*)	Leaf spots first appear at the margin of the leaflets and are dark brown to black and irregular in shape. Lesions produced on the petioles and stems are dark brown and often coalesce and girdle the stems. As the disease progresses entire leaflets may shrivel and die.	Soil should be well-drained. Hot water treatment of seed at 50°C for 15 minutes is effective. Follow crop rotation and destroy infected plant material. spraying with Foltaf (0.2%), Copper Oxychloride (0.3%) satisfactorily control the disease.
Powdery Mildew (*Erysiphe polygoni*)	The symptoms appear as white powdery growth on the leaves and petioles causing the leaves to turn brown and wilt.	Spraying Bavistin (0.1%) or Benlate (0.1%) at an interval of 8-10 days effectively controls the disease.
Watery soft Rot (*Sclerotinia sclerotiorum*)	Symptoms can be identified in the field as characteristic white mold with black sclerotia present on the crown of infected carrots. In storage, a soft, watery rot with white mold and black sclerotia is seen.	Crop rotation, planting on raised beds, winter flooding, rapid cooling prior to storage and meticulous sanitation of all storage components are all necessary to reduce losses from this disease
Black Rot (*Alternaria radicina*):	Symptom is seen as shiny black decay at the crown area and a greenish-black mold on the taproot. The infected tissue is greenish black to jet black due to presence of masses of black spores. Infestation is seen in the field as well as in storage.	Proper field sanitation and practicing rotation help to keep the disease under control. The root surface should be kept dry and stored at 0°C with 95 % relative humidity.
Bacterial soft Rot (*Erwinia carotovora* p.v. *carotovora* and *Erwinia carotovora* p.v.*atroseptica*)	Soft water-soaked, irregular lesions appear on the roots. Initially these lesions are superficial but soon spread to cover the inner tissues. The foliage may remain green until the disease on the tuber advances considerably. The entire plant wilts when complete rotting of the tuber takes place. A foul odour is given out from the decayed roots.	Planting on raised beds in poorly drained areas helps to reduce bacterial infections. The diseased roots should not be stored along with the healthy tubers. The root surface should kept dry and stored at 0°C with 95 % relative humidity.
Carrot Yellows	Virus attack results in yellowing of leaves accompanied by vein clearing. Dormant buds in the crown grow out into chlorotic shoots, which give a witchers broom appearance on the top. Size and quality of roots are reduced and malformed. The disease is transmitted by the six-spotted leaf hopper.	Weed control especially of those acting as alternate host eliminates the disease. Spraying insecticides to control the leaf hopper population helps to reduce the disease attack.
Beetroot		
Leaf Spot (*Cercospora beticola*)	Numerous small circular spots appear on the leaf surface. The spots increase in size, becoming brownish or purplish in color. Individual spots are usually circular but several may coalesce into larger areas of dead tissue.	Destroy affected plants and practice crop rotation. Spraying with Copper Oxychloride (0.3 %) thrice at an interval of 15 days controls the disease effectively.

Contd....

Downy Mildew (*Perenospora schachtti*)	Symptoms appear as irregular greasy greyish areas on the leaves. Under moist conditions, these areas expand rapidly and a white powdery growth appears on the lower surface of the affected leaves. Affected leaf dries and shrivels quickly. The fungus survives on the crop residues in the soil and is also carried by the seed.	Adopt good field sanitation and crop rotation. Seed treatment with Thiram (2.5-3 g/kg of seed) protects the emerging seedlings from the disease attack. Spraying with Dithane Z-78 (0.3 %) thrice at an interval of 15 days is also recommended.
Mosaic :	Transmitted and spread by aphids. Symptoms appear as conspicuous mottling with chlorotic, zonate ring spots on the leaf surface. When these ring spots develop, their centers are usually green. Virus infected plants remain stunted and may lose some leaves.	Destruction of infected plants and controlling the aphid population by spraying Malathion (2ml/litre of water) prevents the spread of the disease.
Curly-top	Transmitted by beet leaf hoppers. External symptoms of curly top virus infection may appear in leaves, stems, flowers, fruits, or roots of infected plants. Infected plant parts may become distorted through curling, twisting, rolling, stunting, etc. Leaves become thickened and leathery. Internal symptoms consist of death of the vascular tissues.	Adopt sanitary measures including the eradication of susceptible weeds. Regulate the time of planting to avoid the main flights of the beet leaf hopper; use barriers of trap crop. Spraying Malathion (2ml/litre of water) controls the population of beet leaf hoppers.

Components of IPM in Vegetable Crops

1. Selection of Crop and Cultural Practices

Agronomic practices like selection of varieties, intercultural operation, sowing/ planting date and trap cropping/intercropping should be suitably modified to make uncongenial condition for pests and to enhance the activity of natural enemies., selection of resistant/ less susceptible varieties, holding good in pest management. Unlike cereals, in vegetables, very less number of resistant/ less susceptible varieties have been developed.

2. Sowing/ Planting Time

Adjustment of sowing and planting date keeping in mind the peak activity of the damaging stage of the pest help in avoiding the critical stage of the crop from pest attack. Early planting of cucurbits in November escapes the attack of red pumpkin beetle . Sowing of okra during second week of June retains less population of borers, thereby incurred maximum healthy yield where as July planted brinjal faces the ravages of shoot and fruit borer.

3. Tillage

Summer ploughing is an effective practice to spoil the soil inhibiting inactive stages of insect. Deep ploughing of the field after the harvest reduces the activity of fruitfly, red pumpkin beetle and cut worm

4. Intercropping

Intercropping of crops with diverse plant geometry obstructs the adults from egg laying. Release of volatile allelo-chemicals from a particular crop deters the adult insect from damaging the other. These effects did not provide suitable conditions for development of micro -climate favouring the multiplication of a particular type of pest. Above all, such planting combination enhances the activity of predators and parasites.

5. Trap Crop

Trap crop will be attractive to the pest than the main crop and by concentrating them in certain part of the field, pest can be trapped and destroyed. Mustard as trap crop along with cabbage has been successfully utilized for the control of diamondback moth, aphid and leaf webber. African marigold in tight bud stage functions as good source to trap the adults of *Helicoverpa armigera* besides it also attract the adults of leafminer for egg laying on the leaves. Maize plants applied with bait-spray trap and kill fruit fly adults when sown in combination with bitter gourd.

6. Biocontrol Agents/ Natural Enemies

India being the home of maximum vegetable crops is a large reservoir of the natural enemies of pests involved with them. But careful studies are required to use the predators/ parasitoids as a tool of pest management. Release of *Trichogramma* spp. (egg parasite), *Chrysoperla carnea* (predator), *Bracon* spp. and *Cotesia plutellae* have shown positive results. The feasibility of bio control agents in vegetable is comparatively less because of the annual nature of crop with short growing period. However, use of bio control agents is limited to some area only. Among the microbial control agents, with the intervention of private companies, Bt has been extensively used but recently resistance development has limited its scope. Use of insect viruses is limited to *Helicoverpa armigera, Spodoptera litura* and *Plutella xylostella*.

Bio-control Agents Recommended in Vegetables

Bioagents	Dose	Target
Trichogramma brassiliensis	2,50,000 parasitised Corcyra eggs/ha (Inundative realese) 50,00parasitised Corcyra eggs/ha(Weekly Inoculative release)	Okra shoot and fruit borer Tomato fruit borer
Chrysoperla carnea	50,000 first instar larvae/ ha (Weekly release)	Okra aphid Cabbage aphid
HNPV	250 LE/ ha (10 days interval)	Tomato fruit borer
SNPV	250 LE/ ha (10 days interval)	*Spodoptera litura*
Bacillus thuringiensis	500 gai/ ha (10 days interval)	Diamondback moth Shoot and fruit borer of brinjal and okra. Tomato fruit borer

7. Judicious Use of Insecticides

Chemical control still remains first choice of farmers. Synthetic insecticides can be judiciously used considering certain aspects like waiting period of chemical, economic threshold level and initiation of control measures through pheromone / light trap catches. Choice of chemicals plays a crucial role in avoiding residue of chemicals. Vegetables with less picking interval should be sprayed with insecticide having less waiting period. Chemicals like plant origin insecticide, growth regulators, attractants and even synthetic pyrethroids can fit well in IPM. Botanicals being non persistant and safe to mammals is ideal in IPM. Neem products have shown efficacy against insects like jassids, aphids and leafminers. Application of crude neem seed kernel extract(4%) controls diamondback moth. Neem products mixed with reduced doses of insecticide proved better against jassids and borers of okra. Endosulfan can be applied with half of recommended dose mixed with neem product. Among the IGRs, moulting inhibitors are extensively used. Teflubenzuron is proved effective against diamondback moth. Another IGR, cyomaizin performs well to control leafminers of tomato and muskmelon. The IGRs are quite safe to non target organisms.

Insect pests of major vegetable crops and their control measures are given elsewhere.

Insect Pests of Major Vegetable Crops and their Control Measures

Crop / Name of pest	Symptom	Control measures
Tomato		
Tomato Fruit Worm (*Heliothis armigera*)	Caterpillar crawls over the leaves and feeds on vegetative parts. Makes holes and burrows. The period of activity commences from October and continues up to March.	Judicious use of synthetic pyrethroids viz., Fenvalerate (50g a.i./ha) or Deltamethrin (10g a.i./ha) is effectively used for controlling fruit borer.
Jassids (*Amrasca biguttula biguttula, Cestius* (*Hishimonus*) *phycitis*)	Both nymphs and adults suck the sap from the lower surface of the leaves. The infested leaf curls upward along the margins, which may turn yellowish and show burntup patches. They also transit mycoplasma disease like little leaf and virus disease like mosaic.	Jassids are controlled by 2-3 sprayings with Malathion (0.1%) or Dichlorvos (0.05%) 20 days after transplanting.
Tobacco Caterpiller (*Spodoptera littoralis*)	The insects are greyish brown and have white marking on upper wings. Young caterpillars feed on tender leaves, shoots and fruits at night.	Spraying Nuvan-100 (0.5 ml/ 2 litres of water)is effective in controlling the disease.
Whitefly (*Bemisia tabaci*)	Milky white, minute flies; nymphs and adults suck the cell sap from the leaves. The affected leaves curl and dry up. The growth of the plant is stunted. Whiteflies act as a vector, transmitting leaf curt virus causing severe loss in yield.	Soil application of Carbofuran (24 kg/ha) as side dressing before transplanting is recommended for the control of white flies. Spraying with Monocrotophos (0.02%) or Phosphamidon (0.02%) at fortnightly intervals starting with the appearance of the attack control the pest.
Mites (*Tetranychus cucurbitae*)	Low relative humidity favours mite multiplication. Both the foliage and the flower buds are seriously affected. Different stages of mite are found in colonies covered by white-silky webs on lower surface of leaves. Nymphs and adults suck cell sap and white patches appear on leaves. Affected leaves become mottled curl, turn brown and fall.	During egg stage and the resting stages, most miticides are ineffective. At high temperatures, it may be necessary to apply these at an interval of two days. Acaricides like Dicofol (0.05%) and Wettable Sulphur (0.3%) give effective control of mites. Destroy severely infested plant parts.
Root-Knot Nematodes (*Meloidogyne* spp.)	The root-knot nematodes invade the roots and stimulate cell proliferation, which results in formation of galls on roots. The aerial symptoms of root-knot nematode infection include stunted plant growth, chlorosis and a tendency to wilt under moisture stress during hot, dry weather.	Use of resistant varieties. Crop rotation with non-host crops such as marigold, maize, onion etc. reduces the nematode population. Application of neem oil cake (1-1.5 t/ha) 15 days after transplanting helps to suppress the nematode population.

Contd....

Brinjal		
Shoot and Fruit Borer (*Leucinodes orbonalis*)	The adult female lays eggs on the ventral surface of the leaves, flower buds and on young fruits. Short pinkish larvae initially bore into the terminal shoots resulting in withering and drying of the shoot. later, it bores into the young fruits by making holes and feeds inside which makes the fruits unfit for consumption.	Affected parts should be clipped along with the insect and destroyed. Spraying the crops with Carbaryl (0.1%) or Cypermethrin (0.5ml/litre of water) at fortnightly intervals starting from 35 days after transplanting controls the pest.
Jassids (*Amrasca biguttula biguttula*), *Cestius* (*Hishimonus*) *phycitis*)	Both nymphs and adults suck the sap from the lower surface of the leaves. The infested leaf curl upward along the margins, which may turn yellowish and show, burnt up patches. They also transit mycoplasma diseases like little leaf and virus disease like mosaic. Fruit setting is adversely affected by the infestation.	Jassids are controlled by spraying Malathion (0.1%) or Dichlorvos (0.05%) 20 days after transplanting.
Red Spider mite (*Tetranychus neocaledonicus, Tetranychus cinnabarinus, Paratetranychus indicus*)	Low relative humidity favours mite multiplication. Different stages of mites are found in colonies covered by white-silky webs on lower surface of leaves. Nymphs and adults suck cell sap and white patches appear on leaves. Affected leaves become mottled, turn brown and fall.	Same as tomato
Mealy Bug (*Centrocccus insolitus*)	Nymphs and adults of mealy bugssuck sap from the leaves, tender shoots, and the fruits. Leaves show characteristic curling symptoms similar to that of a virus. A heavy black sooty mould may develop on the honeydew like droplets secreted by mealy bugs. The infestation may lead to fruit drop or the fruits remain on the shoots in a dried and shriveled condition.	Unlike the adults, the crawlers are free from waxy coating and therefore the crawler stage is the most effective for spraying pesticides. Spraying of insecticides like Dichlorvos (0.02%) or Chlorpyriphos (0.05%) with fish oil rosin soap was found to control the insect population.
Lace Wing Bug (*Urentius hystricellus*):	Attack seen in summer season. Nymphs and dark brown bugs with lace like wings suck the sap from leaves, which turn yellowish and are found covered with insect excreta. Affected leaves ultimately dry up.	Proper crop rotation and spraying with Phosphamidon (0.03%) help in the reduction of the pest population.
Root-Knot Nematodes (*Meloidogyne incognita* and *Meloidogyne javanica*)	Same as tomato	Same as tomato.

Contd....

Chilli		
Aphids :(*Aphis gossypii* & *Myzus persicae*)	Suck the sap from tender shoots, leaves and on the lower surface of the leaves and reduce the vigour of the plant. Secrete sweet substances which attract antsand develop sooty mould. The pods develop black colour due to sooty mould The yields are also reduced by aphids directly and more through the spread of virus diseases acting as vectors indirectly.	Spray with 0.1% Dimethoate or Methyl demeton (2 ml in one litre of water) or Monocrotophos 1.5ml or Acephate (1g in one litre of water). Spray with alternate chemicals at 10 days interval till the end of aphid population checked. The winged forms of aphids rapidly migrate.
Mites (*Polyphagotarsoni - mus latus*)	Nymphs and adults suck sap from leaves. Affected leaves curl downward along the margins of the leaf and attain an inverted boat shape. Leaf petioles elongated and small leaves serrated. In severe cases, fruit wall becomes hard and white strips appear on the fruit.	Spray Phasalone 3ml/litre (Severe conditions) Or Wettable sulphur 3g/litre of water or Dicofol 5 ml/lit of water.
Tobacco caterpillar (*Spodoptera Litura*)	The larvae of second and third instar enter chilli pods by making a hole near calyx and feed on chilli seed. The affected pods drop off or develop white colour on drying.	Spray Chlorpyriphos 2.5ml/litre of water or Quinolphos 2ml/litre of water or Carbaryl 3gm or Monocrotophos 2ml or Thiodicarb 1gm or Acephate 1g/litre of water. Keep Pheremone traps (2/acre) in the field.
Root-Knot Nematodes (*Meloidogyne incognita* and *Meloidogyne javanica*)	Same as tomato	Same as tomato.
Potato		
Aphids (*Myzus persicae, Aphis gossypi, Aphis fabae*)	There are several kinds of aphids that feed on potato plants. Aphids transmit a virus disease and cause severe damage to the plant by sucking the juice from leaves and young stems. The plants may be stunted, with the leaves tightly curled and rolled when the attack is severe.	Aphids are controlled by adjusting the planting dates upto October 15 in north western plains, October 25 in Central plains and upto November 5 in North eastern plains. In the Plains, aphids are controlled by application of Phorate 10 G (1.5kg a.i./ha) followed by spraying with Dimethoate 30 EC (0.03%). The spraying should be repeated every week.
Cut Worms(*Agrotis ipsilon* and *Agrotis segetum*):	The caterpillars feed on young shoots or under ground tubers. In the early crop stages caterpillars cut the stem of the young plants near the ground and feed on the shoots and leaves. After tuber formation, they bore and nibble into the tubers affecting both tuber yield and market value.	Hot weather ploughing in plains and autumn ploughing in hills reduce the population of immature stages. A number of birds, feed on insects that get exposed upon ploughing. Spraying like Chlorpyrifos 20EC (2.5 litres/ha) and drenching the ridges controls the pest.

Contd....

Mites (*Polyphagotarsonemus latus*)	Both adults and nymphs damage the crop by foliage feeding. Mite attack starts from the top potato leaves showing downward curling. At first the lower side of leaves appears 'oily'. Gradually these symptoms can be seen on the entire plant. The infested leaves become short and leathery in appearance with characteristic copper colour deposits on lower side. Under severe mite attack, the infested leaves dry up.	The mite infestation can be avoided by delayed planting towards mid October in Indo-Gangetic plains. Frequent spraying with Dicofol 18.5 EC or Quinalphos 25 EC @ 2.0 litres/ha controls the pest. The first application should be done on appearance of the pest and repeated at 7 to 10 days interval depending on the persistence of mite infestation.
Rootknot Nematode(*Meloidogynae arenaria* and *M. incognita*)	Small galls or knots are formed on potato roots but they often go unnoticed. Heavily infested plants are stunted and exhibit early maturity. Reduction in size and number of tubers reduces the yield and warty 'pimple-like' outgrowths formed on tubers result in qualitative reduction.	Deep ploughing and drying of soil in summer months facilitate drying of infective larva, Follow crop rotation with cereals. Late Planting of autumn crop and early planting of spring crop in North-Western plains reduces nematode damage. Application of Carbofuran (Furadan 3G) @ 3 kg a.i./ha is recommended in two split doses, first at planting and second during earthing up.
Potato Cyst Nematode(*Globodera pallida*)	Small patches of poorly growing plants appear in the field. Such plants show temporary wilting, stunting and premature yellowing symptoms. The size and number of tubers are reduced. Small mustard seed size yellow or white female nematodes are seen sticking to the roots.	Growing non-host crops like radish, garlic, beet and turnip bring down the cyst population. Application of Carbofuran 3G (2 kg a.i./ha) at the time of planting reduces the nematode population.
Potato Tubermoth (*Phthorimaea operculella*)	Potato tubermoth (PTM) larvae mine the younger leaves and feeds on the tubers. Larvae penetrate the leaves and feed within leaf veins or stems of the plant and on tubers in storage by depositing the eggs near the eye buds, causing irregular galleries and 'tunnels' deep inside the tuber.	Follow IPM. Use uninfested seed tubers, follow deep planting, and adopt frequent irrigation wherever possible. Spraying Monocrotophos 40 EC (1.5 litre in 1000 litres of water) on 30 day old crop is effective in controlling the pest. Use of bioagents, viz. *Bacillus thuringiensis* (Bt) and Granulosis virus (GV) is advocated for PTM control in the stores.
Cucurbits		
Red Pumpkin Beetle (*Aulacophora foveicollis*)	The pest attacks cucurbits especially melons at the seedling stage. They make holes in cotyledonary leaves of watermelon. As a result the seedlings in the young stage die.	Spraying Carbaryl (4 g/litre of water) or Metacid (1ml/litre of water) during the seedling stage effectively controls the pest.
Aphids (*Aphis* sp.)	Aphids damage the plants by sucking the leaf sap in young stage, cotyledonary leaves crinkle and in severe case the plants wither. The leaves of fully grown vines turn yellow and plant loses its vigour.	The aphids can be easily controlled by spraying Malathion (0.1 %) or Metasystox (0.1-0.2%) or Rogor (0.1-0.2%) sufficiently early before the attack becomes severe. Usually sprays against aphids and beetles can be combined.

Contd....

Fruit Fly (*Dacus cucurbitae* and *Dacus dorsalis*) :	The fly attack is severe, especially after summer rains when the humidity is high. Maggots of this fly causes severe damage to young developing fruits. The adult fly lays eggs in the flowers. The eggs hatch into maggots, which feed inside the fruits and causes rotting.	The affected fruits should be regularly pinched off and buried in a pit. Spraying of Thiodan @6 ml per 4.5 litres of water during flowering stage partially checks the fly incidence. However, spraying should be done in the evening to prevent the repellent action to pollinating insects like bees from insecticidal sprays. Fix baits with banana fruit pulp/ocimum sap along with furadan in pandals for attracting fruit fly.
Mites	The pest attains serious proportions during summer. Different stages of mite are found in colonies covered by white-silky webs on lower surface of leaves. Nymphs and adults suck cell sap and white patches appear on leaves. Affected leaves become mottled, turn brown and fall.	Acaricides like Dicofol (0.05%) and Wettable Sulphur (0.3%) gives effective control of mites. At high temperatures, it may be necessary to apply these at an interval of two days. Cutting and burning of severely infested plant parts reduce further multiplication.
Okra		
Shoot and Fruit Borer (*Earias vittela* and *E. insulana*)	The incidence of fruit borers usually occurs during humid conditions after the rainfall. The adult female lays eggs individually on leaves, floral buds and on tender fruits. Small brown caterpillars bore into the top shoot and feeds inside the shoot before fruit formation. Later on they bore into the fruits and feed within. Affected fruits become unfit for consumption.	The infested fruits and shoots should be removed regularly and buried deep in the soil. Spraying with Endosulfan 35 EC (2 ml/litre of water) or Quinalphos 25 EC (2 ml/litre of water) or Carbaryl (4 g/litre of water) effectively controls the pest. Before spraying all the affected plant parts should be removed.
Leaf Hopper (*Amrasca biguttula biguttula*)	This pest attacks the crop at its early stage of growth. Small, greenish leaf hoppers; nymphs and adults are found on the under side of the leaves and suck the cell sap from the leaves. As a result the leaves curl upwards along the margins and have a burnt look which extend over the entire leaf area.	Soil application of Carbofuran 3 G (1 kg a.i./ha) at the time of sowing effectively controls the pest. Spraying the crop with Monocrotophos (0.05%) at fortnightly intervals starting with the appearance of the pest provides a good control.
Mites (*Tetranychus* spp.)	Attack is mostly observed during the warm and dry periods of the season. Nymphs and adults suck cell sap and whitish grey patches appear on leaves. Affected leaves become mottled, turn brown and fall	Spraying with Wettable Sulphur 80 WP (2 g/ litre of water) or Dicofol 18.5 EC (2.5 ml/litre of water) effectively control the mites
White Fly (*Bemisia tabaci*)	The milky white minute flies; nymphs and adults suck the cell sap from the leaves. The affected leaves curl and dry. The affected plants show a stunted growth. Also responsible for transmitting yellowvein mosaic virus.	Soil application of Carbofuran (1 kg a.i./ha) at the time of sowing and 4-5 foliar sprays of Dimethoate (0.05%) or Metasystox (0.02%) or Nuvacron (0.05%) at an interval of 10 days gives effective control.

Contd....

Root-Knot Nematode (*Meloidogyne incognita*)	Causes characteristic root knots or galls. The aerial symptoms consist mainly of stunted plant growth and yellowing of leaves. Nematode attack in the seedling stage leads to pre- and post-emergence damage resulting in reduced crop stand.	Follow crop rotation with non-host crops such as cereals, fallowing and deep ploughing 2-3 times in summer months is recommended. Application of Nemagon (30 litres/ha) with irrigation before sowing is recommended to protect the seedling in its early stage of plant growth.
Onion		
Onion Thrips (*Thrips tabaci*):	Adults lay eggs in clusters in leaf tissues. Numerous nymphs and adults are observed between leaf sheaths and stems lacerating the epidermis of leaves and sucking the exuding cell sap. The affected leaves show silvery blotches which later turn into brownish colour. The leaves get distorted from tips downwards and plant ultimately wilt and dry away.	Soil application of Phorate or Carbofuran granules (1kg a.i./ha) is also recommended. Alternate spraying of 0.07% Endosulphan (2ml/litre of water), or 0.07% Nuvacron (2ml/litre of water) or 0.01% Cypermethrin (1ml/litre of water) along with a sticker (Triton or Sandovit) at fortnightly intervals controls the insect.
Head Borer (*Helicoverpa armigera*):	Serious pest of onion seed crop in Northern India. The larvae cuts the pedicel of the flower and feeds on the stalk. Single larva damages many flower stalks.	Spraying Endosulfan (2-3 ml/litre of water) along with sticker (Triton/ Sandovit) gives control.
Onion Maggot (*Hylemia antiqua*):	The flies lay their eggs on old leaves or on soil and larvae enters into soil and damage disc portion of onion bulb. Infested plants turn yellowish brown and finally dry up. The affected bulbs rot in storage.	Regular crop rotation should be followed. Application of Thimet in the soil before transplanting is found effective.
Cabbage & Cauliflower		
Cabbage Diamondback Moth (DBM) (*Plutella xylostella*)	Most serious pests of cabbage and cauliflower under high temperature conditions. Female lays eggs on the leaves either singly or in groups. Caterpillars feed on the leaf epidermis and later make holes in the leaves. Severely affected leaves are completely skeletonised.	Spraying the crop with Malathion (0.1%) or Profenofos (0.25-0.5 kg a.i./ha) gives excellent control of the larvae. Intercropping mustard in the cabbage crop 15 days prior to sowing and 25 days after planting cabbage was effective. As the pest preferred mustard, major pest population is attracted towards it, which can be destroyed by spraying the crop with Dichlorvos (1 ml/ litre of water).
Leaf Webber (*Crocidolomia binotalis*)	Eggs are laid in clusters on the undersurface of the leaves and held together by gelatinous glue. Green caterpillars web up the leaves and live inside the knotted mass. Flowering and pod formation is adversely affected.	Removal and destruction of webbed bunches of leaf help to check the further spread of the disease. Dusting the crop with Carbaryl (4%) or spraying with Malathion (0.05%) is effective.

Contd....

Cabbage Borer (*Hellula undalis*)	The adult female lays eggs on the growing point or on the older leaves. The pale-whitish caterpillars mine the foliage, feed on the shoots and finally bore into the stem. Infested plants are killed or produce side shoots, which do not form heads.	Spraying the crop with Malathion (0.1%) or dusting 4% Carbaryl gives excellent control of the larvae.
Cabbage Butterfly (*Pieris brassicae*)	The larvae feed gregariously on the foliage and burrow into the heads. In case of severe cases the heads become completely unfit for marketing.	The caterpillars should be hand picked and destroyed. Spraying the crop upon the emergence of the pest with Malathion (0.1%) or dusting Carbaryl (0.15%) gives excellent control of the pest.
Aphids (Brevicoryne brassicae, Myzus persicae, Lipaphis erysimi)	Yellowish green nymphs and adults suck cell sap and devitalize plants. Affected parts become discolored and malformed. High humidity favours rapid multiplication of this pest. The aphids are mostly observed on the lower surface of the leaves.	Spraying of Monocrotophos (0.05%) or Malathion (0.1%) at 10-15 days interval contains aphid population effectively. To prevent recurrence of the pest granular insecticides like Phorate @ 1.0 kg a.i./ha should be applied to soil.
Pea		
Pea Aphid (*Acyrthosiphon pisi*)	Attack is seen from January onwards. The green soft-bodied nymphs and the adults attack the young vines and suck the sap of younger parts of plants. The affected plants become stunted and the pods curl, and results in blanking.	The insect can be controlled by spraying Malathion (0.05%) or Dichlorvos (DDVP) (0.05%).
Pod Borer (*Helicoverpa* spp.)	Attack is severe on green pods especially at the stage of maturity. The caterpillars bore into the pods and feed on the seeds.	Handpicking of the caterpillar and pupae during the early stages of infestation reduces the pest damage.
Pea Weevil (*Bruchus pisorum*)	Adults feed in blossoms and lay eggs on young pods. Larvae after hatching from the eggs bores into green seed.	Spraying Malathion (0.05%) or Dichlorvos (DDVP) (0.05%) at the blossoming stage effectively control the pest.
Leaf Miner (*Phytomyza atricornis*)	The larvae attack tender leaves and feed in the epidermal layers of the leaf by making serpentine mines in which air gets trapped and gives them silvery appearance. The affected leaves turn pale yellow, get distorted and crumpled. Such leaves gradually dry and die away.	Spraying the plants with Dichlorvos (0.05%) at the emergence of new leaves is highly beneficial. Application of Phorate 10 G (1 kg a.i./acre) applied one day before planting is effective in reducing the larval population of leaf miner.
Stem Fly (*Ophiomyai phaseoli*)	The maggots bore inside the stem leading to wilting of the plants.	Seed treatment with Carbofuran (4%) protects the newly emerging seedlings from the infection. Spraying with Malathion (0.05%) or Monocrotophos (0.04%) effectively controls the pest.

Contd....

Radish		
Aphids (*Myzus persicae, Brevicoryne brassicae, Lipaphis erysimi*)	Aphids attack both the seedlings as well as the mature crop. The aphids are mostly observed on the lower surface of the leaves. The nymphs and adults suck the cell sap and devitalize plants. Affected parts become discolored and malformed.	Spraying of Monocrotophos (0.05%) or Malathion (0.1%) at 10-15 days interval contains aphid population effectively. To prevent recurrence of the pest granular insecticides like Phorate @ 1.0 kg a.i./ha should be applied to soil.
Mustard Sawfly (*Athalia proxima*)	Attack is seen in the vegetative and flowering stage of the crop. The adult female lays eggs inside the leaf tissue. The grubs after hatching from the eggs feedon the leaves by making small holes.	Hand picking of larvae when the area involved is small, or spraying with Malathion 50 EC (1ml/litre of water) twice at an interval of 10 days is recommended.
Carrot		
Carrot Rust Fly (*Psila rosae*)	The larvae mine the roots, causing holes Heavy maggot feeding is indicated by drooping, discolored foliage. If the roots are severely injured the leaves become rusty brown or even dry.	Follow deep plowing, Crop rotation and destruction of wild hosts.Soil application of Carbofuran at a depth of 10-15 cm effectively controls the pest.
Thrips (*Thrips tabacii*)	Adults and larvae suck the cell sap from the leaves of the plants. White blotches appear on leaves. Tips of leaves wither and turn brown.	Alternate spraying of insecticides like Phosphamidon (0.05%) or Monocrotophos (0.1%) or Malathion (0.05%) offer a good control over the pest.
Nematode *Heterodera carotae, Meloidogyne* sp.)	Nematodes infestation causes substantial yield losses in carrot. Roots become forked and e unmarketable.	Soil treatment with Nemagon is effective in reducing the nematode population.
Beet root		
Beet Leaf Miner (*Pegomyia hyocyami*)	The maggots attack tender leaves and feed in the epidermal layers of the leaf by making serpentine mines in which air gets trapped and gives them silvery appearance. The affected leaves turn pale yellow, get distorted and crumpled. Such leaves gradually dry and die away.	Destroy all fallen leaves and other plant debris after harvesting of roots. Spraying the underside of the leaf with Methyl Demeton (0.03%) and Phosphamidon (0.035%) at the emergence of new leaves is highly beneficial.
Web Worms (*Hymenia* sp. or *Loxostege* sp.)	Eggs are laid in clusters on the undersurface of the leaves and held together by gelatinous glue. Green caterpillars web up the leaves and live inside the knotted mass. Flowering and pod formation is adversely affected.	Removal and destruction of webbed bunches of leaf help to check the further spread of the disease. Spraying with Rogor (1 ml/litre of water) is effective.
Semiloopers (*Plusia* spp.)	The green caterpillars voraciously feed on the foliage damaging green foliage badly.	Hand picking the larvae and spraying the crop with Endosulfan (0.05%) or Carbaryll (0.1%) controls the pest.

REFERENCES

Balappa, S.R. and Hugas, L.B. 2003. An economic evaluation of onion production and its marketing system in Karnataka. *Agricultural Marketting*. 16: 22-25

Dahiya, P.S. 1995. Temporal and spatial overview of marketing costs and margins for potato in India - The implications policy. *Indian Journal of Agricultural Marketing*, Conference Special, 9: 68.

Janick, J. 1979. *Horticultural Science*. W.H. Freeman and Co. San Fransisco.

Lokesh, G.B., Hari Shilpa, G.R. and Chandrakanth, M.G. 2005. An economic analysis of tomato production, Marketting and processing in Karnataka. *Agricultural Marketing* 17: 15-21.

Mishra S.N., Tripathy, S., and Prusty, B.K. 1995. Production and marketing of onion in Kalahandi District of Orissa. *Indian Journal of Agricultural Marketing*, Conference Special, 9 :167.

Nariani, T.K. and Seth, M.L. 1978. *Vegetable Crops in India*, pp. 606-22. Bose, T.K. and Som, M.G. (Eds). Naya Prokash, Calcutta.

Pandey, U.B. 1998. Export of vegetables. *Souvenir: Silver Jubilee National Symposium "Emerging Scenario in Vegetable Research and Development"*. PDVR, Varanasi, 12-14, December 1998,pp 131-138.

Pandey, A.K. and Rai, M. 2005. Popularizing exotic vegetables for nutrition and food diversification. *Indian Horticulture*. 50:19-22.

Prasad, E, Y., Reghuram, P. Satyanarayana, G.and Lakshmana Rao, M . 1980. Economic analysis of chilli farming in Guntur district. *Spices India* 1: 27-30.

Seshadri. V.S. 1986. Cucurbits. (in) *Vegetable Crops in India*. edn. I: Bose, T.K. and Som, M.G.(Eds). Naya Prakash, Calcutta

Singh, D.V., and Kharwal, J.C. 1995. Components vegetable marketing: A case study in Himachal Pradesh. *Indian Journal of Agricultural Marketing*, Conference Special, 9: 65.

Singh, B., Kumar, M. and Singh, A. 2005. Cultivating tomato in green house is lucrative. *Indian Horticulture* 49: 13-14.

Singh, B., Kumar, M. and Sirohi, N.P.S. 2005. Growing parthenocarpic cucumber in green house is ideal. *Indian Horticulture* 49: 22-23..

Singh, P.M., Mathura Rai, Singh, B and Pandey, A.K. 2006. Producing quality vegetable seeds : A remunerative enterprise. *Indian Horticulture* 51 : 31-34.

Singh, Narendra and Netra Pal. 1996. *Bulb and Root Vegetables* (In) Fifty Years of Crop Science Research in India, R. S. Paroda and K. L. Chadha (eds.), Indian Council of Agricultural Research, New Delhi.

Singh, R., and Sikka, B.K. 1992. *Marketing of high value perishable crops in Himachal Pradesh.* Agricultural Economics Research Center, Himachal Pradesh University, Simla, India, 168 p.

Smith, P.G., Villalon,B. and Villa, P.L. 1987. Horticultural classification of peppers grown in the United States. *Hort Science* 22: 11-13.

Vasisht A.K., Singh, C., and Singh, I. 1995. Temporal and spatial changes in the price spread of agricultural commodities in India. *Indian Journal of Agricultural Marketing,* Conference Special 9: 1-1 1.

Wadhwani, M.K. and Bhogal, T.S. 2003. Economics of production, post harvest management and price behaviour of cole crops in Western U.P.-An empirical analysis. *Agricultural Marketing* 16: 10-20.

Yadav, N.K., and Pant, D.C. 1995. Price spread of early grown onion in Alwar District of Rajasthan. *Indian Journal of Agricultural Marketing,* Conference Special 9: 87.

Chapter 3

FLOWER CROPS

Introduction

Flowers are the most fascinating objects in the world and many poems are written glorifying them. They are associated with mankind from the dawn of civilization. Flowers are an integral part of our day to day life and they find a place in worshiping, religious and social functions, wedding, interior decoration and self adornment. Flowers are used to convey the human feelings. To *say it with flowers* is very common in western culture. Emergence of a strong middle class and changing life style has created a great demand for cut flowers in India.

Floriculture is the aesthetic branch of horticulture which deals with cultivation of both traditional and commercial flower crops, either in open field or under protected conditions and growing of ornamentals including potted plants and their marketing.

Demand for cut flowers in developed countries is ever increasing and globally, the export trade in cut flowers is growing at around 8-10% per annum. The climatic condition prevailing in most of the developed countries does not favor year around production of these flowers and many of them import flowers to meet domestic demand. Cost escalation in production prompted many developed nations to shift the base to developing countries, especially to Asia. Floriculture is now emerging as a potential money spinner for many third world countries. It is now recognized as a lucrative profession with much higher potential for returns/unit area than any other horticultural crop. Besides, floriculture improves image of the country, increases tourism, creates employment and attracts investment.

Items of trade under floriculture comprise the following:

1. Cut flower and cut foliage which may be fresh, dried, dyed, bleached or impregnated
2. Live plants/ pot plants and bedding plants
3. Seeds, tuber, bulbs, tissue culture plants and rooted plants of flower crops
4. Value added products such as flower perfumes and natural dyes

Present Status

Internationally, floriculture industry is a billion dollar business. During 2004, the global trade in floriculture products was worth 120 billion USD. Total area under flower crops through out the world is 2.5 lakh ha.

The Major Exporting Countries

Netherlands dominates the world trade in flowers and plants and accounts for a market share of 60 % for cut flowers 50% for potted plants. Being a major market for floriculture, Netherlands both imports and re-exports the flowers. It has 9 auction centers which handle 90 % of the world trade in floriculture products. Two major auction centers are located at Amsterdam and Alsmeer. Other important exporters of flowers and plants are Columbia (10%), Italy (6%), Israel, Ecuador, Kenya, Spain, Zimbabwe and Thailand.

The Major Importing Countries of Cut Flowers

Germany, UK, France Netherlands, Italy, Switzerland, USA, Japan, Hong Kong and Singapore are the major importers. These countries also have domestic units, the flowers of which are consumed locally. The per capita consumption of flowers is the highest in Norway.

The most important flower traded in the international market is still Rose. However, other flowers like chrysanthemum, tulips, carnations, liliums, gerbera, cymbidium, freesia, anthurium, dahlia, poinsettia, gladiolus, gypsophila, lisianthes, alstromeria, etc. are also marketed in large quantities in these countries.

Indian Scene

India occupies 15% of area under flower crops and ranks 2nd next to China, with an area of 73,500 hectares and production of 5.35 lakh tonnes of loose flowers and 700 million stems (25.62 tonnes) of cut flowers. In spite of a long tradition in flower farming and larger area, India's share in the international trade for flowers is negligible, below 1%. Major area is under traditional flowers like marigold, jasmine, aster, rose, chrysanthemum, tuberose, crossandra and most of them are cultivated in open fields. Area under modern flowers like rose, carnation, gerbera, gladiolus, anthurium and orchids are less than 10 %. Traditional floriculture units are mainly based in southern states of Karnataka, Tamil Nadu and Andhra Pradesh. States such as West Bengal, Maharashtra, Rajasthan, Haryana and Delhi also have large areas under floriculture. In loose flower production Tamil Nadu ranks first (24.7%) followed by Karnataka (23.1%), West Bengal (9.3%), Andhra Pradesh (9.0%) and Rajasthan (7.4%).

State Wise Area and Production (1997-1998)

S.No	State	Area	Production (t)
1.	Andhra Pradesh	8420	32900
2.	Karnataka	20780	124290
3.	Tamil Nadu	16745	142333
4.	Maharastra	4786	33250
5.	Haryana	1950	33040
6.	West Bengal	13720	17685
7.	New Delhi	1866	10274
8.	Madhya Pradesh	1334	13127
9.	Others	3900	9354
	Total	**73501**	**416226**

Source: APEDA

Profile of Indian Floriculture

The floricultural activities in India include cultivation and marketing of traditional and cut flowers, indoor plants and landscaping plants, bedding plant industry, plant rental services, floral perfumes, flower seeds, dry flowers and foliage industry.

The traditional flowers are picked individually mostly as buds without the stalk and marketed as loose flowers, on weight basis. Flowers are mostly used for hair adornment and for garland making. Large quantities of flowers are used for decorating marriage halls, temple chariots and arches. The most important traditional flowers are jasmine, scented rose, chrysanthemum, marigold, tuberose and crossandra. Among different traditional flowers, jasmine occupies the major area followed by tuberose, crossandra, rose, marigold and chrysanthemum.

After the policy of economic liberalization since 1991-92, the Govt. of India identified floriculture as a sunrise industry and accorded 100% export oriented status to it. This has given an impetus to the Indian entrepreneurs for establishing floriculture units under controlled climatic conditions. Owing to steady increase in demand of cut flowers, commercial floriculture has emerged as a hi-tech activity taking place under greenhouse.

Several units have come up mostly under corporate sector and now there are more than 230 export-oriented units already under operation and 52 under construction in India covering 323 ha under green house. The area is likely to go beyond 500 ha. Most of the units are located in Bangalore, Pune and Delhi. Almost all of them are capital intensive and made large investments in greenhouse and other infrastructure. More than Rs.3000 crores capital investment has already taken place in the floriculture industry over the past few years. In addition, several entrepreneurs are coming up to start new floriculture projects.

In big cities like Bangalore and Pune, plant rental services to corporate houses, commercial establishments, offices etc., are a thriving business. Supply of tissue culture and other planting materials for bedding plants like chrysanthemum, gerbera, dahlia etc., and flower seed industry are also gaining momentum. Apart from nursery industry, landscaping and home gardening also has become an important part of floricultural activities. Value addition of flower crops by extraction of essential oil and perfumes, cosmetics and natural pigments for food coloring is another emerging area of floriculture industry.

Domestic Market

Even though the floriculture trade is still in infancy in India, the growth rate of Indian floriculture products over the past several years had been impressive at 600 per cent and the internal trade is growing @ 15% annually (Chowdhary, 2003). The cut-flower trade in big cities like Bombay, Calcutta, Bangalore, Madras, Delhi, Pune, Ludhiana, etc. has developed tremendously during the last two decades. The growing of certain flowers and production of planting material like corms, bulbs and pot plants around big cities are fetching good income for farmers. The demand for flowers has lead to opening up of several florists shop in big and small cities for the supply of bouquets, floral baskets and cut flowers for vase decorations.

Export Market

Today India's share in the global floriculture trade is $ 67 million which is only at 0.75 per cent of the total. India's exports are growing between 7 and

8 per cent. During the past 10 years, India's floriculture export has increased 25 times. India has now 6 flower export cities and 4 new export cities are proposed. Ninety percent of the area in India under cut flower is occupied by rose and it is mainly exported to Japan. Apart from roses, other exotic flowers like carnations, chrysanthemums, orchids, gladioli and anthuriums are exported to developed countries mostly in the West and Middle East. Traditional flowers like jasmine and tuberose are exported as fresh flowers to Malaysia, Singapore, Sri Lanka and Gulf countries. Extracting essential oils from flowers is another major industry in India. Jasmine and Tuberose flowers are used for extraction of oils and are exported as high value concrete in the world market.

Prospects of Floriculture in India

India is capable of supplying international quality flowers at highly competitive prices due to a number of favorable factors. India is a large tropical country endowed with rich and diverse agro climatic conditions for floriculture. There is plenty of sunlight, good rains, abundant water supply, good quality soil and water with which India can grow a wide variety of flowers.

During winter months of November to March when flower production is less in the major producing countries, weather in the Indian subcontinent is conducive to top quality production. Demand for Indian flowers from September to February in Europe and Japan is very high. Christmas, New Year and Valentine's Day which offer good market for flowers also come in between these months. In summer, Indian flowers take a back seat in European auctions as the traditional markets are flooded with flowers from temperate countries.

India also has the location advantage. Geographically, the country is located centrally between the markets of Europe and South East Asia. India's proximity to most of the South-East Asian markets like Japan, and Russia, Far East and Middle-East Countries compared to other producing countries in Europe, Africa, South Africa and South America offers tremendous competitive advantage. Flowers reach faster, fresher and safer as these destinations are closer.

Most importantly, India enjoys the advantage of low cost labor compared to other floriculture exporting countries. There is abundant availability of skilled labor at the cheapest cost. Indian floriculture industry takes advantage of the weather factors and is less capital intensive. Hence the production cost is less. There is a well developed research and development system in

Agriculture and there is access to latest technology to grow and dispatch flowers. A strong technical manpower is also available in horticulture.

A few Indian floriculture companies have gained ISO-9002 certification which establishes that the floricultural industry is capable to deliver flowers of international quality. Presence of a number of biotech companies with large tissue culture facility which can propagate planting materials at low cost has facilitated its growth. The new seed policy has already made it feasible to import planting materials of international varieties. The emerging strong domestic market can sustain the industry during lean export period.

Floriculture industry has also attracted large Indian corporate and big business houses that have a seasoned management. Huge computer infrastructure is available for direct sales and logistical support. Many Indian expatriates are living abroad who can be employed in the distribution chain. India has started a Market Facilitation Center in Amsterdam since 2000.

R&D Organizations in Floriculture

Research: Several ICAR institutions such as IARI, New Delhi, IIHR, Bangalore, NRC on orchids, Sikkim, CSIR institutions like NBRI, Lucknow and State Agricultural universities conduct research on flower crops.

National Horticultural Board: NHB Provides funds for establishment of model floriculture units, transfer of technology, financial support through subsidies, trainings, soft loans for infrastructure development etc; Eleven-model floriculture centre units and two large centers and 20 tissue culture units are established under its assistance.

APEDA (Agricultural and Processed Food Products Export Development Authority): This agency is responsible for export promotion and development of floriculture in India, grants subsidies for establishing cold storage, pre cooling units, refrigerated vans and green houses, and air freight subsidy to exports. Direct subsidy up to 50% of the precooling and cold storage units, as well as subsidy for using improved packaging material is given by APEDA.

Refinance assistance is available from NABARD to a number of hi-tech units at reasonable rate of interest. Import duties are reduced on cut flowers, flower seeds, tissue-cultured plants, etc.

The government of India offers tax benefits to new export oriented floriculture companies in the form of income-tax holidays and exemption from certain import duties. Attempts to establish flower auction centers in different intensive flower growing areas to safeguard the growers are attempted and an auction centre has now been established in Bangalore.

Florist's organizations and growers co-operatives are encouraged for marketing and more systematic functioning of the domestic market.

Government of India provides subsidies for air freight, packaging, establishing phytosanitary labs, pre-cooling and cold storage facilities at farms and airports, refrigerated transport, and marketing expos.

Constraints

In spite of an abundant and varied production base, India's export of floricultural product is not encouraging. The low performance is attributed to many constraints.

Floriculture in India is yet to assume the role of an organized industry. High cost of technology and high initial establishment cost have taken the toll on many units and their profitability.

Lack of quality planting materials, inappropriate choice of plant variety and technology, prohibitive cost of imported planting materials and restrictions on plant variety multiplication due to plant breeder's rights and violation of intellectual property rights are a few of the production constraints. Research support for floriculture is poor compared to other horticultural crops.

Poor quality of produce, inadequate surplus for export, poor infrastructure support like post harvest handling facilities specific for individual flowers, lack of market intelligence and unorganized marketing are the other constraints. Foreign investors are showing more interest in China, Ethiopia and other countries than in India.

India also has infra-structural problems like bad interior road, inadequate refrigerated transport and storage facilities, non-availability of space in major airlines and insufficient availability of chartered cargo flights during the peak seasons for export.

Indian floriculture is still subject to import duties in the European Union. Discriminative policies by importing countries on duty tariffs *(8.5% general and 5.5% in winter)* and heavy freight costs are eating into growers' profit margins. Environmental and social standards recommended for the EU like *MPS, EUREPGAP,* Signature of UPOV (Union for Protection of Varieties), respect of patents, trademarks and special licenses of breeders and their registered propagators, payment of royalties, tedious phyto-sanitary certification laws in importing countries, etc., also act as deterrents.

Solutions/ Suggestions

To overcome these problems, attention must be focused on (Ahuja, 2001)

- Establishment of model nurseries for supplying genuine planting materials.
- Changing the traditional flowers including roses which give only low profit margins and growing new cultivars by which the growers can make profits. For example, *Phaleanopis* orchids are the most popular plant sold in Europe. *Phaleanopsis* sales in Europe are increasing at 30%/ year.
- Strengthening focused research on key areas.
- Setting up of centers for training and manpower development for the personnel involved in floriculture
- Ensuring close linkage between governments, research institutions and flower growers
- Establishing co-operative florist organizations at regional level for marketing.
- Assistance for provision of infrastructure and market development, reduction in interest rates by banking institutions and waiver of excise duties on domestic sales
- Establishing specially designed cargo terminals for agricultural produce in international airports, reduction in air freight charges and providing sufficient cargo space in airlines for promoting export.
- Reduction in import duty on planting material and equipment.
- Promotion and use of information technology for trading.

Achievements

The Indian floriculture industry has started showing signs of stabilization. The emerging strong domestic market, establishment of basic infrastructure facilities like cold storage at airports in major growing areas like Bangalore and Delhi and the setting up of an auction house of international status in Bangalore, apart from an increase in freight subsidy, have come as a shot in the arm for growers and exporters.

Packing is crucial in the marketing of flowers. The standards developed by the Indian Institute of Packaging, Mumbai for roses, gladioli, carnation, chrysanthemum, gerbera, orchids and anthurium are being popularized by APEDA to help exporters to maintain quality of floriculture products.

Despite facing several set backs, the Indian Floriculture trade has made a mark on the global front by delivering quality flowers. India, which began exports at the beginning of decade to traditional Dutch markets, has discovered other markets in the past several years. There is also an increasing tendency of the Indian growers to directly send flowers to the European countries rather than routing through Netherlands. Major destinations apart from Europe are Japan, Australia, New Zealand, Singapore, Taiwan and Middle East countries.

Of late, the growers in India have realized the economic potential of flower farming. The advent of protected cultivation, efforts of growers and the achievement of governmental organizations in research and development has definitely aided in the advancement of floriculture. The future of Indian floriculture is extremely bright and its further development lies in identifying and cultivating new flowers, expanding the domestic and export market, and product diversification.

ANTHURIUM

Anthuriums are tropical plants valued for their attractive long lasting flowers and handsome foliage. It is one of the important economic flowers of export potential ranked 11th in international trade. The flowers are used in vase arrangement due to bold effects, bright color and long vase life. Potted anthuriums are ranked 23rd among all potted foliage and flowering plants sold through the Dutch auctions. Containerized anthuriums can be used for interiorscapes and landscapes.

Classification

Anthuriums can be broadly categorized into flowering and foliage types. It belongs to the family Araceae and includes more than 800 species and several varieties. Many of them are partially epiphytic in growth habit. Only few of them are commercially important. *Anthurium andreanum* and *A. scherzerianum* are the most economically viable species. Important flowering species are *A. andreanum, A. scherzerianum, A. amnicola, A. ornatum* and *A. robustum.*

Important foliage species are *A. clarinervium, A. crystallinum, A. grande, A. holtonianum, A. magnificum, A. veitchii, A. warocquianum, A. pedatoradiatum* etc.

A. andreanum (oil cloth flower) is a native of Columbia. It is semi epiphytic in nature, slightly creeping, with aerial roots for anchorage. The plant is vigorous and grows to about 3 feet and produces large leaves that are spirally arranged. It is shy suckering and is suitable for growing in tropical climate. Used as cut flowers and potted plants.

A. scherzerianum is a native of Costa Rica. It is known as flamingo flower because of the peculiarly curved shape of candle. It is a dwarf compact plant with profuse suckering habit. It is suitable for subtropical climate and requires cooler atmospheric temperature for growth. They are mainly used as potted plants.

Physiology: Seedling plants have a juvenile phase which lasts for 12-18 months. There after the plant undergoes alternate vegetative phase and generative phase. Anthurium produces bloom through out the year, and on an average the plant produces one leaf and one bloom arising from the axil of each leaf every month (Prasad *et al.*, 2001). The interval between leaf emergence and bloom emergence is shortened or lengthened with changes in environmental condition and management. More bloom is produced during the actively growing season following the monsoon in Kerala.

The flower in anthurium is a modification of leaf which is colorful and heart shaped known as spathe and the true flowers are present in the spadix or candle which is a pencil like structure arising from the base of the spathe. Hundreds of small bisexual flowers are spirally arranged on the spadix.

Anthurium flowers are seen mainly in 4 basic colors viz., Pink, White, Red and Orange. Varieties with intermediate shades and bicolor known as obakes are also seen. Flower color is due to pigments Cyanidin and Pelargonidin. Two major genes M and O are responsible for color which controls the pigments Cyanidin and Pelargonidin.

Breeding and Varieties

Anthurium breeding work was started in University of Hawaii from the 1980s. They developed several varieties as cut flowers. Some outstanding attributes of UH cultivars include anthracnose resistance, bacterial blight tolerance, attractive foliage, long-lasting flowers, and free-flowering habit etc. Several cultivars were bred for dual purposes (cut/potted).

Selection Criteria for Cut Flower Varieties

1. The plant should be compact, produce leaves at shorter internodes.
2. Profuse suckering habit and flowering through out year with glossy flowers positioned above the foliage
3. The spathe should be bright, heart shaped, showy, puckered (with plenty of blisters) and there should be symmetrical overlapping of basal lobes.

4. Reclining spadix angled to the spathe, short in length than spathe and oriented at an angle below 30°
5. An erect long flower stem about five times the length of the spathe
6. Resistant to common diseases and pests such as bacterial blight caused by *Xanthomonas campestris pv. dieffenbachiae* and anthracnose or black nose, caused by *Colletotrichum gloeosporioides.*

Breeding in India

In India, Anthurium breeding started at IIHR, Bangalore. Four hybrids with desirable characters were evolved (Singh, 1995). Indian varieties are known based on their spathe color. Genuine varieties are difficult to obtain as frequent cross breeding has obliterated varieties. However, there is a popular planter's selection called Agnihotri.

Almost all the international varieties are listed out side the country. They mainly come from the producer countries such as Holland, USA and Mauritius. The predominant anthurium varieties accepted internationally are available from Anthura, Avo Anthurium vogels, Andes Anthurium (Netherlands), University of Hawaii, and Oglesby plant labs (USA). The Hawaiian Anthurium Industry Association is responsible for registering varieties.

Hundreds of varieties are released world wide by different breeders. How ever only some of them are prominent in the cut flower trade. Important varieties from different breeders are given below.

Breeder	Spathe Color	Varieties
Anthura, Holland	Red	Tropical, Cancan, Arizona, Carre, Calisto, Fire, Calore, Montero
	Orange	Casino, Passion, Sunglow, Cheers
	Pink	Spirit, Rosa, Nunzia, Magentis
	White	Acropolis, Carnaval, Bianco, Pierrot, Angel, Moments, Presence, Fantasia
	Obake	Lambada, Sultan, Regina, Amigo, Senator, Cascade, Marquis, President, Monet, Samba, Marshall, Salvo, Cantare, Volare
	Green	Midori, Pistache, Champagne, Laguna, Tequila, Savanna,Verino
	Others	Safari (deep red), Choco (chocolate), Tropic night (chocolate)

AVO Anthurium vogels, Holland	Red	Rosetta, Serge, Gloria, Tinora, Claudia, Nette, Alexis, Diana, Ingrid, Evita, Floriana, Yang, Jerichona
	Orange	Leigh, Gino
	Pink	Lydia, Suha, Patricia, Picasso
	White	Margaretha, Priscilla, Monique, Joy, Laurentian
	Obake	Gabriella, Melanie, Barbara, Eline, Maneula, Brittanie
	Green	Jeanne, Samanta, Simone, Marilene
	Pot varieites	Carre, Champion, Nevada
Andes Anthurium, Holland	Red	Elan, Altipano, Flame, Pumasillo, Lipstick
	Orange	Aymara, Vogue, Salasaga
	White	Cotopaxi, Grace
	Pink	Lucia
	Obake	Titicaca, Jewel, Akapana, Poopo
	Others	Esmeralda(green), Chicas (chocolate)
IIHR, Bangalore	Red	IIHR 243
	Orange	IIHR 26
	White	IIHR 51(fragrant), IIHR 139
Hawaiin varieties	Red	Ozaki, Kaumana, Kohara, Pohoa, Toyamo, Red Hot, Hayashii
	White	Manoa mist, Uniwaii, Chameleon, Jamaica, Frost, Trinidad
	Pink	Abe Pink, Blush, Marian seefurth, Candy Stripe, Lollipop, Lady jane
	Orange	Nitta, Sun burst, Diamond Jubilee

For commercial growing it is advisable to select a few varieties which are in good demand for export market. Deep red and bright orange has greatest demand and now a days bicolor are also gaining importance. The demand in the International market is in the order of Red Anthurium(40%) Pink (15%), White (13%), Orange (12%) and Others (8%).

Propagation

Anthurium is propagated through seed and vegetatively by stem cuttings or by separation of suckers. Plants can be multiplied in large number by micro-propagation techniques from the tender leaf explants.

Seed Propagation

Seeds are produced by hand pollination. Natural seed set is not profuse due to protogyny. Selfing or crossing can be made for seed better set. Developing of new varieties necessitates crossing. Seeds become mature within a period of 4-6 months after pollination. Mature seeds will have a pulpy coating or mucilage when ripe. The pulp is carefully removed without injuring the soft seed. Seeds are sown immediately after extraction. Seeds can be stored up to 1 week, but storage reduces germination. Sowing is done on a medium of clean fine sand or on a moist cotton pad. Seeds germinate with in 10 days. Sprouted seeds on cotton pad are shifted to a sand medium for further growth. Seeds can also be germinated in vitro in nutrient media. Seedlings will take two years or more for flowering. Due to high variability of the seedlings, flower quality cannot be assured.

Vegetative Propagation

Suckers: Anthurium is generally propagated through suckers that arise from the base of the mother plant. They can be transplanted at 4-5 leaf stage with 2-3 good roots in pure sand medium for one month and there after shifted to potting medium.

Suckering varies with varieties. Almost all cut flower varieties are shy suckering. Topping may stimulate suckering. To induce suckering on topped plants, growth regulators such as gibberellic acid (500 mgl^{-1}) or benzyl adenine may be used.

Division

Stem cutting (discs) obtained from fully grown plants of about 5-6 years age can also be utilized for propagation. The thick main stem is cut into 3-5 cm long discs. If the stem is very thick, the discs can be cut vertically. Each bit should have 2-3 nodes and minimum two lateral buds. Stem discs are treated with a fungicide solution and planted on a medium of clean river sand and leaf mould mixture. They develop shoot and root from the axillary buds with in 1-2 months which can be separated and planted independently when they reach 5-6 leaf stage.

Top portion of the stem along with leaves and a few roots that remains after the removal of intermediate stem can also be used as planting material.

Root stumps: After topping and separating the intermediate stem discs from older plants, what remains is the root stump. This can also be planted which produce sprouts, which on rooting can be separated and planted.

Tissue culture: By employing tissue culture technique, uniform plants can be successfully produced. For commercial growing, plants produced by tissue culture techniques are more advantageous as uniform, good quality flowers can be obtained early. In vitro axillary bud/ leaf culture in modified MS media is successful.

Environmental Requirements

Light

Anthurium plants prefer to grow under shade. The tolerable level of light in the tropical region during summer is 20-30 per cent. Bright filtered light is required for better flowering. Excess light causes yellowing and scorching of leaves eventually causing death of the plant. Very low light intensity causes excessive vegetative growth and low flowering.

Temperature: The plant thrives best under day temperature between 25 to 28°C. Day temperature above 35°C is harmful. Night temperature between 18°C to 22°C is ideal. Night temperature is more critical. It should not be less than 15°C.

Humidity: The plant prefers high humidity above 80%. However, continuous higher level of RH above 85% with out proper ventilation causes diseases. Humidity below 60% is unsuitable. In such cases, artificial arrangement to increase humidity through misting/ fogging is required. The plants also require proper aeration.

Growing Structure: It is preferable to grow anthurium in the open, under artificial shade structures for better growth and yield. 75-80% shade is ideal and achieved using shade nets. High rainfall increases the risk of infections to the plants and they should be protected from heavy rains by using poly sheet. Depending on locations and budget, shade house, low cost poly house or climate controlled green house can be employed. Covering the sides with 50% shade net improves the microclimate inside the shade house which may be raised occasionally to facilitate aeration.

Containers and Media

Planting in pots is preferred in the plains. Seedlings and sprouted cuttings of 5-10 cm height are transferred to pots. An ideal pot should be 30 cm diameter at top with 4 large holes at the bottom on sides for sufficient drainage. One seedling can be planted in a pot. But to provide a full appearance, two to three plants can be raised per pot. When plants are young, they are raised in 10 cm pot and as they grow can be shifted to bigger pots of size 15, 20, 25, 30 cm successively.

Media

A lose medium above ground is suitable. Anthurium requires a high organic well aerated medium with good water retention and drainage. An ideal medium should be light in weight and rich in organic matter. Other qualities required are high porosity, low salt concentration and good anchorage. Optimum pH of the media is 5.0 and EC= 0.6 m mhos/cm^2

When anthurium is grown in green houses, artificial media like rock wool, polystyrene granulates and peat moss is used. When grown in shade houses either on beds or pots, the combination of media varies depending on locations. The primary consideration for selecting the components of the media is local availability and suitability. The components of the media commonly used are old and chopped coconut husk (2.5 cm^2 size), saw dust, cocopeat, coir pith compost, charcoal, brick and tile pieces, peat, sugar cane bagasse, tree bark, wood shavings, coffee husk, cured coffee pulp, spent coffee, gravel, leaf mould, FYM, chicken manure etc.

Planting and Management

West cost in South India, Eastern and Western Ghats and North Eastern Hill region are the potential areas for growing anthuriums in India. Wynad, Coorg, Yercaud, Kalimpong are some regions where anthuriums are grown intensively. Anthurium cultivation is done mainly in pots and in beds.

Growing in pots: For terrace gardening and for small scale cultivation, this method is feasible. But for larger units, this will incur a lot of expense. A ten or twelve inch pot with 2-4 holes is preferred. A medium consisting coarse sand, cow dung, coconut husk, charcoal or brick pieces in 2:1:2:0.5 proportion can be used. At bottom one inch layer of coarse sand is spread. Over this, pieces of bricks or charcoal are arranged. The plant along with the roots spread on all sides is kept above this. The roots are then covered with a mixture of chopped coconut husk, sand and cow dung. Initially potting media is filled 1/3rd of the pot and subsequently with the growth of the plant, fresh medium is added to cover the roots gradually. The surface of the media may be raked with out disturbing the roots once in a month to improve aeration. The pot should be supplemented with growing media rich in organic matter such as leaf mould, coco peat (coir pith compost) etc once in three months. Young plants require repotting once a year, and adult 2 or 3 years.

Planting in beds: For large-scale planting, cultivation in beds is preferred at higher altitudes (about 1000m above MSL). Land with gentle slope with out water stagnation is selected. Planting is done on raised beds with a width of 1.2 m, 0.2-0.3 m height and convenient length. A path width of 75 cm can also be given in between beds which may also act as drains.

In Kerala, a 10 cm bottom layer of broken tile and brick pieces above which old and chopped coconut husk mixed with charcoal are provided. After planting, a media comprising organic manure, leaf mould and sand are used on top.

Spacing

Anthuriums are planted at a spacing between 30 to 60 cm depending up on the variety. 30 x 30 cm, 45 x 45 cm, or 60 x 60 cm spacing is given. Usually 3-4 rows are planted per bed widthwise. The roots are spread on the media and another layer of media is provided to cover the roots. Do not plant too deep. The plant population will be 50,000 to 60,000 plants/ha or 6-8 plants/ m^2.

Nutrition

Anthurium plants require manuring through organic and inorganic sources. Organic manures can be provided through fresh cow dung, or neem cake dissolved in 10-15 times of water and fermented for 1-2 weeks. Cow's urine can be diluted 20 times and used. The above solutions can be sprayed on the plants after filtering. Vermicompost, coir pith compost, leaf mould and microbial fertilizers are other options. Organic manures are to be sprayed at least once in a fortnight with the source changed each time in a cycle.

Nitrogen, Potassium, Calcium, Phosphorous, Magnesium and Sulphur are the important elements required for Anthurium nutrition. The nutrients can be provided either through irrigation water or through foliar application.

Young plants may be sprayed with a fertilizer mixture containing more of nitrogen with N:P:K in the ratio of 3:1:1. This may be provided @ 2g/l twice a week. After the plants start flowering, The ratio of N:P:K may be changed to 1:2:2 in the fertilizer mixture and the it may be provided @2-3g/

l twice a week. Commercial fertilizer mixtures available in the market such as 17:17:17, 18:18:18 can also be used. Slow release fertilizers, if used, need be given only once in 4 months. A deficiency of calcium can cause fading of the spathe color which can be corrected through application of lime @5g per plant per month.

For pot grown anthuriums, NPK mixture in the ratio of 1:1:1 @1% (10g/l) may be provided every month.

For green house grown anthuriums, a concentrated solution of NPK mixture 8:12:24 (200g), Potassium sulphate (20g) and magnesium sulphate (40g) is prepared in 100 l of water. This is diluted five times and sprayed on the plants or provided through irrigation. Apart from this, a monthly spray of $CaNO_3$ @ 0.5g /l/m^2 may also be provided.

Irrigation: The plants are to the watered twice daily during summer. Irrigation through drip, microsprinklers, mist or over head sprinklers is the best method. The water should be free from dissolved salt. Sodium and chlorine should be less than 5mmol/l. Slightly acidic water with pH 5.7-6.2 is preferred. Rain water is the best. Irrigation at late evening should be avoided. Reduce quantity during winter. In green houses, irrigation @ 3-5 l/day/m^2 through drip or mist is practiced.

Leaf Pruning

The leaves should not overlap each other. So rigid leaf pruning of older leaves retaining 3-4 young leaves (matured) can be practiced with out any adverse effect on flower production and quality. This will improve health of the plants, and provide better aeration. Healthy cut foliages can also be marketed and adds to the revenue.

Removal of suckers at young stage, cleaning of crown before rains etc is the other operations to be carried out in order to have a better growth and flowering.

Harvesting and Yield

The flowers are harvested with its long stem when 1/3 to 2/3rd of true flowers on the spadix are open, indicated by the change of color. Color change varies with the varieties. The spathe should be completely unfurled and the spadix should be well developed. Flowers are cut either in morning or evening. They are cut with the long stalk using a sharp knife retaining only one or two inches on the plant. The basal ends of the stalk of flowers are kept in water soon after harvest of flowers. A piece of cotton soaked in water is wrapped to the basal cut end to prevent desiccation while in transport. Spathe along with spadix is inserted into a polythene cover of appropriate size, which will prevent bruises on the spathe. The open ends of the polythene cover are secured to prevent movement and bruising of the spathe inside the cover. Till marketing, the flowers can be kept in cold storage at temperatures between 14-17°C.

The yield varies between 6 to 10 flowers per plant/ year which depend on variety, climate and management. Replanting is required once in 6-7 years. The price of spike depends on grade, quality and market. In Indian markets, the price varies from Rs. 5-20 for different grades.

Grading

Flowers having blemishes, black spots, discoloration etc are discarded initially and those having clean and shining spathe are graded based on size of spathe and stalk length. The grades used in important producing countries are given below.

Holland			Hawaii		
Grade	**Flower size (width-cm)**	**Minimum stem length (cm)**	**Grade**	**Flower size (width- inch)**	**Stems/ box (Nos)**
A	6-7	25	Peevee	< 2.5	50
B	7-8	25	Miniature	2.5-3.0	40
C	8-9	30	Small	3.0-4.0	30
D	9-11	35	Medium	4.0-5.0	25
E	11-13	40	Large	5.0-5.5	18
F	13-15	45	Extra large	5.5-6.0	15
G	15 & above	50	Premium	>6.0	8

The grading adopted in India is given below.

S. No	Grade	Size of spathe (cm)	Stem length (cm)
1	Mini	<5.0	20-25
2	Small	5.0-7.0	25-30
3	Medium	7.0-9.0	30-40
4	Big	9.0-12.0	40-60
5	Large	12.0-15.0	50-60
6	Extra large	>15.0	> 60

Packing

Flowers packed in corrugated cardboard boxes lined with polythene sheet and cushioned with news print shredding. The size of the box varies depending on grading and sorting code adopted by different producer countries. In India, boxes of size 60 x 30 x 23 cm are used. Flowers packed in these boxes are further packed in outer cartons known as master boxes. In one master box, six inner boxes can be packed. 11-60 flowers can be packed in an inner box depending on the size of flowers. In Dutch market, different box sizes ranging from 100 x 20 x 10 cm to 100 x 40 x 14 cm are used.

Vaselife

With proper post harvest treatments, the anthurium flowers can last for 2-3 weeks in vase. Various floral preservatives and growth regulators are used in anthurium for prolonging vase life employing pulsing and holding treatments. Pulsing treatments are done immediately after harvest and before shipment to reduce injury and to extend holding period. Pulsing with benzyl adenine (50 mgl^{-1}) for 12 hours, and triadimefon (25 mgl^{-1}) for 8 hrs were effective in extending vase life of anthurium flowers (Valsalakumari and Rajeevan, 2006) Silver nitrate and calcium nitrate treatment of the stem also increases post harvest life by up to 50 percent.

Among holding solutions, BA 25 mg l^{-1} and 8-HQC 30 mg l^{-1} resulted in prolonged vase life up to 27 days (Valsalakumari and Rajeevan, 2006). Apart from pulsing and holding treatments, wax coating, enclosing the inflorescence in polythene film, plugging the cut end of stalks which reduces water loss from the flower also increases post-harvest life.

CARNATION

Carnation is one of the important cut flowers in the international trade. Its array of colors and excellent keeping quality, ability to withstand long distance transportation and ability to rehydrate after prolonged shipping has made it one of the most demanded flowers. They are also suitable for bedding, pots, borders, edging and rock gardens. Carnations in general are grown only under protected cultivation.

Classification: Carnation *(Dianthus caryophyllus)* belongs to family Caryophyllaceae. Even though there are several types of carnations, only perpetual flowering carnations are commercially important. Perpetual flowering carnations can be broadly grouped into standard or sim carnations and spray carnations. Standard carnations produce single large size bloom on an individual longer stalk. In spray carnations, flowers are small and are produced in bunches on branched flower stalks. Standard carnations prefer cooler climate than spray varieties. Spray carnations can be grown in warmer regions.

Varieties

Varieties possess different colors and are selected depending on market demand. The present preference for color is in the order of pink (31%), red (15%), yellow (12%), bicolor (12%), white (9%), mix (8%), purple (5%), orange (3%), salmon (3%) and novelty (2%).

Standard Carnation

Standard Carnation	Red	Scania, Tanga, Killer
	Pink	Nora, Pink Sim, Candy, Calypso
	Yellow	Pallas, Murcia, Tahiti
	White	White Sim, Roma, Candy White
	Others	Charmer, Santiago, Vanessa
Spray Carnation	Red	Rony, Karma, Etna
	Pink	Annelies, Barbara, Silvery Pink
	Yellow	Odeon, Alicetta, Lior
	White	Royalette, Tibet, Excel
	Others	Exquisite, Scarlet Elegance, Kissi
Micro Carnation		Eolo, Pink, Wiko
Mini Spray Carnation		Lima, Onia, Roland

Propagation

Carnations are multiplied vegetatively through cuttings. Terminal cuttings which are healthy, 10-15 cm long with 3-4 nodes weighing around 10 g are ideal for multiplication. Plant bioregulators like IBA or NAA (500 mgl^{-1}) is used for better rooting of cuttings (Gopinath and Khan, 1997). Cuttings can be stored at 0^0C before planting for several weeks. Cuttings are spaced at 5 cm apart and intermittent misting should be done for good rooting. Cuttings normally develop good root system within 21 days. The rooting medium should be sterilized before planting. Drenching with fungicide is ideal to control fungal infection during rooting.

Climate

Carnations are grown in almost every climate. In temperate zones, they are grown in glass houses, in subtropical conditions they are grown in poly houses and open conditions and in tropical climate, under shade nets.

Temperature and light are the major factors, which influence growth and flowering of the plant. Temperature at night is critical in maintaining quality and it should be low. Mild climate with a moderate temperature is ideal for the crop. Optimum temperature during day is 28^{0}C and 16-18^{0}C during night.

Carnation is a facultative long day plant and flowering is more and faster in long days than short days. About 21.51 lux is the minimum natural light intensity required for adequate photosynthesis of carnations. During winter, light intensity should be higher and during summer, the temperatures should be mild.

In India, places having cool climate like Banglore, Coorg, Kodaikanal, Ooty, Yercaud, Pune, Nasik and North Eastern hills are suitable for establishing commercial units for cut flower production.

Soil

Soils with good drainage are suitable for the cultivation of carnations. A rich sandy loam is considered to be the most ideal. Clayey soils should be amended with sand and organic manure. Optimum soil pH is between 5.5 - 6.5. Ideal EC of soil during vegetative phase is 1.2 ms/cm and during the generative phase, it is 1.5-1.7 ms/cm.

Planting and Aftercare

Carnations can be produced during January to April from the planting in August –September (Singh, 2001). Soil should be ploughed up to 50 cm deep. Raised beds with 1 m width and 15-20 cm height and of required length are made. An alley space of 45 cm is left between beds.

Spacings of 12.5 x 15 cm, 15 x 15 cm, and 15 x 20 cm are followed in carnation. Normally, 30-40 plants / m^2 are ideal under shade nets. Shallow planting is adopted in tropical areas. Shading should be given in the beginning of crop for a few days. Care should be taken to maintain humidity by overhead spraying of water to prevent plants from drying after planting.

Support Materials

Flowering stems of carnation has tendency to bend if supports are not provided. They bend and droop instead of growing upright. In a single season crop, 4-5 layers of net may be required to support the stems without drooping to sides. Good support material is metallic wire woven with nylon mesh. At every 2-3 meters, the wire should be supported with poles. Poles at both the ends of bed should be strong and usually made of concrete.

Metallic wire is tied along the bed length using supporting poles. Across the bed, nylon wires are woven like net. For an optimum support, an increasing width of meshes can be used. Bottom net can be of 7.5 x 7.5 cm, then one net each of 10 x 10 cm and 12.5 x 12.5 cm and the upper most can be 15x15 cm.

The nets are made at the right time, should not be early or late. Observe the crop closely and raise the nets one after the other in stages.

Irrigation

Over watering and poor drainage cause stunted growth and root decay. Water logging would cause deprival of oxygen to plants. The growing medium should be evenly moist. During the first three weeks of planting, overhead irrigation can be resorted to prevent the young plants from drying up. Thereafter, irrigation through drip can be provided. Established plants require 6-7 liters of water / m^2/ day.

Pinching

Pinching refers to nipping of terminal bud and encouraging growth of side shoots. This is a mandatory operation for spray carnations. If left unpinched, the main shoot ends up in a flower. Depending upon the need of crop spread a) single, b) one and half and c) double pinches are given. Ideal time for pinching is morning. When the plant attains 6 nodes, the first pinch is given leaving 5-6 leaf pairs on the plant. This is referred as 'single pinch'. This would give rise upto six lateral shoots. It takes 3 weeks from planting of rooted cuttings until first pinching. With a 'one and half pinch', 2-3 of these lateral shoots are pinched again. This can be done about 4 weeks after first pinching (Gill, 2001). For the 'double pinch', all the lateral shoots are pinched off. Fungicide dressing of the pinched plants with Captaf or Carbendazim (0.1%) is recommended.

Disbudding

This is an essential practice in standard carnations to get large sized quality blooms. Disbudding refers to removal of undesirable immature side buds so that the central/terminal flower bud receives maximum food for full development.

Manuring

Regular supply of nutrients is essential through out the duration of crop. A fertilizer dosage of 250g N, 80g P and 200g K, 125g Ca and 400g Mg / m^2/ year is recommended in addition to 5kg of FYM / m^2 for growth and production of quality flowers. The total quantity of fertilizers can be evenly split and applied at fortnightly intervals. Organic manures and basal dosage of fertilizers are incorporated at the time of bed preparation.

Harvest and Post Harvest

After planting, it takes normally 110-120 days to come to peak flowering. A commercial carnation plant is capable of producing 10 to 20 flowers / year. Saleable stems up to 140 numbers / m^{-2} can be harvested.

Standard carnation flowers are cut when they are still in tight bud stage with petals emerging. Spray type carnations are cut when two flowers are open and remaining buds are showing color. The best time of harvesting is in morning. Flowers are harvested every two days and they are graded and bunched.

Harvested flowers are bunched together based on their physical measurements like stem length, flower diameter etc. Standard carnations are bunched into 20 stems / bunch while sprays are bunched into 10 stems per bunch. The bunches are covered with plastic or paper sleeves. After bunching, stem ends of flowers are trimmed and placed immediately in a bucket of preservative solution. The solutions for carnations should be acidic (pH 4.5) with 2-5% sucrose and a biocide not phytotoxic to carnations. The sprays are kept in preservative solution for 3 to 4 hours for conditioning. Commercial preservatives like Chrysal AVB or Florissant 100 @ 0.2% can be used.

Carnation flowers can be stored before marketing. For this, flowers have to be packed in cartons lined with polyethylene pre-cooled without lid. The plastic is then loosely folded on top of stems and the lid is closed. These cartons are stored in cool chambers designed to maintain 5-8^0 C with good air circulation and a constant relative humidity of 90-95%.

Plant Protection

Disorder

Calyx Splitting

Cultivars with too many petals are susceptible to calyx splitting. Petals are deprived of their support which results in bending and shape of the flower is lost. This happens due to boron deficiency. Varying temperature and environmental conditions also influence calyx splitting. Selection of cultivars less prone to splitting, regulation of temperature and maintenance of optimal fertilizer level can minimize these disorders. This can also be reduced by placing a rubber band or use of 6mm wide clear plastic tape around the calyx of flowers which have just started opening.

CHRYSANTHEMUM

Chrysanthemum is a very popular flower crop through out the world. It ranks 2nd / 3rd in international trade, next to rose and close to carnation. Long stem varieties with large flowers are grown for cut flower and exhibition purpose while small flowered varieties are grown for making garlands and religious offerings. Plants are also used in landscaping and exhibition purpose. Erect tall growing types are suitable as a component in border for back ground planting. Dwarf varieties are suitable for front row planting in border, as flower beds and for pot culture.

In India, it ranks 3rd in acreage after Rose and Jasmine. Chrysanthemum performs remarkably well in Indian plains and is mainly cultivated in open field condition for loose flowers. Chrysanthemums are also grown in green houses for cut flower production but only under a few acres. It is mainly cultivated in TamilNadu, Karnataka, Maharashtra and Andra Pradesh and some parts of North India.

Classification

Chrysanthemum (*Dendranthema grandiflora*) is a highly versatile and adaptable plant belonging to family Asteraceae and has wide range of size, type and color of flower. Chysanthemum inflorescence is a composite cymose cluster with a stalk or peduncle and the florets borne on a head or capitulum. Florets which have distinct petal are known as ray florets and contain only female flower parts. Florets with extremely short petals are called disc florets and contain both male and female flower parts.

1.Classification based on ray-disc and arrangement of florets

The flowers are classified based on ray-disc and arrangement of florets into 5 broad groups. (National Chrysanthemum Society, England).

a. *Single:* Have one or more outer pistillate flower (ray) with disc florets at centre.

b. *Anemones:* Similar to singles except disc florets are elongated and tubular forming a cushion. Disc flowers may be same or a color different from ray flowers.

c. *Pompons:* A globular head formed by short uniform ray flowers, the shape is formal, disc flowers are not apparent, further classified into three distinct sizes,

   1. Small buttons (4 cm or less in diameter)
   2. Intermediate (4-6 cm in diameter)
   3. Large (6-10 cm in diameter)

d. Decorative: Have a floret arrangement similar to pompons since they are composed mainly of ray flowers but the outer rows are longer than central flower giving the flowers a flatter appearance or informal shape. Sizes are mostly intermediate and large.

e. Large flowered: Bloom size is more than 10 cm, disc florets are not apparent in most of these forms and are classified to many shapes.

(i) *Incurved double:* Globose and formal with ray flowers similar in size to disc flowers and that curve inward and towards the top.

(ii) *Reflexed double:* Less formal and globose than the incurved double with over lapping ray flowers curved downward except for the ray flowers.

(iii) *Tubular ray flowers:* These are further grouped into

(a) *Spider:* Ray flowers tubular and elongated into the outer rows but short in center. The dropping outer row ray flowers are some times hooked on ends.

(b) *Fuji:* Similar to the spider except that the ray flowers may be shorter, droopless and lack hooks on ends.

(c) *Quill:* Tubular ray flowers, long on outside and short near the center, resembling feather quills. Ends of flowers are open and not flattened.

(d) *Spoon:* Similar to the quill except outer row flowers are open and are flattened, resembling a spoon.

f. Miscellaneous: Novelty types consisting of feathery plume, like hairy ray florets.

2. Classification based on size of flower

Based on flower size chrysanthemums are classified as a) Large flowered b) Small flowered. Further classification of large and small flowered chrysanthemum is based on shape and floret arrangement.

Large flowered: Incurved, Reflexed, Intermediate, Quilled, Spider, Ball, Japanese or Irregular

Small flowered: Korean, Korean Double, Decorate, Anemone, Pompons, Button, Quilled, Spoon, Semi Quilled, Chineraria, Stellate, No stake- No pinching type.

3. Classification based on architecture of growth

a. Standard chrysanthemum: Plants which produce a single large flower with a long stalk, and other buds removed as and when they arise from laterals for increasing flower size.

b. *Spray chrysanthemum:* Flowers are produced in clusters of 7-8 flowers and consist of pompon type flowers. The main apex bud is removed and lateral buds are allowed to develop into more shoots which in turn produce flower clusters.

c. *Pot cherry chrysanthemum:* Small flowered dwarf chrysanthemum with 18-27 cm height are used for bedding or potting. The flowers are produced in clusters with 18-30 flowers.

4. Classification based on duration of crop

Indian varieties are classified depending upon duration as early, medium and late varieties

Early : The varieties bloom in 90 days from final transplanting

Medium : 100-110 days are required to bloom from final transplanting

Late : More than 110 days are required to bloom from final transplanting

5. Classification under green house cultivation

Green house chrysanthemums are classified as Single, (Margaret types), Anemone, Decorative (Dahlia type), Spider, Pompon, Santini, and Pot chrysanthemum.

Varieties

Varieties popular in international market are of single types and these are called margaritas, the color of which differs inside and out side the disc. Flowers exist in different colors. White is the most preferred color for cut chrysanthemum varieties in the international market. Other colors which are in good demand are yellow, purple and pink. Spray types also command a major share in international market.

There are several varieties introduced to India. Some of the varieties are also bred in India. A list of varieties developed by various institutions and for various purposes is given in the following Table.

Propagation: Chrysanthemum is propagated either through root suckers, terminal cuttings or micropropagation. Seeds are also used for raising some annual types and hybrids.

a) *Suckers:* When the stem is cut back after flowering, it induces production of suckers from its base. These can be separated from mother plants and planted in prepared sand beds. Well rooted suckers can be transplanted directly to field.

b) *Terminal cuttings:* Cuttings are taken from healthy stock plants from middle to end of June. Terminal cuttings of 5-7 cm length are taken from flowering shoots. Basal leaves are removed and the laminas of upper leaves are clipped half. Dipping in IBA (2500 ppm) combined with misting improves rooting. The cuttings are planted in sand beds under semi shade and irrigated regularly. They are transplanted to main field when they are properly rooted.

Institution/Type	Variety
IIHR, Bangalore	Arka Swarna, Arka Ravi (hybrids), Indira, Chandrika, Rakhee, Ravikiran, Red Gold, Yellow Gold, Yellow Star, Chandrakantha, Nilima, Pankaj, Ushakiran
NBRI, Lucknow	Agnisikha, Navneet, Shanti, Y2K, Kargill 99, Sadbhavana, Appu, , Bindiya, Birbal sahni, Mother Theresa, Dina, Mini Jessie, Little Darling, Kusum, Indiana
IARI, New Delhi	Poornima, Kirthi, Chandirki
TNAU, Coimbatore	CO1, CO2, MDU1
Introductions Large flowered	Fred Yule, Snow Ball, Star Yellow, Super Giant, Temptation, Sunrise
Small flowered	Flirt, lilith, Sonali, Tara.
Pot chrysanthemum	Julia
Cut flower varieties	
White	Paragon, Tikro, Snowdone white, White Reagan, Cassa cream, Polaris
Yellow	Yellow Paragon, Yellow Spider, Reagan Lemon, Sunney, Sunshine
Red	Rosalis, Dark Splender, Reagan Cherry, Leman's Red , Dark Flamence
Green	Greeno
Pink	Reballet, Reagan, Money Maker
Orange	Reagon Orange, Tigro
Spider	Bartiz, Vesuvio , Yellow Venn

Climate

Light: Flowering of chrysanthemum is very much influenced by quantity and quality of light. Most of the cultivars are photo sensitive and classified as a short day plant (10 hours day light) and require short photo periods for flower bud initiation and development. Under long day condition (light period > 14 hours) they tend to remain vegetative.

When length of dark period exceeds 9.30 hours, flower bud initiation occurs and the bud development takes place when dark period exceeds 10.30 hours. When grown in open, planting time should be adjusted so that cultivars are exposed to long days during vegetative phase and short days for flowering. There are also varieties which are not photosensitive and flower at certain season throughout the year. It can be grown successfully up to an elevation of 1200 MSL.

Temperature

Based on temperature requirement, varieties are classified into three types

Thermo zero cultivars: Varieties which flower at any temperature between 10-27 ^{0}C but most constantly at 16 ^{0}C night temperature.

Thermo positive cultivars: A minimum of 16^{0}C is required for initiation and at 27 ^{0}C there will be rapid initiation but delayed flowering.

Thermo negative cultivars: Bud initiation occurs at low or high temperature between 10 and 27 ^{0}C but continuous high temperature delays bud development.

Soil: Well drained sandy loam soil with a pH of 6.5-7.0 with high organic matter content is suitable. It is a shallow rooted plant and is very sensitive to water logged condition.

Planting

Cultivation of chrysanthemums are done both in open field condition for loose flowers and under green house conditions for cut flowers.

Under protected cultivation, in tropical and sub tropical climate they are grown in poly houses in which temperature is regulated partially by evaporative cooling system. Artificial light and shading are also provided.

In Tamil Nadu, Karnataka and Maharastra, small flowered types are grown commercially in open field conditions. Since chrysanthemum is a short-day plant, it normally flowers during September to December under South Indian conditions. Hence, planting during May- June is recommended.

The field has to be ploughed thrice during March and beds or ridges and furrows are formed. Suckers or rooted cuttings obtained from previous crops are used for planting. Before planting, roots of suckers or cuttings are dipped in a fungicidal solution to protect against wilt. They are planted at a spacing of 30 x 20 cm in beds or on one side of ridges under open conditions (Rangasamy, 1997). Spacing varies from15-30 cm in green houses depending on whether standard or spray types with a population of 12-16 plants / m^2 in standards and 20-25 plants / m^2 in spray types.

Irrigation

The plants need adequate water during active vegetative growth when new leaves are formed. After formation of flower buds, no further leaf is formed and less amount of water is needed. Hence, chrysanthemums are to be irrigated twice a week in first two weeks and subsequently at weekly intervals.

Manuring: Chrysanthemum is a heavy feeder and hence they are to be adequately manured. For open field cultivation, apply 25t of FYM along with 100:160: 80 Kg NPK/ha. Half of N and the entire quantity of P and K are to be applied as basal just before planting. The other half of N is top dressed 30 days after planting the suckers. The same dose can be repeated if a ratoon crop is raised and hoeing should be done once in a month.

For cultivation under protected conditions, fertilizers are provided through irrigation. The nutrient requirement (mgl^{-1}) of cut chrysanthemums is given below.

Age(weeks)	N	P	K	Ca	Mg	B
0-3	120	30	220	78	20	0.33
4-7	130	30	230	78	20	0.33
8-12	150	30	250	78	20	0.33

Special Practices

Staking: Whether grown in pots or field, cut chrysanthemums require staking to keep plants erect and to maintain proper shape of plants and bloom. Supports are prepared mostly from bamboo sticks. Only dwarf varieties do not require staking.

Pinching: This indicates removal of apical bud with an objective to encourage side branches in spray chrysanthemums. Pinching increases number of flowering stems in each plant and it can indirectly control flowering date, number of blooms and quality. In spray chrysanthemum raised through suckers, 2 pinchings are required to encourage lateral growth. This is first done at 4th week after planting and 2nd, 7 weeks after planting. For rooted cuttings, pinching is done 2 weeks after planting. Pinching may be repeated to encourage further branching. In standard chrysanthemum only a single central bloom is desired on a stalk and pinching is not done in such plants. Single pinching is done if 2 flowers are desired and double pinching is done for 4 flowers.

Depending on number of blooms desired, degrees of pinching vary from hard, soft and roll out pinching. In soft pinching the top soft tips of shoot along with 2-3 open leaves are removed. In hard pinching, a longer portion up to hard shoot is removed.

Disbudding: This refers to removal of immature flower buds which are not desired. The objective is to get a few numbers of large flowers. These operations are mostly performed for large flowering and decorative type Chrysanthemums. In standard types, terminal buds are retained and regular disbudding of axillary buds are done to get one large flower. In spray types, large apical buds are removed and the axiliary buds are allowed to develop. Disbudding is done to remove side branches which arrive from axillary buds so that the no of flower is limited and blooms of better size are obtained.

De-shooting: This is practiced to reduce number of branches for improving size and form of flower. Lateral buds and side shoots are removed at their early stage of growth from time to time.

De-suckering: This is mainly practiced in standard chrysanthemums. During vegetative growth phase, suckers develop from base of plants. For proper and vigorous growth of plants, suckers are removed periodically. This is practiced to allow single stem to develop up to a certain height. Without de-suckering, the main plant loses vigor and becomes weak.

Application of growth regulators: Growth regulators can be employed for increasing stem length of flowers as well as for early flowering. Application of GA (5-10 mgl^{-1}) increased stem length of standards while GA (100 mgl^{-1}) was useful in promoting earlier flowering. Spraying of MH (250-1000mgl^{-1}) significantly increased number of flowers per plant (Sharma *et al.*,1990)

Harvesting: Chrysanthemum starts flowering after 3rd month of transplanting. Flowering season varies from region to region under open field cultivation and extends from August to February. Harvesting for loose flowers is done at 4 days intervals and about 15 harvests can be done from a crop. Yield of loose flowers ranges from 8-10 tonnes / ha. Flowers are packed in bamboo baskets and gunny bags. Flowers are heaped on the floor and stored in whole sale market at room temperature. Longevity of flowers is upto 4 days under ambient conditions.

Cut chrysanthemums such as standards and sprays are harvested when 40-50% blooms are fully open. Dwarf varieties are harvested when 80-85% blooms are opened. They are cut with long stem leaving 10 cm on plant. The lower leaves are stripped off and the cut end of stalk is placed in water immediately. Yield of cut flower stems in a hectare for standard types are up to 2.5 lakhs while in cut flower sprays it is upto 1.5 lakhs.

In spray varieties, after harvesting the first crop of flowers, the plants are cut upto ground level and a ratoon crop can be taken after providing organic manures.

Grading: Flowers are graded based on stem length, appearance of flowers, and number of flowers, color and freshness of flowers. The flowers are protected with a cover of transparent polythene sleeve and bunched in 5-10 numbers depending on grade. They can be stored upto 6 weeks at 0.5 ^{0}C.

Vase solution comprising 8HQC (200 mgl^{-1}) + citric acid (75 mgl^{-1}) + sucrose (5%) is useful in extending vase life up to 50 days.

CROSSANDRA

Crossandra flowers are very popular for their bright color, light-weight and keeping quality. Deep orange colored flowers are of great demand for garlands and hair adornments. Crossandra with jasmine adds charm, fragrance and contrast to the garlands. The plant is also used in landscaping as a potted plant and for bedding and border.

The crop is commercially grown in India, Srilanka and other tropical countries for flowers. Karnataka, Tamil Nadu and Andhra Pradesh are the major producers.

Classification and Varieties

Crossandra belongs to family Acanthaceae and important species are *Crossandra undulaefolia, C. guineensis and C. nelotila.* Among these, *C. undulaefolia* is cultivated in India for flower production.

Yellow Orange, Lutea Yellow, Delhi (triploid) and Sebacaulis Red are the important varieties. Two hybrids are released from TNAU. Delhi is a tripolid and others are tetraploids.

Climate and Soil

These are tropical plants, grown around 30 ^{o}C. Crossandra could be cultivated in a wide range of soils. Well drained fertile, red sandy loam soils with pH of 6 to 7.5 are ideal for its cultivation.

Propagation

Propagation is through seeds or stem cuttings rooted under mist. Normally, commercial orange type is propagated by sedsIn triploid varieties like Delhi, cuttings are used. Cuttings should be transplanted when sufficient roots are developed.

Seeds and Sowing

Fresh seeds are sown at a depth of 1 cm during July - October in raised nursery beds 15 cm apart in lines. Watering should be done daily. Seedlings germinate in 10-15 days and are ready for transplanting in 45-60 days when they have four or five pairs of leaves. 4 cents onursery is enough for 1 ha.

Seed rate: Seed rate is 5 kg/ ha

Cultural Practices

Preparation of field: The land is ploughed three or four times. FYM (25t/ ha) is incorporated and mixed well in the soil. Furrows and ridges are prepared 60 cm apart. Dip the roots of seedlings in Carbendazim (1 g/l of water) and plant on the ridge at 60 cm spacing. For Delhi crossandra, a spacing of 45x 40 cm is followed.

Manuring

Apply FYM (25 t/ha) as basal at the time of preparation of land. A fertilizer dose of NPK @ 100:60:60 Kg/ha/year is recommended for the crop. Of this, 33.3: 60: 60 kg/ha N: P_2O_5: K_2O is applied as basal. The crop is top dressed twice with 33.3 kg N / ha each time, the first at 3 months after transplanting and the second 8-9 months after transplanting. Application of fertilizers is to be necessarily followed by irrigation.

For Delhi Crossandra, apply FYM @ 25 t/ha, Gypsum (100 kg/ha) and P & K at 50 and 100 kg/ha as basal. Apply Neem cake @ 250 kg and N 40 kg/ ha 30 days after planting. Another top dressing is given 90 days after planting with N P K @ 40:20:60 kg and repeat this dose at quarterly intervals for of two years.

Iron deficiency is a common problem in crossandra. Application of $FeSO_4$ + 2% urea once in 3 months improves flower yield. Foliar spray of $ZnSO_4$ (0.5%) and $FeSO_4$ (1.0%) recorded highest yield in crossandra (Aruna, 1999). Spraying of ascorbic acid (1000 mgl^{-1}) is beneficial for obtaining higher yield.

The plants are irrigated once in 4-7 days depending on season. Weeding, application of fertilizer and earthing up are combined together for easy maintenance of the crop. Providing of partial shade improves yield of flowers.

Harvest and Yield

The plants start flowering 2-3 months after planting and continue to bear flower throughout the year, and flower production drops during rainy season.

Harvesting commences from third month after planting. Flower opens in sequence from base of spike. Two days are required for complete opening of flower. Fully opened flowers remain fresh on the plant for 3 days. Picking of flowers are done on alternate days. It requires 15-25 days to harvest complete flowers in a spike. Flowers are picked early in the morning by pulling the corolla out of the calyx. Flowers will be available for picking for six months in a year. The crop can be retained in the field for about 3 years including the ratoon crop in third year. About 3- 5 t/ha of flowers can be harvested each year. Delhi Crossandra yields 2800 kg of flowers / ha / year.

After three years the crop has to be replanted. However it is better to replace the crop every year to maintain healthy plants and high yield. About 15000 flowers are required to make a kilogram.

GERBERA

Gerbera is one among the most important cut flowers grown commercially throughout the world. The wide range of color of flowers and the beautiful blooms with long peduncle make them attractive in floral arrangements. The cut flowers also have a reasonably long vase life. In landscaping, gerberas are ideal for beds, borders, pots and rock gardens.

Classification: Gerbera jamesonii (Transvaal Daisy) is a tender, stemless, perennial herb which belongs to Asteraceae family. Flower heads are solitary. Flowers possess a wide range of colors including yellow orange, cream, white, pink, brick red, scarlet, maroon, terracotta and various other intermediate shades. Based on flower heads, they are classified into single, double and semi double cultivars. The double cultivars also have bicolor flowers. Miniature gerberas known as 'germini' which are suitable for smaller flower arrangement are also popular.

Green house Varieties

Flower color	Varieties
Red	Diablo, Ruby Red, Rubian, Sangria, Savannah, Thalassa, Testarossa
Yellow	Cabana, Dana Ellen, Golden Gate, Piton, Solero, Tonnekke, Thalassa

Orange	Careera, Goliath, Mistique, Ornella, Rodis, Sunset, Sunway
Pink	Charmander, Fleurance, Pink Ellegance, Pink Sparklet, Primrose
Lilac	Magnum, Ringo
Cream	Tiramisu, Vital
White	Dalma, Himalaya, Junkfrau, Winter Queen
Magenta	Aquila, Tonga

Propagation: Gerbera is propagated through seeds and clumps and micropropagation. Cuttings can also be rooted and used as planting material. A faster and easier method of propagating *Gerbera* is via division of clumps. The *Gerbera* produces numerous suckers, which can be split into many individual plants. In the process of dividing the plant, try to give each sucker some roots and be as careful as you can so as to minimize any damage to the plant.

In commercial cultivation under green house, tissue culture plants are advised.

Climate and Soil

Light and temperature are the critical environmental factors. High light intensity results in compact growth, shorter stems and discoloration of flowers. In sub tropical and tropical climate, gerbera can be grown under open conditions. But growing under partial shade is recommended for quality blooms. To control light intensity and solar radiation, shade net 50-70% is recommended depending on locations. Poly films may be provided as an additional cover for protecting plants during monsoon. In temperate climate, gerberas are grown under green house.

Ideal temperature for gerbera cultivation is between 23-26 ^{0}C. Flower bud initiation is reduced below 10 ^{0}C and above 35 ^{0}C. Low night temperature favors production of suckers. Optimum humidity for maintaining health of plants is 80-85 %.

Aeration : Gerbera is very sensitive to an oxygen poor environment. Sufficient ventilation space is required on top and sides of growing structure.

Soil: Soil media should be rich, highly porous and well drained. Soil pH should be between 5.5 to 6.5 and should be maintained to get maximum absorption of nutrients.

Planting

Exotic cultivars require more care in their management and they are cultivated in poly house or green house. Gerberas are grown on raised beds. Beds of size 60 cm (width), 45 cm (height) and of convenient length are made. A spacing of 30-40 cm is provided between beds.

The beds should be prepared in such a way that it should be highly porous, well drained and airy. Gravel can be used as the bottom most layer for proper drainage. Coco peat and organic manure are added to improve soil structure and nutrition. Composition of the bed should be such that it consists media, moisture and air in 1:1:1 proportion.

Before planting, sterilization of soil in the beds is necessary to guard against soil borne pathogens such as *Phytopthora* and *Fursarium* and nematode. Soil solarisation by irrigating the beds and covering with plastic sheets for 6-8 weeks is effective. Steam sterilization or fumigation of soil with formaldehyde or methyl bromide (7.5-10l / m^2) can also be done. Chloropicrin can also provide significant control (Duniway, 2002). The beds are covered with plastic sheet for 2 weeks and then the soil is thoroughly flushed with water to avoid traces of bromide.

Spacing: Gerberas are planted at a spacing of 30cm between rows and 30 cm between plants (Gurav *et.al.*, 2005) with a population of 6-8 plants per sq. m. Generally 2 rows are planted / width of bed. While planting, the heart or crown of plant should be in level with soil or slightly above soil level. Deeper planting results in crown rot. Plant the seedlings without disturbing root ball.

The plants are properly shaded after planting till the plants start to grow and the humidity should be kept at 70-80%.

Irrigation

Immediately after planting, irrigate the plants with overhead sprinklers for one month and gradually change to drip irrigation. In hot summer, foggers can be employed to maintain humidity of air. The soil should always be moderately moist without excess and the soil surface should dry out between watering. pH of irrigation water should remain near neutral and the EC should nor exceed 2dS/mThe plants require approximately 700 ml water / day.

Manures and Fertilizers

Organic manures are applied as basal at the time of bed preparation. FYM /cocopeat @75t/ha or 7.5 kg / m^2 is incorporated. For open field cultivation till 2-3 weeks, no fertilizer is applied. From 3-12 weeks, NPK are applied at 20:25:20 g /m^2 at monthly interval and 15:10:30 g NPK/m^2 from 4th month at 15 days interval.

Under polyhouse condition, apply NPK 20:20:20 mixture @ 0.4 g/plant on every alternate days for first three months during vegetative phase to have a better foliage. Once the plants come to flower, apply NPK 15: 8: 35 mixture @ 0.4 g/ plant on alternate days.

Depending on deficiency, micronutrients such as Boron, Calcium Magnesium and Copper can be applied as foliar spray either weekly or fortnightly (0.15%) for better quality blooms.

Harvesting: After planting, the first flowers are produced in 3 months and the plants yield upto 2-3 years. Average yield under green house is 200 flowers / m^2 / year. Under open conditions, flower yield is low @130-150 flowers/ m^2 / year.

Harvesting is done when 2-3 rows of stamens are entirely developed. The flowers are pulled with the long stalk in the morning or late evening. The basal 5-6 cm of stalk is cut and placed in a bucket containing a holding solution of cold water (9-12 ^{0}C) and sodium hypochlorite (1%). The flowers are stored at 80.0% RH. Cut flowers have a vase life of 8-10 days. The use of 6% sugar+ 200 mgl^{-1}8-HQC as apreservative has shown to prolong vase life (Reid, 2004). The flowers are then graded and packed in a box of size 98 cm (l) x 30 cm (w) x 12 cm (h) @50 numbers / box. A good flower will have a diameter of 10-12 cm and stalk length of 50 cm.

GLADIOLUS

Gladiolus is grown for its attractive flower spike having florets of huge form, dazzling color and varying sizes, with long vase life. It is a popular international cut flower crop grown mostly outdoors. The plant is also used in landscaping for bed, rockery, or pot planting.

Classification

Gladiolus belongs to family Iridaceae. In Europe it is classified as *Gladiolus grandiflorus, G. primulinus,* Butterfly, Miniature and Face ups. The American system uses a three digit number to describe a cultivar. In India it is classified as large flowered types, butterfly and miniatures.

Varieties

There are a number of varieties in gladiolus. Theses are either introduced to India or are developed indigenously.

Sources	Varieties
Exotic varieties (Introduced)	Friendship, Spic and Span, Mansoer Red, Dr.Fleming, Peter Pears and White Friendship
IIHR, Bangalore	Apsara, Arathi, Poonam, Nazrara, Sapna, Meera, Shobha, Sindhu, Darshan, Shakthi, Sindur, Tilak, Dheeraj, Sagar, Kumkum and Arka Kesar
IARI, NewDelhi	Mayur, Agnirekha, Suchithra, Sanjivani, Dhanwanthari, Noopur, Neelkanth, Pusa suhagin , Sarang, Swetha, Vandana, Sunayana
NBRI, Lucknow	Arshana, Basnth Bahar, Indrani, Kalima, Manmohan, Manohar, Neel kanth, Muktha, Jwala, Mohini, Manisha, Pithamber, Priyadarshini, Rashmi, Sada Bahar, Sanyuktha, Triloki
Cut flower varieties	White friendship, Priscilla, Pink Friendship, Ben Venuto, Peter Pears, Desico, Fidelio, Sancerre, Eurovision, Hunting song, Teach, Jessica, Nova lux, Maseagni
Varieties based on color	
Pink	Apple Blossom, Friendship, Rose Supreme, Suchithra
Red	Eurovision, Oscar, Patricia, Pusa suhagin, Red fleming
Orange/ Scarlet	Happy End, Hunting Song, Peter Pears
Yellow	Nova lux, True Yellow, Yellow Stone
Blue/ Mauve	Blue lilac, Her majesty, Priscilla, Tropic Sea
White	White Friendship, Snow Princess, Sun rise, White Prosperity

Propagation

Propagation is done through seed and vegetative means. Seed propagation is done for breeding and development of new varieties.

Gladiolus is propagated usually through corms and cormels. Size of corm markedly influences growth and flowering of gladiolus. Medium and large sized corms are preferred for planting, as small corms produce only small flower spikes. Larger corms produce earlier flowering, longer spikes and more number of flowers / spike. Underground corms with 4-5 cm diameter are required for good spikes. Top shaped corms with high crown give better flowers than flat ones (Panwar and Saxena, 2001). Corms can also be divided and used for planting.

Each corm produces a few to several cormels (baby corm) from which also plants come out. However small cormels produce flowers only after 2 or

3 (planting) season. In India, corms produced after flowering of a crop are frequently used as planting material for subsequent crop. However, it is preferred to grow separate crops for cut flower and corm production.

Micropropagation is used for production of large scale disease free, planting materials.

Dormancy

Corms and cormels have a period of dormancy for 3-4 months and 5-6 months respectively. Pre planting treatment with ethylene chlorhydrine (3%) ethrel(1000 mgl^{-1}) or thiourea (1000 mgl^{-1}), GA (100 ppm) can remove dormancy (Seghal, 2001). Soaking corms in hot water gives early sprouting.

Environmental Factors: The development of flower spikes and individual florets are affected by environmental factors like light and temperature.

Light : Even though the plant is not photo sensitive, day length influences growth and flowering in gladiolus. Long photo periods are beneficial in flowering and quality of flowers. Full sunshine is essential for cultivation. It prefers high light intensities and can be grown usually under open conditions. Shading reduces flowering and floret numbers.

Temperature: The plants are tolerant to higher temperature as long as humidity and soil moisture are optimum. A temperature range of 15-25 °C is ideal. Day temperature between 19-25 ^{0}C and night between 15-18 °C are better. Varieties that tolerate higher temperature are Nova lux, Peter Pears, White friendship etc.

Soil

Gladiolus can be grown in a wide range of soils, light sandy to clay loam. Soil rich in organic manure and nutrients are preferred. Deep well-drained acidic soils with a pH of 5.5-6.5 are the best for cultivation. EC should not be more than 2d/sq.m. Soil borne diseases are many and so soil solarisation is desirable.

Planting Season

The planting season depends on locality. Under Bangalore conditions, planting can be done round the year. However, June, late September, October and November are very suitable. In hills, the best period is from February-April. In Pune region, September- November is ideal. In North Indian plains, planting could be done from September to 2nd week of October.

Planting

The land is ploughed two to three times and FYM (25 t/ha) is applied and mixed well with the soil. Ridges/raised beds are made at a width of 1m (4 rows / width).

Bigger corms are selected (4.5-5.0 cm diameter). Corms may be sprouted indoors before planting. Sprouting in sand can also be done. Sprouting can be improved by treating the corms with GA_3 (1000 mgl^{-1}). Sprouted corms will have root like projections on lower side of corms. Before planting, scales should be removed from corms and they are treated with fungicides like Dithane M45 (0.2%) or Bavistin (0.2%).

Spacing

Spacing and depth of planting vary with corm size, cultivar, climate, soil type and irrigation method. The planting should not go deeper. To get more cormel production, shallow planting is recommended. The corms are planted at a depth of 3 - 5 cm. Spacing adopted varies 20-30cm between rows and plants. In green house cultivation, high density planting (40-50 plants / m^2) with a spacing of 20 cm x 10 cm is adopted.

Manuring

Organic manure (25 t/ha) is applied as basal dose at time of preparation of beds. The recommended dosage of fertilizers for the crop is 100:60:60 kg NPK/ha. Of these, 50:60:60 Kg N: P_2O_5: K_2O / ha is applied as basal. The balance of 50Kg N is top dressed 45 days after planting and earthed up. The N fertilization can also be provided in 3-4split doses. The first dose is given just after sprouting of corms followed by one at 4 leaf stage and one when the spike emerges with in foliage. The fourth dose is given one week after spikes come out. This is intended for development of corms and cormels.

Micronutrient sprays are also beneficial for improving flower yield and quality. Spike length, size and number of flowers are promoted by foliar sprays of 0.2% each of ferrous sulphate and zinc sulphate.

Irrigation

The crop has to be irrigated once in two or three days depending upon soil and weather conditions. During warm weather, watering is done twice a week, sufficiently to wet the root.

Staking

The plants may need staking. Individual bamboo stakes or common stakes are given. Butterfly types need no staking.

Mulching

Mulching is done between and across the rows with silt or polythene. This helps to conserve moisture and reduce weed growth. Chopped straw, saw dust etc. can also be utilized for mulching.

Weeding and Earthing up: Manual weed control is mostly practiced. Chemical control using herbicides like Simazine and Attrazine (4 kg /ha) in 100l of water can also be done. Earthing up is done when the plants are about 25 cm high.

Harvesting and Yield

Harvesting depends on corm size, variety etc. Flowering occurs 2-3 months after planting based on variety. Emergence of inflorescence occurs after production of definite number of leaves usually 7-10 leaves on an average. Harvesting is done in the morning. The entire spikes along with two leaves are cut at tight bud stage, when the basal 1-2 buds are showing color. Spikes are harvested leaving 4-5 leaves on the plant for subsequent development of new corms and cormels. Cut spikes are immediately placed in warm water and stored at 6-7^{0}C temperature. While storing, they should be held in upright position.

Nearly 2 lakh flower spikes can be obtained from one hectare. After harvest of flowers, the plants are left undisturbed in the field. When they start yellowing, the plants are uprooted for harvest of corms and cormels. One plant produces at least one marketable spike and a plantable size corm. Up to 2-3 lakh corms /ha, (up to 4 corms /plant) can be harvested.

Post Harvest Management

Pulsing with sucrose (20%) and 8-HQC (200 mgl^{-1}) for 24 hours improves shelf life and subsequent flower opening. Spikes may be stored for 1-2 weeks at 4^{0}C and 85% RH.

Grading

Grading is done based on over all quality, length of spike, number and size of florets in each spike etc. In a good quality inflorescence, flower stem should be strong and straight without branching, florets should be uniformly placed along the stem, open slowly and face in one direction.

Grade	Length of Spike (cm)	No of florets
Fancy	107	16
Special	96	15
Standard	81	12
Utility	81	10

After grading, spikes are bunched and packed in cardboard boxes or vertical hampers, and transported in refrigerated vans. Ethylene absorbents if packed with cut flowers can also reduce respiration and prolong shelf life.

Harvest of Corms

After harvest of flowers, plants are left undisturbed in the field. When the remaining plant parts start yellowing, the plants are uprooted for harvest of corms and cormels. This generally takes 8 weeks from harvest of flowers and 6-7 months from planting to lifting of corms. Cormels are gathered by pulling the plant gently after loosening soil around the plant with a tool.

Curing: At the time of lifting, corms are checked for any disease infection and the affected corms are discarded. After cleaning them of the adhering roots and dirt, they are treated with a fungicidal solution for 30 minutes, dried, graded and stored in trays in shady but well ventilated places. The layers of corms in the tray should not exceed three. They are the best stored in cold storages at 5 ^{0}C and 80-90 % RH. Periodic culling of diseased corms is done.

In the next planting season, the material is taken out of cold store and exposed to outer condition at room temperature for 2-3 weeks which help to break dormancy. Corms can also be soaked in thiourea (500 ppm) for 24 hours for breaking dormancy. Corms are descaled, treated with fungicidal solution (0.2% bavistin) and dried in shade before planting. Cormels are given standard hot water treatment before planting.

JASMINE

Jasmine is an important traditional flower crop cultivated in India. Fragrance of jasmine flowers is known for ages and flowers used for making garlands and bouquets, decorations, personal adorning of hair by women and for religious offerings. The essential oil extracted from flowers is a high value commodity used as starting material for expensive floral perfumes and has high export potential. Jasmine oil also has medicinal properties. Jasmine tea is made by mixing fresh jasmine with green tea and is popular in Japan. The plant is also used in landscaping as climbers, shrubs and rarely as pot plants.

Species and Varieties

In the genus *Jasminum*, 200 species are known out which 40 species are from India. Trailing, climbing and erect growing species and cultivars are available. Flower color ranges from white, yellow and red tinged. Commercially important species of Jasmine are listed below.

S.No	Species	Common name
1	*Jasminum auriculatum*	Mullai
2	*Jasminum grandiflorum*	Pitchi mullai, Spanish Jasmine (or) Jathimalli
3	*Jasminum sambac*	Gundu mullai, Tuscan jasmine (Arabian Jasmine)
4	*Jasminum multiflorum, J. pubescens*	Kakkada mullai

Among these, *Jasminum grandiflorum* and *J. auriculatum* are mostly cultivated in Tamilnadu and Karnataka. *Jasminum sambac* is the most ideal species for cultivation in Kerala. *Jasminum multiflorum* (Kakkada) is grown as a garden plant and the flowers have less fragrance. But it is hardy and less susceptible to pests and diseases and flowers profusely in the northern districts of Kerala.

Species	Varieties
Jasminum auriculatum	Parimullai, CO 1 mulla, CO 2 mulla, Long Point, Long Round, Short Point, Short Round
Jasminum sambac	Ramanathapuram Gundumalli, Motia, Virupakshi, Khoya, Ramabanam, Madanabanam, Single Mogra, Double Mogra, Iruvatchi , Kasthuri Mullai , Oosimalli, Soochimalli
Jasminum grandiflorium	CO1 pitchi , CO2 pitchi , Thimmapuram, Lucknow

Propagation

Various methods like cuttings, layering, suckers, budding, grafting, tissue culture etc., are employed in jasmine. Propagation through cuttings and layering are commonly adopted.

Cuttings: Propagation through cuttings is the easiest method. Several factors like type of cutting, rooting media, season and species influence success of rooting. Terminal/apical five nodal cuttings and semi hardwood cuttings can be used. *Jasminum auriculatum* responds better to rooting of semi-hard wood cuttings. *Jasminum grandiflorum* responds more to terminal cutting. In *Jasminum sambac* terminal and semi-hardwood cuttings can be selected. Single node cuttings with one leaf has been identified as the best planting material for quick, cheap and efficient multiplication of *J. sambac* (Paulos,1980).

Cuttings take longer time to root. Best rooting medium is sand: vermiculate: moss at 1:1:1 ratio (Mukhopadhyay and Bose, 1979). For increasing rooting percentage, the cuttings can be dipped in IBA or NAA (5000 ppm) or IAA (1000 ppm) and planted in coarse sand. Quick dip method is adopted for treatment.

Suckers: Those suckers arising from base of plants can be uprooted and used as planting materials.

Layering: In *Jasminum grandiflorum* and *Jasminum auriculatum* simple and compound layering can be adopted during June to November. The layers will be ready with in 90-120 days. Air layering can be adopted in *Jasminum sambac.*

Soil and Climate

Jasmine prefers a mild tropical climate. For successful cultivation, warm summer and mild winter with sunny days and moderate rainfall are preferred. They give good yield in low rainfall conditions. Jasmine is commercially grown under open field condition.

Jasmine can be planted on a wide range of soils. Well drained rich sandy loam and red loam soil are the best suited. Clayey soil results in increased vegetative growth and reduced flowering.

Planting

Land with proper drainage, irrigation facilities and sunny location are essential. After ploughing the land, pits of size 45 cm^3 are dug at least 1month before planting and exposed to sunlight. A few days before planting, half of the pits are filled with top soil and the balance with a mixture of 15 kg of well rotten FYM, soil and sand. If there is problem of termites, Chlorpyriphos can be drenched. Well rooted healthy and strong plants (2-4 in number) are planted in each pits. Planting is usually done during June-August.

Planting distance depends on species and also on soil and environmental conditions.

Spacing

Species	Spacing
Jasminum sambac	1.2 x 1.2 m
Jasminum auriculatum	1.8 x 1.8 m
Jasminum grandiflorum	2.0 x 1.5 m
Jasminum multiflorum	2.0 x 2.0 m

Pinching

After planting, it is necessary to pinch the newly formed apical bud for inducing lateral growth. This ensures bushy nature of growth and more surface area for flowering later.

Pruning

In jasmine, flowering habit is terminal and axillary. Increasing number of fresh shoots is the criteria to increase flower yield. Pruning at least once a year is a mandatory operation in Jasmine to increase fresh growth, flower bud initiation and differentiation and ultimately flower production. The best season for pruning is between November-January.

The plant is pruned from 2nd year onwards. All the suckers, past season growth including branches which come from the basal portion and which trail on the ground are removed and healthy upright branches are retained. The upright branches are pruned at a height of 45-50 cm from ground level. Cut ends are smeared with Bordeaux paste.

Pruning is a labor intensive operation. Chemical defoliants like Pentachlorophenol (3000ppm), potassium chloride and paraqua dichloride can be used as a substitute for pruning to save labor in larger units (Muthuswamy and Madhava Rao, 1974).

Manuring

Jasmine responds to intensive manuring. Too much of manuring encourages vegetative growth and hampers quality and quantity of blooms.

Apply organic manure (15 kg/ pit) before planting and thereafter every year, after pruning. This is also supplemented with 100 g organic manure every month in the form of groundnut or Neem oil cake and 100 g bone meal. Fertilizer dosages for the different species are given below.

Species	N: P: K dosage (g/pit)	Interval period	Method
J.auriculatum	60 : 120 : 120	Bimonthly	Six equal splits
J.sambac	120: 240: 240	Bimonthly	Six split doses
J.grandiflorum	100 : 150: 100	June- July	
December	Two split doses		

Foliar Nutrition

Foliar spray of 50% of total N requirement was better. Apart from major nutrients, Zinc (0.25%) and Mg (0.5%) before flowering resulted in appreciable increase in yield of flowers. These fertilizers are mixed together and sprayed during January and July. Spraying $FeSO_4$ (0.5%) at monthly intervals reduces iron deficiency and chlorosis.

Irrigation: There should be constant and adequate water supply during peak flowering season for high yield of flowers. Irrigating twice a week in summer from February to May is essential. Usually, plants are irrigated by flooding. After peak flowering is over, irrigation interval can be lengthened. Thereafter irrigation is done once in seven to ten days depending on soil and climatic conditions. Sprinkler or drip irrigation can also be practiced.

Weed control: Regular weeding is necessary. Manual weeding is effective but expensive. Manual weed control is better for smaller units. For larger areas, herbicides like oryzalin, paraquat or gramaxon can be used. Mulching can be done to reduce weeds.

Harvesting

If rooted cuttings and layers are used, first flowering can be expected from 2nd-3rd month. Economic yield starts from 3rd year and stabilized yield is obtained up to 12-15 years. Thereafter the yield declines. *Jasminum sambac* flowers almost through out the year for 9 months. Peak flowering season is from March to October. Flower production will be less after pruning and in heavy rains. Application of Cycocel (1000 ppm) induces early flowering and extended blooming period and also produces high yield (Narayana Gowda and Nagegowda, 1994). Two sprayings with Cycocel, 15 days before and 15 days after pruning are beneficial.

The stage of harvest depends on purpose for which flowers are used. For garlands, fully developed unopened flower buds are picked at tight bud stage early in the morning or previous evening, while for extraction of concrete only fully opened freshly picked flowers are required. Jasmine flower buds are to be harvested between 6- 8 am for improved shelf life. Harvesting after 9.30 am results in the gradual reduction of yield and quality of concrete.

Pluckers collect the flowers in a bag. The Pluckers should maintain personal hygiene and they should carefully handle the flowers. The fresh flowers should not be wrinkled or damaged during harvest and transit. This affects quality and concrete recovery and the entire product may be unmarketable.

Yield: Yield of flowers and jasmine oil vary according to the species and management practices.

Species	Flower yield (t/ha)	Concrete (%)	Oil yield (concrete Kg/ha)
J. sambac	5	0.14 - 0.19	11-15.0
J. auriculatum	5	0.28 - 0.36	28.00
J. grandiflorum	6	0.25 - 0.32	29.00

Grading

There are no standard grades available for jasmine. The flowers may be graded according to the corolla tube length, bud size, shape and freshness.

Post Harvest Handling

Longevity of jasmine flowers are between 24 - 48 hours and vary inherently between species. *Jasminum multiflorum* has a longer shelf while *J. grandiflorum* has the lowest. Different post harvest management practice like precooling, packaging, chemical treatment, cold storage etc., are beneficial in extending shelf life of jasmine flowers. Packing in 300 gauge polythene bags, precooling by dipping in 7.5 ^{0}C cold water and chemical dip in sodium thiosulphate and storage at 5 ^{0}C increased considerably shelf life.

Packing

Corrugated cardboard boxes are the proper packing materials for distant market. Wholesalers pack flowers in bamboo baskets. They are packed to maintain moisture and air circulation in baskets. Water is sprinkled on newspapers covering inside of the basket. The top is covered with paper again and closed with a bamboo basket cover or gunny stack which is stitched at edges.

Processing

Collection of Flowers

For extraction of oil, *Jasminum grandiflorum* is preferred. Fully opened flowers are harvested and used fresh for extraction. Harvesting of flowers is done before 9.30 am. Picking after 11.00 am reduces the yield and quality of concrete especially in hot climates. Flowers are collected in clean baskets (or) cloth bags without damage and contamination. Harvested flowers are kept in cool atmosphere and processed within two hours after harvest.

Method

Usual method of steam distillation is not effective in extraction of jasmine oil. Extraction using a solvent which yields Jasmine concrete is adopted. Jasmine concrete is a wax like substance containing essential oil with some plant waxes, albumin and coloring matter. It is semi solid in consistency with yellowish brown color and fragrance of fresh jasmine flowers. Jasmine concrete is obtained from a two step process involving solvent extraction using food grade hexane and removal of solvent from perfume material by evaporation.

Equipments

1. Extractor (Rotary type of 3 kg capacity)
2. Evaporator (wide mouthed circular stainless steel bowl of 25 litre capacity with air tight lid + hole at the center for outlet of solvent vapor)
3. Condenser (Stainless steel with 12 small pipes fitted parallel for solvent vapor passing and circulating cold water to condense the solvent).
4. Vacuum distillation unit.

Flowers are soaked in Food Grade Hexane (boiling point 70 ^{0}C) (2 litres / kg)of flowers for 30 minutes in a container. Rotate the container slowly for 20 minutes in the rotary type of extractor. The essential oil along with wax and pigments get dissolved in Hexane. Solvent along with its contents is led into the evaporator. Evaporation of solvents occurs at a constant temperature of 75 °C. Vapour of solvent is condensed into liquid for recycling. The remaining mixture comprising essential oil, wax and pigments is distilled in a vacuum distillation unit for complete removal of solvent. Floral concrete is settled in still in the form of molten wax. Around 0.29% yield of concrete can be obtained from fresh flowers. The concrete is cooled and stored in glass or aluminium containers. It can be stored at 10-15 ^{0}C for over a year.

Jasmine absolute: To prepare the absolute from jasmine concrete, it is warmed and stirred with pure ethyl alcohol. During the heating and stirring process, the concrete breaks up into minute globules greatly increasing the surface area of original mass. Since the aromatic molecules are more soluble in alcohol than in wax, an efficient separation of the two takes place. But along with the aromatic molecules a certain amount of wax also becomes dissolved. This can be removed by agitating and freezing the solution at very low temperatures (around -30 ^{0}F). Most of the wax can be precipitated out in this manner. As a final precaution the purified solution is cold filtered leaving only the wax-free material.

The alcohol which is used for dissolving the concrete is recovered from dewaxed extract by gentle distillation under reduced pressure. This process of alcohol removal from the extract is done in several stages until the final removal is done under vacuum. The final product is known as absolute. It is a viscous, clear yellow, brown liquid, soluble in 95% alcohol.

The present world production of jasmine concrete is about 20 tonnes of which India exports 2 tonnes (10%). The international market price for Jasmine concrete and absolute oil is around Rs. 30,000/- and Rs. 60,000/- per kg respectively. Price of Jasmine absolute is 2 times that of concrete. 8 million blossoms of flowers are required to produce 1 kilo of *Jasminum grandiflorum* absolute. The jasmine oil is unique as it blends well with other floral extracts to make high grade perfumes and cosmetics. This is used for high grade perfumes, soaps and cosmetics etc.

India exports Jasmine oil to France and Italy. In India, the essential oil industries are present mainly in Tamil Nadu, Karnataka, Andra Pradesh and Maharashtra. Nearly 50% of world's famous jasmine oil is produced from France and the rest is contributed by Belgium, Netherlands, Italy, Turkey, Morocco and Tunisia.

MARIGOLD

Marigold is a popular annual flower that could be grown on a commercial scale. It has gained popularity on account of its easy cultivation and wide adaptability to varying soil and climatic conditions. Free flowering habit, short duration to produce marketable flowers, wide spectrum of color, shape, size and good keeping quality make marigold an acceptable commercial crop.

The plant is often used in landscaping for bedding, edging and as a component of borders and rockery. They can also be used as potted plants, for hanging baskets and for window box gardening. Flowers are used for both loose and cut purposes. Loose flowers are used for making garlands, Veni and floral carpets. The large yellow to orange flower heads are exploited for manufacturing food colors through pigment extraction. The plant has aromatic foliage and is used for extraction of essential oil and perfumes. The flowers and foliage also have pharmaceutical properties and medicinal principles. Neutraceuticals are derived from them. Flowers are dried and powdered and are used as a component in poultry feed to intensify the yellow colour of the egg yolk and broiler skin.

Classification

Marigold which belongs to family Compositae has many species. The important among them are African marigold (*Tagetes erecta*), French marigold (*T. patula*), Signet marigold (*T. signata/ T.tenuifolia*), wild marigold (*T. minuta*) Sweet scented Marigold (*T.lucida)* Californian Marigold *(T.lacera)* and Shrubby Marigold (*T.lemmonii)* of which African Marigold and French Marigold are used for large scale cultivation in India.

In African marigold, plants are usually tall, grow up to 1 m height, erect and compact with broad leaves. They produce single or double globular flowers of yellow, orange and white color. The plant is suitable for tropical conditions.

In French marigold, plants are dwarf, 20-60 cm high with profuse branching. Flowers are small, produced in clusters with colors ranging from yellow to red. This requires mild climate for luxuriant growth and flowering.

Signet Marigold is of European origin and requires cooler climate for cultivation. They are dwarf, bushy, compact and dome shaped plants, 15-25 cm tall with fine lacy foliage and small single flowers. The leaves contain higher essential oil (4.25 %) of very good quality used in perfumery industry.

Tagetes minuta is the wild marigold found in Himalayan region and North India. The leaves of this species have the highest quality essential oil.

Interspecific hybrids between *T. erecta* and *T. patula* also have been evolved, known as Red and Gold hybrids.

Varieties

Species	Variety
Tagetes erecta	
Tall varieties	Cracker Jack, Guinea Gold, Apricot, Gold Coin, Primrose, Climax, Sun Giant, Yellow Supreme, Giant Double, African Orange, African yellow
Medium tall varieties	Jubilee, Happy Face, Gold Galore, MDU-1, Pusa Narangi Gaintha and Pusa Basanti Gaintha.
Dwarf varieties	Moon Shoot, Mariner, Aztec, Golden Age, Guys and Dolls, Spun Gold, Spun Yellow
F1Hybrids	Appollo, Climax, First Lady, Gold Lady, Inca Series
Tagetes patula	
Single types	Dainty Marietta, Sunny, Tetra, Raffled Red
Double types	Rusty Red
Dwarf varieties	Flame, Flaming Fire, Rusty Red
Extremely Dwarf	Gold, Gypsy, Orange Flame, Sun Kist, Tom Thumb
Large flowered (Super French)	Sparky, Spanish Brocade, Ole, Bonita, Golden Boy, Gypsy Dancer
Signet Marigold	Golden Gem, Lemon Gem, Paprika, Lulu, Pumila, Ursula
Triploid Marigold	Seven Star, Show Boat
Red and Gold hybrids **(Interspecific hybrids)**	Nugget, Show Boat, Red Seven Star

Propagation

Marigold is propagated mainly through seeds and also by cuttings.

Seed Propagation

Seeds are usually used for raising the crop. Crop raised from seeds is tall, vigorous and flowers profusely and is preferred over cuttings.

Cuttings

This method is commonly followed for maintaining purity of varieties. Normally, the presence of adventitious roots along the stem helps in establishment of cuttings. About 10 cm long cuttings taken from shoot tips which are not flowered are planted in sand to strike roots easily and plants thus raised are used for bedding and pot planting. Plant bioregulators like IAA and IBA can be used for improving rooting percentage. (Roy *et al.,* 1983). But this method is not feasible for commercial planting.

Nursery

Seeds of marigold can be sown in pots, seed boxes or on flat or raised nursery beds. Seeds are very light and have 300-350 numbers / g for African marigold varieties. For raising seedlings for one hectare, about 2.5 kg seed is required for African marigold and 800g -1 kg for French marigold.

Nursery beds are thoroughly prepared and mixed with 10 kg of well rotten farmyard manure / m^2. Sow the seeds in rows 7.5 cm apart at shallow depth not more than 2cm. Cover the seeds with fine FYM and irrigate. Drench the nursery with carbendazim (0.1%) against seed/ soil borne pathogens. The seedlings will be ready for transplanting within one month at 4 leaf stage.

Soil

A wide range of soils with good drainage is suitable for cultivation of marigold. African marigold prefers rich moist sandy loam soil. French marigold prefers light soil. A soil that is deep, friable, fertile and well drained having good water holding capacity, and near neutral pH of 7.0-7.5 is the most desirable for its cultivation.

Climate

Marigold requires mild climate for luxuriant growth and profuse flowering. It ceases to grow at high temperature thereby flower quantity and quality is adversely affected. During severe winter including frost, plants and flowers are killed and blackened. African marigold tolerates higher temperatures.

Planting and Management

African marigold is suitable for cultivation in three seasons namely rainy, winter, and summer. Hence, flowers of marigold are available almost throughout the year.

In Kerala, the best season for cultivation of African and French marigold is between September- January.

Land should be well prepared by ploughing 2-3 times. Beds of convenient size are made to facilitate irrigation and other cultural operations. FYM @ 50 t/ha should be incorporated to the soil.

Transplanting of Seedlings

Marigold seedlings establish easily after transplanting in the field without much mortality. One month old seedlings with 3-4 true leaves are used for transplanting. Transplant the seedlings at a spacing of 20-30 x 20-30 cm for French marigold and 30-45 x 30-45 cm for African marigold depending on varieties.

Pinching: It consists of removing terminal portion of the plant 30-45 days after transplanting. Pinching is done for emergence of side branches and to increase total yield. If left without pinching, the plants may grow straight upwards and end up into a terminal flower bud.

Irrigation

Immediately after transplanting, copious irrigation is given. Season of planting determines the frequency of irrigation. If rainfall is normal and well distributed, irrigation is not frequently required. Decide the frequency depending upon soil moisture and weather conditions. In summer, frequent irrigation at the interval of 4-5 days is required.

Weeding

3-4 manual weedings are done during the entire growth period. Weeds have to be removed at monthly intervals.

Manuring

Apart from organic manures, a total dosage of NPK @ 250:60:60 kg/ha is recommended for the crop. Apply a basal dose of fertilizers @ 112.5 kg N, 60 kg P2O5, and 60 kg K_2O / ha at the time of bed preparation. The remaining 112.5 kg N is top dressed at the time of pinching (30-45 days after transplanting). After top dressing, the beds are to be earthed up.

Intercropping

Marigold is often raised as an intercrop in many plantations and orchards. Marigold roots have allelopathic effect and companion planting with other crops reduces infestation by nematodes and insects. The roots exude a substance which spreads in their immediate vicinity killing nematodes (Alam *et al.*,1988). They also help to deter beetles and whiteflies and are a powerful insect repellent.

Harvest and Yield

Marigold plants require 55-60 days to complete vegetative growth and to enter into reproductive phase. Flowers will be ready for harvest in about 10 weeks from the date of transplanting. The plant continues to bear flowers for another 2-21/2 months from the date of first harvest. The flowers are harvested when they attain full size. Plucking of flowers should be done in cool hours of the day, either in morning or evening. Field should be irrigated before so that flowers keep well for longer period after plucking.

Yield of Flowers

Flower yield depends upon season of planting, species, variety, and cultural practices adopted. On an average, a fresh flower yield of 20-22 t/ha during rainy season, 15.0 to 17.5 t/ha in winter and 10-12 t/ha in summer can be obtained.

Seed Production

Marigold is a cross pollinated crop. Hence, proper isolation distance of 1-1.5 km should be given amongst varieties for seed production.

Seed yield varies from 312 to 375 kg/ha in *T.erecta* and 1000-1250 kg/ ha in *T.patula.*

Extraction of Pigments

The fresh flowers are silaged (a sort of fermentation) for 20 days and then pressed to remove water. The resulting cake is dried at dehydration centres and pelletized. After grinding the marigold meal, the oleoresin is produced through solvent extraction methods using hexane.

The orange color is due to presence of carotenoid pigments in flowers. Xanthophylls are the major fraction. Lutein accounts for 80-90% of total xanthophylls content. Currently, the pigment is used as an additive component in poultry feed, in food colouring and as neutraceutical. The major market of marigold oleoresin is Mexico where it is used in poultry feed to change the skin colour of the chicken and the yolk to "orangish".

Lutein is purified from marigold extract using patented processes. Studies in recent years have recommended use of lutein in food as it would help protect against cataracts and macular degeneration, two common age-related eye disorders. Purified extract of marigold petals containing lutein dipalmitate is marketed as an ophthalmologic agent under trade name Adaptinol.

Essential Oil: Leaves and flowers contain odoriferous oil which is used for manufacturing of perfumes and which also has medicinal properties. The oil also has good insect repelling properties.

ORCHIDS

Orchids are the most exotic and fascinating of all flowers captivating in their beauty and complexity. They are the most exquisite and abundant members of the plant kingdom and belong to Orchidaceae, which is the largest family of flowering plants. Their attractive blooms are bewitchingly beautiful and exhibit a wide range of diversity in form, size, shape, color and texture of flowers. Orchids are noted for their long lasting vase life. These features make them highly prized among the ornamentals as a cut flower and as a potted garden plant.

Importance

Orchids are highly priced and occupy 8% of the global floricultural trade. Among the various genera, *Cymbidium* has the highest demand. Cultivation of orchids is taken up as a very profitable occupation by many and the cut flower business has become a flourishing trade locally and globally.

Orchids are also grown for medicinal value, as their tubers are aphrodisiac. Several flavanoides and steroidal compounds are also extracted from certain orchids.

Orchids also constitute an order of royalty in the world of ornamental plants and play a very useful role in balancing the forest ecosystem.

India is blessed with a wealth of orchid flora, and many beautiful species of Orchidaceae are growing naturally in the forests. India is home to about 1,700 species of orchids, of which about 800 are in the North Eastern region of the country. The genus found are *Paphiopedilum, Rhynchostylis, Vanda* etc. Himalayas are not only rich in terms of number of species, but more-importantly many of them rank at the top of the list of ornamentally important ones. Many orchids native to this country are important parent plants for production of several outstanding hybrids in the world. Western Ghat forests in South India also account for about 200 species.

Classification of Commercial Orchids

Most attractive orchids are epiphytic, which produce fasciculate roots by which they fix themselves and adhere to the stem or branches of trees. They also have strong aerial roots, which absorb water and nutrients from the atmosphere. They take nourishment from the minute particles of organic matter in rain water or from the debris that collects around the roots. They always require free moving air. Epiphytic orchids like *Dendrobium, Vanda, Cattleya, Oncidium* etc dominate the cut flower trade.

Based on growth habit, orchids can be classified into monopodials and sympodials. In monopodials, vertical growth takes place only at the tip of the central stem and the plant becomes taller each year. Stems are non-branching, and aerial roots and flower stalks emerge from the central stem. Leaves are arranged in two rows opposite each side of stem, the individual leaves alternating with those in the opposite row. Eg: *Vanda, Arachnis, Phlaenopsis, Aranda, Aranthera.*

Sympodials have a rhizome with horizontal growth which produces thickened pseudo bulbs in clusters. Pseudo bulbs store food and water and have leaves on them. They bloom either from their tips or their most recent growth. New growth originates from the base of previous years growth in the rhizome. A well developed plant consists of a clump of pseudo bulbs of varying age and size. No aerial roots are produced. Eg:- Dendrobium, Cattleya, Oncidium etc.

The management practices adopted for monopodial and sympodial orchids vary.

Propagation

Orchids can be propagated through seed, conventional vegetative means and through micropropagation. Most growers multiply orchid plants by division, through off-shoots cuttings and by air layering.

Seed Propagation

Orchid seeds are mainly used for developing hybrids. The fruit is a capsule and contains thousands of extremely minute seed in it. Each seed contains an undifferentiated embryo devoid of any functional endosperm, the food for embryo.

In nature, the propagation is effected through nourishment of seeds through symbiotic association of mycorrhizal fungi. Seedling propagation is possible only under aseptic conditions. The seeds are to be provided with sterile artificial nutrient medium as in tissue culture, like Vacin and Went, Knudsons, MS media etc. Seedlings produced by embryo culture take 2-5 years for flowering, depending on the genus.

Vegetative Propagation

Stem Cuttings

Monopodial orchids are propagated by stem cuttings. Terminal cuttings 20-40 cm long, with one or two healthy aerial roots are ideal as planting material. The severed pieces, along with its roots form a new plant and the original mother plant continues its growth by another stem arising from an axillary bud on the side. Intermediate cuttings which have developed aerial roots can also be used. But they take longer time to sprout and grow. Off-shoots that arise from nodes of stem also can be cut and potted.

Eg: *Vanda, Arachnis, Ascocentrum, Aranda, Renanthera*

Flower stalk cuttings: This method is useful in *Phalaenopsis, Phaius, calanthe* etc., Treatment with IBA or NAA (4000-5000 mgl^{-1}) enhance root development.

Air-layering: This also is reported as a propagation method in vanda and some other monopodials.

Division : Sympodial orchids are propagated by division of clumps and potted separately into new plant. A plant with a minimum of 3-4 pseudo bulbs along with a portion of rhizome and basal root is ideal for planting. In about 4-5 years growth, a sympodial orchid clump acquires some 8-10 good sized canes and such a plant can be divided to 3-4 individual units having at least 2-3 canes / division. Each division must have at least one old cane of two years' growth, one new shoot and some new roots.

Eg. Dendrobium, Cattleya, Epidendrum, Oncidium, Spathoglottis etc.

Besides, back bulbs or spent canes (shoots that have ceased to produce flowers) before they are shriveled can be severed from the mother plant and placed horizontally over a moist medium to stimulate sprouting of new shoots. When the sprouted shoots produce sufficient roots, they can be separated and planted.

Apart from back bulbs, division of pseudo bulbs, stem, and inflorescence stalk into 10-15 cm length and placing them in a tray of damp sterile media is also in possible in certain genera, for rooting.

Off-Shoots (Keikis): Some of the sympodial types produce off-shoots on the pseudo bulb called keikis. These are miniature plants with roots, arising from nodes of old canes. Keikis when fully grown can be separated and planted independently. Application of cytokinins like BAP ($1gl^{-1}$) stimulates new offshoots in *Dendrobium, Phalaenopsis, Paphiopedilum* etc.

Micropropagation

Meristem culture is very effective in large scale propagation of orchids. T C plants take longer time -18-20 months for flowering.

Flower structure: The flower structure in orchids is typical. Flowers are trimerous, with two whorls of perianth parts. The outer whorl consists of 3 sepals and the inner three petals. Sepals closely resemble petals in structure and color and are not easily distinguishable. Out of the 3 petals, 2 are slightly smaller and the third one is usually larger and the tip is modified into a complex structure known as lip or labellum. The labellum carries androecium and gynoecium fused into a single cylindrical structure known as gymnostemium or column which appears like a hood. Anthers or pollen sac with 2 pollinia are present at tip of the column. Stigma is located below the anther, separated by a flap like rostellum. It is the third petal which is distinctive and different in each orchid and is the distinguishing character in each variety and species and helps to identify the genera and species.

Commercial Orchids: Although there are several thousands of species and a considerably large number of hybrids occurring in nature, only a few groups are grown commercially and are the most important for cut flower production.

Monopodials

a) *Vanda:* Based on shape of leaves, vandas are classified either as terete (pencil like) or strap (flat) leaved. Intermediate types like semi terete and quarter terete types are also seen. Vanda prefers open sunlight for their growth and development.

b) *Phalaenopsis:* These are also known as moth orchids. Flowers are borne on long spikes in large numbers. They prefer low light intensity, temperature and less humidity.

c) *Arachnis:* These are generally known as spider orchids. They have long spikes which are 75 cm or more and carry up to 15 flowers. They prefer bright sunlight and high humidity.

Sympodials

d) *Cattleyas:* They enjoy the position of being the most popular orchids in the world and are used extensively in hybridization. They have a beautiful shape with medium to large size bloom of various colors that last for more than 2 weeks on plant with long vase life when cut. Cattleyas possess a tantalizing fragrance and hence are aptly called Queen of orchids. There are unifoliate and bifoliate cattleyas which produce single leaf and two leaves respectively from each pseudo bulb. They require partial shading under tropical conditions.

e) *Cymbidiums:* Commercially, they are the most sought after orchids, which rank first among orchids. The spikes are arching or cascading types. They prefer a cool climate.

f) *Dendrobiums:* They produce flowers year round and are the most popular genus of Kerala. The color and size of the flowers exhibit a wide range. They prefer partial shade, high humidity and good aeration.

g) *Oncidium:* They are also called as dancing dolls. They also prefer partial shade, high humidity and aeration.

h) *Paphiopedilum :* They are also known as lady's slipper orchids due to their slipper shaped pouches. They prefer low to medium light and cool to intermediate temperature. They are more sought as a potted plant.

Inter generic crosses are common in orchids and bigeneric, trigeneric, tetrageneric and pentagenetic hybrids are available. Some of the promising hybrids are Aranda, Aranthera, Ascocenda, Laeliocattleya, Renantanda (bigeneric), Brassolaeliocattleya, Mokara (trigeneric), Potinara, Robinara (tetrageneric) and Goodlera (pentageneric).

Hybrids of certain Indian orchids like *Vanda coerula* (Blue vanda or Blue moon vanda) and species of *Cymbidium, Dendrobium* and *Paphiopedilum* are considered as the monarchs in the orchid world.

Varieties

The popular genera and hybrids of orchids, suitable for growing under tropical conditions in South India are Arachnis, Aranthera, Phalaenopsis, Vanda, Aranda, Mokara (Monopodials) and Dendrobium, Cattleya, Oncidium, Cymbidium (Sympodials). The important varieties of some of the genera are given below.

Genus	Variety
Cattleya	C. Arctic, C. Snow Flower, C. Nobilis Natalia, C. Hail storm
Phalaenopsis	P. Alice Gloria, P. Sandarana, P. Amabilis, P. Pepper Mint, P. Grace palm
Vanda	Miss Joaquim, Diana, John Club, Isabel, Kasem Delight, Wirat Uchida
Arachnis	Maggie Oei Red Ribbon, Maggie Oei Yellow Ribbon
Aranda	A. Peter Evert, Wendy Scott, Noorah Alsageoff, A. Christine, Luci lay cock
Aranthera	Annie Black, Mohammed Hannefa, James Storie
Renanthera	R. Nancy Chadler and Brooke Chadler
Oncidium	O. Gower Ramsey, O. Taka, O. Ai gold
Dendrobium	Madame Vipa, Ekapol Panda, Sonia 16, Sonia17, Sonia 28 , Queen Sonia, Sonia BomJoe, Earsakul
Purple and white flower	
Purple flower	Madame Pompadour, Renappa, New Wanee, Sabine Red, Juree Red
White	Thailand White, Emma White, Adisak White, Kasem White, Snow White, Fairy White, Savin-5, Pravit White, Dorine
Pink	Sakura Pink, D.D.Pink, New Pink
Yellow	Kasem Gold, Tongchai Gold, Sherifa Fathimah
Striated	Candy Stripe, Kanchana Green

Dendrobium species is a typical tropical orchid species suitable for coastal areas of Kerala where the humidity is high and is the most popular genus.

Climatic Requirements

Unlike many other cultivated ornamentals, orchids are very hardy plants which can withstand quite a lot of adversities. For growing them successfully up to flowering condition, a number of key factors must be satisfied. These are temperature, humidity, light, aeration, watering, containers and media, repotting and division, fertilizer application and plant protection.

As the orchid family is comprised of a large number of genera having characters of extreme diversity, different groups of orchids need different conditions of growth.

Light : Light is the most important single factor that controls healthy growth of plant and their ability to reach flowering. Some orchids need full sunlight to flower while others die if kept in full sun. There are orchids that require full sunlight, those that need intermediate exposure to sunlight and those that need full shade.

Requirements vary from genus to genus. Both terrestrial and epiphytes grow under varying levels of shade. Plants grown under deep shade will have good vegetative growth and poor flowering. Hence shade and light regulations are very important operations for better flowering. This requirement varies with species and varieties. Some species grow in open sunlight.

Optimum requirement in general for orchids is 3000-6000 ft. candles. Vanda and Arachnis require 8000 ft candles and can be grown in open conditions. Cymbidium grows in full sunlight under cooler climate. Light requirement is very low in *Phalaenopsis* (200-300 ft. candles) and so are to be kept in shade.

Intensity of light that falls on the plants can be controlled by proper shading methods. HDPE shade nets of various intensities are now available in the market and those should be used judiciously to provide optimum light /shade condition to the plant. Depending on light requirement, the shade nets are selected. For Dendrobium, 50% shade net is advised.

Temperature

Orchids require proper temperature for good growth and flowering. Accordingly, there are tropical or warm (21-29^{0}C), subtropical or intermediate (18-24^{0}C) and temperate or cool (16-21^{0}C) orchids. Depending on genus, temperature requirement varies. Night temperature should be lower by 10° F than day temperature.

Humidity: This is a critical factor especially for orchids coming from tropics. Since majority is epiphytes, atmosphere saturated with water vapor is ideal. Vanda and Phalaenopsis require high humidity above 80% while Cattleya requires only 40-55%.

When temperature is high, humidity will be correspondingly low. So in warm months of the year, misting or spraying leaves, use of humidifier etc., can be done for maintaining humidity.

Aeration: Orchids require plenty of air and so they are usually not grown enclosed unless provision for air movement is provided. Air circulation facilitates moisture to reach the roots. Air that constantly circulates also keeps the plants in good health as stagnant and still air promote fungal growth. A fresh moist atmosphere with good air circulation is ideal.

Media and Containers

Orchids always require plenty of fresh air near its root. So the media selected should provide a micro climate of an optimum combination of moisture and aeration which favor healthy growth of plant. It should also provide good drainage, support to plant, should not be harmful and available locally.

For terrestrial orchids, a medium containing equal parts of leaf mould, soil and sand would be desirable. Epiphytic orchids thrive well under a wide range of growing media that provide the plants with a rough surface to cling. The most common components of the mixture consist of charcoal, coconut husk pieces, dried tree fern chips, osmunda fiber, wood bark pieces, sphagnum moss, broken pieces of burnt bricks and tiles, perlite, peat moss etc. These are used either alone or in combination depending on climatic condition, genera, method of planting and type and size of planting material. When humidity is low, coconut husks can be used but during excess rain fall, it is advisable to be removed from the media. Larger the plant, thicker the roots and the particle size of media also can be larger.

Planting

The ideal location for growing orchids is in the open condition, under appropriate level of shade nets. In Kerala, it is also grown in interspaces of coconut trees aged above 15 years. UV stabilized HDPE shade nets are available in different grades and a level of 50% shade will be ideal for most of the shade loving orchids. The plants are also to be protected from heavy rains and it is better to provide a cover with UV stabilized poly films during rainy season. This will reduce the incidence of diseases.

In commercial growing, depending on growth habit of genera, two methods are adopted namely, planting in containers and planting on ground/ beds.

Monopodials are usually raised on ground except dwarf ones, since they cannot be managed in pots for large scale commercial cultivation. Sympodials are usually raised in containers. Dwarf monopodials like Phalaenopsis can also be raised in containers. Terrestrial orchids are planted directly in the soil provided with the media.

Planting on the Ground

Monopodial orchids can be grown on the ground in trenches or above soil level.

Select an open site that receives good sunlight and drainage. A raised bed of 60-90 cm width, 15-20 cm height and of convenient length is prepared with its periphery protected with bricks, tiles or laterite to prevent shallowing down of compost/ media. In areas of water logging, the lower layer (10 cm) is loosely filled with broken brick or tile and two upper layers of well-dried coconut husks. Planting is done above the medium with a support for proper anchorage.

Monopodials such as Vanda, Arachnis and their hybrids can be grown in beds in the interspaces of grown up coconut garden without any cover. Terminal cuttings of these monopodial orchids are planted loosely on old coconut husks at a spacing of 20-30 cm between plants and rows in long beds. There can be two or three rows in a bed. If basal cuttings are used, they will sprout within a period of two months.

Wooden/ concrete/ granite pillars are used as standards for supporting these orchids. The poles are fixed in soil on either side of the bed at appropriate intervals and cross wires (GI, coir) are used for connecting them. Each plant is then tied to these cross wires. A mixture of dried, split coconut husk with half rotten compost is placed around the planting. The bed is maintained at its height by regular addition of leaf mould, compost, husk etc.

Planting in Containers

Planting can also be done in pots or other containers. Sympodial orchids are mostly planted in pots. Monopodials which are dwarf in stature like *Ascocenda, Mokara* and *Phalaenopsis* can also be raised in containers.

Clay pots with holes on sides and bottom are used for quick and easy drainage. New pots may be soaked in water overnight to fill the pores thoroughly. Pots with a size between 12-18 cm are used depending on genus and varieties. Size of pot should be proportionate to size of plants.

For planting sympodials like *Dendrobium*, pots are filled with tile bits or freshly broken bricks. Larger pieces are placed at bottom. A bamboo stake should be fixed firmly in the larger peaces of brick on one side of pot. While planting, base of pseudo bulbs should not be buried in potting mixture but should rest on top of it. They are planted in such a way that the oldest shoot touches one side of the container and the youngest pseudobulb faces centre of pot. Both monopodial and sympodial orchids are to be supported properly since the planting is done above the medium. After planting, charcoal can be used above the media. Plastic pots can be used for hanging, since they are light, easy to transport and do not accumulate algae on their sides.

Orchids can also be grown in some type of container or support such as wooden, bamboo and wire baskets filled with small pieces of dried coconut husk and partially burnt charcoal which can be suspended as hanging baskets. Coconut husks, tree fern rafts, pieces of wood etc. can also be used as containers.

Repotting : Repotting is done only if compelling reason exist. Repotting seriously disturbs the plant. This becomes necessary when the plant grows large and overgrows its container or when the potting material deteriorates or when the plant has to be split or divided. This may usually be required in every two to three years. The best time for repotting is when fresh roots emerge at the bases of previous year's growth. For monopodials, repotting or division has to be done when new leaf growth shows at the top and there is new root growth.

Irrigation: Water management is also critical in successful orchid cultivation. Over watering destroys the plants more often than under watering. Avoid continuous wetting of root zone and the potting media should be just evenly moist. Too much water encourages fungal and bacterial growth. Too little water stunts the plant. The frequency and quantity depend on several factors like type of orchid, type and size of pot, potting medium, climate, method of irrigation etc.

Terrestrial orchids require more watering than epiphytes. Plants which are freshly potted should be watered sparingly. Plants grown on brick and charcoal require more water. Materials which retain more water, like coconut fibre, need less watering. Small pots dry out faster than larger ones. Orchids in hanging baskets need more water. In summer, watering should be twice daily, first in the morning and the other before late evening.

The best method to irrigate orchids is through misting or fogging. Microsprinklers can be used for larger units. During summer, spraying water on the floor and walls can be done to increase humidity.

Rain water and deionised water are the best. Orchids prefer near neutral water with pH 6.5-7. Presence of chlorine and dissolved salts in irrigation water is detrimental to plants.

Nutrition

In their natural habitats, orchids grow on no special nutrients, depending only on what they could absorb from t1he atmosphere or surrounding media. But for cut flower culture, orchids require regular nutrition management with both organic and inorganic manures. Since most of the cultivated orchids are epiphytic, foliar feeding is very effective in orchids.

Organic manures like cow dung, cow's urine, groundnut cake, neem cake etc. can be moderately used. It must be diluted in water at 1:10 ratio and kept for a week to allow fermentation. Cow's urine if used, must be diluted 25 times. The supernatant solution is used for foliar spraying once in a week. It is better to change the source of organic manure each week on rotation. Deficiency of certain micronutrients is also taken care of by application of organic manure from different sources.

Monopodial orchids grown on ground can be provided with Farm yard manure @ one kg for one square meter. Two to three applications can be given in a year.

Inorganic manures are also provided through foliar feeding. During initial stages of growth of orchid plant, more nitrogen is required for better vegetative growth. A fertilizer mixture of N: P_2O_5: K_2O in 3:1:1 ratio is recommended during vegetative period. During flowering stage, a mixture with a higher proportion of phosphorus and potassium such as 1:2:2 (N:P:K) is helpful. The usual dose of such mixture is 2-3 g / liter of water, applied twice or thrice a week (Rajeevan, 2006).

A fertilizer mixture available commercially, equally balanced in nitrogen (N), phosphorus (P) and potassium (K) of strength 17:17:17, 18:18:18 or 20:20:20, dissolved in water at the rate of 1-2g / liter applied twice a week gives satisfactory results.

Harvesting and Yield

The stage of harvest influences vase life and marketing. Sprays are harvested when the uppermost 2-3 flowers are yet to open. Harvesting is done either in the morning or evening. They are given a smooth cut at base of stalk close to stem. The yield of sprays in a year varies from 6-10 in Dendrobium and 15-20 for Vanda, Arachnis etc.

Post Harvest Handling

Immediately after harvest, the lower 0.75 cm of peduncle is cut off, and the flower is dipped in a bucket containing lukewarm water.

Treatment with 8-HQC + sucrose (5%) improves flower quality and vase-life of flowers. Aluminium chloride (500 ppm), Ammonium molybdate (100 ppm) or Boric acid (1000 ppm) also lengthens vase life of orchids (Prakash *et al.*, 2001)

When the spikes are packed for marketing, they are taken from bucket and the basal ends are wrapped in wet cotton padding. A protective cover with a piece of polythene sheet is provided around the cotton pad and secured with a rubber band to reduce water loss. They are packed in bundles of five or ten in cartons lined with moist news paper and packed with shredded news paper in between to prevent movement of bundles during transport.

Grading and Marketing

Grading is done based on number of flowers, length of spike, color, etc. In India, most of the orchid flowers are sold in the domestic market in cities like Delhi, Bombay, Bangalore, Chennai and Calcutta.

ROSE

Rose is one of the oldest flowers, grown for over 7000 years. It is a popular garden plant throughout the world and is known as the queen of flowers. Rose is treated as a symbol of love, adoration and innocence. Among flowers, rose is the most favorite gift exchanged on special occasions to dear ones.

Importance and Uses

Traditionally roses are used in landscaping. Rosaries are a part of many large gardens. They are also widely used in arches, pergola, as potted plants and as standards. Certain rose species are good source for extraction of essential oil. Rose oil is used commercially for the preparation of cosmetics, perfumes and flavors which are exported. Rose attar is also extracted from them.

The rose petals are also used to prepare rose water and gulkand. Rose water has food value. Petal preserved in sugar syrup can be eaten. Gulkhand, a preparation obtained by mixing rose petals and sugar is a delicacy. These products are exported to UK, France, Germany and Middle East countries. Rose extract is medicinal and is used as an eye tonic. It is used in Ayurveda according to Charaka Samhitha. Hips or seeds are used as a source of vitamin C.

Loose flowers of roses are used for worship, bouquets and making garlands. Rose ranks first among the cut flower trade at international level. *Rosa hybrida* having long stem is grown for cut flowers under green house conditions and is exported. It is traded worth USD 1.5 billion globally every year.

Classification of Modern Rose

1. *Hybrid Tea Rose:* Most popular among the groups. Bushy plants with blooms single, long stemmed, very attractive and in good shape. Typical

hybrid tea roses have medium to large flowers with many petals forming a central core. They have pleasant fragrance.

2. *Polyanthas:* Medium to dwarf sized plants or climbers which flower in clusters with less number of petals and with perpetual flowering habits.
3. *Floribunda (Hybrid Polyanthas):* This was developed by crossing HT X dwarf polyanthas. The plants flower in clusters, and have perpetual flowering habit. Flower form is inferior to hybrid tea. They produce large quantities of short stems.
4. *Miniatures:* Plants are comparatively dwarf with size 30-45 cm. Flowers are small and are produced in clusters. Used for growing in pots, borders, rockeries and as indoor plants.
5. *Climbers and ramblers:* Climbers have stiff stem, their flowers are larger and continuously blooming. Ramblers have long pliable stem, have large clusters of small flowers and they produce inflorescence once in a year during summer.

Varieties

There are thousands of varieties developed in India and abroad- either by expert nursery men or by breeders. Late Dr. B.P. Pal was a pioneer breeder who carried out research on roses from 1962 onwards at IARI. He authored a book 'Rose in India' and mentioned about 125 varieties in his book.

IARI has the distinction of releasing the largest number of rose varieties in the country. It released 75 varieties comprising 35 HT, 25 Floribundas, 12 Polyanthas, 3 Miniatures and climbers in rose.

Popular varieties developed by research institutions in India are given below:

Institution	Class	Varieties
IARI, New Delhi	Hybrid tea	Dr Homi Baba, Poornima, Kanakangi, Masina, Nayika, Ranjana, Pusa Sonia, Mrinalini, Bhim, Arjun, Jawahar, Raktha Gandha, Priyadarshini, Chithra, Dr. Benjamin Pal, Mother Theresa, Preyasi, Shreyasi, Dr. Bharath Ram, Pusa Bahadur, Puja Garima, Pusa Gaurav, Pusa Priya, Soma, Arjun, Anurag, Manasi, Kavitha, Mohini, Deepika, Dr.S.S. Bhatnagar
	Floribunda	Arunima, Banjaran, Chandama, Chitchor, Deepika, Deepshikha, Delhi Princess,

contd...

contd...

		Kavitha, Lehar, Madhura, Prema, Mohini, Pusa Barahmasi, Pusa Pitamber, Pusa Virangana, Manasi, Surya Kiran, Lahar
	Natural mutants	Chitra, Pusa Abhishek, Climbing Sadabahar, Climbing Dr. Homibhabha, Delhi Pink Pearl, Navsadabahar, Pusa Sadabahar
	Induced mutants	Sharada, Angana, Sukumari, Pusa Christina, Abhisarika, Madhosh.
	Hybrids	Rakthagandha, Arjun, Rakthima, Pusa Gaurav, Pusa Pitamber, Mother Theresa, Priya, Dr.Benjamin Pal, Mohini (with chocolate flower color)
NBRI, Lucknow	Hybrid tea	Kronenberg, Light Pink Prize, Mrinalini Light Pink Mutant, Mrinalini stripe, Pink Montezuma, Salmon Beauty, Summer Holiday Mutant, Winter Holiday Mutant, Sylvia white.
	Floribunda	Angara, Curio, Pink Contempo, Pink Imperator, Sharada, Sukumari, Yellow Contempo, Tangerine Contempo, Twinkle, Salmon Beauty Mutant, Zorina Pink Mutant
	Miniature	Windy City Mutant
	Climbing	Cri Cri
IIHR Bangalore		G.S.Randhawa, Kiran, Nishkanth(root stock)
TNAU, Yercaud	Floribunda	YCD-1, YCD-2, YCD-3

Propagation

Methods adopted include seeds, cuttings, layering, budding, grafting and tissue culture. Commercial method of propagation is by cutting, budding and tissue culture.

Cutting

Scented roses, miniatures, climbers and ramblers and rootstocks are mainly propagated through cuttings. Cuttings are collected from healthier plants having 15-20 cm length and 3-4 nodes. Cuttings are treated with IAA (10 mgl^{-1}) or IBA (100 mgl^{-1}) and planted in mist chamber for easy and quicker rooting.

Budding

Hybrid teas and Floribunda roses are mainly propagated through budding. 'T' budding is the common method. Patch budding is adopted in higher elevations.

Stentling or Top Graft

In this method, root stock is an unrooted cutting, on which the variety is grafted. Grafted plants are raised in the field and graft establishment and rooting take place simultaneously. This procedure saves the time for rooting of root stocks.

Standard Rose

Hybrid teas and floribundas having vigorous growth and spreading habit are budded on straight vertical stem of a suitable root stock at different height from ground level. According to height of budding, they may be full standard (1.0 m) or half standards (40 to 60 cm). Weeping standards are budded at 1.5 m or even higher with scion having cascade branching habit.

Root Stocks

Selection of root stock depends on climatic and soil conditions. Various root stocks are used in different parts of India. In North India, *Rosa indica var. odorata* (Briar root stock) is mostly used (Singh and Saxena, 2001). *Rosa multiflora* is widely used in South India. IIHR, Bangalore has developed a thornless root stock 'Nishkant', a spontaneous mutant of *Rosa multiflora.* Apart from these, *R. bourboniana (*Edward rose), *Rosa canina inermis* (Dog rose) and *Rosa manetti* are also used as root stocks.

Climatic Requirements

Temperature

This affects quality and quantity of flowers. Moderately cool climate with bright sunshine and free ventilation are ideal. Temperature range of 18-30^{0}C (28-30^0 maximum and 18-20^{0}C Minimum) is ideal. Increase in temperature increases the number of flower and reduces their duration. High temperature reduces the flower quality (number of petals gets reduced and color fades easily) and shortens vase life. Stem length also decreases with increased temperature. High temperature also induces long neck (the part of stem between bud and upper most leaf which is less preferred).

Light

Rose is grown in areas of plenty of bright sunlight throughout day. Light requirement is between 10000-12,000 foot-candles. For cut flower production under protected cultivation, 40% shade is given. Growth is slow at low light intensity below 12 hours. Increase in light intensity increases quality and production. UV stabilized Poly films are preferred. Plants grown under low light intensity produce thinner leaves and have lower chlorophyll content of leaves, less flower yield and poor development of color.

Humidity

Humidity up to 75 % is ideal. High humidity induces diseases like powdery mildew. High humidity is a major problem for green house cultivation of roses in tropics. This results in condensation of water on lower surface of leaves and flowers at night which remain wet for some hours. Wet condition results in black spot and powdery mildew, causing leaf drop and root decay.

Soil

Rose prefers a sandy or sandy loam soil with a pH between 6-6.5. Soil should be rich in nutrients, organic matter and with good drainage. Soil with high water table is unsuitable.

Cultivation of Rose

In India, roses are grown under open field conditions for cut flower; loose flower, extraction of rose perfumes and for landscaping. This is mainly for internal consumption.

Planting

The best time of planting is during June-July in South India and September-October in North Indian plains. The initial preparation of rose beds should be started during summer so that soil gets exposure to the sun and air and during monsoon it gets a chance to settle down before planting. Beds of rectangular shape with 20 cm height are thoroughly prepared and made ready. During planting, the soil should not be too wet or too dry. Pits of 30 - 45 cm^3 are made for planting. Care should be taken to keep the bud union about 5.0 cm above soil level. After planting, the bed should be sufficiently irrigated.

Spacing

The spacing under open field cultivation varies depending on different types of roses used for different purposes. The spacing adopted under different situations is given below:

Rose types	Purpose	Spacing
Hybrid tea	Loose flower	45-60 x 45-60 cm
	Cut roses	90 x 60 cm
Floribunda	Loose flower	45-60 x 45-60 cm
	Cut roses	90 x 60 cm
Miniature	Loose flower & landscaping	30 x 30 cm
Bush rose	Loose flower & landscaping	90 X 90 cm
Standard Rose	Landscaping	120 X 120 cm
Climbers and ramblers	Landscaping	200 X 200 cm

High density planting under open conditions with suitable varieties such as Arjun, Sindhoor, Superstar, Montezuma etc., are recommended by IIHR at a spacing of 35 x 25 cm.

Pruning

Pruning in rose is the removal of unwanted and unproductive portions of plant, required for maintenance of productivity, flower quality and plant vigor. The objectives are to open up the bush for proper utilization of solar energy, better air flow, increased number of flowers, improved flower size and long stemmed flowers. It is also done to remove weak, staggered and crisscrossed branches and dead and diseased shoots. Pruning also becomes necessary for rejuvenation of bushes when yield and quality decline.

Time of Pruning

Pruning is done once or twice in a year. Rose is pruned traditionally during winter months of October-November in plains, and in March - April in hills. In Bangalore conditions, pruning is done twice during November and end of June. In Kerala and Tamil Nadu, pruning is recommended during November- December. For cut flower production, pruning is done exactly 45 days prior to the date of requirement of flowers.

Method

The method and intensity of pruning largely depend on cultivar, class of roses and health and vigor and age of plant. HTs withstand hard pruning. Floribundas should be pruned lightly. Polyanthas and miniature need thinning of overcrowded branches and tipping of ends. In climbers, only interlaced or diseased stems are removed.

Regular pruning is required from first year of planting and in subsequent years in Hybrid teas and floribundas. During first year, pruning is done to give definite shape to the plant. In newly budded plants, pinching is practiced to promote axillary branching. When these shoots grow, cut them back at 30-45 cm above bud union so that only 4-6 strong buds remain and allow the lateral shoots to grow. During 2nd year, retain all strong shoots and remove weak and diseased shoots and cut back the strong shoots to 4-5 buds. During 3rd and 4th year, cut back vigorous shoots to half of its growth. Rejuvenation pruning is done after 5-6 years. In this, all the main branches are cut back at 15-20 cm from the base.

The basic rule in pruning is to make slanting cut at an angle of 45^{0} at about 5 mm above a vigorous bud or a strong eye that points to the direction one desires the new shoot to grow. Make the cut always in the healthy region of shoot. Four healthy shoots are retained. Depending on the number of buds retained, pruning may be either light (7-8 buds), moderate (4-5 buds) or hard (1-2 buds).

Bending of Bushes

This is an alternative method to pruning adopted in rose grown under poly houses. This is done by bending of unproductive or unused stem in a slanting or horizontal position.

After pruning, apply Bordeaux paste or other fungicides over cut surfaces to prevent diseases. Pruned bushes should also be supplemented with organic manure at 10kg/ bush.

Pinching: It is the removal of terminal growing portion of stem. This reduces plant height and encourages axillary branching. Maturity of flower bud is delayed by this practice. The blind shoots that appear healthy are pinched hard by 1or 2 nodes to induce flowering

Disbudding: Undesirable buds are removed from time to time keeping only the central bud intact for better quality blooms especially in HTs.

Deshooting: Young vegetative shoots developing from the axil of basal and lateral shoots are removed to encourage growth of terminal shoot. Otherwise, it results in much branched plants that produce smaller flower in HT.

Mulching

Mulching with dry leaves, straw, saw dust etc at 10-15 cm thickness in rose beds during summer reduces weed growth and irrigation requirement of plant. Mulching is done after weed control. Plastic mulches are also used.

Manuring

In new planting, organic manure is provided while preparing beds or in planting pits @ 10 kg/ bush. In subsequent years, basins are formed around plants and manures are applied at 10-15 cm depth after pruning.

A mixture consisting organic and inorganic manures can be made with the following combination and 75 g of this mixture is applied in a month per bush (Sujatha and Narayana Gowda, 1997).

S.No	Manure	Quantity
1	Ground nut cake	5.0 kg
2	Bone meal	6.0 kg
3	Ammonium phosphate	2.0 kg
4	Urea	0.5 kg
5	Super phosphate	2.0 kg
6	Potassium sulphate	1.0 kg

Roses also respond well to foliar feeding. A mixture with the composition consisting urea (2 parts), Dihydrogen ammonium phosphate (1 part) Potassium nitrate (1 part), Potassium phosphate (1 part) can be made and 3g of this mixture is dissolved in 1 liter of water applied at fortnightly interval. Micro nutrient mixture consisting Magnesium sulphate (4 parts), Manganese sulphate (3 parts), chelated iron (2 parts), Zinc sulphate (2 parts) and Borax (1part) sprayed @ 2g/ liter once in two months is reported to be very effective in improving color of flower and foliage.

For high density planting at 30 x 30 cm spacing, a dosage of 300 Kg N, 600 kg P_2O_5, 300 kg K_2O/ ha / year is recommended by IIHR, Bangalore.

Irrigation

The interval between irrigation depends up on soil and climate. Flood irrigation is given at 10 days interval. Plants are watered twice a week during summer. Avoid irrigation immediately before and after pruning. Sprinkler and drip irrigation can also be practiced.

Pot Culture

This is usually adopted for home gardening. Plants are raised in 30 to 35 cm size pots. Pots are filled with potting mixture in the ratio of 3:2:1 Soil: FYM : leaf mould. Repotting is done once in a year or two. Other recommended practices like pinching, deshooting, pruning, disbudding, manuring, irrigation etc., also have to be adopted in pot culture for proper flower production and quality.

Harvesting

In budded plants, flowering starts from 1st year onwards. Economic yield is obtained from 2nd to 10th year. Flowers are harvested when the flower buds are in half open stage. Flower buds showing color having two outer petals unfurled from tip are harvested for local market. For cut flowers, they are harvested at tight bud stage with long stalks.

Yield

Loose flowers: 7.5 t/ha

Cut flowers

1st year : 100-120 flowers/m^2

2nd year: 200-240 flowers/ m^2

3rd year: 300-360 flowers / m^2

Post Harvest Handling

As soon as the flowers are harvested, the stems are lowered into buckets containing clean water with preservatives. The flowers are then transferred to cooling chambers and cooled down to 2- 4^0 C for 5-6 hours. They are graded according to the length of flower stalk which varies from 40-110 cm depending on variety and packed 20/ bunch.

Pre harvest spray of potassium sulphate (200 mgl^{-1}), GA_3 (100 mgl^{-1}) in November and January improved the quality and vase life. Use of Nickel chloride (300 mgl^{-1}) in holding solution increases vase life in roses. Pulsing with silver thiosulphate (1 mM) for 15 minutes along with sucrose (3%) for 16-24 hrs increased vase life. Similarly, D fructose (40 g/l) is also good as holding solution.

Protected Cultivation of Rose

Roses (*Rosa hybrida*) are also grown under poly houses/ green houses for export of cut flowers. In tropics, cultivation of roses round the year requires protection against excess heat, humidity and rains and to maintain productivity and quality. Hence roses are grown under poly houses in controlled climate.

Classification of Green House Roses for Export

Based on stem length, varieties are classified into

1. *Long stemmed rose:* Roses with a stem length 70-120 cm, flowers large, hybrid tea yielding 100 to 150 flowers/ m^2/ year.
2. *Medium stemmed rose:* Stem length between 50 to 70 cm, intermediate in stature consisting HTs and floribundas yielding 220 flowers/ m^2 /year.
3. *Short stemmed rose:* These are also known as sweet heart roses. The varieties consist HTs and floribundas and are small flowered and their stem length ranges from 30 - 50 cm, yields 250 to 350 flowers/ m^2/ year.

4. *Spray Roses:* They produce many blooms/ stem. Spray flowers consist mainly floribunda (bourbons) and their yields are generally low and the stems are heavy.
5. *Miniature roses:* They have stem length between 20 to 40 cm. Flowers are much smaller & are more popular as pot roses. They yield up to 500 flowers/ m^2/year and require more labor for harvesting and packing.

Rose Varieties for Green House/ Poly House Culture

The varieties selected should have high quality in terms of bud size, stem length and lush green leaves and should be free of pests and diseases and their symptoms. Varieties with good color like red, pink, pastel and bicolors which are high yielding, mildly fragrant and with longer vase life of more than 16 days, are selected. They should preferably be thornless for easy handling, take up water easily with no bent neck and which open perfectly after harvest and resistant to mechanical damage during transport.

At present more than 100 popular varieties of cut roses are grown all over the world and new varieties are added every year. In India, Dutch and Israeli varieties are popular in green houses. Some of the varieties grown are given below.

Aloha, Alsmeer Gold, Ambiance, Baronesse, Bellisima, Bianca, Burkila, Delilah, Esprit, Fantasia, Femma, First Red, Flintstone, Gabrielle, Golden Gate, Grand Gala, Gypsy, Hollywood, Jaguar, Jazz, Kiss, Konfetti , Laguna, Lambada, Mercedez, Miracle, Noblesse, Papillion, Passion, Prophyta, Ravel, Rodeo, Romantic Curiosa, Sacha, Saffire, Sphinx, Sublime, Susanne, Timeless, Versillea and Vivaldi.

Green House Requirement

For tropical region, saw tooth design of green houses is better in terms of effective climate control. Plenty of full sunlight with out shading the canopy level of plants throughout the year is required. Poly films specially made for growing roses are available. Usually 40% shade is provided during summer. The ideal temperature in green house is between 17- 18^{0}C during night and 24- 26^{0}C during day. A constant relative humidity between 70-80% can be maintained. Enrichment with CO_2 above 1000 ppm enhances productivity.

Planting Materials

a) *Rooted cuttings:* Certain varieties can be propagated through this method. They have shorter productive life span and show rapid decline in production.

b) *Tissue culture plants:* TC propagated plants are small and take longer time to establish. This is achieved through micro propagation of shoot apices, leaf blades, axillary buds etc.

c) *Budded plants:* 6-18 months old budded plants are used. These plants start flowering in four months of planting. Root stocks used are similar to that of open field condition.

Media: In Holland and Israel, rose is cultivated in soil less media such as Rock wool, perlite, sponge, coco peat, thermofoam, volcanic ash, pumis, vermiculite etc., In India, common medium used is cocopeat:perlite: vermiculite mixture in 8:1:1 ratio prepared in double layer tray of 20 cm depth. 10-12% humus soil is also used along with artificial media.

Planting

Green house roses are planted on raised beds 25-30 cm high or flat beds of width between1.0-1.6 m and 30-40 m long. The beds are provided with an alley space of 60-75 cm between them and after every 4-5 beds, a bay width of 6.4 -8.0 m is provided for easy movement. Roses are planted during May-June or in August- September in pits prepared on beds.

Spacing: The plants are usually raised in two rows/ width with spacing of 17-20 cm in rows and 40 cm between rows. The planting density varies from 7-8 plants/ m^2 (70000 to 80000 plants/ha).

After Care

Staking: supports are provided by posts erected at an interval of 3m or either side of the beds cross tied with GI wire or UV stabilized plastic ropes and the plants are fastened to them for support.

Disbudding: It is the removal of terminal buds one month after planting when they reach pea size leaving the leaves intact, to encourage lateral shoot production.

Cutting under: After a month of disbudding, the shoots are cut down leaving 3-4 true leaves and allowed for flower production. Cutting under is also practiced to reduce the height of plants.

Deshooting: Axillary buds that appear between flower and leaf should be removed regularly. Dead shoots are also removed frequently as they may harbor fungus.

Bending of bushes: Instead of pruning, bending of bushes is practiced in many polyhouses. It is first done in 3 months old plants. Bending induces cytokinin which results in lateral shoots and increases surface area and frame work for the plant. After this, unproductive and thin shoots are regularly bent.

Irrigation

Irrigation is carried out either through mist, micro sprinkler or drip system depending on soil or media used. Rose plants under protected condition require approximately 1 ltr of water / plant / day. The water quality must be good and salt water is unsuitable. EC should not exceed 1800 m mhos/cm. A drainage line is also provided below the beds for disposal of excess water.

Nutrition

Organic manure is added so that top 30 cm of soil has 30% of organic matter content. A dose of 25 kg of manure/ m^2/year takes care of this requirement. Soil analysis is critical in deciding the precise dosage of fertilizer requirement. Before planting, fertilizers can be incorporated in the bed itself as basal dose. When the plants are established, nutrition is provided through fertigation or fertilizer injection.

The fertilizer concentrations required for fertigation of mature roses in green house during summer and winter season are as follows (Chin and Sirohi, 2001).

S.No	Nutrient	Concentration (mgl^{-1})	
		Summer	winter
1	Nitrogen	100-120	150-160
2	Ammonium	20-25	20-25
3	Nitrate	80-100	130-140
4	Phosphorus	20-25	30-35
5	Potassium	120-140	160-170
6	Calcium	60-80	90-100
7	Magnesium	30-35	35-40
8	Sulphur	40-50	40-50
9	Iron	1.0-1.1	1.3-1.4
10	Manganese	0.5-0.6	0.6-0.7
11	Zinc	0.25-0.30	0.30-0.35
12	Copper	0.10-0.12	0.10-0.12
13	Boron	0.25-0.30	0.25-0.30
14	Molybdenum	0.05-0.06	0.05-0.06

Harvesting

After planting in May- June, first flower can be harvested in October. Flowers are harvested early in the morning leaving 1-2 mature leaves with 5 leaflets left on the plant. For local markets harvesting is done when the outer petal starts unfurling. For export, they are harvested before opening at tight bud stage when the bud develops a pin hole at center. Prematurely harvested bud bends by neck even if placed in water.

Yield

Yield varies depending on type of roses grown and for hybrid teas and floribundas may yield15-30 cut flowers/ year/ plant (150-200 nos /m^2/year)

Post Harvest Treatment

Harvested flowers are placed in buckets containing clean water treated with a mild bactericide immediately after harvest. Flowers are taken for pre cooling to remove field heat at 10^0 C for 18-24 hours before transit.

Grading

After precooling, they are graded based on stem length and quality. Graded flowers are bundled into 20 flowers according to stem length, wrapped in corrugated cardboard sleeves and boxes of standard size and stored at 2^0C in preservative for minimum period of 12 hrs before packing for shipment. Chemical preservatives like 8-HQC (150 mgl^{-1}) and Silver nitrate (1mM) or commercial preservatives like Florisant 600, Chrysal RVB-1%, Krislite VB etc., can be used for extending vase life.

Physiological Disorders

Water shoot: These are vigorous shoots which arise from below bud joint, grow erect and produce succulent and dark leaves and fail to produce flowers

Blindness: It is the condition in which the terminal flower bud fail to develop and abscises.

Bull head: formation of malformed flowers. Have more number of petals and petalloides.

TUBEROSE

Among the commercially grown flowers in India, tuberose occupies prime position because of its popularity as a cut flower, loose flower as well its potential in perfume industry. It is a bulbous flower known for its prettiness, elegance and delightful fragrance. It has good economic potential for loose/ cut flower trade and essential oil industry. The flowers are white and waxy and are used mainly for making garlands, bouquets, bridal makeup and other floral arrangements. The flowers borne on long spike stand long distance transport remain fresh for long and are excellent for table decoration. Tuberose flowers are also used in the perfume industry for extraction of high value floral concrete. The plants are also used in landscaping as a component in beds and boarders and as potted plants.

Classification

Tuberose is botanically known as *Polianthes tuberosa* Linn. which belongs to family Amaryllidaceae. On the basis of number of rows of petals, tuberose is classified into three groups of cultivars:

1. *Single:* Cultivars having flowers with only a single row of corolla segments. Flower is pure white and more fragrant than double. Flowers are extensively used for essential oil extraction and also for loose flowers. Concrete content is observed 0.08 to 0.11 %.
2. *Semi-double:* White flowers with 2-3 rows of corolla segments on straight spikes used as cut flowers.
3. *Double:* Flowers with more than three rows of corolla segments on long and sturdy spikes. Flowers are white and also tinged with pinkish red. They not bear seeds. It does not open well and is not commercially viable as a single cultivar. It is used as cut flower as well as loose flower and for extraction of essential oil. Concrete recovery is 0.06%.

Type	Varieties
Single	Sringar, Calcutta Single, Mexican Single, Rajat Rekha, Prajwal.
Semi double	Semi Double, Vaibhav
Double	Suvasini, Svarna Rekha, Calcutta Double, Pearl

Propagation

Tuberose is propagated through seeds, bulbs, bulblets, and through tissue culture. It is grown generally by division of bulbs each containing a bud and basal plate. Bulb size influences flowering. Large bulbs flower early and produce longer spike and more flowers (Nagaraju, 1992). Spindle shaped bulbs with a diameter of 2 to 3 cm are preferred to conical or flat bulls for planting. About 1.25 *to 1.50* lakh bulbs (800 to 900 kg) are required for planting one hectare.

Climate and Soil

Tuberose is adapted to mild tropical and subtropical conditions. It prefers open sunny location and should not be grown under shade which may drastically reduce the flower yield. Long days with 16 hour day length promote vegetative growth, precocity in flowering and length of spike. Optimum temperature for proper growth and development is 20 to 30 ^{0}C. Double cultivars prefer cooler climate compared to single.

Tuberose is grown in wide range of soils from light sandy to clay soils including those affected by salinity and alkalinity conditions. Porous and well-drained sandy loam soils with a pH of 6.5-7.5 are the best suited for tuberose cultivation.

Planting and Management

Under mild tropical conditions like that of Bangalore, tuberose can be raised round the year. In North India, it can be planted from March to May. In Kerala it is planted in July-August.

An open field having plenty of sunlight and adequate drainage is selected. Land is prepared well by ploughing two or three times. The land is ploughed 30 cm deep, to a fine tilth and raised beds with 1.0-1.2m width are taken. Decomposed organic matter or FYM @ 30 t/ha should be thoroughly incorporated into the bed.

The bulbs preferably of size 2-5 cm or above are planted at a depth of 4-8 cm. Planting single bulb / per hill is better over clump planting. In single cultivars, planting up to 3 bulbs / hill for one year crop and 2 bulbs / hill for more than 1 year crop is recommended. In 'double' 2 bulbs / hill is better, for 1 year crop.

Spacing : Spacing varies with soils. A spacing of 30 cm x 30 cm is recommended by TNAU. The spacing recommended for Kerala is 20 x25 cm (KAU) and for Karnataka 20 x 20 cm (IIHR).

Manuring

Tuberose is a heavy feeder and responds well to application of organic and inorganic nutrients. A fertilizer dosage of 200 kg N, 400 kg P_2O_5 and 200 kg K_2O kg/ha N:P:K is recommended for Tamil Nadu. In Kerala, NPK (100:50:50 Kg/ha) is recommended. Of these, half N, complete P and K are applied at the time of planting. Remaining N is applied when the flower spikes start to appear. Under Bangalore conditions, NPK @ 15:40:40 g /m^2 is recommended for 'double' cultivars and 40:40:40 g NPK/ m^2 for single. NPK

200:200:200 kg/ha is recommended for commercial cultivationby AICRP on floriculture. The fertilizers can be provided in three splits, as basal, 60 days after planting and 90-120 days after planting.

Weed Control

Tuberose field can be kept almost weed free by pre-emergent application of gramaxone (3.0 l/ha) followed by post-emergent sprays thrice at an interval of 40 days in between rows of crop. Pre-planting application of atrazine (3.0 kg/ha) was also recommended for weed control in tuberose and more production of good quality flowers.

Irrigation

The filed is irrigated before planting. Further irrigation is given only when the bulbs germinate. Thereafter, a heavy irrigation is provided once in 10 days or at weekly intervals during summer.

Effect of Growth Regulators

The effects of pre-planting treatment of bulbs with GA_3, ethrel or thiourea promoted early appearance of flower and the number of flower spikes but reduced number of bulbs / plant. Treatment with GA_3 (200ppm) produced the highest number of longer spikes with maximum number of florets.

Harvest and Yield

Flowering of tuberose starts 80 to 100 days after planting. Tuberose flowers all the year round. The peak flowering is between July and October. For cut flower purpose, the whole spike is cut from the base when the basal 1or 2 pairs of florets are opened. Spikes are harvested preferably in the morning before sunrise or late in the evening by clipping with a sharp knife that gives a clean cut. About 4-6 cm basal portion has to be left to allow growth of bulb. The lower portions of cut spikes immediately after harvest, are to be immersed in water for prolonging life of spikes. Yield of spikes varies from 2-4 lakh spikes/ ha.

For loose flower purpose and essential oil extraction, individual flowers at full maturity stage are plucked from spikes early in the morning. Loose flowers are usually packed in bamboo basket which can hold about 10-15 kg of flowers and sold on weight basis. The yield of fresh loose flowers / year / hactare in "Single" tuberose varies from 10 to 15 tonnes / hactare in the first year for the plant crop, 15 and 20 tonnes per hectare in the second year for the first ratoon and 5 to 7.5 tonnes per hectare in the third year for the second ratoon. Out of the approximate total yield of 30,000 kg of loose flowers from one hectare, in three years, 27.5 kg of 'concrete' can be obtained. This concrete in turn will yield about 5.50 kg of absolute.

Ratoon Crop

After the harvest of main crop, flower stalks are headed back and the plot is manured and irrigated. The preceding year's crop is allowed to grow undisturbed by following usual cultivation practices. Two or three ratoon crops can be taken from single planting. However yield and quality of flower are poor in ratoon crop (Singh and Uma, 1995). If the bulbs are not uprooted and replanted after three or four ratoons, the spikes and flowers tend to become smaller and unattractive. For cut flower purpose, fresh crop is advised for every year.

Grading

The flower spikes are graded according to total length, length of stalk, number of flowers/ spike and weight of spikes. Straight and strong stem of uniform length and uniform stage of development are preferred. Flowers should be free from bruises, diseases and pests. Florets are graded according to their size for loose flowers.

Packing and Transport

For room decoration, long spikes are preferred and are bunched in bundles. Each bundle contains 50-100 spikes. These bundles are packed in rectangular bamboo baskets lined with Hessian cloth. For long distance transport, it is better to pack them in corrugated cardboard boxes.

Holding Solutions

Citric acid (300mgl^{-1}) + Sucrose (2%) as holding treatment is recommended for tuberose cut spikes.

Curing and Storage of Bulbs

Bulbs reach maturity at cessation of flowering when leaves become yellow and dry during winter (February-March) in North India. At this stage, irrigation is withheld and the soil is allowed to dry. The leaves are cut off at ground level and bulbs are dug out. After digging, the bulbs are lifted out and cleaned thoroughly. The offsets are then separated out and used as seed-stock for next season. The bulbs are then graded based on size into mature (> 1.5 cm diameter) and immature (< 1.5 cm diameter). Cleaned and graded bulbs are placed on shelves to dry or cure. The bulbs must be stirred or have their position changed every a few days to prevent fungal attack and rotting. The bulbs have dormancy and they will have to be stored for about 4.5 weeks before planting. Storing the bulbs for 30 days at 10°C results in good germination.

Extraction

Tuberose concrete and absolute are prepared in the same way as jasmine concrete and absolute. Concrete is extracted from fresh flowers harvested before sunrise before petals open, using solvents such as hexane followed by distillation to recover solvent. Single cultivars are suitable for concrete production. Recovery of concrete from flower ranges from 0.08 -0.14%. 1150 kg flower is required for making 1kg concrete. Good quality concrete contain 30-35% absolute.

Disease and Pest Management of Major Flower Crops

Anthurium

Pest	Scientific Name	Control
Nematodes	*Meloidogyne incognita* *Radopholus similis* *Pratylenchus*	Media sterilization Maintaining hygiene Nemacure15G 5-7g/m^2
Aphids	*Myzus persicae* *Aphis gossypii*	Dimethoate (2 mll^{-1})
Mites	*Tetranychus urticae* *Polyphagotarsonemus latus* *Tarsonemus pallides*	Abamectin Kelthane
Thrips	*Thrips tabaci*	Dimethoate (2 mll^{-1})
Slugs and snails	*Levicantis alti* *Limax sp*	Traps using beer or potato slices metaldehyde

Disease	Causal organism	Control
Bacterial blight/ Anthurium decline	*Xanthomonas campestris* *pv. Dieffenbachiae*	Sanitation streptomycin sulfate oxy tetracycline
Anthracnose	*Colletotrichum* *gloeosporioides*	Mancozeb (0.25%) Carbendazim (0.1%)
Root rot	*Pythium sp.* *Phtophthora sp.*	Mancozeb (0.25%) Metalaxyl (0.25%)
Black rot	*Rhizoctonia solani*	Benlate (0.1%)
Mosaic	Virus	Control of aphids Removal of affected plants

Carnation

Pests	Name	Control
Aphids	*Myzus persicae*	Dimethoate (2ml/l), endosulphan (2.5 ml/l)
Thrips	*Thrips tabaci*	Dimethoate (2ml/l)
Red Spider Mite	*Tetranychus urticae*	Kelthane (2.5ml/l), dicofol (1 ml/l)
Nematodes	*Meloidogyne* spp.	Carbofuran 3g (0.5Kg ai/ha)

Disease	Organism	Control
Fusarium wilt	*Fusarium oxysporum*	Benlate (0.2%), Ridomil (0.2%)
Rust	*Uromyces dianthii*	Mancozeb (0.3%)
Grey mould	*Botrytis cinerea*	Benlate (0.2%)
But rot	*Rhizoctonia solani*	Benlate (0.2%)
Mosaic	Carnation mosaic virus	Removal of affected pants

Chrysanthemum

Pest	Scientific name	Control
Leaf miner	*Chromatomyia syngenesiae*	Triazophos (2 $ml l^{-1}$)
Spotted spider mite	*Phytonemus pallidus*	Abamectin (o.4 $ml l^{-1}$)
White fly	*Aleurodicus dispersus*	Triazophos (1.5 $ml l^{-1}$)
Thrips	*Thrips nigropilosus*	Dimethoate (2 $ml l^{-1}$)
Aphids	*Myzus persicae*	Dimethoate (2 $ml l^{-1}$)
Nematodes	*Pratylenchus penetrans*	Carbofuran (15 kg/ha)

Disease	Causal organism	Control
Dry rot / crown rot	*Rhizoctonia solani*	Rhizolex (0.4 $ml l^{-1}$)
Wilt	*Verticilium sp*	Soil sterilization with propanocarb (2-3g/m^2)
Fusarium Wilt	*Fusarium oxysporum*	Carbendazim (0.5 ml/lit)
Root rot	*Pythium sp* *Phytophthora sp*	Soil sterilization with propanocarb (-3g/m^2) Mancozeb (1.5g l^{-1})
Leaf spot	*Septoria chrysanthemella*	Mancozeb (1.5g l^{-1})
Grey mould	*Botrytis cineria*	Polyoxin (0.3g l^{-1})

Black rot	*Ascochyta sp*	Mancozeb (1.5g l^{-1})
Stem rot	*Fusarium solani*	Dithane M45
Bacterial disease	*Pseudomonas cichorii*	Soil sterilization with copper hydroxide (0.3%)

Crossandra

Pests	Scientific Name	Control
White flies	*Bemecia tabaci*	Phosalone (0.07%)
Nematodes		
Root knot Nematode Leision nematode Needle nematode	*Meloidogyne incognita* *Pratylenchus dellatteri* *Longidorum atriconcy*	Phorate (1g per plant), carbofuran 3 G at 1 kg a.i./ha
Aphids	*Aphis gossipy*	Spray dimethoate 30 EC 2 ml/lit.

Disease	Causal organism	Control
Wilt	*Fusarium solani*	Carbendazim (0.5 ml/lit)
Wilt	*Rhizoctonia spp*	Benlate (0.1%)

Gerbera

Pest	Scientific name	Control
Aphids	*Aphis gossypii*	Dimethoate (0.05%)
Thrips	*Thrips tabaci*	Monocrotophos (0.05%)
White fly	*Bemisia tabaci*	Phosalone (0.07%)
Spider mites	*Tetranychus urticae*	Dicifol (0.1%)
Leaf miner	*Liriomyza trifolii*	Triazophos (2 mll^{-1})
Nematode	*Meloidogyne incognita*	Carbofuran (0.5 kg ai/ha)

Disease	Causal organism	Control
Crown rot	*Phytophthora cryptogea*	Metalaxyl (0.2%) soil drenching
Root rot	*Pythium* sp.	Copper oxychloride (0.4%) drenching
Fusarium rot	*Fusarium oxysporum*	Bordeux mixture (1.0%)
Botrytis blight	*Botrytis cinerea*	Carbendazim (0.1%)
Powdery mildew	*Erysiphe cichoracearum*	Dinocap (0.05%)
Leaf spot	*Alternaria alternate*	Zineb (0.2%)

Gladiolus

Pest	Scientific name	Control
Thrips	*Taeniothrips simplex*	Dimethoate 30EC (2ml/l)
Aphids	*Aphis gossypii, Myzus persicae*	Dimethoate 30EC (2ml/l)
Cut worms	*Agrotis segetum*	Carbaryl or Malathion (0.1%)
Mite	*Rhizoglyphus echinopus*	Dicofol (1 ml/l)

Diseases	Causal organism	Control
Fusarium corm rot/wilt	*Fusarium oxysporum*	Selection of tolerant variety- Dhiraj
Botrytis blight	*Botrytis gladiolorum*	Dithane M45 (0.2%)
Curvularia blight	*Curvularia sp*	Benomyl (0.2%), Captan (0.03%)

Jasmine

Pest	Scientific name	Control
Blossom midge	*Contarinia maculipennis*	Monocrotophos (0.05%)
Bud Worm	*Hendecasis duplifacialis*	Carbofuran (40g/plant)
Thrips	*Thrips orientalis*	Dimethoate 2 ml/l
Jasmine Moth	*Palpita unionalis*	Monocrotophos (0.05%)
The Gallery Worm	*Elasmopolpus jasrninoghagus*	Malathion (0.2%)
Root-Knot Nematode	*Meloidogyne incognita*	Neem cake (1 t/ha) or Carbofuran (2.5kg/ha)

Disease	Scientific Name	Control
Leaf blight	*Cercospora jasminicola Alternaria jasmini*	0.2% mancozeb or 0.1% benomyl
Fusarium wilt	*Fusarium spp*	1 % Bordeaux mixture.
Rust	*Urornyces hobsoni*	0.2% Zineb
Wilt	*Fusarium solani*	Drenching with 1% Bordeaux mixture.

Marigold

Pest	Scientific Name	Control
Aphids	*Neotoxoptera oliveri*	Dimethoate 0.03%
Mites	*Tetranychus urticae*	Dicofol (0.1%)
Leaf hoppers	*Amrasca biguttula biguttula*	Quinalphos (0.05%)
Flower thrips	*Megalurothrips distalis*	Dimethoate (2 mll^{-1})

Disease	Causal organism	Control
Wilt	*Phytophthora cryptogea*	Improve drainage
Stem rot	*Sclerotinia sclerotiorum*	Drenching with copper oxy chloride
Powdery mildew	*Oidium* sp., *Leveillula taurica*	Sulfex (3 g/l)
Leaf Spot and Blight	*Cercospora and Septoria* sp	Dithane M 45 (0.2%)
Flower Bud Rot	*Alternaria dianthi*	Mancozeb (2g/litre)

Orchids

Pest	Scientific Name	Control
Flower Thrips	*Megalurothrips distalis*	Dimethoate (2 mll^{-1})
Aphids	*Macrosiphum luteum*	Dimethoate (2 mll^{-1})
Spider mite	*Tetranychus urticae*	Kelthane (2.5ml/l), dicofol (1 ml/l)
Brown Soft scale	*Coccus hesperidium*	Carbaryl (0.2%)
Mealy bugs	*Pseudococcus microcirculus*	Dimethoate (2 mll^{-1})
Snails and slugs	*Levicantis alti* *Limax sp*	Hand picking, Dimethoate (2mll^{-1}) or Carbaryl (0.2%)

Diseases	Causal organism	Control
Heart rot	*Phytophthora sp*	Metalaxyl (0.25%)
Leaf spot	*Colletotrichum gloeosporioides*	Mancozeb (0.3%)
Leaf blight	*Pythium*	Carbendazim (1g/l)
Collar rot Orchid wilt	*Sclerotium rolfsii*	Drenching with Zineb (0.3%) or Bavistin (0.5-1%)
Mosaic and ring spot	Virus	Destroy the affected plants

Rose

Pest	Scientific name	Control
Aphids	*Macrosiphum rosae*	Dimethoate 0.05%
Thrips	*Frankliniella tritici* *Thrips maginis*	Dimethoate 0.05%
Bugs	*Philaenus spumarius*	Quinalphos 0.05%
Midges	*Dasyneura rhodophaga*	Quinalphos0.05%, Imidachloprid 0.006%
Mite	*Tetranychus urticae*	Dicofol 0.1%
Red scale	*Aonidiella aurantii*	Dimethoate 0.05%.

Disease	Causal organism	Control
Powdery mildew	*Sphaerotheca pannosa*	Wettable Sulphur 0.2-0.5%
Dieback and stem blight	*Diplodia rosarum*	Carbendazim 1-1.5 g/l
Black spot	*Diplocarpon rosae*	Zineb 0.2-0.4%
Rust	*Pragmidium sp.*	Mancozeb 02-0.4%
Leaf spot	*Alternaria alternate*	Zineb 0.2-0.4%
Rose wilt	*Marmor flaccumfasciens*	Control of aphid vector and removal of affected plants

Tuberose

Pests	Scientific name	Control
Thrips	*Frankliniella bondari*	Endosulfan ($2ml^{-1}$) or Malathion ($1.5\ ml^{-1}$)
Bud borer	*Helicoverpa armigera*	Endosulfan (0.2%) or Monocrotophos (0.2%)
Aphids	*Aphis gossypii*	Malathion (0.1%)
Nematodes	*Meloidogyne incognita,* *Rotylenchulus reniformis,* *Aphelencoids besseyi*	Carbofuran (2-5 kg/ha), Neem application (1 tonne/ha)
Weevils	*Myllocerus sp*	Applying BHC dust (10%) in the soil

Disease	Causal Organism	Control
Stem rot	*Sclerotium rolfsii*	Drenching with Zineb (0.3%) or Bavistin (0.5-1%). Soil application of Brassicol (20%) @ 30 Kg/ha
Leaf blight	*Botrytis elliptica*	Ammonical copper (2%) or Greeno (0.5%)
Alternaria leaf spot	*Alternaria polyanth*	Mancozeb (0.2%) or Iprodione (0.2%) at 10 days interval
Flower-bud rot	*Erwinia* sp.	Streptomycine (0.01%) spray

REFERENCES

Ahuja, A.K. 2001. *Floriculture- current status, vision and action plan*. TIFAC, New Delhi.

Alam, M.M., Khan, A.M. and Saxena, S.K. 1988. Management of plant parasitic nematodes by different cropping sequences. *Indian Jurnal of Plant Pathology* 6:102-109.

Aruna, 1999. *Yield and physiological responses to Nitrogen levels and micronutrients in crossandra (Crossandra infuundibuliformis)*. Ph.D (Hort) thesis. TNAU, Coimbatore.

Chin, A. and Sirohi, N.P.S. 2001. Advances in hi-tech cultivation of roses. (*In*) *Adavances in Floriculture* Eds. M.L.Chowdhary K.P. Singh and C.T.S. Hussain. Division of Floriculture and Landscaping, IARI, New Delhi.pp.70-72

Choudhary, M.L. 2001. Floriculture in India: An insight. (*In*) *Commercial flower production*. Eds. S.S. Sindhu, and M.L. Choudhary, Division of floriculture and landscaping, IARI, New Delhi. pp.1-7.

Duniway, J.M. 2002. Status of chemical alternatives to methyl bromide for preplant fumigation of soil. *Phytopathology* 92:1337-1343

Gill, A.P.S. 2001. Carnation growing in tropics. (*In*) *Advances in Floriculture*. Eds. M.L.Chowdhary K.P. Singh and C.T.S. Hussain. Division of Floriculture and Landscaping, IARI, New Delhi.pp.73-77

Gopinath, G. and Khan, M.M. 1997. Protected cultivation of carnations. (*In*) *Progressive Horticulture*. Eds. I.S. Yadav and M.L.Choudhary. The House of Sarpans (Media), Bangalore pp.34-40

Gurav, S.B., Singh, B.R. Desai, U.T., Katwate, S.M., Kakade, D.S. and Dhane, A.V. 2005. Effect of spacing on yield and quality of gerbera under polyhouse. *Journal of Ornamental Horticulture* 8:62-64.

Mukhopadhyay, T.P. and Bose, T.K. 1979. *Punjab Horticulture Journal*. 19:98-101.

Muthuswamy, S. and Madhava Rao, V.N. 1974. *Indian Perfumer* 18:1-4.

Narayanagowda, J.V. and Nagegowda, V. 1994. *Floriculture- Technology, trade and trends-* Eds. Prakash, J. and Bhandary, K.R. Oxford and IBH Publishing Co., New Delhi, pp.145-147

Nagaraju, H.T., 1992. National Seminar on Commercial Floriculture in India- Present and Potential. pp:109.

Panwar, R.S. and Saxena, N.K. 2001. Gladiolus cultivation. (*In*) *Commercial Flower Production* Eds. Sindhu, S.S. and Choudhary, M.L. Division of Floriculture and Landscaping, IARI, NewDelhi.pp.35-40

Paulos, D. 1980. National seminar on Production technology of commercial flower crops. TNAU, Coimbatore. pp.17-18.

Prakash, D., Prasad, K.V., Choudhary, M.L., Priya Marie, C. and Bhushan, B. 2001 The art of growing orchids: A diverse crop.(*In*) *Advances in Floriculture* Eds. M.L Choudhary, K. P. Singh and C.T.S. Hussain. Division of Floriculture and Landscaping, IARI, New Delhi.pp.103-115

Prasad, K.V., Prakash, D., , M.L. and Pandey, R. 2001. The art of growing anthuriums. (*In*) *Advances in Floriculture* Eds. M.L Choudhary, K. P. Singh and S. Hussain. Division of Floriculture and Landscaping, IARI, New Delhi.pp.116-126

Rajeevan, P.K. 2006. Production technology of Orchids. (*In*) *Reflections on Indian Flori-business.* Eds.P.K. Rajeevan., P.K.Valsalakumari., C.K.Geetha, and P.K.Sudha Devi. Dept. of Pomology and Floriculture, College of Horticulture, Vellanikkara. pp.109-123

Rangasamy, P. 1997. Production technology of Chrysanthemum (*In*) *Progressive Horticulture.* Eds. I.S. Yadav and M.L. Choudhary, The House of Sarpan (Media) Bangalore pp.71-79.

Reid, M.S. 2004. Produce facts-Gerbera, Transvaal Daisy- Recommendation for maintaining post harvest quality, post harvest technology. Research & Information Center, University of California, Davis, USA

Roy, B.N., Basu, R.N. and Bose, T.K. 1973. *Plant Sciience.* 5:117-122.

Seghal, O.P. 2001. Advances in production technology of gladiolus. (*In*) *Compendium of lectures on Recent Advances in Floriculture Research.* Eds. M.L. Choudhary, A.P. Singh, K.V. Prasad and K.P. Singh. Division of floriculture and Landscaping IARI New Delhi.pp11-23.

Singh, F.1994. New strategies in tissue culture propagation of *Anthurium andreanum.* Paper presented at the First International Seminar on Anthuriums, Thiruvananthapuram, Kerala, India during 8-9 May 1994.

Singh, K.P. and Uma, S.1995. Studies on ratoon crop in tuberose cv. 'Single' and 'Double' *Indian Perfumer* 39:158-160.

Singh, H.P. 2001. Commercial cultivation of carnation under North Indian condition. (*In*) *Commercial Flower Production*. Eds. S.S Sindhu and M.L.Chowdhary. Division of Floriculture and Landscaping, IARI, New Delhi.pp.26-34

Singh, A.P. and Saxena, N.K. 2001. Package of practices for growing roses. (*In*) *Commercial Flower Production*. Eds. S.S. Sindhu and M.L.Chowdhary. Division of Floriculture and Landscaping, IARI, New Delhi.pp.17-25

Sharma, H.G., Verma, L.S. and Tiwary, B.T. 1990. Effect of foliar application of some plant growth regulators on growth and flowering of Chrysanthemum. *Orissa Journal of Horticulture* 23:1-2.

Sujatha, K. and Narayana gowda, J.V. 1997. Production technology of rose. (*In*) *Progressive Horticulture*. Eds. I.S. Yadav and M.L.Choudhary. The House of Sarpans (Media), Bangalore pp.55-70

Valsalakumari, P.K. and Rajeevan, P.K. 2006. Production technology of anthurium.(*In*) *Reflections on Indian Flori-business*. . Eds.P.K. Rajeevan., P.K.Valsalakumari., C.K.Geetha, and P.K.Sudha Devi. Dept. of Pomology and Floriculture, College of Horticulture , Vellanikkara.pp.124-145.

PART - II

Chapter 4

PLANTATION CROPS

Introduction

Plantation crops are high value industrial crops with great economic importance and play vital role in trade. It contributes to environment protection and sustains a large number of agrobased industries. It has great potential for utilization of waste lands in varied agro-ecosystems like rainfed, dryland, hilly, arid and coastal, providing higher employment opportunities, nutritional security, eco-friendly, high potential for foreign exchange earnings and above all providing livelihood security to people. They contribute 25% of AGDP. Major plantation crops grown in India are coconut,cardamom,cinchona, arecanut, oilpalm, cashew, palmyra, rubber, cocoa, tea and coffee. These crops grown between 20° N and 20° S of equator occupy nearly 3.2 million ha with a production of 9.8 million tonnes. All these crops are perennial in nature and demand labour round the year. All plantation crops enter international trade and all of them have a long juvenile period. Coconut, arecanut and cashew are cultivated by individual land owners who are not organized to the same extent as planters of tea, coffee and rubber. Coconut and arecanut are cultivated from time immemorial. Tea, coffee and rubber have comparatively large plantations which started fairly recently with arrival of Europeans. Since independence, these crops showed increase in area and production. Changes in area and production are neither uniform among crops nor over the periods as both are very much influenced by climatic and seasonal changes, especially occurrence of droughts, cyclones, disease and pest incidence and market prices. Plantation crops are among the oldest organized industries in India and continue to be the mainstay of agrarian economies in many states of the country. They provide direct and indirect employment to more than 30 million people in the country.

India is the leading producer of most of these plantation crops, its position fluctuates from year to year. In addition to primary products from these crops which are largely exported, the by-products from many of these crops also have considerable industrial and commercial uses. Coir, wood and shell products from coconut, leaf sheath in arecanut, cashewnut shell liquid (CNSL) and fenni from cashew are a few of the by-products which made considerable impact in industrial field.

Most of the plantation crops introduced into India received little attention in scientific management and research till 1950's. Stagnation of yield is mainly due to lack of systematic replanting programme.

Coconut

Coconut occupies prime position in the cultural, social and economic lives of millions of people across the world. The crop husbandry and allied activities provide livelihood and food security to more than 10 million people in India. Coconut oil, the main commercial product, determines prices of coconut and its products. The recurring phenomenon of price fluctuations, seasonal variations in production, availability of other vegetable oils like palm oil and soybean oil are the root causes for the crisis of the industry. The increasing trend in area and production of coconut in the country with the regress in the consumption of coconut oil in both edible and non-edible sector on account of cheap substitutes necessitated development of broad based processing and cultivation technologies for sustainable growth of the industry.

Coconut is grown in 93 countries in the world. Indonesia is the leading producer with an area of 3.88 million ha.(32.2%) producing 19.358 billion nuts followed by Philippines(26.65%) and India(15.91%). Indonesia (31.94%), India(19.88%), Philippines (22.54%) and Sri Lanka together account for 78.27% of global production. Productivity of coconut in India in terms of nuts/ha is the highest among major coconut growing countries . Indonesia has the highest productivity in terms of out put of oil/ha. The other countries contributing significantly to the world production of coconut are Bangladesh, Thailand, Vietnam, Papua New Guinea, Brazil, Myanmar, Mozambique and Tanzania.

In India, area and production of coconut generally showed increasing trends. There was an overall increase of 187% in area and 326% in production during 1995-96 compared to 1950-51. The production showed an increase of

303% over a period of 46 years. More than 91% of area under coconut is in the four Southern states viz. Kerala, Tamil Nadu, Karnataka and Andhra Pradesh sharing 92%of India's production. Even though area and production of coconut in the country is increasing, growth in productivity is at a low pace.

Traditionally coconut industry in India was centered on copra making, extraction of coconut oil and coir manufacturing. Economic globalization has made integrated various regional markets into a world market. As a result, various new coconut products from elsewhere in the world have found a prominent place in the product profile of the food chain markets in the country. To make the industry competitive significant changes have to be made in domestic market in terms of product developments and deep market integration. Emphasis has to be given for value addition in coconut through product diversification and by-product utilization and for evolving technologies for development of new value added products in tune with those of other leading countries.

Arecanut

Arecanut is one of the most important commercial crops in South East Asia. It finds a place in all religious, social and cultural functions. The nut is cultivated in palm plantations and the tree and the nut have a never-ending list of uses like chewing purpose, vegetable, medicine, stimulant, timber, fuel wood, clothing, wrapping, lubricant and tannin. When the nut is chewed along with betel leaf, it increases the stimulating effect, though excessive consumption enhances risk of cancer. Areca is taken up from the Malayan language that means 'cluster of nuts'.

World production of arecanut sums up to 0.64 million t., Asian contribution being the maximum. India is the leading country in production contributing to 51% of world's total production followed by China(25%) . Major producers are India, China, Myanmar, Bangladesh, Indonesia, Thailand and Malaysia. Regarding area under cultivation of betel nut, the world figures have reached 0.476 million ha. India has 60% of the total area under arecanut in the world.

India dominates production of arecanut producing almost half of the nuts produced in the world. It has also the maximum area covered under betel nut as compared to other nut producing countries of the world, and earns about Rs.45 million annually by exporting arecanut in various forms. Area and production of arecanut in India showed an increasing trend during the past 40 years. The area under arecanut in India was 0.106 million ha during 1955-56 which increased to 0.236 million ha during 1994-95 recording an increase of 123% over a period of 40 years. Production for the corresponding period increased from 0.081 to 0.272 million t. (236% increase). The productivity remained almost constant from 1957 to1974 (around 850kg/ha) and slightly increased thereafter during the next five years to around 950 kg/ha, and further increased and reached the peak level of 1200 kg/ha. Yellow leaf disease (YLD) is the most serious malady affecting arecanut in Kerala and Karnataka.

Arecanut production in India has now reached a level of self-sufficiency and the present production is more than adequate to meet the internal demand. There is steady decline in price of arecanut mainly due to increased production. In the absence of availability of an effective alternative use of arecanut and limited export potential, the national policy has to be for concentrated developmental efforts in more favourable areas of arecanut. This can be done through investment oriented programmes with all input components like high yielding variety, optimum fertilizer application, irrigation and plant protection measures. Inter and mixed croppings in arecanut gardens are advocated to augment income from existing gardens.

Oil Palm

Oil palm is produced in 42 countries worldwide on about 12 million ha. Cultivation of oil palm expanded in recent years such that it is now second only to soybean as a major source of the world supply of oils and fats. Palm oil shares 40% of the world's trade in Edible Oil. Over the past 30 years, the worldwide area planted to oil palm increased by more than 150%. Most of this increase took place in South East Asia, with spectacular production increases in Malaysia and Indonesia There are several reasons for this rapid expansion. Crude palm oil and kernel oil prices have been strong, due to the rapid increase in consumption of dietary oils and fats in the developing economies of China and India. This encouraged investors to develop plantations on large areas of suitable land found in peninsular Malaysia and the islands of Sumatra in Indonesia and Borneo, where part belongs to Malaysia (Sabah and Sarawak) and part to Indonesia (Kalimantan).

Malaysia is the largest palm oil producer in the world accounting 50 % of global production of which 85% exported. The share of Indonesia in global production of palm oil is 30 % of which 40% is exported. Global demand for palm oil is likely to increase from the present level of 22.5 to 40 mt in 2020. This forces the major producing countries in South East Asia to establish new plantations, a significant share of these by converting natural forests. Oil palm was introduced to India a century ago as ornamental palm in the

Botanical Garden of Calcutta and isolated trees are found growing from Kanyakumari to Tripura. Oil palm, gained commercial plantation status during seventies, grown largely in Kollam district of Kerala and Little Andaman Islands. During later half of eighties only, potential of its cultivation as a commercial oilseed crop under irrigation was realised. Its cultivation is spreading very fast in Andhra Pradesh, Tamil Nadu, Karnataka and Maharashtra in the identified areas. Production of palm oil in India is negligible (0.05 mt). India is the world's largest importer of palm oil, followed by EU, China and Pakistan. Currently, India imports around 3 mt of palm oil annually to meet one-fourth of its edible oil requirements. These imports are from East Asian countries, particularly from Indonesia and Malaysia.

The oil palm has the distinction of being the most productive of all oil crops with an average yield in major producing countries (3-4 t. of mesocarp (palm) oil/ha/year). By contrast, yields of most competing oil crops are typically less than one t./ha/year. This means that productivity of oil palm is at least 3-8 times more than most oil seed crops. Thus, only 7 m.ha of oil palm are required to supply 20% of the world demand for oil and fats (1.09 billion t.), compared to 80 m. ha. of oilseeds needed to supply another 24% of this demand.

Cashew

Cashew is produced in almost 32 countries of the world and now it has become the number one crop in the world taking over almonds. It is generally found in the tropical zone of the planet as it thrives on a hot and dry climate. The world production figures of this crop hover around 22.5 lakh t./ year. The major raw cashew producing countries are Vietnam, India, Brazil, Nigeria, Indonesia, Tanzania, Cote de Ivoire, Guinea, Mozambique, Benin, Thailand, Malaysia, Kenya,Ghana, Philippines, Madagascar, Sri Lanka ,Senegal, Burkina Faso and El Salvador. Regarding production of cashew kernels in the world, India is the top most producer. Prior to 2002, India was also the leading producer of raw cashews in the world but in that year Vietnam took over the lead from India. Production trend of cashew has ever been rising and it doubled since 1994. World's total area under cultivation of cashew crop is around 35.1 lakh ha, India is at the top in list of the countries regarding area utilized in production of cashew. It is a high ratio of area as compared to crop yield due to low productivity, the world's average yield being just 700 kg / ha.

Cashew is one of the important plantation crops introduced to India during the 16th century. The potential of this crop in international trade was first realized by India in early 1900s through export of cashew kernels. As cashew was considered suitable for afforestation and soil conservation , cashew plantations drew till recently very little attention, and cashew was always planted in waste lands and marginal lands with poor fertility status. India has always been a major player in the production of cashew. It is the second largest producer of raw cashew in the world but conquers 1st place among the largest producing countries of cashew kernels and also in the maximum area covered (7.70 lakh ha currently). Indian production of cashew contributes to around 4.6 lakh t. / annum. The country provides around 55% supply of cashew kernels in the world. Country exports cashew kernels since 1900s. Over the years, both export earnings and quality of kernels are increasing. As the domestic production is only 50%, India imports raw nuts from African countries to meet demand of cashew processing industries.

The pace of industrial growth is slowed down due to shortage of raw nuts and decline in productivity from over aging and senility of plantations . To bridge the gap of industrial demand and present production, steps are needed to increase production two folds. This can be achieved by adopting scientific inputs of high yielding varieties and production practices. New areas for expansion of cashew are restricted and even if it is possible, such areas are not superior as the present plantations . Nearly 70% of existing area are constituted of non-pedigree planting materials. Genetic upgradation with high yielding varieties by replanting should be resorted in future strategy.

The changing global scenario of production of cashew at an internationally competitive price by reducing cost of production and increasing the production/ unit area demand research programmes are to be reoriented in India. Technologies developed in research centres have to be effectively transferred to farmers fields to achieve higher yields.

Natural Rubber

The invention of tyres and internal combustion engines during 19'th century and rapid development in automobile sector, paved the way for spectacular growth of rubber industry and natural rubber achieved the status of a strategic raw material. India ranks third in production and fourth in consumption of natural rubber in the world. Country's relative share in global output of NR progressively increased from an insignificant 0.8% during 1950 to 9.0% during 2005. The higher efficiency of rubber over other major crops in the country is endorsed by the fact that this plantation industry accounts for 1.0% of India's GDP though rubber occupies only 0.4% of the gross cropped area in the country. The achievements of Indian rubber plantation industry during post-independence period are mainly attributable to the protectionist policy pursued by the Government of India under various five-year plans, aimed at attainment of self-sufficiency in production and import substitution.

Due to limited availability of land suitable for rubber cultivation, scope for increasing production through extending area of cultivation is limited. Productivity improvement through short-term and long-term measures received prime importance in the strategy adopted for increasing production. Emergence of free trade regime in NR sector since April 2001 necessitated a paradigm shift in policy orientation to make the industry sustainable. Focus needs to be on improving its efficiency by exploring measures for cost - reduction and the key instrument in attaining cost efficiency is technology. The efficiency of Indian NR market facilitated investments in production sector. Average holding size less than 0.50 ha.,cultivation of rubber has become a subsistence activity and today smallholdings account for 88 % in area and 91% of production. Small growers are receptive to technological innovations. As a result, the productivity progressively increased from 298/kg/ha during 1951-52 to 1796 kg/ha. during 2005-06.

Importance of cost competitiveness in a free trade regime and importance of technology as a tool merit special mention. Cost of production of NR in India is the highest among major producing countries despite having the highest productivity. This is because of the agroclimatic conditions in the regions identified as suitable for rubber in the country are approximately favourable for growing rubber as compared to the ideal conditions prevailing in Thailand, Indonesia, and Malaysia. Rubber holding size in India is relatively smaller with the average size below 0.50 ha. compared to 2-3 ha. in the first three major producing countries.This is another cost-pushing factor in India. The undulating terrain necessitates soil conservation measures involving additional investment. Technology has to be used as a tool for addressing additional cost factor arising from above three factors by the industry to improve its efficiency.

In efforts towards reducing cost of production, various research findings leading to optimization of agro-management practices and development of improved crop harvesting technologies need attention. With the release of two improved latex timber clones in RRII 400 series, rubber plantation industry reached the threshold of another technological revolution. In this context, it is important to note that about 50,000 ha.of existing yielding area have already attained replanting age and another 125,000 ha.would be attaining replanting age during eleventh plan period (2007-12). This anticipated replanting boom provides opportunity for growers to upgrade the technology by adopting RRII 400 series. The diagnosis of performance of the Indian rubber plantation industry has identified technological intervention as the most important single contributing factor for dynamic growth attained during post independent phase. Though the achievements made so far are impressive, India's disadvantageous position as the costliest region in NR production has to be addressed. A feasible approach in addressing this issue is to bridge the gap between the potential productivity and the realized productivity.

Tea

Production of tea, after stagnating between 1997-98 and 2004-05 at around 830-850 million kg, increased by over 12 % in 2005-06 and now accounts for 27.74% of the global tea production in the world. This position is retained over the last century. With a cultivated area of around 5.12 lakh ha. India accounts for one-fifth of the global tea area. India is also the largest consumer of tea and accounts for 22% of global tea consumption. Annual consumption is estimated over 700 million kg. Tea industry provides direct employment to more than one million workers of whom a sizeable number are woman belonging to weaker sections of society. In addition, a significant number of persons derive their livelihood from ancillary activities associated with production, value addition and marketing of tea. Substantial foreign exchange earnings with negligible import content and contribution to the state and central exchequers are the other significant economic features of the Indian tea industry. Tea industry is one of the most environment friendly industries since it is mostly cultivated in remote rural areas.

The foundation for tea industry was laid by entrepreneurs a long time ago in an organized manner. More than 80 % of the production in India is now accounted by corporate sector of the industry. Currently small scale sector gained prominence and accounted nearly 20% of the all India tea production. Growers holding tea area up to 10-12 ha. are small tea growers Because of the scattered nature ,the growers find it difficult to obtain proper technical help on good crop husbandry practices. Despite technological advancement in tea science, productivity of small holdings is way behind that of the organized sector. In order to bridge extension gap and effectively meet the developmental needs of small growers, it is proposed to set up a Small Growers Development wing within the Tea Development Directorate of Tea Board, which will be responsible for development and regulation of the small sector.The Government has approved setting up of a Special Purpose Tea Fund (SPTF) under the Tea Board for funding re plantation and rejuvenation activities aimed at improving the age profile of tea plantations.

Coffee

The coffee industry has prime place among plantation crops in India. At present, coffee is mainly cultivated in the traditional states of Karnataka, Kerala and Tamil Nadu and to a small extent in non-traditional areas of Andhra Pradesh, Orissa, and seven North Eastern States. It is only in India that both commercially important species viz., arabica and robusta are cultivated in almost equal proportions. The area under coffee showed an increasing trend through out the period from 1950-51 onwards. At present (2005 - 06), arabica coffee is cultivated in an area of 1. 70 lakh ha with a production of 1.03 lakh t. while robusta is grown in 1.85 lakh ha with a production of 1.72 lakh t.. Indian coffee industry is characterised by marked cultural dualism between subsistence oriented small growers and market oriented large and corporate producers. Small growers with less than 10 ha land account for 98.4% of the 1,75,475 holdings covering approx. 71.84% of land under coffee and contribute around 60% towards the country's production. Indian coffee industry provides direct employment to nearly four lakh workers in the plantations, besides providing indirect employment to thousands of people in processing and trade. India exports nearly 70% of its coffee to destinations all over the world, and the remaining quantity is consumed in the domestic market. During 2005-06 India exported 1.57 lakh t. of coffee worth Rs.1147 Crores.

Indian coffee by contributing 4% to global production enjoys a niche market. Farm productivity and production slumped in many producing countries during the crisis years of 2002-2005, perhaps due to which the demand has finally caught up with supply and prices started firming up.

Coffee beans stand at the 3rd place in the list of legally traded products in the world and are considered a very important traded commodity. The world production of coffee is around 6 million t. annually. Coffee is produced in 70 countries of the world. Major coffee producing countries are Brazil (33.16%), Columbia (11.65%) ,Vietnam (10.61%), Indonesia (5.97%), Mexico (4.59%), India (4.60%), Guatemala (rest of the countries have around 28%) Ethiopia, Uganda, Honduras, Cote d'Ivoire, Costa Rica, Peru, Papua New Guinea, Puerto Rico and Castalia . The top three producing countries account for over 50% of total production and hence control the world coffee market. The level of production of these countries is on a declining stage except for Columbia, Uganda and Peru. The major consuming and importing countries of coffee are United States, Canada, Japan, Germany, Italy, United kingdom, Poland and Spain. Other than the above mentioned countries it is also imported by African countries. The net imports of coffee figures around 4.5 million t. The export scenario is that the major share of exports in the world is also held by Brazil and followed by other leading producing countries.

Cocoa

Global production of cocoa is concentrated effectively in the regions between 10° North and 10° South of the Equator. West Africa, which supplies approximately 70% of global output, is the most important cocoa production region in terms of economics. The most important producing countries are the Ivory Coast, Ghana, Nigeria and Cameroon. Among non-African countries producing significant amounts of cocoa are Indonesia, Brazil, Ecuador, the Dominican Republic and Malaysia. As a result of their importance in terms of world production, West African producers are seeking to increase their cocoa output. Cocoa growing represents an important source of revenue for a large number of small farm-owners. Most plantations are family farms of 2-10 ha. This production is particularly significant in national economic terms because local demand for cocoa is relatively weak and therefore almost all production is for export. In Africa, cocoa beans are generally harvested in September and October, although the season can continue until January or March.

Traditionally, cocoa is cultivated in producing countries and sold for export in the form of beans. Importing countries then process the beans, transforming the raw goods into finished or semi-finished products (cocoa butter, cocoa liquor, cocoa powder.). In recent years, in an effort to increase the value of exports, some producer countries, like the Ivory Coast, Ghana, Nigeria and Brazil, developed their own facilities for grinding beans. World consumption is estimated 2 800 000 t./ year. The largest cocoa importers are

Europe (more than 1.2 million t./ year) and the United States (0.4 million t./ year). The largest importers are Holland, USA, Germany, Britain and Brazil. European countries are the largest consumers of cocoa and chocolate. Each country has its own preferences and style of chocolate, the popularity of different products varying according to national taste. On average, the Swiss munches approximately 10.55 kg chocolate / person / year. As a nation, Great Britain consumes more than 500,000 t. chocolate / year. In France, the average consumption / person is 6.8 kg / year, with the New Year and Easter celebrations being the most important occasions for tasting and offering chocolate gifts. Eastern European countries are important emerging new markets and would remain so for foreseeable future. Since 1990s, Asia developed into a major chocolate market. Japan has experienced a significant increase in chocolate consumption during the past decade. Demand in China increased to 9,000 t. in 2000, an increase of more than 90% from the previous year! Among cocoa producing countries, Brazil saw an increase of 10% in its annual consumption / person since 1993.

In India, though cocoa was the most compatible as a mixed crop in coconut and arecanut plantations, area expansion programme received a setback during the last decade mainly due to marketing problems and the pricing policy. The trend is now steadily changing due to establishment of processing factories including one in the co-operative sector offering an effective competition to the multinational company which was operating alone earlier. Cocoa cultivation as a mixed crop in coconut and arecanut plantations of Kerala, Karnataka and Andhra Pradesh is now popular and the production is expected to meet the existing demand in near future.

Constraints in Increasing Production of Plantation Crops

The production and productivity of crops like tea, rubber and arecanut have increased substantially, though these remained stagnant or even declined in crops like coconut and arecanut. Technologies to increase production are available, a break through is not achieved in transferring them to farmers in most of these plantation crops. Long pre-bearing period and time lag between input application and response are also among the constraints for adopting improved technologies in most of plantation crops. Most of the holdings in many plantation crops are small and marginal and the performance of large and small holdings differs substantially. Contrary to the assumption that small plantations are more productive because of the personalized attention, larger plantations under corporate sector show better productivity. Small holder farmers fail to give appropriate attention to the crop in time because of long pre-bearing period, uncertainity of the returns and slow response to improved management practices. Whenever there is slump in price for a particular commodity, small farmers divert their resources to another crop, more remunerative at that time. In perennial crops such irregular input management will reflect on the yield of subsequent years. Being perennial crops, regular replanting, under planting and also removal of senile trees or unproductive plants are seldom attended to. Sometimes closer planting results in plant canopy competition and root competition and thereby reduced productivity.

Marginal and sub-marginal lands, improper input management, dependence on monsoon rain, uneven distribution of rains, long dry spells coupled with high temperature in summer resulting in drought condition affect productivity of these crops. Inadequate availability of quality planting material and control of pests and diseases are yet another constraints in production. Though most of the pests and diseases can be controlled by proper and timely application of plant protection chemicals and adoption of other proven means it could not be done by all farmers simultaneously due to large number of small holdings and lack of collective approach to tackle the problem.

Though plantation crops can be compared to industry with regard to the investments and returns, there are differences also. In industry, the sons or close relatives take over the management from the founder. They are educated and trained to get expertise in specific areas related to concerned industry. In farming, less educated in most cases take to farming by force without much aptitudc. There is reluctance on the part of such people to change over to improved crop production technology. This is one the reasons for low productivity and returns from plantation crops. Unless plantation crops farming are considered as business and managed with adequate and timely inputs, it will not be a profitable venture. Lack of credit facilities on easy terms, deficiencies in input supply, uncertain markets and crop prices, lack of necessary institutional infrastructure and socio-economic and cultural problems are a few of the other constraints.

Future Strategies for Increasing Yields

In India, production and productivity of many plantation crops remained static or even showed a downward trend during the last two decades. Taking into account the future demand, it is necessary to develop production technologies suited to small and marginal holdings to make them a viable production unit.

The concept of production can be based on energy on one side and income on the other side. Since conversion of energy into biomass is the primary function in any production activity, a critical analysis is worth-while. More studies are to be undertaken to elucidate the mechanism so that productivity / unit land / unit time can be increased.

There is need for increasing harvest indices along with biomass production. Photosynthetic rates are very important. Energy conversion efficiency of the crop is very low being 1-2%. If this is increased, there is possibility of increasing dry matter production. Such increase could be achieved by (i)Increased rate of growth (ii)Improved canopy structure and subsequent increase in light penetration resulting in assimilation rates (iii) Improved partitioning of assimilates etc.(iv) Ways to reduce dark respiration in tropical trees in which 75-85% of fixed carbon are lost by respiration.

Since moisture and nutrients have strong interaction and also the fact that moisture availability becomes limiting at least during certain periods of the year, research to identify and develops stress tolerant varieties should receive priority in plantations. It is necessary that the production components like land, air space, inputs and time be considered together. It is also essential that instead of one crop / unit area and time more crops are planned to be raised either in sequence or together. A clear under standing of optimal exploitation levels of the above basic components by a crop community enables maximization of production through this pathway. Perennial plantation crops offer better scope for yield maximization, provided selection is made on the basis of canopy and root architecture and space to make better use of incoming solar energy at different strata of crop canopies and producing much higher biomass and higher income. Palm-based multi species cropping system wherein compatible annual, biannual and perennial crops are cultivated to utilize the horizontal and vertical space, land and sunlight continuously for production of biomass like food, fuel, fodder etc. having economic value and diverse end uses relevant to the farmer and society. Such a cropping system provides more employment opportunities and income, spread over time. Research efforts on product diversification and bye-product utilization also need to be undertaken. Harvest and post harvest technologies should receive attention to avoid wastages and also to improve the quality of products.

ARECANUT

Arecanut palm *(Areca catechu L.)* is cultivated primarily for its kernel obtained from the fruit which is chewed in its tender, ripe or processed form. Arecanut belongs to family Palmae. It is native of Malayan Archipelago, Philippines and other East Indian Islands. Commercial cultivation is confined only in India, Bangladesh and Sri Lanka. Kerala, Karnataka and Assam account for more than 90% of the total area and production in our country. Arecanut is an essential ingredient of 'gutka' and 'pan masala.' It is consumed both as a raw/ripe nut (adaka or kacha tamul), dried ripe nut (chali supari) and as semi-mature, cut and processed `Bateldike' or `Kalipak'. There are over 150 trade types, differing in maturity, processing conditions and varying in their taste characteristics as per market conditions prevailing at different centers of the country.

Soil and Climate

The arecanut palm is capable of growing under a variety of climatic and soil conditions. It grows well from almost sea level up to an altitude of 1000 m amsl in areas receiving abundant and well distributed rainfall or under irrigated conditions. Though it grows at altitudes up to 1,000 m amsl, at higher altitudes it is not at all productive. The cultivation is mostly confined to 28° North and South of the Equator. It is unable to withstand extremes of temperature and wide diurnal variations. Arecanut grows well within the temperature range of 14 to 36°C and is adversely affected by temperatures below 10°C and above 40°C. It can be grown in areas receiving annual rainfall of 750 to 4,500 mm.

Areca thrives well in a variety of soils, the laterite soils of the West Coast, the red loamy soils of the Mettupalayam (Tamil Nadu), the alluvial soils of Assam and West Bengal and the loam of Orissa. The foremost factor that has to be considered in establishing an arecanut plantation is that the site selected should have adequate facilities for irrigation. The soil also should be deep and well drained, without a high water-table. Being highly susceptible to a sun-scorch, the areca palms need adequate protection from exposure to the South-Western sun.

Varieties

There are a few local varieties known by the name of place where they are grown are given below:

Released Varieties

Variety	Country of in troduction	Year	Mean yield ripe nut (kg/palm/yr)	Chali yield (kg/ palm/ Yr)	Special features
Varieties from exotic collections					
Mangala CPCRI	China	1973	8.82	2.02	Semi tall, more no. of female flowers/inflorescence, high yielding, early bearing, quick stabilization of yield, semi tall, more leaf lets than local cultivars. Leaves dark green with crinkling at the tip, good chewing and marketing quality. Fruits - medium size
Sumangala CPCRI	Indonesia	1985	12.93	3.28	Tall palm, Flowering 4-5yrs, colour of nut deep yellow - orange, oblong to round
Sreemangala CPCRI	Singapore	1985	12.82	3.10	Tall palm, Flowering 4-5 yrs, Ripe nuts oblong to round, deep yellow
Varieties from indigenous collections					
Mohit Nagar CPCRI	W. Bengal	1991	15.08	3.67	Greater uniformity, Bunches well spaced, nuts loosely arranged, Early stabilization of yield when compared to Sumangala & Sreemangala
Samrudhi (CARI-Portblair)	A & N islands	1995	18.89	4.34	Tall palm, consistent high yield, well spaced bunches, round & bold nuts recommended for A & N islands
SAS-1 Sirsi Arecanut Selection UAS, Bangalore	Sirsi		24.03	4.60	Tall, regular bearer, Nuts round, closely arranged, uniformsized nuts for hilly areas of Karnataka - tender and ripe nut

Promising Cultivars

Cultivar/Variety	Growth habit	Shape & size of nut	Yield Chali (kg/palm/yr.)	Recommended Agro climatic area
SK Local	Tall	Round and bold	2.00	Northern Kerala, Kasaragod & DK,Karnataka
Thirthahalli	Tall	Small and elongated	2.60	Malnad areas of Karnataka
Sagar	Tall	Small and round	2.25	Shimoga and Uttara Kannada of Karnataka
Sree Vardhan	Semi-tall	Round and medium	2.00	Coastal Maharashtra and Karnataka
Hirehalli Dwarf	Dwarf mutant	Round to oval medium	1.20	Hirehalli area of Karnataka

Nursery Management

Arecanut is propagated only by seeds. Being a perennial crop, adequate care should be taken in selecting the planting material. There are four steps in selection and raising of arecanut seedlings viz., selection of mother palms, selection of seed nuts, germination and raising the seedlings and selection of seedlings.

Selection of Mother Palms

Select genetically superior planting-material for which mother-palms possessing characters of high heritability, correlated with high yield are to be identified. The criteria for selection of mother palms are; early bearing, regular bearing habit, large number of leaves on the crown, shorter internodes and high fruit set. The selection of seed nuts may be commenced only after the stabilization of yield of palm, generally takes 4 to 5 years from the first bearing. Only fully ripe nuts should be selected as seed nuts, discarding underdeveloped and malformed ones.

Selection of Seed Nut

Fully ripened nuts having weight of above 35 g should be selected. The nuts selected should float vertically with calyx-end pointing upwards when allowed to float on water. These nuts produce vigorous seedlings of greater vigour.

Primary and Secondary Nurseries

The selected seed nuts are sown soon after harvest, 5 to 6 cm apart, in beds of sand under partial shade, with their stalk-ends pointing upwards. Sand is spread over the nuts just to cover them. The nuts are irrigated daily. Germination starts about 40 days after sowing and the sprouts are ready for transplanting when they are about three months old. Nursery-beds of 150 cm width and of convenient length are prepared for transplanting sprouts. The sprouts are to be transplanted at a spacing of 30 X 30 cm with onset of monsoon. A basal dose of well-decomposed cattle manure at the rate of 5 tonnes/ha may be applied to the secondary nursery. Proper shade should be provided to the nursery, shade may be either of coconut or arecanut leaves spread over a pandal or by planting some fast growing green manures or banana. The seedlings would be ready for transplanting when they are 12 to 18 months old. The nursery should be watered regularly in summer and proper drainage should be provided in rainy season. The nursery should be kept clean by periodical weeding.

Selection of Seedlings

Select good seedlings for transplanting in main field when they are 12-18 months old. Selection of seedlings can be based on selection index. Multiplying leaf number by 40 and subtracting the seedling height gives the selection index. Select seedlings with higher selection index values.

Planting and Management

Selection of Site and Layout

The crop thrives well in humid areas protected well against hot sun and heavy wind. Since the areca palm does not withstand either water logging or drought, the site selected should have proper drainage and adequate source of water for irrigation. Arecanut palm cannot withstand extreme temperature and exposure to direct sun. The site selected should have protection from southern and western sides. The soil depth and depth of water table are other two parameters to be considered while selecting the site. The soil should be deep (preferably not less than two meters) and water table should be sufficiently low for better root development. Aligning the rows in north-south direction with a deviation of 35° towards south-west lowers incidence of sun-scorch.

Spacing

This depends on rooting pattern of the crop along with fertility and depth of soil. The studies conducted at different places with different spacing revealed that a spacing of 2.7 X 2.7 m is optimum for arecanut. This was justified by root distribution studies also.

Depth of Planting

In well-drained soils and in fields where proper drainage can be provided, deep planting is preferred. Deeper planting provides a firm anchorage and larger volume of space for root development. In areas where water table is high, shallow planting is preferred. Thus in well-drained soils, planting at a depth of 90 cm and in heavy soils planting at a depth of 60 cm are recommended.

Season of Planting

In areas where South-West monsoon is severe, planting in September-October is recommended. In other areas planting can also be done in May-June.

Planting

Pits of 90 x 90 x 90 cm are dug and the pits are filled with a mixture of top soil, powdered cowdung and sand to a height of 50 to 60cm from the bottom. The seedlings are planted in the centre of pit, covered with soil to collar level and pressed around.

Drainage

For better growth and development of plants proper drainage is essential. The number of drainage channels depend upon soil type. In light soils, number of channels may be less and in heavy soils channels should be dug in each row for proper draining of excess water. Channels should be at least 15 - 30 cm deeper than the depth at which seedlings are planted.

Shading

The palms are highly susceptible for sun scorching. The seedlings should be given protection against direct exposure to sun. This may be done by either covering plants with areca or coconut leaves or by raising crops like banana in between two rows of arecanut. Sun scorching is mostly seen during October - January. During this period even the stems of young palms have to be protected. For this, quick growing shade plant can be planted on Southern and Western sides of garden.

Cultural Operations

Main cultural operations, such as light forking or digging to break the irrigation crust and to control weeds, are performed towards close of monsoon, generally in October-November. In slopes, prevent soil erosion by terracing. Sow seeds of green manure-cum-cover crops such as *Mimosa invisa, Stylosanthes gracilis* and *Calapagonium muconoides* in April-May with the onset of pre-monsoon rains. Cut and apply them to palms in September-October.

Nutrient Management

A steady and high yield depends much on adequate availability of plant nutrients in soil. This is more important in a perennial crop. Since almost all the arecanut-growing areas are in heavy-rainfall tracts, the soils are liable to leaching and erosion thus making them poor in major plant nutrients and organic matter. The annual application of 100 g of N, 40 g of P_2O_5 and 140 g K_2O in the form of fertilizers and 12 kg of each of green manure and compost or cattle manure/bearing palm is recommended. The fertilizers may be applied in two split doses during September-October and February, and green manure and compost in single dose in September-October. Under rain-fed conditions, application of second dose of fertilizers may be given in March or April after adequate summer showers. For young palms, a full dose of green manure and compost or cattle-dung may be applied from the first year of planting along with one-third of the quantity of fertilizer, followed by two-thirds of fertilizer in second year and full dose from third year onwards. Manuring during September-October is done in basins of 0.75to 1 m radius, made around the palm to a depth of 15 to 20 cm. Second dose of fertilizer can be applied to base of each palm all around and mixed with soil by a light forking. Soils of arecanut-growing areas are mostly acidic. Application of lime at the rate of 500 g / palm once in two or three years corrects soil acidity, increases availability of plant nutrients, reduces phosphorus fixation in soil and enhances the microbiological activity and improves the soil properties. Application has to be completed at least three weeks before manuring in September-October.

Organic Matter Recycling

On an average, 5.5 to 6.0 tonnes of wastes are available from one ha of areca garden /year. This can be effectively used as organic source of nutrients for areca palms. But direct application of these wastes in the garden will take long time for decomposition and will not meet the nutrient demand of the crop immediately. Hence, these materials can be composted using earthworms effectively and used as organic manure in areca gardens. To prepare vermicompost, areca wastes are chopped into small pieces of 10 cm and heaped. The heap should be mixed with cow dung slurry @ 10 kg / 100 kg of waste and kept for two weeks with sprinkling water daily. Then the chopped material is arranged in beds of one-meter width and convenient length. Cement tanks or trenches can be used for this purpose. A layer of 10-15 cm waste material is alternated with 2 cm layer of cowdung over which earthworms are released at the rate of 1000 numbers/square meter. The wastes are converted into fine granular, odourless vermicompost within 60 days. During this period, the earthworm population is doubled. About 8 kg /palm/year of vermicompost meets the crops nutrient demand in terms of nitrogen. The two species of earthworms *Eudrilus eugeniae* and *Eisenia foetida* can be used.

Water Management

Irrigation and Drainage

Arecanut cannot withstand drought for a long time. Being a perennial crop, once affected by water stress, it may require two-three years to regain the normal vigour and yield. The death of palms due to moisture stress is also not uncommon.

The palms should be irrigated once in four to seven days depending on soil type and climatic factors. . In Southern Kerala, where arecanut is mainly grown under rain-fed conditions, it is found that manuring along with irrigation gives three times more yield than manuring alone. In west coast, where major area of arecanut gardens are irrigated, watering the garden once in seven or eight days during November-December, once in six days during January-February and once in three to five days during March-April-May is recommended. In each irrigation, give about 175 litres of water/palm. Where there is shortage of water follow drip irrigation. Application of organic mulch to the garden helps conservation of soil moisture.

Construct drainage channels (25-30 cm deep from bottom of pits) between rows and drain out water during periods of heavy rainfall to prevent water logging.

Fertigation

Application of nutrients through irrigation water is called as fertigation. This procedure can be profitably followed in arecanut.

Advantages of Fertigation

1. Nutrients can be supplied as per crop growth, stages and demand of crop.
2. Nutrients can be supplied uniformly to all plants.
3. Since nutrients are applied in the area of highest root activity and at optimum soil moisture, the nutrient availability and nutrient absorption by roots will be improved.
4. As the method is simple it saves time, labour and energy.
5. Only a small quantity of fertilizer will be available in soil. This avoids losses of nutrients through leaching and erosion.
6. Saving in quantity of fertilizers to the extent of 20-40%.
7. Fertilizer use efficiency will be increased due to easy availability, reduced losses and increased yield.
8. This method can be used effectively to grow crops in soils, which have low nutrient and water holding capacity.

Criteria for Applying Fertilizers Through Drip

- Should avoid corrosion
- Should be soluble
- Should not react with salts or other chemicals in irrigation water.

Fertigation Methods

1. By using main pump
2. By using a separate small pump
3. By using a fertilizer tank placed above main irrigation tank
4. Ventury

Sources of Nutrients

Nitrogen fertilizers	:	Urea, ammonium nitrate and calcium nitrate
Phosphatic fertilizers	:	Mono ammonium phosphate
Potassic fertilizers	:	Potassium nitrate, potassium sulphate and mono-potassium phosphate

Quantity and Interval

Studies at CPCRI, kasaragod showed that in the initial stages of areca garden only 75% of recommended dose of fertilizer is sufficient when fertilizer was given through drip irrigation. The fertilizer should be split into ten parts and should be applied once in 20 days from November to May.

Arecanut Based Cropping System

Arecanut cultivated with a spacing of 2.7 x 2.7 m provides ample scope for cultivation of various annual, biennial and perennial crops in inter spaces. This practice of growing inter/mixed crops in the arecanut garden is called as mixed/intercropping.

Experiments conducted at CPCRI Regional Station, Vittal and other centres indicated that some precautions to be observed while practicing inter/ mixed cropping. The crop chosen should not have any problem of soil exhaustion and pest build-up. At the same time, it should have maximum production potential and give maximum returns/ unit input to farmers. Crops, which love shade or those which withstand the canopy of arecanut palms and heavy dripping during monsoon are to be preferred. The choice of intercrops also depends upon locality.

A large number of annuals, biennials and perennial crops tried in areca gardens to evolve suitable crop combinations at CPCRI Regional Station Vittal and the Research Centres spread over Karnataka, Kerala, Assam and West Bengal. Crops like banana, pepper, cocoa, elephant foot yam, citrus, betelvine, pineapple etc. were found suitable for inter/mixed cropping in arecanut. However, it is found that as the age of the garden advances, only a few crops

Crop Husbandry for Mixed Crops in Arecanut Garden

Crop	Spacing (m x m)	Population /ha	Fertiliser (g/palm or tree/year)			FYM (kg/palm)	Age of arecanut at which inter/mixed crop can be planted
			N (Urea)	P2O5(RP)#	K2O (MOP)$		
Arecanut	2.7x 2.7	1300	100 (220)	40 (200)	140 (235)	10	-
Black Pepper*	2.7x 2.7	1300	100 (220)	40 (200)	140 (235)	..	6-8 years
Banana†	2.7x 5.4	650	160 (350)	160 (800)	320 (535)	20	Before or with planting
Cocoa†	2.7x 5.4	650	100 (220)	40 (200)	140 (235)	..	4 years
Lemon†	2.7x 5.4	650	300 (655)	250 (1250)	500 (835)	20	4 years
Betelvine*	2.7x 2.7	1300	100 (220)	40 (200)	140 (235)	10	6-8 years

* Planted at the base of the palms; † Planted at the centre of four palms.
# RP – Rock phosphate; $ MOP – Muriate of potash

Cost of Cultivation/ ha of Arecanut

Sl. No.	Components of different cost concepts	Cost/ha (Rs.)	(%)distribution of cost
1	Hired human labour	7672	42.12
2	Animal labour	87	0.48
3	Machine labour	420	2.31
4	Seed/ seedlings	15	0.08
5	Farmyard manure and chemical fertilizers	5488	30.13
6	Plant protection	206	1.13
7	Land tax and irrigation cess	75	0.41
8	Repair and maintenance charges of implements, machinery and building.	1477	8.11
9	Interest on working capital	1514	8.31
10	Other expenses	1259	6.91
11	Total cost 'A' (1-10)	18213	100.00
12	Interest on fixed capital	1451	
13	Cost 'B1' (11+12)	19664	
14	Interest on land value	141560	
15	Cost 'B' (13+14)	161224	
16	Imputed value of household labour	3356	
17	Cost 'C' (13+14)	164580	

Source: Department of Economics and Statistics, Kerala.

can be grown profitably as mixed crop viz., pepper, cocoa, banana, lime and betelvine. The intercrops, apart from giving additional income, create additional employment opportunities.

Harvesting

The pre-bearing age of palm ranges 5 to 8 years. Though inflorescence initiation is observed in every leaf axil, there is absorption of inflorescence to about 5%. Colour of fruit during its growth changes from green to different shades of yellow and red during ripening. In some places, tender nuts and mature nuts are harvested. Tender nuts are harvested from July to December and ripe nuts from December to March or from May to July. Three or four plucking are done during the whole season.

Harvesting of arecanut is done by skilled climbers who cover about 50 - 100 palms a day. Mechanical arecanut palm climbing device is available for easy climbing of palm by any unskilled persons. Stage of harvesting depends on type of product to be prepared. There are two main types viz., kalipak, prepared out of immature green nuts, and chali prepared from ripe nuts. In each case, maturity at which fruit is harvested and season of harvest affects quality of processed nut considerably.

Preservation of Fruits

Fully ripe arecanuts after preserving are used throughout the year. These are quite popular in Kerala and Assam. In Assam, fresh fruits are preserved in thick layers of mud. This product is known as Bura Tamul. In Kerala, fresh fruits are stored by steeping in water, resulting in discolouration of outer husk and foul smell due to bacterial attack but the inner core is well preserved. This product is known as Neettadakka. Use of a solution containing sodium benzoate (0.1 %) and potassium metabisulphate (0.2%) acidified to a pH of 3.5-4.0 with hydrochloric acid for steeping ripe nuts after initial heat blanching is suggested to eliminate the foul smell and improve quality of nuts.

Dried Ripe Nuts

The most popular trade form of arecanut is dried whole nut known as Chali or Kottapak. Ripe nuts are dried in the sun for 35-40 days, dehusked and marketed as whole nuts. The optimum moisture content is around 12%. Inadequate drying results in fungal infection and in a poor quality product. Depending on size, different grades of decrease order are Moti, Srivardhan, Jamnagar and Jini. The main producing states of Chali are Kerala, Karnataka, Assam and Maharashtra. To facilitate drying and dehusking, the fruits are cut longitudinally into 2 halves and sun-dried for 10 days. The kernels are scooped out and given a final drying. This product, known as Pareha, is popular in Kerala and Karnataka. Mechanical driers are also used to make Chali. Drying takes 60-70 hours over a period of 7-8 days at 45° - 75°C. The dehusking can be done using a manually-operated arecanut dehusker developed by CPCRI, Kasaragod. About 40 kg chali can be made within a period of 8 hours.

Kalipak

It is another important processed product of arecanut. Kerala and Karnataka are main producers of Kalipak. Tender nuts of 6-7 months are dehusked, cut into pieces, boiled with water or a diluted extract from previous boiling, coated with kali and dried. Kali coating can be repeated 3-4 times to get a glossy appearance. Kali is the concentrated extract obtained after 3-4 batches of boiling. The kalipak is known by different names depending on number, shape and size of cuts. Api or unde (without any cuts), batlu (transverse cut into halves), choor (several longitudinal cuts), podi (both longitudinal and transverse cuts) and erazel (transverse thin slices). Iylon is another variety made from green nuts which are cut transversely into 5-6 discs and without kali coating. During preparation of kalipak, the tannin content is reduced substantially. A well-dried product with a dark brown colour, glossy appearance, crisp chewing feel, well-toned astringency and absence of over-mature nuts are rated superior.

Scented Supari

It is made both from chali and kalipak. Chali supari is more popular. The dried nuts are broken into bits, blended with spices and flavour mixture and packed in butter paper. Instead of raw spices, essential oils are used for easy blending, with coconut gratings to avoid microbial growth. Saccharin is occasionally used for sweetening. Rose essence and menthol are commonly used for flavouring.

Uses of Arecanut Constituents and Byproducts

- *Tannins:* Polyphenols or tannins form the major constituent of the nut. Tender nut contains tannins 30-37 %, while ripe nuts have only 16-22 %. The areca tannins obtained as byproduct from tender nut processing can be utilized for dyeing clothes, leather, rope, for making black writing ink along with ferrous sulphate, as an adhesive in plywood manufacture and even as food colour.

- *Fat:* The nuts contain 8-12 % fat, which is extractable with organic solvents like hexane or chloroform. The fat is rich in myristic acid and can be a good source of myristic acid and its derivatives. Refining of fat with alkali makes it edible. It can be mixed with cocoa fat for confectionery products and as an extender to cocoa butter for various products. It can be a substitute for vanaspati in preparations of sweets and biscuits.

- *Alkaloids:* Arecanut has 1.5% of alkaloids such as arecoline, arecaidine guvacine, isoguvacine and arecolidine. Of them, arecoline alone accounts for 0.24 %. They possess antihelminthic property and are effective against tapeworms and roundworms. It is also used as a CNS depressant drug. It has anti-bacterial property and inhibits growth of *Escherichia coli, Staphylococcus typhi* and *Staphylococcus aureus*. The blood sugar lowering effect of arecoline is mentioned in Ayurveda.
- *Areca husk:* The husk of arecanut constitutes 60-80% of total weight of fresh nut. Several processes are developed for utilization of areca husk for making hard boards, plastic boards and brown wrapping paper. The husk fibre extracted by soaking in water for 3 weeks and beating with a mallet can be used for thick boards, fluffy cushions and non-woven fabrics. It can be a good source of furfural and xylose for industrial applications.
- *Areca stem and leaf sheath:* Arecanut stem is a useful building material. Its timber can be used for making a variety of elegant utility articles like rulers, shelves and waste-paper baskets, due to its hardness. Hollow stem can be used as drainage and irrigation pipes. A process is developed at CPCRI, Kasaragod, for making plyboards from leaf sheath for tea chest, sint case and file board manufacture. The CFTRI, Mysore, developed a machine for making cups and throw-away plates which can substitute the paper plates.

Curing of Areca Nut in Meghalaya

People of Meghalaya are fond of taking fresh fruit instead of processed one. But fresh fruits are not available round the year as the harvesting periods in the areas are November to December and March to April. The Khasi cures the arecanut in flowing water. The technique has been developed by trial and error method. In olden days a tribal family used to store arecanuts in earthen pots of water for maintaining moisture and freshness. The water used to be changed every week. A typical curing of Areca nut consists of following components.

1. *Curing pond:* A pond for curing of Areca nut is constructed of stones and masonry. It is normally in the size of 15m length, 10m width and

2.5m depth. Such a pond can take 150 baskets of nuts. The size and shape of a pond vary as per requirement and land scape of the area; however the depth remains the same.

2. *Water flowing Channel:* The perennial stream is the main requisite for the location of a pond. In the typical arrangement, water flow is diverted from the upstream of the nearby pond at one end and outlet is kept diagonally opposite of the inlet. The water level maintained 2.35 m from the bottom with the continuous flow.
3. *Holding basket:* The holding baskets are locally (Nongpriang village) prepared of bamboo by matting of size of 1m diameter at top and 2.65 m height with the square bottom of 0.65m. The matting is done with 1 cm bamboo strips by keeping the 3 cm perforation. Each basket has a capacity to hold 1200 kg of areca nut which is equivalent to 12 gunny bags of areca nut.
4. *Raising basket:* A small bamboo basket is used for raising cured fruit from the pond. This basket is matted in the size of 36 cm diameter at top, 30 cm high with the square base of 16 cm.

Method

The pond is at first fragmented into the square of 90 x 90 cm each at the top with frames made of strong bamboo. Thus holding baskets of 1 m diameter when inserted into these squares could be reduced to that size. The bamboo frames hold the individual baskets and keep them vertical. These baskets are filled with freshly harvested areca nut. Fruits normally float on water. The baskets are therefore covered with jute clothes and stone weights are kept over them to ensure proper curing. Once the areca nuts are charged into the holding baskets, then they are not disturbed till the curing is completed. No special care is required except maintenance of water level of pond and water flow. The minimum water flow is assured (36 l/min) to prevent fruits from being spoiled. Excess water flow is good for keeping fruit fresh.

The criteria for adequate curing are: (i) Yellow colour turns prominently bright yellow and (ii) the husk gets freed from kernels. The normal curing time is around 4 to 5 months. Fruits are then taken out from the pond. The fruits are sometimes cured for one and half months and disposed as and when demand for them arises. Once the fruits are raised from one holding basket then that basket should be completely raised out otherwise rest of fruits will be spoiled. The loss in curing is generally 15-30% of total weight. The loss can be minimized by keeping same lot in one basket. The units of fresh fruits and nuts in one kg amount to 42 and 75 in number respectively. The dry nuts in one kg however amount to 135 in number. Curing improves colour, taste and freshness of the nut. Cured kernels are less astringent and are better to chew and taste.

Production of Mushroom from the waste of Arecanut

In India, cultivation of mushrooms is limited to three species viz., white button mushroom, paddy straw mushroom and oyster mushroom. Oyster mushrooms are the ideal ones for the arecanut sector due to its ability to utilize lignin rich arecanut wastes and the climatic conditions prevailing in plantations are also ideal for its growth. Oyster mushrooms, known as wood fungi, are endowed with ligninolytic and cellulolytic properties to utilize a wide range of agricultural residues as substrates for growth and fruit body production. Paddy straw is the most widely used substrate for its cultivation. But its increasing cost and decreasing availability are factors, which prompted research workers to look for alternate substrates for oyster mushroom cultivation. Conditions have been standardized for cultivation of oyster mushrooms using arecanut leaf sheath and bunch waste. Steps in oyster mushrooms cultivation include development of spawn, substrate preparation, spawning, incubation for spawn running and opening and maintenance of beds for cropping.

Spawn, the vegetative seed of the fungus, can be obtained either from research institutions or can be prepared with adequate training. An efficient and stable strain of Pleurotus isolated from sporocarps should be used for spawn preparation using grains like wheat, sorghum, maize, jowar or paddy straw as substrates. Pasteurization of substrates is necessary to avoid contamination and to obtain higher yield. Steam sterilization, hot water treatment and chemical sterilization are the effective methods. Steam sterilization at 1.02 kg/cm^2 pressure in an autoclave for one hour is an efficient method of sterilization. Hot water dip of the substrate at 80°C for 60-120 minutes is another method, which can be easily adopted. Chemical sterilization method involves treatment with formalin and bavistin in substrate soaking water for 18 hours. Arecanut bunch waste and leafstalk are pasteurized by soaking in a solution of formalin (500 ppm) + bavistin (25 ppm)

Polyethylene bag method is the commonly followed method of cultivation. Polyethylene bags (150-200 gauge) of 66 x 45 cm size are punched to facilitate cross ventilation. Spawning is done by multilayered technique using 3 % spawn. For spawn running and cropping, the filled up bags are incubated in cool dark place. In 15 to 20 days white thread like mycelium covers the entire substrate and the whole mass turns into a solid cylindrical structure. At this stage, the polyethylene bags are ripped open and incubated for cropping by hanging or by stacking on the shelves of mushroom house with watering daily twice after two days of opening of beds.

Low cost mushroom sheds can be built with coconut/areca stem and plaited coconut leaves inside arecanut plantations. Multilayer rack can be prepared with coconut/ areca stem inside the shed to keep the spawned substrate for spawn running and cropping. Ventilators with insect proof nets are to be provided on all sides of the shed. Mushroom production is 69% and 49.8 % in a cropping period of 47 to 52 days in arecanut bunch waste and leaf sheath, respectively.

Marketting

Demand and Supply Analysis

Arecanut.is an important commercial crop in India. Arecanut is the seed of the arecanut palm. It plays a prominent role in the religious, social and cultural functions and economic life of people in India. The economic produce is the fruit called "betel nut" and is used mainly for masticatory purposes. There are two varieties of arecanut, called white Supari and Red Supari. White variety Supari is prepared by harvesting fully ripe arecanut and by sun drying for 40 to 50 days. Red variety Supari is prepared by harvesting the tender (green) arecanut, boiling it and peeling off the husk.

Arecanut has uses in ayurvedic and veterinary medicines. It is estimated that nearly ten million people depend on arecanut industry for their livelihood in India. The quality, variety and types of arecanut vary from one place to other. Consumption of arecanut in the country steadily increased from 0.114 million tones in 1956-57 to 0.336 million tones in 1999-2000. The projected demand of arecanut by 2020 is 0.617 million tones. The arecanut is location specific, but the consumption is wide spread throughout the country. The per capita availability of arecanut in India is 1.58 g /adult/day.

Global Scenario

Besides India, China, Bangladesh, Indonesia, Myanmar and Thailand are the other important arecanut producers. The current production of arecanut in the world is about 0.64 million tonnes from an area of 0.476 million hectares. India ranks first in both area (58%) and production (53%) of arecanut. The current world productivity of arecanut is 1.287 tones/ha. China ranks first in arecanut productivity with 3.752 tones/ha. India ranks fourth in terms of productivity (1.189 tones/ha).

Arecanut has limited export potential. However, a small quantity of it is exported in its processed form or other products like panmasala, scented supari and gutka. During 2000-01, India exported 1502 million tones of arecanut. The quantity of import varied from 545 tonnes to 10823 tones during 1994-95 to 1997-1998. Arecanut imported during 2000-01 was reduced to 805 tones from Srilanka, Indonesia, Bangladesh Singapore and Myanmar.

State wise Area and Production of Arecanut During 2002 - 03

States	**Area ('000ha)**	**Production ('000 tones)**
Andhra Pradesh	0.3	0.7
Assam	73.7	51.6
Goa	1.6	2.5
Karnataka	119.1	162.8
Kerala	93.2	84.7
Maharashtra	2.2	4.4
Meghalaya	11.2	14.2
Mizoram	1.0	1.8
Tamil Nadu	3.7	12.7
Tripura	3.2	6.8
West Bengal	9.1	16.6
Pondicherry	0.1	0.1
Andaman	4.4	7.2
All India	322.7	366.0

Source: CPCRI, Kasaragod

Indian Scenario

India continues to dominate the world in area, production and productivity of arecanut and has achieved self sufficiency in arecanut production (0.27 million tones). Most of the production is domestically consumed. Total area under arecanut cultivation is around 3.2 lakh hectares and the total production is around 3.4 lakh tones. Arecanut is mainly grown in Kerala, Karnataka and Assam. Kerala and Karnataka account for about 70% of country's production. Tamil Nadu, Maharashtra, A.P, West Bengal and Orissa are the other important producing states. Out of the total production, the percentage of white variety is around 60% and that of red variety 40%. Arecanut consumption in the country is around 3.3 lakh tones. Raw arecanut is used as supari while processed nuts are used in gutka industry.

Arecanut production in India has now almost reached a level of self-sufficiency. Uses for Arecanut other than chewing are negligible. Its export prospects are also very much limited. Therefore, the present policy is not to expand area under arecanut, but to adopt intensive cultivation and take up replanting of aged and unproductive gardens. Inter and mixed cropping in arecanut gardens is advocated to augment income from existing arecanut garden.

Arecanut has limited export potential. A small quantity of it is exported in its processed form or other products like panmasala, scented supari and gutka.

Price

Arecanut prices, which showed a steady increase in almost all markets of India from 1948-49, had an unprecedented fall in 1972-73. Within India the habit of chewing is slowly declining (Das, 1985). The price variations have become almost a regular feature in the market, which widely affect the poor growers who entirely depend upon income from areca for their livelihood.

Major Indian Markets

Campco is the main nodal agency for procuring arecanut. Mumbai, Ahmedadad, Indore, Jaipur, Delhi, Nagpur, Patna, Calcutta, Cuttack, Bangalore, Rajkot, and Chennai are the important marketing centres of arecanut in India.

Market Influencing Factors

- Weather conditions during growth period.
- Demand from overseas consuming countries like Nepal and Pakistan.
- Quantum and timing of imports from Indonesia, Thailand and Burma.
- Government intervention through MSP purchase.
- Carryover stocks.
- Growth of Gutka industry

Price Spread Analysis of Arecanut

Studies showed that 80% of the farmers dispose the produce immediately after harvest and they are small cultivators. Rest 20% who dispose the produce when the prices in the market are favorable, are large farmers. It was observed that indebtedness and lack of proper infrastructure facilities for storage compelled the small farmers to dispose the produce at the earliest. There are four predominant marketing channels in the arecanut trade. Producer's share in consumer price was the highest in the channel-4 where co-operative society markets the produce through it's own sales depot at major consuming centers. The producer's share in consumer price was the lowest for channel-3, which is the longest marketing channel with six levels of market functionaries. Marketing efficiency analysis through composite index method revealed channel-4 as the most efficient trade channel with the lowest composite index. Producer's share in consumer price ranged from 63.53 to 68.10% among the channels. The analysis of marketing efficiency, highlighted channel–4 as the most efficient marketing channel with the lowest composite index.

CASHEW

Cashew (*Anacardium occidentale* L.) is a tropical evergreen tree belonging to family Anacardiaceae. The cashew tree, native to Brazil, was introduced to Mozambique and then India in the sixteenth century by the Portuguese, as a means of controlling coastal erosion. The first introduction of cashew in India was made in Goa from where it spread to other parts of the country. In the beginning, it was mainly considered as a crop for afforestation and to check erosions. Because of its adaptive ability in wide range of agro climatic conditions, it became a crop of high economy and attained the status of an export oriented commodity bringing considerable foreign exchange to the country.

The nuts, apple and other by products of this crop are of commercial importance. The cashew kernels are used in confectionery and dessert. The shells contain a high quality oil known as cashew nut shell liquid (CNSL) which has wide industrial uses. Cashew apple is eaten fresh or mixed in fruit salads and a drink is prepared from the juice. Cashew can be distilled to produce alcoholic drink (Fenny).

Soil and Climate

It is well adapted to a wide variety of soils from coastal sands to degraded laterites. Cashew grows well in soils which are very poor in nutrients. In Sri Lanka and Philippines, it is grown in alluvial soils and in volcanic soils in Indonesia, Fiji and Philippines. It is grown abundantly in loose sandy soils particularly in South India, East Africa and Brazil. Cashew prefers well-drained aerated sandy loam with adequate moisture.

Cashew is essentially a tropical crop, grows the best in the warm, moist and typically tropical climate. Its distribution is restricted to altitudes below 700 m where the temperature does not fall below 20°C for prolonged periods, although it may be found growing at elevation up to 1200 m. It is the best adopted to the coastal regions. The cashew is hardy and drought resistant, but it is damaged by frost. Cashew is grown in areas with rainfall ranging from 600 - 4500 mm/annum.

Cashew is a sun-loving tree and does not tolerate excessive shade. It tolerates temperature of more than 36°C for a shorter period but the most favourable temperature lies between 24 to 28° C.

The climatic factors influence the cashew

(i) Dry spell during flowering and fruit setting ensures better harvest.

(ii) Cloudy weather during flowering enhances scorching of flowers due to tea mosquito infestation.

(iii) Heavy rains during flowering and fruit set damage production.

(iv) High temperature (39-42°C) during stage of fruit set development causes fruit drop.

Varieties

Selection of suitable cashew varieties for the specific region and appropriate package of practices determines the final yield. More than 42 varieties having exportable grade of cashew kernels are released from different research institutes in India.

Softwood Grafting and Nursery Management in Cashew

Different methods of propagation viz., epicotyl grafting, soft wood grafting, veneer grafting, side grafting, layering etc., tried in cashew with varying degrees of success. Among them, soft wood grafting is the best for commercial multiplication of cashew.

I. Establishment of Scion Bank

i. Production of Scion Sticks

- The Scion Bank should be established with recommended varieties of the region.
- The plants should be planted in a closer spacing of 4x4m to get continuous supply of scions.
- It should be fertilized with recommended quantity of NPK during monsoon and tender shoots should be protected by timely sprays of insecticides.
- The flower panicles should be cut off to get more number of scions.

- The plants in the scion bank should be maintained by heading back the top to a height of 1.5 m from ground level and by pruning the dried branches.
- Pruning of trees may be carried out annually during September –October in the scion bank

Released Cashew Hybrids

Research Station	Hybrid	Parentage	Yield (kg/ tree)	Nut wt. (g)	Kernel wt. (g)	Shell -ing (%)	Kernal grade
CRS, Bapatla	BPP-1	T.No.1 x T.No.273	10.0	5.0	1.3	27.5	W 400
	BPP-2	T.No.1 x T.No.273	11.0	4.0	1.0	25.7	W 450
	BPP-8	T.No.1 x T.No.39	14.5	8.2	2.3	29.0	W 210
RFRS, Vengurla	Vengurla-3	Ansur-1xVetore-56	14.4	9.1	2.4	27.0	W 210
	Vengurla-4	Midnapore Red x Vetore-56	17.2	7.7	2.4	31.0	W 210
	Vengurla-5	Ansur Early x Mysore Kotekar 1/61	16.6	4.5	1.3	30.0	W 400
	Vengurla-6	Vetore 56 x Ansur-1	13.8	8.0	2.2	28.0	W 210
	Vengurla-7	Vengurla-3 x M 10/4	18.5	10.0	2.9	30.5	W 180
	Vengurla-8		15.7	11.5	2.8	28	W180
CRS, Madakkathara	Dhana	ALGD-1 x K-30-1	17.5	9.5	2.2	28.0	W 210
	Kanaka	BLA 139-1 x H-3-13	19.0	6.8	2.1	31.0	W 210
	Amrutha	BLA 139-1 x H-3-13	18.3	7.2	2.2	31.6	W 210
	Priyanka	BLA 139-1 x K-30-1	16.9	10.8	2.8	26.5	W 180
	Damodar	(A-1xH-3-13)	13.36	8.2	2.0	27.3	W 240
	Raghava	ALGD-1-1xK-30-1	14.65	9.2	2.27	26.6	W 210
	Poornima	K-30-1x BLA 139-1	14.1	7.8	2.6	31.0	W210
CRS, Anakkayam	Dharasree	Kottarakkara x Brazil	15.02	7.8	2.4	30.5	W 240
	Anakha	T.No. 20 x K-30-1	13.7	10.0	2.9	29.0	W 180
	Akshaya	H-4-7 x K-30-1	11.8	11.0	3.12	28.4	W 180

ii Selection of Scion and Precuring

- Select non-flowered, 3-5 months old lateral shoots of current season's growth

- The selected shoots should be about 10-12 cm long, straight, uniformly round with pencil thickness, brown coloured having dormant plumpy terminal buds.
- The top 4-5 leaves should be dark green in colour indicating proper maturity of scion sticks.
- The selected scion sticks should be pre cured by clipping off leaf blades, leaving petiole.
- The scion sticks can be detached from the mother tree after 8-10 days, before they sprout and utilized for grafting.

Released Cashew Selections

Research Station	Variety	Source	Yield kg/tree	Nut wt. (g)	Kernal grade
NRCC, Puttur	NRCC Sel.1	VTH 107/31	10.0	7.6	W 210
	NRCC Sel.2	VTH 40/11	9.0	9.2	W 210
ARS, Ullal	Ullal-1	8/46	16.0	6.7	W 210
	Ullal-2	3/67	9.0	6.0	W 240
	Ullal-3	5/37	14.7	7.0	W 210
	Ullal-4	2/77	9.5	7.2	W 210
	UN-50	2/27	10.5	9.0	W 180
ARS, Chintamani	Chintamani-1	8.46	7.2	6.9	W 210
CRS, Bhubaneswar	Bhubaneswar-1	WBDC-5	10.5	4.6	W 320
CRS, Bapatla	BPP-3	3/3	11.0	4.8	W 400
	BPP-4	9/8	10.5	6.0	W 400
	BPP-5	T.No.1 of Bapatla	11.0	5.2	W400
	BPP-6	T.No.56	10.5	5.2	W 400
RRS, Jhargram	Jhargram-1	T.No.16	8.5	5.0	W 320
RFRS, Vengurla	Vengurla-1	Ansur-1	19.0	6.2	W 240
	Vengurla-2	WBDC-6	24.0	4.3	W 320
RRS, Vridhachalam	VRI-1	M 10/4	7.2	5.0	W 320
	VRI-2	M 44/3	7.2	5.0	W 320
	VRI-3	M 26/2	10.0	7.2	W 280
CRS, Madakkathara & CRS, Anakkayam	A-1	BLA 139-1	12.0	6.0	W 280
	MDK-1	BLA 39-4	13.8	6.2	W 280
	MDK-2	NDR-2-1	17.0	7.3	W 240
	Sulabha	K-10-2	22.0	9.8	W 210

iii Collection of Scion Stick

- The pre cured scion sticks should be separated early in the morning to avoid desiccation and its length should not be less than 10cm.
- Soon after separation from the mother tree, scion sticks should be dipped in water and placed in a polythene bag of 100 guage thickness and brought to nursery shed for grafting.
- The scion sticks wrapped in sphagnum moss cloth and placed in a polythene bag of 100 guage can be kept for 3-4 days and used for grafting, if necessary.

II. Raising of Root Stock Seedlings

i.Selection of Seednuts

- Collect fresh seednuts during peak period of harvest in February-April and sun dried for 2-3 days.
- Medium sized nuts of 6-7g should be selected to get uniform and vigorously growing seedlings required for soft wood grafting. Fresh seed nuts are used for raising root stocks. More than one year old seed nuts may be avoided.
- Prepare potting mixture in the proportion of one part red soil, one part sand and one part compost in heavy rainfall areas and one part red soil, one part compost in low rainfall areas and mixed with little quantity of rock phosphate (10g/2Kg of potting mixture). Incorporation of Azospirillum+PSB+AMF@10G/bag obtained 100% germination.
- Use high density polythene bags of 25 x15 cm size and 300 gauge thickness for filling potting mixture.
- Punch about 30-40 holes of 0.5 cm diameter uniformly on polythene bags for ensuring good drainage.
- Fill the bags properly up to the brim with potting mixture and arrange bags in beds of 1000(10x 100) number in each.

ii.Sowing of Seeds

- The seed nuts should be soaked in water for 12-24 hours before sowing to get good germination.

- Sow the seeds in the centre of bag, stalk-end upwards, with a depth of not more than 2.5 cm and cover with a little soil.
- Water the polythene bags immediately after sowing and daily during drier months. Excess irrigation should be avoided.
- The seednuts usually germinate within 15-20 days after sowing.
- Seed bed may be mulched with paddy straw till germination takes place and partial shade may be provided during summer.
- To control pests while seeds germinate, malathion (5%) dust, or spraying of chlorpyriphos (Durmet 20 EC:0.05%) should be applied

iii.Maintenance of Seedlings in Nursery

- Daily watering is required during summer.
- During rainy season, to avoid water stagnation and disease occurrence seedlings may be covered by polythene sheet.
- To get vigorous seedlings with single main stem, the side shoots arising from leaf axils should be removed frequently.
- To prevent penetration of roots into the ground when seedlings are kept in same place for three months and have to be shifted and arranged in beds of different place.
- To control tea mosquito and other sucking insect pests, ekalux / monocrotophos @ 1.5 ml / litre of water may be sprayed.
- Select 45-60 days old seedlings as root stocks for grafting.

III. Soft Wood Grafting Technique

i.Preparation of Root Stock

- Disinfect the grafting knives by dipping them in fungicide solution.
- Root stock seedling brought from nursery beds should be kept in the nursery shed under shade for grafting.
- Remove leaves by a sharp grafting knife leaving two pairs of bottom leaves.
- At a height of 15-20 cm from ground level, a transverse cut is made on root stock and the terminal shoot is removed.
- A cleft of 6-7 cm deep is made in middle of decapitated stem by giving a longitudinal cut.

- A little portion of wood is removed from inner sides of cleft at the top, so that after grafting, the joint will be perfect.

ii.Preparation of Scion

- Select a matching scion stick of same thickness as that of the root stock of 10-12cm long by cutting off the excess portion at the bottom.
- Shape the cut end of the scion into a wedge of 6-7 cm long by chopping off the bark and little portion of wood from two opposite sides.
- While preparing the wedge, the gum on cut surface should not be disturbed and cut surfaces should not be soiled by touching with fingers.

iii.Grafting

- Wedge of scion is inserted into cleft of root stock to see that cambial layers of both the root stock and the scion come in perfect contact with each other.
- The graft joint is secured firmly with 2.5cm wide and 30cm long polythene strip of 100 gauge thickness.
- A long and narrow HD polythene bag of 15 x 12.5cm size and 100 gauge thickness is inserted on the grafted plant, tied at the bottom with a single knot to protect the scion stick from drying up.
- In places with high humidity, the polythene caps need not be used
- The freshly grafted plants are left in the nursery shed for about two weeks to encourage sprouting of terminal buds.
- After two weeks, the polythene caps are removed gently and the grafts are shifted to open condition in nursery.
- Within 3-4 weeks, 70-80% of grafts will sprout.
- Graft ready for planting within 5-6 months

IV. Management of Cashew Grafts in the Nursery

- Cashew grafts should be watered daily during dry periods.
- During rainy season, excess water should be removed.
- Plants in the nursery should be shifted at least once in a month to avoid penetration of roots in the ground.

- During summer (January-May) grafted plants should be protected by providing partial shade by erecting pandal of dry coconut fronds or nylon nets.
- Bordeaux mixture spray (1%) may be given at 10 days interval during rainy season to control fungal infection of tender seedlings and grafted plants.
- Ekalux / monocrotophos (1.5ml/litre of water) may be sprayed to control sucking insects and leaf eating caterpillars as and when the damage is seen.
- The flower panicles produced by the graft should be removed as and when observed in nursery.
- The polythene strip should be removed from graft joint after four months of grafting.
- When scion leaves turn bronze colour, the graft can be exposed to more light

Planting and Management

Method and Season of Planting

The square system or triangular system of planting is practiced. The ideal time for planting is usually during monsoon seasons (June-July or Sept-October). If irrigation facilities are available, planting can be done throughout the year except winter months.

Normally cashew grafts are planted in pits of 60 cm^3 . The pits should be filled with a mixture of top soil and organic manure to 3/4 of the pit capacity. The grafts are planted after carefully removing the polythene bag. Care should be taken to see that the graft joint remains at least 5 cm above ground level .The polythene tape around the graft union need to be removed carefully. Staking should be done immediately after planting to protect grafts from wind damage. Mulch basins of plants with organic waste materials during early years.

Planting Materials

Selection of planting material is the most important in cashew cultivation. Cashew is highly cross pollinated and vegetative propagation is recommended mainly on commercial scale to produce true to type planting materials. Softwood grafting is the only method, commercially feasible and practically successful in cashew.

Spacing

The normal spacing recommended is 7.5 x 7.5 m to 8 x 8 m and spacing may be reduced to 4x 4m depending on type of soil and managerial capacity.

High Density Planting

A close spacing of 4 x 4m or 5 x 5m or 6 x 4m and maintained up to 9 years with proper pruning and training. Later the excess plants can be thinned out to provide a final spacing of 8x 8m or 10 x 10m or 6 x 8m depending upon canopy expansion rate. This enables high returns during initial years.

Nutrient Management

Cashew is a nitrogen lover and N and K nutrition are of great significance in enhancing production and productivity of cashew (Abdul Salam, 1997). According to Beena *et al.* (1995), the nutrient up take through a kg of nut and its apple is 64.1g of N, 2.05g P, 24.7g k, 4.1g Ca, 1.57 g S, 525.7 mg of Fe, 63.6 mg Mn, 87.8 mg Zn and 26.5 mg Cu.

Spray application of urea 2% with any of the recommended pesticide twice at flushing and panicle initiation ensure high fruit set and effective pest control. In almost all the cashew growing states of India, FYM is recommended @50-100kg/ tree to sustain cashew yield. Manures and fertilizers promote growth of plants and advance onset of flowering in young trees.

A fertilizer dose of 750 g N, 325 g P205 and 750 g K2O/ plant/ year is recommended. The ideal period for fertilizer application is immediately after cessation of heavy rains and with available soil moisture. During 1st, 2nd and 3rd year of planting 1/3rd, 2/3rd and full doze of fertilizers should be applied and 3rd year onwards full quantity is to be applied. It is applied either as a single dose during September-October or in two equal splits during June-July and Sept-Oct.

Broadcast the fertilizers over the entire tree basin (15 cm deep) in a radial distance of 2 to 3 m within the canopy spread, leaving half a metre from the tree trunk and incorporate by light raking.

Vermi Composting

Apple residue and leaf litter can be effectively utilized for production of vermicompost. According to NRCC (2004) two tones of organically recyclable biomass could be converted to 1.3 tonnes of vermicompost with net recovery of 65%. The nutrient composition of the compost was analysed as 1.65% N, 0.43% P and 0.59% K compared to vermicompost prepared from other sources. *Eudrillus* sp. (1000 worms/ tonne after 14 days of initial composting) could convert recyclable cashew biomass into vermicompost within 3 months.

Weeding

Chemical, manual or mechanical weeding should be done depending upon type and intensity of weed growth during June-July or Aug-Sept. Glyphosate (post emergent) application at 6 to 7 ml/ litre of water (0.8 kg a.i./ha) during June - July also effectively controls weeds.

Mulching

Mulching the tree basins helps in conservation of soil moisture and prevents soil erosion. Mulching with organic matter or residues inhibits weed growth and reduces surface evaporation during summer and also regulates soil temperature.

Irrigation and Water Management

In India, cashew is grown mainly under rainfed condition. However, protective irrigation especially during summer (January-March) at fortnightly intervals @ 200 liters/plant improves fruit set, fruit retention and thereby increasing nut yield.

Soil and Water Conservation Measures

To eliminate subsoil moisture stress during summer in West Coast soil, water conservation measures like terracing base of plant, opening catch pits, terracing with crescent bund and opening pits in middle of four plants across the slope or opening trenches of 1.0 m width and 0.5 m depth and of convenient length between two rows of cashew and burying coconut husk can be adopted to withhold water for longer period particularly during pre and post monsoon showers. In East Coast, intercept bunds of 100 x 50 x 30 cm size are prepared in cashew plantations. This bunding, apart from breaking velocity of run off water, it helps to impound rain water.

Under sloppy areas, soil and water conservation practices can be done by making trenches of 30 cm width, 60 cm depth and of convenient length may be taken in between rows along the contour. Forming terrace at 2m radius around the plant at the base within three years of planting is beneficial in conserving soil moisture. This will not only conserve soil and moisture but will also enable to enhance growth of cashew.

General Pruning

In older cashew plantations, removal of dried or dead wood, criss-cross branches, water shoots etc. should be attended to at least once in 2-3 years. This allows proper growth of canopy and receipt of adequate sunlight on all branches. Pruning of cashew plants should be done during May / June.

Training and Pruning

Training and pruning are important horticulture practices to be employed to make better frame work of cashew plant. It helps to control growth and make it easy for cultural practices. The sprouts arising from root stock portion of cashew graft should be removed frequently during first 3-4 years of planting. The plant is allowed to grow by maintaining a single stem upto 0.75-1.00m from ground level. As an orchard management technique to improve sanitation, removal of water shoots, lower branches, crisscross branches and dry branches are beneficial to enhance flowering and yield.

Intercropping

Intercropping received little attention in cashew. Depending on soil and climatic conditions and local situations, annual tubers and spices like tapioca, turmeric, ginger etc, can be grown as inter crops. When once plants become sufficiently big, pepper can be taken as mixed crop.

Top Working

Top working is a technique evolved to rejuvenate unproductive and senile cashew trees. Top working can successfully rejuvenate poor yielders in the age group of 5-20 years. The unproductive trees are to be beheaded at a height of 0.75 to 1.00 m from ground level. The stem should be cut with a saw to avoid stump splitting. The best season for beheading trees is May-September. Soon after beheading, the stumps and cut portions should be given a swabbing with copper oxychloride and carbaryl 50 WP (50 g each/ litre of water). Sprouts emerge 30-45 days after beheading. Sprouting will be profuse in young trees. New, 20-25 days old shoots should be grafted with scions of high yielding varieties using softwood grafting technique. To ensure at least six or seven successful grafts, 10-15 graftings are to be done on new shoots of every tree. The best season for grafting is July-November. Thinning of extra shoots arising from stumps should be done to obtain better growth of grafts. Removal of sprouts below graft joint and removal of polythene strip from graft joint should be done. Top working is simple and can be done by farmers after getting proper training.

The top worked trees start yielding right from second year after top working. Thus Precocity can be considered as one of the best advantages of this technique. The major disadvantage associated with top working is the huge casualty of trees due to stem borer attack. Intensive care and management to ward off stem borer are essential. As such, adoption of top working on a larger scale would be difficult.

Cost of Cultivation (200 Plants/ha)

1. Labour Requirement (Mandays and Cost)

S. No.	Items of operation	1st year	2nd year	3rd year	Total
i	Land cleaning & Peg marking (200 pits)	42	2	1	45
ii	Digging 200 pits	16	-	-	16
iii	Pit filling & Planting - stacking, mulching & shading	9	2	-	11
iv	Weeding twice a year	17	10	5	32
v	Soil conservation measures	30	30	-	60
vi	Application of manures and fertilizers	4	6	10	20
vii	Plant protection operations	4	4	6	14
viii	Fencing, watch and ward	24	14	14	52
ix	Miscellaneous operations like removal of sprouts, deblossoming, transportation of inputs, pruning and training	4	5	4	13
	Total	150	73	40	263
	Wages @ Rs.100/- per manday	15000	7300	4000	26300

Source: Directorate of cashew and cocoa development, Kochi.

2. Cost of Inputs

a) Conventional Methods

(in rupees)

S. No.	Items of operation	1st year	2nd year	3rd year	Total
i	200 grafts @ Rs.20/- per graft	4000	-	-	4000
ii	Gap filling @ 10% (20 grafts)	-	400	-	400
iii	Manures and fertilizers	825	1650	2550	5025
iv	Plant protection chemicals	375	750	1125	2250
	Total 2(a)	5200	2800	3675	11675

S. No.	Items of operation	1st year	2nd year	3rd year	Total
i	Cost for soil conservation measures	3000	3000	-	6000
ii	Cost for irrigation	5000	5000	5000	15000
Cost of cultivation under conventional methods excluding labour Total [2(a) +i + ii]		13200	10800	8675	32675
Financial assistance limited to Rs.30,000/- and disbursed @ 50:30:20		15000	9000	6000	30000

b) Organic Cultivation

S. No.	Items of operation	1st year	2nd year	3rd year	Total
i	200 grafts @ Rs.20/- per graft	4000	-	-	4000
ii	Gap filling @ 10% (20 grafts)	-	400	-	400
iii	Organic manures/bio-fertilizers	5000	5000	4000	14000
iv	Organic pesticides	500	1000	1500	3000
	Total 2(b)	9500	6400	5500	21400
Total cost for organic cultivation excluding labour [2(a) +iii + iv]		17500	14400	10500	42400
Financial assistance limited to Rs.40,000/- and disbursed @ 50:30:20		20000	12000	8000	40000

Source: Directorate of Cashew and Cocoa Development, Kochi.

Harvesting and Yield

Flower panicles emerging from the graft during 1st and 2nd year should be removed to allow the plant to put good vegetative growth and better framework. Economic bearing in cashew commences after 3rd year of planting. The ripened fruits fall down and nuts from fallen fruits have to be collected. Nuts can be dried in sun for 2 to 3 days on cement floor and can be stored in gunny bags. Yield starting from 1 kg in 3rd-4th year, goes on increasing as the canopy develops and one can expect more than 10 kg of nuts in 8 to 10 years old plants depending on management.

Processing at Orchard Level

The processing of cashew in the orchard is mainly confined to removal of raw nuts from cashew apple and drying. Harvest only fully matured nuts. Usually, nuts are picked when they fall off from the tree. Best quality nuts are obtained, where fresh fallen fruits are collected. The apples are removed and the nuts are sun dried for 2 - 3 days to reduce moisture from 25% to below 9 %. It is very essential to dry the nuts after harvest to prevent spoilage during subsequent storage. This helps the kernel to retain their quality, particular the flavour. However, if cashew apples are used for processing, it is better to harvest them from the tree without damaging apples. The mature nuts will sink in water while the immature/unfilled ones will float. This test could be used to find out whether the nuts are mature or not

Processing of Rawnuts at Factories

The nuts which are required to be processed at factory should be dried again for 1-2 days to reduce and maintain moisture level of 7-8%. The processing of raw nuts involves roasting, shelling, drying, peeling, grading and packing. The steps involved in commercial processing of cashew nuts are detailed below :-

Roasting

Roasting of raw nuts is done to separate adhering shell from kernel. There are three types of roasting viz, drum roasting, oil bath roasting and steam roasting.

(a) *Drum roasting:* This is one of the oldest and more widely used methods. The nuts are fed into a rotating red hot drum which will ignite the shell maintaining its temperature because of burning of shell liquid. The drum is kept in rotation for 3-4 minutes and roasted nuts are discharged from lower end of drum and immediately covered by ash after sprayed with a little water, to absorb oil on the surface. This facilitates removal of remaining oil on shell. Due to draw backs of method is superseded by oil bath roasting.

(b) *Oil bath roasting:* In this method, conditioned nuts are passed through CNSL bath heated to 170-200°C by conveyer buckets for 1-2 minutes during which period, the shell gets heated rupturing the wall and releasing oil into bath. The oil is recovered by continuous overflow arrangement. The roasted nuts are centrifuged to remove adhering oil, cooled and shelled by hand and leg operated shelling machines. The kernel with the adhering testa is scooped out using a sharp needle.

(c) *Steam roasting:* The raw nuts are steam cooked at about 120-140 lbs/sq inch pressure. Shell oil can be extracted in later stages by crushing. The nuts are shelled by hand and leg operated shelling machines.

A general comparison of above three methods would show that the oil bath method and steam roasting systems require more initial investment and higher maintenance costs: the drum roasting method being the cheapest. From the point of view of environmental pollution, the safest is steam roasting method.

Shelling

Cashew nuts after roasting and cooling are to be shelled to remove kernels. One has to be very careful while shelling nuts. Hands are to be protected from Cashew Nut Shell Liquid (CNSL) which is highly corrosive. Hand gloves should be used while shelling. For the same reason, it is advisable to dust the nuts with wood ash. Commercial processing units use foot operated shell cutters (mechanical device) for shelling. This device consists of a pair of blade (knives) shaped in the counter of half a nut which could be operated by foot. The blades cut through shell all around the nut, leaving the kernel untouched. After shelling kernels and shell pieces are separated manually. The nuts have to be grouped into various sizes, each size matching a pair or blades of appropriate size.

Mechanical Shelling

Several pieces of equipments are designed to remove shells from cashew nuts. The main challenge with mechanical shelling is to remove kernel without damage or contamination from the CNSL. This challenge is exacerbated by irregular shape of nut and wide variation in size of nut. The most successful mechanized decorticators work on nuts that have been conditioned by hot oil process, which makes the shell brittle and easier to break.

Development of a Cashew Nut Sheller in India

A mechanical sheller was designed in India by researchers at the Post Harvest Technology Centre, Indian Institute of Technology Kharagpur. The sheller uses principles of compression and shear, taking into account physical and mechanical properties of cashew nuts. The machine consists of four compartments - power supply, transmission, feeding, shelling and discharging.

The feeding section consists of a hopper and horizontal screw conveyor for positive feed of roasted nuts to shelling section. The design criteria for the hopper and screw conveyor use size, bulk density, coefficient of friction and angle of repose of roasted cashew nuts. A flat plate sliding gate is used to control feed rate of cashews into sheller.

Pre-Grading

Pre-grading can be done before or after drying kernels and greatly reduce the work involved in final grading. Pre-grading can be done mechanically for large-scale processes, separating mainly the whole from the broken kernels and sometimes separating the different size groups of whole kernel.

Drying

Kernels after shelling will have moisture content of more than 6%. Drying of these kernels is necessary to prevent fungus attack during subsequent storage and to facilitate peeling of testa. The kernels are to be dried to moisture content of about 4- 5 %. This is done by drying the kernels in hot chambers at 70 - 80 °C in perforated trays for about 6 - 8 hours. Uniform drying could be achieved with a cross flow drier using forced hot air circulation through kernel layers. To ensure uniform drying, the position of trays has to be changed frequently, as scorching may occur at hotter places. Excess drying of kernels leads to kernels becoming very brittle resulting in higher breakage. After drying, the kernels are kept in the moist chamber for 24 hours which facilitate easy removal of testa (peeling) and minimize broken kernels.

Peeling

This process involves removal of testa (seed coat) from kernel. Peeling is done using a sharp knife or bamboo piece. Care has to be taken while removing testa. If kernels are scorched more, it results in poor quality kernels.

Grading

Kernels are graded according to size manually. In the International Market bold whole kernels fetch premium price. The grading standards developed in India refer to white whole (undamaged) kernels and indicate number of kernels/lb of weight. The largest kernels come in the grade W 210 (440-460/kg) and the smallest of the seven grades is W 500 (1000-1100/kg). Generally, Brazilian kernels have a relatively high proportion of large wholes, with the extra attraction of the 180's grade, known as Special Large Wholes. The white whole kernels are priced according to size. Further classification refers to broken kernels, butts, splits, pieces, small pieces and whether kernels are white or scorched.

Packaging

As far as possible, packaging material used should be eco-friendly and recyclable and containers are hermetically sealed after filling carbon dioxide.

The normal packaging used for export of kernels is air-tight cans of 25lbs (11.34 kg) weight capacity. The packaging material needs to be impermeable, since cashew kernels are subject to rancidity and go stale very quickly.

Grades of Cashew

Grades of cashew					
White wholes	Scorched wholes	Dessert wholes	White pieces	Scorched pieces	Dessert pieces
WW 180	SW-180	SWS (Scorched whole seconds)	B (Butts)	SB (Scorched Buts)	SPS (Scorched pieces seconds)
WW 210	SW-210		S (splits)		
WW 240	SW-240		LWP (Large white pieces)	SS (Scorched splits)	
WW 320	SW-320	DW (Dessert wholes)	SWP (Small white pieces)	SP (Scorched pieces)	DP (Dessert pieces)
WW 450	SW-450				
WW 500	SW-500		Baby bits (BB)	SSP (scorched small pieces)	

Source: Cashew Export Promotion Council of India

After filling and weighing, the cap should be soldered in preparation for vita pack process. This consists of removing all the air from the can and substituting it with carbon dioxide (CO2). The advantages of packing cashew kernels in carbon dioxide are two-fold. Some processors do not have vacuum pumps, and displace the air in the can by feeding in carbon dioxide through a small hole in the bottom of a side of the can. The carbon dioxide valve is turned off when all the air has been replaced. Holes in the can are then sealed, with the hole at the bottom of the side of the can being sealed first, and the one at the top last.

Infestation

The most important defence against infestation of any type is cleanliness, and is essential in the rooms used for drying, peeling, grading, conditioning, and packaging. Floors and walls must be sound and free from cracks. They should be white-washed on a regular basis.

The speed of operations between drying and packaging must be stressed as this reduces critical period when attacks may occur, to a minimum. The equipment used must also be thoroughly cleaned on a regular basis as insects may breed in hidden crevices and gaps.

The kernels are most vulnerable when dry, being both brittle and susceptible to insect infestation. Therefore, at this stage, they must be handled with care and moved to next stage of peeling as quickly as possible.

Hygiene and Safety

Successful business activity depends upon adequate food safety and hygienic practices. The main ways in which a producer can harm consumers are by selling food that:

- contains poisonous materials
- contains bacteria or moulds or the poisons they produce
- contains glass or other contaminants that could cause harm if eaten

Safe food can be produced by careful attention to hygiene and by use of proper quality control.

Proper hygiene means careful attention to cleanliness of processing equipment and personal hygiene of food handlers.

Proper quality control means careful attention to selection of good quality raw materials; correct processing conditions, like temperature and time of heating; preventing contaminating materials such as dirt, metal and stones from becoming mixed with food; and use of suitable packaging materials to protect food after processing. These factors ensure that only wholesome food is produced without contaminants. Any bacteria in the raw materials will be destroyed, or controlled at a safe level, and prevented from growing and multiplying.

Uses of Cashew Apple

In cashew-producing countries, the nut is only one of the products enjoyed by the local population. The cashew "apple" or false fruit is an edible food rich in vitamin C. It can be dried, canned as a preserve, or eaten fresh from the tree. It can also be squeezed for fresh juice, which can then be fermented into cashew wine, a very popular drink in West Africa. In parts of India, it is used to distill a cashew liquor referred to as feni. In some parts of South America, local inhabitants regard the apple, rather than the nut kernel, as a delicacy. In Brazil, the apple is used to manufacture jams, and soft and alcoholic drinks.

The cashew apple is very rich in vitamin C (262 mg/100 ml of juice) and contains five times more vitamin C than an orange. A glass of cashew apple

Overview of Cashew Nut Processing

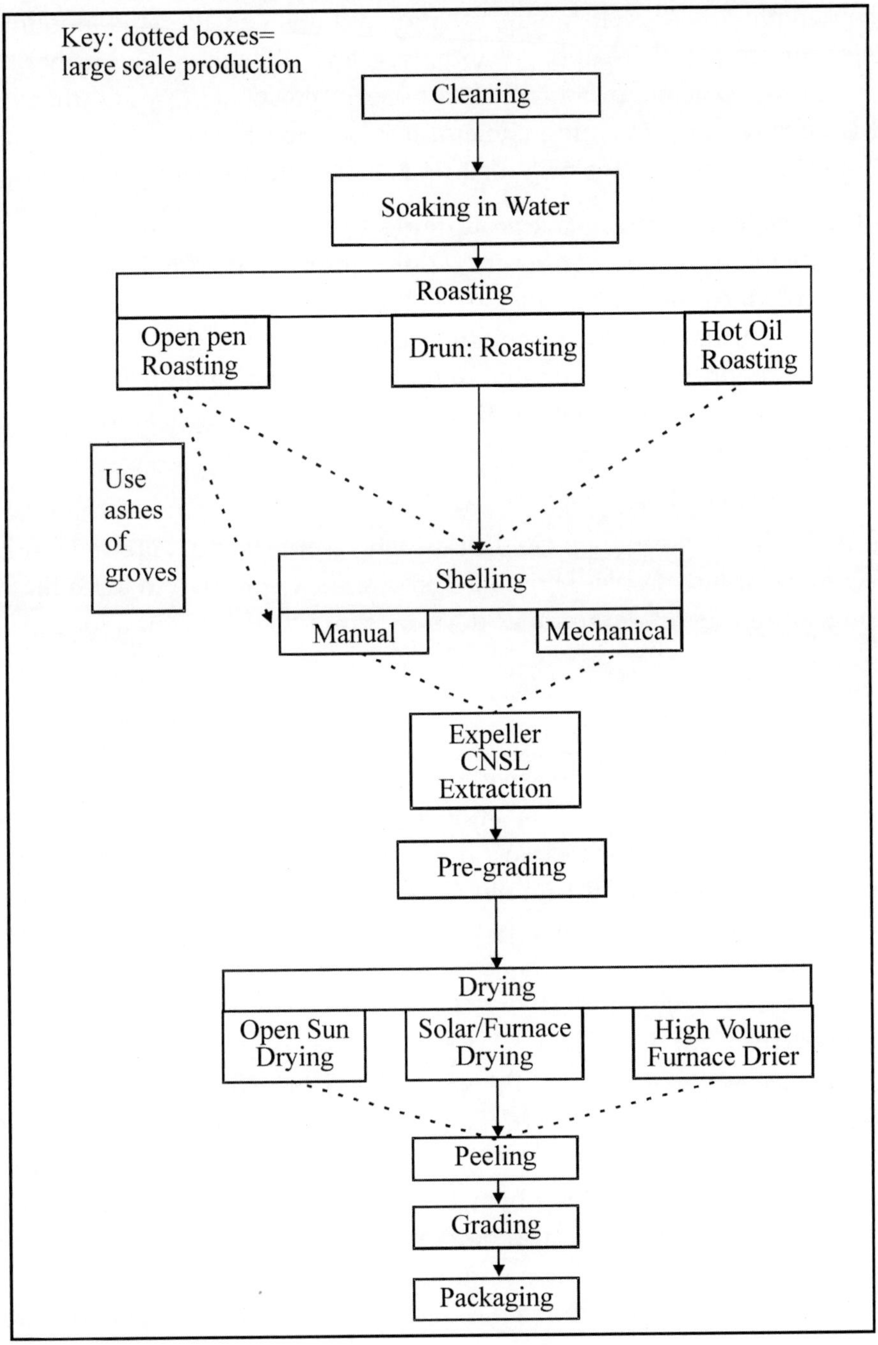

Source: www.fao.org

juice meets an adult individual's daily vitamin C (30 mg) requirement. The cashew apple is also rich in sugars, and contains considerable amounts of tannins and minerals, mainly calcium, iron and phosphorous. Furthermore, the fruit has medicinal properties. It is used for curing scurvy and diarrhoea, and is effective in preventing cholera. It is applied for cure of neurological pain and rheumatism. It is also regarded as a first-class source of energy.

Cashew fruit can be made suitable for consumption by removing undesirable tannins and processing apples into value-added products, like juices, syrups, canned fruits, pickles, jams, chutneys, candy and toffee. The recommended methods for removing astringent properties of cashew apple include steaming the fruit for five minutes before washing it in cold water, boiling the fruit in salt water for five minutes, or adding gelatin solution to the expressed juice.

The fruit should be picked from the tree by hand to avoid bruising the delicate flesh. They are then carefully washed and the nuts are removed for processing. Cashew apples should be processed within two to three hours of picking, since they undergo rapid deterioration when kept for a longer time.

By-Products

Cashew Apple

Cashew apple refers to the swollen peduncle to which the nut is attached. Cashew apple, though very juicy and sweet, is not normally eaten because of its astringent and acrid principles. It is very rich in Vitamin C and contains four to five times as much Vitamin C as citrus. Sound cashew apple can be used for juice extraction. Juice is extracted either by using screw press or basket press.

Cashew Apple Juice

The juice can be extracted with screw press, basket press or simple hand pressing. The extracted juice is strained through muslin cloth which is clarified by adding 1.4 g of PVP (Poly Vinyl Pyrolidine)/ litre of juice and stir the mixture for two minutes after which it is strained again through muslin cloth. Add sugar according to taste and boil the juice. The finished product may be chemically preserved by using sodium benzoate at the rate of 0.8 g./100 °ml. The mixture should be poured into well sterilized bottles, air tight with crown cork and store in a cool dry place.

Cashew Syrup

Extraction of juice and removal of astringency are done in the same way in the pretreatment of juice. Add sugar at the rate of 1 to 1.25 kg for every litre of juice. 20-22 g citric acid /litre and 0.08% as sodium benzoate are added to the juice. Dissolve Sodium Benzoate in a small quantity of water before adding to the mixture. Mix all ingredients thoroughly and keep it as such for three to five hours so that clear syrup forms a separate layer which can be easily siphoned. Bottling can be done as described for juice.

Cashew Apple Jam

Cashew apple must be thoroughly cleaned by washing with water. Immerse the apple in salt solution (3%) for three days to reduce tannin content, after which the fruits are steamed for 15 to 20 minutes at 0.7 to 1.05 kg steam pressure. Then the apples are crushed and mixed with 750 g.sugar/kg of apple and boil it. A pinch of citric acid is added towards the end of cooling process to improve the taste. Store it in well sterilized jam bottles.

Cashew Apple Candy

Just as in Jam, cleaned apples are first immersed in a 3% salt solution. Next day drain out the salt water and steep the fruits in fresh salt solution and repeat the process for a third time.

Remove the salt water and add potassium metabisulphite (625 mg/kg) and keep in this solution for another two to three days. Apples are then thoroughly washed in water. Keep them in a perforated crate made of aluminium or stainless steel and blanched in boiling water for five minutes, followed by steaming in a pressure cooker for five minutes, at 0.35 kg pressure. The apple should not be very soft.

Candy processing is then carried out as usual, starting with a 300 brix syrup containing 0.1% citric acid and 500 mg. potassium metabisulphite/ kg of apple. Pour the syrup over apples until they are completely sub-merged. Keep the fruits immersed in the syrup by placing stainless steel, wooden or glass disc. Cover the container with lid and keep it as such for one day. Next day the syrup is taken out and sugar is added to same syrup for raising concentration upto 350 brix. Syrup is again boiled for about ten minutes and poured back over the apples. Repeat the process daily, raising the sugar content by 50 brix each time for the coming three days and then raise the sugar content by 100 brix each time for the 6th and 7th day ie. the final strength of syrup is 700 brix. Keep the apple for eight to ten days in syrup for complete absorption of sugar. Remove the syrup and dry up the apple which can be stored in screw capped glass jars in a cool dry place.

Cashew wine

Cashew wine is made in many countries throughout Asia and Latin America. It is a light yellow alcoholic drink, with an alcohol content of 6-12 %.

Processing

Cashew apples are cut into slices to ensure a rapid rate of juice extraction when they are crushed in the juice press. The fruit juice is sterilized in stainless steel pans at a temperature of 85°C to eliminate any wild yeasts. The juice is filtered and treated with either sodium or potassium meta bisulphite, to destroy or inhibit the growth of undesirable types of micro-organisms such as acetic acid bacteria, wild yeasts and moulds.

Wine yeast (*Saccharomyces cerevisiae - var ellipsoideus*) should be added. Once the yeast has been added, the juice is thoroughly stirred and allowed to ferment for about two weeks. The wine is separated from sediment and clarified by mixing fining agents, such as gelatin, pectin or casein, with the wine. Filtration is carried out with filter-aids such as fullers earth. The filtered wine is transferred to wooden vats.

The wine is pasteurized at 50-60°C. The temperature should be controlled, so that it does not rise exceed 70°C, since alcohol vaporizes at a temperature of 75-78°C. The wine is then stored in wooden vats and subjected to ageing. At least six months should be allowed for ageing. If necessary, the wine should be clarified again before bottling. During ageing and subsequent maturing in bottles, many reactions, including oxidation, occur. The formation of traces of esters and aldehydes, together with the tannin and acids already present enhance the taste, aroma and preservative properties of the wine. The product is packaged in glass bottles with corks and should be kept out of direct sunlight.

Cashew Nut Shell Liquid - A Versatile Industrial Raw Material

Cashew nut Shell Liquid (CNSL) is a valuable raw material obtained as a by-product during isolation of cashew kernel. The cashew nut shell contains a viscous and dark liquid, known as cashew nut shell liquid (CNSL), which is extremely caustic. It is contained in the thin honeycomb structure between the soft outer skin of the nut and the harder inner shell. The CSNL content of the raw nut varies between 20 to 25 %.

Uses of Cashew Nut Shell Liquid (CNSL)

Cashew nut shell liquid (CNSL) is an important and versatile industrial raw material. It is used as raw material for phenolic resins and friction powder

Flow Diagram - Cashew Wine

Selection	Mature, sound cashew apples
↓	
Slicing	To increase extraction of juice
↓	
Crushing	
↓	
Sterilization	At 85° C
↓	
Filtering	
↓	
Inoculation	With S. cerevisiae
↓	
Fermentation	Two weeks
↓	
Filtering	
↓	
Pasteurization	50-60° C
↓	
Ageing	In wooden vats for 6 months

Source: www. Fao.org

for the automotive industry (brake linings and clutch disks). In drum-brake lining compounds, cashew resins are used as fillers, and may also be used as binders. In disc pads, the role of cashew resin is restricted to the use of friction dust as a filler. The advantage of cashew resins compared with synthetic phenolic resins is that they are more economical and produce a softer material, which gives a quieter braking action

CNSL is also used in mouldings, acid-resistant paints, foundry resins, varnishes, enamels and black lacquers for decorating vases, and as insecticides and fungicides. In tropical medicine, CNSL used in treating leprosy, elephantiasis, psoriasis, ringworm, warts and corns.

Like cashew nuts, CNSL also has an excellent international market and its imports have reached almost US$10 million annually, corresponding to sale of raw liquid. However, the exporting country would earn much more foreign currency if manufactured products were exported.

After extracting the CNSL, the cashew nut shells can be burned to provide heat for decorticating operation or can be used in manufacture of agglomerates. Together with the testa, it may be used either in manufacture of dyestuff or to provide durability to hammocks and fishing lines.

Marketing

Area, Production and Productivity

According to FAO statistics, India has the largest area under cashew. Of the total world area of 3098 thousand hectares under cashew in 2005, approximately 24% is in India. Brazil, the native land of cashew accounted for only 22.24% of the area. In India, it is grown in Kerala, Karnataka, Goa and Maharashtra along the west coast and Tamil Nadu, Andhra Pradesh, Orissa and West Bengal along the east coast. To a limited extent it is being cultivated in Chattisgarh, North Eastern States (Assam, Manipur, Tripura, Meghalaya and Nagaland) and Andaman & Nicobar Islands.

India is the second largest producer of cashewnuts *i.e.,* 460 thousand tonnes in 2005, which constituted 19.68% of the global output of cashew in 2005.

Global Cashew Production

In the early 1970s, majority of global cashew production (68 %) took place in African countries, in particular, Mozambique and Tanzania. Over the following thirty years, production trends shifted, with Asian countries emerging as the world leaders in cashew production. Today, India commands

Area, Production and Productivity in India and Kerala

Years	India			Kerala		
	Area harvested ('000 ha)	Quantity produced ('000 tonnes)	Yield (tonnes/ha)	Area harvested ('000 ha)	Quantity produced ('000 tonnes)	Yield (tonnes /ha)
2001	720.00	450	0.63	92.10	66.20	0.72
2002	740.00	460	0.62	89.70	65.80	0.73
2003	730.00	470	0.64	88.50	66.10	0.75
2004	730.00	460	0.63	88.40	65.20	0.74
2005	730.00	460	0.63	86.10	63.70	0.74

Source: FAOSTAT, Statistics for Planning 2001; 2005

about 40% of the international market in cashew production. Other Asian countries, particularly Vietnam and Indonesia, are begining to expand their production capacities. Currently, the four main cashew producing regions are India, Brazil, Nigeria and Tanzania.

World production of cashew nuts grew rapidly during the 1950s and 1960s, reaching a peak of 624 000 tonnes of raw nuts in 1973. Three countries - India, Mozambique and Tanzania accounted for majority of this production, while smaller industries had developed in Brazil, Kenya and several other African countries. In 1975-76, there was a sharp decline in world production, which continued to 1980s. This decline was largely due to decreased

Cashew Production (tonnes) in Asian Countries (1961 to 2000)

Country	2000	1998	1995	1990	1985	1980	1975	1970	1965	1961
Bangladesh	10	10	20	80	60	20	10			
China	2000	2000	923	1000	6500	6000	4800	3250	2400	2000
India	440000	440000	321640	285590	221330	180266	144254	123319	100000	85000
Indonesia	69027	69027	74995	29907	21114	9074	9122			
Malaysia	13000	13000	13000	10000	6000	5200	11200	4300	3600	3600
Sri Lanka	15000	15000	13000	10000	9500	8260	4330	2770	8800	4240
Thailand	22000	21000	18000	11850	9500	4500	1800	1300	800	500
Vietnam	41200	54000	52800	26000	9000	5600	3500	2100	700	700
Total Asia	**602237**	**614037**	**494378**	**374427**	**283004**	**218920**	**179016**	**137039**	**116300**	**96040**

Source: FAO, 2000

production in Mozambique and Tanzania, since production in India during that period remained static. During the late 1980s and continuing into the 1990s, production picked up and continued to increase gradually. In 2000, world cashew production exceeded 1.2 million tonnes. Asian and African countries produced 0.6 million and 0.4 million tonnes respectively.

Cashew producing countries

India is the second largest producer of raw cashew nuts in the world in 2005 with 20 % share after Vietnam (37%). The other important producers in 2005 were Nigeria (9%), Brazil (8%) and Indonesia (5%).

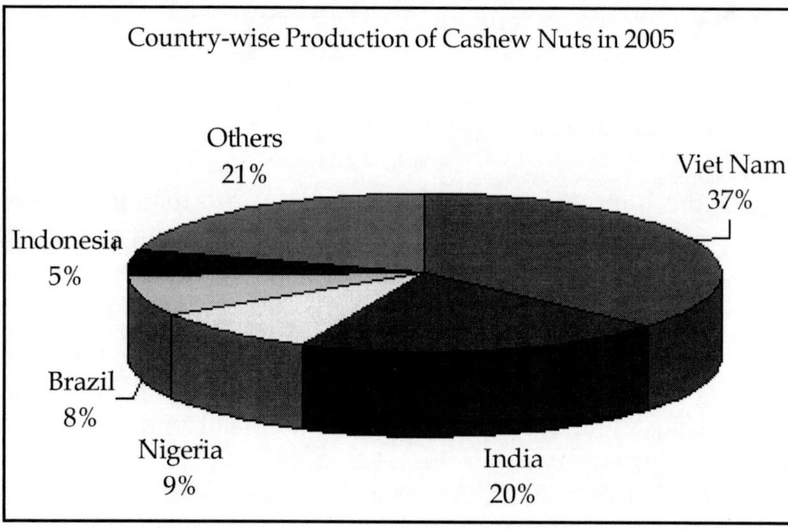

Source:http://faostat.fao.org

Cashew is produced in almost 32 countries of the world and now it has become the number one crop in the world taking over almonds. It is generally found in the tropical zone of the planet as it thrives on a hot and dry climate. The world production figures of this crop cover around 22.5 lakh tones/ year. The major raw cashew producing countries along with their production figures in 2005 are Vietnam (827000 tons), India (460000 tons),Brazil (251268 tonnes), Nigeria (213000 tonnes), Indonesia (122000 tonnes), Tanzania (100000 tonnes), Cote de Ivoire (90000 tonnes), Guinea (81000 tonnes), Mozambique (58000 tonnes), Benin (40000 tonnes), Thailand (24000 tonnes), Malaysia (13000 tonnes), Kenya (10000 tonnes), Ghana (7500 tonnes), Madagascar (6500 tonnes), Sri. Lanka (6200 tonnes), Senegal (4500 tonnes), Burkina Faso (3500 tonnes) and El Salvador (2600 tons).

Regarding production of cashew kernels in the world, prior to 2002 India was the leading producer of raw cashews in the world but in that year Vietnam took over the lead from India. The production trend of cashew has ever been rising and it has doubled since 1994. World's total area under cultivation of cashew crop is around 35.1 lakh hectares, India is at the top in list of countries regarding the maximum area utilized in production of cashew. It is considered to be a high ratio of area as compared to the yield of the crop due to low productivity, the world's average yield being just 700 lb/acre of land.

Indian Scenario

India is the largest producer and exporter of cashew in the world accounting for over 60%of the global cashew kernel supply. Among Indian states, Maharashtra comes first in production followed by Andhra Pradesh, Orissa and Kerala. In India, cashew is grown in Kerala, Karnataka, Goa, Maharashtra, Andhra Pradesh, Orissa and Tamil Nadu. Kerala accounts for approximately 12% of cashew production in India. Kerala has a centralized procurement system and much of the processing is done on an outworker basis by families. Although Kerala only accounts for about 12.5% of the total area under cashew, yields in this state are relatively low when compared to those in other states. Estimated cashew productivity in Kerala is 900-950 kg raw cashew nuts/ha.

India annually produces 3-3.5 lakh tones of RCN and imports 3.5-4.5 lakh tones of RCN which together account for an annual supply of 6-8 million cartons. India annually exports 4-4.5 million cartons of cashew kernels. India is the third largest consumer in the world next to the US and EU .It is estimated that India consumes above 3 million cartons. However, the share of unorganized sector is more in domestic consumption. It is estimated that the Indian cashew industry employs around 300,000 people.

India's Export of Cashew Kernels

Export of cashew kernels from India during 2005-06 was 114,143 tones valued at Rs2514.86 crores. This was 12666 tones valued at Rs.2709.24 crores during 2004-05. There was a decrease in 9.9% in quantity and 7.2% in value. The average unit export price realized during 2005-06 was Rs.220.33/ kg compared to Rs.213.88/ kg during 2004-05, an increase of 3.0%.

During 2005-06, USA continued to be India's major buyer of cashew kernels importing 43150 tones valued at Rs. 958.3 crores and this comprised of 37.8% of our total exports in quantity and 38.10% in value terms.

Country-wise Export of Cashew Kernels from India

Countries	2003-2004		2004-2005		2005-2006	
	Qty (t)	Value (Rs.Cr)	Qty (t)	Value (Rs.Cr)	Qty (t)	Value (Rs.Cr)
U.S.A	48504	881.55	61546	1288.49	43149	958.33
Netherlands	12237	215.03	15693	345.67	18736	408.82
U.A.E	6239	102.42	6690	148.23	8274	184.41
U.K	5392	95.61	7108	158.34	6238	140.88
Japan	5522	101.95	5047	111.21	4685	110.2
France	2444	43.61	3329	66.49	3470	75.79
Saudi Arabia	2011	32.95	2998	61.55	2827	59.61
Spain	2198	41.11	2870	64.8	2648	61.09
Belgium	1769	33.88	2497	53.18	2378	54.48
Germany	1574	26.11	1966	42.39	1991	42.21
Others	12938	230.27	16925	368.9	19747	419.05
Total	100828	1804.43	126667	2709.24	114143	2514.86

Source : Various Custom Houses, * : D.G.C.I & S, Calcutta

Country-wise Export of Cashew Kernels from India During 2005-06

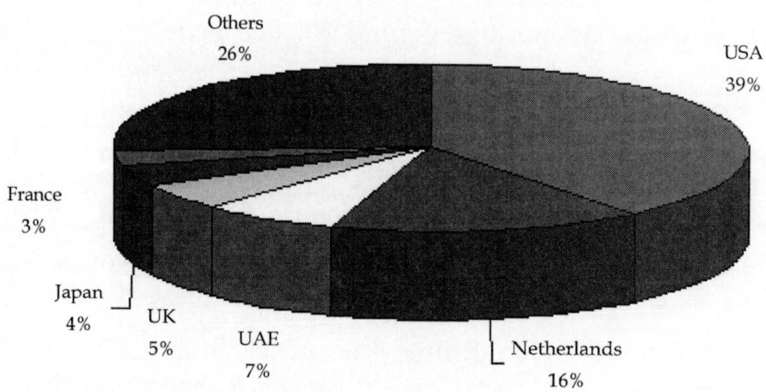

Source : Cashew Export Promotion Council of India

Netherlands was the second largest importer of cashew kernels from India with an intake of 18736 tones valued at Rs405.82 crores representing 16.41% in value terms. UAE overtook UK during 2005-06. Exports to Japan, France, Saudi Arabia and Germany registered an increase. Export to UK, Russia and Norway decreased. Out of the total quantity of cashew kernels exported during 2005-06, 39.38% were to American Zone and European Union accounted for 37.93%. There was a decrease of 28.97% in exports to American Zone and increase of 6.06% to European Zone and 19.08% to West Asian Zone compared to previous year. The above pie diagram shows country-wise export of cashew kernels from India during 2005-06.

Export of cashew nut shell liquid

The export of cashew nut shell liquid from India during 2005-06 declined to 6405 tonnes valued Rs.7.09 crores compared to 7474 tones valued at Rs.7.91 crores during 2004-05. While the quantity exported declined by 14.30% export value declined by 10.37% over the previous year. The unit value realised was Rs.11.07/Kg in 2005-06 registering an increase of 4.63% over previous year. USA accounted for 91% of the quantity and 84 % of the value during 2005-06. The shares of Indonesia, Spain, China during 2004-05 and Mexico were negligible during 2005-06, which accounted for 8.63 % of India's exports during 2004-05. However the share of Korean republic rose from 1.67% to 8.18 % during this period.

Country- wise Export of Cashew Nut Shell Liquid from India

Countries	2004-05				2005-06			
	Qty (tones)	Value (Rs.cr.)	% of Qty	% of Value	Qty (tones)	Value (Rs.cr.)	% of Qty	% of Value
USA	6550	6.46	87.64	81.67	5834	5.97	91.09	84.22
Indonesia	384	0.50	5.14	6.32	0	0.00	-	-
Spain	213	0.33	2.85	4.17	0	0.00	-	-
Japan	154	0.24	2.06	3.03	31	0.09	0.49	1.26
Korea Rep	125	0.31	1.67	3.92	524	1.00	8.18	14.32
China	32	0.04	0.43	0.52	0	0.00	-	-
Mexico	16	0.03	0.21	0.37	6405	0.00	-	-
Zimbabwe	-	0	-	0.00	16	0.03	0.25	1.26
Total	7474	7.91	100	100	6405	7.09	100.00	100.00

Source: Cashew Export Promotion Council

Export of Cashew Kernel, Cashewnut Shell Liquid (CNSL) & Import of Raw Cashewnut During 1990-91 to 2005-2006

Year	Cashew Kernel		C N S L		Raw nut	
	Quantity	Value	Quantity	Value	Quantity	Value
1990-1991	49874	44224	5658	556	82639	13400
1991-1992	47738	66909	4542	402	106080	26668
1992-1993	53436	74549	4258	381	134985	37633
1993-1994	69884	104602	3625	290	191322	48270
1994-1995	77000	124602	3807	244	228109	69094
1995-1996	70334	124050	760	145	222819	76008
1996-1997	68663	128550	1735	277	192285	64060
1997-1998	76593	139610	4446	717	224968	74400
1998-1999	75026	160990	1572	326	181009	68024
1999-2000	92461	245145	764	184	200584	95352
2000-2001	89155	204975	2246	3894	249318	96080
2001-2002	97550	177680	1814	419	356566	96001
2002-2003	104137	193302	7215	925	400659	123657
2003-2004	100828	180442	6926	703	452898	140093
2004-2005	126667	270924	7474	791	578884	218324
2005-2006	114143	251486	6405	709	565400	216295

Source: Cashew Export Promotion Council (Quantity in tones); Value in Rs.Lakhs

India's Import

Import of raw cashew nut into India during 2005-06 was 565400 tones valued at Rs.2162.95 Crores against 578844 tones valued at Rs.2190.94 Crores during 2004-05. This was 2.3% less in quantity and 1.3% less in value compared to previous year. The average import price was Rs.38.26/ kg during 2005-06 against Rs. 37.85 during 2004-05. Ivory Cost continued to be major importer of rawnut into India and Guinea Bissau emerged as second leaving behind the former second Tanzania to third place. Benin supplied 59695 tonnes and became fourth among major importers of raw cashew nuts to India.

Kerala's share in raw cashew nut imports rose marginally from 61% to 65% between 2000-01 and 2003-04. However the share of Kerala fell to 49% in 2004-05 which went up further to 54.25% in 2005-06. Hence lion share of imported raw cashew was utilized by processing units in Kerala. Low production forced many cashew-processing factories to remain closed for a considerable time in recent years. This, points to the need for increasing indigenous production of raw cashew nuts to make the country self sufficient.

Import of Raw Cashew Nuts: Kerala and India

Year	India	*Kerala's share (%)	India	*Kerala's share (%)
	Qty (tones)	Qty (tones)	Value (Rs.crores)	Value (Rs.crores)
2000-01	249318	61.17	960.85	57.52
2001-02	355556	53.88	950.01	52.88
2002-03	400659	62.38	1236.60	62.46
2003-04	452398	65.10	1400.90	64.91
2004-05	578884	48.91	2183.26	48.32
2005-06	565400	54.25	2162.95	52.46

*Import through Cochin Port, **Various Houses, P-Provisional
Source: The Cashew Export Promotion Council of India

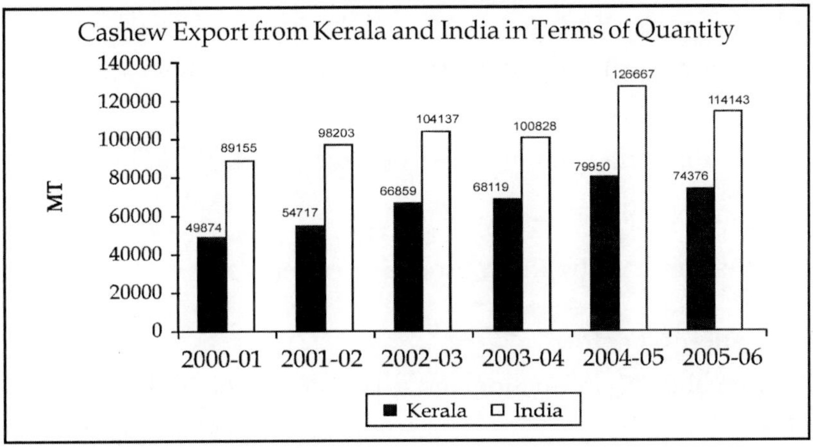

Source : Economic Review 2006

Unit Price of Cashew Kernels

The unit price of cashew kernels rose to Rs 213.89/ Kg during 2004-05 compared to Rs.178.96/ Kg during 2003-04, which shows an increase of 19.52 %. The unit price of cashew nut shell liquid in India was Rs.10.58 /Kg during 2004-05.

Wholesale Prices of Raw Cashew Nuts
Pricing Trend of Raw Cashew Nut in Different States

Year	Kerala	Karnataka	Andhra Pradesh	Tamil Nadu	Goa
2000	42.00	47.00	34.50	36.20	46.50
2001	26.75	29.90	28.25	29.60	33.50
2002	25.00	28.00	27.75	29.14	34.10
2003	29.50	33.00	30.00	34.60	35.50
2004	30.00	30.00	28.70	36.25	35.50
2005	37.88	35.20	38.10	47.75	45.20

Source: www.dacnet.nic.in (Price Rs./Kg.)

Major Trading Centers of Cashew

The major trading centers of cashew in India are Kollam, Manglore and Kochi.

Market Influencing Factors

- Production fluctuations in the major raw cashew nut producing countries.
- Indirect domestic demand due to presence of large unorganized sector
- Government policies
- Market share held by the competitive countries

Price Influencing Factors

- The annual RCN production and exports from the major RCN exporting nations of Africa.
- The demand for cashew kernels from the export markets can be estimated with fair amount of accuracy. The domestic demand and processing can be determined only indirectly as share of unorganized sector is more.
- As cashew is a plantation crop, requiring significant initial investments, the Govt. policies regard to production, processing, exports and imports across all the producing nations have a bearing on prices.

Demand Forecast

The relatively high price elasticity of demand for cashews in Europe suggests considerable opportunity for increasing demand through price competition. Therefore, as an alternative to the base case demand projection, the 2010 projection was also made under the assumption that real cashew kernel prices declined by 25% over the 1998–2010 period. Under this alternative, world cashew kernel demand would grow by 5.9% a year over the forecast period, and cashew demand would increase to 279,338 tones by 2010 compared with 4.6% growth under the base case.

Projected Cashew Kernel Import Demand to 2010

Country or region	Demand in 1998 (tons)	Real per capita GDP growth rate 1998-2010 (%)	Population growth 1998-2010 (%)	Price income (%)	Demand in 2010 (tons)	Growth rate 1998-2010 (%)
France	6,250	2.58	0.24	-0.596.51	21,404	10.80
Germany	12,454	3.01	-0.22	-1.323.23	23,690	5.51
Japan	5,532	2.29	-0.03	- .191.95	8,868	4.01
Netherlands	10,667	2.94	0.25	-1.442.94	24,429	7.15
UK	6,604	2.58	0.02	-0.391.97	11,283	4.56
United States	70,368	2.41	0.69	-0.361.06	103,216	3.24
Sub total	111,875				192,890	4.64
Rest of world	28,000				48,223	4.64
World	139,844				241,113	4.64

Note: The average annual growth rate for the subtotal of six countries from 1998 to 2010 is applied to the rest of the world to obtain world demand. Real prices of cashews are assumed constant over the forecast period.
Source: Real per capita GDP and population growth rates are from the World Bank (2001).

COCOA

Cocoa (*Theobroma cacao* L.) a native of Amazon base of South America entered India in the early half of 20th century. Bulk of it is produced in the tropical areas of the African continent. There are over 20 species in the genus but *Theobroma cacao* is the only one cultivated widely. It is one of the most important beverage crops, third place after tea and coffee. Cocoa bean is the primary raw material for confectioneries, beverages, chocolates and other edible products. The commercial sector of cocoa in India hardly takes place in a major way in international export trade. Majority of the processed cocoa products are consumed within India. The tropical diversified congenial climate available in India provides immense scope for its cultivation.

Soil and Climate

Cocoa grows well in regions where temperature ranges 30-32^{0}C mean maximum and 18-21^{0}C mean minimum but around 25^{0}C is favourable. It can't be grown commercially in areas where the minimum temperature falls below 10^{0}C and annual average temperature is less than 21^{0}C. Average annual rainfall of 1250-3000 mm. and preferably between 1500-2000mm. with a dry season of not more than 3 months with less than 100mm. rain/month is ideal, but the quantity is less important than distribution. Rainfall can be supplemented with irrigation during dry months.

Cocoa grows successfully in a wide range of soils. Well drained, rich soils acidic to neutral in reaction are the best suited. Cocoa trees are more sensitive to moisture stress than other tropical crops. In addition, cocoa trees are sensitive to water logging. The best soil for cocoa is forest soil rich in humus. Clay loams and sandy loams are suitable. The basic requisite is that the soil should have a minimum soil depth of 1.5m.

Varieties

There are three major varietal groups, namely, Criollo, Forestero and Trinitario. Among these, Forestero is the one commercially grown all over the world. It is high yielding resistant to pests and diseases and more tolerant to drought compared to Criollo. Important varieties developed are furnished separately.

Types of cocoa

T. cacao
1. *T. cacao* sub sp. cacao → criollo
2. *T. cacao* sub sp. sphaerocarpum → Forastero and Trinitario

Nursery Management

Cocoa can be propagated through seeds and by vegetative means.

Seed Propagation

For raising seedlings, seeds of mature pods are taken from high yielding mother plants. The mother plants selected should yield more than 100 pods/ year and should have medium or large green pods with an average dry bean weight of not less than one gram. A more suitable procedure for planting

Characters	Criollo	Forastero	Trinitario
1. Fresh bean colour (Fresh nib colour)	Whitish or light brown	Purple or deep purple	White to nearly black
2. Flavour of chocalate	Mild and delicate	Strong	Medium
3. No. of beans/pod	20-30	30 or more	30 or more
4. Pod shape	Big, warty, beaked pods with prominent constriction and deep furrows	Smaller pods with smooth surface, less prominent beans and constriction	Highly variable, surface ranging from complete smoothness to heavy sculpturing
5. Immature pod colour	Red	Green	White, green red or purple
6. Mature pod colour	Yellow	Yellow	Yellow, orange or red
7. Pod husk texture	Soft	Hard	Medium
8. Reaction to biotic & abiotic stress	Highly susceptible	Tolerant	Susceptible/tolerant
9. Vigour of plant	Less vigorous	More vigorous	Less vigorous to more vigorous
10. Easyness to cultivate	Difficult	Easy	Easy

Released Varieties

Name of variety	Characters	Mean no. of pods	Mean pod wt. (g)	No. of beans/ pod	Mean yield wet bean (kg/plant)	Mean yield dry bean kg/plant
CCRP-1	Selection, green changes to yellow, self incompatable	56.2	384.7	46.2	6.2	2.5
CCRP-4	Selection, large purple pod changes to deep yellow. Self incompatable	66.2	402.1	45.4	8.3	3.9
CCRP-5	Selection, pods are large, elliptical green changes to yellow, self incompatable	37.9	425.0	45.2	4.3	1.7
CCRP-6	Selection, pods are big, green to yellow, self incompatable	50.1	895.0	48.0	11.3	3.1
CCRP-7	Selection, pods large, elongated green to yellow. Self incompatable	78.1	526.7	46.9	9.7	4.0

Cocoa - Hybrids

Name of variety	Characters	Pod weight (g)	Mean pod yield	Wet bean wt./pod (g)	Bean No.
1. CCRP-8	CCRP-1 x CCRP-7S.I	389.3	90.4	126.3	48.8
2. CCRP-9	CCRP-1 x CCRP-4S.I	370.4	105.7	84.9	36.7
3.CCRP-10	CCRP-3 x G.VI 68	332.5	79.6	102.5	41.5

good quality seedling will be to collect hybrid seeds from biclonal or polyclonal seed gardens involving superior self- incompatible parents.

Seeds generally lose viability after seven days of harvest. Seeds are to be sown immediately after extraction from fruits. Viability can be extended for some more days by storing seeds in moist charcoal and then packed in polythene bags.

Potting Mixture and Time of Sowing

The normal potting mixture with farm yard manure, sand and soil in equal proportions is good for raising cocoa seedlings. Though cocoa seeds germinate at any time of year, the best period of sowing seeds in nursery is December- January so that four to six months old seedlings become available for field planting by the onset of monsoon in traditional areas.

Method of Sowing

Seeds are to be sown with hilum end facing downwards or are sown flat. Seeds should not be placed too deep in soil. Seeds germinate in a week's time but the process continues for another week. Generally 90% of seeds germinate. Regular watering is essential to keep the soil moist. Over watering should be avoided to check outbreak of diseases.

Selection of Seedlings for Field Planting

Four to six months old seedlings are generally used for field planting. Since seedling vigour and final yield are closely related, seedlings for field planting should be selected based on seedling vigour. Seedling vigour can be estimated based on height of seedlings and stem girth.

Vegetative Propagation

Large scale production of superior planting material is done in cocoa through vegetative means like budding and grafting of which budding is the easiest. The different budding methods feasible are 'T', inverted 'T', patch and modified Forkert. The new method of micro budding is also may be followed.

Root stock: Seedlings of about 60-90 days are generally used as root stock. While selecting root stock, care should be taken to see that both root stock and scion are of same thickness and physiological age.

Scion selection: Bud wood from chupons are collected from high yielding, disease resistant elite plants. Shoots having bark and just hardened leaves are selected as bud wood. Scions are procured by cutting off laminae and after 10 days when the petioles have fallen off, these scion shoots are cut and used for budding immediately.

Budding: Patch budding is adopted. Remove a patch of about 2.5 cm. long and 0.5 cm wide with a single vigorous bud on it. Bark of same size is removed from root stock and bud patch is inserted. It is then tied with grafting tape. The patch selected should have bud, visible to naked eye but it should not have signs of proliferation. About three weeks after budding, grafting tape is removed. If there is successful bud union, a vertical cut is made half way through stem above the bud and the stock portion is snapped back. Such snapped root stock portion is cut and removed only after the bud has grown sufficiently with at least two leaves hardened. After about four to six months, they are ready for field planting. Care should be taken to remove new sprouts from root stock portion.

Planting and management

Lay out and planting

Cocoa is usually planted under coconut and arecanut plantations in India. Shade levels under coconut canopy are highly variable depending mainly on spacing of coconut, extend of canopy development and age of palms. It is estimated that light infiltration through coconut canopy ranges 30-80%. Based on this, a general recommendation is made.

- Coconut plantation that will let in more light through the canopy may be chosen for raising cocoa.
- If the light infiltration is over 50%, it may be beneficial to provide additional shade using temporary shade plants like banana.

Spacing

Spacing adopted for cocoa in major producing countries is highly variable. A close spacing from 1.7x1.7m to 2.7x2.7m was the optimum. Closer spacing was advantageous in early years, especially for unshaded cocoa. The ultimate spacing and population level to be followed should be decided taking into account factors contributing to extent of canopy development, variety and type of management.

In countries like Philippines, Papua New Guinea and Malaysia, where cocoa is cultivated along with coconut, but spacing followed is close one (2 x 2 m), two rows of cocoa in between rows of coconut. In Indian condition, closer planting may not be beneficial. Coconut planted at normal spacing of 7.5 x 7.5 m, there may be a row of cocoa in between two rows of coconut.

For arriving a recommendation in coconut, rooting patterns of the crops may be taken into account. About 75% of line roots are concentrated in 2m round the base, lateral spread is 1.2 to 1.5m. Hence distance between cocoa and coconut may have to be 3.2 to 3.5 m to avoid over lapping and cocoa to cocoa is 3m. In arecanut gardens too, cocoa can be planted as intercrop. The spacing of arecanut should not be less than 2.7 x 2.7 m.

Shade Management

Cocoa is a plant that has originated under shade and this crop is often cultivated under shade. Cocoa needs shade for its early establishment, young cocoa plants grow the best with 50% full sunlight. As the tree grows, its shade requirement is reduced.

Cocoa is planted as pure, mixed crop or intercrop. When planted as a pure crop, Dadap (*Erythnina lithosperma*) is planted at 3 x 3 m spacing to provide shade. Dadap needs pruning every year. For more permanent shade, *Albizzia stipulata* can be planted adopting 9 x 9 m or 12 x 12 m spacing.This requires 4 to 6 years to develop proper canopy to provide sufficient shade. Protection from north east winds by planting wind-breaks is also necessary. Cocoa can be planted as intercrop in coconut gardens provided spacing of coconut is sufficient to provide enough shade and soil is suited to cocoa.

Planting Method

Digging pits of 50 cm^3 and filling with a mixture of surface soil and organic manures. The planting hole should be at least the same size as to hold the basket or polythene bag in which seedlings are raised. Planting should coincide with the onset of monsoon but in places where irrigation facilities are available planting can be done throughout the year.

Post Planting Care

Immediately after planting, basins are mulched with organic materials and provide mulch during summer to conserve moisture. Remove shoots emerging from the lower portions of bud patch periodically. During first three to four years of planting, it is essential to keep field free from weeds.

Nutrition Management

Application of organic manures will be useful in early establishment period. It may not be necessary after about three to five years as cocoa litter will be the rich source of organic matter. For production of 1kg dry beans, plants remove N P K @43.8, 8.04 and 64.29 g respectively. Under average management, annual application of 100g N, 40g P2O5 and 140g K2O/ plant /year in two equal split doses is recommended. During first year of planting, plants may be given 1/3rd of above dose, while the second and third year 2/3rd and full dose of fertilizers applied. While applying manures and fertilizers, care should be taken to open only shallow basins around plants (radius of 1.5m for adult cocoa) and to avoid serious damage to surface feeding root systems. The radius of the basins should be proportionately smaller for young cocoa. Providing adequate irrigation helps in increasing yield by about 30 % both in mono and in mixed crop. Irrigation could beneficially be given once in a week in dry months.

Pruning and Training

Pruning is an important continuous operation and is often taken as a sign of good management in cocoa. Cocoa grows naturally in tiers. Under good management and adequate shade, the plant jorquette at a height of 1-1.5m.The number of fans arising from jorquette varies from 3-5. If growth is unrestricted, new chupons arise on the main stem below the first jorquette after first tier has fully developed and one or more of stem grow and jorquette to produce second tier. Such growth continues upward in tiers. As growth advances and as new tiers are formed, the lower tier gradually becomes weak and a large number of leaves shed. This type of growth is not preferred as such a plant will be difficult to manage. It is desirable to restrict growth to one tier at 1-1.5 m height. When jorquetting takes place at lower levels, this may be raised by nipping off all fan branches and allowing one chupon to develop and grow further to jorquette at desired height. Further vertical growth is arrested through periodical removal of chupons. A second tier is allowed when the first tier is damaged .Drooping fan shoots are to be cut and removed once a year when the crop load was low (Dec-Jan).Pruning and shaping of grown up plants may be done gradually without much shock to bearing plant. Similarly overlapping branches are also removed for facilitating uniform light penetration of every part of canopy.

Rejuvenation of Senile Gardens

Top working is a method to convert old poor yielding cocoa plants to high yielders. This technique helps in rejuvenation of old and unproductive cocoa plantations. A poor yielding cocoa tree of any age can be converted to a high yielder by simple procedure of top working. The method is similar to budding on seedlings. The tree to be topworked is snapped back just below the jorquette (1-1.5 m above the ground) after cutting half way through the width. Patch budding is done on three or four newly formed most vigorous chupon shoots and rest of chupons are removed. Budding should be done only when shoots attain pencil thickness and their leaves are hardened. The bud wood is taken only from fan shoots of high yielding trees. Patch budding can be easily done on these shoots by removing a patch of bark of 2.5 cm

Cost of Cultivation – Cocoa (500 Plants / ha)

1. Labour requirement (Mandays and cost)

S. No.	Items of operation	1st year	2nd year	3rd year	Total
i	Land cleaning & Peg marking (500 pits)	45	-	-	45
ii	Digging 500 pits	30	5	1	36
iii	Pit filling & Planting - Stacking, mulching & shading	15	3	2	20
iv	Weeding once a year	3	4	5	12
v	Pruning and training	10	15	15	40
vi	Soil working	5	10	10	25
vii	Application of manures and fertilizers	6	20	25	51
viii	Plant protection operations	8	18	18	44
ix	Irrigation (gets as and when due for coconut)	-	-	-	-
x	fencing, watch and ward	24	24	24	72
xi	Miscellaneous operations like removal of sprouts, deblossoming, transportation of inputs	6	12	13	31
	Total	152	111	113	376
	Wages @ Rs.100/- man day	15200	11100	11300	37600

Source: Directorate of Cashew and Cocoa Development

length and 0.5 cm width and inserting a patch of bud of similar size. This scion is fixed in position and protected by tying with a polythene tape. Three weeks after budding, the tape is cut off and the stock portion of seedling above bud union is snapped back. The snapped portion is removed only after at least two hardened leaves develop from the bud. When sufficient shoots are hardened, canopy of mother tree can be totally removed. Top working can be done during all seasons. Still, it may be more convenient if this operation is done in a rain-free period in irrigated gardens. For rain fed situations, this may preferably be done after the receipt of pre- monsoon showers. Top worked trees grow much faster than budded plants of same age especially because of presence of an established root system. They start yielding heavily from second year onwards while a budded cocoa plant of same age may take five years for the same.

2. Cost of Inputs (in rupees)

a) Conventional method

S. No.	Items of operation	1st year	2nd year	3rd year	Total
I	500 grafts @ Rs.21/- graft	10500	-	-	10500
ii	Gap filling @ 21% (100 grafts)	-	2100	-	2100
iii	Manures and fertilizers	2500	3900	4900	11300
iv	Plant protection chemicals	1150	1150	1150	3450
V	Irrigation (fuel for the pumpsets available for the irrigation of coconut)	900	900	900	2700
	Total 2(a)	15050	8050	6950	30050
Total cost excluding labour charges		15050	8050	6950	30050 (30000 rd)
Financial assistance limited to 75% of Rs.30,000/ ha, excluding labour charges and disbursed @ 50:30:20		11250	6750	4500	22500
Subsidy / plant		23	13	9	45

Source: Directorate of Cashew and Cocoa Development

b) Organic cultivation

S. No.	Items of operation	1st year	2nd year	3rd year	Total
I	500 grafts @ Rs.21/- graft	10500	-	-	10500
Ii	Gap filling @ 21% (100 grafts)	-	2100	-	2100
Iii	Organic manures/bio-fertilizers	7000	7000	5000	19000
Iv	Organic pesticides	2000	2000	2000	6000
V	Irrigation (fuel charges)	900	900	900	2700
	Total 2(b)	20040	12000	7900	40300
Total cost for excluding labour charges		20400	12000	7900	40300 (40000 rd)
Financial assistance limited to 75% Rs.40,000/- ha, excluding labour charges and disbursed @ 50:30:20		15000	9000	6000	30000
Subsidy / plant		30	18	12	60

Source: Directorate of Cashew and Cocoa Development

COCOA as Inter Crop in Oil Palm or Coconut (0.4 hectare)

S.No	Item of Expenditure	*Year 1*	*Year 2*	*Year 3*	*Year 4*	*Total*
A.	**MATERIAL**					
1.	Planting Material including 10% casualty replacement in second year	1000	100			**1100**
2	Manure and fertilizer	852	1245	1680	1680	**5457**
3	Plant Protection	300	300	300	300	**1200**
	Sub total	**2152**	**1645**	**1980**	**1980**	**7757**
B	**LABOUR**					
1	Land Preparation					
2	Digging of Pits @ Rs. 5/pit	1000				**1000**
3	Manure and fertilizer application (4mandays)	200	200	200	200	**800**
4	Pesticide application (2 mandays)	100	100	100	100	**400**
5	Irrigation (35 units)	1750	1750	1750	1750	
6	Interculture operations (10 mandays including pruning)	500	500	750	750	**2500**
7	Harvesting @ 10 mandays in 3 rd year & 15 mandays in 4 th,5 th year and 20 mandays from 6 th year onwards			500	750	**1250**
	Sub total	**3550**	**2550**	**3300**	**3550**	**12950**
	Contingencies (5 %)	**285**	**210**	**264**	**277**	**1036**
	GRAND TOTAL (rounded off)	**6000**	**4400**	**5500**	**5800**	**21700**

Harvesting

Development of pod takes 5-6 months from fertilizing the flower to full ripening. Harvesting involves removing ripe pods from trees and opening them to extract wet beans. As they ripen, pods change colours, green pods becoming orange, yellow and red pods turning orange. Each pod will have 25-45 beans embedded in white pulp (Mucilage). Generally cocoa gives two main crops in a year during September – January and April-June, though off-season crops may be seen almost all through the year especially under irrigated condition.

Only ripe pods are harvested without damaging flower cushions by cutting the stalk with knife. The harvesting is done at regular intervals of 10-15 days. The damaged, unripe and infested pods are separated to ensure better quality of beans after processing. The harvested pods should be kept for minimum period of two days before opening for fermentation. However, pod should not be kept beyond four days.

Processing of Cocoa Beans

Curing is the process by which cocoa beans are prepared for the market which requires beans of good flavour, potential and good keeping qualities. The curing process involves fermentation followed by drying. Fermentation involves keeping mass of cocoa beans well insulated so that heat is retained, while at the same time air is allowed to pass through the mass. The process lasts up to 7 days and followed immediately by drying. Cocoa bean mass under the process of fermentation has to be overturned regularly to maintain uniform specified temperature all over the mass.

Fermentation

Fermentation of Cocoa beans is essential to remove mucilaginous pulp, to develop flavour and aroma precursors, reduce bitterness, kill germ of seed and to loosen the testa. Among the various methods adopted for fermentation in different cocoa producing countries, Heap, Box, Tray and Basket methods are the standard methods.

Heap Method

This method involves keeping a mass of not less than 50 kg. of wet beans over a layer of banana leaves. The banana leaves are spread over a few sticks to keep them a little raised over the ground level to facilitate flow of sweating.

The leaves are folded and kept over a heap of beans and a few wooden pieces kept over it to keep leaves in position. The heaps are dismantled and beans mixed on third and fifth days. It needs about six days for completion of fermentation and beans can be taken out for drying on seventh day.

Even though the minimum quantity of beans required for effective fermentation is 50 kg, a further increase in quantity of beans in a heap will be beneficial. Heaps of more than about 500 kg. may be difficult to handle.

Tray Method

Wooden trays of size 90 x 60 x 13 cm with battens or reapers fixed at the bottom with gaps in between, are filled with beans. Each tray contains about 45 kg. wet beans. Six such trays are stacked one over the other and an empty tray is kept at the bottom to allow for drainage of sweating. After stacking, beans of the top most tray are kept covered with banana leaves.

After 24 hours of setting the stack of trays is kept covered with gunny sacking to conserve heat that develops. There is no need for mixing beans and fermentation will be completed in four days. On fifth day, beans can be taken out for drying. The minimum number of trays required to be stacked is about six but as many as 12 trays can be used simultaneously.

Box Method

Wooden boxes of 1.2 x 0.95 x 0.75 m with holes at bottom and sides of the box are filled with wet beans. These boxes can hold one tonne of wet beans. The beans are to be stirred on alternate days. As quantity of beans is high, this is the best done by changing beans from one box to another at time of mixing. This would necessitate having a minimum of three boxes. The duration of fermentation is commonly for 3-5 days.

Drying

The fermented beans can be dried either in sun or by artificial means. Sun drying can be done in thin layers of 2 - 3 cm. depth and stirring from time to time. Under normal condition, drying is completed in 4-5 days. While in mechanical driers, care must be taken to avoid exposure of beans to smoke and fumes. The dried beans after cooling maintaining 6 -8% moisture should be cleaned before storage.

Storage

Under conditions prevailing in cocoa growing belt of India, humidity is too high that the beans absorb moisture and attain equilibrium moisture which is not safe for long term storage. Dried beans with moisture content of 6-8% are packed in polythene lined gunny bags. Some special conditions are to be provided in store to maintain quality of cured beans. Beans should be cleaned of flat and broken beans before storage. The store should be sufficiently ventilated and bags are to be kept on a wooden platform with air space of about 15-20 cm below the wooden planks set over the floor. The humidity should not exceed 80% prevent mould development and pest incidence in beans. Before storage, the store can be cleaned and insect free by application of pesticides in advance. Pesticides should neither be applied nor be kept with beans inside the store .As cocoa can absorb and retain permanently any odour in its surroundings, other food stuffs should not be kept with cocoa.

Quality

There are two types of cocoa beans. More than 90% of cocoa produced annually are bulk (or basic) cocoa. Bulk cocoa mostly come from Africa and Brazil, mainly of the forastero variety. Fine and flavor cocoa have distinctive aroma and flavor characteristics, sought after by chocolate manufacturers. They represent only 5% of world cocoa production.

The International Cocoa Standards require cocoa of merchantable quality to be fermented, thoroughly dry, free from smoky beans, abnormal or foreign odors and any evidence of adulteration. It must be reasonably free from living insects, broken beans, fragments and pieces of shell and foreign matter and reasonably uniform in size.

Throughout the world, standards against which all cocoa are measured are those of Ghana cocoa. Cocoa is graded on the basis of count of defective beans in cut test. Defective beans should not exceed following limits:

Grade I

Mouldy beans, maximum 3% by count;

Slaty beans, maximum 3% by count;

Insect-damaged, germinated or flat beans, total maximum 3% by count.

Grade II

Mouldy beans, maximum 4% by count;

Slaty beans, maximum 8% by count;

Insect-damaged, germinated or flat beans, total maximum 6% by count.

Chocolate Making Process

Chocolate making begins when cocoa beans arrive from growing countries. It is now up to the chocolate maker or chocolatier to unlock the flavor treasures, the beans conceal and create fine chocolate. Production techniques change as chocolate makers introduce new technology. But the stages of chocolate making remain the same: roasting, grinding and refining, conching and tempering. The secret to good chocolate is to carry out these tasks with patience, know-how and love, combined with sublime chocolate recipes, the required attention Chocolate making receives when cocoa beans arrive from growing countries.

Roasting

Roasting, like cocoa seed fermentation, is a crucial stage in establishing final flavor of chocolate. The ultimate quality of the chocolate depends as much on the correct choice of roasting conditions as it does on the characteristics of the raw cocoa beans.

There are several roasting processes:

Pre-Roasting

The cocoa beans are quickly heated with hot air or infra-red radiant heaters to separate shells from 'nibs', or bean meat. Beans are roasted at temperatures between 100 and 150°C for between 20 and 40 minutes.

Direct Roasting (without Pre-Roasting)

The beans, still in their shells, are roasted and then shelled all at once. This older and more traditional method allows flavor to develop properly. As a guide, this type of roasting is carried out at between 150 and 160°C for forty to fifty minutes.

Both techniques are used today and each has its advocates. Use of pre-roasting is more productive, but some varieties of beans, when shelled, can be damaged by a too-violent change in temperature during roasting.

Roasting dries and browns cocoa beans and develops their flavor. This is the first important stage in the manufacturer's 'customization' of product.

Earlier, fermentation produced the flavor precursors: reducing sugars, glucose, fructose, and amino-acids. Even though roasting can only reinforce results of good fermentation, it can only spoil very high-quality beans if not done correctly.

Shelling

After or during roasting (i.e. following pre-roasting), the beans are shelled. If the aim is quality, as little shell as possible should be left out. Shelling is carried out in several stages: milling, sifting and winnowing. Each of these is important and shelling is complete when there are very a few residual shells and grains are regular in size.

The shelled beans can be made into fine, smooth chocolate.

Milling

The roasted, shelled and crushed beans are milled to reduce them to even finer particles. It is important to take care when milling, friction generated heat causes smoky or burnt off-flavors and some of the already existing flavors can be damaged or lost.

Refining

Refining converts the milled particles into liquid cocoa mass created as the cocoa butter in the beans binds dry particles.

Conching

Conching involves powerful machines, 'conches', stiring the chocolate in a controlled way.This is a crucial stage of chocolate making. Even when perfectly milled and partly homogenized, the product that emerges from mill does not yet deserve to be called chocolate.Conching is usually done in two stages:

Dry conching: the chocolate is slowly stirred at a temperature of around 80°C to rid it of any residual moisture and add viscosity.

Liquid conching: follows immediately after dry conching - in the same conch and without stopping the machine - affecting texture and creaminess. (Sometimes it is necessary to add a little cocoa butter).

Tempering

Tempering is art of transforming liquid, or semi-liquid, chocolate into a solid. It seems simple but is a very delicate process. The chocolate is heated to a specific temperature until the cocoa butter crystals have melted completely, and then cooling it at a carefully selected temperature. Properly tempered chocolate is smooth and glossy, snaps when broken and produces smooth texture in mouth and delivers flavor and aroma.

Summary of the Process of Transforming Cocoa Beans into Chocolate

Step 1. The cocoa beans are cleaned to remove all extraneous material.

Step 2. The beans are roasted to bring out the chocolate flavour and colour. The temperature, time and degree of moisture involved in roasting depend on type of beans used and the sort of chocolate or product required from the process.

Step 3. A winnowing machine is used to remove shells from beans to leave just the cocoa nibs.

Step 4. The cocoa nibs undergo alkalisation, usually with potassium carbonate, to develop flavour and colour.

Step 5. The nibs are then milled to create cocoa liquor (cocoa particles suspended in cocoa butter). The temperature and degree of milling vary according to type of nib used and product required.

Step 6. Manufacturers generally use more than one type of bean in their products and therefore the different beans have to be blended together to required formula.

Step 7. The cocoa liquor is pressed to extract cocoa butter leaving a solid mass called cocoa presscake. Amount of butter extracted from liquor is controlled by manufacturer to produce presscake with different proportions of fat.

Step 8. The processing now takes two different directions. The cocoa butter is used in manufacture of chocolate. The cocoa presscake is broken into small pieces to form kibbled presscake which is then pulverised to form cocoa powder.

Step 9. Cocoa liquor is used to produce chocolate through addition of cocoa butter. Other ingredients sugar, milk, emulsifying agents and cocoa butter equivalents are also added and mixed. Proportions of different ingredients depend on the type of chocolate being made.

Step 10. The mixture then undergoes a refining process by travelling through a series of rollers until a smooth paste is formed. Refining improves texture of chocolate.

Step 11. The next process, conching, further develops flavour and texture. Conching is a kneading or smoothing process. The speed, duration and temperature of the kneading affect flavour. An alternative to conching is an emulsifying process using a machine that works like an egg beater.

Step 12. The mixture is then tempered or passed through a heating, cooling and reheating process. This prevents discolouration and fat bloom in the product by preventing certain crystalline formations of cocoa butter developing.

Step 13. The mixture is then put into moulds or used for enrobing fillings and cooled in a cooling chamber.

Step 14. The chocolate is then packaged for distribution to retail outlets

Health Benefits of Cocoa Consumption

Chocolate and cocoa contain a high level of flavonoids, specifically epicatechin having beneficial cardiovascular effects on health. Ingestion of flavonol-rich cocoa is associated with acute elevation of circulating NO, enhanced flow-mediated vasodilation, and augmented microcirculation. Prolonged intake of flavonol-rich cocoa is linked to cardio-vascular health benefits.

Marketing

Kerala was the leading State in promoting cocoa cultivation. Massive area coverage was possible through distribution of cocoa seedlings. Perhaps Cadbury India Ltd. was the only industrial unit during the period of massive expansion of area under cocoa. There was an attractive price for cocoa pods and beans prevalent till 1980's. This favourable situation, coupled with large scale distribution of planting materials could bring about an enviable area coverage recording 29,000 ha under cocoa by 1980-81. Being a crop subjected to the monopolistic exploitation of the available industrial unit, however paved ways for fall in price in 1981-82 and 1982-83. Inadequate marketing network and the fall in price developed a sense of insecurity among the planting communities, which detrimentally affected its expansion besides attributing to a neglectful approach by the plantation community. The entry of CAMPCO towards the marketing scenario from 1990's, though created a favourable atmosphere, the services rendered towards procurement of cocoa were far below the requirement. As a result, expansion of cocoa came to a standstill in spite of favouring the growers with a better price.

From 1997-98 onwards the non-traditional tracts of Karnataka and other States like Andhra Pradesh and Tamil Nadu started developing cocoa. As on day, with the implementation of 8th Five Year Plan programmes through distribution of high yielding varieties in the form of clones and hybrid seedlings, the area under cocoa is 17,800 ha. with a meager production of 10,200 tones compared to total world production of 27 lakh tones

Import of Cocoa Beans

Year	Quantity (tonnes)	Value (Rs. in Lakhs)
1999-2000	1422	1034.96
2000-2001	2027	1144.66
2001-2002	2149	1154.60
2002-2003	1213	1074.08
2003-2004	1337	1142.00
2004-2005	4177	3951.00

Source: Directorate of Cashew and Cocoa Development, Cochin

Cocoa Producing Countries

Cocoa cultivation initially spread into the immediate vicinity of original habitat of the cocoa tree from Brazil and Mexico to Central America and the Caribbeans. In sixteenth century, the Spanish introduced cocoa to the Indonesian archipelago, and to West Africa, which now account for more than half of the yearly world crop. In Southeast Asia, production has also markedly increased.

Cocoa is grown in more than thirty-five countries. The cultivated area covers between 3.5 and 4.5 million hectares. This area yields an annual production of approximately 2.7 million tones of cocoa beans.

The largest production area for cocoa is West Africa where about 60% of the world's cocoa is grown. The four major West African cocoa producers are the Ivory Coast, Ghana, Nigeria and Cameroon. The Ivory Coast produces about 43% of the world's cocoa. The next largest producer is Ghana with about 14% of the world's output. Nigeria produces about 6 %of the world's cocoa. Outside of West Africa, the major producers of cocoa are Indonesia, Brazil, Malaysia, Ecuador, and the Dominican Republic. Cocoa producers like Ghana and Indonesia are making efforts to increase cocoa production while producers like Malaysia have been switching to other crops. Ghana has an ongoing problem with black pod disease and smuggling of the crop into neighboring Ivory Coast. Brazil was once one of the largest producers but had problems with witches' broom disease.

Global Production

Cocoa is grown principally in West Africa, Central and South America and Asia. In order of annual production size, the eight largest cocoa-producing countries at present are Côte d'Ivoire, Ghana, Indonesia, Nigeria, Cameroon, Brazil, Ecuador and Malaysia. These countries represent 90% of world production.

World cocoa production is projected to grow at a rate of 2.2% a year, from 1998 - 2000 to 2010, compared with the 1.7 % growth during the previous decade, and reach 3.7 million tonnes. During the same period, Africa's share in the global production is expected to decrease slightly from 69% to 68 %, while the share of the Far East is projected to remain at 18% and that of Latin America and the Caribbean at 14%.

Africa to remains the world's leading cocoa producing area over the next decade. Production in Côte d'Ivoire, the world's largest cocoa bean producer, should grow by 2.3% a year from 1.2 million tonnes of the base period to 1.6 million tonnes in 2010, and account for 44% of global cocoa production due mainly to the increased foreign direct investment followed by the market liberalization. Yields in Cote d'Ivoire are well below levels seen in Asia partly because of less use of agricultural inputs. The recent surge in world cocoa prices has made it easier for growers to use more inputs. If this trend continue, volume of cocoa produced in Cote d'Ivoire could increase further. Output in Ghana, the second largest cocoa bean producer in Africa, would grow from 410 000 tonnes in 1998 - 2000 to 490 000 tonnes in 2010, an annual average growth rate of 1.6%. The corresponding growth rate for the previous decade was 3.3%. The lower projected growth rate over the next decade would result from the outbreak of diseases (such as swollen shoot virus, black pod and mirids), increased competition at the world market and low export prices. Over the same period, Nigeria and Cameroon are projected to increase outputs by 1.4% and 0.3 %, respectively.

Cocoa production in Latin America is projected to increase from 397 000 tonnes during the base period to 520 000 tonnes in 2010, an annual growth rate of 2.5%. Outputs in Brazil, the largest cocoa bean producer in the region, and Colombia, the third largest, are expected to fall, but an increase in outputs in other cocoa producing countries in the region would more than offset the decline. Output in Brazil is projected to increase by 2.2% annually and reach 180 000 tonnes by 2010. The production and yields of cocoa beans in Brazil have decreased during previous decade because of the detrimental production loss caused by witches' broom disease. During the same period, output in Ecuador, the second largest cocoa bean producer in Latin America, would increase by 0.8% annually and reach 94 000 tonnes. Ecuador has successfully used a new variety resistant to the witches' broom disease, which had also affected their cocoa production areas. Growth is expected to be only slight because of the increasing costs of production and lower returns to growers. In Colombia, outputs are projected to fall by 3.1%/ annum. On the other hand, outputs in Dominican Republic and Mexico are expected to grow by 1.8 % and 0.5%, respectively.

In the Far East, production grew rapidly over the past two decades, and this growth is likely to continue. Production in the Far East is projected to grow by 2.7%/ year from 509 000 tonnes during the base period to 680 000 tonnes in 2010 reflecting the expected improvement in yields. The Far East is expected to replace Latin America and the Caribbean as the second largest cocoa producing region by 2010. Most of the production growth in Asia would come from Indonesia, the world's third largest cocoa bean producer after Cote d'Ivoire and Ghana. Production in Indonesia is projected to grow by 3.5% annually to 574 000 tonnes in 2010 and account for 16% of the global production by 2010, compared to 14% in 1998 - 2000. In Indonesia, the Government policies encouraged expansion of production, and most of the increases during the last two decades were bulk cocoa from hybrid trees. While the expansion of production area in Indonesia has slowed since the late 1990s, yields in the country are still the highest among major cocoa producing countries. A close link between the world market prices and the producer prices in Indonesia also contributed to the country's high yields. Since the growers earn a high proportion of the market prices, they can invest in inputs, which in turn results in improvement in yields. Production in Malaysia, where expansions of urban areas and real estate development have reduced cocoa producing areas, is projected to fall by 1.7 % annually and reach 43 000 by 2010. The downward trend has been observed since the early 1990s when the outbreak of disease coincided with the deterioration of country's macro-economic conditions. In addition, farmers switched to production from cocoa to more lucrative crops, such as palm oil, in response to the fall in world cocoa prices during 1990s. Therefore, outputs in Malaysia are unlikely to rebound to the level achieved a couple of decades ago. In the early 1970s production was concentrated on Ghana, Nigeria, Cote d'Ivoire and Brazil, but it has now expanded to areas such as Pacific region, where countries like Indonesia showed spectacular growth rates in production.

Global Consumption

Although cocoa is largely produced in developing countries, it is mostly consumed in industrialized countries. For cocoa, buyers in consuming countries are processors and chocolate manufacturers. A few multinational companies dominate both processing and chocolate manufacturing.

In 2010, world grindings of cocoa beans, a proxy for world cocoa consumption, would amount to 3.6 million tones, reflecting an average annual increase of 2.1% from 2.8 million tones during the base period. Consumption continues to be concentrated in developed countries, which are expected to account for 64% of world cocoa consumption in 2010. Consumption in these countries is projected to increase at an annual rate of 2.2 % from 1.8 million tones during the base period to 2.3 million tones in 2010.

World Production of Cocoa Beans, in Thousand Tones

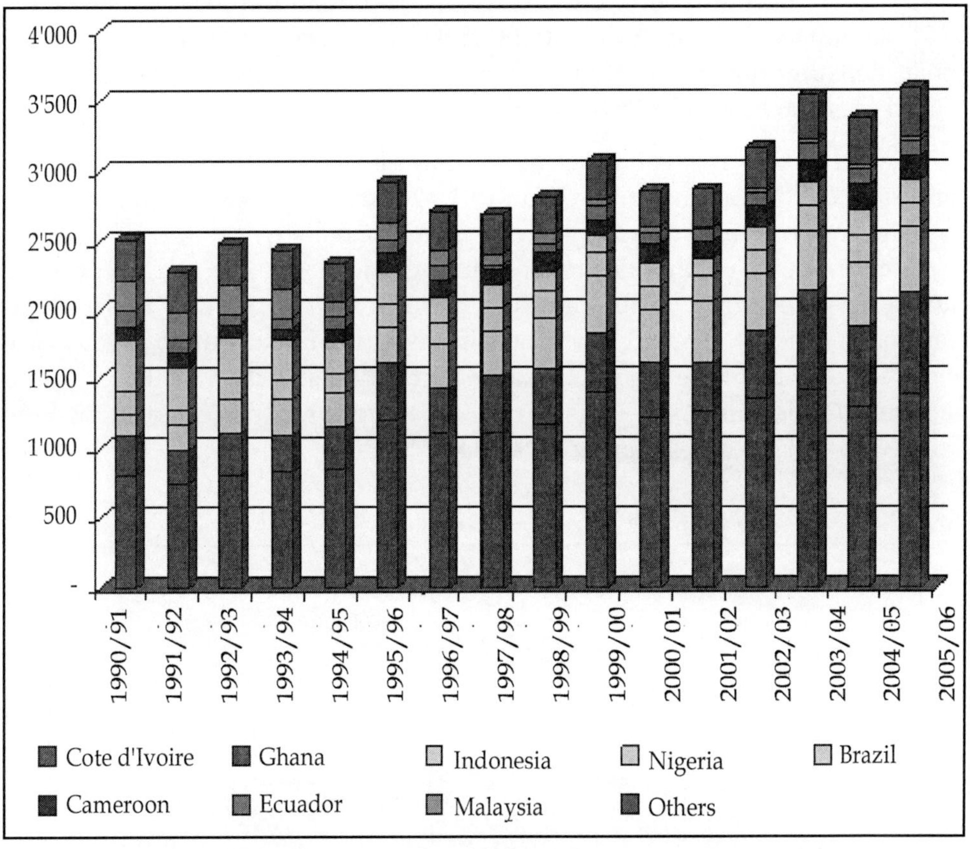

Source: UNCTAD based on the data from International Cocoa Organization, quaterly bulletin of cocoa statistics

Consumption in Europe is projected to grow by 1.7% per annum and reach 1.4 million tones. Europe is likely to continue to be the world's largest cocoa consuming area, accounting for 40% of global cocoa consumption in 2010. Chocolate and cocoa based products in the EU are currently governed by a 2000 directive which authorises the replacement of cocoa butter by less expensive cocoa butter substitutes up to 5 %of the total weight of finished product. Under the Directive, chocolate products that contain vegetable fats other than cocoa butter may be marketed in the EU provided that their labelling is supplemented by a statement. Member countries have until August 2003 to implement national laws to enforce this directive. Consumption in North America, the world's second largest cocoa consuming area, is likely to grow by 3.6% per annum and reach 703 000 tonnes. In the former Soviet Union / CIS, consumption is expected to grow by 0.8 % per annum from 65 000 tones to 71 000 tonnes, reflecting expected increase in income in these countries. In Japan, consumption is expected to increase from 48 000 tones during the base period to 56 000 tones in 2010.

Consumption in developing countries as a group is expected to 1.3 million tonnes by 2010, an annual growth rate of 1.8%. Africa, where capital formation for grindings has grown rapidly over the past decade, will remain the largest consuming region in this group, accounting for 35% of the consumption of developing countries. The share of consumption in Latin America and Caribbean, where the relative cost for grindings are higher compared to Africa, is expected to decrease from 32% to 28 %. In the Far East, where per capita consumption is still small, the share in consumption is projected to increase from 31 % during the same period to 34 % by 2010.

Share of Main Consuming Countries in 2004/05

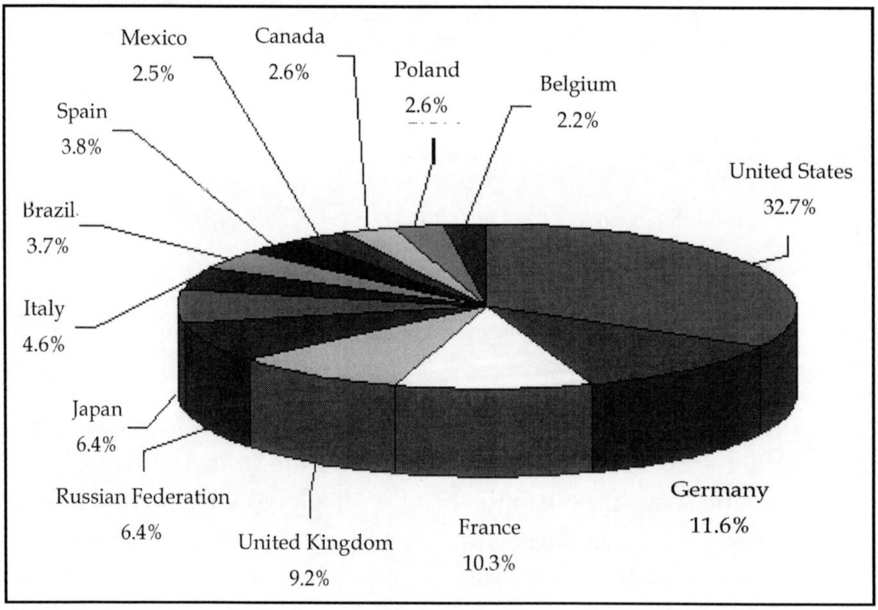

Source: UNCTAD based on the data from International Cocoa Organization, quaterly bulletin of cocoa statistics

Taking into account the data for the last 30 years it is possible to see that with the exception of eight years, there has always been a surplus of production. The following graph presents the supply and demand balance for the last thirty years, considering grindings as primary demand of cocoa beans.

World cocoa bean production, grindings and supply / demand balance in thousand tonnes from 1960/61 to 2005/06 (forecasts)

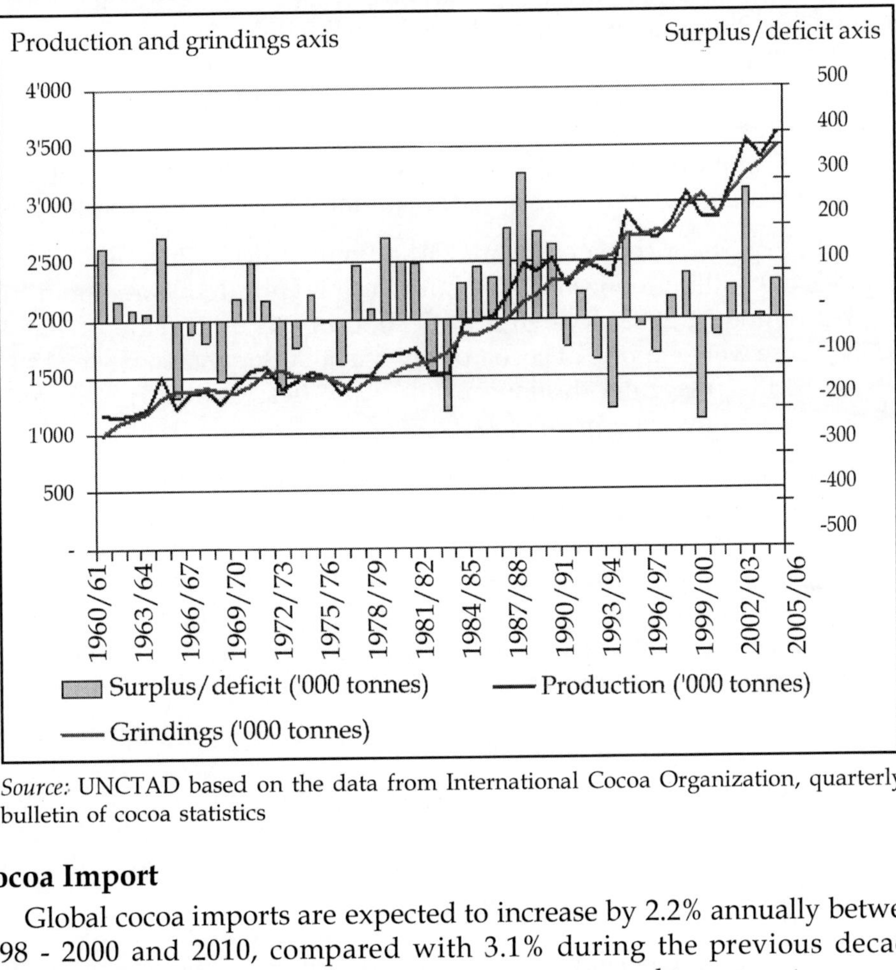

Source: UNCTAD based on the data from International Cocoa Organization, quarterly bulletin of cocoa statistics

Cocoa Import

Global cocoa imports are expected to increase by 2.2% annually between 1998 - 2000 and 2010, compared with 3.1% during the previous decade. Imports in developed counties as a group are expected to grow at an annual rate of 2.5% to 2.6 million tonnes. Europe should continue to be the main consumer of cocoa, accounting for 65 % of global cocoa imports in 2010. In North America, imports are projected to grow by 0.3 % per year, to reach 505000 tonnes by 2010. Shipments to the countries of the former Soviet Union/

the Commonwealth of Independent States (CIS) are likely to decrease slightly by 1.1% per annum. In Japan, imports are expected to increase by 1.4 % per year from 48 000 tonnes in 1998 - 2000 to 56 000 tonnes in 2010. Imports in developing countries as a group are projected to remain unchanged and would account for 11.3% of world cocoa imports.

The International Cocoa Market

Total production of cocoa beans in crop year 2003/04 was an all-time record of 3.5 million tonnes, although output fell back to 3.376 million tonnes in 2004/05. The forecast output for 2005/06 in August 2006 was 3.504 million tonnes. The largest producers were Côte d'Ivoire and Ghana, which grew 39% and 19% of the world total respectively, followed by Indonesia, Nigeria, Brazil and Cameroon, with 18%, 11%, 7% and 5% respectively. Together, ACP countries thus account for over three-quarters of world production.

World exports of cocoa beans in 2004 amounted to 2.725 million tonnes, up from 2.409 million tonnes in 2003, but worth US$4.176 billion, down from US$4.392 billion in 2003, following a decline in the unit value of 16%. In addition there were exports of cocoa paste, cocoa butter and cocoa husks and shells. A further marginal decline in unit value (less than 1%) followed in 2005 down to an average price of US$0.6977 per lb.

The world price of cocoa is quoted on international exchanges, the principal of which are London and New York, and is determined by supply and demand. The world price of cocoa rose seven-fold in 1970s to over US$3,500/ tonne, stimulating the entry of Malaysia and Indonesia into production. Since then production has doubled and prices have halved, in real terms the price in 2002 was only 19% of 1980 price. Both London and New York operate cocoa futures and options on cocoa futures.

Prices

Cocoa prices mainly respond to cocoa supply and demand factors. International prices tend to follow a long-term pattern linked to the cocoa cycle, estimated to be of over 20 years. During cocoa boom periods, there is a supply surplus that will result in falling and then stagnating prices. Consequently, low prices due to overproduction generally have a negative impact on harvesting, encouraging farmers to switch to other crops, a factor which again permits world prices to rise. The cocoa cycle is thus characterised by boom and bust effects.

Prices experienced an important increase in the 1970s, which encouraged production in countries such as Malaysia and Indonesia. However, since the beginning of the 1980s prices have declined. In spite of a modest recovery in the mid 1990s, international cocoa prices are low compared to those prevailing in the 1970s.

The Futures Market

The world price of cocoa is determined by futures market. Without having to conduct personal meetings, buyers and sellers are able to conclude contracts about future deliveries of cocoa beans on the so called futures market. At the time of negotiation, the cocoa beans in question usually have not even begun to grow, and so this provides an opportunity for a lively and exciting world-wide trade. The contracts are registered with the Clearing House. Via this agency, a delivery commitment can be nullified by completion of a sales contract for the same quantity and month of delivery. The Clearing House settles the differences in price. The most important futures markets are in London and New York. Price trends can be closely followed anywhere in the world.

The projections employed dynamic time series models to analyse world cocoa economy. Projections were based on assumption of normal weather conditions, and a continuation of past trends in yields, planted areas, population and income growth. Adjustments were made to reflect current policies and future market prospects. The forecasting models captured cycles and trends in the world cocoa beans market to a satisfactory level, compared with the 14 % during previous decade.

COCONUT

The coconut palm (*Cocos nucifera* Linn.) is the most useful palm in the world. Every part of the tree is useful. It is much attached to the emotions of the people in the Southeast Asia that it forms a part of the mythology and culture and is auspicious in various ceremonies. Coconut tree is ranked among one of the 10 most useful trees of the world and is often termed as 'kalpavriksha', the "tree of life".

Copra obtained by drying the kernel of coconut is the richest source of vegetable oil containing 65 to 70% oil. Several other products are derived from coconut palm and they too are used in many applications. Copra is used to extract coconut oil and coconut meal in the ratio of 3:2. The products like hair oil, soaps, shampoos and medicines require coconut oil as an important ingredient. Also the kernel is quite popular in a vast number of cuisines.

Climate and Soil

The coconut palm grows under varying climatic and soil conditions. It is essentially a tropical plant growing between 20°N to 20°S latitudes. A well distributed rainfall of about 2000 mm / year is ideal for proper growth and maximum production.

Coconut is grown under different soil types like loamy, laterite, coastal sandy, alluvial, clayey and reclaimed soils of marshy low lands. Ideal soil conditions for better growth and performance of the palm are proper drainage, good water-holding capacity, presence of water table within 3m and absence of rock or any hard substratum within 2m of the surface.

Coconut palm is a tropical plant that requires hot and humid climate to prosper. It tolerates salinity in water and soil and that is why it is generally found in the coastal areas. The soil type that is appropriate for production of coconut palm is sandy soil. Consistent rainfall and ample sunlight are also key factors for effective growth of coconut palm. Plantation is difficult to be established. Propagation of the trees is done through nuts.

Coconut Types

Coconut palm is categorized into two basic types –tall and dwarf.

Tall Variety

Tall coconut is a slow growing variety. The tree produces nuts 6 to 10 years after planting, which is comparatively late as compared to dwarf. The tall cultivars grown extensively are the West Coast Tall and East Coast Tall. The tree remains productive even at 80 to 120 years and produces good quality copra, oil and other products. Nuts of palm get mature in 12 months duration.

Dwarf Variety

The dwarf or short variety of trees are fast growing and produce fruits in 4 to 5 years. Dwarf palm has a relatively short productive age. The fruit may be of variable colors such as yellow, red, green and orange. The dwarf variety is shorter in stature and its life span is short as compared to tall. Tall x Dwarf (TxD), Dwarf x Tall (DxT) are the two important hybrids.

There are 10 different combination of hybrids, developed by Kerala Agricultural University, Tamil Nadu Agriculture University and CPCRI and released for commercial cultivation. They are high yielders under good management. Laccadive Ordinary, Andaman Ordinary, Philippines, Java, Cochin-China, Kappadam etc. are the other tall cultivars.

Released Varieties and Cultivars of Coconut

Characters of Released Hybrids

Hybrids	Nut yield (palm/ year)	Mean copra (content / nut (g))	Mean copra content / palm/ year (kg)	Copra yield/ ha (t)	Oil content (%)	State for which recomm-ended
Chandra Sankara (COD x WCT)	116	215	25	4.4	68	Kerala Karnataka Tamilnadu
Kera Sankara (WCT x COD)	108	187	21	3.5	68	Kerala Karnataka
Chandra Laksha (LO x COD)	109	195	21	3.7	69	Kerala Karnataka
Laksha Ganga (LO x GB)	108	195	21	3.7	70	Kerala
Ananda Ganga (AO x GB)	95	216	21	3.6	68	Kerala
Kera Ganga (WCT x GB)	100	201	21	3.5	69	Kerala
Kera Sree (WCT x MYD)	112	216	24	4.2	66	Kerala
Kera Sowbagya (WCT x SS Apricot)	130	195	25	4.3	65	Kerala
VHC-1 (ECT x DG)	98	135	13	2.3	70	Tamilnadu
VHC-2 (ECT x MYD)	107	152	16	2.9	69	Tamilnadu
VHC-3 (ECT x MOD)	156	162	25.3	3.95	70.2	Tamilnadu
Godhavari Ganga (ECT x GB)	140	150	21	3.7	68	Andhra Pradesh

Characters of Released Cultivars

Cultivar (Tall)	Mean		Yield /		
	Palm / year		Copra yield		
	No. of nuts	% over WCT	Mean nut (g)	Mean/palm / year (kg)	% over WCT
Lakshadweep Ordinary (Chandrakalpa)	97	21.2	195	18.9	32.3
Benawali Green Round (Pratap)	151	88.8	152	22.7	57.6
Philippines Ordinary (Double Century)	110	37.5	198	21.8	57.9
West Coast Tall (WCT)	80	-	180	14.4	-
Chavakkad Orange Dwarf (COD)	For tender coconut, 350 ml water 7.1% TSS				

Production of Quality Planting Materials

Nursery Technique

Selection of Site for Nursery

Nursery sites should be well drained with light textured soil and with adequate but not too much shade. In open areas, provide shade during summer. Prepare beds of 1.5 m width and of convenient length with 75 cm space between beds. In areas, where drainage is poor, prepare raised beds. Before planting, examine seed nuts and discard those without nut water and rotten kernels. Sow the nuts in the nursery after commencement of southwest monsoon during May-June.

Spacing of Nuts

Plant seed nuts at a spacing of 30 cm (between rows) x 30 cm (between nuts) with four or five rows/ bed.

Method of Planting Seed Nuts

Plant seed nuts in the beds 25-30 cm deep and cover with soil so that top portion of husk alone is visible. The nuts may be planted either horizontally with the widest of segments at top or vertically with stalk-end up. Vertical planting is preferable on account of convenience in transporting and lesser risk of seedling injury.

Care and Management of Nursery

Provide protective fencing to nursery if it is located in open areas. If the soil is sandy, provide mulching immediately after cessation of monsoon rain. Irrigate the nursery once in two days during summer. Keep nursery beds free of weeds by periodic weeding. If termite is noticed, remove soil in affected area up to a depth of 15 cm and dust soil and nuts with carbaryl or chlorpyrifos. Repeat if attack persists. Periodically spray the plants with Bordeaux mixture(1%)or any other copper fungicide to prevent fungal infection.

Selection of Seedlings

Remove seed nuts, which do not germinate within six months after sowing and those with dead sprouts. Select only good quality seedlings (9-12 months old) by a rigorous selection based on the following characteristics.

- Early germination, rapid growth and seedling vigour.
- Six to eight leaves for 10-12 months old seedlings and at least four leaves for 9 months old seedlings.

- Collar girth of 10-12 cm.
- Early splitting of leaves.

The recovery of quality seedlings will be about 60-65%. Since early germination is one of the criteria for selection of seedlings, the storing and sowing of seed nuts should be in lots rather than in a staggered manner.

Removal of Seedlings

Remove seedlings from nursery by lifting with spade and cutting roots. Keep the seedlings in shade and do not expose to sun. Plant seedlings as early as possible after removal from nursery. Never allow lifting seedlings from soil by pulling leaves or stem.

D X T Hybrid Production

The following guidelines are suggested for large-scale production of D x T hybrid seedlings. Assisted pollination should be done to get maximum hybrid nut production. As far as possible use prepotent palms as parents in the hybridization programmes.

Selection of Mother Palms

Palms with the following phenotypic character combination may be selected for hybridization work.

- Nuts without ridges and having yellow, orange or red colour.
- Palms with overlapping female and male phases
- Small crown and canopy compared to that of tall palm.
- Narrow stem without any bulging at the base with close leaf scars

Hybridization

- Use mixed pollen from identified tall palms.
- Emasculate the inflorescence by cutting the male flowers with scissors and stripping if necessary within 5-7 days of opening the spathe.
- Cloth bags made of very close mesh should be used for covering the inflorescence.
- Hairy caterpillar larvae cause serious damage by boring into the female flowers and developing buttons through stigmatic ends. The damage is more serious under bagged conditions. Spraying carbaryl suspension (0.15%) mixed with fish oil(1%) soap or sandovit ensures protection from the larvae. Spraying may be done prior to pre-pollination bagging.

- Dusting of pollen-talc mixture in 1:9 proportion using pollen dispensers is recommended.
- Assisted pollination for at least 3-5 days on each inflorescence till last female flower becomes receptive and fully pollinated.
- Remove bags after the seventh day of pollination of last female flower.

The nuts should be harvested before it is tree-ripe and sown immediately in the bed without storage. Nursery beds should be mulched or shaded and watered regularly and adequately.

Selection Criteria for Planting Materials

Selection of Mother Palms

Coconut is propagated through seedlings raised from seed nuts collected from superior palms referred as mother palms. Select mother palms having the following characters:

- Regular bearing habit and yielding not less than 80 nuts/ year.
- Age 20 years or more (5 years after reaching full bearing capacity). If the mother palms are the progeny of elite planting material and give consistently higher yields for a period of not less than 6 years, seed nuts can be collected from such palms. There is no need for insisting 20 years as minimum age for mother palms in such conditions.
- More than 30 fully opened leaves with short strong petioles and wide leaf base firmly attached to stem.
- Bearing at least 12 bunches of nuts with strong bunch stalks.
- Bearing nuts of medium size and oblong shape.
- Husked nuts should weigh not less than 600 g.
- Mean copra content of 150 g/ nut or more.

Avoid palms which (i) have long, thin and pendulous inflorescence stalks (ii) produce long, narrow, small sized or barren nuts (iii) show shedding of immature nuts in large numbers and (iv) are grown under favorable environmental conditions.

Collection and Storage of Seed Nuts

Collect mature nuts (above 11 month old) during December to May. Lowering of bunches by means of ropes may be done when the palms are tall and ground is hard. Discard nuts, which show improper development or other undesirable features. Store seeds in shade for a minimum period of 60 days prior to sowing in nursery. For storing, arrange seed nuts with stalk-end up over an 8 cm layer of sand in a shed and cover with sand to prevent drying of nut water. Up to five layers of nuts can be arranged one over the other. The nuts can also be stored in plots, provided the soil is sandy and ground is sufficiently shaded. In nuts harvested in May, heap them in partial shade, till husk is well dried and then sow them in the nursery.

Planting Material

Coconut is propagated through seedlings raised from selected seed nuts. Generally 9 to 12 months old seedlings are used for planting. Select seedlings, which have 6-8 leaves and 10-12 cm collar girth when they are 9-12 months old. Early splitting of leaves is another criteria for selection of coconut seedling.

Selection of Seedlings

Select only vigorously growing seedlings characterized by early germination, good girth at collar, more number of functional leaves and early splitting of leaflets, at the age of about nine to twelve months.

Planting and Management

Use 9-12 months old seedlings for planting, in pits dug to a dimension of 1x1x1m and filled with a mixture of topsoil, farmyard manure or compost in equal proportions and 30 to 60 g of carbaryl (5%). Add sand to this mixture if the soil is heavy black clay. Give a spacing of 8 x 8 m for hybrids and tall varieties in main field. Give protection to newly planted seedlings against grazing by cattle and provide shade in hot weather.

Site Selection

Select sites with deep (not less than 1.5 m depth) well drained soil. Shallow soils with underlying hard rock, low lying areas subject to water stagnation and clayey soils are to be avoided. Proper supply of moisture either through well distributed rainfall or through irrigation should be ensured before planting.

Plantation Lay-out

Before preparing lay-out of a plantation, a few important factors will be considered like: selection of farm type, monoculture or multiple cropping; selection of planting material talls, hybrids or dwarfs; planting system: square, rectangular or triangular; crop combination, mono cropping, mixed cropping, intercropping or mixed farming and type of intercrops, annuals or perennials, and their degree of shade tolerance.

Water availability, soil texture and depth are other factors playing an important role in the decision on planting system and crop combination. The decision on farm type is the most important for entire farming period, with long-lasting effects, as coconut palms have a very long productive life, especially if well managed.

The lay-out of a new plantation should be planned before clearing land. For this purpose, the land will have to be lined first. To reap maximum benefit from sunlight on flat land, lining should be carried out in a North -South and East-West direction. When planting in hedge systems, the wide lanes should run in the East-West direction and narrow lanes North-South. On slopes, lining will be carried out on contour, permitting erosion control measures like contour bunds, etc. After clearing and burning the debris, land has to be staked, a drainage and a road system can be laid out. Stakes of about 1.5-2 m should be used to mark the planting sites. A wooden frame may be used to mark size of planting holes. This may be done with the stake in the middle or with the stake in one corner.

Preparation of Land and Planting

On slopes and in areas of undulating terrain, prepare land by contour terracing or bunding. In low-lying areas, mounds are to be formed at planting site to a height of at least 1m above water level. In reclaimed 'kayal' areas, seedlings are planted on field bunds.

In loamy soils with low water table, a pit size of 1x1x1m is recommended. In laterite soils with underlying rocks, take larger pits of size 1.2 x 1.2 x 1.2m. In sandy soils the size need not exceed 0.75 x 0.75 x 0.75 m.

Spacing and Systems of Planting

Spacing depends upon planting system and soil type. In general, following spacings are recommended under different planting system in sandy and laterite soils.

Planting System	Spacing	Approximate Number of Plants/ ha
Triangular	7.6 m	198
Square	7.6 to 9 m	>170-120
Single hedge	5 m in the rows 9 m between the rows	220
Double hedge	5 x 5 m in rows 9 m between pairs of rows	280

Low Density Planting

As the yield curve for different planting densities is rather flat between densities of 100-180 palms/ ha, a wider spacing is preferred where multiple cropping is planned. It provides more land, light and water for intercrops. Planting in a square or rectangular system facilitates mechanized cultivation and crop management. In climates with long dry seasons, spacing at 9 x 9m in a triangular system, or 10 x 10m in a square system will not give very different yields, but the latter improves light conditions for the intercrop considerably. Selection of coconut variety to be planted is also of importance when intercropping is considered. When high-yielding hybrids are preferred, tall x tall hybrids may be selected instead of dwarf x tall hybrids.

Hedge Row or Double Hedge Row Planting

The hedge row system is based on narrow spacing between palms in the row, and large spacing between rows. The double hedge row system is based on alternation of a narrow spacing between two rows with a very wide spacing. The narrower the spacing in the row, more the palms tend to bend outwards, reducing effect of wide inter-row spacing. With a hedge system, intercropping is possible during entire life-time of palms, with a reduction of space when palms are young, and low leaves may hamper cultivation of other crops close to palm. The system to be chosen depends very much on crop(s) planned for intercropping.

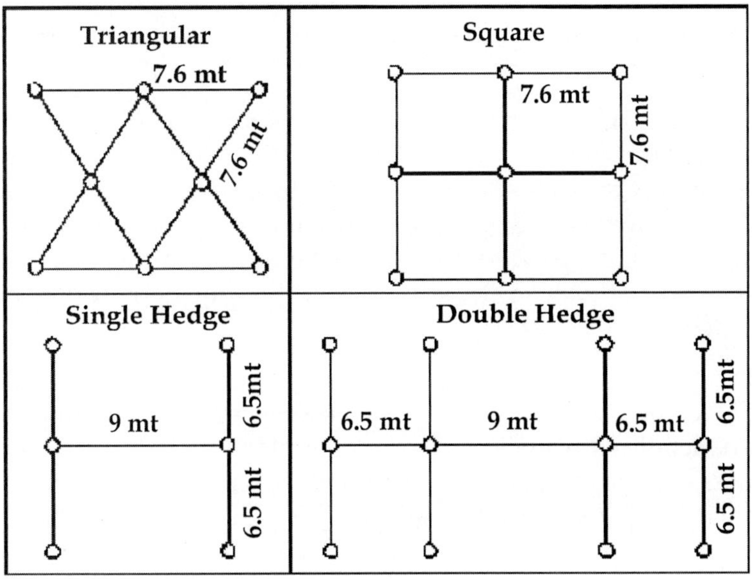

Time of Planting

Seedlings can be transplanted in the beginning of south west monsoon. If irrigation facilities are available, it is advisable to take up planting at least a month before onset of monsoon so that seedlings get well established before heavy rains. Planting can also be taken up before onset of north-east monsoon. In low-lying areas subject to inundation during monsoon period, transplanting may be done after the cessation of monsoon

Planting

Before planting, the pits are filled up with top soil and powdered cow dung/ compost up to a depth of 50 to 60 cm. Then take a small pit inside this, so as to accommodate the nut attached to the seedling. Plant the seedling inside this pit and fill up with sand. Press the sand well so as to avoid water stagnation. If there is chance for white-ant attack apply Sevidol 8G (5gm.) inside the small pit before planting.

In laterite areas apply 2 kg common salt/ pit for improving physical condition of soil. Burying 25 to 30 coconut husks/pit in layers will be useful for moisture conservation.

Care of Young Palms

Shading and Irrigation

The transplanted seedlings should be shaded and irrigated adequately during the summer. Also provide staking so that winds may not uproot the young seedlings. For the first two years after planting, irrigate the seedling twice a week during dry summer months. Shading is a must to transplanted seedlings.

Regular manuring from the first year of planting is essential to achieve higher productivity. For coconut 20-50kg organic manure should be applied / palm/ year with onset of south west monsoon, when soil moisture content is high. Different forms of organic manures like compost, farm yard manure, bone meal, fish meal, blood meal, neem cake, groundnut cake etc. could be made use for this purpose. In addition, following Fertilizer Schedule is recommended.

Apply fertilizer, in two equal split doses one in June-July and again in September-October. The second dose can be given even in December - January if irrigation facilities are available. For young plants 1/3, 2/3 may be given in 1st, 2nd year after planting and the full dose from 3rd year onwards. This is to be applied between 0.5 to 2.0 meters, after making 2 meter basin around the trunk.

Fertilizer Requirement of Young Palms in Relation to that of Adult Palms

Time after planting	Time of application	
	(Proportion of adults palm dose)	
	June-July	Sept-Oct.
3 months (1/10th of full dose)		1/10
1 year (1/3rd of full dose)	1/9	2/9
2 year (2/3rd of full dose)	2/9	4/9
3 year onwards (full dose)	3/9	6/9

Nutrients Recommended for Adult Palms

		Quantity,(kg/palm/annum)		
		N	P_2O_5	K_2O
1	General recommendation 1. Average management 2. (b) Good management	0.34 0.50	0.17 0.32	0.68 1.20
	For reclaimed clayey soils (as in Kuttanad-Kerala)	0.25	0.35	0.90
3	Red loam so ils (Southern Kerala)	0.68	0.23	0.90
4	Hybrids & high yielding palms (a) For irrigated areas (b) For rain fed conditions	1.00 0.50	0.50 0.32	2.00 1.20

In addition to the above dose of fertilizers, two to three kg of finally ground dolomite lime stone or 0.5 kg Magnesium sulphate/ palm/ year is also recommended for use in acidic soils, light sandy soils and in root (wilt) affected tracts in Kerala. The dolomite may be broadcasted prior to onset of monsoon in basins and forked in and should not be applied along with other fertilizers. There is however no harm in applying magnesium sulphate along with other fertilizers.

Weeding and Interculture

Keep the pits free of weeds by periodical weeding. Remove the soil covering the collar of seedlings. As the seedlings grow and form stem, fill up pits gradually by cutting sides. Proper intercultivation provides control of weeds and creates soil mulch. Any tillage system (ploughing, digging, raking or forming mounds) that provides soil mulch and control weeds may be followed depending upon local conditions. For laterite, sandy and red sandy loam soils give two ploughings or diggings in May-June and September-October and one raking in January. In areas, where surface run off is more, form mounds in September-October and level them in November-December.

Drought Management in Coconut Gardens

Coconut produces nuts round the year. Therefore, adequate supply of water is essential for its unhindered growth. Soil moisture is essential for absorption of nutrients by roots. Moisture stress leads to stunted growth, drooping of leaves, immature nut fall and decreased yield. Importance may be given on following aspects so as to ward off stress.

Husk Burial

Burying fresh or dried coconut husks around the palm is a very beneficial practice particularly for moisture retention especially in drought prone areas. The husk can be buried either in linear trenches taken 3 m away from the trunk between rows of palms or in circular trenches taken around the palm at a distance of 2 m from the trunk. The trenches may be dug at 0.5 m wide and at same depth. The husks are to be placed in layers with concave surface facing upwards and covered with soil. The beneficial effects of husk burial will last 5-7 years.

Mulching

Mulching is an effective method of conserving soil moisture. Mulch the coconut basins with green / dry leaves at close of northeast monsoon (October-November). Mulching also adds organic matter to soil and reduces soil temperature. Do not disturb soil in coconut garden during summer. In level lands, during rainy seasons, excess water may be conserved in small trenches dug out in the plantation. In sloppy areas, land may be terraced and trenches dug across. This facilitates maximum percolation of rainwater and water storage. For moisture conservation, lowermost 3-5 leaves may be cut and removed. Provide adequate shade for transplanted seedlings for 1-2 years. To minimize heat load on stem, application of lime solution on trunk up to a height of 2-3 m at start of summer season is recommended.

Green Manure and Cover Crop

This helps to increase organic matter content of soil and also prevents soil erosion in coconut gardens. Following green manure/cover crops are recommended for cultivation in coconut gardens.

- *Crotalaria juncea*
- *Tephrosia purpurea*
- *Gliricidia maculata*
- *Calapagonium muconoides*
- *Mimosa invisa*

Sow the green manure / cover crops during April-May with onset of pre-monsoon showers. The green manure crops should be ploughed in and incorporated in the soil during August-September.

Water Management

Irrigation

Soil moisture very often limits coconut production in those areas where long spell of dry weather prevail or where the rainfall is scanty and ill-distributed. Irrigate the palms during summer in basins around the palm. Irrigation requirement varies according to the soil type and climatic condition. Generally, an adult palm requires 600 to 800 litres of water once in four to seven days. Irrigate in basins of 1.8m radius and 10-20 cm depth. In coastal sandy soils, sea water can be used for irrigating adult palms. Do not irrigate seedlings and very young palms upto 2 year with sea water. In irrigated gardens interruption of irrigation would lead to serious set-back in yield and general condition of palms. Once started irrigation, should be continued regularly and systematically. Drip irrigation is the best suited method of irrigation for coconut. It saves water, labour and energy.

Conventional Basin Irrigation

Irrigate the palms during summer in basins around palms as shown below:

Parameters	Soil texture			
Sandy	**Sandy loam**	**Loam**	**Silty**	**clay**
Available soil moisture (cm/m)	8	12	17	21
Quantity of water / irrigation / palm in litres in a basin of 1.8 m radius	600	900	1300	1600
Frequency of irrigation (days)				
All areas in Kerala except north eastern portion of Thrissur and Palakkad districts	3-4	5	7-8	9
North eastern portion of Thrissur and Palakkad districts	2-3	3-4	5-6	6-7

Drip Irrigation

Drip irrigation is the most suitable system of irrigation to coconut. Some of the major advantages of drip irrigation are: it saves water, enhances plant growth and yield, saves energy and labour, most suited for soils having low water holding capacity and undulating terrain, reduces weed growth and improves efficiency of fertilizers. For coconut, generally, three to four drippers are given/ palm. The water requirement for an adult palm is 40 to 50 litres/ day.

Mixed / Inter / Multi-species Cropping System in Coconut Garden

Schedule for inter-mixed cropping may be drawn up based on canopy size and orientation of palms. A variety of intercrops like pineapple, banana, elephant-foot yam, groundnut, chillies, sweet potato, tapioca and different vegetables can be raised in coconut garden. In older plantation, cocoa, cinnamon, pepper, clove and nutmeg etc. can be grown as mixed crops. These inter/mixed crops are to be adequately and separately manured in addition to manures applied to coconut palm.

Mixed Farming

Mixed farming by raising fodder grasses like Hybrid Napier or Guinea grass along with leguminous fodders like Stylosanthes is profitable. Raising above crops in one ha.of coconut garden can support three to four diary animals. Animals supply large quantities of cattle manure which when applied to the soil improve its fertility status. This sort of mixed farming improves the yield of palm.

Coconut-based Multi-storey Cropping System

A multi-storey cropping system is more complex, to accommodate two or more intercrops of different heights, canopy patterns and rooting systems, to maximize use of available sunlight, nutrients, moisture and land area under coconut. The fundamental objective is to increase productivity of coconut land. The coconut palm serves as the 'top floor', whereas perennials such as coffee, bananas, papaya etc., form mid-storey crops, and short-growing crops like vegetables and pineapples, form the ground floor.

Multi-storey cropping systems require more management skills, labour and other inputs than most systems. In Philippines, India, Indonesia, Sri Lanka and many Pacific Island countries, multi-storey cropping systems are widely practised. The basic principle related to multi-storey cropping systems is crop compatibility, combining different crop heights and rooting systems .In India, a crop combination of coconut + pepper + cocoa or cinnamon + pineapple met the general criteria of a multi-storey cropping system. While coconut formed the top floor, pepper grown along coconut trunk formed the second floor, cinnamon grown to 2 or more metre high formed the first floor, and pineapple with its shallow root systems represented ground floor. Studies at CPCRI, Kasaragod showed coconut + black pepper + cocoa + pineapple combination as the most promising crop combination. It appears that suitability of a multi-storey crop combination depends on farmers' perception and other technical/management requirements.

Organic Farming Practices for Coconut

The tall variety of coconut, widely cultivated in coconut growing regions uses nutrients to the level of 56 kg N, 12 kg P, 70 kg K, 34 kg Ca and 12.5 kg Mg/ palm/ year from a limited area of soil. As the palm is perennial with a continuous flowering habit, it requires replacement of nutrients continuously to maintain productivity.

Most of coconut production comes from small and marginal farmers. Only limited percentage of farmers use chemical fertilizers and as a result, vast majority of palms remain unfertilized or under-fertilized. Continuous use of chemical fertilizers result in increased soil acidity, imbalance of major and micronutrients and decrease in soil biological properties. The situation demands application of low cost organic nutrient sources, which helps in improving soil fertility and productivity. Research conducted at CPCRI resulted in development of a number of viable agrotechniques suitable for organic cultivation of coconut.

Availability of waste biomass from a well managed coconut garden with 175 trees/ha was estimated 14 tonnes/ha/yr in the form of leaves, spathe, bunch waste and husk. A considerable portion of husk is used for extraction of coir fibre. The by-product of coir processing factories, coir dust is usually dumped without any use. It is estimated that 7.5 million tones of coir dust is available in India from the various coir defibering units. The total availability of waste biomass from 1.81 million hectare of coconut plantation in India is estimated 14.36 million tones annually. The natural decomposition of these wastes and the nutrient release are very slow due to high lignin content and nature of lignocellulose complex of coconut waste materials. Substantial saving in terms of fertilizer input is possible through effective recycling of waste biomass.

The organic wastes can be directly utilized as mulch or converted into composts by employing earthworms or microbial cultures. Addition of nitrogen through biological nitrogen fixation can be favoured by growing green manure legumes and use of biofertilizers based on asymbiotic and associative symbiotic nitrogen fixing bacteria.

Vermicomposting of Coconut Palm Wastes

One hectare of coconut plantation produces about 8000 kg dry coconut leaves/ year. As use of this biomass as fuel and thatching material has declined over years, its disposal in a meaningful way has become a problem in many coconut gardens. Studies conducted at CPCRI, Kasaragod revealed that coconut plantation wastes could be effectively converted into rich vermicompost using a local species of epigeic earthworm belonging to *Eudrilus* sp. The local strain of *Eudrilus* is different from *Eudrilus eugeniae*, the African night crawler, which has its origin in tropical and subtropical Africa. Much faster multiplication of this worm on coconut palm wastes and coir pith was also observed compared to African one.

The local species of *Eudrilus* can be multiplied in a 1:1 mixture of cowdung and decaying leaves taken in a cement tank or wooden box or plastic bucket with proper drainage facilities. The nucleus culture of earthworms is to be introduced into above mixture at the rate of 50 / 10 kg of organic wastes and properly mulched with grass, straw or wet gunny bag. The unit should be kept in shade. Sufficient moisture level should be maintained by occasional sprinkling of water. Within 1-2 months, the earthworms multiply to 300 times, which can be used for large scale composting.

Vermicomposting

A low cost technology was standardized for vermicomposting the biomass from coconut palms left exposed to action of weather in the field for about 3 months (Prabhu *et al.*, 1998; Thomas *et al.*, 2001). The weathered wastes obtained during rainy seasons may be preferred. This waste can be used without chopping, thus saving a lot of labour. The coconut wastes used for oyster mushroom cultivation were also suitable for vermicomposting. These organic wastes are to be treated with cowdung at the rate of 10% by weight in the form of slurry and must be allowed to undergo a preliminary decomposition for about 2-3 weeks. The earthworms at the rate of 1000 worms/ tone of coconut leaves are to be introduced. The compost bed should be mulched properly using any locally available plant material or gunny bags and has to be protected from direct sun light. Watering is to be done to maintain enough moisture. As full leaves are used for composting, compact mass is not formed, thus allowing free movement of air in the bed. In about 60-75 days, compost will be ready, leaving behind only mid ribs of the leaves. This method was tested in cement tanks and under the roof of a shed by heaping. Composting was faster in heaps and the earthworms were seen feeding and casting from all sides, may be due to better aeration and exposure. On an average, 70 % recovery of vermicompost was obtained. The same technology for vermicomposting was also tested in large pits taken in the inter spaces of four coconut palms in sandy loam and coastal sandy soils and was found to work well. As composting is done in the field itself, lot of labour required for transportation of the biomass and compost can be saved. This technology may be tried even in plantations with very limited irrigation facilities, as only a limited number of pits or trenches need to be watered.

The average nutrient composition of the vermicompost recovered was : N % (1.8), P % (0.216), K % (0.16), organic carbon % (17.84), and C/N (9.95). Total microbial counts and beneficial microbial population were also more in the compost compared to base material. The C/N ratio of the organic matter ingested by earthworm decreases and bound nutrients are converted into easily available forms.

Vermicomposting of Coirpith

Coir pith is yet another organic material resistant to decomposition, available in huge quantities from coir processing units and can improve soil physical properties and moisture holding capacity to a great extent. But this can be applied to soil only after composting and efforts were made to prepare vemicompost from this waste also using the local strain of *Eudrilus* sp. Coir pith alone was not able to support growth of earthworms. Coir pith treated with lime and rock phosphate @ 0.5% each and incubated for three weeks has to be mixed with cowdung @ 10% and fresh vermicompost @ 10%. This mixture has to be layered with uncut coconut leaves @ 20% to facilitate aeration in the bed. The earthworm *Eudrilus* sp. has to be introduced at the rate of 1000/tonne of organic materials and the bed should be mulched and protected from direct sunlight. Moisture has to be maintained at 50% by regular irrigation. Earthworms form burrows in bed and vermicastings appear as surface casts. A granular vermicompost with 1.2% nitrogen and 16.7:1 C:N ratio can be obtained in two months. Preparation of vermicompost from coir pith on a large scale, would bring back both economic and environmental benefits.

Composting of Lignin Rich Biomass Using Microbes

Coir pith is a lignocellulosic waste biomass which accumulates around coir processing factories as a waste material. Though coir pith has a number of beneficial properties, its direct utilization as manure is not advisable as it contains large amounts of lignin and phytotoxic polyphenols and less of nitrogen. Technologies are available for bioconversion of coir pith having a C:N ratio of 100:1 to acceptable organic manure with help of biopolymer degrading micro-organisms. The composting of coir pith utilizing mushroom fungus, *Pleurotus safor caju* has been reported (Nagarajan *et al.*, 1988). For composting of one tonne of coir pitch, five kg urea and five bottles of *Pleurotus* are required.

Studies conducted at CPCRI resulted in the isolation of efficient strains of fungi with high lignocellulose degradation capabilities, from naturally decomposing coconut wastes. *Marasmiellus troyanus* and *Trochoderma* inoculations were effective in production of quality compost from coir pith. Large scale composting is standardized with amendment of lime and rock phosphate at 0.5% level each and *Glyricidia* leaves (10%) and fungal inoculation at 0.2% level (Thomas *et al.*, 2001). The raw coir pith with a C:N ratio of 100% is converted to an acceptable manure with a C:N ratio of 17:1 within a period of 40-45 days.

Coconut Wastes as Mulch in Plantations

Coconut leaves, husk and coir pith could be utilized as mulches to reduce loss of soil moisture and create conditions for proper root growth and proliferation of soil flora and fauna. Decomposition of the mulches after a period of time result in enrichment of soil organic matter pool. Burial of husk in trenches in between rows of palms is highly effective for moisture conservation in coconut gardens. Husk burying is done at beginning of monsoon in linear trenches of 1.5 to 2 m wide and about 0.3 to 0.5 m deep between rows of palms with concave side of husk facing upwards. Each layer is covered with soil. The husks are also used as a surface mulch around base of palms. Husks in a single layer are placed with convex side up around the palm upto a radius of 2 m from the base. Coconut husks are also important sources of potash which becomes available to palms over a period of time.

Basin Management with Leguminous Cover Crops

The technique for utilization of leguminous cover crops as green manures to supply biologically fixed nitrogen and easily decomposable biomass to coconut was standardized at CPCRI (Thomas and Shantaram, 1984). It involves cultivation of leguminous creepers such as *Pueraria phaseoloides*, *Mimosa invisa* and *Calopogonium mucunoides* in coconut basins during monsoon period from June to October and incorporation of legume biomass in respective basins. During a growth period of 140-150 days, the promising legumes generated 15-28 kg of biomass and 102-197 g of nitrogen in the basin of a coconut palm. A field experiment conducted in a coconut plantation in an acidic laterite soil type revealed feasibility of substituting upto 50% of fertilizer nitrogen with nitrogen contributed by leguminous green manures.

Growing Glyricidia as Green Manure Crop

Generation of large quantities of nitrogen rich biomass is also possible through cultivation of fast growing leguminous tree crop, *Glyricidia* in the coconut plantations. It can also be grown in littoral sandy soils where no other green manure can establish. The planting density and pruning frequency to obtain higher biomass yield from *Glyricidia sepium* when grown as green manure crops in coconut plantation in littoral sandy soil was standardized.

Biofertilizers

There are also indications that coconut palms can benefit from the use of beneficial micro-organisms as biofertilizers. Formulations containing living cells of beneficial micro-organisms multiplied in suitable carriers are biofertilizers which can form low cost ecofriendly inputs for crop production. The group of micro-organisms responsible for biological nitrogen fixation, phosphorus mobilization and production of plant growth promoting substances are found closely associated with coconut palms (Thomas *et al.*, 1991). The diazotrophic bacteria isolated from coconut roots include different species of *Azospirillum, Herbaspirillum* sp., *Azoarcus* sp., *Burkholderia* sp., *Arthrobacter* sp., *Pseudomonas* sp. and *Bacillus* sp. (Prabhu and Thomas, 1999, Thomas *et al.*, 2001), *Beijerinckia indica* was the predominant asymbiotic N_2 fixer in coconut soils. Phosphate solubilizing microbes of coconut soil include *Pseudomonas* sp., *Bacillus* sp. and *Micrococcus* sp. (Thomas and Shantaram, 1986). Efficient strains of nitrogen fixing bacteria and phosphate solubilisers were used for preparation of biofertilizers employing locally available materials vermicompost and coir pith as carrier materials. Inoculation trials indicated that biofertilizers were effective in enhancing vigour of coconut seedlings raised in polybags with 1:1 coir pith soil mixture as substrate (Thomas *et al.*, 2001).

Systems Approach for Sustaining Soil Productivity

Farming / cropping systems are designed with coconut as base crop with a large number of component crops and even animal enterprises to utilize natural resources to the maximum extent and achieve maximum productivity/ unit area of plantations (Bavappa *et al.*, 1986). These systems create favourable conditions and microclimate for multiplication and activity of a variety of beneficial organisms (Bopaiah, 1988). As a number of component crops are involved, soil resources are utilized to the maximum extent. As the biomass production/ unit area is very high, soil health and coconut yield can be sustained even in the absence of external input when the available organic wastes are recycled (Khan *et al.*, 2002).

On Farm Biomass Recycling in Coconut based Cropping System

In the premier coconut growing state of Kerala (India), many crop combinations with coconut have evolved over years to mainly satisfy home needs. This approach has changed over years and commercial inter / mixed cropping ventures are in vogue in the southern states of Andhra Pradesh, Karnataka and Tamil Nadu. Elephant foot yam, banana and cocoa are major crops. There is a renewed interest among farmers to grow vanilla, patcholi besides medicinal plants. Such coconut-based cropping/ farming systems also generate huge amount of biomass. Studies in India and else where, showed that a major part of these can be recycled in an effective manner for ecological sustenance. The biomass production by coconut-pineapple cropping system was 4.3 times that of pure coconut stands. This combination is one of the optimum cultivation patterns in tropical areas (Peng-Fang Ren *et al.*, 1996). Studies at CPCRI, Kasaragod revealed that cocoa as a mixed crop in coconut garden added to the soil 818 and 1985 kg (oven dry wt.)/ hectare/ year of biomass through shed leaves and pruning under single and double hedges systems respectively. As much as 50 kg N, 4.85 kg P and 29.1 kg K / hectare could be returned to the soil every year under the souble hedge cocoa system (Varghese *et al.*, 1978). Addition of large amount of organic refuses by the cocoa to improved the soil physico-chemical characteristics and thereby positively influenced the yield of coconut.

The total removable biomass (in terms of fronds, bunch waste, nuts in case of coconut, fallen leaves and flower buds in clove, above ground biomass for banana and fruits, crown and leaves in case of pineapple) from the system was 19.1 t/ha in the control plot and ranged from 20.11 to 27.55 t/ha under graded levels of fertilizer. In this the contribution from coconut alone was approximately 85%. Availability of biomass for recycling in the system was around 12.7 to 18.2 t/ hectare under various treatments. This biomass, if recycled ,can enhance productivity and sustenance of system in terms of nutrient need besides economic benefits.

From available data in coconut based farming system, it is apparent that system is highly dynamic with complex interactions. The high biomass production and its recycling triggers favourable biological activities which over long period of time can stabilize self generating power of the system (Bavappa, 2000).

In general, it may be noted that the benefit accrued in a cropping / farming system is not only in terms of savings in the cost of inorganic fertilizer, but most importantly in terms of soil health benefits (environment benefit?) which has not been quantified till date.

Cost of Cultivation/ Hectare of Coconut

S. No.	Components of different cost concepts	Cost per hectare (Rs.)	% distribution of cost 'A'
1	Hired human labour	8284	52.03
2	Animal labour	23	0.14
3	Machine labour	161	1.01
4	Seed/ seedlings	27	0.17
5	Farmyard manure and chemical fertilizers	4538	26.5
6	Plant protection	32	0.2
7	Land tax and irrigation cess	83	0.52
8	Repair and maintenance charges	624	3.92
9	Interest on working capital	1363	8.69
10	Other expenses	768	4.82
11	Total cost 'A' (1-10)	15923	100.00
12	Interest on fixed capital	1745	
13	Cost 'B1' (11+12)	17688	
14	Interest on land value	185581	
15	Cost 'B' (13+14)	203249	
16	Imputed value of household labour	1851	
17	Cost 'C' (13+14)	205100	

Source: Dept. of Economics & Statistics, Kerala

Percentage Distribution of Hired Human Labour Hours to the Total Human Hours

Sex	Holding Size Class			
	Small	Medium	Large	All sizes
Male	56.34	59.61	64.66	56.3
Female	8.99	30.55	14.7	13.78
Total	65.33	90.16	79.36	70.08

Source: Dept. of Economics & Statistics, Kerala

Value of Output / Hectare

Output	Value (Rs.)
Product	27989
By-product	2711
Total	30700

Source: Dept. of Economics & Statistics, Kerala

Harvesting and Processing

Twelve months old nuts are harvested for seed copra making. For tender nut purposes, 7 to 8 months old nuts are harvested. In tall, the nuts harvested for seed purpose can be stored for 2 to 3 months period before sowing, whereas in dwarfs and hybrids, nuts should be sown with in a period of 10 –15 days of harvest. On an average, we can have eight harvests, though the coconut palm produces inflorescence every month.

Products derived from the coconut palm include kernel (copra, oil, copra cake, copra meal, desiccated coconut, milk or cream) and husk (a series of white fibre and brown fibre products) and shell (charcoal, activated carbon, flour) which enter international trade. Fresh coconuts and seed nuts are also exported from certain countries.

Coconut Products

1. Copra
2. Coconut Oil
3. Raw Kernel
4. Coconut Cake
5. Tender Coconut Water
6. Coconut Toddy
7. Coconut Shell based Products
8. Coconut Wood based Products
9. Coconut Leaves
10. Coir Pith

Copra

Two forms of copra are made in the country namely milling copra and edible copra. Milling copra is made in the form of cups used for oil extraction. It is used as a dry fruit, consumed as such or with jaggery especially during winter, also used in social and religious functions throughout the country. Edible copra is shredded and used as a garnish in various dishes. Copra contains 65-72% oil. Sun drying takes about 6-8 days. Now a days, indirect hot air driers are being used to make milling copra. The drying time is 20-24 hours. Different types of dryers including electrical and solar dryers for small holders are available.

In drying process, the moisture content has to be reduced from 50-55% to 5-6%. Dried copra remains without any appreciable deterioration during summer but it gains moisture during storage during rainy months and undergo deterioration. Copra stored in alkalthene lined gunny bags remain unspoiled for a period of 6 months without any deterioration.

Edible Copra

Edible copra is made in the form of cups and balls.

Ball Copra

Fully ripe nuts of 12-14 months are stored in upper floor of a specially made store, usually a two storied brick and mortar building. The floor and four sides of upper story being made of wooden bars are spaced 6-9cm. The nuts are frequently stirred and smoked by a slow fire set under the platform using coconut palm wastes. During the period of storage, water inside the nuts gets dried up and kernel gets detached from shell and raffles on shaking. The entire process takes 8-12 months. Small sized nuts are the best compared to large sized one.

Edible Cup Copra

Nuts stored as above for a shorter period are used for making edible cup copra. The copra balls are cut into halves and they are sun dried for a few days.

Characteristic quality of copra

Items	Grade 1	Grade 2	Grade 3
Moisture content (%) by weight, max.	5	6	7
Oil content (on dry basis)% by weight, min.	64	62	60
Free fatty. acid (as lauric acid)(%) by weight, max.	1	2	5
Aflatoxin B1 content, mg/1000 kg	Not found	Not found	Not found
Mould pieces(%) by weight, max.	None	5	10
Impurities(%) by weight, max.	0.5	1	2
Charred or black pieces(%) by weight, max.	None	4	8
Colour of the expelled oil, max.	4	7	11

Source: http://www.foodmarketexchange.com

The Wet Meat or Kernel

The Kernel or endosperm of matured coconut is an important food in all the coconut growing countries. Apart from being used for culinary purposes, milk or cream obtained by squeezing grated kernel goes into preparations of commercial importance.

Nutritive Value of Coconut at Different Stages of Maturity / 100 g

	Coconut Water	Tender Coconut	Coconut Milk	Fresh	Dry
Moisture (%)	93.8	90.8	42.8	36.3	4.3
Calories	24	41	430	444	662
Protein (g)	1.4	0.9	3.4	4.5	6.8
Fat (g)	0.1	1.4	41	41.6	62.3
Carbohydrate (g)	4.4	6.3	11.9	13	18.4

Coconut Oil

Coconut oil is obtained by extracting oil in copra. The residue is generally called copra cake, and copra meal. Coconut oil is used for food and for industrial purposes. Edible oil, by international standards should have less than 0.1% free fatty acid content.

Coconut oil is used in the country as a cooking fat, hair oil, body oil and industrial oil. Coconut oil is made from fully dried copra having maximum moisture content of 6%. Steam cooking of copra is also practised by a few millers to enhance quality and aroma of oil. Coconut oil is marketed in bulk and in packs ranging from sachets containing 5 ml. to 15kg tins. The branded coconut oil in small packs is mainly marketed as hair oil and body oil. There are several brands known for their superior grade which have export market throughout the world. India has unbeatable quality advantage in this sector. Refined coconut oil is also manufactured in the country for industrial uses. Refined coconut oil is mainly used in the manufacture of biscuits, chocolates and other confectionery items, ice cream, pharmaceutical products and costly paints. Generally, filtered coconut oil is used for cooking and toiletry purposes.

Coconut Processing Technology

Coconut oil processing methods or technologies are classified into two major types: dry and wet processes. The oil extraction technology which starts with copra as raw material is termed as dry process, while the method that uses fresh coconuts as starting material is called wet process.

Oil extraction from copra is carried out by two methods: mechanically by pressing; or by use of solvents. Mechanical and solvent extraction use standard technology, developed in the vegetable oil industry. For copra with high oil content, mechanical extraction is efficient and economical and for further oil extraction from copra cake, the solvent extraction method is more suitable. Oil extraction involves five basic steps copra storage; preparation of copra; oil extraction - full press, prepress - solvent, and full solvent methods; processing of extracted oil; processing of cake or meal and storage of products.

Full Press Method

In full press method of oil extraction, maximum pressure is applied to the material and sufficient time is allowed for oil to escape. Temperature of material should not be allowed to rise to a level where oil darkens due to overheating.

Pre Press–solvent Method

In prepress-solvent process, oil is partially extracted by preliminary low pressure mechanical extraction, and then subjected to solvent extraction to remove most of residual oil. Oil content of initial copra material is around 70%; after the mechanical prepress extraction, residual oil content in pre pressed cake is 16 - 20% for optimum operation.

Full Solvent Method

In solvent extraction, oil in material is leached with a solvent (where the oil dissolves in solvent) whereas insoluble meal is retained unaffected. Extent of extraction depends upon kind of solvent, temperature of solvent, ratio of solvent to meal, number of extraction stages, shape of particles, porosity of material, and contact duration. The most common solvent used is hexane.

In full solvent process, the prepared copra is first subjected to a first extraction (percolation), using the weak micella from the second extractor as starting solvent, and producing the strong micella for oil recovery. The extracted meal is then flaked and subjected to a second extraction (immersion) which uses fresh solvent as starting solvent and produces the weak micella solvent for the first extractor. The solvent with the extracted oil is removed in a Desolventizer-Toaster (DT) which is either of the multi-decked vertical design with steam heated pans and paddle scrapers, or the horizontal barrel type with rotating conveying paddles attached to a horizontal shaft, and steam jacketed walls. These are equipped with condensers for solvent recovery and scrubbers to remove dust entrained in vapours.

The solvent with micella is recovered first by an evaporator, usually of the falling film type, where hexane is distilled off by indirect steam heating, and subsequently by a stripping column where traces of hexane in the oil are stripped by steam under vacuum. The water-hexane vapours from the stripper are condensed and collected in a water-hexane separator where hexane is separated by gravity from the water. It is decanted and reused in the extraction. Vent vapours of hexane from the extractors and condensers are recovered in a vent recovery system where hexane is absorbed by mineral oil. Hexane is recovered from mineral oil by stripping, and recycled to extractor.

Processing of Extracted Oil

Oil from the expellers contains substantial quantities of solids. The oil is cleaned first by settling and screening and subsequently by filtration. The screening equipment is a rectangular steel tank equipped with a continuous drag chain conveyor with scraper blades, which scoop settled solids and lift them over a fine screen for drainage at one end of screening tank, after which they are conveyed back to expellers to be mixed with copra. The filtering equipment generally cosists of a plate and frame filter press with canvass filtering media.

Processing of Cake or Meal

Copra cake leaving the expellers has a temperature of about 110 °C and is cooled in a cake cooler. After cooling, it is ground to fine particles by hammer mills or disc mills. The ground cake is bagged for local use or pelletized for export.

Storage of Products

Oil is stored in open in vertical cylindrical steel tanks. These tanks have conical covers and manholes at the top and bottom for cleaning. Capacity of these tanks depend upon capacity of plant, frequency and volumes of withdrawals, and marketing factors. The unpelletized copra cake should be bagged in woven sacks and stacked about 10cm high on wooden pallets.

Coconut Oil Quality

Typical quality specifications of coconut oil milled from good quality copra from Sri Lanka and Papua New Guinea and hot-air dried copra from the Pacific would have 1.0% maximum free fatty acid, 0.2% maximum moisture, and colour not deeper than 5. Typical free fatty acid value of oil extracted from copra from the Philippines and Indonesia will be about 3% and this is suitable for use as industrial-grade oil. The moisture level of stored coconut oil must be within reasonable limits as moisture acts as a catalyst to split oil into fatty acid and glycerine.

Properties of Coconut Oil

Fatty Acid

Lauric	:	49.5 %
Myristic	:	19.5
Palmitic	:	8.5

Dietary Oil

Coconut oil is an excellent cooking/ frying oil. Most of the saturated fats in coconut oil belong to short and medium chain fatty acids. It contains extremely low linoleic acid.

Cosmetic Oil

The natural flavour is due to presence of volatile materials like ketones and lactones. It is a skin friendly oil and has therapeutic applications. The unique feature of coconut oil is its low unsaturation, low viscosity, water white colour and unsaponifiable matter consisting of phytosterols, tocopherols and squalene. It is resistant to rancidity if kept free from moisture.

Coconut oil is a preferred base for making preparations for skin diseases of children. Coconut oil is preferred to be used as hair oil due to its low viscosity and pleasant aroma.

Industrial Oil

Coconut oil forms an important raw material for production of toilet soaps, liquid soaps, shaving cream and natural shampoos due to its high lauric acid content. Coconut oil soaps are known for its excellent lathering property even in hard and saline water. It gives hardness and stability than any other oil.

Coconut oil finds extensive use in food industry due to its characteristics like easy melting behaviour, resistance to oxidative rancidity, pleasing flavour and good digestibility. Coconut oil is preferred as a source of fat in preparations of filled milk, infant milk powder, ice-cream and confectionery and bakery products. Coconut oil when used for surface spray for biscuits gives a glossy appearance and provides a moisture barrier. It can be used for manufacturing margarine and shortenings.

Virgin Coconut Oil

Oil obtained from coconut milk is called virgin coconut oil. In traditional method the milk extracted from grated coconut kernel is boiled to get oil. In modern method (wet processing), kernel is made into a viscous slurry and milk is separated by pressing. The milk is sieved and concentrated. Cream is separated from milk and cream is stirred vigorously to get virgin coconut oil by a process called phase inversion. It is superior in quality for use as hair oil and baby oil due to its pleasing aroma and purity.

By Partially Mechanized Method

Coconut kernel is grated using a rotary grater, crushed in a bridge press to extract milk. The extracted milk is boiled in an open pan until oil separation takes place. The residue is used for making sweets, chutney and powder.

Health Benefits of Coconut Oil

Coconut oil used for centuries as a vital source of food for health and general well being in traditional communities of tropical regions. Over many decades coconut oil received bad publicity due to its saturated fat content, but research showed that not all saturated fats are alike and coconut oil is unique in its structuaral make up.

Convenience Products from Raw Kernel

Desiccated coconut (DC), Coconut Cream, Coconut Milk and Spray Dried Coconut Milk Powder are the convenience products manufactured in the country. Desiccated coconut is used as a substitute to grated raw coconut in various food preparations. Desiccated coconut is marketed in bulk in small packs. Defatted desiccated coconut is also available in the country.

Processed coconut cream/ coconut milk are used in various food preparations as a substitute to milk extracted from raw kernel in traditional method. They are available in cans and tetra packs. Spray drying is the best method for preservation of coconut milk. The product has advantages like less storage space, bulk packaging possible at low cost and long shelf life. Spray dried coconut powder is manufactured by one unit in the country.

Desiccated Coconut

Desiccated coconut (DC) is the white kernel of fresh mature coconuts, shredded and dried down to about 2.5% moisture content under strict hygienic conditions. It is used for human consumption, and it retains the original oil and protein of a fresh nut. Four standard grades based on particle size are produced: extra fine; fine medium and coarse.

Processing

The fresh matured coconuts are de-husked and de-shelled. The de-shelling is done by a special knife to get the kernel in the form of a ball, which is pared using scrapers to remove the testa. The pared kernel balls are then cut open to drain off water and then washed thoroughly in fresh water to remove invert sugars from inner surface of kernel. The kernel is then ground into a fine mass using hammer or pin mills. The ground mass is blanched with live steam for about 20 minutes to bring down the microbiological counts. The blanched mass is then dried in a hot air drier at a temperature of 80-90°C for about 10 hours to bring down moisture content to below 3 %. The hot air drier is provided with a drying chamber consisting of a series of trays, which hold the feed. Hot air is blown into this chamber from an external source through blowers. The dried mass is tested for moisture, free fatty acid and microbiological counts. The product is packed in polythene pouches. The brown testa obtained during paring of kernel can be dried and the oil extracted, known as paring oil.

Physical Requirements: Colour - Natural white; Foreign matter - Shall be free from all foreign matter; Taste and smell - Shall be sweet, pleasant and free from cheesy, smoky, soapy, sour or other undesirable flavours.

Size/grade: There are 4 grades. The sieve analysis for each grade based on sieve aperture and BSS sieve is given below:

(a) Coursegrade

4.76 mm 100% pass through;
3.35 mm 5-15% max retained;
2.00 mm 8-15% max pass through;
1.40 mm 12-2.5% max pass through.

(b) Mediumgrade

0.80 mm 6-100% pass through;
2.00 mm 8-15% max retained;
1.40 mm 12-15% max pass through;
1.00 mm 16-2.5% max pass through;
-Free fatty acids (FFA) 0.3% max (as lauric).

(c) Finegrade

1.88 mm 10-100% pass through
1.40 mm 12-15% max retained;

(d) Super fine grade

1.00 mm 16-100% pass through;

Special grades and cuts may be manufactured to meet customers' specifications, provided they comply with other requirements.

Chemical Requirements

- Moisture 3% max for above 4 grades and 3.5% max for special grades.
- Oil content 68% min.

Coconut Milk or Cream

Coconut milk or cream refers to oil-protein-water emulsion obtained by squeezing grated fresh coconut kernel. It is the processed milk extracted from fresh matured coconuts. This is an instant product, which can either be used directly or diluted with water to make various preparations like curries, sweets, desserts, puddings, etc. It can also be used in the manufacture of bakery products and for flavouring food stuffs. Processed and packed coconut cream has a shelf life of six months and once opened it should be stored in refrigerator for subsequent uses. The technology was developed by the RRL, Thiruvananthapuram.

The dehusked nuts are split into halves and deshelled to separate the kernel. These two operations usually are done manually. Kernel is washed and then blanched by immersing in hot water at 80°C for 10 minutes. The next step is comminution of kernel into small gratings using a hammer mill. The gratings are subjected to pressing using continuous screw press to extract the milk. The coconut milk thus obtained is filtered by passing through a vibratory screen. Food additives like emulsifiers and stabilizers are added to milk to obtain a stable consistency and texture. For this purpose, permitted emulsifiers and stabilizers are mixed with hot water separately and mixed thoroughly. This is added to coconut milk and then subjected to emulsification using a mechanical impeller emulsifier. The emulsified milk assumes a creamy consistency. The coconut cream is then pasteurized at 95°C for 10 minutes in a plate heat exchanger. The pasteurized coconut cream is hot filled in cans using a mechanical volumetric filling machine followed by steam exhausting. The cans are seamed using an automatic can seamer. The seamed cans are sterilized in a rotary retort at 15 psi for 20 minutes. The cans are then cooled in running water.

The residue obtained after extraction of milk is dried in a hot air dryer to 3% moisture level. The residue forms a base for making coconut burfi, coconut cookies, curry and chutney powders. Other by-products like coconut water and coconut shell could be processed into value added products. Vinegar and Nata-de-coco manufactured from coconut water. Shell charcoal and shell powder can be manufactured from coconut shells. The utilization of byproducts would improve economic feasibility of the process.

Instant Coconut Milk Powder

Instant coconut milk powder, is a spray- dried product similar to dairy milk powder. Spray dried coconut milk powder, if dissolved in water, results in coconut milk which can be used in place of fresh coconut milk for food preparations/ beverages in households and food industries by dissolving it in water This is packed in triple laminate pouches and has a shelf- life of over 6 months. The major producers can be found in the Philippines, Sri Lanka and Malaysia. A by-product of this industry is de-fatted coconut (residue) which is a low-priced substitute or diluent for desiccated coconut.

Coconut Sap and Products

Fresh (unfermented) sap (toddy), obtained from unopened spadix is sweet due to sugar content, and consumed as a beverage. The sap undergoes alcoholic fermentation due to yeast in the environment. When fermenting, the alcohol content rises and finally the sap turns sour due to acidification of alcohol. If the sap is collected for its sugar, fermentation is inhibited by applying lime to the earthernware pots.

Coconut Syrup and Sugar

Unfermented sap, when gently boiled, yields a syrup or treacle or honey (65% total sugar min.) which is a traditional food item in coconut producing countries like India, Sri Lanka, Indonesia, Thailand and Malaysia. By boiling, volume of liquid is reduced to about 1/6th of its original volume, at which stage, a viscous syrup is obtained. It is golden yellow to reddish brown due to caramelization of some sugar. It is consumed just like golden syrup from sugar cane.

Coconut sugar is obtained by further boiling, reducing the liquid to about 1/7th of its original volume. This sugar is also called palm sugar (jaggary). It contains a minimum of 70% non-reducing sugars and a maximum of 13% reducing sugars. Palm sugar is also used in various traditional food preparations.

Fermented Coconut Sap and Spirit

The fermented sap (sometimes called coconut wine) is consumed as an alcoholic beverage in coconut producing areas, contains about 7.5% alcohol.

Alcohol can be made by distillation of fermented toddy. The alcohol may be diluted as desired for production of alcoholic beverages. This arrack contains about 32% alcohol. It has a typical brown colour, characteristic of other hard liquor such as whisky.

Coconut Vinegar

Technology is developed by CFTRI, Mysore for the production of vinegar from matured coconut water using vinegar generators. Vinegar has extensive use as a preservative in pickle industry and flavouring agent in food processing sector. Natural vinegar enjoys export market in place of synthetic vinegar

The matured coconut water consisting of about 3% sugar content is concentrated to 10% level by fortifying with sugar. The fortified coconut water is then fermented by inoculating the solution with yeast,(*Sacharomyces cerviseae)*.

After alcoholic fermentation for 4 to 5 days, the clear liquid is siphoned off and inoculated with mother vinegar containing *Acetobacter* bacteria. The alcoholic ferment obtained is then fed into a vinegar generator, where the feed is uniformly sprayed over the surface of porous packing medium (corn cobs). Here the alcoholic ferment is oxidised to acetic acid. The product is run out from packing medium by gravity flow into the receiving vat from where it is recycled into the vinegar generator and the process of acetification is repeated until a strength of 4% is attained. This acetified vinegar is then aged before bottling.

Coconut Water Based Products

Tender Coconut

The tender coconut is valued both for sweet water and delicious gelatinous kernel. Tender coconut water, which is a refreshing drink, possesses many therapeutic properties.The drink is at its optimum level of acceptability when nuts are of 6-7 months of maturity. It has caloric value of 17.4/ 100g.

Packed Tender Coconut Water

Tender coconut water can be processed and packed for marketing as a nutritious soft drink. The Defense Food Research Laboratory (DFRL), Mysore developed a technology for preservation and packing of tender coconut water in aluminium cans and pouches. Shelf life of the product is three months under ambient condition and six months under refrigerated condition. The CDB, Cochin has transferred the technology to seven firms and three have commercialized the technology.

Snow Ball Tender Coconut

It is the whole round soft kernel pulled out from shell with the water intact in it. Eight months old coconuts are used for the purpose. It can be packed in LDPE film of 200 gauge, which can be stored for 15 days at about 15°C under refrigerated condition without affecting its keeping quality (Technology developed by CPCRI,Kasaragod).

Comparison of Mature and Tender Coconut Water

	Mature Coconut Water	**Tender Coconut Water**
Total solids(%)	5.4	6.5
Reducing sugars(%)	0.2	4.4
Minerals (%)	0.5	0.6
Protein (%)	0.1	0.01
Fat(%)	0.1	0.01
Acidity mg	60.0	120.0
pH	5.2	4.5
Potassium mg	247.0	290.0
Sodium mg	48.0	42.0
Calcium mg	40.0	44.0
Magnesium mg	15.0	10.0
Phosphorous mg	6.3	9.2
Iron mg	79.0	106.0
Copper mg	26.0	26.0

Source: Satyavati Krishnankutty (1987)

Amino Acid Composition of Coconut Water (% of Total Protein)

Alanine	2.41
Arginine	10.75
Aspartic acid	3.60
Cystine	0.97 - 1.17
Glutamic acid	9.76 - 14.5
Histidine	1.95 - 2.05
Leucine	1.95 - 4.18
Lysine	1.95 - 4.57
Proline	1.21 - 4.12
Phenylalanine	1.23
Serine	0.59 - 0.91
Tyrosine	2.83 - 3.00

Source: Pradera et al, (1942)

Vitamins of B complex in Coconut Water

Nicotinic acid	0.64 microgram / ml
Pantothenic acid	0.52 ,,
Biotin	0.02 ,,
Riboflavin	< 0.01 ,,
Folic acid	0.003 ,,
Thiamine	Trace ,,
Pyridoxine	Trace ,,

Nata de Coco

This is a gelatinous dessert produced by the bacteria *Acetobacter aceti* in coconut water, and the process was developed in Philippines. The 'nata', when formed, is cooked in thick sugar syrup and often served with fruit. It is composed mainly of polysaccharides, probably dextrose and to be cellulous in nature.

Initially coconut water is strained mixed with sugar and glacial acetic acid in stipulated proportions. Boil for ten minutes and cool. Add culture solution and distribute the mixture in wide mouthed glass or plastic jars, cover the jar with a paper or a thin cloth to protect from dust. It is then kept aside undisturbed for two to three weeks. After this period, the white jelly like thick surface growth is harvested, washed thoroughly to remove all acids and sliced into cubes. It is then immersed in flavoured sugar solution, again boiled and packed in glass jars or retortable pouches, sterilized and sealed.

Coconut Jam

Coconut jam is a traditional high-sugar coconut food product in the Philippines, commonly used as a dessert, bread spread, and rice cake topping (Gonzalez, 1991). It is light to dark brown in colour, thick yet spreadable in consistency and with a rich creamy coconut flavour. It is traditionally prepared by cooking sweetened coconut milk to a very thick consistency at low heat with constant stirring.

Twelve mature coconuts are required to prepare 6 standard jam bottles of 450g each. In this process, milk is extracted from grated coconut and water mixed in equal parts. A small portion (1/2 cup or 125 g) of coconut milk is set aside and citric acid is added at a rate of 0.25% of total weight of coconut milk. Bulk of coconut milk is mixed with brown sugar and glucose, at 10.25% and 5.5% of weight of coconut milk, respectively. The mixture is boiled over low heat with constant stirring for about 20 minutes. The mixture is strained to remove suspended matter, and then boiled over high heat with constant stirring. Just before the mixture thickens, remaining coconut milk and citric acid mix are added, and cooking is continued over low heat until the mixture thickens. The hot product is poured into sterilized bottles and sealed hermetically. Coconut jam is a creamy, free-flowing, sweetened product, with a moisture content of about 25%, fat 4%, protein 6%, and total solids 75%.

White Soft Cheese from Coconut and Cow Milk

White soft cheese made from skim milk (SM) extended with coconut milk (Sanchez and Rasco, 1983), at the Institute of Food Science and Technology, University of the Philippines, Los Banos, Laguna. The mixture of 40% SM and 60% coconut milk yielded white soft cheese comparable to that made from 100% cow milk in terms of flavour, aroma, texture, and general acceptability. (The mixture of 50% SM and 50% coconut milk, though less economical, also produced an equally good soft cheese.

Coconut Shells Products

Activated Carbon

Activated carbon is a non-graphite form of carbon which could be produced from any carbonaceous material like coal, lignite, wood, paddy husk, coir pith and coconut shell. Activated carbon manufactured from coconut shell is superior to those obtained from other sources mainly because of small macropores structure which renders it more effective for adsorption of gas/ vapour and for removal of colour and odour of compounds.

The activated carbon is extensively used in refining and bleaching of vegetable oils and chemical solutions, water purification, recovery of solvents and other vapours, recovery of gold, in gas masks for protection against toxic gases, and in filters for providing adequate protection against war gases/ nuclear fall outs.

Steam and chemical activations are the two commonly used processes for manufacture of activated carbon. Coconut shell based activated carbon units are adopting steam activation process to produce good quality activated carbon.

The process of activation is carried out in two stages. Firstly the coconut shell is converted into shell charcoal by carbonization process which is usually carried out in mud-pits, brick kilns and metallic portable kilns. The coconut shell charcoal is activated by reaction with steam at a temperature of 900 - 1100°C under controlled atmosphere in a rotary kiln. The reaction between steam and charcoal takes place at the internal surface area, creating more sites for adsorption. The temperature factor, in the process of activation, is very important. Below 900°C, the reaction becomes too slow and is very uneconomical. Above 1100°C, the reaction becomes diffusion controlled and therefore takes place on outer surface of charcoal resulting in loss of charcoal.

Coconut Shell Powder

Coconut shell powder is manufactured from matured coconut shells. The manufacture of coconut shell powder is not an organised industry in India. The product finds extensive use in plywood and laminated board industry as a phenolic extruder and as a filler in synthetic resin glues, mosquito coils and agarbathis. Coconut shell powder is preferred to other alternate materials available in the market like bark powder, furfural and peanut shell powder because of its uniformity in quality and chemical composition, better properties in respect of water absorption and resistance to fungal attack.

The product is manufactured in sizes ranging from 80-200 mesh. Keeping in view of the vast industrial uses, the demand for coconut shell powder appears to be promising.

Coconut shells free from contamination of coir pith, are broken into small pieces and fed into a pulveriser. The powder from the pulveriser is fed into a cyclone and the parallel product is collected in bag filters. The shell powder is then fed into a vibrating sieving machine and packed according to mesh size requirements for various end uses.

Coconut Wood Based Products

Coconut wood is traditionally used for making roofing components like rafter joints, purlins, beams, banisters, etc. It is also used for the construction of temporary sheds, cow sheds, workshop buildings, farm buildings and small rural bridges. Because of its distinct grain characteristics, it is ideal for making wall panels, furniture, door, and windows and show pieces.

There are two general methods for preserving coconut wood.

1. Pressure treatment
2. Non-pressure treatment

1. High pressure sap displacement treatment (HPSD) system works by forcing out the sap from living sap wood of a freshly felled tree using water borne preservative solution.
2. Non-pressure treatments are brushing or spraying, dipping, soaking/ steeping, dip diffusion, double diffusion and hot and cold bath treatment.

Coconut Leaves

Coconut leaves are plaited and used for thatching houses and sheds in rural areas. It is also used for thatching 'honeymoon huts' in towns and cities. Technology is available with Regional Research Laboratory, Thiruvananthapuram, Kerala for extending life of leaf thatch from one year to four years. Plaited coconut leaves are also used for making baskets, headgears and for erection of temporary fences. Plaiting of coconut leaves is a cottage industry in traditional coconut growing states. Midribs of leaves are used to make brooms of different types which are used for cleaning rough grounds and floors. Brooms of midribs of coconut leaves are manufactured on a commercial scale in Tamil Nadu and Karnataka.

Coir

It is a versatile natural fibre extracted from mesocarp tissue,or husk of nut. Generally fibre is of golden color when cleaned after removing from coconut husk and hence the name " The Golden Fibre". The husk contains 20 to 30% fibre of varying length. After grinding the husk, the long fibres are removed and used for various industrial purposes, like rope, floor and outdoor mats, aquarium filters, cordage, and garden mulch. The remaining materials, composed of short and medium-length fibres as well as pith tissue, are commonly referred as waste-grade coir. The waste grade coir is screened to remove part or all of the fibre, and the remaining product is referred to as coir pith.

Structure of Coir Fibre

The individual fibre cells are narrow and hollow, with thick walls made of cellulose. They are pale when immature but later become hardened and yellowed as a layer of lignin, is deposited on their walls. Mature brown coir fibres contain more lignin and less cellulose compared to fibres like flax and cotton and so are stronger but less flexible. They are made up of small threads, each less than 1.3 mm long and 10 to 20 micrometres in diameter. White fibre is smoother and finer, but weaker. The coir fibre is relatively waterproof and is the only natural fibre resistant to damage by salt water.

Green coconuts, harvested after about six to twelve months on the plant, contain pliable white fibres. Brown fibre is obtained by harvesting fully mature coconuts when the nutritious layer surrounding the seed is ready to be processed into copra and desiccated coconut. The fibrous layer of fruit is then separated from hard shell (manually) by driving fruit down onto a spike to split it (De-husking). Machines are now available which crush the whole fruit to give loose fibres.

Brown Flibre

The fibrous husks are soaked in pits or in nets in a slow moving body of water to swell and soften fibres. The long bristle fibres are separated from shorter mattress fibres underneath the skin of nut, a process known as wet-milling. The mattress fibres are sifted to remove dirt and other rubbish, dried and packed into bales. Some mattress fibre is allowed to retain more moisture so that it retains its elasticity for 'twisted' fibre production. The coir fibre is elastic enough to twist without break and it holds a curl as though permanently waved. Twisting is done by simply making a rope using machine or by hand. The longer bristle fibre is washed in clean water and then dried before being tied into bundles or hunks. It may then be cleaned and 'hackled' by steel combs to straighten fibres and remove any shorter fibre pieces. Coir bristle fibre can also be bleached and dyed to obtain hanks of different colours.

White Fibre

The immature husks are suspended in a river or water-filled pit for up to ten months. During this time, micro-organisms break down the plant tissues surrounding the fibres to loosen them - a process known as retting. Segments of husk are then beaten by hand to separate out long fibres, which are subsequently dried and cleaned. Cleaned fibre is ready for spinning into yarn using a simple one-handed system or a spinning wheel.

Uses / Applications

Brown coir is used in brushes, doormats, mattresses and sacking. A small amount is also made into twine. Pads of curled brown coir fibre, made by needle-felting (a machine technique that mats the fibres together) are shaped and cut to fill mattresses and for use in erosion control on river banks and hillsides. A major proportion of brown coir pads are sprayed with rubber latex which bonds the fibres together (rubberized coir) to be used as upholstery padding for automobile industry in Europe. The material is also used for insulation and packaging. The major use of white coir is in rope manufacture. Mats of woven coir fibre are made from finer grades of bristle and white fibre using hand or mechanical looms. Coir is recommended as substitute for milled peat moss because it is free of bacteria and fungal spores.

Importance of Coir

Moth-proof; resistant to fungi and rot.

Provides excellent insulation against temperature and sound.

Not easily combustible.

Flame-retardant.

Unaffected by moisture and dampness.

Tough and durable.

Resilient; springs back to shape even after constant use.

Totally static free

Easy to clean.

Physical Properties of Coir Fibre

Ultimate length		0.6 mm
Diameter/width		16 micron
Single fibre	Length	6 to 8 inches
	Density	1.4 g/cc
	Tenacity	10 g/ tex
Breaking Elongation		30%
Moisture regain at 65% RH		10.5%
Swelling in water		5% in diameter

Chemical Properties (Composition) of Coir Fibre

Water solubles	5.25%
Pectin & related compounds	3.30%
Hemi-Cellulose	0.25%
Cellulose	43.44%
Lignin	45.84%
Ash	2.22%

In most areas coir is a by-product of copra production, and the husks are left on the fields as mulch or used as fertiliser due to high potash content. For production of light coloured fibre of spinnable quality, green husk of 10 to 12 months old coconuts is ideally suitable. India and Sri Lanka are the main areas where the fibres from the husk (termed 'coir') are extracted by traditional methods for commercial production of a variety of products (brushes and brooms, ropes and yarns for nets and bags and mats and padding for mattresses). World wide, only a small part of the fibres available are currently used for these purposes.

Fibre Extraction

The processes of fibre extraction are varied, and depend on effectiveness of the wet processing such as bleaching and dyeing of coir and also varied end uses.

Traditional Fibre Extraction

The traditional production of fibres from husks is a laborious and time-consuming process. The husks are processed by various retting techniques generally in ponds of brackish waters (for three to six months) or in backwaters or lagoons. This requires 10-12 months of anaerobic (bacterial) fermentation.By retting, the husks are softened and can be decorticated and the fibre is extracted by beating, which is usually done by hand. After hackling, washing and drying (in the shade) the fibres are loosened manually and cleaned. The remaining residual pith - which was previously considered a waste problem - has recently found new profitable markets as a peat moss substitute for horticultural production. Traditional practices of this kind yield the highest quality of (white) fibre for spinning and weaving. Retted fibres from green husks are the most suitable fibres for dyeing and bleaching. For the production of more coarse brown yarns, shorter periods of retting may be applied. These find an increasing outlet in geo-textile applications.

Mechanical Extraction

Alternatively, either de-fibering or decorticating equipment processes husks after five days of immersion in water tanks. Crushing the husk in a breaker opens the fibres. By using revolving "drums", the coarse long fibres are separated from short woody parts and pith. The stronger fibres are washed, cleaned, dried, hackled and combed.

Green Decortication and Microbial Treatments

New environment friendly methods for fibre production are of interest. These can be locally exploited on relative small-scale, and have the potential to produce a more constant quality of fibres. Novel developments by the Central Coir Research Institute, Alappuzha, under Coir Board of India, using a bio-technological approach with specific microbial enzymes reduced the retting time substantially to three to five days. High quality fibre production has been maintained.

Similar protocols can be developed to enhance properties of fibres in regard to surface properties like smoothness and porosity. By using specific (microbial) lignolytic enzymes (laccase/phenoloxidase), the fibre surface can be bleached (or activated to react more easily with the dyes).

Coir Pith

Coir pith, a waste product obtained during extraction of coir fibre from husk is very light, highly compressible and highly hygroscopic. It is used as a soil conditioner, surface mulch/ rooting medium and desiccant. Composted coir pith is an excellent organic manure for indoor plants and for horticulture crops. Several firms are manufacturing composted coir pith in the country. Compressed coir pith in the form of briquettes for easy transportation is also manufactured in the country.

Product Specification

pH Value	6.5 - 7.5
Methylene Value adsorption mg / gm	190 - 350
Adsorption capacity at % by mass (min)	45
Moisture (max.)	5%
Ash (max)	5%
Hardiness	90

Marketing

Currently India stands third in the list of maximum coconut producing countries of the world. The coconut produce in India was 9.5million tones i.e. 12.8billion nuts in 2005. The area under coconut sums up to 1.78 million ha. in India. States in India indulging in production with their annual average production figures are Kerala (5.7 billion nuts), Tamil Nadu (3.24 billion nuts), Karnataka (1.20 billion nuts), Andhra Pradesh (1.19 billion nuts), West Bengal (0.31 billion nuts), Orissa (0.27 billion nuts), Maharashtra (0.27 billion nuts) and Assam (0.15 billion nuts).

Kerala is the largest producer of coconuts in India contributing to around 45% share in the country's produce. Coconut oil is obtained from copra with 65 to 70% recovery rate.

Domestic Demand

At present India produces about 10 lakh tonnes of copra annually of which about 2 lakh tonnes are of confectionery or edible grade and 8 lakh

State wise Area and Production of Coconut in India

	2003-2004 (Revised) @			2004-2005 (Final)		
States/ Union Territories	AREA ('000 ha)	Production (Million nuts)	Productivity (Nuts/ha)	AREA ('000 ha)	Production (Million nuts)	Productivity (Nuts/ha)
Andhra Pradesh	104.0	1195	11490	104.0	1199.3	11532
Assam *	21.3	154.3	7244	21.3	154.3	7244
Goa	25.1	122.2	4869	25.2	123.5	4901
Gujarat	16.0	111.7	6981	16.4	138.3	8433
Karnataka	376.0	1529.1	4067	385.4	1209.6	3139
Kerala	906.2	5484.0	6052	897.8	5727.0	6379
Maharashtra*	18.0	273.4	15189	18.0	273.4	15189
Nagaland	0.9	1.2	1333	0.9	1.2	1333
Orissa	55.4	243.4	4394	50.8	274.8	5409
Tamil Nadu	352.7	2560.5	7260	357.1	3243.5	9083
Tripura *	3.3	7.0	2121	3.3	7.0	2121
West Bengal	24.4	317.5	13012	24.3	310.9	12794
A &N Islands	25.4	95.2	3748	25.6	87.1	3402
Lakshadweep*	2.7	53.0	19630	2.7	53.0	19630
Pondicherry	2.3	30.7	13348	2.2	30.0	13636
All India	**1933.7**	**12178.2**	**6298**	**1935.0**	**12832.9**	**6632**

Source: Directorate of Economics & Statistics, Ministry of Agriculture, Govt. of India.
@ Revised / Final estimates for Gujarat, Kerala.

tonnes milling grade. The edible copra produced in Malabar region is specially preferred in India and abroad because of its attractive colour and sweetness.

In recent years, use of coconut oil by the industry is declining because of its high price in the country. The toilet soap industry in India uses coconut oil only if its price is below Rs. 35/kg. If it is more, there are substitutes available which are much cheaper to use than coconut oil. At present, India produces about 11614 tones of coconut oil annually, of which about 50% is absorbed by the toiletry sector (in hair oils, toilet soaps, shampoos, etc), 35 % by the households for cooking purposes and 15% is used by confectionery and ice-cream industry. Earlier the toilet soap industry used to consume 23% of the coconut oil produced in the country but now its consumption has dropped to just 5%. The use of coconut oil as hair oil in India, however, has been rising.

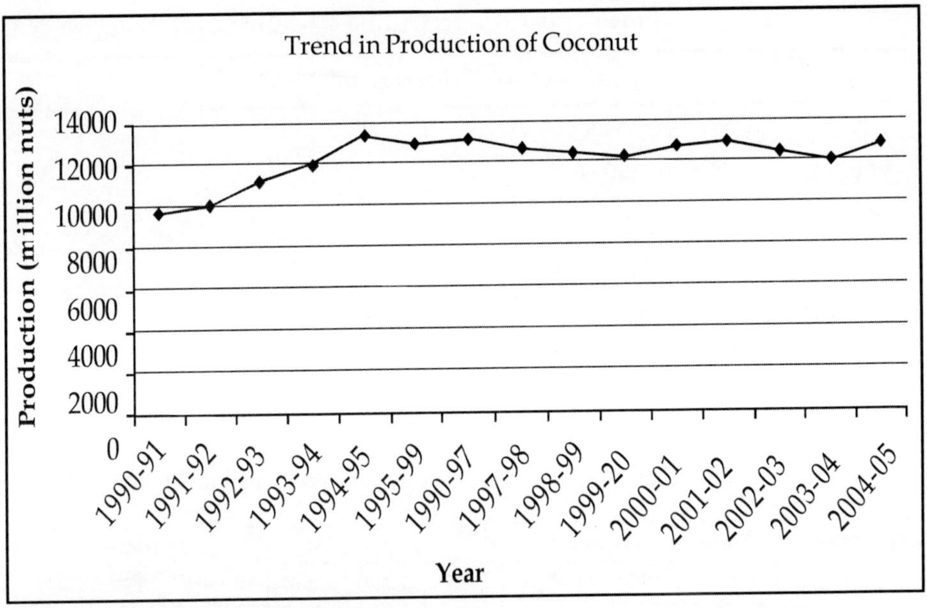

http://coconutboard.nic.in/stat.htm

Price Analysis of Coconut

Coconut oil prices are on the decline and one of the main reasons for this price fall was the uncontrolled import of palm oil and other edible oils which are good substitutes to coconut oil.

International prices of copra and coconut oil are far below domestic prices. The difference in coconut oil price between domestic and international makes a negative impact on domestic market. In the international market, crude coconut oil is available at around Rs 21,000 a tonne and when it was refined it could be sold at Rs 25,000 per tonne. The ruling domestic price of coconut oil ranged between Rs 45,000 and Rs 48,000 a tonne in May 2006.

Since price of coconut oil in the international market is much lower than domestic price, quality and attractiveness of consumer packs are important factors to compete in world market.

The regional analysis reveals that the labour cost of cutting, de-shelling and drying one load (25,000 shells) of coconut in Kangeyam in Tamil Nadu was only Rs 1,250 and hence the units were taking coconut for processing compared to Kerala. Data on average farm price of coconut grown in Kerala show increase in 2004-05 compared to previous year in respect of coconut.

Average Farm Price of Coconut During 2000-01 to 2004-05

Year	Average Farm Price (Rs/100 Nos)
2000-01	281.43
2001-02	340.64
2002-03	475.63
2003-04	582.73
2004-05	635

Source: Directorate of Economics and Statistics

Annual Average Retail Prices of Coconut with Husk in Kerala 2003 to 2006

Commodities	Unit	Prices at the end of December (Rs)					(%) Variation over previous year		
		2003	2004	2005*	2005	2006*	2004	2005	2006*
Coconut oil	Kg	70.86	71	64	63.80	56	0.2	-10	-13
Coconut with husk	100 Nos	689.04	676	628	625.8	555	-2	-7	-12

Export Pattern

India's Share in World Trade

Though India is among the leading producers of coconut in the world, its relative share in the international trade of coconut products is insignificant compared to other major producing countries like Philippines, Indonesia and Sri Lanka.

India trades in major coconut products like copra, coconut oil, desiccated coconut, copra meal (coconut cake) and coir. However, compared to production of coconuts, the quantity traded is quite low. Moreover compared to other crops in Kerala, coconut products are relatively less traded.

The total value of India's exports of coconut products increased from Rs. 25.30 crores in 2001-02 to Rs.44.74 crores in 2004-05, which is quite insignificant, given the area under the crop and level of production. Of the 11 important items related to coconuts exported by India, coconut oil accounted 61.2% of the value of exports in 2004-05. Shell charcoal and dried coconut contributed 13.6% and 5.7% respectively.

Export of Coconut Products from India (Qty. in tonnes, value Rs. in Lakhs)

Item	2003-04		2004-05		2005-06 (Upto September 2005)	
	Qty.	Val.	Qty.	Val.	Qty.	Val.
Coconut (fresh)	610.32	70.08	935.34	100.73	595.19	49.96
Coconut (dried)	594.61	229.07	583.80	256.28	387.15	153.49
Desiccated coconut	332.07	68.56	431.93	49.28	170.23	25.3
Other coconuts excluding fresh/dried	196.98	55.12	796.42	168.46	305.92	93.94
(a) Coconut oil (crude)	224.33	101.46	434.61	204.72	51.36	47.15
(b) Coconut oil (refined)	5789.24	2482.93	5519.75	2739.69	2180.14	1121.55
Other residues of coconut or copra	196.00	16.26	3049.76	210.75	110.11	6.37
Oilcake (defatted/expellers)	285.62	32.94	1229.43	57.05	142.70	8.25
Coconut shell (raw)	266.36	83.16	301.61	75.50	347.05	95.20
Shell charcoal	9351.00	1109.12	2976.00	610.44	17830.00	290.40
Shell hukah	73.71	12.73	4.40	1.37	1.93	0.67
Total		4261.43		4474.27		1892.28

Source : Director General of Commercial Intelligence and Statistics, Kolkata (Quoted by Coconut Development Board)

India exported 6014 tonnes of coconut oil in 2003-04 which declined marginally to 5954 tonnes in 2004-05. Value of exports rose from Rs. 25.89 to Rs 29.44 crores during this period.

India accounts for 0.6 to 0.7% of the trade in copra at global level. The aggregate export of copra was 1.15 lakhs tonnes at the gobal level valued $45 million in 2004. India's export in 2004 was only 761 tones valued at $0.285 million equivalent to Rs 128.25 crores .India exported 583.80 tones of copra in 2004-05 valued at Rs 2.56 crores. Between April and September 2005, India exported 387 tones worth Rs. 1.54 crores only.

Desiccated Coconuts

India exports very small quantity of desiccated coconuts which rose from 144 tonnes in 2000 to 432 tonnes in 2004. This accounts for a very insignificant proportion (0.15%) of the total world export of desiccated coconut of 2.83 lakhs tonnes in 2004. In value terms, India's export share was around 0.05%.India exported 432 tonnes of desicated coconut in 2004-05 valued at

Rs 49 lakhs. During the first six month of 2005-06, India exported 170 tonnes valued at Rs 25 lakhs.

Import Pattern

India had a trade deficit in coconut trade as imports were much higher. Imports increased from Rs.78.81 crores to Rs.104.99 crores between 2002-03 and 2004-05. The trade deficit grew from Rs.16.05 to Rs.63.64 crores during this period.

Import of Coconut Products into India (Qty. in tonnes, value Rs. in Lakhs)

Item	2003-04		2004-05		2005-06 (upto September 2005)	
	Qty.	Val.	Qty.	Val.	Qty.	Val.
Coconut fresh	7.00	0.42	15	1.18	0	0
Coconut (dried)	11.00	3.70	416.15	120.92	0.00	0.00
Other coconut excluding fresh/dried	0.50	0.73	654.21	67.61	1.92	0.36
Desiccated coconut	3049.36	827.35	8208.01	2795.35	625.94	275.17
(a) Coconut oil (crude)	37.25	10.02	1285.35	416.01	107.00	34.45
(b) Coconut oil (refined)	13722.71	2969.18	11427.14	3532.33	1524.00	464.01
Other residues of coconut or copra	1324.82	96.90	2532.00	188.00	2570.25	173.37
Oilcake (solvent extracted and expeller variety)	69262.80	3973.40	50651.68	3314.08	42285.52	2251.52
Coconut shell charcoal	1.00	0.23	563.00	63.71	1086.00	132.56
Total		7881.93		10499.19		3331.44

Source : Director General of Commercial Intelligence and Statistics, Kolkota (Quoted by Coconut Development Board)

Import of Coconut Oil

India imports substantial quantity of coconut oil. The world imports of coconut oil ranged 2.0 to 2.26 million tones, of which India's share ranged 0.6 to 1.5% during five years from 2000 to 2004.

Although India's import of coconut oil declined from 13760 tonnes in 2003-04 to 12712 tonnes in 2004-05, the value of exports rose from Rs.29.79 crores to Rs. 39.48 crores during this period.

Heavy import of palm oil and other types of edible oil resulted in depressing domestic oil prices and it has become a major threat to domestic coconut oil industry. India imports 4.4 to 5.6 million tonnes of edible oil annually of which the share of palm oil varied from 60 to 80% in different years. The share of palm oil in total imports increased from 70.3% in 1996-97 to 79.4 % in 2003-04.

The price of imported edible oil has been lower compared to the domestic coconut oil price by 53 – 65% over the last 6 years.

Edible oil Import and Consumption

Year	Domestic production (million tonnes)	Import			Consumption (million tonnes)	Domestic price of coconut oil in Cochin (Rs/Kg)
		Quantity (Million tonnes)	Value (Rs. Crores)	Unit value (Rs/Kg)		
1999-00	20.72	4.2	8046.05	19.16	10.21	52.49
2000-01	18.44	4.18	5976.53	14.30	9.68	30.83
2001-02	20.66	4.32	6464.97	14.97	10.47	32.50
2002-03	14.84	4.37	8744.88	20.01	9.09	48.29
2003-04	25.29	5.29	11683.24	22.09	12.4	60.11
2004-05	26.10	4.54	10755.65	23.69	12.2	68.11

Source : (i) Ministry of Finance, Government of India, Report of the Committee on Rationalisation of Customs and Excise Duties on Edible Oils and Oilseeds (ii) Coconut Development Board, Coconut Statistics 2005

Import of Copra

The world import of copra declined from 2.09 lakh tones to 1.83 lakh tones between 2001 and 2004. India's share in world copra imports rose to 1% in 2003 which declined to 0.60% in 2004. The value of imports also declined from 1 % in 2003 to 0.50% in 2004. While 416 tones of dried coconuts was imported in 2004-2005, 2532 tonnes of other residues of coconut or copra was also imported. This excluded 50652 tonnes of oil cakes (expeller varieties) imported during the year. The value of dried coconut imported was Rs. 68 lakhs and other residues of coconut or copra was Rs.25.70 crores. The import of oil cake was to the tune of Rs. 33.14 crores in 2004-05.

Import of Coconut Cake

India accounts for 17 to 18% of the world import of coconut cake. The import of coconut oil cake into India declined from 69263 tonnes in 2003-04 to 50,652 tonnes in 2004-05. The value of imports also fell from Rs. 39.73 crores to Rs. 33.14 crores. The first six months of 2005-06 witnessed import of 42286 tonnes valued at Rs. 22.51 crores showing a rising trends in imports.

Import of Desiccated Coconut

Import of desiccated coconut, refined coconut oil and oil cake are much higher than the export. Import of desiccated coconut meat and other coconut products, including charcoal, are on the rise as coconut productivity continues to be lower and coconut-based industries are reluctant to go in for modernization.

India had insignificant level of import till 2003. The year 2004 witnessed sharp increase in imports to the tune of 3.60% of the quantity due to India-Sri Lanka FTA. Because of duty readjustments, the year 2005 witnessed much less imports. India's share import value was 3.0% in 2004, which got reversed in 2005.

Import of Desiccated Coconut (Qty. MT)

Year	World (Qty. MT)	World (Value 1000$)	India (Qty. MT)	India (Value 1000$)	Sri Lanka (Qty. MT)	Sri Lanka (Value 1000$)	(%) Share in World (Qty. MT)
							India
2000	240988	227939	19	15	104	121	0.00
2001	216191	158934	30	6	39	59	0.00
2002	230835	190056	24	9	145	146	0.00
2003	224818	198305	3049	1777	0	0	1.40
2004	230920	203798	8208	6170	10	11	3.60

Source: Computed from http://faostat.fao.org

India imported substantial quantity of desiccated coconut in 2003 and 2004, especially from Sri Lanka. The imports rose from 3049 tones in 2003-04 to 8208 tones in 2004-05. Due to certain duty readjustments, the imports could be reduced to 626 tones in 2005-06.

Global Coir Trade

Currently, the global annual production of coir fiber is about 3.5 lakh tones. Yet, even in the world's top two producers, India and Sri Lanka, which account for 90% of global coir fiber production, combined, this renewable resource is underutilized; local coir mills process only a fraction of the available husks, which accrue more or less year round as a waste during coconut processing.

Traditional uses for the resilient and durable coir fiber include rope and twine, brooms and brushes, doormats, rugs, mattresses and other upholstery, often in the form of rubberized coir pads. In the 1980s and 90s, global exports of coir fiber fell by almost half, as Western consumers shifted to synthetic foam and fibers. Since 1990, domestic demand in India more than doubled global production benefiting exclusively the Indian coir industry. Finally, since 2001, a rising Chinese demand for coir, an expanding market for coir-based erosion control products, and the spread of coir pith as a peat moss substitute in horticulture have further pushed up global production and prices. In turn, other coconut growing countries, including Philippines, Thailand and Vietnam are now expanding their production and export of coir fibre. These changes are also reflected in international trade in coir. Historically, Sri Lanka was the world's largest exporter of various fibre grades, whereas India exports largely value added products - yarn, mats, and rugs. While in 1990 about 80% of global production was exported, growth of the Indian domestic market dropped the export below 40%. Global trade volume for coir fiber, value added products - yarn, mats, rugs - and coir pith now stands at $140 million /year with India and Sri Lanka respectively accounting for $70 and $60 million.

COFFEE

The coffee plant is a woody perennial evergreen dicotyledon that belongs to Rubiaceae family. It has a main vertical trunk (orthotropic) and primary, secondary, and tertiary horizontal branches (plagiotropic). Two main species of coffee cultivated today are *Coffea arabica* known as Arabica coffee accounts for 75-80% of the world's production and *Coffea canephora*, known as Robusta coffee, is more robust than the Arabica plants, but produces an inferior tasting beverage with a higher caffeine content. The coffee plant can grow to heights of 10 meters if not pruned, but producing countries will maintain the coffee at a height reasonable for easy harvesting.

Indian coffee is the most extraordinary of beverages, offering intriguing subtlety and stimulating intensity. India is the only country that grows all of its coffee under shade. Typically mild and not too acidic, these coffees possess an exotic full –bodied taste and a fine aroma. India has consistently produced and exported a remarkable variety of high-quality coffees for over one hundred and fifty years.

Soil and Climate

The soil should be deep, friable, and rich in organic matter with a pH of 6.0- 6.5. The soil and climatic requirements for Arabica and Robusta types are as follows:

Requirements	Arabica	Robusta
Elevation	1000-1500 m above msl	500-1000 m above msl
Annual rainfall	1600-2500 mm	1000-2000 mm
Blossom rain	March-April (2.5 - 4.0 cm)	Feb-March (2.0 - 4.0 cm)
Backing rain	April-May (5.0 – 7.5 cm)	April-May (5.0 - 7.5 cm)
Shade	Medium to light	Uniform thin
Temperature	15-25°C,cool,equable	20-30°C,hot,humid
RH	70-80%	80-90%
Soils	Deep, friable, rich in organic matter, well drained and slightly acidic (pH 6.0-6.5)	Same as Arabica
Slopes	Gentle to moderate	Gentle slopes to fairly level
Aspect	North,East and North East aspect	Same as Arabica

India's coffee growing regions have diverse climatic conditions, which are well suited for cultivation of different varieties of coffee. Some regions with high elevations are ideally suited for growing Arabica while those with warm humid conditions are the best suited for Robustas.

Coffee Types

Three species of coffee trees exist: *Robusta, Arabica* and *Liberica*. *Robusta* and *Liberica* are used in commercial operations because they have a higher yield than the *Arabica* varieties. Beans of *Robusta* and *Liberica* species are flavored more harshly than the *Arabica.*

Arabica varieties are the most important type of coffee beans. Their production is almost solely to produce coffee for gourmet coffee market. *Arabica* trees produce less fruit than other two varieties. In general, *Arabica* trees require specific environmental conditions. They do not grow in wide climatic conditions as the *Robusta* and *Liberica. Arabica* tree is more fragile in nature and necessitates hand picking.

Important Varieties

Kents : Kents is the earliest variety of Arabica, selected by an English planter of same name during 1920s. This variety remained popular with planting community till 1940s, because it was less susceptible to rust. Today, it is grown in a few areas but it is still known for its exceptional cup quality.

S.795 : Most popular Arabica selection released during 1940s with high yields, bold beans, superior quality and relative tolerance to leaf rust. This selection was developed using 'Kents' Arabica, known for its high quality. Even today, the S.795 is a favourite for planters and is a widely cultivated Arabica variety. S.795 has a balanced cup with subtle flavour notes of Mocca.

Cauvery : Popularly known as Catimor, Cauvery is a descendant of a cross between 'Caturra' and 'Hybrido-de-Timor'. Caturra is a natural mutant of the famous Bourbon variety. Cauvery inherited the high yielding and superior quality attributes of Caturra and resistance of 'Hybrido-de-Timor'.

Sln. 9 : Selection 9 is a derivative of a cross between an Ethiopian Arabica collection, 'Tafarikela', and 'Hybrido-de-Timor'. Sln.9 inherited all the superior cup quality traits of Tafarikela. This variety has won the Fine Cup Award for the best Arabica at the 'Flavour of India - Cupping Competition 2002' organised by Coffee Board of India

Typica : This is the base from which many coffee varieties are developed. Like other Coffea, Arabica varieties are developed from it. Typica coffee plants have a conical shape with a main vertical trunk and secondary verticals that grow at a slight slant. Typica is a tall plant reaching 3.5-4 m in height. The lateral branches form 50-70° angles with the vertical stem. Typica coffee has a very low production, but has an excellent cup quality.

Bourbon : Bourbon coffee plants produce 20-30% more coffee than Typica, but has a smaller harvest than most coffee varieties. Bourbon has less of a conical shape than Typica coffee plants, but has more secondary branches. The angle between secondary branches and the main stem are smaller, and the branch points on the main stem are closely spaced. The leaves are broad and wavy on edges. The fruit is relatively small and dense. The cherries mature quickly and are at a risk of falling off during high winds or rains. The best results for Bourbon coffee are realized between 1,200-2,200m. Cup quality is excellent and similar to Typica.

Caturra : Caturra is a mutant of Bourbon coffee discovered in Brazil. It is a mutant with high production and good quality, but requires extensive care and fertilization. It is short with a thick core and has many secondary branches. It has large leaves with wavy borders similar to Coffee Bourbon. It adapts well to almost any environment, but does the best between 500-1825m with annual precipitation between 2,500-3,500 mm. At higher altitudes , quality increases, but production decreases.

Catuai : Catuai is a high yielding coffee plant resulting from a cross between Mundo Novo and Caturra. The plant is relatively short, and the lateral branches form close angles with primary branches. The fruit does not fall off the branch easily, which is favorable with areas with strong winds or rain. Catuai also needs sufficient fertilization and care.

Pache comum : Pache comum is a mutant of Typica coffee first observed on the farm El Brito, Santa Cruz Naranjo, Santa Rosa and Guatemala. Many consider the cup to be smooth or flat. This coffee variety adapts well between 1,160-1,830m.

Pache colis : Pache colis was found in Mataquescuintla, Guatemala in a farm consisting of Caturra and Pache comum. The coffee fruits are very large and the leaves are roughly textured. Pache colis provides resistance to phoma. It has secondary and tertiary branching, and typically grows to 0.8-1.25 m. It adapts well to altitudes of 1000-2000m with temperatures between 20-21°C.

Catimor : Catimor is a cross between Timor coffee (resistant to rust) and Caturra coffee. It was created in Portugal in 1959. Maturation is early and production is very high with yields equal to or greater than the yield of other commercial coffee varieties. For this reason, the method of fertilization and shade must be monitored very closely. The Catimor T-8667 descendants are relatively small in stature, but have large coffee fruits and seeds. The Catimor line T-5269 is strong and adapts well to lower regions between 600-1000m with annual rainfall over 3,000 mm. T-5175 is very productive and robust, but can have problems at either very high or very low altitudes. At low altitudes, there is almost no difference in cup quality between Catimor and the other commercial coffee varieties, but at elevations greater than1,200m Bourbon, Caturra, and Catuai have a better cup quality.

Mundo Novo : Natural hybrid between Typica coffee and Bourbon coffee. The plant was first found in Brazil. The plant is strong and resistant to berrry disease. Mundo Novo has high production, but matures slightly later than other kinds of coffee. It does well between 1,150-1,830 with an annual rainfall of 1,200-1,800 mm.

Maragogype : This coffee variety is a mutant of Typica coffee and discovered in Brazil. The Maragogype coffee plant large and taller than either

Bourbon or Typica. Production is low, but the seeds are very large. Maragogype adapts the best between 660-810m. The cup characteristics are highly appreciated in certain coffee markets.

Amarello : This coffee variety, as its name indicates, produces a yellow fruit. It is not widely planted.

Blue mountain : Blue mountain is a famous coffee variety favored for its resistance to coffee berry disease and ability to thrive in high altitudes. It was first grown in Jamaica and is now grown in Kona, Hawaii. Blue mountain coffee, however, cannot adapt to all climates and maintain its high quality flavor profile.

Propagation

Growing Seeds in Seed Beds

Coffee, generally propagated through seeds and in Robusta, clonal propagation was successful. To a limited extent, grafted plants are also being planted. From selected and certified seed blocks, healthy and fully matured fruits of normal appearance with three quarters of ripening are harvested from the marked plants during November-December in Arabica and January-February in Robusta. Discarding the floats, sound fruits are pulped and sieved to remove defective beans. The beans are then mixed with wood ash @ 0.75 kg/ kg of seeds and dried under shade stirring from time to time to facilitate uniform drying. To protect the seeds against any microbial infection, the seeds are treated with either carbendazim @ 1 g or with vitavax @ 0.66 g/ kg of coffee seed.

Primary Nursery Beds

Seeds are sown in raised seed beds (15 cm above the ground level) provided with proper drainage prepared out of soil, compost and sand at 6:2:1 proportion. A bed of 4 x 3 m will be sufficient for 1.5 kg of seeds, if sown 1.0 to 1.5 cm apart in rows, with the flat side of seed towards the soil. A thin layer of soil is spread after sowing and covered with dry straw to a thickness of about 5 cm to ensure uniform temperature and to regulate moisture retention. Sowing is to be taken up in December-January for Arabica and in February-March for Robusta. Watering of the seed beds is to be done twice a day in the initial week and thereafter regulated. The seeds sprout in about 40 days when the straw mulch is removed. The primary beds are provided with pandal covered with coir mats or dry leaves.

Secondary Nursery Beds

In some areas, seedlings from germination beds are transplanted to secondary nursery beds of same soil composition as that of germination beds. Transplanting is done at button stage. Seedlings are planted 30 cm apart. If the taproot are bent or excessively grown, it is nipped off while transplanting. The beds are mulched and watered at regular intervals. Watering should be done during early morning hours.

Poly Bag

Seedlings from germination beds are transplanted to polythene bags in Feb-March when they are at the 'button' stage. Polythene bags of 23 x 15 cm and 150 gauge thickness with adequate number of holes of 3 mm in the bottom half of the bag, are preferred. The bags are filled with potting mixture of 6 parts of sieved jungle soil, 2 parts of well rotten sieved cattle manure and 1 part of fine sand. The soil is filled into the nursery baskets and pressed firm. Nursery baskets are arranged conveniently in rows of 10 within a rectangular frame with bamboo reapers. These frames are held in positions with bamboo or wooden props driven into the ground at suitable distance.

Coffee seedlings at button stage are transplanted into nursery baskets. The seedlings are gently lifted from the germination beds with minimum injury to roots. Prior to transplanting, the nursery basket is watered and a vertical hole of 5 cm deep is made in the soil at centre of basket. At the time of transplanting, it is preferable to slightly nip off the taproot of the seedling. The shoot portion of transplanted seedling should be at the same height above soil level as it was in germination bed. Transplanting is done preferably in early morning hours or late in afternoon. Seedlings uprooted from primary bed should not be stored for a long time but transplanted immediately.

Regular watering and aftercare of the seedlings should follow. Excess moisture and watering in the afternoon should be avoided as it may induce damping off.

After Care of Seedlings

Seedlings are to be manured once in two months with urea dissolved in water or supernatant solution of fermented cow dung slurry. For an area of 1 m^2, 20 g urea dissolved in 4.5 litres of water is sufficient. Adequate protection is to be given against nursery diseases and pests. Overhead shade in the nursery has to be thinned and finally removed after onset of monsoon (this is not applicable to northeastern areas). The seedlings grow vigorously if watered judiciously and protected against afternoon sun.

Clonal Propagation

In Robusta, which is highly cross-pollinated, clonal propagation is more adaptable. In Arabica, the stabilization of desirable characters in the selected plants could be easily maintained by adopting clonal or vegetative propagation method. The vertical (orthotropic) shoots are marked after harvest, single node green wood (semi-hardwood) cutting of 10 cm length and 3 to 6 months old are planted in polythene bags with medium of jungle soil, sand and cattle manure in the proportion of 6:3:1. The bags with cuttings are arranged in a propagation chamber made of a trench of size 2 x 1 x 0.5 m covered over with a thick polythene sheet (500 gauge) spread over a framework of bamboos.

A trench could accommodate about 108 filled up bags of size 22 x 15 cm. Pre planting treatment of the base of cuttings with IBA (indole butyric acid) (5000 ppm) enhances early rooting. Under South Indian conditions, cuttings collected during June-July recorded the highest% of rooting. Cuttings root in 3-4 months after planting. Rooted cuttings should be hardened by keeping them under shade for about two months and then can be transplanted into the field.

Establishment of Coffee Plantations

Selection of Site

In choosing a site for a new plantation, due consideration should be given to altitude, rainfall, exposure to wind, slope of land, sources of water and approach etc. Arabica coffee grows well at an elevation of 1000-1500 m amsl, while robusta coffee comes up well at lower altitudes of 500 - 1000 m amsl. Locations with gentle to moderate slopes covered with a good canopy of evergreen trees are to be preferred. Southern and western aspects should be avoided especially at lower elevations. Incase unavoidable, such areas should be provided with more shade to protect coffee from afternoon sun. In wind prone areas, wind belts consisting of tall trees like silver oak, tree coffee etc. should be raised.

Choice of Varieties

The varieties selected for coffee production must be well adapted to local conditions and tolerant/resistant to pests/diseases. In Arabica, varieties with wider adaptability like S.795, Sln.5-B, Sln.6 and Sln.9 may be preferred, while in robustas improved varieties like S.274 and CxR may be selected.

Land Preparation

Clean felling of trees is not advocated when land is prepared for planting coffee. Selective retention of evergreen trees providing filter shade at a spacing of 10-13m is desirable. The land is divided into blocks of convenient size by laying out footpaths and roads in between. The ground level bushy growth is cleared by uprooting and *in situ* burning. Land preparation should be completed well in advance of commencement of South - West monsoon (June).

Soil Conservation

The loss of top soil is negligible when the land is covered by a two tier shade canopy comprising of lower tier of temporary shade trees like dadap (*Erythrina lithosperma*) and top canopy of permanent shade trees. The soil erosion attains serious dimension on steep slopes without proper shade coverage. In such fields, there is need to prevent erosion of top fertile soil by appropriate soil conservation measures like contour planting and terracing.

In established plantations, opening of cradle pits (60 cm wide, 30-36 cm deep) across the slope in between coffee rows in a staggered manner, would not only help in prevention of soil erosion but also in better harvesting of rain water and conservation of soil moisture for dry months. These pits also act as *in situ* compost pits for leaf litter, weed biomass and shade tree loppings. These pits are to be renovated every year just before onset of South - West monsoon.

Line Marking

In each block, the spots for planting of coffee and shade trees should be marked at recommended spacing soon after land preparation. The following spacing is suggested for different coffee varieties.

If it is a jungle, only selective felling of trees is done maintaining the trees, which are desirable at appropriate spacing. The under growth may be cleared to enable line marking with a base line and opening of pits. The entire plot may be conveniently divided into blocks with roads and footpaths.

In April, pits of 45 x 45 x 45 cm may be opened at appropriate spacing for different coffee cultivars as described below.

Tall Arabica like S 795, S 288	:	2.1 x 2.1 m
Semi-dwarfs like Cauvery	:	1.8 x 1.8 m
Dwarfs like S 7 (San Ramon)	:	1.5 x 1.5 m
Hybrids like Congensis x Robusta (CxR)	:	2.5 x 2.5 m
Robusta selections like S 274, BR series	:	3.0 x 3.0 m

The pits after digging will be kept open for weathering for a couple of months until monsoon. In June, the pits are covered with topsoil and staked. In poor soils, 250 g of FYM or compost / pit may be added before filling.

Planting of Shade Trees

It is advisable to plant temporary shade trees like dadap at closer spacing initially, for providing optimum shade to young coffee plants at a spacing of 5-6.5m. In large open spaces, evergreen permanent shade trees such as *Ficus* sp., *Albizzia* sp., *Artocarpus* etc. should be planted at suitable intervals (10-13m). For planting shade trees, pits should be taken out during pre-monsoon period and filled with top soil after exposing for about a fortnight. Planting of shade trees should preferably be completed before onset of South - West monsoon.

Shade Management

India cultivates all of its coffee under a well-defined two-tier mixed shade canopy, comprising evergreen leguminous trees. Nearly 50 different types of shade trees are found in coffee plantations. Shade trees prevent soil erosion on a sloping terrain; they enrich the soil by recycling nutrients from deeper layers, protect the coffee plant from seasonal fluctuations in temperature, and play host to diverse flora and fauna. Coffee plantations in India are essential spice worlds too: a wide variety of spices and fruit crops like pepper, cardamom, vanilla, orange and banana grown along with coffee plants.

Shade plays a vital role in maintaining the ecosystem and required micro-climate in coffee plantations. A two-tier mixed shade canopy consisting of a lower canopy of temporary shade trees like dadap (*Erythrina lithosperma*) and a top canopy of permanent shade trees belonging to *Ficus spp.* and *Albizzia spp.* are recommended for coffee. The optimum shade requirement would be 50% for arabica coffee and 30% for robusta coffee. By maintaining optimum shade, incidence of pests such as white stem borer and green scale and diseases such as leaf rust and black rot could be brought down substantially in arabica coffee. Similarly, the shot hole borer incidence in robusta could be minimized by maintaining optimum shade. Temporary shade plants like dadap should be lopped just before onset of South West monsoon to facilitate better light penetration and aeration during monsoon. The newly sprouted branches should be thinned out towards end of South West monsoon to retain an umbrella shaped frame for providing shade during post monsoon period. The canopy of permanent shade trees should be regulated once in two to three years to provide optimum shade.

Planting of Coffee

Planting of coffee seedlings should be taken up during August-September towards end of heavy monsoon rains. At the time of planting, it is advisable to add about 50g of rock phosphate to each pit, for encouraging root growth and better establishment of plants. In cockchafer infested fields, neem cake @ 250g / pit is advocated.

After Care of Young Plantations

During the year of planting, following operations help in better establishment of coffee and shade plants. After planting, the coffee seedlings should be provided with staking and mulching to protect against wind damage and to conserve soil moisture for ensuing dry period. Towards the commencement of dry period, the young plants in open area should be protected by erecting temporary shade huts with jungle tree twigs. Stems of young dadap plants should be coated with lime solution to prevent sun scorching.

Maintenance of New Plantations

Green Manuring for Soil Enrichment

In newly planted fields, green manure crops like cow pea and horse gram could be cultivated for two or three years to build up soil fertility. These crops should be grown during June-September, to prevent competition for soil moisture. These green manure crops contribute around 6-10 tonnes of dry matter/ha and also effectively suppress weed growth in early years. As most of these crops are leguminous in nature, they fix nitrogen from atmosphere. The green manure crops should be cut before flowering and incorporated into soil to improve soil fertility.

Nutrient Management

The following practices would be essential for meeting nutrient requirement of young coffee holdings. Correction of soil pH using agricultural lime or dolomite, based on soil test values, at least once in 2-3 years is practiced. Application of farmyard manure or compost prepared on the farm @ 1250 kg/ha/ year.

Weed Management

Weeds should be controlled by manual (slashing) weeding or by scuffling. However in slopes, only hand weeding should be followed. Mulching with weed slashings and shade tree leaf litter etc. would also be beneficial in suppressing weed growth especially grasses. Cultivation of cover crops like cowpea, horse gram etc., in vacant spots would help in suppression of weed growth also.

Fertilizer Recommendation for Coffee (N:P2O5:K2O, kg/ha)

	Time of application			
	Pre-blossom (March)	**Post-blossom, pre-monsoon (May)**	**Mid-monsoon (August)**	**Post-monsoon (October)**
Arabica				
1st year	15:10:15	15:10:15		15:10:15
2nd and 3rd year	20:15:20	20:15:20		20:15:20
4th year	30:20:30	20:20:20		30:20:30
Bearing coffee 5 years and above: for less than 1 t/ha crop	40:30:40	40:30:40	20:0:0	40:30:40
For 1 t/ha and above	40:30:40	40:30:40	40:30:40	40:30:40
Robusta				
For less than 1t/ha crop	40:30:40			40:30:40
For 1 t/ha and above	40:30:40	40:30:40		40:30:40

Weeds pose a serious problem especially in new coffee clearings. Grasses need to be controlled in the initial years itself. The following measures are suggested for controlling weeds. In new clearings, cultural practices like cover digging (36cm deep) during the year of planting and scuffling (12-18cm) for the next two to three years carried out during post-monsoon season, would not only bring down weed growth but also help in conservation of soil moisture. However, in sloping terrain, avoid soil digging to prevent soil erosion. In such areas, adopt only slash weeding. Cultivation of green manure crops/cover crops and mulching with weed slashings and shade tree leaf litter etc. would also help in smothering of weeds. Once the coffee bushes cover up, the weed growth would naturally get suppressed and manual slash weeding alone would be sufficient.

Training and Pruning

The young coffee plants should be trained to provide proper shape to bushes and to improve efficiency of operations like spraying, harvesting etc. at later stages. Generally, single stem system of training is recommended for coffee grown under shade. In this system, the plant height is restricted by topping (capping) at prescribed heights. The tall arabica varieties are topped at two stages (two-tier system) while the dwarf arabicas as well as the robustas are capped at single level (single tier system). The prescribed topping heights for different coffee varieties are:

Tall arabicas - 1st topping at 0.9m. 2nd topping at 1.5 to 1.7m (second topping is done after harvesting 4-5 crops, when the spread of lower canopy is complete). Dwarf arabicas - Single topping at 1-1.7m depending on soil fertility, wind proneness etc. Robustas - Single topping at 1.5 to1.7m. In topping operation, the terminal portion of main stem is decapitated at the prescribed height, by providing a slant cut. In arabicas, one of the top most primary branches is also cut near to base to prevent splitting of main stem due to crop load. After topping, all the new suckers produced on the main stem are to be removed periodically. Apart from periodic removal of suckers, the young plants require very little pruning.

Pruning

Coffee being an evergreen plant requires regular annual pruning, to create a balance between vegetative and cropping wood. The coffee plants trained on a capped single stem system require regular pruning once in a year especially in arabica coffee. Pruning should be done immediately after harvesting of crop. Only in case of die- back affected bushes, pruning should be delayed till receipt of a few showers and recovery of plants with new vegetative growth. In pruning operation, certain criteria are recommended.

- Never cut primary branches to their base, as they can not regenerate and the lost primaries can not be replaced easily
- The primary branches which have exhausted all cropping nodes, should be cut back by leaving some nodes from base,
- Remove secondaries and tertiaries, which have produced two to three crops to encourage new branches
- Remove all criss-cross branches, and those growing upwards (gormandisers), downwards and inwards towards the main stem
- Remove all the suckers
- Remove all the diseased, damaged and lean and lanky branches.

In robusta, it has a self pruning habit by shedding of laterals after 3 to 4 crops. Also, it exhibits more determined growth habit in which the production of secondaries and tertiaries are pre-determined. Regular pruning is not required. Removal of shot hole borer affected twigs, gormandisers, suckers etc. should be carried out three to four times a year. In irrigated fields, light pruning every year after harvest is suggested to regulate cropping wood.

Handling, Centering and Desuckering

Handling is thinning out of young flush that arises after main pruning. Handling should be carried out at least once or twice in a year depending on growth characteristics of varieties. The first round of handling is done during June-July and in case required, a second round during September. In handling operation the following criteria are suggested.

- Retain one healthy young flush on each side of node on primary branch and remove remaining flushes. In essence, each primary branch should have four to six secondaries.
- Remove all new growth arising on primary branches within 15cm radius of the main stem (centering), to allow aeration of the main stem.
- The suckers should be removed all the time, at least three to four times a year (desuckering).

Drought Management

Wherever water is available, overhead irrigation by sprinkler system is adopted to a greater advantage during November-January to keep the soil moisture level and in February-April, for ensuring blossom and retention, if necessary. Plants affected by drought limit vegetative growth, show floral abnormalities and poor fruit set resulting in reduced yield. Therefore, drought tolerance is an important aspect of coffee productivity. Arabica coffee is more tolerant to drought than Robusta. For inducing tolerance in Robusta, the following nutrient solution can be sprayed @ 1 litre/ plant. (Urea 1 kg, Superphosphate 1 kg, Muriate of potash 750 g, Zinc sulphate 1 kg in 200 litres of water).

Spraying Schedule

1st spray: 45 days after the last rainfall (usually the 2nd fortnight of January)

2nd spray: 30-45 days after first spray

Foliar application of anti-transpirants like Ralli Dhan 110 @ 200 ml in 200 litres of water (0.1%) is also useful for drought management in coffee.

Pest Management

No serious pest is observed in young coffee plantations except for sporadic incidence of some foliar and soil borne pests. The damage by sucking pests like mealy bug, green scale and foliar pests like leaf miner and grasshoppers could be avoided by spraying neem kernel extract, plant based extracts and other permitted products. Application of neem cake @ 250g/ plant can be effective against soil borne pests like cockchafers (root grubs) and nematodes.

Disease Management

Young coffee plants are usually free from major diseases. However, in exposed areas, brown eye spot disease may cause defoliation. Providing adequate shade against exposure, mulching to conserve moisture and spraying with Bordeaux mixture(1%) can take care of this minor disease.

Inter Cropping

Cultivation of short duration crops like ginger, elephant foot yam, pineapple, banana, papaya etc can be adopted to augment income during the pre-bearing stage of coffee.

Mixed Cropping and Diversified Farming

Planting of compatible crops like pepper and orange in the coffee blocks at 6.o to 9 m distance and associate crops like cardamom, arecanut etc., in valleys is a common feature among coffee estates in India. Of late, high value crops like vanilla are gaining popularity in coffee areas.

The nutritional requirement of coffee and its' associate crops should be met by proper recycling of all by-products available within the farm. For this purpose, all the produces from various crops should be processed within the farm and the crop residues should be recycled after composting, to minimize net loss of nutrients from the farm. It is also advisable to go for diversification, by maintaining subsidiary activities like apiculture and dairy farming. For maintaining dairy cattle, appropriate areas should be demarcated for pastures especially in slopes and marginal lands, which are generally not very ideal for coffee. Similarly, areas unsuitable for coffee planting like scrubby, rocky jungle may be left out undisturbed as a natural habitat for birds and natural enemies of pests.

Composting of on-farm Organic Wastes

The by-products of coffee processing namely coffee pulp and cherry husk are rated higher than cattle manure in terms of their manurial value and soil conditioning properties. It is estimated that for every tonne of clean coffee produced on the estate, around one tonne of dry pulp or cherry husk is produced. These by-products when composted and recycled would contribute approximately 17 kg of nitrogen, 1.0 kg of phosphorus, 52 kg of potassium,

1.0 kg of sulphur and 5.0 kg of magnesium. Thus the micronutrient requirement of coffee is also taken care of. The coffee processing by-products should be composted along with weed biomass, within the estate by following aerobic "Heap method".

Harvesting and Processing

Coffee is harvested during dry season when the coffee cherries are bright red, glossy, and firm. Ripe cherries are either picked by hand, stripped from the tree with both unripe and overripe beans, or all the beans are collected using a harvesting machine. These processes are called selective picking, stripping, and mechanical harvesting, respectively. To maximize the amount of ripe coffee harvested it is necessary to selectively pick the beans from the ripe tree by hand and leave behind unripe, green beans to be harvested at a later time. About 12-20 kg of export ready coffee will be produced from every 100 kg of coffee cherries harvested. Once coffee is picked, it can be prepared either by a "dry" method, which produces natural coffee, or by the "wet" method, which produces washed coffee.

The 10 Steps To Coffee

There are 10 steps through which a coffee bean passes from seed to cup.

- Harvesting
- Processing
- Drying
- Hulling
- Polishing
- Grading and sorting
- Exporting the beans
- Tasting
- Roasting
- Grinding

Harvesting the Beans

After three to four years, when they reach maturity, coffee trees bear fruits in lines or clusters along branches of trees. Fruit turns red when it is ready to be harvested. Coffee beans are actually the seeds of these ripened cherries. Most arabica cherries ripen after 6-8 months; robusta beans take 9 to11 months to ripen. Beneath the cherries' red skin (exocarp) is a fleshy pulp (mesocarp), a slimy layer (parenchyma), and a parchment-like covering of bean (endocarp). Inside these layers are usually two beans, which are covered by a thin membrane or coat. This membrane or seed skin (spermoderm), is referred to in the coffee trade as the "silver skin".

Harvest times vary according to geographical zone, but typically there is only one harvest a year. North of the Equator, harvest takes place between September and March and South of the Equator, the main harvest occurs in April or May, although it may last until August. In countries where wet and dry seasons are not clearly defined like Colombia and Kenya, there may be two flowerings a year, therefore permitting a main and a secondary crop. Equatorial countries can harvest fruit all year round. The average coffee tree produces enough beans to make about 0.45 kg to 0.7 kg of roasted coffee/ year. It takes approximately 9,000 handpicked green coffee beans to make a kg of coffee.

Harvesting Methods

There are four different methods of harvesting coffee.

Stripping method. This form of harvesting is done by hand, removes all of the cherries, flowers, green cherries, and deeply over ripened cherries. This method produces poor results because of mixing of good cherries with bad ones. However, it is still practiced in some parts of Africa and Brazil.

Method uses a comb to brush the trees. This method does remove all ripe cherries leaving the unripe cherries and green leaves still connected to the branches of the tree. This is a time-consuming process that could be worth , if the plantation owner were being paid a premium for the quality that he was producing. (Coffee in Latin America, 1963).

Mechanical harvesting. This process uses a vibrator fixed to the trunk of the tree, shaking the ripe cherries loose so they fall to the ground where they can be reached with ease. The other mechanized harvesting tools are rotating brushes connected to the side of tractors. This process damages the tree, nipping off green cherries, flowers, and leaves at the same time.

Hand picking. Selective picking involves making several passes among the coffee trees at intervals of eight to 10 days so that only the fully ripe berries are taken. Selective picking is more expensive and, when used, is used only for arabica beans.

Processing the Beans

Processing beans, or preparing them for roasting, is done in two ways: (1) the dry method, or (2) the wet method.

Dry Method

This is the simplest, cheapest and most traditional method of processing coffee. The harvested cherries are spread over a concrete, brick or matting surface, ideally in sunlight, and raked at regular intervals to prevent fermentation. If it rains or if the temperature falls, the cherries have to be covered for protection. After about 7-10 days, when the moisture content of each cherry will have fallen to about 11%, the cherries are dry. The outer shell will have become dark brown and brittle, and the beans can be heard rattling around inside their husk. The dried cherries are then stored in silos, where the beans continue to lose moisture.

Wet Method

This method requires greater investment and more care than the dry method, but it causes less damage and helps to preserve the intrinsic qualities of the bean. The main difference between the two methods is that the wet method uses a procedure to remove the pulp from the bean within 12-24 hours of harvesting, instead of allowing the cherries to air dry. Using a pulping machine, the beans are separated from skin and pulp, which are washed away with water. The lighter, immature beans are separated from heavier, mature beans through specially designed washing channels or by an Aagaard pre-grader, a system involving shaking the beans through a strainer into a tank of water. Next, the beans are stored in fermentation tanks for 12 to 48 hours, during which time enzymes work to naturally separate the slimy layer (parenchyma) from the parchment-like covering (endocarp). When the process is complete, the endocarp has a "pebbly" feel. Fermentation is usually carried out in concrete tanks, which vary considerably in size. The length of fermentation process is based on the condition of beans and the climate's conditions. In regions of low altitude, fermentation time is short. At higher altitude, fermentation takes up to 48 hours.

Drying

The endocarp must then be dried to retain about 11% moisture so the beans can be stored in a stable condition. The endocarp can be dried by the sun or by mechanical dryers. If drying is left to the sun, the beans are spread on concrete floors or on drying tables, and turned regularly. After 7 to 15 days, the beans are known as "parchment coffee," and will ideally remain in this form until immediately before export.

Hulling

In wet processed coffee, hulling is used to remove the hull or dried parchment layer immediately surrounding the bean. Hulling dry processed coffee refers to removing husks or whole of dried outer coverings of original cherries.

Polishing

Polishing beans is an optional process that is not always done. During the polishing process, any silver skin that remains on beans after hulling is removed in a polishing machine. While polished beans are considered superior to unpolished ones, in reality there is little difference between the two.

Grading and Sorting

Although coffee beans are of a fairly uniform size and proportion, they are graded first by size and then by density (the elephant bean is the exception). Typically the size of beans is expressed on a scale of 10 to 20. Beans are sized by being passed through a screener. Beans are also sorted by using an air jet to separate heavy and light beans (the pneumatic method). Over-fermented or un hulled beans are removed. This is usually done by hand as the beans move along on a belt, but other methods include electronic color sorting. Different countries grade their beans according to different systems. In general, there are six export grades, the top grade being SHB (strictly hard bean) or strictly high grown, which means that the coffee beans are produced at a minimum altitude 1,300m above sea level.

Exporting

Approximately 7 million tons of "green coffee" (the term used to describe coffee beans before they are roasted) are produced each year. Beans are often kept and transported in coarse fiber bags made of jute or sisal. Beans are also shipped in bulk using bulk containers with plastic liners. The shipments are then sent to warehouses or directly to the roaster.

Tasting Coffee

The tasting of coffee is referred to as "cupping." Cupping coffee is a process by which coffee brew is slurped by an expert for evaluating the brew and determining characteristics. The tasting of coffee is a rigorous and disciplined process. The taster (also known as the liquor) first assesses green beans for appearance. The green beans are then roasted in a small laboratory roaster and tested for flavor and aroma. After the coffee has been infused in water, the taster "noses" the brew. After three minutes, the brew is lightly stirred and smelled again. The resulting foam is removed and the tasting begins. The cupper takes a spoonful of coffee into his/her mouth and "chews" it around before spitting it out. The procedure is repeated with all of the samples, and notes are made as each brew is sampled.

Following 10 criteria are used to describe and categorize coffee:

Type	-	robusta, washed, arabica
Taste	-	strictly soft, harsh
Body	-	lacking, too heavy
Acidity	-	some, too much at the top
Age	-	old to fresh
Defects	-	sour, grassy, musty
Cup	-	roast, watery, burned, old
Overall assessment	-	neutral, spicy, hard
Aroma	-	weak to strong
Fullness	-	slight to considerable

Roasting Coffee

Roasting is the heat treatment which transforms the green beans into the aromatic brown nuggets that we buy whole or ground. Most roasters use air temperatures of about 550°F. The beans are kept moving during the roasting process to keep them from burning. When the beans reach a temperature of about 400°F, they start to turn a darker brown, and the oils (called coffee essence, coffee oil or caffeol) start to emerge. This process, called pyrolosis, is the heart of roasting, as it produces the flavor and aroma of coffee we drink. When the beans are removed from roaster, they are cooled by air or water. Roasting is performed in individual importing countries, because roasted beans must reach the consumer quickly to get full coffee flavor.

Grinding Coffee

The objective of grinding coffee is to get the most flavor from beans. This is done by infusing ground beans in hot water. Generally, faster the infusion, finer the grind.

Speciality Coffee

Specialty coffees are high quality coffees that differ from normal coffee with relevance to visual quality or cup or both. Specialty Coffees are getting increasingly popular in the world coffee market.

Processing of Specialty Coffees

Utmost attention is paid while preparing majority of specialty coffees both at primary and secondary processing levels.

Flow chart showing key wet and dry processing steps for coffee cherries and final processing steps for green coffee.

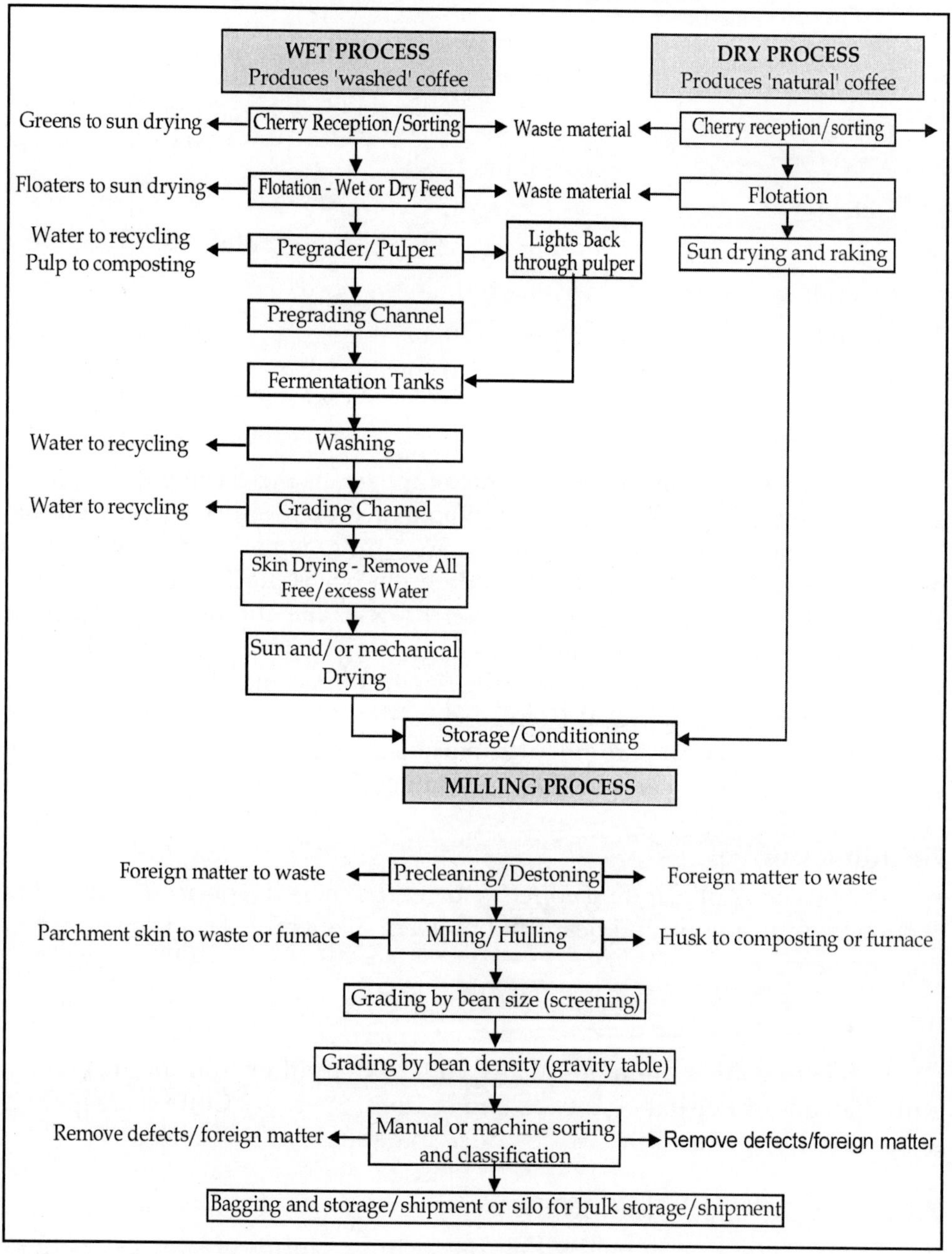

Source: International trade centre,UNCTAD and WTO 2002
Coffee: An exporters guide Geneva, Switzerland

Primary Processing: Selective harvesting is done with special attention to careful hand picking of just ripened berries. Natural fermentation is a must for flavour development and a high standard of quality. Soaking washed beans under fresh water over night is essential for development of colour and flavour in the bean. Slow drying is absolutely necessary under natural lights.

Secondary processing: Grading of coffee beans at curing level involving separation of beans according to size, shape and density followed by meticulous garbing improves the quality.

Today, consumers around the world demand the best coffee. Planters in producing countries are now developing finer strains of coffee to custom-requirement of gourmet coffee drinker.

Specialty coffee is distinguished by its clear origin, distinction made by careful cultivation practices, careful routine of plucking, special processing and of course, special branding and special handling and appearance thereafter.

Consumers to appreciate gourmet coffees in their discerning cups; As a result, specialty coffees fetch better prices, attract premiums and custom. These coffees have a personality, aura, craving, and branding of their own. Specialty branding that brings in bigger bucks for the grower. The world of coffee drinkers is going the specialty way. It makes sense to grow and sell specialty coffee. Specialty coffees are broadly classified into five categories based on their unique features:

Decaffeinated Coffee

Coffees with caffeine removed artificially are known as decaffeinated coffees. There is a prominent segment of consumers preferring decaffeinated coffees owing to health consciousness.

High Grown Coffees

The coffees grown at higher elevations i.e. 1250m and above possess a distinct flavour and acidity in cup due to slower development of beans. The coffees are of high quality with dense beans.

Estate Coffee (Single Origin Coffees)

Production of good quality coffee by highlighting special features of estate can be classified as estate branded coffees. This may be elevation, variety, cultural practices, and special processing techniques. Estate coffees are those that originate on single form with common soils and with distinct identity in terms of flavour and aroma characteristics. Estate coffees are generally more expensive and actually define the specialty coffee market.

Variety Coffee

Some of the varieties are known to possess good inherent quality due to genetic make up of plant. Fruits from such varieties can be harvested and processed separately to retain unique quality. Some of the Indian selections like Kents, Agaro, Cioccie, CxR are unique in their cup quality. Separate processing of these varieties help in processing their intrinsic quality.

India offers several varieties of specialty coffees, popular in the West. Continuous research by Indian Scientists has helped to identify better strains that will make finer coffees with added flavour profiles, in both Arabica and Robusta varieties.

Monsooned Coffee

Monsooned coffee has a story to tell. The "Monsooning" of coffee first happened quite by accident in the deep of sailing ships - a shipload of coffee bound for Europe acquired a mellow yet unique taste en route, with the coffee beans 'swelling' due to the moisture in the air. A new kind of coffee was born - Monsooned Coffee.

Even today, India offers the same golden quality Monsooned Coffee. Prepared by the unique natural elements and the special process of today, the Monsooned Coffee still has the monsooned flavour, mellow taste and golden look. Consumers in Scandinavian countries love it for its special colour and flavour.

Main Grades

Monsooned Malabar AA

Monsooned Basanally

Monsooned Robusta AA

Mysore Nuggets Extra Bold

This coffee is a premium coffee that represents the best quality coffee from India. The beans are very large, uniform blush green in colour with a clean polished appearance. In cup, the coffee exhibits full aroma, medium to good body, good acidity and fine flavour with a hint of spice.Arabica Plantation Coffee (Washed coffee) is grown in the Mysore, Coorg, Biligiris and Shevaroys regions.

Robusta Kaapi Royale

This coffee is prepared from Robusta Parchment AB from the regions of Mysore, Coorg, Wynad, Shevaroys, Pulneys and Barbabudans. The beans appear bold, round with pointed ends and gray to bluish gray in colour. This cup ensures full body, soft, smooth and mellow flavour.

The Quality Control division of the Coffee Board has set specifications for processing, grading and garbling of specialty coffees to ensure quality of these coffees.

Organic Coffees in India

Organic coffee means a coffee grown under holistic agricultural production management system that includes onfarm composting of farm resources, mulching, shade regulation and eco friendly pest and disease control measures. It completely excludes the use of agro chemicals with an aim of preserving the environment and to obtain residue free product. To provide authenticity to the product as organic, it has to be certified by an accredited certification agency.

Scope for Organic Coffee Production in India

There is a lot of scope for production of organic coffee in India, due to natural advantages existing in coffee, which are as follows:

- Coffee is mainly grown in deep fertile jungle soil under well-defined mixed shade consisting leguminous and non-leguminous trees.
- Majority of coffee holdings are dominated by small and tribal growers with low or zero inputs or synthetic inputs with sustainable yields, which are ideally suited for conversion to organic farming without any serious setbacks in yields
- Majority of small and tribal holdings are following traditional farming practices such as cattle manure, composting, manual weeding, recycling of organic wastes, which are prerequisites for organic coffee production.
- Coffee in principle is mainly grown under high degree of mixed cropping under shade with sustainable income.

Role of Coffee Board in Promotion of Organic Coffee Production

Coffee Board, initiated steps to promote organic coffee production in the country by implementing a Project "Promotion of Organic Coffee Production in India" during the IX plan period (1997-2002) with the objectives of standardizing package of practices for production of organic coffee, to create awareness among the growers by organizing trainings/workshops/seminars/education campaign and to identify potential zones for production of organic coffee in coffee growing regions of India.

Role of Coffee Board Under National Programme for Organic Production

Coffee Board represents the 'Standing-cum-Accreditation Committee' since 1994 and currently a member of reconstituted "National Steering Committee (NSC)" as well. Under the NPOP, the Coffee Board is identified as an "Accreditation Agency (AA)" for development and promotion of Organic coffee in the country.

As Accreditation agency, the Board accredited three inspection and Certification agencies under National Programme for Organic Production (NPOP) viz., M/s. IMO, Bangalore, M/s. SKAL, Bangalore, and M/s. APOF, Bangalore.

Marketing

Coffee is a beverage obtained from coffee plant's fruit called cherry. *Coffea arabica* and *Coffea robusta* are the two most commonly cultivated species of coffee plant having economic significance. Arabica accounts for about 70% of the world's coffee production. Robusta coffee trees represent about 30 % of the world's market.

Indian Scenario

India contributes only around 4% of world production but Indian coffee has created a niche for itself in international market, particularly the Indian Robusta and the Arabica Coffee are highly preferred for their good blending quality. In India, coffee plantation occupies an area of around 3.5 lakh hectares providing rural employment pre-dominantly in Karnataka, Kerala and Tamil Nadu, which contribute about 99% of total coffee production. There are 1,78,308 coffee holdings, out of which 1,75,475 fall within the small growers' category and remaining 2,833 holdings fall under large holdings (above 10 hectares category). The small growers account for about 71.8% of the total area.

Production

Coffee is probably the most traded agricultural product in the World. India was the seventh largest producer of coffee in the world. India produced 2.76 lakh tones of coffee in 2004-05, which accounted for 4.07% of the world output. India improved her share to 4.66% in 2005-06 with an estimated production of 2.81 lakh tones.

The crop year of Indian coffee lies between October to September. Coffee growing regions in India can be grouped under three distinct categories - Traditional areas representing Southern states of Karnataka, Kerala and Tamil Nadu and Non-traditional areas comprising Andhra Pradesh and Orissa in the Eastern Ghats of the country and the Northeastern region comprising the 'Seven Sister' states of Assam, Manipur, Meghalaya, Mizoram, Tripura, Nagaland and Arunachal Pradesh.

In Karnataka, Chikmagalur, Coorg and Hassan are the major coffee producing areas producing around 76,300 tones and 1,19,975 tones annually of Arabica and Robusta coffee respectively .In Kerala, Wyanad, Idukky and Nelliampathy are the major coffee producing areas producing around 1,375 tones and 55,450 tones annually of Arabica and Robusta coffee respectively. In Tamil Nadu, Pulney, Nilgiris, Shevroy (Salem) and Anamalais (Coimbatore) are the major coffee producing areas producing around 14,375 tones and 4,450 tones annually of Arabica and Robusta coffee respectively.

Among non-traditional areas, Andhra Pradesh, Orissa and North Eastern Region produce around 1,950 tones and 125 tones annually of Arabica and Robusta coffee respectively. India annually exports 2,01,498 tones of coffee to around 44 countries across globe with Italy (26.5%), Russian Federation (14.61%), Germany (8.31%), Belgium (5.55%) and Spain (5.11%) being the top 5 destinations for Indian coffee. In 2005, estimated domestic consumption was 80200 tones.

Global Scenario

After petroleum, coffee is the world's most important traded commodity. Coffee supply and demand world wide are subject to considerable fluctuations year to year. These fluctuations are caused by a variety of factors like economic interests of the producing and purchasing companies besides climatic factors. Till 1989, the International Coffee Organisation (ICO) controlled global coffee trade. After 1989, coffee became a free-trade commodity. The present coffee producing belt around the globe encompasses approximately 70 countries. About 7-million tones of green coffee are shipped worldwide each year. Coffee is grown on about 10 million ha of land in countries in Africa, Asia, and Latin America .The three major growing regions are Africa, South America and Asia. Brazil, Columbia and Vietnam are the world's top 3 coffee producing countries. The estimated value of the export of green coffee is 10 billion US dollar.

Export

Total global coffee export in 2004 was 61.23 lakh tones of which 92.45% was green coffee and the rest roasted coffee. Brazil was the top coffee exporter with 23% share followed by Vietnam (16%), Columbia (9%), Indonesia (5.62%) and Germany (5.21%). India has 11th rank with 2.3% of global coffee export. Brazil holds first place in export of green coffee with 25 % while Germany who holds first place in the export of roasted coffee with 21%. India exported 3 % of green coffee and 0.01% of roasted coffee and is ranked 11th and 69th positions respectively.

Global Export of Green and Roasted Coffee in 2004

Countries	Quantity (Lakh tones)	Percentage
Brazil	14.13	23.08
Viet Nam	9.75	15.92
Colombia	5.76	9.40
Indonesia	3.44	5.62
Germany	3.19	5.21
Guatemala	2.09	3.41
Peru	1.91	3.12
Honduras	1.58	2.58
Uganda	1.53	2.50
Côste d'Ivoire	1.42	2.32
India	1.41	2.30
Other 129 countries	15.02	24.53
World	61.23	100.00

Source: http://faostat.fao.org

Brazil and Vietnam together accounted for 39% of the global export. Columbia (9%), Indonesia (6%) and Germany (5%) had a total share of 20%. The other 129 countries constituted 41% of the export.

India's Export

The export of coffee from India ranged from 1.53 to 1.70 lakh tonnes during 2000-01 to 2006-07. It declined from 1.65 lakh tonnes valued at Rs.

803.48 crores in 2000-01 to 1.58 lakh tonnes valued at Rs. 1124.75 crores in 2005-06. 2003-04 witnessed rise in exports to 1.70 lakh tonnes. The major item of export was coffee neither roasted nor decaffeinated with 99.81 % 2005-06. The other coffee items included (i) not roasted but decaffeinated coffee (ii) roasted not decaffeinated coffee (iii) roasted decaffeinated coffee, and (iv) other coffee.

Global Export of Green and Roasted Coffee (2004)

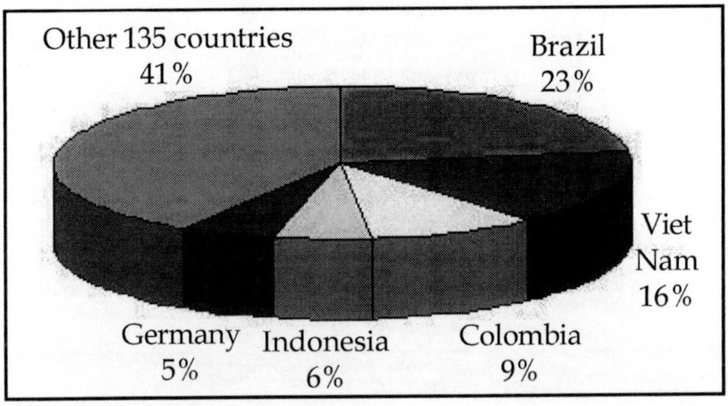

Source: http://faostat.fao.org

Export of Coffee from India During 2000-01 to 2006-07

Years	Quantity ('000 tones)			Value (Rs. Crores)		
	Coffee (Neither roasted nor decaffeinated)	**Other coffee**	**Total**	**Coffee (Neither roasted nor decaffeinated)**	**Other coffee**	**Total**
2000-01	159.16 (96.17)	6.33 (3.83)	165.49 (100.00)	779.73 (97.04)	23.75 (2.96)	803.48 (100.00)
2001-02	149.03 (97.11)	4.43 (2.89)	153.46 (100.00)	674.28 (97.03)	20.67 (2.97)	694.95 (100.00)
2002-03	164.57 (98.08)	3.23 (1.92)	167.80 (100.00)	692.81 (97.84)	15.33 (2.16)	708.13 (100.00)
2003-04	167.24 (98.14)	3.18 (1.86)	170.42 (100.00)	730.38 (97.86)	15.97 (2.14)	746.35 (100.00)
2004-05	147.48 (98.85)	1.724 (1.15)	149.21 (100.00)	753.73 (98.93)	8.19 (1.07)	761.92 (100.00)
2005-06	157.21 (99.81)	0.29 (0.19)	157.50 (100.00)	1122.37 (99.79)	2.38 (0.21)	1124.75 (100.00)
2006-07 (Apr-June)	63.84 (99.87)	0.082 (0.13)	63.92 (100.00)	477.01 (99.79)	1.03 (0.21)	478.04 (100.00)

Note: Figures in parenthesis are percentages to total
Source: http://dgft.delhi.nic.in/

The following figure depicts the variations in annual export of coffee during 2000-01 to 2005-06.

Export of Coffee from India During 2001-02 to 2005-06

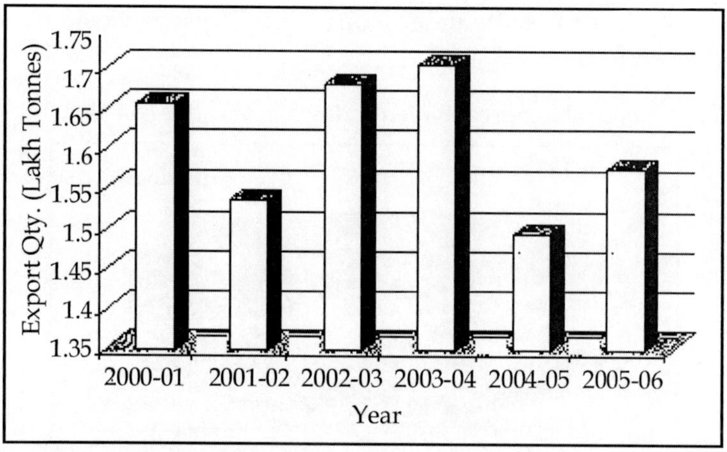

Source: http://dgft.delhi.nic.in/

Indian coffee is exported to 74 countries. The major destination of India's coffee export was Europe. Among them, Italy was the leading importer with 53.36 thousand tonnes accounting for 33.88% of the coffee export in 2005-06, followed by Germany (12.09%), Belgium (7.32%), Spain (6.60%) and Russia (5.30%).

Direction of Export of Coffee from India (2005-06)

Importing Countries	'000 Tones	%
Italy	53.36	33.88
Germany	19.05	12.09
Belgium	11.53	7.32
Spain	10.39	6.60
Russia	8.35	5.30
Slovenia	7.60	4.82
Greece	4.49	2.85
Japan	3.94	2.50
Portugal	3.59	2.28
Switzerland	3.28	2.08
Others (64 Countries)	31.93	20.27
World	157.50	100.00

Source: http://dgft.delhi.nic.in/

Import

Global import of green coffee accounted 54.63 lakh tonnes valued at Rs. 33529 crores in 2004. The leading importer of green coffee was USA with 22.68% share, followed by Germany (17.55%), Japan (7.34%) and Italy (3.95%). India was in 34th rank with an import of 0.20 lakh tonnes constituting 0.38% of the global imports.

Global Import of Green Coffee (2004)

Importing countries	Quantity		Value	
	Lakh tonnes	%	Rs. Crores	%
USA	12.39	22.68	8890.96	26.52
Germany	9.59	17.55	5496.87	16.39
Japan	4.01	7.34	2857.99	8.52
Italy	3.95	7.23	2199.11	6.56
France	2.40	4.39	1412.20	4.21
Spain	2.26	4.13	1094.35	3.26
Belgium	1.94	3.54	1236.31	3.69
Netherlands	1.39	2.54	986.76	2.94
Canada	1.37	2.50	1005.31	3.00
Algeria	1.29	2.36	505.03	1.51
Other 157 countries	14.05	25.72	7844.08	23.39
World	54.63	100.00	33528.99	100.00

Source : http://faostat.fao.org

USA was the leading importer. Germany, Japan, Italy, France and Spain altogether constituted 22.21% of the import. The other 161 countries accounted for 37% of the global import.

India's Import

The import of coffee into India shows a steady increase from 4.12 thousand tonnes in 2000-01 valued at Rs.17.63 crores to 38.19 thousand tonnes valued at Rs.166.84 crores in 2005-06. The coffee export from India since 2002-03 is almost exclusively confined to coffee neither roasted nor decaffeinated. The import of coffee has been steadily increasing since 2002-03.

Import of Coffee into India

Year	Quantity (Thousand tonnes)				Value (Rs. Crores)			
	Coffee neither roasted nor decaffe-inated	Roasted not decaffe-inated coffee	Other	Total	Coffee neither roasted nor decaffe-inated	Roasted not decaffe-inated coffee	Other	Total
2000-01	0.79 (19.22)	0.93 (22.71)	2.39 (58.07)	4.11 (100.00)	3.13 (7.61)	3.62 (8.80)	10.89 (26.48)	17.63 (42.89)
2001-02	1.14 (62.32)	0.02 (0.87)	0.67 (36.81)	1.83 (100.00)	4.27 (23.33)	0.34 (1.87)	4.16 (22.69)	8.77 (47.89)
2002-03	5.13 (94.02)	0.00 (0.06)	0.32 (5.92)	5.46 (100.00)	11.23 (20.56)	0.22 (0.41)	1.29 (2.35)	12.74 (23.32)
2003-04	8.76 (99.85)	0.01 (0.06)	0.01 (0.09)	8.77 (100.00)	21.33 (24.32)	0.25 (0.28)	0.18 (0.21)	21.76 (24.81)
2004-05	22.19 (99.95)	0.01 (0.03)	0.01 (0.03)	22.20 (100.00)	63.36 (28.54)	0.30 (0.13)	0.20 (0.09)	63.85 (28.76)
2005-06	38.17 (99.95)	0.01 (0.02)	0.01 (0.03)	38.19 (100.00)	166.34 (43.55)	0.26 (0.07)	0.24 (0.06)	166.84 (43.68)
2006-07 (Apr-Jun)	6.29 (99.93)	0.00 (0.04)	0.00 (0.03)	6.29 (100.00)	33.26 (52.86)	0.09 (0.15)	0.03 (0.04)	33.38 (53.05)

Source: http://dgft.delhi.nic.in/

Import of Coffee into India in Quantity

Although India has been importing coffee from 25 countries, Indonesia and Vietnam alone accounted for 97% of quantity and 96% of value. The share of both Vietnam and Indonesia were more or less same at around 48%.

Kerala's Production

Kerala is the second largest producer of coffee in India next to Karnataka. Of the 3.80 lakh hectares of planted area at the national level in 2005-06, Kerala accounted for 22.3% compared to 58.5% by Karnataka. Kerala produced 60175 tonnes of coffee in 2005-06, which constituted the 21.35% of national production.

Kerala's Export

The export of coffee through Cochin port was 0.69 lakh tonnes in 2005-06 although production in Kerala was at 0.60 lakh tonnes. This is due to the fact that good part of exports from Karnataka has been routed through Cochin port. The share of export through Cochin port at national level decreased from 55.56% in 2002-03 to 34.16% in 2005-06. Kerala's export is estimated 40,000 to 44,000 tonnes, which constituted 67 to 73% of the production in the state.

Export of Coffee

Year	Kerala (through Cochin port)		India		Kerala's share (%)	
	Quantity (Lakh Tonnes)	Value (Rs. Crores)	Quantity (Lakh Tonnes)	Value (Rs. Crores)	Quantity	Value
2002-03	1.15	387.37	2.07	1051	55.56	36.86
2003-04	1.19	443.42	2.33	1158	51.07	38.29
2004-05	1.15	462.97	2.11	1224	54.50	37.82
2005-06	0.69	603.23	2.02	1510	34.16	39.95

Source: (i) Economic Review 2006 (ii) Economic Survey 2006-07

Kerala's Import

Import of coffee through Cochin port declined from 52.69 tonnes in 2004-05 to 16.73 tonnes in 2005-06. Among the coffee products imported, other coffee was the principal item imported in 2004-05 with a share of 86.9 %. Rest was in the form of instant coffee. The entire import of coffee in 2005-06 was instant coffee.

Import of Coffee through Cochin Port

Items	2004-05	2005-06
Instant coffee	6.90	16.73
Other coffee	45.79	-
Total coffee	52.69	16.73

Source: Consolidated Annual Reports (Various issues), The Cochin Chamber of Commerce and Industry

Factors Affecting Coffee Prices

- The size and availability of coffee stocks worldwide
- Seasonal factors have a significant influence on price of coffee

- Change in coffee consumption pattern governed by factors population size of coffee consumers, standard of living and cultural acceptability
- Availability and price of coffee substitutes like tea, cocoa and coffee complements like sugar also play a very important role in determining coffee prices

Prices

The Coffee prices in India are largely influenced by New York Exchange for Arabicas and London Exchange for Robustas. The international and domestic prices have improved during 2005-06 and the average Indian Coffee Trade Association (ICTA) price for January to December 2005 was Rs104.44/ kg for Arabica and Rs 53.66/kg for Robusta.

Marketing System

Consequent upon the implementation of 100% Free Sale Quota announced by the Govt. of India during 1995, the Coffee Board is no longer involved in pooling and marketing of coffee. However, the Marketing Intelligence Unit of the Board continues to collect and compile daily market information and disseminates information/data on prices, supply, demand and other factors that are important for market analysis. The Unit published a Data Base on a monthly basis, coordinates the implementation of Price Stabilization Fund of the Govt. of India, the schemes on Promotion of Indian Coffee Exports to Russia and CIS Countries, and also advised the Board on World Trade Organisation (WTO)/Trade related issues.

Coffee Forecasting

A dynamic model was used to make coffee projections. It covers the major exporting and importing countries of green coffee. Supply, demand and stock functions were estimated for each of major exporting and importing countries. The model performs dynamic simulation forward in time and generates forecasts on the basis of assumptions for future behaviour of GDP, consumer price indices and exchange rates. For each year in the future, the International Coffee Organization's composite price is solved to achieve equilibrium between supply and demand for green coffee. This model was developed to provide forecasts for green coffee production, consumption and trade, assuming that coffee is treated as a homogeneous commodity with no distinction between Arabica and Robusta varieties.

Production

World coffee production is projected to grow by 0.5% annually from 1998 - 2000 to 2010, compared to 1.9% of the previous decade. Global output is expected to reach 7.0 million tonnes by 2010 compared to 6.7 million tonnes in 1998 - 2000. The world's largest coffee producing region is likely to continue as Latin America and the Caribbean, although projected annual growth rate for the region is expected to decrease from 1.7 % in the previous decade to 0.4% annually during the projection period. Its output is projected at 4.0 million tonnes by 2010, compared to 4.2 million tonnes in 1998 - 2000. Coffee production in Brazil in 2010 is expected to decrease to 1.3 million tonnes compared to 2.1 million tonnes in 1998 - 2000. In Brazil, improved prices from the mid - 1990s stimulated planting and replanting after a period of decline when growers responded to lower prices by reducing use of agricultural inputs and uprooting plants in marginal areas. In Colombia, based on age profile of coffee areas, output is projected to grow at an annual rate of 0.7% to 2010 to reach 747 000 tonnes, compared to 699 000 tonnes in 1998 - 2000. Some plantings took place during the 1990s in response to the surge in demand for Colombian Milds, which fetch premium prices over other Arabicas.

In Central America, output of Mexico in 2010 is expected to reach 273,000 tonnes, more or less the same as the base period. In Guatemala, the projected annual growth rate of 1.7% would take production to 348 000 tonnes by 2010. A growth rate of 3.9 percent for El Salvador is likely to bring their output to 165 000 tonnes by 2010, while Costa Rica should experience an increase of 4.2% that brings output to 194 000 tonnes.

In Africa, coffee production is expected to increase by 1.5 % annually from the base period to 2010, mostly reflecting increases in yields rather than an expansion in area. Output is anticipated to increase from 961 000 tonnes in 1998 - 2000 to 1.1 million tonnes by 2010. Production in Ethiopia, the largest Arabica coffee producing country in Africa, is expected to expand by 1.6% annually to reach 207 000 tonnes by 2010. Coffee output in Côte d'Ivoire is expected to increase by 3.8%/ annum, which would likely bring its output to 217 000 tonnes by 2010. The output in Uganda is projected to increase at a rate of 0.7% annually from 1998 - 2000 to 2010. Output may rise to 222 000 tonnes by 2010 from 207 000 tonnes in 1998 - 2000, through replanting and higher yields. Kenya is projected to expand output by 1.1% annually during the projection period to arrive at 88 000 tonnes.

Production in Asia is projected to grow by 2.1% annually to reach 1.7 million tonnes by 2010. Much of the expansion is expected to occur in

Indonesia, the largest producing country in the region. Its coffee production expanded rapidly during the 1970s, slowed in the 1990s and is projected to expand at a growth rate of 1.7% annually to 2010 when output is likely to reach 654 000 tonnes. Also, in India, output is projected to rise at 3.1% annually to reach 409 000 tonnes by 2010. An increase of 2.0%/ annum is expected in Viet Nam, where output could reach 561 000 tonnes by 2010. An annual increase of 0.7 percent is expected for Thailand, where output is projected to reach 59 000 tonnes by 2010.

In Oceania, Papua New Guinea is the only significant producing country. Its production has been relatively stable during 1980s and its output in 2010 is estimated 150000 tones.

Consumption

World consumption of coffee is projected to increase by 0.4 % annually from 6.7 million tonnes in 1998 - 2000 to 6.9 million tonnes in 2010. Coffee consumption in developing countries is projected to grow from 1.7 million tonnes in 1998 - 2000 to 1.9 million tonnes in 2010, at an annual rate of 1.3 %, while their share in the world market is expected to increase from 26% in the base period to 28% in 2010. The projected higher growth rate for developing countries compared to developed countries is due mainly to higher income and population growth in developing countries, with increased coffee consumption continuing to be concentrated in the major coffee producing countries.

Developed countries, including countries in transition, are likely to continue to account for the larger, though slightly declining, share of world coffee consumption. In the base period, their share of consumption was 74%, nearly 5 million tonnes, compared with 72% projected for 2010. Coffee consumption in developed countries is projected to grow by 0.1% annually to 5.0 million tonnes by 2010. In Europe, demand for coffee is projected to increase by 0.4%/year to 3.1 million tonnes by 2010. The European Community (EC) is projected to account for 2.2 million tonnes, or 68% of total consumption in Europe. Demand is expected to rise slightly in the EC, but growth in consumption in the rest of Europe, excluding the former Soviet Union/CIS, is expected to show a slight decline. Growth in the former Soviet Union/CIS is expected to be more or less the same as in the base period. In North America demand is projected to decrease by 1.0%/ year, mainly reflecting income and population growth in the region.

Trade

In 2010, global coffee net-exports is projected to reach 5.5 million tonnes. Latin America and the Caribbean, with an export of 2.9 million tonnes, is expected to continue as leading exporting region, although there will be a decline in the net-exports of 0.5% annually. By contrast, in Africa there will be a net export increase at a rate of 1.6 % annually, reaching 1.0 million tonnes and accounting for 18% share of global exports. In Asia, export availabilities are expected to grow to 1.5 million tonnes in 2010, accounting for 27 % of world coffee exports. Export availabilities from Oceania are estimated to increase by 7.3%, reaching 150 000 tonnes, about 3.0% of global export availabilities.

World coffee imports are expected to increase by 0.2% annually during the projection period to reach 5.5 million tonnes by 2010. This compares with average imports of 5.4 million tonnes in 1998 - 2000. Imports by developing countries are projected to reach 421 000 tonnes in 2010, accounting for less than 8% of the world's total and similar to their share in 1998 - 2000. Reflecting the slower growth of consumption, import requirements of the developed countries are projected to grow at an annual rate of 0.1%, reaching 5.1 million tonnes by 2010 and accounting for 92 % of the global total. Import demand by North America is projected to decline moderately to 1.54 million tonnes by 2010. Imports into Europe are projected to decrease marginally to 2.96 million tonnes by 2010. Imports to Japan are projected to grow at 1.6 % annually reaching 460 000 tonnes. Growth in import demand by the former Soviet Union/CIS, where consumption in soluble form has grown but no processing firm has been established in the area, is expected to remain low at less than 1% / annum during the projection period.

Issues and Uncertainties

The result of the projections indicates that global green coffee demand and supply would continue to grow, although at a rate slower than in the previous decade, and be almost in balance at around 7 million tonnes by 2010. The projections indicate that several major changes would take place in the world coffee market to 2010. First, most production growth would come from Asia and Africa, instead of Latin America where most coffee had been produced. Second, the growth of consumption would be faster in developing countries than in developed countries, in contrast to trend over previous decade. Part of the growth in consumption in developing countries would come from increase within producing countries, and partially because of this, international trade would grow slower. This scenario, however, is subject to sudden and substantial changes in world coffee economy.

Recent price crises had an important implication for world coffee economy. The price crisis, which has adversely and seriously affected incomes of all coffee producers, hit some producers more severely than others due to differences in various economic factors such as production cost and exchange rates. These variations may change the relative competitiveness among exporters, and could therefore alter the pattern of world coffee trade. In addition, various international initiatives are expected to take place as exporters underline importance of promoting higher quality coffee with the aim of improving prices through boosting consumption. All these factors may affect the demand and supply conditions in the world coffee markets to 2010 although the price would continue to be the primary determinant.

OIL PALM

Oil Palm *(Elaeis guineensis)* originated in the tropical rain forest of West Africa. The main belt runs through southern latitudes of Cameroon, Cote d'Ivoire, Ghana, Liberia, Nigeria, Sierra Leone, Togo and into the equatorial region of Angola and the Congo. Processing of oil palm fruits for edible oil was practiced in Africa for thousands of years. The oil produced, highly coloured and flavoured, is an essential ingredient in much of the traditional West African cuisine. The oil palm comprises two species of Arecaceae, or palm family. They are used in commercial agriculture in the production of palm oil. The African Oil Palm *Elaeis guineensis* is native to West Africa, occurring between Angola and Gambia, while the American Oil Palm *Elaeis oleifera* is native to tropical Central America and South America . During the 14 to 17 centuries some palm fruits were taken to the Americas and from there to the Far East. The plant thrived better in the Far East, thus providing the largest commercial production of an economic crop far removed from its centre of origin.

Oil palm popularly known as red oil palm is the highest oil producer among perennial oil yielding crops. It produces two distinct oils viz., palm oil (extracted from mesocarp of fresh fruits) and palm kernel oil (from kernel). Palm oil has excellent health attributes. It is rich in vitamins A and E and is cholesterol free. Palm oil can be used in formulation of margarine and cooking fat like vanaspathi. It can be used in manufacture of biscuits, ice creams, soaps, detergents, and shampoos and also as frying fat. Palm kernel oil has a variety of industrial uses.

Soil and Climate

Oil palm is a humid tropical palm which thrives well in localities where annual temperature ranges 19-33° C with an evenly distributed rainfall of 2500 to 4000 mm/ annum, humidity more than 80% and sunshine hours of not less than 5hr/day. It can be grown upto 900m from sea level. It is grown mostly within 10 degree N or S, of the Equator. It grows commercially in latitudes of 17 degrees from the Equator, but yields are reduced at such sites. It is a water-loving crop and requires adequate irrigation. The crop responds well to drip irrigation and yields increase by at least 20%.

It grows on a wide range of soils, provided good drainage and pH between 4 and 7.It tolerates periodic flooding or a high water table; many soils are alluvial in nature. Deep well-drained medium loam soil rich in humus is the most suitable for oil palm cultivation. Highly alkaline, highly saline, water logged and coastal sandy soils should be avoided. At least one metre depth of soil is necessary.

Varieties

Broadly, there are three varieties viz., Dura, Piscifera and Tenera. Tenera, a hybrid of Dura and Piscifera is characterized by thin shell and medium to high mesocarp (65-90%) and high oil content (16-20%). It is a commercially cultivated variety.

Propagation

Oil palm is propagated by seed, using F1 hybrid seed from controlled crosses that produce tenera types (Dura xPiscifera). Seed is produced by companies specializing in oil palm breeding.

Seed Processing and Germination

Full ripened bunches are harvested, fruits are separated and seeds are extracted by retting. For retting, fruits are soaked in water for about 6-7 days, fruits are removed from water and the mesocarp is removed by pounding fruits with sand in a wooden mortar. Seeds are treated with 0.1% Hemisan and then dried in shade for 20 min. Such seeds can be stored in polythene bags to a period of six months without losing viability.

Seed germination is poor and takes several months for germination, results in sporadic germination. Special treatment is required to improve germination.

First seeds are soaked in water for 5 days by changing water daily. Then seeds are spread out and dried in shade for 24 h. Seeds are then to be kept in polythene bag, air tightly closed and kept in a germinator at 40°C for 80 days. Seeds are taken out and again soaked in water for 5 days with daily change of water and dried in shade for 2 h. The seeds are again kept in polythene bag at room temperature. Seeds start germinating within 10-12 days with 90-95% germination.

Nursery Technique

A nursery is essential for growing germinated seeds to produce strong, healthy and viable seedlings for field planting. The old system of raising seedlings in soil beds replaced by poly bag nursery. It can be of single or double stage system. In single stage system sprouts are directly placed in bigger poly bags and nurtured till planting, where as in double stage system, the sprouts are first planted in small sized poly bags and later they are planted in bigger bags and maintained till field planting.

Nursery Planting

One stage poly bag nursery is the best method. Black polythene bag (500 gauge) of size 40x35 cm with perforated holes are used. It is filled with potting mixture (sand, soil, cowdung) in equal proportion. The precise stage of planting sprouts is 10-14 days from germination. Sprouted seeds are placed at a depth of 2.5 cm with plumule pointing upwards. Shade is provided during early stages.

Irrigation twice a day and good mulching during summer are desirable. A fertilizer mixture of NPK in the ratio of 18:46:0 is applied at 3,5,8 and 12 g/ plant during 2nd, 3rd, 4th and 5th month respectively. Afterwards NPK 15:15:15 at 15,20,25,30,35,40 g/plant is applied during 6th, 7th, 8th, 9th, 10th & 11th months, 5-15 g MgSO4 and 0.5-1.5 g borax are applied according to need after 6th month. A few supplementary applications of urea may be required if nitrogen deficiency is noticed.

Site selection, Layout and Planting

Planting

Oil palm is planted in pits of size 60 x60x60cm, in triangular system at a spacing of 9 x 9 x 9 m accommodating 143 plants/ ha. Planting can be done in any season. The best period is June to December. Seedlings of 10-14 months age are the best suited for planting. Rock phosphate (250g) and phorate(50g) are applied and mixed with soil at the base of pit. Place seedlings in pit along with soil after removing polythene bag. Fill up the pit with soil and press all around firmly. Irrigation should be given immediately after planting. Staking is given in wind prone areas.

Nutrition

The oil palm has a high demand for nutrients. Nutrients that are removed continuously through the harvested FFB need replacing, if soil nutrient reserves are not to become depleted.

Fertilizer Recommended for Oil Palm (kg/palm/year)

Year	FYM	Urea	SSP	MOP	MgSO4
1	50	0.870	1.250	0.667	0.125
2	50	1.740	2.500	1.333	0.250
3	50	2.610	3.750	2.000	0.500
4 onwards	50	3.000	5.000	4.000	0.500

In gardens, where 20-25 t FFB/ha is obtained by 6th year, an additional 20% of the recommended dose can be applied to maintain the productivity. Borax @ 100g/palm/yr is recommended when boron deficiency symptoms are noticed. Fertilizer can be applied in two equal splits doses during July-August and December-January. For the newly planted crop, the dose of fertilizer may be given three months after planting. Along with second dose of fertilizer, 50-100 kg FYM or 100 kg green manure and 5 kg neem cake/palm is applied.

Past studies showed that empty fruits bunch (EFB) mulching significantly improved oil palm yield (Hamdan *et al.*1998). EFB generally contains 0.80 % N, 0.22 % P_2O_5, 2.90 % K_2O, and 0.30 % MgO on a dry weight basis. Yield improvement ranging from 5 % to 23 % is achieved depending on soil type. Hamdan *et al.* (1998) showed that nutrients from 60 t/ha/year of EFB, without any inorganic fertilizer, were sufficient to support palm growth. Nevertheless, because of the high C/N ratio, a lower rate of EFB application is made with supplements of inorganic fertilizer. This approach maintains soil productivity through better soil structure and reduces fertilizer cost for immature palms by 58 %, and by 5 % for mature palms.

Intercropping

Oil palm is a wide spaced perennial crop with a long juvenile period of three years. During initial stages of plantation in oil palm i.e., up to 3rd year, some of the light feeder inter crops like pulses, cereals, vegetables, grasses etc., can be grown.

Cover Cropping

A good cover crop should be established by the time, palms are to be field planted. *Pueraria phaseoloides, Pueraria javanica, Calapagonium muconoides and Centrosema pubescens* are commonly used.

Seeds at the rate of 2-3 kg /ha are treated (soaked in hot water 50-60°C for 2 h.) are sown about 2.5cm deep and covered. Five rows 1.2 m apart can be sown in between two rows of palms. Regular hand weeding at two weeks interval and NPK Mg mixture (15:15:6:4 @25-30 kg/ha) is given as a starter one week and three weeks after germination. Again at 3 months and 6 months after sowing, rock phosphate @60 kg/ha is broadcasted. For the next two years, application of rock phosphate @25kg/ha is necessary.

After Care and Management

i) *Basin management*: During first year of planting, basins of 1m radius are to be taken around the palm, removing the soil from inside so that the soil will not accumulate at the collar region. Basin must be widened to 2m radius during 2nd , 3m from 3rd year onwards. Sunhemp seeds can be sown in the basin during summer to reduce weed growth and add nitrogen to the basin

ii) *Weeding*: Active root zone can be kept weed free by hand weeding or spraying herbicide. Herbicide mixtures of paraquat with atrazine, monuron and diuron sprayed on ground twice a year control weeds effectively.

iii) *Mulching*: It can be done with dried leaves, empty bunches and male inflorescence.

Irrigation

Oil palm requires adequate irrigation, as it is a fast growing crop with high productivity and biomass production. For yielding palms of above three years age, a minimum of 150 litres/ day is required. In older plantations the requirement goes up to 200 litres/ day. The crop responds well to drip or micro sprinkler irrigation particularly when water is limited. If drip is installed, four drippers have to be placed for each palm. If each dripper discharges 8litre/hour, 4-5 hours of irrigation are sufficient to discharge 160 litres/ day. Drip irrigation increases productivity by 15-20%, reduces wastage of water, and requires less power/fuel/irrigation compared to conventional irrigation methods. It is important to note that any physiological stress shifts sex ratio in favour of male flowers and consequently the productivity is reduced.

Pruning

Frond pruning has an influence on yield. If the leaves are allowed to break of its own normally it breaks off 30-60cm from the trunk. These remaining frond bases interfere with harvesting and pollination. Each palm produces 24 leaves annually. Pruning helps in easy visual assessment of fruit ripeness, harvest, minimizing the loss of loose fruits which are retained in the frond bases. Excessive pruning results in decreased yield. During immature period, pruning should be confined to the removal of senescent and useless fronds lying close to the soil surface. When inflorescence starts developing, two whorls of fronds are left below the lowest developing inflorescence. This system continues up to 10th year and from 11th year a single frond whorl below the lower most bunch only be retained. Pruning is organized in such a way that about 32-35 top leaves are left undisturbed on the adult palms.

Ablation

The practice of removing young male and female inflorescences and young bunches before starting economic yield is termed as ablation. It usually starts about 14th months after field planting and continues until about 26th month, when about 70% of the palms are producing inflorescence. Usually 4-5 weeks intervals are given between two successive ablation rounds.

Yield

Oil palm starts bearing from 3rd year onwards and its economic life varies from 30 to 35 years. The yield of oil palm varies according to age and management. Under average management conditions, in a mature plantation (8 to 9 years old), yield of 20-25 tonnes of fresh fruit bunches (FFBs)/ha is expected. Under good maintenance and management, yield up to 30-40 tonnes of FFBs /ha is possible.

Cost of Cultivation

Oil Palm (one hectare) - Without Drip Irrigation System

Varieties	**Tenera, a hybrid of Dura and Piscifera**
Spacing	9 x 9 m (triangular system)
Planting density	143

S. No	Item of expenditure	Units	Year 1	Year 2	Year 3	Year 4	Total
I	**Material**						
1	Planting material including 10% casualty replacement in second year	Seedlings	7865	770			**8635**
2	Manure and fertilizer		3259	5088	7007	9281	**24635**
3	Drip irrigation system for mono crop						
4	Plant protection		600	600	1200	1200	**3600**
5	Intercrop		0	0			**0**
6	Fencing		0				**0**
	Subtotal		**11724**	**6458**	**8207**	**10481**	**36870**
II	**Labour**						
1	Land preparation	LS	1750				**1750**
2	Digging,filling and planting@Rs. 15/plant		2145				**2145**
3	Manure and fertilizer application	4	240	240	240	240	**960**
4	Pesticide application	2	120	120	120	120	**480**
5	Irrigation	5	300	300	300	300	**1200**
6	Interculture operations	10	600	600	600	600	**2400**
7	Harvesting (@ Rs. 150 per ton)	150				600	**600**
8	Watch and ward					1000	**1000**
	Subtotal		**5155**	**1260**	**1260**	**2860**	**10535**
III	**Contigencies**	5%	844	386	473	667	**2370**
	Total (rounded off)		**17700**	**8100**	**9900**	**14000**	**49700**

*Source:*NABARD

Principles of Preservation and Processing Methods

The general principles of preservation include destruction of enzymes in the raw material and removal of micro-organisms by heat (sterilization) during processing. Elimination of as much water as possible from the oil is done to prevent microbial growth (bacterial activity, or disease-causing germs) during storage. The oil therefore has a long shelf life due to its low moisture content. Proper packaging and storage of extracted oil are required to slow down chemical deterioration (rancidity).

Unit cost capitalised upto 4th year : 49700
(without drip irrigation system)

Oil Palm (one hectare) - With Drip Irrigation System

Type of drip system	Unit cost for oil Palm	Drip cost	Total Cost
With 16 mm online drip	49700	23850	73550
With 12 mm online drip	49700	21990	71690
With 16 mm microjet	49700	26090	75790
With 12 mm microjet	49700	24230	73930
Average Cost with Micro irrigation system	49700	24040	73740
Rounded off	49700	24000	73700

*Source:*NABARD

Palm Oil Processing-unit Operations

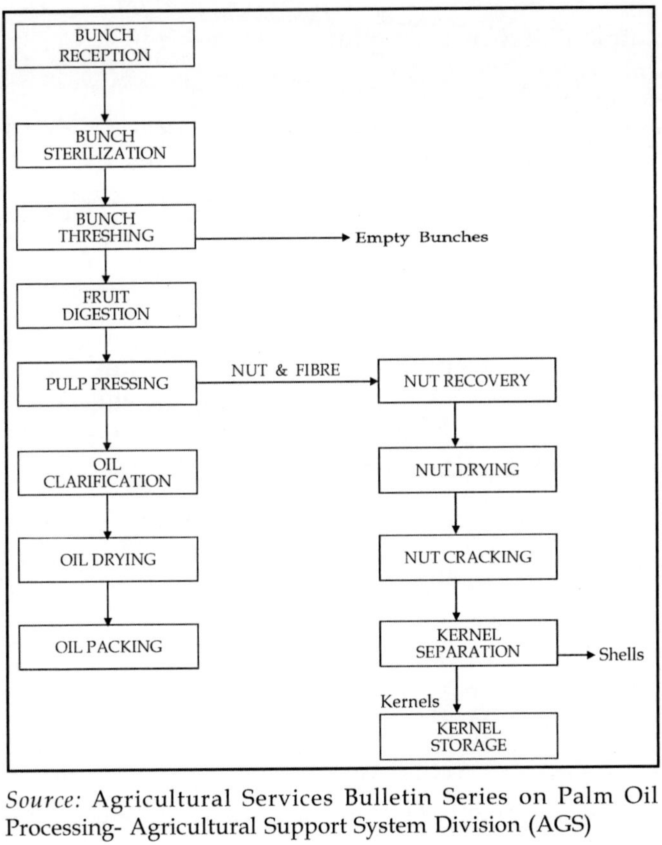

Source: Agricultural Services Bulletin Series on Palm Oil Processing- Agricultural Support System Division (AGS)

Palm Oil Processing Unit Operations

The fresh fruit bunch consists of fruit embedded in spikelets growing on a main stem. Manual threshing is achieved by cutting the fruit-laden spikelets from bunch stem with an axe or machete and then separating the fruit from the spikelets by hand. Most small-scale processors do not have the capacity to generate steam for sterilization. Therefore, the threshed fruits are cooked in water. Whole bunches which include spikelets absorb a lot of water in the cooking process. High-pressure steam is more effective in heating bunches without losing much water. Therefore, small-scale operations thresh bunches before the fruits are cooked, while high-pressure sterilization systems thresh bunches after heating to loosen the fruits.

Sterilization of Bunches

Sterilization or cooking means use of high-temperature wet-heat treatment of loose fruit. Cooking normally uses hot water and sterilization uses pressurized steam. The cooking action serves several purposes.

- Heat treatment destroys oil-splitting enzymes and arrests hydrolysis and autoxidation.
- Wet heat weakens fruit stem and makes it easy to remove fruit from bunches on shaking or tumbling in the threshing machine.
- Heat helps to solidify proteins in which oil-bearing cells are microscopically dispersed. The protein solidification (coagulation) allows oil-bearing cells to come together and flow more easily on application of pressure.
- Fruit cooking weakens pulp structure, softening and making it easier to detach fibrous material and its contents during digestion process. The high heat is enough to disrupt partially oil-containing cells in mesocarp and permit oil to be released more readily.
- The moisture introduced by steam acts chemically to break down gums and resins. The gums and resins cause oil to foam during frying. Some of the gums and resins are soluble in water. Others can be made soluble in water, when broken down by wet steam (hydrolysis), so that they can be removed during oil clarification. Starches present in fruit are hydrolyzed and removed in this way.
- When high-pressure steam is used for sterilization, the heat causes moisture in nuts to expand. When pressure is reduced, contraction of nut leads to detachment of kernel from shell wall, thus loosening kernels

within their shells. Detachment of kernel from the shell wall greatly facilitates nut cracking operations later. Sterilization (cooking) is one of the most important operations in oil processing, ensuring success of several other phases.

- During sterilization, it is important to ensure evacuation of air from the sterilizer. Air not only acts as a barrier to heat transfer, but oil oxidation increases considerably at high temperature; hence oxidation risks are high during sterilization. Over-sterilization can also lead to poor bleach ability of resultant oil. Sterilization is also the chief factor responsible for discolouration of palm kernels, leading to poor bleach ability of extracted oil and reduction of protein value of press cake.

Digestion of Fruit

Digestion is the process of releasing palm oil in the fruit through rupture or breaking down of the oil-bearing cells. The digester commonly used consists of a steam-heated cylindrical vessel fitted with a central rotating shaft carrying a number of beater (stirring) arms. Through action of rotating beater arms, the fruit is pounded. Pounding, or digesting fruit at high temperature, helps to reduce viscosity of oil, destroys the fruits' outer covering (exocarp), and completes disruption of oil cells already begun in the sterilization phase. Unfortunately, for reasons related to cost and maintenance, most small-scale digesters do not have the heat insulation and steam injections that help to maintain their contents at elevated temperature during this operation.

Pressing (Extracting Palm Oil)

There are two distinct methods of extracting oil from digested material. One system uses mechanical presses and is called the 'dry' method. The other called the 'wet' method uses hot water to leach out the oil.

In the 'dry' method, objective of extraction stage is to squeeze the oil out of a mixture of oil, moisture, fibre and nuts by applying mechanical pressure on digested mash. There are a large number of different types of presses but the principle of operation is same for each. The presses may be designed for batch (small amount of materials operated upon for a time period) or continuous operations.

Batch Presses

In batch operations, material is placed in a heavy metal 'cage' and a metal plunger is used to press the material. The main differences in batch press designs are as follows: a) the method used to move the plunger and apply the pressure; b) amount of pressure in the press; and c) size of cage.

Continuous Systems

Early centrifuges and hydraulic presses have now given way to specially designed screw-presses similar to those used for other oilseeds. These consist of a cylindrical perforated cage through which runs a closely fitting screw. Digested fruit is continuously conveyed through cage towards an outlet restricted by a cone, which creates pressure to expel oil through cage perforations (drilled holes). Oil-bearing cells that are not ruptured in the digester remain unopened if a hydraulic or centrifugal extraction system is employed. Screw presses, due to turbulence and kneading action exerted on fruit mass in the press cage, can effectively break open unopened oil cells and release more oil. These presses act as an additional digester and are efficient in oil extraction.

Clarification and Drying of Oil

The main point of clarification is to separate oil from its impurities. The fluid coming out of the press is a mixture of palm oil, water, cell debris, fibrous materials and 'non-oily solids'. Because of non-oily solids, the mixture is very thick (viscous). Hot water is therefore added to the press output mixture to thin it. The dilution (addition of water) provides a barrier causing the heavy solids to fall to bottom of container while the lighter oil droplets flow through watery mixture to the top when heat is applied to break emulsion (oil suspended in water with aid of gums and resins). Water is added in a ratio of 3:1.

The diluted mixture is passed through a screen to remove coarse fibre. The screened mixture is boiled from one or two hours and then allowed to settle by gravity in the large tank so that the palm oil, being lighter than water, will separate and rise to the top. The clear oil is decanted into a reception tank. This clarified oil still contains traces of water and dirt. To prevent increasing FFA through autocatalytic hydrolysis of the oil, moisture content of oil must be reduced to 0.15 to 0.25%. Re-heating the decanted oil in a cooking pot and carefully skimming off the dried oil from any engrained dirt removes residual moisture. Continuous clarifiers consist of three compartments to treat the crude mixture, dry decanted oil and hold finished oil in an outer shell as a heat exchanger.

Oil Storage

In large-scale mills, the purified and dried oil is transferred to a tank for storage prior to dispatch from the mill. Since the rate of oxidation of oil increases with temperature of storage, oil is normally maintained around 50°C, using hot water or low-pressure steam-heating coils, to prevent solidification and fractionation. Iron contamination from storage tank occurs if the tank is not lined with a suitable protective coating.

Small-scale mills simply pack the dried oil in used petroleum oil drums or plastic drums and store the drums at ambient temperature.

Kernel Recovery

The residue from press consists of a mixture of fibre and palm nuts. The nuts are separated from fibre by hand in small-scale operations. The sorted fibre is covered and allowed to heat, using its own internal exothermic reactions, for about two or three days. The fibre is then pressed in spindle presses to recover a second grade (technical) oil that is used normally in soap-making. The nuts are usually dried and sold to other operators who process them into palm kernel oil

In large-scale kernel recovery process, the nuts contained in the press cake are separated from the fibre in a depericarper. They are then dried and cracked in centrifugal crackers to release the kernels. The kernels are normally separated from the shells using a combination of winnowing and hydrocyclones. The kernels are then dried in silos to a moisture content of about 7% before packing.

During the nut cracking process some of the kernels are broken. The rate of FFA increase is much faster in broken kernels than in whole kernels. Breakage of kernels should therefore be kept as low as possible, given other processing considerations.

Summary of Unit Operations

Unit operation	**Purpose**
Fruit fermentation	To loosen fruit base from spikelets and to allow ripening processes
Bunch chopping	To facilitate manual removal of fruit
Fruit sorting	To remove and sort fruit from spikelets
Fruit boiling	To sterilize and stop enzymatic spoilage, coagulate protein and expose microscopic oil cells
Fruit digestion	To rupture oil-bearing cells to allow oil flow during extraction while separating fibre from nuts
Mash pressing	To release fluid palm oil using applied pressure on ruptured cellular contents

Contd...

Oil purification	To boil mixture of oil and water to remove water-soluble gums and resins in the oil, dry decanted oil by further heating
Fibre-nut separation	To separate de-oiled fibre from palm nuts.
Second Pressing	To recover residual oil for use as soap stock
Nut drying	To sun dry nuts for later cracking

Source: Agricultural Services Bulletin Series on Palm Oil Processing- Agricultural Support System Division (AGS)

Schematics of Processing Models and Associated Machinery

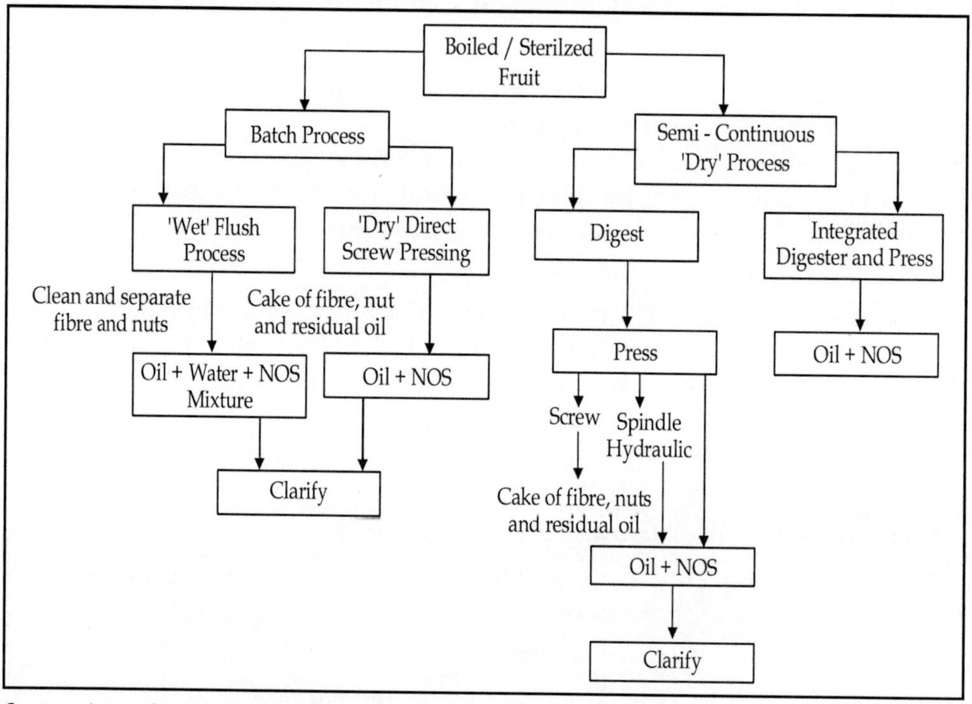

Source: Agricultural Services Bulletin Series on Palm Oil Processing- Agricultural Support System Division (AGS)

Process Technology/Capital Investment Considerations

The wide array of machinery options makes it possible for a processor to start operations with a manual spindle-press used to pound the palm fruit. Another may start with a single motorised vertical wet process digester. Further up the investment scale are those who can afford the combination of horizontal digester and screw-press or combination of horizontal digester and hydraulic press along with associated sterilizers, threshers, and oil clarifiers. Another combination that is yet to be tried is horizontal motorised screw-press in combination with a second stage vertical flushing digester for maximum palm oil extraction and fibre/nut separation.

Type of unit	Key machines	Rated capacity (kg FFB/hr)	Extraction efficiency (%)	Capital investment (US$)
Single batch unit				
Dry	Spindle	100-200	55	150-200
	Hydraulic	200-300	67-74	5 000-7 000
	Screw	250-400	77.4	1 500-6 000
Wet	Vertical digester	500-800	80-90	1 500-2 500
Dry	Motorised horizontal digester (only)	500-1000	55	2 500-3 000
Dual separate units				
Dry	Digester + Spindle presses	200-300	60-70	3 000-5 000
	Digester + hydraulic press	400-800	67-78	7 000-10 000
Semi-continuous combined units	Motorised digester + hydraulic press	500-850	70-87	10 000
Dry	hydraulic + spindle-press			-15 000
	Digester + screw-press	500-850	76-90	12 000-15 000

Source: Agricultural Services Bulletin Series on Palm Oil Processing- Agricultural Support System Division (AGS)

Palm Kernel Oil Extraction

Mechanical Extraction

Mechanical extraction processes are suitable for both small and large capacity operations. The three basic steps in these processes are (a) kernel pre-treatment, (b) screw-pressing, and (c) oil clarification.

Mechanical Extraction of Palm Kernel Oil

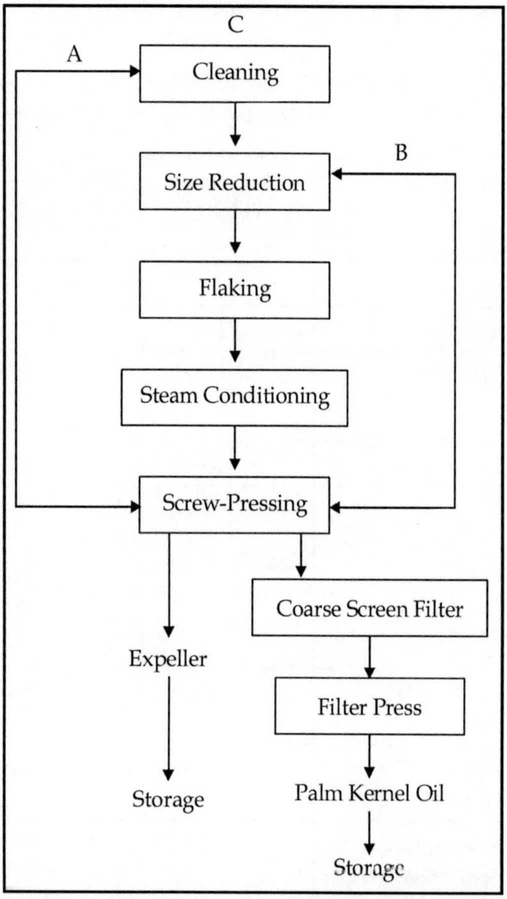

Source: Agricultural Services Bulletin Series on Palm Oil Processing- Agricultural Support System Division (AGS)

Line (A) is for direct screw-pressing without kernel pre-treatment; Line (B) is for partial kernel pre-treatment followed by screw-pressing; and Line (C) is for complete pre-treatment followed by screw-pressing.

Kernel Pre-treatment

Proper kernel pre-treatment is necessary for efficient extraction of oil from kernels. The feed kernels must first be cleaned of foreign materials that may cause damage to the screw-presses, increasing maintenance costs and down time, and contamination of products. Magnetic separators commonly used to remove metal debris, while vibrating screens are used to sieve sand, stones or other undesirable materials.

A swinging hammer grinder, breaker rolls or a combination of both are used to break the kernels into small fragments. This process increases the surface area of kernels, thus facilitating flaking. The kernel fragments subsequently are subjected to flaking in a roller mill. A large roller mill consists of up to five rollers mounted vertically above one another, each revolving at 200-300 rpm. The thickness of kernel cakes is progressively reduced as it travels from the top roller to the bottom. This progressive rolling initiates rupturing of cell walls. The flakes that leave the bottom nip are from 0.25 to 0.4 mm thick.

The kernel flakes are then conveyed to a stack cooker for steam conditioning, the purpose of which is to adjust moisture content of meal to an optimum level, rupture cell walls (initiated by rolling), reduce viscosity of oil, coagulate the protein in the meal to facilitate separation of oil from protein materials.

The meal flows from the top compartment down to fifth compartment in series. At each stage a mechanical stirrer agitates the meal. Steam trays heat cookers, and live steam may be injected into each compartment when necessary. The important variables are temperature, retention time and moisture content. In the palm kernel, meals are normally cooked to a moisture content of 3% at 104-110°C.

Screw-pressing

The properly cooked meal is then fed to the screw-press, which consists of an interrupted helical thread (worm) which revolves within a stationary perforated cylinder called the cage or barrel. The meal is forced through the barrel by the action of revolving worms. The volume axially displaced by the worm diminishes from the feeding end to the discharge end, thus compressing the meal as it passes through the barrel.

The expelled oil drains through the perforation of the lining bars of the barrel, while the de-oiled cake is discharged through an annular orifice. To prevent extreme temperatures that could damage oil and cake quality, the worm-shaft is always cooled with circulating water while the barrel is cooled externally by recycling some cooled oil.

Oil Clarification

The expelled oil invariably contains a certain quantity of 'fines and foots' that need to be removed. Oil from the presses is drained to a reservoir. It is then either pumped to a decanter or revolving coarse screen to remove a large part of solid impurities. The oil is then pumped to a filter press to remove remaining solids and fines to produce clear oil prior to storage. The cakes discharged from the presses are conveyed for bagging or bulk storage.

Not all crushers use same procedure for mechanical extraction of kernel oil. There are three variations: direct screw-pressing, partial pre-treatment, and complete pre-treatment.

Direct Screw-pressing

Some mills crush kernels directly in presses without any pre-treatment. Double pressing is usually required to ensure efficient oil extraction. The screw-presses used normally are less than 10 tonnes/ unit/ day.

Partial Pre-treatment

The kernels are first broken down to smaller fragments by grinding prior to screw-pressing. In some cases, cooking is also carried out.

Complete Pre-treatment

The full pre-treatment processes described earlier are carried out prior to screw-pressing. Plants with larger capacities (50-500 tonnes/day) choose complete pre-treatment and the equipment is usually imported from Europe. FATECO and Faith Engineering now offer complete line for small-scale operators.

Solvent Extraction

Solvent extraction processes can be divided into three main unit operations: kernel pre-treatment, oil extraction, and solvent recovery from the oil and meal. For the purposes of small-scale operations, it is sufficient to mention solvent extraction process as an alternative for high capacity mills. However, the process is not recommended for small enterprises.

Uses of Palm Oil

Palm oil's unique composition makes it versatile in its application in food manufacturing and in the chemical, cosmetic and pharmaceutical industries. Its semi-solid physical properties are needed in many food preparation. Its non-cholesterol quality and digestibility make it popular as source of energy, while its technical and economic superiority makes it preferable as base material in the manufacture of various non-edible products.

In food industries, palm oil is the choice for manufacturing solid fat products. Palmoil, olein and stearin are popularly used worldwide in making margarine, shortenings and confectionery, and in frying snack foods.

Palm oil is cost-effective as it does not go through expensive hydrogenation process. Its high content of natural antioxidants and its stability at high temperatures make it excellent as a deep frying medium. It also gives fried products a longer shelf life, while its bland taste brings out natural flavors of food.

Palm oil is also used in the manufacture of soaps, detergents and other surfactants. It is a good raw material for producing oleochemicals, fatty acids, fatty alcohols, glycerol and other derivatives for manufacture of cosmetics, pharmaceuticals, house hold and industrial products.

Oleochemicals manufactured from palm oil and palm kernel oil are now popular for manufacture of environmentally friendly detergents as they are readily biodegradable.

Dietary Value/100 Gram Edible Portion

	Oil palm fruit	**Palm oil**
Water (%)	26	0.5
Calories	540	878
Protein (%)	1.9	0
Fat (%)	58.4	99.1
Carbohydrates (%)	12.5	0.4
Crude Fiber (%)	3.2	0
	% of US RDA	
Vitamin A	3.5	2.3
22 Thiamin, B1	13.3	2.0
Riboflavin, B2	5.6	Trace
Niacin	7.0	---
Vitamin C	26.7	---
Calcium	10.3	0.9
Phosphorus	5.9	1.0
Iron	45	55
Sodium	2.4	---
Potassium	5.3	---

Source: Agricultural Services Bulletin Series on Palm Oil Processing- Agricultural Support System Division (AGS)

Crop oil Quality and Nutritional Value

Palm oil has wide range of uses, both food and non-food, 80% of oil produced serves as an important source of vegetable fat for an increasing number of people. The crude oil comprises of approximately half saturated (mainly palmitic) and half unsaturated (mainly oleic) fatty acids. Saturated fats have in the past been widely regarded as being undesirable dietary components due to their association with high cholesterol levels and heart disease. There are various forms of cholesterol, not all bad, and recent studies showed that intake of palm oil raises levels of high-density lipoprotein (HDL, 'good' cholesterol) at the expense of low-density lipoprotein (LDL, 'bad' cholesterol.).

Unsaturated oils soybean are usually hydrogenated to render them less liquid and more suitable for manufacturing of margarines and other products. This essentially converts them into saturated fats while at the same time generating *trans*-fatty acids that do not occur naturally. As such, *trans* isomers impose a risk of heart disease. There is now a preference for *trans*-free oils such as palm oil. Palm oil is therefore promoted both as a'balanced' oil (having equal saturated and non-saturated content) and one that is *trans*-free (Wahle and James 1993; Ascherio 2002).

Palm oil is rich in carotenoids, from which it derives its deep red colour, and the major component of its glycerides is the saturated fatty acid palmitic; hence it is a viscous semi-solid, even at tropical ambients, and a solid fat in temperate climates. Vitamin E is a free radical scavenger, with anti-cancer properties. Red palm oil, which is the product of a new technology that aims to preserve these minor components, has a further role to play in a healthy diet. More recently, it is shown that the oil palm is a rich source of phenolic antioxidants, a minor fraction of which enters the oil phase and confers further health benefits. Palm oil is now accepted worldwide as a healthy dietary ingredient.Palm oil is high in vitamin K and dietary Magnesium. Palm oil contains 43% saturated fats, 43 % monounsaturated and 13 % polyunsaturated fats.

Standards for Oil Palm Fibre

The oil palm industry in Malaysia started 80 years ago. Today it is the largest in agricultural plantation sector, exceeding rubber plantation by more than double in area planted.

In terms of hectare, the total area under oil palm cultivation is over 2.65 million hectares, producing over 8 million tones of oil annually. The oil consists of only 10% of the total biomass produced in the plantation. The remainder consists of huge amount of lignocellulosic materials such as oil palm fronds, trunks and empty fruit bunches. The projection figures of these residues are as follows:

- 7.0 million tonnes of oil palm trunks
- 26.2 million tonnes of oil palm fronds
- 23% of Empty Fruit Bunch (EFB)/ tonne of Fresh Fruit Bunch (FFB) processed in oil palm mill.

Recent report shows that the mesocarp fibre and shell are used as boiler fire source to produce steam and generate power. Empty fruit bunches are mainly incinerated to produce bunch ash to be distributed back to the field as fertiliser.

The conventional method of burning these residues often creates environmental problems in that it generates severe air pollution and is prohibited by Environment Protection Act. In abiding by the regulations, these residues are expensive to dispose. An alternative way of optimising usage of oil palm residues fibre is based on production of value-added products.

The empty fruit bunch fibre (EFB) was identified as the first of the series of standards on oil palm fibres. The EFB has the highest fibre yield and is the only material commercially utilised for fibre extraction but there are good potentials for the exploitation of the other two materials (oil palm fronds and trunks). The critical parameters affecting the quality of the end product of oil palm fibre include the fibre length, moisture content, oil content and impurities.

For the purpose of commercialisation, the standard recommends one grade of empty fruit bunch fibre (EFB). In quantifying the percentage proportion of fibre length, the standard has established following numerical values:

Fibre Length	**% Proportion (Oven dry weight basis)**
1) >100 mm	30
2) >50 - 100 mm	35
3) <50 mm	35

Characteristically, EFB fibres are clean, biodegradable and compatible than many other fibres from other wood species. EFB fibres are suitable for manufacture of mattress, car seat, insulation, composite panel product and particle board. The choice and selection of fibre length are very much dependent on nature of manufactured product. Long fibre length is suitable for mattress and short fibre makes excellent particle board.

Inherent characteristics such as high moisture content will have a detrimental effect on oil palm residues. Degradation and infection will easily take place thus leading to deterioration in properties of fibres. To prevent this from happening, a maximum moisture content of 15% (on wet basis) was set. This value is comparable to moisture content for all wood-based fibres.

In addition, the standard has also identified impurities as a critical element in producing high quality fibres. Impurities by definition means calyxes, spikes, aggregates or fibre strands, any dust or fine particles and parenchyma of EFB. Since it is practically impossible to have fibre free impurities, the maximum level of impurities allowed for in the standard is 15%. Nevertheless with future improvement in the process technologies, this limit could be further reduced.

Presence of a significant percentage of residual oil posed a major problem in that it reacts with moisture content, giving rise to rancidity and ultimately fungal growth. The control limit for oil content shall not exceed 3%.

Other inherent characteristics like salt content, odour, mass, etc. were taken into consideration during the process of formulating the standard. Since these parameters do not affect the overall quality of the EFB fibre significantly, it was therefore excluded from the general requirements of the standard.

Marketing

Crude palm oil and palm kernel oil are edible vegetable oils and now have a wide range of markets in the food and oleochemical industries. In addition, palm oil has a very healthy component of human diet.

The oil palm remains a formidable competitor to other vegetable oil crops in terms of oil yield/ha and resource use efficiency due to its unrivalled ability to transform solar energy into vegetable oil. The oil yield from properly maintained oil palm is over six times more than oil yields from commercially grown rapeseed. Additionally, the energy balance expressed by the ratio of energy output to input is wider for oil palm than other commercially grown oil crops. These characteristics undoubtedly favour oil palm as a renewable energy source in future. With a share of over 54% (18 million tons), palm oil dominates global vegetable oil export trade. India, China, Pakistan, Middle-East countries and European Union are the major importers of the commodity.

Indian Scenario

India is amongst the largest producer and consumer of vegetable oils. In the face of huge imports of edible oils by mid 1980s, the government made a concerted effort to make oilseeds more attractive to growers, through a combination of mission mode programmes. High prices of oils caused expansion of processing capacity and increase in oilseeds production by over 70 % in six years. India became almost self-sufficient in edible oil (up to 98%) and the imports of edible oil was reduced to 0.1 million tones in1992-93 from 1.5 million tones in 1986-87. Removal of all quantitative restrictions (QR) on imports and regulation through tariff in 1994 under WTO rules fundamentally changed the import regime of India's edible oil and from 0.1 million tonnes in 1992-93 the country's import reached 4.3 million tonnes in 2002-03. The corresponding export earnings however, much lower due to depressed prices in international market (due to quality consideration). There is a shift away from oilseed cultivation in India due to decline in relative profitability. The country's demand for edible oils is expected double from the current level in the next 12 years. India's future self sufficiency under WTO regime is examined and factors discussed include international oil prices and consequent impact on domestic availability and demand, importing of oilseeds instead of oil for local crushing, fiscal incentives for motivating farmer to shift to oilseed cultivation; ICAR- led research initiatives for increasing productivity of oilseeds, and agricultural reforms and modernization of oilseed industry to reach world standards.

India, imports nearly 3 million tones of palm oil annually (mainly from Malaysia, followed by Indonesia). This implies that the country depends on palm oil imports of over 25% of its annual edible oil requirement. There is sharp rise in the imports of palm oil into the country during the post 1998 period.

Imports of Palm Oil into India

	RBD Palmolein	**CPO**	**Crude Oil**	**CPKO**
02-03 (Nov.- May)	11,862	1,089,995	1,005,347	46,311
01-02 (Nov.-Oct.)	118,895	1,891,538	920,015	
00-01 (Nov.-Oct.)	1,516,688	1,404,723	32,877	
99-00 (Nov.-Oct.)	2,213,777	828,359		
98-99 (Nov.-Oct.)	2,676,712			

RBD: Refined Bleached Deodourised ; CPO: Crude Palm Oil ;CPKO: Crude Palm Kernel Oil.
Source: National Commodity and Derivatives Exchange Limited.

Composition of Indian palm oil import basket has also undergone an observable change during past 5 years. Crude oil has almost replaced refined oil imports owing to differential duty between the two.

Duty Structure for Various forms of Palm Oil Imported to India.

Description of goods	Import Duty(%)	Tariff Value * (US$/tonne)
Crude Palm Oil	65.00	390
RBD Palm Oil	70.00	426
RBD Palmolein	70.00	434
Crude Palmolein	65.00	415

Source: National Commodity and Derivatives Exchange Limited.

Tariff Value/Base import price refers to price on which Government of India levies (calculates) import duty payable by a palm oil importer at a given point of time. This was introduced by the Government during 2005 with the objective of curtailing large-scale under invoicing that was being done by a few unscrupulous importers. The value keeps changing in line with fluctuation in CPO prices in Malaysia.

Palm oil imported into India is used in various forms- consumed directly as 'palm oil' after refining, used in the manufacture of Vanaspati, for blending with other vegetable oil, crude oil and kernel oil for industrial purposes. It is imported mainly through the ports of Kandla, Kolkata, Mangalore, Mundra, Mumbai and Chennai.

Imported crude oil/olein is traded at cities of the importing ports in the form of RBD palmolein. RBD palm oil (60-70%) of palm oil is imported by organised players in the industry by large refiners/corporates who own brands). Since 100% of the palm oil traded in the country is 'imported' from Malaysia and Indonesia, the prices in the main trading centres of Mumbai, Chennai, Kandla and Kakinada. depend mainly on the futures prices of CPO at MDEX.

'Volatility' is at 'its highest' in palm oil. Being the sole oil dominating international edible oil trade, it almost 'makes' the price of vegetable oils traded across the globe; more so for India owing to its high dependence. As mentioned earlier, palm oil factors in all the fundamental and technical factors influencing oilseed complex World over, making the prices highly volatile in line with frequent fluctuation in supply/demand complex. It becomes quite essential for any farmer or processor or trader dealing with oilseed/oil/oil meal, brokers, speculators and all others involved in the supply and value chain of oilseed complex to hedge against their price risk by participating in futures trading.

The Price Trend and Volatility of CPO and RBDPO at the spot markets of Kandla at Kakinada enunciates the vitality of futures trading for hedging price risk in CPO and RBD Plamolein in specific, and other oils/oilseeds in general. The volatility also stands as a testimony to the 'trading opportunity' that the commodities (CPO and RBD palmolein) extend. With a share of over 54% (18 million tonnes), palm oil dominates the global vegetable oil export trade. India, China, Pakistan, the Middle-East countries and the European Union are the major importers of the commodity.

India in World Palmoil Industry

(Rounded figs.)	India	World	% Share
	(million t)		
Annual Oil Production	0.05	26	0.2
Annual Oil Imports	3	22	13.6
Annual Oil Consumption	3	26	11.5

Global Scenario

Palm oil with an annual production of 25-27 million tones is the second most produced oil in the world. Malaysia (14.8 million tones) and Indonesia (15 million tones) are the major producers. They together account for 85% of production. Around 80% (21-23 million tones) of global production is exported. Malaysia and Indonesia with 12-12.5 million tonnes and 6-7 million tones respectively are major exporters. India, China and EU are the major importers.

Oilpalm Area and Output

	Malaysia		Indonesia		World	
	2005	2006	2005	2006	2005	2006
Mature area (million ha)	3.65	3.80	3.92	4.03	9.15	9.76
Yield (t/ha)	4.09	3.97	3.52	3.54	3.65	3.64
Output (million t)	14.96	15.10	13.80	14.70	33.42	35.48

Source: Oil World (various issues) GAPKI, MPOB

World Oil Palm Production

Top Producing Countries				
	2002-2003	**2003-2004**	**2004-2005**	**2005-2006***
World	27.64	29.05	34.03	34.80*
Key Producers				
Malysia	13.18	13.40	15.19	14.80
Indonesia	9.80	10.60	14.00	15.00
Thailand	0.82	0.84	0.76	0.80
Nigeria	0.64	0.78	0.79	0.80

Current estimates in million t.
Source: Oil World (various issues) GAPKI, MPOB

Development of Palm Oil Sector

Global palm oil production and trade rose steeply and continuously from 1970s onward. A comparison of the rise - over the last ten years - in world palm oil production and trade with that of other oils and fats, evidences the above average growth rates achieved by palm oil. Consequently, the share of palm oil in global vegetable oil production has more than doubled over the last twenty years. This development is accompanied by a strong rise in the share of palm oil in global vegetable oil trade ; also the share of palm oil in global consumption grew significantly. It is interesting to compare the developments in palm oil with those of its main competitor, soybean oil. Regarding production, since 2005, global palm oil output is matching that of soy oil, whereas for trade, global shipments of palm oil surpassed those of soy oil in the mid 1970s and today palm oil exports exceed soy oil shipments by almost two times.

Another outstanding feature of palm oil is its very high trade-to-production ratio, which refers to the amount of global production that enters trade. The ratio for palm oil by far exceeds that observed for other oils and fats, oilseeds and other crops in general. Malaysia and Indonesia - account for roughly 90% of both global production of and global trade in palm oil. There are a number of factors which explain remarkable expansion of palm oil during the past decades:

- Palm oil production costs are low, compared to other oil crops. Comprehensive analysis conducted by LMC shows that of all vegetable oils, palm oil has the lowest/ unit production cost, followed by soybean oil, with 20% higher costs; the ranking continues with sunflower oil, coconut oil and finally rapeseed oil, which has the highest production costs. It should be noted here that labour costs weigh strongly on palm oil's overall production costs; in fact, with regard to labour productivity, performance of palm oil is relatively weak.

- The oil palm industry has benefited from a favourable economic environment and policy setting in two main producing countries, government presence tends to be relatively strong and from the beginning industry enjoyed (and continues to do so) indirect and direct protection in various forms; furthermore, the industry is successful in attracting domestic and foreign investors, and a high level of efficiency can be noted throughout the sector. From the industry's inception, following key areas received priority attention: research and development, general infrastructure and shipping facilities, processing for domestic value addition, vertical integration and sector-financed marketing and trade promotion programmes.
- The high level of market concentration (i.e. only two main producing countries and a limited number of big players in each country) facilitated the control and continuous advancement and modernization of production, trade and technological development.
- Compared to other oil crops, palm oil milling is relatively independent from byproduct markets. High-value palm kernel oil is traded on a separate market (the market for lauric oils) and palm kernel meal production and trade are relatively small.

The success story of palm oil also is associated with a number of drawbacks like: (i) a strong concentration of agricultural resources as opposed to sector diversification; (ii) a high level of vulnerability to weather changes; (iii) a strong dependence on export markets; and (iv) increasing concerns regarding social and environmental sustainability of production expansion.

Not withstanding these constraints, the palm oil sector is a highly performing industry and palm oil enjoys a strong market position. Palm oil is the most price competitive vegetable oil in global market for the last twenty years. Vis-à-vis its main competitor - soybean oil - palm oil is traded at a considerable discount in most years. As a result, today palm oil is consumed world-wide; it is the vegetable oil with the highest level of market penetration, and many nations (including China and India) depend heavily on palm oil imports.

Factors that explain the interest of the global market place in palm oil include the following: (i) a high level of substitutability with other soft oils; (ii) a high melting point and a low content of trans fatty acids, which are of special interest to food industry; and (iii) the fact, that a number of negative health attributes originally associated with palm oil were proven wrong. A closer look at past and current price trends reveals that the industry witnessed a long term, steady decline in the world price of palm .

New demand for vegetable oil for biodiesel production had a major influence on recent strengthening of prices; and the biofuel driven surge in the price of rapeseed and its oil has lifted vegetable oil prices in general. Although strengthening along with other oils, palm oil remained the lowest priced vegetable oil. As far as relationship between world palm and soybean oil prices is concerned, attention needs to be drawn to the following

- Impact of biodiesel demand on the two markets differs: biodiesel-driven soybean crushing inevitably generates surplus meal supplies, which lead to downward pressure on meal prices and thus reduced crush margins for soybeans; palm oil production is not subjected to such mechanisms due to the limited amount of meal by-products generated.
- Trade policies play a role: the discount of palm oil over soybean oil is also determined by India's tariff policies. India is a major buyer of both oils, but maintains a strong tariff preference for soybean oil. To make up for the disadvantage, palm oil must be sold at a certain discount. Development of palm oil price discount at global level is therefore closely dependant on changes in India's import tariffs.

The dominant position of palm oil in total vegetable oil output is expected to remain unchanged and may strengthen slightly. The pattern of trade, share of palm oil is anticipated to grow further, approaching 70% in 2015. With regard to two leading producers of palm oil, in or around 2009, Indonesia is to overtake Malaysia production-wise; and towards 2015 Indonesia is also anticipated to become the world's leading exporter of palm oil.

Important World Palm Oil Markets

- Bursa Malaysian Derivatives (BMD) is the largest future's market for crude palmoil.
- Malaysian and Indonesian FOB prices set the mood in physical market.

Factors Influencing Palm Oil Prices

- The supply demand and price scenario of competitive oils, viz., soybean oil.

- The supply-demand scenario of all oils and oilseeds in consuming centers, viz., India, China and EU. This in turn is manifested as imports from these countries.
- The palmoil production cycle: April - December is peak production period.
- Import regulations imposed in importing countries.

World Production of 17 Oils & Fats : 1995 - 2005 ('000 T)

Oils/Fats	1995	1996	1997	1998	1999	2000	2001	2002	2003	2004	2005
Palm Oil	15,211	16,286	17,946	16,920	20,625	21,867	23,984	25,392	28,111	30,909	33,326
Palm Kernel Oil	1,945	2,083	2,230	2,191	2,559	2,698	2,947	3,042	3,339	3,568	3,906
Soyabean Oil	20,404	20,322	21,033	24,008	24,794	25,563	27,828	29,861	31,288	30,713	33,287
Cottonseed Oil	3,905	4,119	4,077	4,059	3,893	3,850	4,052	4,234	3,995	4,417	5,033
Groundnut Oil	4,423	4,563	4,512	4,498	4,697	4,539	5,141	5,181	4,511	4,746	4,509
Sunflower Oil	8,556	9,006	9,162	8,407	9,308	9,745	8,200	7,624	8,962	9,402	9,681
Rapeseed Oil	10,955	11,479	11,828	12,290	13,247	14,502	13,730	13,307	12,660	14,904	16,027
Corn Oil	1,855	1,834	1,855	1,874	1,935	1,966	1,962	2,016	2,015	2,015	2,099
Coconut Oil	3,350	2,867	3,313	3,153	2,399	3,261	3,499	3,145	3,286	3,037	3,143
Olive Oil	1,888	2,042	2,701	2,588	2,475	2,540	2,761	2,718	2,903	3,055	2,903
Castor Oil	483	479	442	441	435	497	515	438	430	492	539
Sesame Oil	589	668	715	709	686	705	747	822	782	828	823
Linseed Oil	701	666	695	692	734	705	648	585	597	623	607
Total Vegetable Oil	**74,265**	**76,414**	**80,509**	**81,830**	**87,787**	**92,438**	**96,014**	**98,365**	**102,879**	**108,709**	**115,883**
Butter	5,677	5,648	5,688	5,765	5,885	5,967	6,010	6,188	6,274	6,351	6,521
Tallow	7,507	7,500	7,591	7,806	8,171	8,202	7,693	8,073	8,029	8,239	8,196
Fish Oil	1,285	1,336	1,184	886	1,413	1,411	1,131	934	989	1,077	1,054
Lard	5,692	5,936	6,141	6,520	6,619	6,739	6,780	7,006	7,210	7,363	7,545
Total Animal Oils/Fat	20,161	20,420	20,604	20,977	22,088	22,319	21,614	22,201	22,502	23,030	23,316
GRAND TOTAL	**94,426**	**96,834**	**101,113**	**102,807**	**109,875**	**114,757**	**117,628**	**120,566**	**125,381**	**131,739**	**139,199**

Source : i) Oil World Annual (1999 - 2005) & Oil World Weekly (16 December, 2005) ii) MPOB - For data on Malaysian palm oil and palm kernel oil

RUBBER

Many plant species produce natural rubber. Considerations of quality and economics, however limit the source of natural rubber to one species, namely *Hevea brasiliensis.*

It is a native of Amazon basin and introduced from there to countries in the tropical belts of Asia and Africa during late 19th century. It is the most far reaching and successful of introductions in plant history resulting in plantations over 9.3 million ha. 95% of it across the globe in Asia.

Hevea brasiliensis, also known as Para rubber tree after the Brazilian port of Para, is a quick growing, fairly sturdy, perennial tree of a height of 25 to 30 metres. It has a straight trunk and thick, somewhat soft, light brownish gray bark. The young plant shows characteristic growth pattern of alternating period of rapid elongation and consolidated development. The leaves are trifoliate with long stalks. The tree is deciduous in habit and winters from December to February in India. Refoliation is quick and copious flowering follows. Flowers are small but appearing in large clusters. Fruits are three lobed, each holding three seeds, quite like castor seeds in appearance but much larger in size. The seeds are oil bearing.

The rubber tree live for a hundred years or even more. But its economic life period in plantations, on general considerations is, only around 32 years, 7 years of immature phase and 25 years of productive phase.

Climate and Soil

Traditionally a rainfed crop, a well distributed rainfall of 2000 to 3000 mm without any marked dry season and with 125 to 150 rainy days/ annum. Maximum temperature of about 29-34°C and minimum of about 20°C or more with a monthly mean of 25-28°C.Temperature more than 35°C is harmful resulting in leaf injury and drying of leaf margins. Bright sunshine amounting to about 2000 h/annum at the rate of 6 h/ day through all the months. An increase in sunshine duration towards end of rainy season is often associated with an increase in latex production. Winds have tremendous influence on performance of rubber. In addition to mechanical effects, cold and dry winds affect physiological processes.

Soil should be well drained, have a minimum depth of one metre without any intervening hardpan or impenetrable layer. In marshy areas and waterlogged conditions growth of rubber is always found very poor. Generally, soil in the rubber tract is highly weathered and are mostly lateritic. Sedimentary types and non-lateritic red and alluvial soils are also seen in a few non-traditional areas. Red soil found in some areas is characterized by reddish to brown colour and fine loamy texture.

Clones

The planting materials approved by the Rubber Board are classified into three categories I, II and III. Category I comprises of materials approved for large scale planting. However, it is strongly recommended that these may be used to cover only 50% of total area. Rest of the area are with category II clones (35%) and category III clones (15%).

Categories of Clones

Category I	RRII 105, PB 260, RRII 414 and RRII 430 (RRIM 600 and GT 1 in non traditional areas)
Category II	RRIM 600, GT 1, RRII 5, RRII 203, PB 28/59, PB 217, RRII 417 and RRII 422
Category III	RRII 50, RRII 51, RRII 52, RRII 118, RRII 176, RRII 208, RRII 300, RRII 429, PR 107, PR 255, PR 261, PB 86, PB 5/51, PB 235, PB 255, PB 280, PB 311, PB 312, PB 314, PB 330, RRIM 605, RRIM 701, RRIM 703, RRIM 712, RRIC 100, RRIC 102, RRIC 130, KRS 163, IRCA 111, IRCA 130, SCATC 88-13, SCATC 93-114, Haiken 1, BPM 24 and Polyclonal seeds

Clones Released from RRII

Clone	Parentage	Status
RRII 105	Tjir 1 x Gl 1	Category I
RRII 414	RRII 105 x RRIC 100	Category I
RRII 430	RRII 105 x RRIC 100	Category I
RRII 5	Primary clone	Category II
RRII 203	PB 86 x Mil 3/2	Category II
RRII 417	RRII 105 X RRIC 100	Category II
RRII 422	RRII 105 X RRIC 100	Category II
RRII 50	Primary clone	Category III
RRII 51	Primary clone	Category III
RRII 52	Primary clone	Category III
RRII 118	Mil 3/2 x Hil 28	Category III
RRII 176	Mil 3/2 x PB 5/60	Category III
RRII 208	Mil 3/2 x AVROS 255	Category III
RRII 300	Tjir 1 x PR 107	Category III
RRII 429	RRII 105 X RRIC 100	Category III

Planting Material Production and Nursery Management

(a) By Clonal Seeds

Seeds of clones are termed clonal seeds. There are different kinds of clonal seeds - monoclonal, polyclonal, legitimate and illegitimate. The seeds are collected and seedlings are raised during July-September. Hybrid polyclonal seeds collected from approved polyclonal seed gardens are the seed materials recommended for planting.

(b) By Budded Plants

Depending on colour and age of bud and age of stock plant used, three types of budding are mainly recognized. They are (1) Brown (conventional) budding, (2) Green budding, and (3) Young budding. In the first method, older buds having brown colour are used while in the other two, green tender buds are utilized. Depending on part of stock where budding is carried out, budding is classified into: (1) Base budding, (2) Crown budding, (3) Over budding, and (4) High budding. Base budding is carried out at base of stock plant.

Brown Budding

Brown budding is generally carried out by grafting brown coloured buds taken from bud wood of about one year's growth onto stock plants of 10 months or more growth. Vigorously growing healthy stocks having a girth of 7.5 cm at the collar region are ideal for budding. Stocks should be budded when the bark peels off very easily.

Green Budding

Both stock and budwood for green budding are young. Vigorous seedlings five to seven months old having a girth of about 2.5 cm at base, with brown bark up to a height of about 15 cm, can be used for this purpose. The stock plants require about four to five months to attain this size.

Buds are collected from six to eight weeks old budwood, also known as bud shoots or bud sticks. Buds are found above scale leaves of shoots alone are used for budding. These buds are green in colour and hence the name green budding. Green budwood is obtained from budwood plants (source bushes).

Budding can be carried out at any time of the year. Too dry or very wet weather is unsuitable. Generally more success is obtained during rainy season than in summer.

Young Budding

This is a kind of green budding carried out on very young plants less than two months old. Stocks are raised in small bags of size 33 x 15 cm. When seven to eight weeks old, they are green budded. Four weeks after budding, plants are cut back leaving a snag of 20 to 25 cm length. Buds on the snag are nicked or the shoots coming from them pruned off promptly. When the scion develops two or three whorls of leaves, plants are transplanted to field.

This technique has certain advantages like plants could be produced within seven months after planting of germinated seeds in bags. The cost of production is also slightly less compared to normal method.

Crown Budding

High yielding clones show many undesirable characters like susceptibility to wind and diseases. An undesirable crown can be replaced by a desirable one through crown budding. The tree produced by crown budding is a three-part-tree comprising root system of stock plant, trunk of one clone and crown of another clone. Thus the desirable characters of trunk of one clone and crown of another clone are combined.

Crown budding is ideally carried out when scion of budded plant has attained a height of 2.4 to 3 m. One to two years are usually required for the plants to attain such a growth. Budding is carried out at a height of 210 to 240 cm on the inter-whorl region below top whorl of leaves. It should be done only when the top flush of leaves are fully expanded and hardened. Stem tissue should be green or dark green at the time of budding. For crown budding, green budding technique is followed.

Successfully budded plants are cut back leaving a snag of about 5 cm. Treating cut ends of the stem with some wound dressing compound and trunk shoots arise should be pruned regularly at fortnightly intervals. This should be continued until the crown bud sprouts and the crown shoot grows to a length of about 2.5 cm. After that, two or three trunk shoots arising about 15 to 45 cm below the crown shoot are allowed to grow. These shoots should be spaced apart and be on different sides of stem as far as possible.

Over Budding

Budwood plants are sometimes budded at higher levels for converting an existing budwood nursery of a clone to another clone without replanting. This is termed as over budding. The method adopted is to carry out budding at the basal portion of the brown budwood before harvesting it. The budwood is harvested after the new bud is successfully attached to the plant, by cutting above the budded portion. Since the budding is carried out on a well-established plant, the scion emerging from new bud grows vigorously producing more budwood compared to a newly established budwood plant. Thus it is a quick and economic method for converting budwood nursery of one clone to that of another.

Nursery Management

Rubber seeds lose viability very rapidly if left in the field. The seeds are therefore picked up daily during seed fall season and quickly transported to nurseries for germination and planting. Planting materials for establishing rubber plantations are generated in ground and polybag nurseries.

Ground Nursery

Ground nurseries are established for production of budded stumps, stumped buddings and budwood.

Selection of Site

The site should have good accessibility for supervision and transport of materials.

A good soil depth of at least 75 cm is essential

Loamy soils are ideal

A well-drained level area is ideal. Undulating lands are also suitable if adequate soil conservation measures are adopted

Contour terracing is done where the slope is in excess of 2%.

Waterlogged areas should be avoided and water table should be sufficiently low to allow sufficient soil depth for root development

Shade-free areas are preferred

Land with a history of intensive cropping needs proper build up of nutrient status to satisfactory level.

Preparation of Nursery Beds

Dig to a depth of 60-75 cm. Stones, stumps, roots, etc. present in the soil are removed and soil is brought to a fine tilth. Beds should be 90 to 120 cm wide and of convenient length. In the level lands, raised beds are made with footpaths of about 45 cm width between beds. In undulating lands, beds are prepared along contours, one below other. At the time of preparation of nursery beds, 25 kg of compost or well-rotten cattle manure and 3.5 kg of powdered rock phosphate (20% P_2O_5) are incorporated for every 100 m^2 of nursery bed. When nurseries are established in newly cleared forest areas, rich in organic matter, compost or cattle manure need not be applied during first year. Similarly when the same area is repeatedly used as a nursery, rock phosphate need be applied only once in three years. Drainage and pathways should be provided appropriately.

Seed Germination Beds

A well-drained area with moderate shade is the ideal site for germination beds. Level beds of 90 cm width and convenient length are prepared with walking space in between. The beds should be raised 10-15 cm above soil surface to avoid water-logging. A free-draining friable material like river sand, spread above the bed to a thickness of 5 cm, is used as medium for germination. Seeds are washed thoroughly to remove charcoal and other packing debris and spread over the bed in a single layer touching one another and pressed gently into the sand. To prevent loss of too much moisture from rooting medium, the beds are covered with a thin layer of gunny bag, coir matting or similar material. A high level of moisture is maintained in the bed by evenly sprinkling water early in the morning and late in the evenings. Germination of seeds starts within 6-7 days after sowing. The beds should be inspected daily and the germinated seeds picked up and collected in a bucket containing water as soon as the radicle emerges for planting in the nursery beds or main field. Seeds, which do not germinate within two to three weeks, should be discarded.

Planting in Nursery

Small holes enough to accommodate the seeds in a horizontal position and approximately 5 cm deep are made. The seeds are carefully placed in the holes with the radicle pointing downwards and covered with soil. The sprouted seeds should be planted when the young root is less than 2 cm long.

The common spacing adopted are:

Seedling stumps: 30 x 30 cm

Green-budded stumps: 23 x 23 cm

Brown-budded stumps: 30 x 30 cm or staggered pairs of rows 60 cm apart and 23 cm between plants

Stumped buddings: 60 x 60 cm, 90 x 30 cm, 90 x 60 cm or 90 x 90 cm

Soil core plants: 35 x 35 cm, 38 x 30 cm or 60 x 60 cm.

Budwood nursery: 90 x 60 cm or 120 x 60 cm

Nursery for Seedlings and Budded Stumps

The nursery beds should always be kept free of weeds. Three rounds of weeding are needed. The first weeding is done just before application of first dose of fertilizers and second weeding before second dose. The third round of weeding is done just before commencement of budding during May or June. The first round of manual weeding can be replaced with application of pre-emergence herbicides. After the final preparation of nursery beds, diuron (2.5 kg / ha) in 700 l water is sprayed on beds and germinated seeds planted five days later.

Mulching is an important operation to be followed in seedling nurseries before beginning of summer season and after the second round of fertilizer application. A single round of good mulching in December is adequate. Black polythene sheets properly anchored to soil to prevent them from being blown away by wind can also be used for mulching. Spreading a thin layer of soil above the sheet is an effective way to achieve this.

During dry period, which usually extends from December to April, the nurseries should be irrigated. In large nurseries, overhead sprinkler irrigation systems are ideal. Daily watering is preferred during initial weeks. Later, the frequency of irrigation can be reduced to once in two or three days.

Budwood Nursery

Buds required for bud grafting are collected from budwood obtained from plants raised specifically for this purpose. Budwood nurseries are of two types: Brown budwood nursery (produces brown buds) and Green budwood nursery (produces green buds).

After cleaning and levelling, soil is first dug to a depth of 45-60 cm. Planting can be done with polybag plants, budded stumps or seed at stake followed by budding. For green bud shoot nursery, the spacing is 1 x 1 m or 80 x 90 cm. Proper fertilizer application may be carried out to ensure good growth. Other agronomic practices like irrigation, mulching, weeding, shading, protection against diseases and pests are followed in a similar manner as for seedling nursery.

During first year of planting, only one shoot is allowed to grow. About 1 m of brown budwood can be obtained from this after one year. From second year, two or three shoots are allowed to develop on a plant depending on spacing adopted. To remove the leaves present in the brown-coloured budwood, the leaflets are first removed by clipping tip of leaf stalk. After about one week, the leaf stalk dries and falls off. Budwood is then harvested by sawing off, leaving about 15 cm at the base. From this portion, shoots develop in the subsequent season.

Green bud shoot plants are shaped from brown budwood plants. For this, a well-established brown budwood plant is first cut back at a height of about 75 cm. A number of shoots emerge below the cut end. Among these, only 3-5 most vigorous ones are retained and the rest removed. When these shoots have grown and produced brown wood to a length of about 5 cm, they are pruned at the point where the brown colour ends to produce more branches. Two to three most vigorous branches are retained on each shoot and others cut off. When these secondary branches develop brown colour at the basal 5 cm, they are again pruned. New branches arise from these and give the budwood plant a bushy appearance. For producing green shoots, all branches of a green bud shoot plant (also called source bush) are pruned. The new branches arising are harvested when one whorl of leaves develops. The harvested budwood is cut into pieces of convenient length before being taken to nursery beds for budding.

Polybag Nursery

Planting materials in polybags can be prepared by two different methods.

Budded stumps can be planted in polybag and the scion allowed to develop till they are ready for planting in the field.

Germinated seeds are planted in polybags and bud-grafted when 5-6 month old.

The roots of budded stumps can be treated with indole butyric acid (IBA), a hormone that enhances root growth. Dipping root in cowdung slurry before planting enhances root development.

Planting in Polybags

Polythene bags of size 55-60 cm length and 25-30 cm width and 400 gauge thickness holding about 8-10 kg soil, are usually used for raising plants upto two to three whorl stage. For producing plants of 6-7 whorls, larger bags of 65 x 35 cm size and 500 gauge thickness holding about 23 kg soil should be used.

Regular cultural operations like manuring, watering, weeding, shading and plant protection are adopted. Application of N-P-K-Mg 10-10-4-1.5 mixture is done at monthly intervals. During first month, 10 g / bag which is gradually increased to 30 g in four months time.

Watering should be done soon after manuring. During dry periods, irrigation should be done regularly. Watering can be done manually in small nurseries while sprinklers or drip irrigation system is more economical in large nurseries. Too much watering should be avoided to prevent water logging. During summer, partial shade may be provided to plants by erecting overhead shade. Appropriate prophylactic and curative measures may be taken against diseases and pests.

Land Preparation and Planting

June-July is the ideal period for planting rubber in South India, all the pre-planting operations should be completed before onset of monsoon.

Clearing

Land is cleared, a light burning after felling trees and drying facilitate planting operations and slow down regeneration of weeds. Excessive cleaning and burning cause destruction of soil and expose soil to erosion. Replanting old areas has to be thought of when yield falls and cultivation becomes uneconomic. The old trees may be slaughter tapped with yield stimulant application before felling.

Lining

Rubber is planted by adopting square or rectangular planting system.Square planting is suitable for level and near level lands. Rectangular system can be adopted in flat lands and slopes and the lines should be oriented in the East-West direction to intercept maximum sunlight. Contour lining is done in undulating and hilly areas where the slope exceeds 8 %. Here the planting points are marked as lines passing through points of same elevation.

Planting Density

Planting materials and terrains	Spacing	Plants/ha
Buildings-Hilly areas	6.7x3.4 m	445
Flat areas- square	4.9x4.9m	420
Triangular	4.9x4.9m	470
Seedlings-Hilly areas	6.1x3.0m	539
Flat areas	4.6x4.6m	479

Higher initial stand is recommended for allowing proper thinning out. During the period of immaturity and initial years of tapping, selective thinning should be carried out to reduce the stand/ha to 310 by the 10th year of tapping.

Soil and Water Conservation

On hilly and undulating terrain, cutting of terraces along the contour is a recommended practice to conserve moisture and prevent erosion. The soil on the hill side is cut from a distance of 60-75 cm in front of the planting row and thrown back so that terraces formed will have a width of 1.25 to 1.5 m and an inward drop of 20-30 cm. Steps of uncut earth are left out at intervals along the terraces to check lateral flow of water. Silt pits (trenches of about 120 cm length, 45 cm width and 60 cm depth) are also made along the contour at suitable intervals to check erosion and to conserve water. Pits can be taken at the rate of 200-250/ ha depending on degree of slope. Construction of stone pitched contour bunds is another method to check erosion in steep slopes.

Pitting and Refilling

The standard size of pit is 75 x 75 x 75cm. Size of pits varies depending upon the planting material to be used. Stumped buddings need comparatively deeper and larger pits. Smaller pits are sufficient for small and medium sized polybag plants.

In deep, loose and friable soils, pits are dug wider at the top and tapering towards bottom or depth is reduced to 60 cm with a central crow bar hole of 15 cm or more depth for taproot. But in hard, stony and compact soils, the pits should be widened.

While digging, topsoil is kept on one side and subsoil on another side. Filling should be done with the top fertile soil as far as possible. The organic manure and phosphatic fertilizers applied to pits should be mixed with top 20 cm soil in the pit. The pits should be filled to about 5 cm above ground level. A peg is placed in the centre of pit to locate the planting point.

Types of Planting

Continuous wet weather during June-July is ideal for planting rubber.

Seed at Stake Planting

The system of planting seeds is called seed at stake planting and it is followed by field budding. Two or three germinated seeds may be planted in a line or triangle. The weaker plants should be thinned out later, allowing most vigorous one to grow on which field budding is carried out at appropriate stage.

Seedling Stumps and Budded Stumps

It is advisable to plant stumps soon after pulling out from nursery beds. While planting budded stumps, bud patch should be just above ground level. If a considerable part of the stock below the union is left above ground level, it may maximize effect of 'elephant foot'. If kept below soil, the union is likely to be infected by soil organisms.

Polybag Planting

A planting hole slightly bigger than size of bag is made .Bottom of bag is completely cut and then the bag along with plant is inserted into planting hole. While planting, scion of polybag plants should be directed towards north-east to minimize adverse effect of direct sunlight on bud patch.

Aftercare and Management

Various cultural operations like establishment of ground cover, intercropping, weed management, vacancy filling, pruning, mulching, induction of branching and thinning should be judiciously followed to reduce immaturity period and to obtain optimum stand of healthy trees at minimum cost under proper management.

Pruning

After planting, plants should be inspected at regular intervals and false shoots sprouting from the stock should be removed and only the vigorous bud shoot is allowed to develop. Any side shoot developing upto 2.5 m from ground level should also be removed.

Intercropping

Growing of intercrops necessitates soil disturbing tillage operations of various kinds. This will predispose the topsoil to erosion losses in steep and undulating lands. The growing of intercrops, therefore has to be restricted to level lands and gentle slopes. Even in such lands it should be ensured that leguminous cover crops are established side by side with intercrops or immediately after intercropping is stopped.

Common Intercrops Cultivated in Rubber Plantations

Banana

During the first year, 1200 plants/ha. can be grown in a double row system at 2 x 2 m spacing. The stand should be restricted to 600 during the second year in a single row system. Non-ratoon types like Nendran is suitable as intercrop during the initial two years. Ratoon types like Palayankodan and Poovan should be planted in a single row and only one sucker should be retained during second year. The pseudostem, leaves and mother rhizome left after harvest can be used for mulching the rubber plant.

Pineapple

Pineapple can be grown as intercrop during initial four years. Two or three rows of pineapple suckers are planted along inter-rows of rubber in trenches during year of planting of rubber. In sloppy lands, the trenches should be taken across the slope parallel to the contour terraces to reduce soil erosion. Proper manuring using organic and chemical fertilizers is necessary to ensure high yield and to prevent depletion of soil fertility. Pineapple, once planted, will give fruits for a period of about four years.

Ginger and Turmeric

Since thorough digging of the soil is required for the cultivation of these crops, they may be grown only on level and near level lands. Heavy application of cattle manure or compost and organic mulches are important in the cultivation of these crops. They can be grown for the initial two years.

Vegetables

Vegetables like cowpea, cucumber, ladies finger and amaranth can be grown during the initial 2-3 years with proper addition of organic manures.

Medicinal Plants

Shade tolerant medicinal plants like *Strobilanthus haenianus* (Karimkurinji), *Adathoda vasica* (Valiya adalodakam) and *Plumbago rosea* (Chuvanna Koduveli) are found suitable for cultivation in rubber plantations.

Establishment and Management of Cover Crops

Cover crops are established and maintained in rubber plantations for the purpose of conserving soil and improve soil fertility. Leguminous ground cover enhances growth and yield of rubber. Most Common leguminous cover crops grown are *Pueraria phaseoloides, Mucuna bracteata, Calopogonium mucunoides,* and *Centrosema pubescens.*

Establishment of Cover Crop

Cover crops are generally established from seeds. *Pueraria phaseoloides* and *Mucuna bracteata* can also be propagated by stem cuttings. About 420 beds of 1.2 x 1.0 m/ ha.are prepared immediately after pre-monsoon rains. If cuttings are used, fresh cuttings 0.75-1.0m long should be planted when frequent rains are available during June or July. On the other hand, if seeds are used, they should be sown in prepared patches during May after the pre-monsoon rains.

Seeds of cover crops have very hard seed coat, which delays or inhibits germination. Therefore pre-sowing treatment is done to ensure uniformity and higher percentage of germination.

*Acid treatment:*Seeds are treated with concentrated sulphuric acid for a period of 10 min for *P. phaseoloides,* 30 min for *M. bracteata* and 20-30 min for *C. mucunoides.*

Hot water treatment: This method is suitable for seeds of *P.phaseoloides* and *C.mucunoides*. Hot water treatment is carried out by pouring water at 60-80°C over the seeds in a container till all the seeds are submerged, for a period of 4-6 hours before sowing.

The pretreated seeds are mixed with equal quantity of rock phosphate and sown in rows or in equidistant patches between plant rows. Powdered rock phosphate (20% P_2O_5) @ 150 kg/ha is applied in two equal splits, the first one month after sowing and the second two months after first application. In areas, where soils are known to be deficient in available potassium, application of a mixture of 150 kg of rock phosphate and 50 kg of muriate of potash is recommended. Cover crop should not be allowed to grow in a circle of about 2 m diameter around the plants or along the contour terraces for 3 to 4 years to prevent its competition with the young rubber.

Mulching, Shading and Whitewashing

Mulching or covering the plant basin is done with dry leaves, cover crop cuttings, grass cuttings and paddy straw. Dried African payal (Salvinia sp.) also can be effectively used as mulch at the rate of 5 kg/sq.m (sun dried material).November is the ideal time for mulching to protect plant from adverse effect of drought.

Plaited coconut leaves or used gunny bags can be used for shading purpose. The brown bark of young plants can be protected from scorching sun by white washing main stem of plant from second year of planting. This may be continued till canopy of plants develops and partially shades the plantation. Whitewashing can be done using lime or china clay.

Induction of Branches

To achieve a high rate of girth increase, rubber plant should produce branches at a height of about 2.5 to 3.0 m from ground. In high branching trees, girth increment is poor compared to low branching trees. Some clones like RRIM 600 and GT1show a tendency for high branching. In such cases branching has to be induced by encouraging a few lateral buds to develop. Double blade ring cut device and the leaf cap method can be utilized for this purpose.

Weed Control

Maintenance of a luxuriant ground cover in early phase of plantation can minimize weed infestation to a great extent. Manual weeding involves slashing, scraping and uprooting. As manual clean weeding exposes soils to erosion, mulching should be done immediately after weeding.

Weeds can be controlled with use of herbicides or weedicides. There are two main types of herbicides, the pre-emergent and post-emergent herbicides.

Pre-emergent Herbicides

1. Diuron (Karmex, Klass): 2.5 kg in 700 litres
2. Simazine (Princep, Parintop): 3 kg in 700 litres
3. Alachlor (Lasso): 5 litres in 600 litres

Post-emergent Herbicides

1. Paraquat (Gramoxone) and 2,4,D (Fernoxone): 2.25 litres or 1.25 kg in 500-600 litres (4 rounds at an interval of 6-8 weeks)
2. Glyphosate (Glycel, Weed off or Round up): 2 litres in 400 litres (2-3 rounds at 3 months interval and spot application of 0.5 to 2.5 litres Gramoxone/ ha).

Manuring/Fertilizer Application

The nutrient requirements of rubber vary considerably during three important stages of growth - the nursery, immature and mature stages.

Seedling Nursery

Basal dressing: Incorporate 2.5t of compost or well rotted cattle manure and 350 kg of powdered rock phosphate (18-20% P2O5) for every effective ha, i.e., 25 kg of compost and 3.5 kg of rock phosphate/ 100 m^2 of nursery bed.

Top Dressing: Apply 2500 kg of 10-10-4-1.5 N-P-K-Mg mixture/effective ha, i.e., 25 kg/ 100 m^2 of the nursery bed 6-8 weeks after planting. Again apply 550 kg of urea/ effective ha, i.e., 5.5 kg/100 m^2, six to eight weeks after first top dressing.

The first application of recommended quantity of N-P-K-Mg mixture should be made during September-October, i.e., 6-8 weeks after planting the germinated seeds in nursery beds.

Budwood Nursery

1. Incorporate 150 kg of powdered rock phosphate/ha, i.e. 1.5 kg/ 100 sq. m of nursery bed as a basal dressing at the time of preparing nursery bed.
2. For the first crop of bud-wood, apply 250 g of 10-10-4-1.5 N-P-K-Mg mixture/ plant in two equal split doses. Apply the first split of 125 g / plant two to three months after planting the budded stump or cutting back if budding is carried out *in situ.* The second dose of 125g / plant should be applied 8-9 months after planting.
3. For the second and subsequent crops of bud-wood from the nursery, apply 125 g of 10-10-4-1.5 N-P-K-Mg mixture/ plant in one single dose two to three months after cutting back.

The fertiliser application should be made during September–October and March-April either in a band 8 cm away from base of plant or in between two rows and lightly forked with a hand rake.

Immature Rubber

(a) Up to the Fourth Year

Pit manuring: Incorporate 12 kg of compost or well rotted cattle manure and 175 g of rock phosphate in every pit at the time of filling to provide good soil conditions for development of a good root system. In newly cleared forest areas, it is enough to apply only 175 g rock phosphate, well mixed with the top 20 cm soil in pit as surface soil in general will be rich in organic matter.

In Kanyakumari district of Tamilnadu, Thrissur, Palakkad, Malappuram, Kozhikode, Waynad, Kannur and Kasargod districts of Kerala and Karnataka, Goa, Maharashtra and N. E. India, available Mg status of soil is high and application of 12-12-6 N-P-K mixture is recommended. In all other regions, 10-10-4-1.5 N-P-K-Mg mixture may be applied.

Quantity and the Schedule of Fertilizer Application During the First Four Years

Year of planting	Months after planting	Time of application	Qty of mixture/ plant (g)		Qty of mixture/ ha with 440-450 plant points (kg)	
			10-10-4-1.5 (NPKMg)	12-12-6(NPK)	10-10-4-1.5 (NPKMg)	12-12-6 (NPK)
1st year	3 months	Sep-Oct	225	190	100	85
2nd year	9 months	Apr-May	450	380	200	170
Do	15 months	Sep-Oct	450	380	200	170
3rd year	21 months	Apr-May	550	480	250	215
Do	27 months	Sep-Oct	550	480	250	215
4th year	33 months	Apr-May	450	380	200	170
Do	39 months	Sep-Oct	450	380	200	170

1) In first two years, it is advantageous to apply 50% of phosphorus in water-soluble form for better growth of rubber plants.

Fertilizers should be evenly distributed over a circular band of about 30 cm around base of young plant, leaving about 7 cm from the base all around and slightly forked into the top 5-8 cm of soil. The plant bases should then be immediately mulched.

(b) From Fifth Year of Planting to Maturity

The fertilizer recommendation from 5th year of planting till tapping for a well maintained field where mulching is practiced and leguminous ground cover established and maintained, is to apply NPK:30:30:30 kg/ha in two split doses, one in April-May and other in Sept-Oct.

This can be supplied by:

1. Mixing straight fertilizers, 65 kg urea, 150 kg rock phosphate and 50 kg muriate of potash
2. 300 kg/ha of 10-10-10 or 250 kg/ha of 12-12-12 NPK mixture
3. Any of the complex fertilizers of grades 15-15-15 or 17-17-17 or 19-19-19 NPK and quantities of these should be 200,175 or 160 kg/ha respectively
4. 10-26-26 NPK complex (115 kg) mixed with urea (40 kg) or ammonium phosphate sulphate (ammophos 20-20) (150 kg) with muriate of potash (50 kg) or 65 kg diammonium phosphate (DAP) mixed with 40 kg urea and 50 kg muriate of potash

Diammonium phosphate is recommended for soils having pH 6.0 or above which is very rare.

In areas where no legume ground covers were established and no mulching practiced during the initial years application of NPK: 60: 40: 24 kg/ha. This can be supplied through 400 kg of 15-10-6 mixture applied in two split doses of 200 kg each during 5th and subsequent years till the plants become ready for tapping.

Mature Rubber Under Tapping

For rubber under tapping, NPK at the rate of 35:35:35 kg/ha.is recommended. Fertilisers should be applied in square or rectangular patches in between rows, each patch serving four trees. Once the canopy of the plants closes, say, 5 to 6 years after planting, light forking to incorporate fertilisers into the topsoil is necessary.

In areas where the legume ground cover is present or where the legume cover has died out leaving thick mulch, it is enough to broadcast fertilizers between two rows of rubber trees. Deep pocket placement of fertilizers and application too close to base of trees should be avoided.

Discriminatory Fertilizer Recommendation

Fertility status of individual fields can be assessed by soil and leaf analyses and the fertilizers are recommended based on individual case history of field.

- Soil may be analyzed before starting rubber cultivation to ascertain soil fertility and to obtain fertilizer schedule to be adopted during initial four years
- During fourth year of planting, soil and leaf samples may be analyzed to obtain fertilizer recommendation for fifth year
- Leaf sample may be analyzed every year while soil samples need be analyzed only once in three years.

Harvesting

Tapping and Stimulation

Latex is obtained from bark of rubber tree by tapping. Tapping is a process of controlled wounding during which thin shavings of bark are removed. The aim of tapping is to cut open latex vessels in case of trees tapped for first time or to remove coagulum which blocks cut ends of latex vessels in case of trees under regular tapping.

Bark

An inner layer of soft bark, an intermediate layer of hard bark, and an outer protective layer of cork cells can be distinguished in bark of rubber tree. The vessels are concentrated in soft bark arranged in a series of concentric rings of interconnecting vessels.

Marking

Panels are marked on trees selected for tapping, using a template and marking knife. After deciding position of panel, a vertical line, called front channel line, is drawn. On this line, opening height is marked. Another vertical line, called back channel line, is marked on half spiral point above opening height. With aid of template placed between these two lines, at opening height, ensuring a high left to low right, line for tapping cut and a few guidelines are marked through the grooves. The vertical front end of template should be kept on front channel of tree, parallel to it and free end of template is on trunk towards left side. After marking guidelines, spout and cup hanger are fixed. Subsequent guidelines are marked every year before commencing of tapping.

Slope and Direction of Tapping Cut

Tapping cut of budded trees should have a slope of about 30° to the horizontal. For seedling trees, cuts need to have a slope of only about 25°, since the bark is fairly thick.

The latex vessels in the bark run at an angle of 3-5° to the right and therefore a cut from high left to low right will open greater number of latex vessels.

Latex Flow

Hevea latex in the latex vessels of tapped trees contains 30-40 % rubber in the form of particles. The best yield is obtained by tapping to a depth of less than 1 mm close to cambium.

Tapping Implements

The main implements used for tapping and collection of latex are knives, spouts, cup hangers, collection cups, collection buckets and scrap baskets.

Types of tapping knives

'Michie Golledge' knife

'Jebong' knife, commonly used in Malaysia, is more suitable for speedy and easier tapping but with a slightly higher bark consumption.

Modified gouge knife is in use mainly for controlled upward tapping (CUT).

Spout

A metal piece made of galvanized iron (GI) or tin sheet (5 x 3 cm size) bent all along into a V-shape.

Cup Hangers

Used to keep the collection cups in place below the spouts. They are made of GI wire and tied on the trees with the help of plastic thread.

Collection Cups

Coconut shells are widely used. These are now almost completely replaced by higher capacity (400–900 ml) and cleaner plastic cups.

Scrap Baskets

Used for collection of field coagulum or scraps such as tree lace and cup scrap.

GI Buckets

Commonly used for carrying away latex from field.

Time of Tapping

It is necessary to commence tapping early in the morning, since late tapping reduces exudation of latex due to increased transpiration by trees.

Tapping Task

Normal tapping task (Number of trees allotted to a tapper for a day's tapping) in India varies from 300 to 400 trees.

Tapping Systems

- *Budded trees:* Half spiral alternate daily (1/2S d/2) system
- *Seedlings:* Half spiral third daily (1/2S d/3) system
- *Medium yielding clones (RRIM 600, GT 1, PB 28/59 etc):* Alternate daily tapping

- *High yielding clones (RRII 105, PB 217):* Low frequency tapping systems (d/3, d/4 & d/6) with stimulation may be practiced

Panel Notation

Panel notation indicates panel position and renewal succession of panel. This helps to describe tapping. Common panel notations are B0-1 (A) (First basal panel of virgin bark), B0-2 (B) (Second basal panel of virgin bark), BI-1 (C) (First renewed bark of B0-1), BI-2 (D) (First renewed bark of B0-2), BII-1 (E) (Second renewed bark of B0-1) and BII-2 (F) (Second renewed bark of B0-2).

First, second, third and fourth high panels of virgin bark above the standard height of opening are indicated by HO-1, HO-2, HO-3 and HO-4 respectively. Renewed bark is usually not tapped in high panels.

Intensive Tapping

Intensive tapping is generally done on old rubber trees for a few years prior to their removal.

High Level Tapping

When tapping of renewed bark on basal panels becomes uneconomic, new cuts are opened at higher levels, 180 cm from bud union or even higher. The tapper uses a small ladder to reach the cut.

Controlled Upward Tapping

Controlled upward tapping (CUT) is practiced for longer exploitation of the virgin bark above basal panel. In CUT, instead of using ladder, a long handled modified gouge knife is used for upward tapping from ground.

The tapping cut can be 1/4 spiral and its angle 45°. Length of handle used for tapping is 120 cm for tapping upto 40 cm height of high panel and 180 cm for tapping beyond that height.

Rainguarding

About 35-40 additional tapping days could be obtained every year by rainguarding trees by fixing polythene (LDPE) rain-guard to trunk above tapping cut. Systematic application of panel protectants at frequent intervals is necessary because chances of bark rot diseases are high. Raingurding is recommended only in areas where yield is 675 kg/ha/annum or more and 25 or more tapping days are annually lost by rain.

Four types of rainguards, viz. polythene skirt, 'Tapping Shade', 'Guardian Rainguard' and 'Tapping Shield' are recommended and popularized.

Tapping Rest

During refoliation and flowering of rubber trees, yield will be comparatively poor and normally trees are given about four weeks rest if the soil is very dry and yield is uneconomic.

Yield Stimulation

Certain chemicals induce ethylene formation in plant tissue while certain other groups of chemicals generate ethylene directly by decomposition. Between two types of chemicals tested, Ethephon containing 2-chloro-ethyl phosphonic acid as active ingredient, has proved very potent yield stimulant.

Under low frequency tapping systems, trees are stimulated from first year of tapping using ethephon (2.5%) by panel application method to achieve sustainable yield. Use of ethephon is also recommended for trees tapped under d/2 frequency (panel C) and trees tapped in panel D (BI-2) for short-term increase in production. In newly opened trees or on trees tapping resumed after annual rest, ethephon at 5% (active ingredient) concentration may be applied once (over the lace after two tappings).

The application is done with a brush just above the tapping cut to a width of 1 cm, on cut surface of bark over the lace or after removing it. The 10% formulation of ethephon available in the market can be diluted to 2.5 % or 5% with palm oil, petroleum jelly, coconut oil or even with water. The water-mixed formulation should not be kept for more than two hours. Continuous application of ethephon is not recommended for periods of more than three years at a stretch, under d/2 frequency of tapping.

Puncture Tapping

In puncture tapping, an ethephon-based stimulant is applied to a strip of scraped bark, usually vertical, 1.5-2.0 cm wide and 50-100 cm long. On each tapping day, 4-10 punctures are made, at equal spacing in the stimulated bark and latex is collected.

A vertical strip of 50 to 60 cm length with 6 punctures/ tapping, stimulation with 2.5% ethephon and a third daily or lower frequency tapping appear suitable for most clones. At monthly intervals, a new strip adjacent to old strip is prepared, stimulated and puncture tapped. Trees can be puncture tapped when they attain a girth of 40-45 cm, one year before normal opening.

Precautions

1. Puncture should avoid cambium, but should be deep enough
2. Puncture should be distributed uniformly along stimulated portion
3. Needle should have a diameter of 1 mm with blunt tip
4. Ten applications of Ethephon (2.5 % active ingredient) / year will be required

Slaughter Tapping

Highly intensive tapping prior to replanting is the last stage in the tapping cycle and is called slaughter tapping. It is done at higher levels, sometimes even on branches, with help of ladders to extract as much latex as possible from available bark.

Bee Keeping in Rubber Plantations

The rubber tree is a prolific source of honey, obtained from the extra floral nectaries at tip of petiole, where leaflets join. Honey flow period is between January and March.

Cost of production of rubber per kilogramme, 1998-99 to 2004-05

S. No.	Item	1998-99	1999-00	2000-01	2001-02	2002-03	2003-04	2004-05
1	Cultivation cost							
A	Cost of labour, labour wages inclusive of salaries supplements, of-field supervisors	5.01	5.45	5.58	6.11	6.78	11.53	6.78
B	Cost of materials - manure, dolomite, tractor/truck, sprays, chemicals, operati ng costs, field tools, other costs of maintaining rubber	1.52	1.54	1.78	3.14	2.76	3.14	3.26
C	Tapping/collecting cost - tapping/ collecting labour inclusive of supplements, drivers inclusive of supplements, tapping supervisors salaries, truck/tractors sup plements, tapping supervisors salaries, truck/tractors	15.09	13.14	13.95	16.57	16.78	22.1	27.93
	Group total	**21.62**	**20.13**	**21.31**	**25.82**	**26.31**	**36.77**	**37.97**

Contd...

2	Cost of maintaining							
	Young rubber - manure, dolomite, sprays, chemicals, plants from nursery, baskets, field tools, tractor/truck operating costs, other materials costs	5.82	6.72	4.91	5.04	5.23	5.7	6.41
	Group total	**5.82**	**6.72**	**4.91**	**5.04**	**5.23**	**5.7**	**6.41**
3	Factory cost							
A	Wages and salaries - total wages of labour including supplements, total salaries of rubber makers, mechanics, other factory staff	1.89	2.52	2.67	2.38	2.64	1.81	1.79
B	Factory materials - chemicals, firewood, fuel, diesel, furnace oil, electricity packing materials, acids, other sundries, machinery and general factory maintenance	3.72	3.22	5.30	3.86	3.87	3.02	2.68
	Group total	**5.61**	**5.74**	**7.97**	**6.24**	**6.51**	**4.83**	**4.47**
4	Marketing and agency house expenses transport to Colombo, storage and handling, brokerage, agency commission and other marketing costs	0.55	0.37	0.73	0.79	0.68	0.75	0.80
	Group total	**0.55**	**0.37**	**0.73**	**0.79**	**0.68**	**0.75**	**0.80**
5	Cost of estate maintenance							
A	Cost labour, wages of labour including supplements	1.34	0.90	1.10	0.84	0.52	1.41	3.65
B	Other equipments - lines, latrines, other buildings, roads, water supply, estate boundaries, and other general estate maintenance	1.32	1.39	2.27	1.65	2.02	2.87	3.33
	Group total	**2.66**	**2.29**	**3.37**	**2.49**	**2.54**	**4.28**	**6.98**
6	Estate general charges salaries, allowances, director fees, visiting charges, holiday wages, feeding and schooling, sanitation, medical maternity, all insurance, E.P.F. contribution, concessionary grants, office supplies, rent depreciation auditors' fees bonus to staff workers,	12.81	16.47	14.76	15.36	18.48	13.65	15.93

Contd...

	compensation and other general charges							
	Group total	**12.81**	**16.47**	**14.76**	**15.36**	**18.48**	**13.65**	**15.93**
	Total cost per kilogramme of made rubber	**49.07**	**51.72**	**53.05**	**55.74**	**59.76**	**65.98**	**72.56**

1. The cost of production of rubber estimates are based on a survey carried out jointly by the Department of Census and Statistics and the Rubber Development Department
2. The costs figures given above are, all inclusive cost up to the point of the Colombo auctions for shipping

Crop Collection

The main crop from a rubber plantation is latex, a milky white dispersion of rubber in water, harvested by tapping process. Two to three hours after tapping, the latex collected in the cup is transferred to a clean bucket. About 70-80% of crop from a rubber plantation is in the form of latex.

The latex which gets solidified in the tapping panel (tree lace) and the collection cups (cup lump) also form part of the crop and are collected by the tapper in a basket just prior to tapping. The latex spilt and/or overflowed to the ground (earth scrap) when gets dried up is also collected as scrap once in a month. These are collectively called field coagulum.

Latex and field coagulum are highly susceptible to bacterial action and therefore it is essential to process these into forms that will allow safe storage and marketing.

General Composition of Latex

Rubber	**30-40%**	**Resins**	**1-2%**
Proteins	2.0 - 2.5%	Sugars	1.0 - 1.5%
Ash	0.7 - 0.9%	Water	55-65%

Latex can be Processed into any of the Following Forms

1. Preserved field latex and latex concentrate
2. Sheet rubber
3. Block rubber
4. Crepe rubber

Field coagulum can be processed only into crepe rubber or block rubber.

Latex Preservation and Concentration

Preserved Field Latex

Field latex is preserved using suitable preservative for long-term storage. The processing of preserved field latex consists essentially of adding the preservative (usually ammonia, minimum (1%) to the sieved latex, bulking, settling, blending and packing. Field latex can also be preserved with LATZ (Low ammonia - TMTD - zinc oxide) system.

Latex Concentrate

Two important methods of processing latex into preserved latex concentrate are commercially practised.

1) Concentration by Creaming

Mix a creaming agent like ammonium alginate or tamarind seed powder with properly preserved field latex and allow the latex to separate into two layers, an upper layer of concentrated latex and a lower layer of serum containing very little rubber. The lower layer of serum is removed, leaving the latex concentrate having about 50-55% DRC (dry rubber content), which is often tested, packed and marketed.

2) Concentration by Centrifugation

Separate preserved field latex into two fractions, one containing the concentrated latex of more than 60% dry rubber and the other containing 4-8% dry rubber (skim latex).

Skim latex is generally coagulated with sulphuric acid, made into crepe, dried and marketed as skim rubber, which is a low-grade rubber.

Ribbed Smoked Sheet (RSS)

Latex is coagulated in suitable containers into thin slabs of coagulum and rolled through a set of smooth rollers followed by a grooved set and dried to obtain sheet rubber. Depending upon drying method, sheet rubbers are classified into two:

- Ribbed smoked sheets
- Air dried sheets (Pale amber unsmoked sheets).

The latex collected is brought to processing centre before pre-coagulation sets in. In cases where the latex is prone to pre-coagulation, an anticoagulant is used.

Latex brought to the centre is strained through 40 and 60 mesh stainless steel sieves. The volume of latex is measured with a standard vessel and a calibrated rod. The dry rubber content (DRC) can be estimated with a metrolac, which is a special type of hydrometer calibrated to directly read the DRC.

Latex is diluted in bulking tanks to a standard consistency of 1/2 kg of dry rubber for every 4 litre of diluted latex (12.5% DRC). The diluted latex is allowed to stand in bulking tank for a fixed time (usually 15 to 20 minutes) for the heavy dirt particles to sediment. The diluted latex is drawn out from bulking tank without disturbing sediment layer of impurities into coagulation pans or tanks. Four litres of latex are usually transferred to each pan.

Coagulation

Formic acid or acetic acid is generally used for coagulation. Only diluted acid should be used for coagulation and should be thoroughly mixed with latex.

Acid Requirement for Coagulation of 4 Litres of Diluted Latex Containing ½ kg Rubber

	Acetic acid	**Formic acid**
For the next day sheeting	3 ml diluted to 300 ml with water	1.5 ml diluted to 300 ml with water
For the same day sheeting	4 ml diluted to 400 ml with wate	2 ml diluted to 400 ml with water

Catalyst AC (100 ml of a 5% solution) and sulphuric acid are also used by growers. 300 ml of a 0.5% solution of the sulphuric acid is required for same day sheeting and 250 ml for next day sheeting. Coagulum from latex often shows a tendency for surface darkening. To prevent this, a small quantity of sodium bisulphite (1.2 g/ kg DRC), dissolved in water may be added to the diluted latex before coagulation. After coagulation, the coagulum is removed from the pan or tank and thoroughly washed in running water. They are rolled either in a sheeting battery or smooth rollers to a thickness of 3 mm and finally passed through grooved roller. While sheeting, the coagulum is continuously washed. The sheets are again washed in running water in a tank.

Mould growth on sheet rubber can be prevented by treating freshly machined sheet in a dilute solution of para-nitrophenol (PNP). The concentration of para-nitrophenol is 0.05 to 0.1% in water (100 litres of solution for treating 100 sheets). The wet sheets are allowed to drip on reapers arranged in a well-ventilated dripping shed.

Smoking and Smoke House

The sheets after two or three hours of dripping in shade are placed in the smoke house where the temperature is maintained between 40-60°C. In the smoke house, sheets are dried gradually whereby blisters are avoided. In addition, the creosotic substances present in the smoke prevent mould growth on smoked sheets.

Smoke sheets on first day at a low temperature (40-45°C). Sheets can be dried by placing them on first day on reapers at bottom region of smoke house and at higher regions on subsequent days of smoking.

Four days of smoking is generally sufficient under normal conditions, but during rainy season five to six days are required for satisfactory drying of sheets.

The smoke house is a chamber in which, sheets can be placed on reapers, a furnace outside the chamber and a flue duct connecting furnace to chamber. Smoke houses are of two types, those in which furnace is inside drying chamber and those in which furnace is outside (the ground-floor type and the tunnel-type).

Other Methods of Drying Rubber Sheets

(1) Solar-cum-smoke drying, (2) Sun drying and (3) Air drying:

In solar-cum-smoke drying, hot air from solar collectors is blown into drying chamber in which sheets are placed on reapers placed on trolleys. The system also contains a furnace for burning firewood which acts as a back-up heat source for maintaining inside temperature during night and also on cloudy days.

Storage

Sheets after grading are packed in 50 kg bales. The grades are marked on bales and marketed.

There are six international grade descriptions for sheet rubber. They are RSS 1X, RSS 1, RSS 2, RSS 3 ,RSS 4 and RSS 5.

Crepe Rubber

Crepe rubbers are processed from fresh latex coagulum, field coagulum or cuttings of RSS. When any of these materials, after necessary preliminary treatments, is passed through a set of crepe making machines, crinkly, lace-like rubber is obtained. This, when dried, is called crepe rubber. Crepe rubbers are classified into latex crepe and field coagulum crepe, depending on starting material used.

Latex Crepe

Pale latex crepe (PLC) and sole crepe are two latex grade crepe rubbers. Latex to be used for production of PLC and sole crepe shall be free from yellow pigments as far as possible. Also coagulum shall not undergo enzymic discolouration.

1. Pale Latex Crepe

Starting Material: Fresh Field Latex: It is produced from latex of which the yellow colouring pigment is removed by either fractional coagulation or by bleaching or a combination of the two. The latex is sieved, diluted, bulked and treated with sodium bisulphate to avoid enzymic discolouration. Fractional coagulation is then effected by adding acetic or oxalic acid to remove undesired colouring matter present in latex. The coagulum containing colouring matter in latex due to partial coagulation is removed by sieving. In bleaching process, colouring matter is bleached using certain bleaching agents like tolyl mercaptan or its sodium/potassium salts. After fractional coagulation and or bleaching, the final coagulation is done using formic or oxalic acid. The coagulum is then passed through a creping battery. The thin crepe is then dried, graded and packed. Pale latex crepe is generally graded into four groups viz; IX, 1, 2 and 3. Different grades of PLC are used for high quality products like pharmaceutical articles, light coloured and transparent goods, adhesives, tapes, tubes and derivatives like chlorinated rubber.

2. Sole Crepe

Dried PLC is cut into proper sizes and lamination is done with hand. Sole crepe is marketed in different sizes and thickness. Thickness of various types of sole crepes, specified by consumers is 1.5, 3.2, 4.7, 6.3 and 9.5 mm and size of laminated piece is generally 90 x 30 cm. A variation of 0.4 mm in thickness is allowed. The sole crepe after sizing is packed in bales. Maximum weight of crepe in each bale shall be 101.7 kg and minimum 72.6 kg. Sole crepe is used primarily for production of translucent shoe soling materials.

Field Coagulum Crepe: Unit operations involved are soaking of coagulum in water, crepe making, drying, grading and packing. The crepe prepared from field coagulum materials fall into five types.

i. *Estate Brown Crepe (EBC) - Starting material:* Cup lumps and other higher grades of field coagulum. The pre processing consists of soaking coagulum in water for 24 h and washing the cup lumps and scraps to remove dirt. The washed materials are passed through a creping battery. The crepe thus obtained is air-dried, graded and packed. Three different grades of EBC produced are IX, 2X and 3X.

ii. *Thin Brown Crepe (Remills) - Starting material:* Wet slab coagulum, unsmoked sheets and cup Lumps. The materials are washed and milled in the creping battery. The crepe obtained is air-dried, graded and packed. Four different grades are produced;1, 2, 3 and 4.

iii. *Thick Blanket Crepe - Starting material:* Wet slabs, unsmoked sheets or other high grade Scraps. The processing consists of milling the material in creping battery, air drying and packing. Four different grades are viz, 1,2, 3 and 4.

iv. *Flat Bark Crepe - Starting material:* All types of low grades scrap including earth Scrap. The details of processing are same as that of estate brown crepe.

v. *Smoked Blanket Crepe - Starting material:* Smoked rubber derived exclusively from ribbed smoked sheet or cuttings. The processing consists of milling materials in the creping battery, air drying and packing. The crepe from field coagulum grade is packed in bales.

Technically Specified Rubber (TSR) : Technically specified rubber (TSR) is a modern marketable form of natural rubber produced from latex and field coagulum. The production of TSR started only during 1960's, prior to which raw natural rubber was marketed in conventional forms like sheet and crepe for which only visual grading was done. From 1950's onwards, well packed and scientifically characterized synthetic rubbers became commercially available in bulk quantities. To compete with synthetic alternatives, TSR was developed. TSR is produced in almost all NR producing countries. In India, TSR production started in 1973. It is marketed with quality certification from the Bureau of Indian Standards, under the name Indian Standard Natural Rubber (ISNR).

Production

Several commercial processes are used for production of TSR. Unit operations in the process of manufacture are 1) Size reduction, 2) Dewatering, 3) Dirt removal, 4) Drying, 5) Blending and 6) Grading. The important parts of machinery being used for said operations are slab cutter, coagulum crusher, pre-breaker, macerator/crepe roller, hammer mill/shredder driers and baling press. TSR bales produced are packed in low-density polyethylene (LDPE) bags. Each bale shall have a weight of 25kg and will have an identification number like lot/batch number marked on cover.

Specification and Grading

In India, six grades of TSR are being produced. The Bureau of Indian Standards (BIS) specifications for each grade of rubber are given below.

Physical and Chemical Requirements for Raw Natural Rubber

S. No	Characteristic(2)	Requirements for						
		ISNR 3CV(3)	ISNR 3L(4)	ISNR 5(5)	ISNR 10(6)	ISNR 20(7)	ISNR 50(8)	Method of test Ref. to (9)
a)	Dirt content, percent by mass, Max	0.03	0.03	0.05	0.10	0.20	0.50	IS 3660 (Part 1)1985*
b)	Volatile matter percent by mass, Max	0.80	0.80	0.80	0.80	0.80	0.80	IS 3660 (Part 2)1985*
c)	Ash percent by mass, Max 0.50	0.50	0.50	0.60	0.75	1.00	1.50	IS 3660 (Part 3)1988*
d)	Nitrogen, percent by mass, Max	0.60	0.60	0.60	0.60	0.60	0.60	NR: 11 of IS 3660 (Part 52): 1968*
e)	Initial plasticity (Po)	As agreed to between the purchaser and the supplier	30 (Min)	30 (Min)	30 (Min)	30 (Min)	30 (Min)	IS 3660 (Part 11): 1989*
f)	Plasticity Retention Index (PRI), Min.	60	60	60	50	40	30	IS 3660 (Part 12): 1989*
g)	Colour (Lovibond scale), Max	---	6.0	---	---	---	---	NR: 14 of IS 3660 (Part 54): 1979*
h)	Mooney viscosity ML (1+4) at 100°C	60±5	---	---	---	---	---	IS 3660 (Part 7): 1988*
i)	Accelerated storage hardening DP Max	8	---	---	---	---	---	NR: 15 of IS 3660 (Part 54): 1979*

Marketing
Indian Scenario

Rubber is grown in Kerala, adjoining Kanyakumari district of Tamil Nadu, Tripura, Assam, Megahalaya, Nagaland, Mizoram, Manipur, Goa, Andaman and Nicobar Islands and Coastal Karnataka. Rubber plantations are spread over 5.81 lakh ha in the country. The production sector of the country is dominated by small holdings, which accounts for 91% of production and 88% of area with an average holding size of 0.5 ha. There are nearly 1 million producers and about 0.7 million people engaged in plantation sector as workforce either directly or indirectly.

India is the fourth largest producer of natural rubber (NR), next to Thailand, Indonesia and Malaysia sharing about 8.6% of world output. The productivity of 1745 Kg/ha in India is the highest among major producing countries in the world. The country is also the fourth largest consumer of natural rubber. A large chunk of its domestic demand (nearly 97 %) is met from domestic production. Rubber production in India is around 6-7 lakh tones. Kerala accounts for 90% of India's rubber production. The other producer is Karnataka. RSS (Ribbed Smoked Sheets) account for 72% of production and 45% of imports. Block Rubber accounts for 10% of production and 40% of imports. The tyre industry, consumes 52 % of rubber produced in the country.

Natural Rubber-Consumption, Production and Yield

Year	Consumption ('000 t)	Production ('000 t)	Yield (kg/ha)
1997-98	572	583	1549
1998-99	592	605	1563
1999-00	628	622	1576
2000-01	631	630	1576
2001-02	638	631	1576
2002-03	695	649	1592
2003-04	719	711	1663
2004-05	755	749	1705
2005-06 *	799	780	1745

*Anticipated *Source:* Ministry of Commerce and Industry.

Kerala's Production Pattern

Kerala accounts for 83 % of area under rubber in the country. The coverage under the crop in 2005-06 was 4.94 lakh ha, and production of 7.39

lakh tones indicating 7% increase over the previous year. The increasing trend in productivity continued during 2005-06. It was 1190 kg/ ha in 1998-99, which rose to 1495 kg. during 2005-06. In terms of tapping area, productivity recorded was 1765 kg/ha during 2004-05. (Source: Economic Review 2006).

Consumption of Rubber in Kerala (in t) During 2000-01 to 2003-04

Year	Kerala			
	Natural rubber	Synthetic rubber	Reclaimed rubber	Total
2000-01	88221	32978	6773	127972
2001-02	100163	32114	6583	138860
2002-03	126100	36438	6861	169399
2003-04	124630	46921	7367	178918

Source: Rubber Board Statistics

Global Scenario

Thailand, Indonesia, India, China, Malaysia and Vietnam are the major producers of rubber in the World. The global production fluctuates between 6-8 million t, with a production of 7.9 million t in 2003, of which Asian countries produced 6.76 million t. On the consumption front, global NR consumption is 7.89 million t in 2003, of which 1.9 million t was consumed in India and China alone. The total synthetic rubber consumption in 2003 was 1.13 million t. Around 60 % of global rubber production is used by transportation sector. In this sector, natural or synthetic rubber cannot be used individually and has to be blended.

Export Pattern

The trade in natural rubber is slightly less than one eighth of global output. The total export of natural rubber in 2004 was 10.52 lakh t, out of total production of 89.76 lakh t produced from 82.99 lakh hectares. The world export doubled from 5.17 lakh t in 2000 to 10.52 lakh t in 2004.

Thailand emerged as the world's leader in rubber exports since 2000. Her share in world export rose steadily from 62.5% in 2000 to 81.2% in 2004. At the same time, the share of Malaysia declined steadily from 17.8% in 2000 to 7% in 2004. Vietnam, which produces 4.5% of world output, is another potential exporter. Asian countries account for 90% of the natural rubber exports. Hence if rubber is included for free trade or preferential tariff, farmers in Kerala will have lot of problems.

Quantity of Export of Natural Rubber from Different Countries (Qty in t)

Year	India	Sri Lanka	Indonesia	Malaysia	Myanmar	Thailand	World
2000	554	196	9095	91803	20100	323024	517109
2001	585	355	10375	80424	25000	410458	590234
2002	3299	974	8637	77974	22600	438889	621553
2003	8627	392	12526	76866	30000	491926	701866
2004	5403	2014	11755	73331	30000	853403	1051562

Source: http://faostat.fao.org

Export of Natural Rubber by Different Countries (%)

Year	India	Sri Lanka	Indonesia	Malaysia	Myanmar	Thailand	World
2000	0.1	0.0	1.8	17.8	3.9	62.5	100.0
2001	0.1	0.1	1.8	13.6	4.2	69.5	100.0
2002	0.5	0.2	1.4	12.5	3.6	70.6	100.0
2003	1.2	0.1	1.8	11.0	4.3	70.1	100.0
2004	0.5	0.2	1.1	7.0	2.9	81.2	100.0

Source: http://faostat.fao.org

India's export went up by ten fold from 554 t in 2000 to 5403 t in 2004. This was higher during 2003 (8627 t). As a result, India's share in the export of natural rubber appreciated from 0.1% in 2000 to 1.2% in 2003. However it declined to 0.5% in 2004.

Value of Export of Natural Rubber by Different Countries (%)

Year	India	Sri Lanka	Indonesia	Malaysia	Myanmar	Thailand	World
2000	0.2	0.0	1.7	20.0	2.4	53.7	100.0
2001	0.1	0.1	1.7	16.2	2.6	61.2	100.0
2002	0.7	0.2	1.2	14.9	2.1	64.1	100.0
2003	1.4	0.1	1.4	12.1	1.9	67.5	100.0
2004	0.7	0.2	1.4	11.2	1.5	71.6	100.0

Source: http://faostat.fao.org

The aggregate value of natural rubber exports in 2004 was $995 million compared to $436 million in 2000. Thailand alone contributed 71.6 % of the value of exports in 2004. The next significant exporter was Malaysia. Her export share declined from 20% in 2000 to 11.2% in 2004. While India's share was only 0.2% in 2000, it went up to 1.4% in 2003. However, it declined by half in 2004 to 0.7%.

Import Pattern

The total quantity imported was 26229 t in 2002-03 which increased to 44199 tonnes in 2003-04. The annual growth of domestic natural rubber industry for the period 1996-97 to 2003-04, showed an average increase of 3.3 % in production whereas consumption increased at 3.1%/ annum while import of natural rubber increased by 10.6%.

Import of Rubber in India (Qty in tonnes)

Years	Natural Rubber		Synthetic Rubber		Total
	Qty	%	Qty	%	
2000-01	8970	7.73	106923	92.26	115893
2001-02	49590	30.8	111323	69.18	160913
2002-03	26229	17.40	124475	82.59	150704
2003-04	44199	20.27	173784	79.72	217983

Source: Government of Kerala, Economic Review 2004, p.S31

Import of Natural and Synthetic Rubber into India

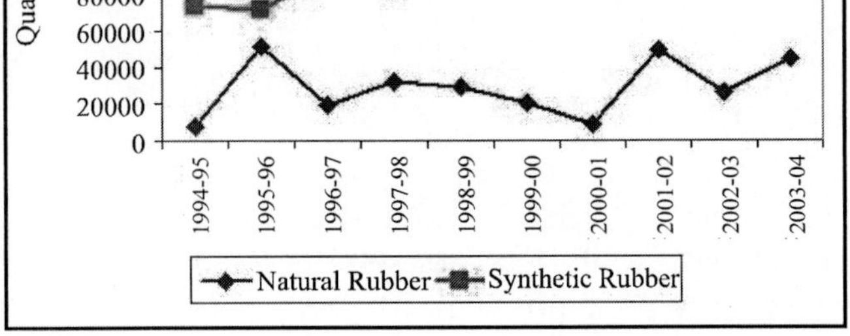

Source: Government of Kerala, Economic Review 2004, p. S31

Value of Natural Rubber Imported from Selected Countries into India (Rs lakh)

Countries	2001-02	2002-03	2003-04P
China	7.67	-	22.14
Indonesia	1816.86	1671.59	3307.66
Malaysia	5668.86	2772.85	10621.83
Myanmar	355.36	20.91	104.87
Singapore	620.30	351.29	128.08
Sri Lanka	222.53	179.88	204.02
Thailand	5023.64	4019.96	6326.31
Vietnam	400.77	647.61	846.50
Others	330.48	265.95	452.51
Total	**14446.47**	**9930.04**	**22013.92**

P: Provisional. *Source*: Rubber Statistical News, Rubber Board Kottayam.

Data on import of rubber and rubber products revealed a growing share of the major product groups, viz. synthetic rubber and allied products. The table below summarizes four products concentration in import of rubber and rubber products.

Price Variations

World rubber prices continue to show some strength in near future largely due to high oil price and continuing global economic growth. If global economic growth, especially in developed countries like the EU, Japan, and the United States and developing countries like China and India continues at the current pace, further price strengthening could be expected. The magnitude of the growth in rubber consumption depends considerably on the growth in China. It would, however, be limited in the long run by the good potential to increase supply by more intensive tapping and by increasing yield in major producing countries.

Market Prices

The higher price in the international market is reflected in domestic market also. The average price of RSS4 in domestic market at Kottayam was Rs. 55.70/ kg. in 2004-05 which increased to Rs. 66.99 in 2005-06. The international price of RSS3, equivalent of RSS4 of India, increased from Rs. 57.51 to Rs. 73.98 in the corresponding period. The price of RSS 4 in Kottayam

Trend in Price of Rubber During 2001-2005

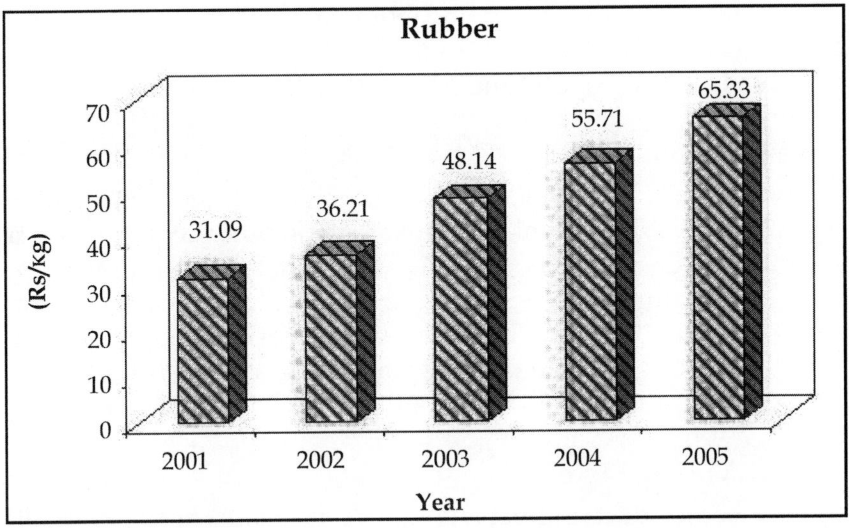

Source: Economic Review

reached Rs.106.92 during June 2006 and then declined to Rs.87.09 in October 2006. The Indian price (RSS4) which was higher than international price till June 2006 started moving below the international price during remaining months of the year owing to sharp increase in international price.

Average Farm Price of Rubber During 2001 to 2006

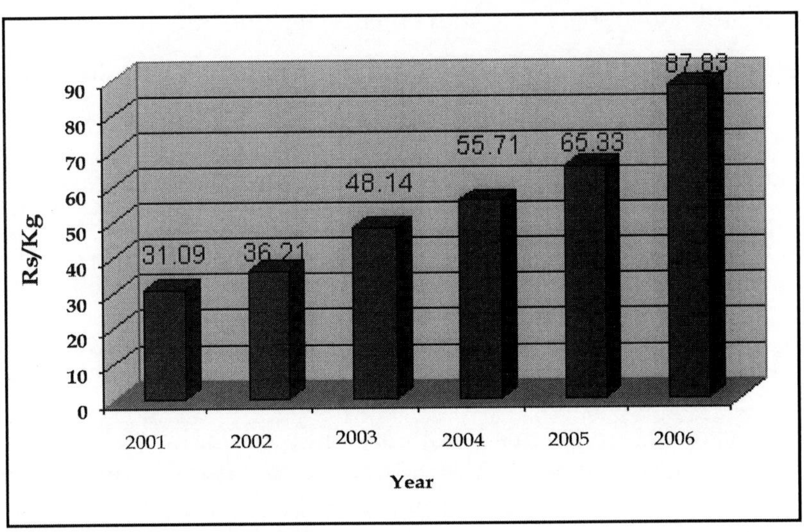

Source : Economic Review 2005

Major World Markets

Tokyo Commodity Exchange, Singapore Commodity Exchange and Osaka Mercantile Exchange are the major exchanges undertaking futures trading of rubber. Kuala Lampur, London and New York are the major physical markets.

Marketing System

Kottayam, Kochi, Kozhikode and Kannur in Kerala are the major primary markets in India. Marketing link to about a million producers and 4800 rubber goods manufacturers is provided to 10175 licensed dealers and 136 processors operating in the country. The marketing network of primary Co-operative Marketing Societies and the Trading and Processing Companies jointly promoted by the Board and the Rubber Producers' Societies ensures the highest farm gate price for Indian NR among all the NR producing countries.

Market Influencing Factors

- The rubber production fluctuates between months and is normally low during rainy season.
- Growth in industrial production: automobile industry.
- Ratio of utilization of domestic production and imported rubber by tyre manufacturers.
- Government policies have a profound influence on rubber prices. These include subsidies, restrictions on ports etc.
- International rubber price movements, have a slow influence.
- Stockists and speculators also play a significant role in influencing prices

Forecasts for the World Rubber Economy

Natural rubber and different types of synthetic rubbers are used in many different end-products. The most important is the tyre sector taking about half the total elastomer consumption. The other category, general rubber goods, includes hoses, belting, footwear, surgical goods, and rubberized cloth.

The 1990s witnessed several important changes in the global natural rubber economy. World consumption of natural rubber rose by about 32%, while that of synthetic rubber increased by only 4% and total elastomers by 14 %. The major reason for this is collapse of Soviet Union, which consumed a massive amount of rubber, with less than 10 % natural rubber. Also, there was a significant change in the relative share in utilization of natural rubber between developed and developing regions. Consumption of natural rubber in the developing countries rose by over 4 % / annum against less than 2 5% in the developed countries. Consequently, share of the developing countries in global consumption of natural rubber rose from 42% to 48 % between 1988-1990 and 1998-2000. On the other hand, requirements in the tyre industry led to a decline in share of natural rubber in total rubber consumption in most Asian developing countries.

Most natural rubber production is traded internationally. The share of exports in production declined from 86 % in the late 1970s to 80% in the late 1980s and to around 70 % in the late 1990s. This decline was partly the result of increased domestic utilization by the exporting countries themselves, and rapid growth in production in some net importing countries like China and India. World exports increased by over 1.6 % / annum between 1988/90 and 1998/2000. The decade also saw demise of International Natural Rubber Organization, another failed attempt to influence markets through managing stocks. The sharp decline in world market prices and fluctuations in exchange rates made the price support programme too expensive to continue.

Global Production

World natural rubber production was 6.8 million t during 1998-2000. It is projected to reach 7.9 million t in 2010. The annual growth rate would be 1.3% in current decade, which is significantly below 2.9 % during past decade.

Rubber is produced entirely in developing countries and Asia is the largest producing region. It is projected that total production in Asia would touch 6.8 million t. by 2010, more than 85 % of world output, with an annual growth rate of 1%. Compared with the fast growth of last decade of 2.7%, slower growth is largely due to further contraction in Malaysia and Sri Lanka, which have made efforts to diversify to more remunerative crops. On the other hand, expansion would be less rapid in almost all other countries except Viet Nam. Compared to 2001, production in Thailand would even decline. There are a number of reasons for this. First, a large part of the area was planted in 1980s and is now due for replanting and thus will to be out of production for around six years. Second, the rate of replanting natural rubber is declining as farmers and land shift to other employment because of low prices seen in recent years. Finally, there is not much new planting. Nevertheless, Thailand is expected to retain its position as the world's largest supplier to produce 2.2 million t. by 2010. In Indonesia, production should grow by 2.2 % annually to reach 2.05 million t. by 2010. While production in other major producing countries like China and India experience little growth, Viet Nam and the Philippines are likely to see their production grow annually by 8.4 and 5.9 % respectively, to double their output by 2010. Africa is expected to increase its production to nearly 0.5 million t. with an annual growth rate of 2.2 %, which is slightly higher than 1.8 % in the previous decade.

Latin America experienced more rapid expansion in production than other regions. In the past decade, it grew by 8 % annually, but growth to declines in the next decade to around 5 % annually. Brazil is the largest producing country in Latin America. Favourable agronomic conditions, abundant land resources and lower labour costs are major factors driving growth in production. However, given the very low production base of only 31000 t. in 1988-90, Brazil would still be a small producing county in global terms with output of 0.12 million t. by 2010.

Demand

Demand for natural rubber is part of the total elastomer demand for tyres and other rubber products. More than 60 % of natural rubber is used for tyres, which is the major driving force behind changes in natural rubber demand. The share of natural rubber in the tyre sector was about 50 % in 2000 and is likely remain around this level in the next decade. Technological requirements are the overwhelming determinants of the share of natural rubber in tyre production and no significant new developments are foreseen that would allow synthetic rubber to replace natural rubber in tyres.

In the general rubber goods (non-tyre) sector, the share of natural rubber is projected to decline from 29 % in 2000 to about 22 % in 2010. This decline reflects partly a continuation of a shift to speciality rubbers and other materials, coupled with a projected shortage of natural rubber. Natural rubber can more easily be replaced in many non-tyre applications than in tyres.

Overall, natural rubber will constitute about 36% of total elastomer consumption by 2010 declining from 40 % in 2000.

World demand for natural rubber in 2010 is projected at 7.9 million t., a 1.3% annual increase in the current decade, considerably lower than in the 1990s. Demand is projected to grow the fastest in developing countries, at about 2.5% annually. Above average economic growth in developing countries as well as the in demand for domestic processing of rubber products would contribute to the higher growth in natural rubber demand.

Asia accounts for more than 86 % of total developing countries' demand, among which China is the largest consuming country with the highest growth rate. It is projected that China would grow by 5.1% annually to reach nearly 1.6 million t. by 2010, reflecting largely its rapidly growing demand for tyres. Demand in Indonesia is expected to grow by more than 5 % annually, driven by development of domestic processing industries and demand for tyres. India and the Republic of Korea are also major consuming countries in this region. Although their annual growth rates are likely to be only around 2 %, demand is expected to exceed 0.7 million t. in India and nearly 0.4 million t. in Korea by 2010. Most other Asian countries would also see their demand increase. Demand in countries in Latin America would remain at around the current level, largely due to slow economic growth and hence slow growth in demand for vehicles.

Demand in developed countries would to remain at the current level, the result of sharply increased demand in Eastern European countries and the former Soviet Union offsetting reduced demand in the United States and the EU. It is projected that demand in the United States will reach 1.09 million t by 2010, slightly lower than the average of 1.16 million t in 1998-2000. Major car making countries in the EU including Germany, France, Italy and the United Kingdom would see demand weaken slightly in the next decade. Following a period of stagnation since early 1990s, demand would pick up in Eastern Europe and the former Soviet Union as their economies grow. Japan would see its demand decline slightly, at an annual rate of about 1%. As a result, total consumption in developed countries is projected to be about 3.5 million t, about the same as the average consumption level during 1998-2000.

Trade

During 1998-2000, 70 % of natural rubber was traded in the world market. Two significant developments in trade occurred in the last decade. Net exports of natural rubber increased less rapidly than production. While production grew at 2.9 % annually, the annual growth rate of net exports was only 1.6%, which was largely due to weak demand in developed countries and strong increases in domestic consumption in producing countries. Also, while new exporting countries, especially Viet Nam, expanded their exports rapidly, exports from a few traditional exporting countries waned gradually. Malaysia, once the largest exporter in the world, saw its exports drop from about 1.4 million tonnes in the late 1980s to less than 0.2 million tonnes in 2001, while Sri Lanka's exports declined from nearly 0.1 million tonnes to about 0.03 million tonnes. On the other hand, Viet Nam exported almost 10 times more in 2001 than in 1988-90, with an annual growth rate of 21.6%, a trend that is expected to continue in the current decade.

World net exports in 2010 are projected to grow by 1.3% annually to reach 5.5 million t., 15 % above the average of 1998-2000, with the bulk of the increase from Indonesia, Viet Nam and some smaller Asian countries. Exports from Indonesia should grow by 2.1% to reach about 1.9 million t, and exports from Viet Nam are expected to reach 0.5 million t by 2010, with an annual growth of 8.1 %. Thailand would experience little change in exports, largely reflecting its slow increase in production and higher domestic demand. Exports from Malaysia would continue to decline at an annual rate of 9%. As a result, it is expected to export only 0.12 million t by 2010, just 8.5 % of its 1988-90 level. Net exports from Sri Lanka are likely to have disappeared by 2010 because of declining production and increasing domestic consumption. Export availability in Africa and Latin America would continue to grow. However, their shares in the world market are still small. By 2010, exports from Africa and Latin America would be 0.38 and 0.03 million t, respectively.

Developed countries are the largest importers in world market. Although import demand in developed countries increased steadily last decade, higher growth in import requirements from developing countries was the major determinant of imports. It is projected that growth in imports will be essentially flat in developed countries while developing countries as a whole would grow by 3.5 % annually in current decade. Economic growth in developing countries which induces greater demand for cars and hence tyres is the major factor responsible for higher growth rate. Notable growth is expected in China, which would import more than 1.1 million t.by 2010 with an annual growth rate of 8 % next decade. Higher import demand from China accounts for almost all import growth in developing countries.

Issues and Uncertainties

The last decade witnessed considerable changes in world natural rubber suppliers. Emergence of new low-cost producing countries forced a few high-cost countries to shift away from natural rubber production.

On the demand side, with economic growth developing countries have become increasingly important importers in the world market. These trends are expected to continue in the current decade.

Natural rubber prices declined steadily after 1996 and reached a historical low in 2000. While prices have recovered somewhat since then, they are still significantly lower than in 1996. Demand is likely to grow steadily in the current decade, leading to some tendency towards higher prices. However, price increases tend to be dampened because they promote increased production on one hand, and increased use of synthetic rubber on other. In addition, existing excess stocks may take a few years to be cleared.

TEA

The tea plant is an evergreen shrub or small tree that reaches 10 m if unpruned. It is adapted to sub-tropical areas. Tea belongs to family Camelliaceae and all the cultivated tea plants belong to two distinct species, viz., *Camellia sinensis* (L). O. Kuntze, the short leaved 'China' plants and *Camellia assamica* (Masters) Wight, the broad leaved 'Assam' plants. The 'Cambod' variety, a subspecies of the latter, is named C. *assamica lasiocalyx* (Planchon exWatt) Wight. The 'China', 'Assam' and 'Cambod' and a large number of their hybrids are seen in many tea fields. The tea plant is an evergreen of *Camellia* family, native to China, Tibet and North India. The small leaf variety, known as *Camellia sinensis,* thrives in cool, high mountain regions of Central China and Japan. The broad leaf variety, known as *Camellia assamica*, grows the best in the moist, tropical climates in Northeast India and the Szechuan and Yunnan provinces of China. The plant produces dark green, shiny leaves and small, white blossoms.

Tea was originally cultivated in China, and most of the world's tea still comes from Asia, with lesser quantities from Africa and South America. Sri Lanka and India are now the largest producers, but sizeable quantities also come from China, Japan, the former USSR, Indonesia and Pakistan. The Islamic Republic of Iran, Turkey, Viet Nam and Malaysia are small-scale growers. Since Second World War, the area under tea cultivation in Africa is expanding rapidly, particularly in Kenya, Mozambique, Congo, Malawi, Uganda and the United Republic of Tanzania. Mauritius, Rwanda, Cameroon, Zambia and Zimbabwe also have small acreages. The main South American producers are Argentina, Brazil and Peru.

Climate

Tea growing tracts of South India, extending along the Western Ghats, vary in their elevation from 300 to 2,300 m above MSL and experience an annual rainfall from 90-750 cm. These plantations, with their adjoining forest ecosystem contribute greatly to maintenance of terrestrial ecology by providing extensive land cover and minimizing soil erosion.

Tea grows the best in areas with a maximum temperature of 16-32°C and a well distributed rainfall of 150 cm/annum. Relative humidity should be around 80% most of the time and should never be less than 40%. The area should not be prone to frost. The soil should be acidic (around pH 5.0) having good drainage facility.

The average annual rainfall in North East India ranges 2000-4000 mm. More than the total amount, distribution of rainfall matters a lot for sustained high yield of tea throughout the season.

Temperature affects tea yield by influencing rate of photosynthesis and controlling growth and dormancy. In general, the ambient temperature within 13°C and 28-32°C are conducive for growth of tea.

Day length influences growth and dormancy in tea bushes. When days of less than 11hr 15 min duration last for at least six weeks, tea bushes become dormant. Hence the length of growing season decreases with increasing distance from the equator. Seasonal dormancy appears from around 18° North and South latitudes. In the Northeast India (25°-27°N latitude), tea bushes remain dormant during winter for about 3 months on account of combined effects of short days and low temperature.

Soil

Tea grows well on high and well drained soils having good depth, acidic pH in the range 4.5 to 5.5 and more than 2% organic matter.

Varieties

Clone

Until middle of last century, tea was grown essentially from seed throughout the world. Development of a simple, cheap and rapid method of vegetative propagation of tea plant in third and fourth decades of last century at Tocklai, Assam made it possible to release elite plants and multiply them vegetatively as clones for economical planting. Tocklai released the first tea clone in the world in 1949.The total number at the end of 2002 being 30 excluding 151 clones selected in commercial tea estates. Two new TV clones are in the pipeline, which are under multilocational trials. Following are clones so far released by Tocklai Tea Research station, Assam for commercial cultivation:

Standard clones: TV2, TV3, TV4, TV5, TV6, TV7, TV10, TV11, TV12, TV13, TV14, TV16, TV17, TV20, TV24, TV27, AND TV28 have above average yield and quality with high yield potential of about 3000-3500 kg made tea/ha.

Yield clones: TV8, TV9, TV15, TV18, TV19, TV22, TV23, TV25, TV26, TV29 and TV30 have average quality but high yield with yield potential of 4000 kg made tea/ha and above.

Quality clones: TV1 and TV21 have high quality but average yield with yield potential of about 2500-2800 kg made tea/ha.

So far Upasi Tea Research Foundation released 29 clones (vegetatively propagated cultivars). Among these, clones like UPASI-3, UPASI-8, UPASI-17, UPASI-25, UPASI-27, UPASI-28 and TRF-1 are high yielders. Clones like

UPASI-2, UPASI-6, UPASI-9, UPASI-20 and UPASI-26 are moderate yielders but are tolerant to drought. Certain SriLankan clones TRI-2024, TRI-2025 and TRI 2026 and a few estate clones CR-6017, C-1, ATK-1 and SA-6 are also widely used for planting in South India. Current clonal selection work in the Anamallais, Nelliampathy and Nilgiris resulted in selection of 140 bushes based on visual examination.

Biclonal Seed Stocks

The biclonal seed stocks are UPASI: BSS-1 to BSS-5. These seed stocks germinate early in the nursery and establish well in the field. They are excellent for infilling, replanting and extension planting. The stocks are moderate yielders but are tolerant to drought.The biclonal seed stocks released by Tocklai till the end of 2002 are TS 378, TS 379, TS 397, TS 449, TS 450, TS 462, TS 463, TS 464, TS 491, TS 506,TS 520, TS 557, TS 569 and TS 589.

Production of Planting Materials- Propagation Methods

Tea Seed Bari-Source of Seed

Tea seeds are produced in seed baries (orchards) planted specially for the purpose where the parents (selected generative clones) of approved stocks are established either planting directly or by grafting on any vigorous jat/ clonal root stocks. Tea seeds are produced over the entire surface of a seed tree. In general, spreading type of plants require about 5.5-6m space between them, while the erect type and Chinery plants may be planted at a closer spacing of 3.5-4.5m and 3-3.5m. The planting of seed bearers may be either square or triangular design. Number of plants varies 330 to 1283/ha, depending on spacing and design of plantings used. When both parents are seed bearers, they are planted alternately in each row. If only one of the parents is a seed bearer, then the planting pattern is such that the male parent (pollen donor) is surrounded by a few seed bearers (female).Seed baries can be raised on any soil suitable for tea cultivation. A well drained, fertile sandy loam to loam soil is the most ideal.

Tea is a cross pollinated plant. Pollination is caused by pserphid, a small insect having a short flying range. It is therefore, necessary to isolate seed baries from other teas to avoid undesirable out crossing. A minimum distance of 10m should separate seed baries from main tea fields.

Planting, after care and manuring are similar as that of young teas. Mature trees should be manured with NPK 2:1:2 mixture @ 100 kg / ha/year in spring. Pruning is not recommended except to remove unwanted shoots to clear congestion which facilitates aeration and light penetration. For higher seed production, 30-40% sunlight should reach the ground in a diffused manner. If there were large height difference between clonal pairs (parents), balancing the height by lopping would facilitate efficient pollination. Old seed baries can be rejuvenated by heavy pruning as in tea under plucking. The major part of flower bud initiation (75%) takes place during March-June and about 97% of flower bud formation is completed by end-September. Blossoming generally starts from early-October and ends by January with a peak period in November-December. Most of the seeds are produced from this peak flowering after 10 months. The seeds ripen by next October. The November-May period is very crucial for fruit development and moisture deficit in this period affects seed yield adversely.

The tea fruit is allowed to dehisce on the tree and the dropped seeds are collected from ground. As these seeds lose viability rapidly, the normal practice is to collect them everyday. Tea seeds vary in size from 10 to 22 mm dia.

Grading and Sorting

After collection, seeds are passed through a rotary type shifter to eliminate very small seeds. Remaining seeds are allowed to soak 2-3 hours in a trough of water. The seeds which sink (sinkers) are then spread on a plastic sheet or on concrete floor and examined for pest, disease and mechanical damage. The suspended or floating seeds (floaters) are discarded.

Storage of Seeds

Tea seed is stored for a minimum period in a cool place covered by a layer of semi-moist sand or sub-soil. Rapid drying during storage causes loss of viability.

Packing

After sorting, tea seeds are packed as soon as possible. For long distance transport, the seeds should be packed in wooden boxes in units of 20 kg. Seeds are spread in layers along with packing material and covering each layer with a thin sheet of tough paper. When the box is full, a sheet of paper is laid on top and the lid is nailed down. The different packing materials used are moist sand, sub-soil, powdered charcoal, ash and their mixtures. The moisture content of packing material varies from 10-30%. If sandy loam sub-soil is used, the optimum moisture content would be 10-12% while for powdered charcoal it varies 25-30%. Depending on size of grader used, one kg of graded and sorted seed contains 300-500 seeds.

Seed Propagation

Sandy loam soil is the best for raising nursery on beds. The nursery site should be on virgin soil. In absence of virgin soil, uprooted tea areas or area with poor soil may be used after the site has been put under green manure crops for a period of two years. The plants need lopping 2-3 times a year and the loppings should be left insitu for decomposition. Alternatively, the area may be heavily dressed with cattle manure @ 25 tonnes/ha. The area should be criss-crossed, ploughed and harrowed about a month before the beds are ready for sowing. Beds should not be prepared during rains. The seed beds may be of any convenient length but should not be wider than 1.5m for convenience of cultural operations. In North East India, the beds are generally prepared flat 10-15 cm above the surrounding ground level. Each nursery bed is enclosed by a drain 20 cm deep and 20 cm wide, which is linked up with outlet drain. Single super phosphate (SSP) should be applied to the soil @ 50g/m2 before sowing seeds. A nursery bed measuring 30 x 1.5 m or 37.5 x 1.2 m accommodates about 1300 seeds when sown at 20 x 20cm (triangular) spacing.

Tea seeds are planted on nursery beds in rows at a depth of 1.5 cm with the "eye" facing downwards or parallel to the ground surface. If the seedlings are to be lifted with the soil (a cylinder of earth about 15 cm in diameter and 20-25 cm in depth around the roots) for transplanting in the field later, the seeds are sown with a spacing of 20 cm. This spacing can be reduced to 15 cm or so for preparing seedling for stump or carrot planting where lifting of plants with soil is not required. The nursery beds can be of any convenient length up to 120 cm wide.

Polythene Sleeve Nursery

When land is insufficient, seedlings are raised in polythene sleeves (17 cm lay flat wide, 20-30 cm long and 150 gauge thick). Soil can be brought from any convenient site but must be tested for acidity and eelworm population. It should be mixed with super phosphate @500 g/m^3 of soil and the sleeves filled with soil 3-4 weeks before sowing. Seeds are sown at 0.5-1cm depth with the "eye" facing downwards and covering seeds with a thin layer of soil. The sleeves can be placed in a permanent nursery site where irrigation facilities exist. If irrigation facilities are not available, sleeves can be placed under usual overhead shade or green crop shade. Vigorous seedlings can be planted in the field in April-May when they are 5-6 months old. For planting 1-year old seedlings, seeds should be sown in 30 cm long sleeves. It will be necessary to shift the sleeves once or twice after mid-July to prevent taproot from penetrating the ground. For planting one hectare, 40-60 kg tea seeds are required depending upon the spacing.

Pre-germination of the seeds in moist sand kept in germinating pits with mulch cover is necessary to raise good nursery. They are ready to be transferred to the sleeves as the seeds crack. Seeds that do not germinate even after 2 months should be discarded.

After sowing, the bed or the sleeves can be kept under mulch when the nursery is raised under green crop shade. Frequent examination is required to ensure scrapping of the mulch around the germinating seeds so that it does not cause any obstruction. Excess watering should be avoided. Only hand weeding should be done carefully and proper care against pests and diseases should be ensured to protect plants.

Vegetative Propagation

The nursery is raised on a high land with adequate drainage and sunlight which is convenient for irrigation and frequent inspection. The soil should be neither too sandy nor too clayey with pH 4.5-5.0 and free from eelworm. If the soil is too acidic (below pH 4.5), it is treated with slaked lime @ 2t/ha. Sub-acidic soils (pH above 5.1) can be amended with aluminium sulphate solution (2%) 2-3 times. Soil should be tested before planting cuttings.

Beds under low shade should be north -south oriented. Shade is provided with the help of bamboo lath frame about 22 cm over the propagation beds, supported by sidewalls. The bamboo lath frame should be raised gradually with increase in height of plants. It is necessary to prepare beds at least 6-8 weeks before planting cuttings. The period of heavy rain should be avoided for preparation of beds as clods are difficult to break when the soil is wet. The bed is generally 120 cm wide and of any convenient length. Each bed is enclosed by 45 cm deep and 30 cm wide drain, which should run into one outlet drain. Application of super phosphate @500g/m^3 of soil used for filling sleeves or @50g/m^2 of bed by incorporating in the soil to 15-20 cm depth helps in root development.

Polythene Sleeve Nursery

Clones can also be planted directly into sleeves (15 cm lay flat x 22cm long; 150 gauge thick). Smaller sleeves produce weaker plants. Before filling sleeves, the soil should be sieved through a No.4 wire mesh to eliminate undecomposed organic matter and pebbles. The sleeves are filled with sieved soil and the filled sleeves should be kept covered preferably under shade for 6-8 weeks prior to planting cuttings.

Type of Cuttings

Cuttings should be taken from primaries with dormant apical buds (banji state) for higher success. In some clones, lateral shoots are produced from lower portion of the primaries quite early. The unbranched portion of primaries, in such cases, should be taken first followed by laterals when they mature after a few weeks. A good cutting should have a fully mature healthy undamaged leaf, dormant or just swelling bud and a hard green stem of about 2.5 cm length below and about 0.5 cm above the leaf.

Planting of Cuttings

With a dibber (small bamboo sticks slightly thinner than the stem), 2cm deep holes are made 5 cm apart on the bed. Cuttings are then planted firmly by holding the cutting at the node with the leaf pointing away from the hand in a semi erect position and inserting the stem on holes upright. In north south oriented beds, tip of leaf should point towards north and in east west aligned beds, the tip should point towards west. The mother leaf should be more or less in an upright position after insertion into holes. There should be no air pockets in interspaces of cutting and soil. Dipping cuttings in 0.1 % zinc sulphate (0.1%) solution before planting helps root initiation. The soil moisture should be maintained at field capacity. The cuttings should be transferred to sleeves just after initiation of roots. Shade should be such that cuts about 50% of the mid-day sun light are falling on nursery bed. In the low lath frame shade, the bamboo frames should be raised on the east after the cuttings have produced 4-5 new leaves. Where thatch is used in overhead shade, automatic thinning occurs in time. When bamboo lath is used, gradual removal of roof (alternate one) ensures adequate hardening of plants. Young plants need protection from pests and diseases by high volume spray of suitable pesticide.

Nutrient Management in Nursery

When the plants have produced 3-4 leaves, they are manured with a mixture of NPK (2:1:2) and sand at 1:9 ratio. One teaspoonful of this mixture should be applied/ plant at 3-4 weeks interval during growing season. If required, foliar application of nutrients can be started when cuttings have produced 6-8 leaves during April-May and August-November.

Composition of Tea Nursery Mixture is as follows

Ammonium phosphate (20:20)	:	60 parts
Potassium sulphate	:	24 parts

(or) Muriate of potash	:	20 parts
Magnesium sulphate	:	16 parts

Stock Solution

Dissolve 30 g of the mixture in 10 litres of water. This can be sprayed over 2 m^2 (450 plants) at weekly interval.

Land Preparation for New Planting

The vegetation is cleared one season ahead by ring barking trees followed by uprooting the stumps along with roots and leaving the litters to decompose *in situ*. The pits so formed are filled up with excavated soil and the ground is levelled with minimum soil disturbance. The drains are then laid out and excavated soil spread thinly over the ground. Following this, staking for tea plants is made and seeds of green crop like *Crotolaria anagyroides* are sown between tea lines for providing quick ground cover and temporary shade to young tea plants. Tea is planted as soon as land preparation is completed. In general, planting is completed by end of May/June in drought hit areas and by September to early November in other places. The newly planted tea fields are then adequately mulched. Between clearing land and planting, there should be minimum time interval.

Land Preparation for Replanting on Uprooted Land

Ploughing and cross ploughing should be undertaken when the soil is moist . If the soil is highly acidic, 2-3 tonnes of powdered slake lime or dolomite is broadcast before ploughing. After levelling the land, drains are dug and the area put under rehabilitation crops for at least two years. After rehabilitation period is over, the grass is cut at ground level leaving the lopping in situ. The stumps are removed with minimum soil disturbance. The drains are deepened or re-graded and the area is replanted with vigorous clones or seedlings.

Planting

Correctly planted tea plants establish in the field quickly, grow vigorously and come into full bearing earlier.

Time of Planting

Planting can be done in April-June and September-October or October-November with adequate irrigation. Periods of heavy rains should be avoided.

Type of Plants Used for Planting

Only healthy plants 40-60 cm high with at least 12 good mature leaves and of pencil (0.5 cm) thickness (at collar) should be taken for planting in field. In general, 9 to 12 months old plants attain this stage. Sub-standard plants should be discarded. Before plants are removed from nursery, they should be hardened by gradual exposure to full sun.

Type of Planting

There are two types of planting, i.e., pit planting and trench planting.

Pit Planting

This method is followed when spacing between plants is wide enough to allow digging of individual pits of proper size and without much difficulty. Pits should be about 45 cm wide and 45 cm deep, circular and straight walled. The excavated soil is conditioned by mixing with 4-5 kg well-decomposed cattle manure or 150-200g well-decomposed oil cakes and returning soil into pits. No other manure is used except 30 g rock phosphate and 30g SSP at the time of planting.

Trench Planting

This method is adopted for closer spacing and in heavy soils. Trenches 30 cm wide and 45 cm deep are dug along rows.

Method of Planting

Two methods of planting, are bheti planting and stump planting.

Bheti Planting

The plants are lifted along with the soil and the roots intact from the nursery bed. This is convenient with plants grown in ploythene sleeves, which reduce difficulty in transportation, reduce root damage and give a very high percentage of survival. The polythene is removed carefully by slitting the tube and the bheti is held in the pit half-filled with conditioned soil in such a manner so that top of bheti is flushed with ground surface. Rock phosphate(30g) is added at bottom of bheti and the pit is filled with soil with adequate ramming. At about 5 cm depth, 30g SSP(30g) is added around the bheti and the pit is filled up to the collar of plant with soil.

Stump Planting

The lifted plants from nursery bed without having any soil around the roots are cut off 15-20 cm from collar and excess roots trimmed off before putting them into pits. This method is generally followed with overgrown nursery plants and has the advantage of easy transport and reduced chances of withering after planting.

Style of Planting

Up and down	:	1.2 x 1.2 m (6800 plants/ha)
Contour planting, single hedge	:	1.2 x 0.75 m (10,800 plants/ha)
Contour planting, double hedge	:	1.35 x 0.75 x 0.75 m (13200 plants/ha)

The double hedge planting accommodates more number of plants/ unit area gives early high yield, better soil conservation and less weed growth. About 14000-16000 (up to 17000 in hilly areas) plants/ hectare are ideal bush population with spacing of 105-110 cm between rows and 60-75 cm between plants.

Shade Management

The best permanent shade tree for tea plantation in South India is silver oak (*Grevillea robusta*).

Planting of Silver Oak

Silver oak seeds should be sown within 6 months after collection. Seed should be sown in raised beds of 1 m width and of convenient length using sandy loam soil with a pH around 6.0. Seeds should be covered with thin layer of sand / ash. Germination takes place in 2-3 weeks. Use 6-9 months old seedlings for planting. Plant along tea rows at a spacing of 6 x 6 m (275 plants/ha). Apply a mixture of 100 g rock phosphate and 400 g dolomite/ pit and thoroughly mix with soil prior to planting. Apply NK mixture @ 100 g/ tree twice a year; rock phosphate at 250 g/tree and borated lime 1.1 kg (1 kg dolomite + 100 g boric acid) during alternate years.

Shade Regulation

Tea requires only sparse shade. Retain optimum stand of shade based on growth of tree, altitude of garden and aspect of field (south and west slopes require more shade). Thin out shade initially to 12 x 6 m after 8-10 years of planting and if required, further thinning may be done to 12 x 12 m at later stages (12 years from planting). Always thin out shade prior to pruning.

Pollarding

Cutting the main stem with the objective of developing lateral branches is pollarding. Commence pollarding when trees attain a girth of around 50 cm at elbow level. Pollarding depends on altitude (8 m height for higher altitude, 9 m for low elevation). Leave one branch in each direction and 3 to 4 tiers of branches below pollarding height.

Annual Lopping

Cutting the erect growing branches on laterals is lopping, which should be done before onset of monsoon and lop only erect branches and retain laterals.

Shade Removal

Useful age of Grevillea is 40-60 years. Remove old trees after establishing new shade.

Temporary Shade

For frost prone areas: *Acacia mearnsii*

For mid elevations: *Indigofera teysmanii*

For higher altitude: *Sesbania cinerescens, Crotalaria agathiflora* and *Acacia elata*

Planting at 3 x 3 m spacing is adopted. Temporary shade should be removed 3 years after establishment of Grevilliea.

Nutrient Management in Main Field

Manuring of young tea commences 2 months after planting. The ratio and source of nutrients vary according to soil reaction (pH).

Fertilizer Application for Young Tea in Soils with pH Below 4.5

Age	**$N:K_2O:MgO$ kg/ha/annum**	**No. of splits**
1st year	180: 270: 30	5
2nd year	240: 360: 40	6
3rd year	300: 450: 50	6
4th year & above	300: 300: 50	6

Apply phosphorus at 90 kg/ha every year in one application. Quantity of fertilizer/ bush may be calculated assuming a population of 13,000/ ha.

Fertilizer application for soils with pH between 4.5 and 5.5

Age	N:K_2O kg/ha/annum	No. of splits
1st year	180: 270	5
2nd year	240: 360	6
3rd year	340: 450	6
4th year & above	300: 300	6

Apply phosphorus at 90 kg/ha every year in one application.

Fertilizer application for soils with pH above 5.5

Age	N:P_2O_5:K_2O kg/ha/annum	No. of splits
1st year	180:60:180	5
2nd year	240:80:240	6
3rd year onwards up to 1st pruning	300:100:300	6

The rate of fertilizer application for mature tea varies with yield and soil test values while the N:K_2O ratio varies with stage of pruning.

Method of Application

Apply the recommended quantity of mixtures along drip circle of plants. In semi-circular furrow taken above the plant on the slope, using a Kokra-eyebrow method, apply fertilizers when there is adequate soil moisture and when the fields are free from weeds. Punch holes of 15-22 cm depth in soil on either side of plants and place the rock phosphate.

Time of Manuring

As a general rule, the best time for fertilizer application is after the first rain in spring has moistened the soil to a depth of 45 cm and there is some new growth in case of unpruned tea.

Method of Manuring

It is recommended to apply N and K fertilizers in two splits if the dose exceeds 100 kg/ha. This ensures adequate supply of nutrients throughout the growing period and higher utilization efficiency. The 60% dose of N and K should be applied in March-April (1st split) and the second split (i.e. the remaining 40%) should be applied in August-September. The entire quantity of phosphatic fertilizer should be applied in the beginning of season along with first split of N and K. In very light texture soil, the number of splits may be increased.

Micronutrients

Among various micronutrients tried, only zinc showed consistent response and is recommended particularly in unpruned / skiffed teas. Foliar application of zinc sulphate at 1-2% concentrations (w/v) was beneficial and produced up to an average of 10% increase in yield. The present recommendation is not to exceed 12.5 kg zinc sulphate/ hectare/ year, which is to be applied in 4-6 sprays (1-2% concentrations) in early (March onwards; 2-3 sprays) and late season (September onwards; 2-3 sprays). Indiscriminate use of zinc sulphate sprays should be avoided as it may lead to high levels in made tea.

Foliar Nutrition

Foliar nutrition is beneficial under stress condition or coinciding with physiological changes in the bush. NPK mixture 2:1:2 or 2:1:3 where potash status is low, (0.5-1 %) can be sprayed during this period.

Foliar sprays of potash(1%)MOP (w/v) during moisture stress period enhance the water use efficiency of bushes. MOP (1%) and $MgSO_4$ (1%) sprays from Jan to March help to ameliorate ill effect of moisture stress in UP teas.

Irrigation in Tea

During the period from October-February, rainfall is scanty and the quantity varies from 5 to 10% of the annual rainfall. In this period, average evaporation exceeds average precipitation by 8 to 357 mm. Therefore, conservation and supplementation of soil moisture by irrigation becomes necessary during this period of moisture stress.

Irrigation in Young Tea

In planning irrigation, first priority should be given to the newly planted young tea. Among tea areas those having coarse texture soil and in the hills especially the south facing slopes should also get priority.

Mature Tea

Sprinkler irrigation is the most widely accepted method in tea gardens. Drip irrigation is confined to seed baris. In general, depending upon rainfall received in October, irrigation should commence from November and continue till March/April. The first application in November can be a little more than the estimated field irrigation requirement followed by five more applications, each at an interval of three weeks. In severely drought prone years, irrigation in April may be necessary but after the rainfall in April exceeds 75-125 mm, irrigation can be discontinued. As in case of young tea, irrigation schedule should not be interrupted except after heavy rainfall exceeding 38 mm and then for not longer than 2-3 days.

Drainage

The objective of drainage is to remove excess moisture in the root zone within a specified time so that the soil returns to field capacity without any harm to the tea plants. Developing outlet augments the rate of outflow and thereby reduces water logging. Without a proper outlet, drainage planning can not be successful and thus for an efficient drainage layout, development of outlet should be taken up first with the ultimate objective of lowering water table below root zone depth. Studies at Tocklai indicate that improved drainage increases tea yields to the extent of 30-35% over a period of time.

Young Tea Management

After planting, the area should be mulched adequately with green vegetative matter leaving a 10 cm distance from collar of plants uncovered. Sowing of green crops, in the young tea fields between alternate tea rows provides shade and generates mulching materials. This is particularly useful in drought hit areas. The green crop has to be periodically lopped and loppings left *in situ* to decompose.

The young plants should also be protected from pests and diseases using high volume sprays of suitable pesticides. If tea plants are affected by red rust, apart from spraying with copper fungicide as a control measure, it is also necessary to look for correction of drainage, shade, soil pH, and damage by drought.

Plants like *Indigofera teysmanii, Melia azadirach, Gliricidia sepium* ., should be planted for providing temporary shade. They will be ready to provide shade within one year of planting and should be kept lopped periodically. *Indigofera teysmanii* is the most common species used for temporary shade and is propagated by means of seed, air-layering or stem cutting.

Soon after planting, the ground is cheeled and mulched heavily or sprayed with a pre-emergent herbicide (simazine or oxyfluorfen). Subsequently hand weeding in rings of 15 cm radius around the collar of the plant is done.

Herbicides spraying in young tea field is done with great care and using a protective shield.

Manuring of Young Tea

After planting, as soon as the plants have produced two or three new leaves, they can be foliar sprayed with NPK or NK mixture at fortnightly intervals till the plants have produced 4-5 new leaves. Well-established plants can be manured with YTD mixture (NPK 10:5:10 or 10:5:15 depending up on soil available potash status) at rates depending upon spread of bush as determined by age. During first four years of formative stage, plants are ring manured upto 4 splits while on 5th and 6th year; manures are applied in 2 splits. A general guide line is as follows:

Young Tea Manuring During Formative Stage

Year of planting	Nitrogen (kg/ha)	YTD manure (Kg/ha)	No. of splits	Application method
0 year	20-40	200-400	2-3	Ring
+ 1 year	80-100	800-1000	4	Ring
+ 2 year	100-120	1000-1200	4	Ring
+ 3 year	120-140	1200-1400	4	Ring
+ 4 year	140-150	1400-1500	2	Strip
+ 5 year	140-150	1400-1500	2	Strip

Foliar Spraying of Nutrients

Sprayings of zinc sulphate and magnesium sulphate along with urea and MOP, depending upon season, cultivars and growth of plants are useful.

Bush Frame

In young tea plants, formation of a proper frame is achieved by following three operations:

a) Decentering, lung pruning, or debudding, b) Formative pruning 20-26 months after planting c) Final Frame formation pruning.

The objective of first two operations are to suppress centrally dominating apical growth and distribute vigour for even radial spread so that at the time of final frame formation, there is uniformity in distribution and thickness of branches at the pruning level.

The permanent frame is formed at a height, that ensures optimum coverage of ground and convenience in harvesting, by pluckers. The height of permanent frame will be influenced by spacing and cultivars of tea and it should normally be at 35-45 cm from ground level.

Decentering

The main stem is removed with a pruning knife at a height of about 20 cm from ground. The operation should be done after the plants make one flush of growth following transplanting in field and when they are at banji (dormant) stage.

Lung Pruning or Thumb Pruning

The seedling is held between thumb and index finger at about 20 cm height and stem is half broken in such a way that tissues on one side are left intact for movement of water, nutrients and photosynthates. The broken portion should be bent towards ground facing either south or west depending on row direction and should be removed after branches below it produce a flush of growth and go banji. Lung pruning can also be done by using a small pruning knife. After development of side laterals, the broken portion is removed with a knife. After 6-8 weeks of decentering or lung/thumb pruning, any leader primary should be head backed, depending on thickness, to 25-30 cm from ground.

Debudding

Buds from leaf axils are removed up to the normal level of 20 cm from the ground. Two weeks before debudding the top two and a bud should be nipped off from plants in the nursery. This helps the buds below to swell. Four to five days before plants are taken out to the field, they can be debudded in nursery.

Frame Forming Pruning

The frame forming prune should be given at a height 5-10 cm below pre-determined height of permanent frame. Thicker branches should be pruned down on merit (head back) and any congestion at centre removed.

Permanent Frame

Bush Architecture

An ideal bush frame is one which will have a three-tier branching system in proportion of 1:3:7 at decentering/debudding, frame forming prune and permanent frame forming prune levels, respectively.

Number and Thickness of Branches

On an average, there should be 26 to 32 branches on permanent frame. 70 % of these branches should have a thickness of 0.5-1.0 cm. It should have a density of one stick / 100 sq. cm of frame area. The permanent frame should cover 36-40% of ground area. Judicious pruning at time of formative prune is important to achieve this objective.

Pruning and Skiffing

Pruning is one of the most important operations, next to plucking, which directly determines productivity of tea bushes. It is a necessary evil in the sense that it has to be carried out periodically otherwise results in huge crop loss. If pruning is delayed, age of wood from pruning increases, size and weight of growing shoots on plucking surface decreases.

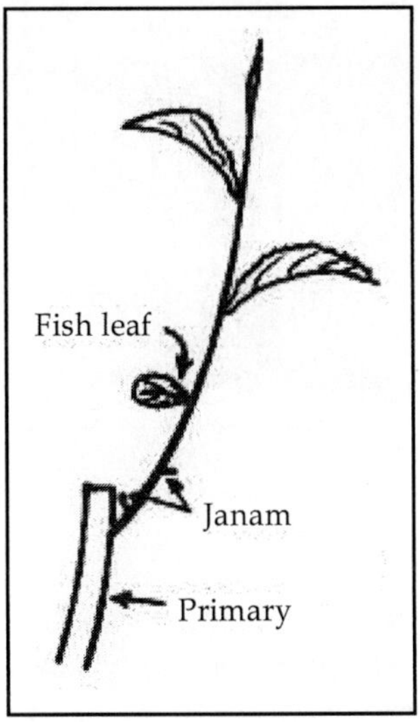

Types of Pruning

Light Prune

Tea bushes are usually pruned every 3 or 4 years at 4 -5 cm above the last pruning cut. This type of pruning is called light prune (LP). The time period from one light prune year to another is called one pruning cycle and LP is thus, natural sequence given at the end of a pruning cycle.

Height Reduction Prune and Medium Prune

When tea bushes grow tall and plucking becomes difficult, they are brought down to an optimum height by height reduction prune (HRP) at 60-70 cm, or medium prune (MP) at 45-60 cm above ground. Both HRP and MP help in rejuvenating tea bushes that have become old and their yields started declining. MP removes the knots and unproductive excess woods and facilitate consolidation by infilling of vacancies. Such sections are brought back to normal 3-4 year pruning cycles in about 4-5 years.

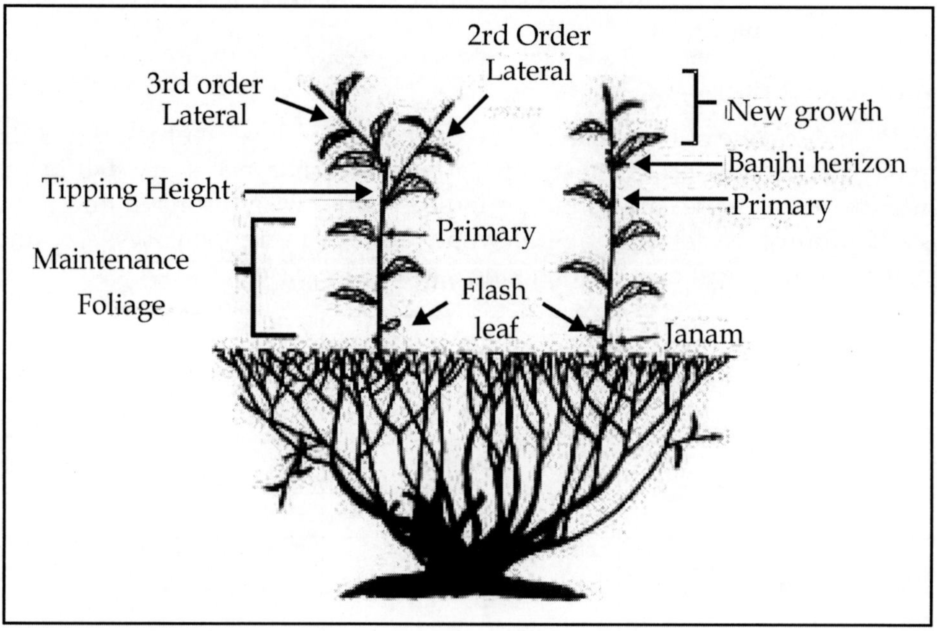

(*Source:* Swaminathan et al., 1990)

Heavy Prune

Heavy prune (HP) is given at 15-45 cm for complete renewal of frame. In collar prune, all above ground parts of the tea bushes are cut down and this operation is carried out only when the root system is strong enough to withstand shock and initiate new growth. In practice, however, very low pruning is generally avoided now a days as it results in heavy mortality, particularly in Assam jats growing in poorly shaded sandy soil. In between two successive prune (LP) years, tea bushes are given lighter forms of cuts which are termed as deep skiff (DS), medium skiff (MS), light skiff (LS), level of skiff (LOS) or untouched which is called unpruned (UP).

Deep Skiff

Deep skiffing (DS) of tea bushes is done normally between 12-15 cm above the last LP mark.The DS helps to regulate crop distribution and to reduce ill effects of drought, excessive creep and height of plucking table.

Medium Skiff

Medium skiff (MS) is normally given at 5 cm over last Deep skiff mark. The objective of MS is to regulate crop distribution, reduce ill effects of drought, reduce incidence of excessive banji formation and reduce height of plucking table.

Level of Skiff and Light Skiff

Level of Skiff (LOS) is given 4-6 cm above tipping mark mainly to level plucking surface. Light Skiff (LS) is usually given up to 1 cm above previous tipping height.

Time of Pruning and Skiffing

In general, pruning is carried out when tea bush is dormant and there is good reserve of root starch. In drought areas of Northeast India, pruning is done in mid-December to mid-January avoiding shoot initiation during drought. Medium pruning can be carried out in mid-December to end of January. Skiffing is performed at anytime between mid-December and end-January; deeper the cut earlier the skiffing time. In Darjeeling, pruning can be started as early as November.

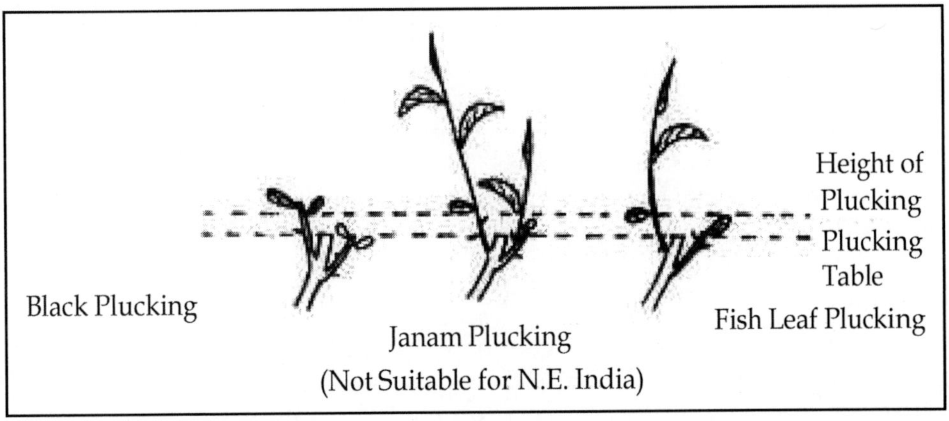

Resting Before Pruning

For LP, generally resting is not required. If there is inadequate foliage, the tea bushes are given rest by stopping plucking from 3 weeks before pruning. Very weak tea plants and those due for MP should be rested 5-8 weeks prior to pruning. Vigorously growing well-shaded tea bushes are pruned first. For MP/HP in the year of prune, additional doses of potash and phosphate are also applied and in inadequate shaded areas, temporary shade of *Indigofera teysmanii* at 3-4 m apart should be planted a year before pruning.

Pruning and Bush Sanitation

After pruning at predetermined height, bushes are sprayed with an aqueous solution of copper fungicide on same day. This reduces chances of damage by Poria, Aglaospora etc. Knife cleaning out (KCO) can be started any time after prune, and should be completed before bud-break starts. KCO involves removal of knots and of dead, diseased and unproductive wood. All large cuts should be disposed within 24 hours. In MP, some of the branches may have to be cut below general pruning height to get clean wood. To prevent chances of sun scorch, the bushes may be covered with pruning litter immediately after prune. At times they may also require lime-caustic-wash.

The caustic wash is prepared using 6 kg washing soda and 2kg quick lime in 100 litres of water and is used after keeping preparation overnight.

Crop Loss Associated with Prune/skiff Compared to Unprune

Type of skiff/skiff	(%) loss of crop
MP	60-70
LP	30-35
DS	10-15
MS	5-10

Medium or rejuvenation pruning is given only to those sections, which show potential for improving after the prune, should be considered for heavy prune.

Pruning Cycle

In a pruning cycle, all forms of pruning/skiffings are introduced sequentially for distribution of crop and renewal of wood producing crop/ quality. In comparison to annual prune, which causes a heavy rush only during the main season, well-balanced extended pruning cycles provide an even distribution of crop resulting in an even flow to the factory throughout the season. The choice of pruning cycles depends on a number of factors, viz., crop, labour, quality and market requirements, pests and diseases, climate and soil, bush height, age, vigour and kind of tea. In the plains of Northeast India under normal growing conditions 3-4 year pruning cycle and for Darjeeling depending upon elevation, a 4-5 year pruning cycle is suitable.

Normally Followed Pruning Cycles are

1. Well shaded, young and vigorously growing tea areas	: LP-UP-UP
2. Normal tea growing areas	: LP-UP-DS-UP
3. Droughty areas	: LP-UP-DS/LP-DS-UP
4. Darjeeling (low elevation)	: LP-UP-DS-UP
5. Darjeeling (mid and high elevation)	: LP-UP-UP-DS-UP

Primary objective of leaving bushes unprune or giving LS/LOS is to have more early and total crop. Therefore, conditions favourable to achieve that objective must be ensured before hand. The first two cycles shown above are crop oriented and the 3-year cycle of LP-UP-DS or LP-DS-UP is quality oriented.

Approximate Gain in Crop Over Annual Prune

DS	10 - 15%
MS	15 - 20%
LS	20 - 25%
LOS	25 - 30%
UP	30 - 35%

Plucking of Tea

Plucking in tea is synonymous with harvesting in other crops. The tender apical portions of shoots consisting of 2-3 leaves and the terminal buds are nipped off in plucking. The plucked shoots are manufactured to produce tea.

Plucking Systems

Three plucking systems are present.

- Janam plucking
- Fish leaf plucking
- Single leaf plucking

In theory, only janam should be left behind on stubs of plucked shoots when following janam plucking system, but in practice some fish leaves are also left on the bushes along with janams. This cannot be avoided for sake of maintaining a flat plucking surface. Similarly, fish leaf plucking and single leaf plucking systems are generally a mixture of the two, with predominance of one or other.

In any plucking system, the harvested leaf should not be coarse as coarse leaf is harmful for quality of product. Secondly, shoots below plucking surface should be left alone.

Janam plucking is the system followed in North East India, this system has proved superior to other systems. Other countries claim superiority of one or other two remaining systems under their cultural conditions.

Due to janam plucking, the major bulk of tea produced in North East India is derived from tiny janam axil buds.

Shoot Production

Light pruning stimulates dormant buds on the short 2.5-3 cm pieces of stems known as sticks, to produce a new set of leafy branches, called primaries. When the primaries grow to a pre-set height above bush frame, these are tipped (broken back) parallel to ground surface. The average height of fish leaf on primaries is the usual height of tipping in N.E. India. All leaves left on bushes below the tipping level are known as maintenance foliage. Tipping stimulates buds on axils of the leaves borne by primaries, but the stimulus becomes weaker at increasing distance from the point of tipping. As a result, the top most point axillary buds of all tipped primaries produce lateral shoots, which declines to 50% in second axillary buds and 25% in the third.

The lateral shoots are plucked when they produce two or more leaves above the tipping level. Plucking of these lateral shoots of first order originating from axil buds on primaries stimulates growth of second order laterals which when plucked produce third order laterals and thus production of successive orders of laterals continue until end of plucking season. In N.E. India, up to 8th order laterals have been produced in a year.

Maintenance Foliage

Photosynthetic capacity develops gradually in a young, expanding tea leaf and it does not attain full capacity before attaining more than half its final size. It therefore, follows that the young shoots harvested for manufacture of tea grow largely at the expense of metabolites supplied by the mature maintenance foliage remaining below the plucking table.

Maintenance leaves on a tipped and plucked primary remain for 18 months, although its efficiency as a photosynthesising organ goes on diminishing after 6 months.

Standard of Plucking

Standard of plucking denotes type of shoot harvested. Depending on length of plucking round, or type of shoots harvested, there may be five standards of plucking as follows:

Plucking Standard

Name of System	Shoot size	Breaking back	Plucking Round (days)	(%) crop gained/loss over standard plucking
Fine	1 + B; small 2 + B	Done	5/6	- 11.3
Standard	Large 1 + B; all 2 + B small 3 + B & single banjis	Done	7	Base
Medium	All 2 + B; 3 + B; single & double banjis	Not done	7/8	+ 0.5
Coarse	3 + B or larger shoots all banjis	Not done	8 or more	+ 28.2 to 38.4
Black	All 1 + B; 2 + B & single banjis	Done	6/7	- 5.0

To maintain a balance between quality and yield, 75% fine and 25% coarse leaf in the harvest are ideal.

Plucking Rounds

The time interval between two successive pluckings is called plucking round. Plucking round may be extended from 4 to 14 days, but to keep a balance between crop and quality, normally 6-8 days plucking round is practised in North East India depending on growth rate and quality of tea, one desires to produce. The time required for unfolding of successive leaves from a growing bud varies from 3 to 6 days depending on climatic variation. This is called leaf period. The mean leaf period of seed jat plants of N.E. India is 4 days during main flushing season and the leaf should be plucked a day

Type of Shoots Available on Different Plucking Round

Type of shoot left in the bush	Type of shoots harvested after		
	4 days	8 days	12 days
Only bud	1 + B	2 + B	3 + B
1 + B	2 + B	3 + B	4 + B*
2 + B	3 + B	4 + B*	5 + B*

* After 4+B or 5+B shoots go banji

earlier than twice of leaf period (2 x leaf period - 1 = 7 days). The type of shoots left out during previous plucking round determines size of harvestable shoots in next round, as is evident from table.

Creep and Breaking Back

It is advised to pluck close to janam and not allow undue rise (creep) of the plucking table, since the latter results in crop loss. Under normal conditions, the creep should not exceed following limits:

Type of Prune	By end July	By end August	By end November
LP	2.5 cm	3.5 cm	5 - 6 cm
DS	2.5 cm	3 cm	4.5 - 5 cm
MS	2.5 cm	3 cm	4 cm
UP	2.5 cm	3 cm	4 cm

Under a good plucking system, breaking back is not required. However, this becomes necessary when rounds are very long and supervision is improper as otherwise it results in crop loss.

Banji Shoots

Leaving banji shoots on the surface until they come through results in loss of crop, which could be as high as 90 kg/ha. This loss is attributed to uneven surface resulting in inefficient plucking.

Organic Farming

The objective of organic tea cultivation is to have an ecologically sustainable plantation, aimed at conservation of ecology and natural habit without polluting soil, air and water and yet maintaining sustainable tea production. Tea is produced in the absence of synthesized chemicals like pesticides, fungicides, herbicides, growth regulators and concentrated fertilizers. Naturally occurring, mined products and bulky and concentrated organic manures are used for resistant cultivars, with regulation of micro-climate or by the introduction of biological control agents / the use of biological products, naturally extracted without the use of inorganic solvents.

Requirements

Selection of Site

The area needs to be sufficiently isolated to ensure that there is no possibility of any pollutants or contaminants flowing or drifting into it from any known or unknown sources. A detailed history of a period of about ten years, fully documented to give details of external inputs during pre conversion period, should be maintained to facilitate inspection of organic tea cultivation.

Conversion Period

The minimum conversion period should be three years from the last usage of synthetic agrochemicals. One can start marketing the tea as "in conversion organic tea" only after the lapse of one year from the start of conversion.

Soil

The minimum depth of soil profile should be 1.5 to 2.0 metres and organic matter status should be medium to high level depending on elevation and rainfall of the area.

Water

A perennial source of water free from pollutants is required in the estate for large scale compost preparation, which is essential for organic tea cultivation.

Livestock

Livestock is an integral feature of organic farming for cowdung, sheep, goat, poultry and pig manure.

Manual Workers

Additional workers are required for organic tea cultivation in comparison to conventional tea cultivation. The extra workers are required for trenching, compost preparation, weed control, application of bulky and concentrated organic manures and shade regulation.

Cultural Practices

Soil Water Conservation Measures

Trenching and mulching conserve rain water and make moisture available to plants on a sustained basis. Tea pruning, leaf litter and compost should be buried in trenches. Mulching can be done with Guatemala grass (*Tripsacum laxum*), shade tree loppings and other plant materials. Guatemala grass can be planted in large vacant patches with the twin objectives of rehabilitating the soil and raising mulch material.

Soil Reaction

The soil pH should be maintained at about 5.0 by application of agricultural lime or dolomite lime once in a pruning cycle.

Manuring

Manuring aims at making available all essential nutrients deficient in soil at optimum quantities and to return nutrients removed, by using the manures of organic origin demanded by the tea bushes for a sustainable productivity level. Compost, oil cakes and rock phosphate are main inputs to substitute removal of nutrients. Nitrogen is also supplied by regular lopping of low and medium shade trees and leguminous trees (*Gliricidia sepium, Gliricidia maculata*) which can be grown along roadsides and other vacant patches. Bone meal, fish meal and other manures of organic origin can also be used. Wood ash can also be applied.

Manuring of Organic tea-pruned Year (for 2500 - 4000 kg made tea/ha)

Manures	Quantity	Nutrients (kg/ha)		
		N	P	K
Compost*	10 t/ha	199	14	72
Neem cake ** (1 split)	2.5 t/ha	121	9	26
Castor cake *** (1 split)	2.5 t/ha	121	9	28
Rock phosphate (20% P_2O_5 (1 split)	400 kg/ha	-	35	-

Manuring of Organic tea-other years (for 2500 - 4000 kg made tea/ha)

Manure	Quantity	Nutrients (kg/ha)		
		N	P	K
Compost*	5 t/ha	242	18	52
Neem cake ** (1 split)	5 t/ha	242	18	56
Castor cake *** (1 split)	500 t/ha	-	-	28
Rock phosphate (20% P_2O_5 (1 split))	400 kg/ha	-	35	-

* Compost (1.99 % N, 0.14 % P, 0.72 % K, moisture less than 10.0 %)

** Neem cake (4.85 % N, 0.36 % P, 1.03 % K)

***Castor cake (4.82 % N, 0.35 % P, 1.11 % K)

Rock phosphate may be applied in the pruned after and third years as placement based on soil test values. Soil available P should be maintained at above 22 ppm.

Wood ash may be applied only in mature tea fields except in pruned year, preferably in dry periods by broadcasting 15 cm away from collar. Organic manures should not be broadcasted in tea fields. It should be applied in staggered trenches. The dimension of trench should be 2 x 0.3 x 0.45 m, 2 m apart between trenches.

Shade Management

Adequate attention to maintenance of optimum stand of shade should be given in organic tea cultivation and should be regulated properly by annual lopping.

Weed Control

Weeds should be controlled manually by hand pulling during dry period and slashing during monsoon; uprooted and slashed weeds may be retained in field.

Pests and Diseases Management

Pests and diseases control in organic tea cultivation are primarily preventive rather than curative. If insects and diseases occur, non-toxic biological methods are applied. Within a balanced ecological system, the pests and diseases are controlled by use of resistant clones, balanced nutrient supply, parasites and predators, pheromones, herbal sprays and by appropriate cultural operations. Certain caterpillars like flushworm, leaf rollers etc., can be controlled to a certain extent by manually removing the infested shoots during plucking.

Blister blight disease can be controlled by use of resistant clones and by modifying micro-climate by thinning of shade trees.

Post-harvest and Manufacturing Practices

Manufacturing Practices

For manufacturing, only mechanical and physical processes are allowed with natural fermentation. It can be manufactured as orthodox, CTC, Oolong or green tea.

Isolation of Manufacturing Facility

Manufacturing of organic tea should be carried out in a separate factory to eliminate all changes and possibilities of coming into contact with conventional tea.

Storage and Packing

There should be separate store for organic tea where no fumigation, insecticides or fungicides are used. Vacuum, steam or high pressure water cleaning is permitted. Organic tea should be packed in plywood chests or biodegradable packing materials on the same day of production and the organic quality grade should be clearly indicated on each chest or container along with the invoice number of dispatch.

Transportation and Shipment

The chests of organic tea should be transported separately and there should not be any chance of it coming into contact with the conventional tea. Before shipment to destination, it should be stored in an isolated place, away from the conventional tea.

Inspection and Certification

Soil Analysis

Soil analyses of zero to 30 cm depth should be necessarily done each year, regularly, for macro and micronutrients and heavy metal status. This will reveals fertilizer usage and contamination from sources other than intentional application such as wind drift from neighboring plantations or water migrating from plantations above.

Leaf Analysis

Tissue analysis of mother leaf should also be carried out, annually, for pesticide and fungicide residues and for the levels of macro and micronutrients and heavy metals. This reveals mineral nutrient status and pesticide and fungicide usage, if any.

Marketable Tea Analysis

The marketable organic tea should be analysed twice a year, pre- and post- monsoon, for pesticide residues, flavour and quality.

Book-Keeping

Book keeping should be properly maintained to make an audit trial of inputs and marketable tea supplied to consumers. The invoice number should be mentioned on tea chests, in the factory book and in the documents. With invoice number, origin of the tea can be traced back to when it was manufactured, the quantity manufactured and the site and quantity of leaves plucked in the field with the factory diary and field book.

Inspection by Independent Agency

Inspection by representatives of an independent agency unannounced at least once a year is required by field visits to check organic growing techniques, taking samples of leaf and marketable tea at random and also to check maintenance of book-keeping system.

Standards for Organic Manure

Fertilizer Advisory Development and Information Network for Asia and Pacific (FADINAP), Bangkok formulated following standards for organic manure:

pH (1%): 6.5 - 7.5 more or less neutral
Organic matter should not be less than 20.0 %
Carbon: Nitrogen ratio (C : N ratio) -10:1 to 15:1
Moisture content should not exceed 25.0%
Pathogens*: absent

*Pathogens and heavy metals originate mainly from sewage sludge and urban garbage.

The above specifications are also applicable for organic manures suitable for tea plantations. In addition, electrical conductivity (EC) is also considered to judge their suitability for use in tea fields. EC (1%) up to 0.50 dSm^{-1} is recommended for young tea EC (1%) up to 1.5 dSm^{-1} recommended for mature tea only. The minimum rate of organic manures suggested for application in tea fields is 5 t/ha and the maximum rate is 20 to 25 t/ha to be applied in two equal splits. Organic manures should not be broadcast in tea fields; it should be applied in staggered trenches across the slope once in every two/three rows depending on gradient.

Yield Estimate

	Year					
	1	**2**	**3**	**4**	**5**	**6 onwards**
High yielding	Nil	1,800	4,200	6,500	9,000	11,000
Low yielding	Nil	700	1,800	4,200	6,500	9,000

(Figs. in kg green leaf/ha)

Sale Price

Keeping in view the present rates of price of green leaf prevailing in different parts of the country, an average sale price of Rs.12/Kg considered for this model.

Income : (Rs. / ha)

Particulars	Year					
	1	2	3	4	5	6 onwards
I. High Yielding Areas :						
Gross income	-	21,600	50,400	78,000	108,000	132,000
Total cost (including plucking cost)	136,900	30,180	32,010	31,830	38,040	42,840
Net Income	**-136,900**	**-8,580**	**18,390**	**46,170**	**69,960**	**89,160**
II. Low Yielding Areas :						
Gross income	0	8,400	21,600	50,400	78,000	108,000
Total cost (including plucking cost)	136,900	27,540	26,250	26,310	32,040	38,040
Net Income	**-136,900**	**-19,140**	**-4,650**	**24,090**	**45,960**	**69,960**

Estimated Cost for Development of Tea Plantation by Small Growers

Unit Size : 1.0 ha, Plant Population : 15000/ha

A. Itemwise mandays (m.d.) requirements

S. No.	Item	No. of Mandays				
		Year 1	Year 2	Year 3	Year 4	Year 5
1	Clearing/Levelling land	125	-	-	-	-
2	Soil/Land preparation	50	-	-	-	-
3	Drainage	65	-	-	-	-
4	Surveying	10	-	-	-	-
5	Organic matter application	30	-	-	-	-
6	Planting/Infilling tea					
i)	Pitting @125 holes/m.d	120	12	6	-	-
ii)	Filling pits @200 holes/m.d.	75	8	4	-	-
iii)	Transportation @300 plants/m.d.	50	5	3	-	-

Contd...

iv)	Planting @75 plants/m.d. (Including staking)	200	20	10	-	-
v)	Fertiliser application @1000 plants/m.d.	15	2	1	-	-
7	Planting Shade trees :		-	-	-	-
	Permanent	10				
	Temporary	20	-	-	-	-
8	Mulching @300 plants/m.d. per round	150	50	-	-	-
9	Pruning/Tipping	10	20	30	40	40
10	Manual cultivation @30 m.d. round / chemical weed control	60	120	90	90	90
11	Application of plant protection chemicals @4 m.d./round	12	12	12	16	16
12	Application of fertilisers @4 m.d./round	12	12	12	12	12
13	Shade maintenance @5 m.d./round	15	10	10	10	6
14	Maintenance of drainage fencing etc.	-	10	10	10	10
	Total mandays	**1,029**	**281**	**188**	**178**	**174**

B. Investment in Mandays & Material (Figs. in Rs.)

S. No.	Items	Year				
		1	**2**	**3**	**4**	**5 onwards**
A	Mandays (No.)	1,029	281	188	178	174
	Cost of labour @ Rs.60.00 per manday	61,740	16,860	11,280	10,680	10,440
B	Materials :					
	1. Organic matter @Rs.2.5kg/pit i.e.37.5 tonnes/ha. costing Rs.300 per tonne	11,250	-	-	-	-
	2. Mulching materials (collection from neighbouring areas/ forests) (25 to 30 t/ha)	-	-	-	-	-
	3. Tea planting materials @ Rs.3.00 each	45,000	4,500	2,250	-	-

Contd...

	4. Shade tree planting materials (including 10% for gap filling) : i. Permanent sp. 490 @Rs.200 each ii. Temporary sp. 1730 @Rs.1.00 each	980 1730	- -	- -	- -	- -
	5. Fertilizers	1,500	3,000	6,000	3,000	3,000
	6. Plant Protection Chemicals @Rs.300.00/lit.	1,500	1,500	1,800	1,800	2,100
	7. Weedicides		-	600	750	900
	Total	136,900	25,860	21,930	16,230	16,440

Unit Cost : Rs. 1,84,690.00/Ha (Capitalised Upto The 3rd Year)

Financial Analysis of Small Tea Growers' Scheme (High Yielding Area)

YEAR	**1**	**2**	**3**	**4**	**5**	**6-15**
Devlpmnt.cost	136900	25860	21930	16230	16440	164400
Plucking cost	NIL	4320	10080	15600	21600	264000
Total cost	136900	30180	32010	31830	38040	428400
Benefit	0	21600	50400	78000	108000	1320000
Net benefit	(136,900)	(8,580)	18,390	46,170	69,960	891,600
NPV of cost	385232					
NPV of benefit	718436					
NPW at 15% df	333204					
BCR	1.86 : 1.00					
IRR	> 50%					

Financial Analysis of Small Tea Growers' Scheme (Low Yielding Area)

YEAR	**1**	**2**	**3**	**4**	**5**	**6-15**
Devlpmnt. Cost	136900	25860	21930	16230	16440	164400
Plucking cost	NIL	1680	4320	10080	15600	216000
Total cost	136900	27540	26250	26310	32040	380400
Benefit	0	8400	21600	50400	78000	1080000
Net benefit	-136900	-19140	-4650	24090	45960	699600
NPV of Cost	352557					
NPV of Benefit	555064					
NPW at 15% DF	202507					
BCR	1.57: 1.00					
IRR	40%					

Harvesting and Processing

Young tea fields and the fields immediately after pruning are harvested by hand plucking. Fields which have crossed more than 15 months from pruning are harvested with the help of a hand held shear. Shear harvesting increases productivity of workers. In recent years, hand held motorized harvesters were evaluated and found useful to achieve high labour productivity. Both one man and two men operated Kawasaki and Ochiai motorized harvesters are useful in tea fields planted on moderate and gentle slopes.

Ten Commandments for Plucking

1. Harvest two to three leaves and a bud and / or single and two-leaf banjis.
2. Pluck mother leaf during January-March.
3. Pluck new level during rest of month.
4. Pluck at 7-10 days interval during high cropping months
5. Pluck at 12-15 days interval during low cropping months.
6. Removal of banjis and breaking-back should be a part of plucking operation.
7. Do not pluck below the level.
8. Leave immature shoots.
9. Shear-harvest during rush periods.
10. Cut lanes in older fields.

Tea Processing

It is the method in which the leaves and flushes from *Camellia sinensis* is transformed into dried leaves for brewing tea. The types of tea are distinguished by processing, they undergo. In its most general form, tea processing involves oxidising leaves, stopping oxidation, forming tea and drying it. Of these steps, degree of oxidation plays a significant role of determining final flavour of tea, with curing and leaf breakage contributing to a lesser amount flavour.

Picking: Tea leaves and flushes, which include a terminal bud and 2 young leaves, are plucked from tea bushes twice a year during early spring and early summer or late spring. Autumn or winter pickings of tea flushes are much less common, though they occur when climate permits. Picking is done by hand when a higher quality tea is needed, or where labour costs are not prohibitive. Hand-picking is done by pulling flush with a snap of wrist and does not involve twisting or pinching the flush, since doing the latter reduces quality of leaves. Tea flushes and leaves can also be picked by machine, though there will be more broken leaves and partial flushes.

Wilting: The tea leaves wilts soon after picking, with a gradual onset of enzymatic oxidation. Wilting is used to remove excess water from the leaves and allows a very light amount of oxidation. The leaves can be either put under sun or left in a cool breezy room to pull moisture out from the leaves.

Bruising: To promote and quicken oxidation, leaves may be bruised by tumbling in baskets or by being kneaded or rolled-over by heavy wheels. This also releases some of the leaf juices, which may aid in oxidation and change the taste profile of tea.

Oxidation: For teas that require oxidation, leaves are left on their own in a closed room where they turn progressively darker. This process is referred to as fermentation in tea industry, although no true fermentation happens since the process is not driven by microorganisms. The tea producer may choose when the oxidation should be stopped. For light oolong teas, this may be anywhere from 5-40% oxidation, in darker oolong teas 60-70%, and in black teas 100% oxidation.

Kill-green: Kill-green is done to stop tea leaf oxidation at a desired level. This process is accomplished by moderately heating tea leaves, thus deactivating their oxidative enzymes, without destroying flavour of tea. Traditionally, the tea leaves are panned in a wok or steamed, but with advancements in technology, kill-green is sometimes done by baking or "panning" in a rolling drum. In CTC black teas, kill-green is done simultaneously with drying.

Yellowing: Unique to yellow teas, warm and damp tea leaves from after kill-green are allowed to be lightly heated in a closed container, which causes the previously green leaves to yellow.

*Shaping:*The damp tea leaves are then rolled to be formed into wrinkle strips. This is typically done by placing the damp leaves in large cloth bags, which are then kneaded by hand or machine to form the strips. This rolling action also causes some of the sap and juices inside the leaves to ooze out, which further enchances taste of tea. The strips of tea can then be formed into other shapes, such as being rolled into spirals, kneaded and rolled into pellets, or tied into balls and other elaborate shapes.

Drying: Drying is done to "finish" the tea for sale. This can be done in a myriad of ways including panning, sunning, air drying, or baking. Baking is usually the most common. Great care must be taken not to over-cook the leaves.

Curing: While not always required, some teas required additional aging, secondary-fermentation, or baking to reach their drinking potential. As well, flavoured teas are manufactured by spraying with aromas and flavours or by storing them with their flavorants.

Type Specific Processing

Tea is traditionally classified based on the degree or period of "fermentation" the leaves have undergone.

White Tea

Young leaves (new growth buds) that have undergone almost no oxidation; the buds may be shielded from sunlight to prevent formation of chlorophyll. White tea is produced in lesser quantities than most other styles, and can be correspondingly more expensive than tea from the same plant processed by other methods.

Green Tea

The oxidation process is stopped after a minimal amount of oxidation by application of heat, either with steam ,a traditional Japanese method, or by dry cooking in hot pans, the traditional Chinese method. Tea leaves may be left to dry as separate leaves or they may be rolled into small pellets to make gun-powder tea. This process is time consuming and is typically done with pekoes of higher quality. The tea is processed within one to two days of harvesting.

Oolong

Oxidation is stopped somewhere between the standards for green tea and black tea. The oxidation process takes two to three days. In Chinese, semi-oxidized teas are collectively grouped as blue tea (literally: blue-green tea), while the term "oolong" is used specifically as names for certain semi-oxidized teas.

Black Tea / Red tea

Black tea is the most common form of tea in South Asia(Sri Lanka,India,Pakistan,Bangladesh *etc.*) and in the last century many African countries includingKenya,Burundi,Rwanda,Malawi,and Zinbabwe. The literal translation of the Chinese word is red tea, which is used by some tea lovers. The Chinese call it red tea because the actual tea liquid is red. Westerners call it black tea because the tea leaves used to brew it are usually black. However, red tea may also refer to rooibos, an increasingly popular South American tisane. The oxidation process will take between two weeks and one month. Black tea is further classified as either orthodox or as CTC (Crush,Tear,Curl), a production method developed about 1932). Unblended black teas are also identified by the estate they come from, their year and the flush (first, second or autumn). Orthodox processed black teas are further graded according to the post-production leaf quality by the Orange Pekoe system, while CTC teas use a different grading system.

Post–fermented Tea

Teas that undergo a second oxidation, are collectively referred to as secondary or post-fermentation teas in English. In Chinese they are categorized as Dark tea or black tea. This is not to be confused with the English term Black tea, known in Chinese as red tea. Pu-erh, also known as Póu léi (Polee) in Cantonese is the most common type of post-fermetation tea in the market.

Yellow Tea

Either used as a name of high-quality tea served at the Imperial court, or of special tea processed similarly to green tea, but with a slower drying phase.

Kukicha

Also called winter tea, kukicha is made from twigs and old leaves pruned from the tea plant during its dormant season and dry roasted over a fire.

Green Tea Production

Green tea comes from the same plant as black tea, but the fermentation process is prevented by heat treatment immediately after withering.

Withering: This process is only carried out where necessary. The necessity and duration of withering varies widely according to the desired type of tea.

Steaming / dry heat treatment: This destroys the plants own enzymes so that the leaf will retain its green color instead of turning "black".

Rolling: Rolling is performed manually or by machine depending on the type of tea. In some cases the leaf is rolled into artistic shapes following a tradition which dates back thousands of years.

Drying: For this purpose, the leaves are either stacked in hot-air rack driers or exposed to the natural heat of the sun.

Sorting: Green tea is available in the same familiar grades - leaf, broken, Fannings or dust - as black tea, depending on the production specification. Green tea is a strongly alkaline drink which protects the body from hyperacidity.

Oolong Production

Oolong is a semi-fermented tea. Its secret lies in the fermentation of the leaf's outer edges, while the heart of the leaf remains unfermented. It is mainly consumed in Taiwan and China.

White Tea Production

The production of white tea is different from green tea. White tea leaves come from a special varietal tea bush called Narcissus or Big White bushes. For white tea, the little buds that form on the plant are covered with silver hairs that give the young leaves a white appearance. According to the different standards of picking and selecting, white teas can be classified as Silver Needle, White Peony, Tribute Eyebrow, and Noble Long Life Eyebrow. All of these white teas are widely produced in China and are available in America. The highest-quality white teas are Silver Needle and White Peony. Silver Needle is carefully hand selected from the tender fleshy sprouts of the "Big White" or the "Narcissus" tea bush. If the buds are selected with two leaves intact, then the resulting selection will be made into White Peony tea. The leaves and other material left over from the selection of Silver Needle and White Peony will be processed into Noble, Long Life Eyebrow.

White Tea leaves are not steamed or pan-fired as is the case in green tea but rather the leaves are withered, hand selected and sun dried. The special nature of white tea's color, leaf shape and hair fragrance is mainly created during the withering stage. If mechanical drying is required the leaves are baked (not fired) at temperatures less that 40°C. Only special 'two leaves and a bud' are selected. These leaves must show a very light green almost gray white color and be covered with velvet peach fuzz down. The ideal is a leaf or two being wrapped around a newly developing shoot. These shoots are plucked and segregated from the rest of the leaf being plucked. These leaves are then naturally withered and the painstaking process on final manual selection occurs. "Three Whites" is the primary requirement of the fresh tea material. This means the buds, the first leaves and the second leaves must be covered with tiny white hairs.

A tea maker's ability to balance solar and indoor withering of white tea is the major determining factor of quality. There are many nuances of white tea production that are dependent on the region and climate where the tea is made, but the major stages in the process are selective picking from specific varietals, withering, careful hand selection, and bake drying.

Black Tea Production

There are three different methods of producing black tea: Orthodox production / CTC production / LTP production.

The Orthodox Production Method

This production method consists of five stages: withering, rolling, fermentation, drying, and sorting.

The CTC Production Method

CTC stands for crushing, tearing and curling. Both the CTC and LTP methods are mainly used for the finer end of the scale, i.e. fanning and dust grades. These teas are usually destined for teabag production. The withered leaf is often cut to a uniform size by machine. Then the leaves are fed into the CTC machine where they are crushed, torn and curled in a single operation by metal rollers. The extracted cell sap is collected and added to the leaves again. The crushed leaves are then fermented, dried and sorted.

The CTC method is mainly used in Indian regions.

The LTP Method

The third method of producing black tea is the LTP method, named after the inventor of the relevant machine, the Lawrie Tea Processor. In this method, the withered leaves are often leveled before being processed in the LTP machine. Here they are virtually torn to pieces by blades rotating at high speed. This is followed by the usual fermentation, drying and sorting procedures.

Tea Grading and Sorting

The leaf grades result exclusively from the last stage of production, the sorting stage. There are 4 basic groups in orthodox production: Leaf, Broken, Fannings and Dust. These categories specify and indicate the different leaf sizes and associated strengths.

Professional tea tasters - brokers, agents, and buyers - are the most particular tea drinkers in the world. Experienced tasters can compare as many as a hundred teas, and can quickly judge which teas will blend well with others. Tea tasters have an extensive vocabulary to describe the teas they are tasting. Like wine tasters, tea tasters seek complex flavor characteristics - flowery, woody, malty, pungent - and especially prize mellowness (lack of harshness).

Here are some terms used by professional tea tasters to described infused tea:

- *Body* - The weight and qaulity of the tea on the tongue. Can be described as wispy, light, medium, or full.
- *Brassy* - Strong and bitter; caused by underwithering of black tea.
- *Bright* - Lively, fresh, and high quality.
- *Brisk* - Opposite of "flat"; pungent without being undesirably high in tannin content.
- *Burnt* - An off flavor caused by overfiring.
- *Chocolaty* - A desirable flavor quality of fine Darjeelings.
- *Coarse* - Bitter or overly acid; attributable to improper processing.
- *Dull* - Muddy looking; the opposite of "bright" or "brisk."
- *Earthy* - May be inherent to the leaf, or caused by damp storage.
- *Flat* - Off, stale taste; usually a property of old teas.
- *Full-bodied* - An ideal combination of strength and color.
- *Green* - When said of black tea, refers to immaturity of character due to under oxidation or under withering.
- *Harsh* - Very rough in flavor; associated with underwithered teas.
- *Heavy* - Low in briskness and very full-bodied.
- *Light* - Lacking strength and depth of color.
- *Malty* - Subtle, underlying flavor; a desirable quality in Assam teas.
- *Mellow* - Smooth, easy, pleasant.
- *Metallic* - A sharp, coppery flavor.
- *Point* - Used to describe a leaf with desirable brightness and acidity.
- *Pungent* - Pleasantly astringent; a good combination or briskness, strength, and brightness.

- *Self-drinking* - A tea that can be drunk alone, without blending with other leaves.
- *Smooth* - Rounded in flavor, pleasant on the palate.
- *Soft* - The opposite of "brisk"; caused by inefficient oxidation or firing.
- *Vegetative* - Grassy flavor, a desirable characteristic of some green teas.

Storing Tea

Although each type of tea has a different shelf life, it is best to use tea within 6 to 12 months of purchase. Green teas perish the quickest and begin to deteriorate within a year of harvest. Oolong and black teas retain their characteristics for several years.

Keep your loose-leaf teas stored in a cool, dark place to preserve their freshness. Avoid exposing the tea to light. Always store your teas in opaque (non-clear) jars with airtight lids.

Another factor contributing to a tea's longetivity is the way in which the tea leaf is rolled. Tea leaves rolled into pellets (Gun powder and Imperial green teas) or twisted (like the black Yunnan) last longer than an open, flat leaf, because less of their surface area is exposed to air. Keep it away from heat, moisture, and, of course, other strongly scented teas or spices.

Tea Byproducts

Tea Seed Oil

Tea seeds are used as a source of supplementary protein and edible oil for human consumption, apart from many other industrial applications. The prospects of tea seed oil are immense. It can be a substitute for any edible oil; it can also be used as a cheaper alternative to olive oil which is mostly imported. The cosmetic industry may use it for manufacturing hair lotion, soap etc. The thiosaponin has some medicinal value. It acts as an adjuvant, anti-inflammatory agent. Tea oil cake and waste are used as fodder and bio-manuring.

Other Products from Tea

Instant Tea

People are accustomed to seeing and using different kinds of instant coffee, but the development of instant teas has lagged far behind that of instant coffee market. In the case of instant tea, the liquor (tea brew) itself is further concentrated and dried to form a powdered or granular product. Hot water soluble and cold water soluble instant teas are found in the market.

Criteria for the selection of the raw material include processing requirements of the market for which the tea is intended and the commercial restrictions placed upon manufacture. Instant tea can be made from black tea leaf, (partially) fermented, un dried leaf, oolong or green tea. Manufacturers in tea producing countries use fermented un dried leaf as a basis for instant black tea production, as this does not have to pass through the auctions and is therefore cheaper.

The process of the production of instant tea consists of the following operations, selection of raw materials, extraction, aroma stripping, cream processing, concentration and drying.

Extraction

Commercial extraction techniques are designed to optimize two parameters; yield of the solids extracted from the leaf and concentration of extract solution. Counter - current extraction is commonly used as a suitable technique to achieve these targets and is used in both batch and continuous forms. In a typical continuous counter - current extraction process tea leaves are fed into the lower end of the extractor and are conveyed upwards by the rotation of the screw while the extracting water flows in the reverse direction. The screws are perforated to allow water to flow more easily through the body of the tea and temperature of the extractor is controlled by the series of heaters on the body of the unit. Particle size of the leaf has to be carefully controlled to avoid too many fine particles which may pack together causing channeling or overflow of the water.

Yield of solids from the extraction system is commonly between 25 and 35% use of hydrogen peroxide, ammonia or ammonium carbonate, extraction at reduced pH 2.0 - 3.0. Use of enzymes such as tannase have been found to increase the yields up to 42% solids.

Aroma Stripping

Tea aroma is an important criteria for the acceptability of tea, considerable efforts have been made to preserve the aroma in the instant tea product. Most process deal with the problem by stripping the aroma of the extract before concentrating and treating it separately.

The aroma is removed by either passing the stripping gas through extract or spraying extract into a moving stream of gas. The aroma is not condensed but is immediately added back to material which has already been concentrated. The use of inert gas is considered advantageous as it prevents the oxidation and hence deterioration of the aroma.

Tea Cream Processing

When a solution of tea cools down it becomes opaque and changes color to pale brown due to a phenomenon known as 'creaming'. The cream essentially consists of caffeine, theaflavins and thearubigins and teas can be characterized by the amount of cream formed on cooling.

In the manufacture of instant tea, tea cream is processed either by maintaining the temperature above 65° C or by removal by cooling and precipitation or centrifugation. Since tea cream contains many flavor compounds removal by precipitation decreases the quality of the tea and incidentally, decreases the yield of solids from the leaf, so solubilisation by maintaining the temperature is a preferred method.

Cold soluble tea, is required to be clear and bright when reconstituted and the cream causes an unacceptable haze, hence solubilisation of the cream is crucial. The cream is separated from tea extract and is oxidized at 70° C at a pH of 9 - 10 with molecular oxygen. Use of enzyme tannase has also been documented in dissolving the tea cream. Cold soluble teas are produced by allowing the tannins in the cream to react with an aqueous solution of natural plant catechins extracted from other plants such as green tea.

Concentration

The liquor from the extraction process is too dilute to pass directly to a drier. It is uneconomic to dry from such low solids but the powder produced by both spray and freeze drying is of such low density. The liquor has to be concentrated to 20 - 40 % solids before drying.

Historically concentration was carried out by evaporation under reduced pressure. Tea extract is easily damaged by exposure to high temperatures during concentration and drying. Most manufacturers supply units which separate the volatiles and condense them prior to addition back into concentrate before drying. Techniques such as reverse osmosis and freeze concentration have also been documented.

Drying

Drum spray and freeze drying have all been proposed for instant tea, but spray drying is most commonly employed because it is cheaper in terms of both capital and running costs. Spray dried tea is rarely agglomerated and the most important criteria for the product are; taste, color, bulk density and moisture content. The bulk density is controlled by changing the solids concentration of the feed stock and by careful choice of atomizing system.

Developments in the drying technologies have allowed the ratio of inlet to outlet temperature to change from 185° C / 95° C to 240° C / 70° C. This still allows the powder to experience not more than 70° C (decreasing the risk of flavor loss) and yet be dried within a few seconds to a moisture content of less than 5%. Moisture content of spray dried tea is generally between 3 and 5%.

Speciality and Herbal Teas

These are classified according to the raw material used and the type of manufacture adopted. Orthodox processed tea leaves produce "particular" teas like semi-fermented oolongs and reprocessed, scented and compressed teas. Delicately plucked and processed teas make numerous "prominent" teas. The raw material is taken from the plants which are well grown in an excellent habitat with advantageous ecological conditions and are made skillfully and exquisitely with delicate flush buds and leaves of the tea plants belonging to fine varieties. Herbal teas are made by processing veritable teas together with medicinal herbs, which are used to complement the curative properties of the former for many ailments. A few speciality teas are White teas, Yellow teas and Tea extract.

Use as Dye

Wood, silk or synthetic yarn type fibres are dyed in tea extract solution (1:20) containing 10% of tea extract powder with metal-mordant at 60°C for six hours. After washing the dyed yarn with water, the yarn weight increased by 3-5%. By metal-mordants, the colour of yarn changes to pale green, yellow-green, yellow, brown and copper-yellow. It is found that despite repeated washing or exposure to sunlight for the fading test, the colour of yarn is maintained at least for one year.

Use as a Detergent

Oolong tea extract has been developed as one of the effective substitutes for (CFC) chlorofluoro carbon used to wash away oil on chips used in display coated liquid crystal. The advantages of oolong tea extract as detergent are (i) it is an organic material and therefore ecofriendly and the waste water is easily disposed of (ii) it is a safe detergent which is non toxic and non-inflammable (iii) the extract forms a high quality detergent for washing oily substances. Manure produced from a mixture of green tea waste and fowl dropping is sold in the Japanese manure market and is used widely in kitchen gardens for flowers and vegetables.

Health Benefits of Green Tea

Current research shows tea contains specific antioxidants and health promoting ingredients, lowering the risk of heart disease, stroke and certain types of cancer. (American Health Foundation).

It's no surprise that the health benefits of tea have been scientifically and medically upheld by independent research in several countries. Even before modern science formalized its findings, the rest of the tea-drinking world already knew it. Tea is good food. The properties contained in tea have played a significant role in benefiting patients with heart disease, stroke, atherosclerosis, high blood pressure, diabetes and certain types of cancer. We know that antioxidants destroy free radicals, which wreak havoc to the human body, and tea is rich in antioxidants. Tea also contains vitamins including carotenoids and vitamin K, and mineral components including potassium, aluminum, manganese and fluoride. As if this weren't enough, the antibiotic effects of tea have proven helpful to oral and bone health and have also been used as a treatment for anemia. A recent study in Britain found that tea may inhibit certain brain enzymes linked to Alzheimer's disease.

A serving of tea contains less than half the caffeine of coffee. For those who are caffeine sensitive, naturally caffeine-free Rooibos is an excellent alternative. In addition to the health properties of tea, Rooibos increases iron absorption and relieves stomach and digestive discomfort. Hot or over ice, tea tastes great—and most importantly it's good for health.

Chemical Composition of Fresh Tea Shoot

Compounds %	Dry weight	Contribution
Total Polyphenols	25 - 30	Astringency
Flavanols		
(-) Epigallocatechin gallate	8 - 12	
(-) Epicatechin gallate	3 - 6	
(-) Epigallo catechin	3 - 6	
(-) Epicatechin	1 - 3	
(+) Catechin	1 - 2	
(+) Gallocatechin	3 - 4	
Flavonols and flavonol glycosides	3 - 4	

Contd...

Leuco anthocyanins	2 – 3	
Polyphenolic acids and depsides	3 – 4	
Caffeine	3 - 4	Briskness
Theobromine	0.2	
Theophylline	0.5	
Amino acids	4 – 5	Brothyness
Organic acids	0.5 – 0.6	
Monosaccharides	4 – 5	
Polysaccharides	14 - 22	
Cellulose and Hemicellulose	4 – 7	
Pectins	5 – 6	
Lignin	5 – 6	
Protein	14 – 17	
Lipids	3 – 5	
Chlorophylls and other pigments	0.5 - 0.6	Colour and Appearance
Ash (minerals)	5 – 6	
Volatiles	0.01 – 0.02	Aroma

Average Chemical Composition of Orthodox and CT teas made from Assam Clones

Composition	Orthodox teas	CTC teas
Water soluble solids %	39.52	41.12
Theaflavins %	0.59	1.30
Thearubigin %	6.5	18.00
Total Lipid	3.11	3.68
Carotenoids (¾g/g)	215	176
Flavour volatiles	18.40	8.20
Total fibre %	19.35	18.93
Crude fibre%	11.70	11.12
Chlorophyll a mg/g	1.38	0.48
Chlorophyll b mg/g	0.77	0.58

(Sum of total area of GC peaks compared to internal standard compounds)

Biochemical Compounds Responsible for Colour

Compounds	Colour
Theaflavins	Yellowish brown
Thearubigins	Reddish brown
Flavonol glycosides	Light yellow
Pheophorbide	Brownish
Pheophytin	Blackish
Carotene	Yellow

Biochemical Compounds Responsible for Taste

Compounds	Taste
Polyphenol	Astringent
Amino acids	Brothy
Caffeine	Bitter
Theaflavins	Astringent
Thearubigin	Ashy and slight astringent

Biochemical Compounds Responsible for Flavour

Compounds	Flavour
Linalool, Linalool oxide	Sweet
Geraniol, Phenylacetaldehyde	Floral
Nerolidol, Benzaldehyde, Methyl salicylate, Phenyl ethanol	Fruity
Trans-2-Hexenal, n-Hexanal, Cis-3-Hexenol, Grassy, b-Ionone	Fresh flavour

Principal Components of Black Tea Beverage

Components	Concentration (g/100g)
Catechins	3
Theaflavins	3
Thearubigins	12
Flavanols	6
Phenolic acids and Depsides	10
Amino acids	13
Methylxanthines	8
Carbohydrates	10
Protein	0.8
Mineral matter	8
Volatiles	0.05

(Components measured in wt % of extract solids)

Marketing

Indian Scenario

India was the leading producer and consumer of tea in the world till 2003. But China emerged as the leading producer, which accounted for 27.37% of global production in 2005 followed by India with 24.18%. The production in India is estimated at 8.31 lakh tonnes. The tea production in the world increased from 29.56 lakh tonnes in 2000 to 34.36 lakh tonnes in 2005. The corresponding increase in India's tea production was from 8.37 lakh tonnes in 2000 to 8.51 lakh tonnes in 2004. There was a marginal reduction in tea output in 2005 to 8.31 lakh tonnes.

Tea is grown in 16 States in India, of which Assam, West Bengal, Tamil Nadu and Kerala account for about 96% of the total tea production. The teas originating from Darjeeling, Assam and Nilgiris are well known for their distinctive quality world over. While tea exports contribute a significant amount of foreign exchange into the country, tea also contributes revenue to the national exchequer by way of cess, sales tax, agricultural and corporate income tax, etc. More than two million people derive their livelihood from ancillary activities associated with the industry. The tea industry provides direct employment to more than a million workers, of which a sizeable number are women.

Area, Production and Productivity of Tea in India

Years	**Area harvested ('000 ha)**	**Quantity produced ('000 tonnes)**	**Yield (tonnes/ha)**
2001-02	438.00	854.00	1.95
2002-03	511.94	837.60	1.64
2003-04	512.00	850.49	1.66
2004-05	512.00	830.74	1.62
2005-06 (P)	512.00	832.88	1.62

Area, Production and Productivity of Tea-Kerala

Years	**Area harvested ('000 ha)**	**Quantity produced ('000 Tonnes)**	**Yield/hectare (Tonnes/ha)**
2001-02	36.90	66.09	1.79
2002-03	37.07	55.35	1.49
2003-04	38.33	57.55	1.50
2004-05 (P)	38.26	57.57	1.51

Government of Kerala (2006): Economic Review 2005

Global Production and Consumption

Tea is grown in 36 tropical and semi-tropical countries, 19 of them ACP countries. The six largest producing countries - India, China, Kenya, Sri Lanka, Turkey and Indonesia (in that order) - account for 80% of world output. Less than half of production is exported, as India and China, in particular, are major consumers as well as producers. World tea production increased at an annual growth rate of 2.8% between 1970 and 2000 expanding from 1.27 million tones to 2.97 million tones. Most of the growth was due to the increase in productivity rather than an expansion in area. Major producers are: India, which accounts for about 29 % of global production; China, 23%; Sri Lanka, 9%; Kenya, 8%; and Indonesia, 6 %. Black tea accounts for more than 70 % of world tea production and green tea, 22%. Over the past decade, production of black tea has increased at 1% per annum, while that for green tea at 2.5 %.

Global tea production has grown by around 2% a year since 1993-95, consumption in Western countries has grown by barely 1%. In a number of developing countries, but however, consumption has kept pace with, or exceeded production. Global production grew by 110,000 tonnes or over 3% to 3.47 million tonnes in 2005, largely as a result of favourable weather conditions; however there have been forecasts of a fall of 45-50,000 tonnes in 2006. In 2005 India accounted for 26.7% of world output, followed by China, 25.2%, Kenya 9.5% and Sri Lanka, 9.1%.

Main world producers and leading ACP producers	2005 (tonnes)
India	927,984
China	875,000
Kenya	328,498
Sri Lanka	317,196
Turkey	205,500*
Indonesia	160,000
Malawi	37,978
Uganda	35,000*
Tanzania	31,342
Rwanda	15,484*
Zimbabwe	14,884
World total	3,470,000

Sources: FAO; *Kenya Times*, September 26th 2006 * 2004

Tea is consumed in both producing and importing countries. India is the largest producer and consumer of tea and accounts for more than 20% of global tea consumption, while the United Kingdom, the Russian Federation, the CIS countries, Pakistan and Japan each account for about 5 % of the total tea consumed globally. Over the past decade, world tea consumption has increased by 1% annually.

Export Pattern

Global Exports

World tea exports reached 16.15 lakh tonnes in 2004. Sri Lanka was the largest exporter followed by China. Sri Lanka accounted 2.99 lakh tonnes (18.51%) valued at 3291.33 Rs. (22.38%). India is the fourth major tea exporter with 1.75 lakh tonnes (10.82%) earning a foreign exchange of 1697.25 crores Rs. (11.54%).

Global Tea Export (2004)

Country	**Quantity**		**Value**	
	(lakh tonnes)	**%**	**Rs. Crores**	**%**
Sri Lanka	2.99	18.51	3291.33	22.38
China	2.85	17.68	2088.98	14.20
Kenya	2.84	17.61	2083.59	14.17
India	1.75	10.82	1697.25	11.54
Viet Nam	0.99	6.16	429.32	2.92
Indonesia	0.99	6.10	521.29	3.54
Argentina	0.68	4.20	182.03	1.24
Uganda	0.37	2.28	167.40	1.14
Malawi	0.33	2.03	177.19	1.20
UK	0.29	1.77	1163.76	7.91
Other Countries	2.07	12.84	2904.21	19.75
World	16.15	100.00	14706.35	100.00

Source: http://faostat.fao.org

India's Export

India exported 1.58 lakh tonnes of tea valued at Rs.1632.09 crores in 2005-06 The figure shows the export pattern of tea from India which shows a declining pattern over the years from 2001-02 to 2005-06.

Market Share of Major Tea Exporting Countries (2004)

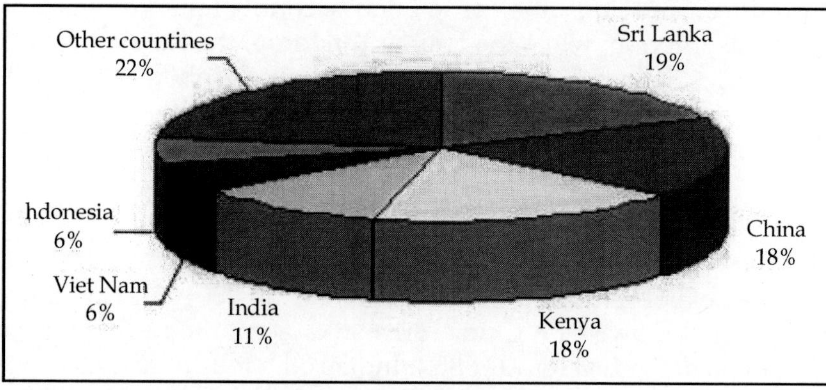

Export of Tea from India

Year	Quantity (lakh tonnes)	Value (Rs. Crores)
2001-02	1.80	1719.22
2002-03	1.83	1652.07
2003-04	1.78	1637.35
2004-05	1.83	1840.30
2005-06	1.58	1632.09

Source: CMIE

Russia was the largest importer of Indian tea (17.5%) followed by UK (15.5%) and UAE (15.2%).

Country-wise export of tea shows that Russia is the major destination with 0.28 lakh tonnes in 2005-06 (17.48%) with 13.52% of total value, followed by UK and UAE with 15 % and USA (7.31%) which together accounted for lionshare of India's exports. In value terms also these four countries acounted for more than half of the Indian tea export. In USA and Germany, percentage share in value terms is higher than percentage share in quantity.

Country wise Export of Tea from India (2005-06)

Importing Country	Quantity		Value	
	Lakh tonnes)	%	Rs. Crores	%
Russia	0.28	17.48	220.70	13.52
UK	0.25	15.80	222.20	13.61
UAE	0.24	15.16	239.57	14.68

Contd...

USA	0.12	7.31	146.91	9.00
Kazakhstan	0.10	6.18	96.53	5.91
Iran	0.08	4.99	78.77	4.83
Germany	0.06	3.84	112.18	6.87
Pakistan	0.05	3.07	24.13	1.48
Australia	0.05	2.88	81.69	5.01
Saudi Arabia	0.04	2.72	39.26	2.41
Other countries	0.33	20.59	370.14	22.68
World	1.58	100.00	1632.09	100.00

Source: CMIE

The following pie chart shows the countrywise export of tea from India in 2005-06. The top four importers share accounted 55.75% of Indian tea export.

Country-wise Export of Tea from India in Quantity Terms (2005-06)

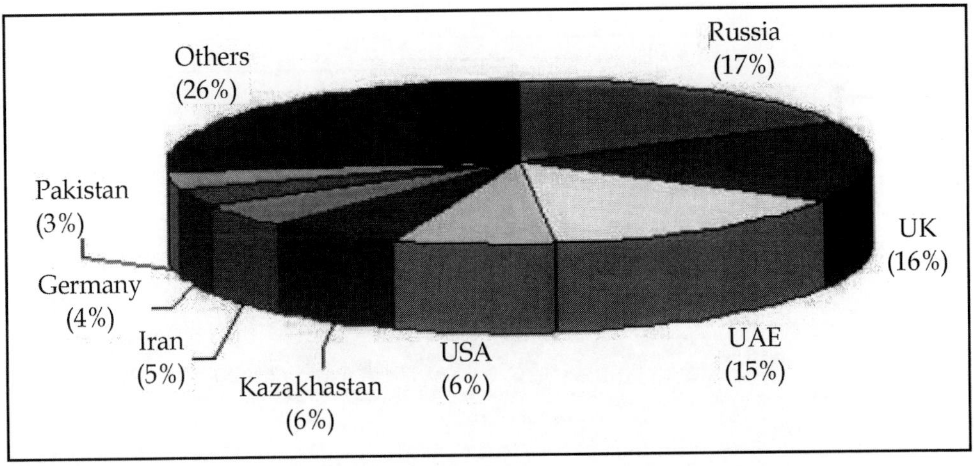

Source: CMIE

Organic Tea

Total world production of organic tea is estimated at around 9000 tonnes in 2001, which represents less than 1% share of global tea production. India produced around 3150 tonnes of organic tea in 2001 which accounts for one-third of the world production. India is not only the pioneer but also one of the leading producers of organic tea in the world.

World Net Imports

World net tea imports was to the tune of 14.05 lakh tonnes in 2004. The major importing country of tea was Russian Federation with 1.72 lakh tonnes with 12.25 per cent of total world import. The other major importing countries with more than 1 lakh tonnes are are UK (11.13%) and Pakistan (8.26%). Other countries other than the major top 10 importers contributed 43.8 per cent of the volume of world import of tea and 47.9% of the value.

Global Tea Import (2004)

Importing Countries	**Quantity**		**Value**	
	Lakh tonnes	%	**Rs. crores**	%
Russian Federation	1.72	12.25	1312.68	9.54
UK	1.56	11.13	1312.09	9.54
Pakistan	1.16	8.25	854.58	6.21
USA	0.99	7.08	920.56	6.69
UAE	0.66	4.68	773.10	5.62
Japan	0.56	4.00	869.36	6.32
Morocco	0.46	3.25	327.81	2.38
Germany	0.43	3.09	527.07	3.83
China	0.35	2.47	262.13	1.91
Others	6.15	43.80	6594.66	47.95
World	14.05	100.00	13754.04	100.00

Source: http://faostat.fao.org

Global Import Share of Tea (2004)

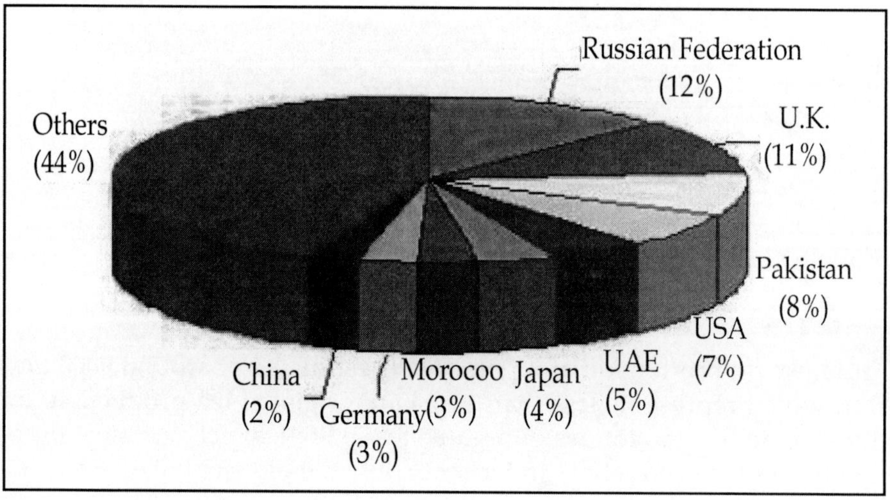

Source: http://faostat.fao.org

India's Import

India's import of tea has increased from 0.10 lakh tonnes valued at Rs. 67.01 crores in 2001-02 to 0.19 lakh tonnes valued at Rs.108.16 crores in 2005-06. The years 2002-03 and 2004-05 showed high import with 0.24 lakh and 0.32 lakh tonnes respectively.

Import of Tea into India (Quantity)

Of the total import of tea, 25.02 per cent was from Nepal. The other major importing countries which crossed thousand tonnes in 2004 were Vietnam (22.66%), Kenya (19.77%), Indonesia (10.45%) and Argentina (5.59%).

Country Wise Import of Tea into India (2005-06)

Country wise import of Tea into India (2005-06)				
Exporting Country	**Quantity**		**Value**	
	'000 tonnes	**%**	**Rs. Crores**	**%**
Nepal	4.72	25.02	25.98	24.02
Viet Nam	4.27	22.66	14.75	13.64
Kenya	3.73	19.77	29.07	26.88
Indonesia	1.97	10.45	11.23	10.38
Argentina	1.05	5.59	4.85	4.49
China	0.79	4.17	4.76	4.40
Malawi	0.69	3.64	4.05	3.74
Papua New Guinea	0.58	3.06	3.37	3.12
Malaysia	0.37	1.97	1.54	1.42
Sri Lanka	0.30	1.58	3.90	3.61
Others	0.39	2.09	4.66	4.31
World	18.85	100.00	108.16	100.00

Source: CMIE

The import pattern of tea into India in quantity terms in 2005-06, which shows that Nepal (25.02%), Vietnam (22.66%), Kenya (19.77%) and Indonesia (10.45%) were the major countries.

Kerala's Export

India's tea production in 2005-06 was 9.31 lakh tonnes of which 56384 tonnes (6.06%) was contributed by Kerala. A declining trend in production in the state was observed for during last three years from 57553 tonnes in

Import Share of Tea into India in Quantity Terms (2005-06)

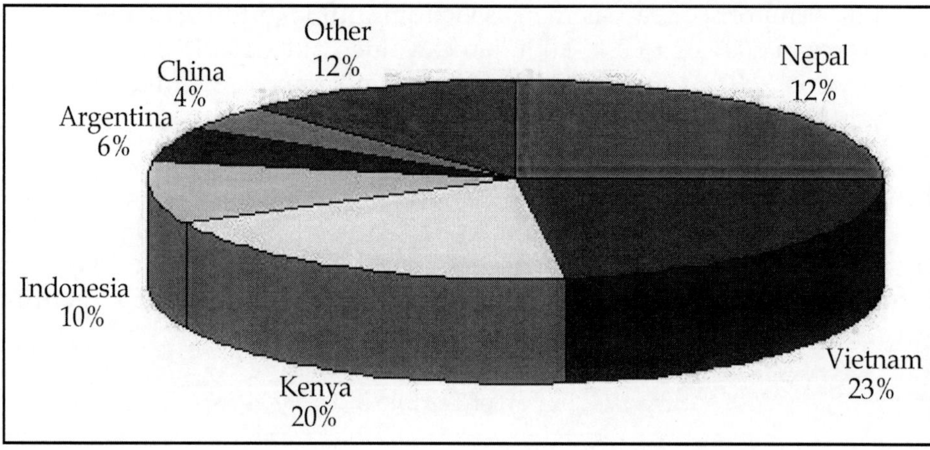

Source: CMIE

2003-04 to 56384 tonnes in 2005-06. The export of tea from Kochi declined from 94839 tonnes in 2000-01 to 31985 tonnes in 2005-06 although it had reached an all time record of 103544 tonnes in 2002-03. A marginal increase was reported in 2002-03 at the tune of 103544 tonnes. In value terms it declined from Rs. 198.8 crores in 2000-01 to Rs. 253.48 crores in 2004-05.

Export of Tea from Kochi Port

Year	Quantity (Tonne)	Value (Rs. Crore)
2000-01	94839	198.80
2001-02	96155	205.48
2002-03	103544	325.23
2003-04	86794	269.93
2004-05	86589	96.20
2005-06	31985	253.48

Source: Economic Review 2006

The figure shows the export pattern of tea from Kochi port for the period 2000-01 to 2005-06, which shows a decline trend over the years.

Marketing System

The primary marketing of tea is conducted through public tea auction centers licensed by the Tea Board which are located at Kolkata, Guwahati,

Export of Tea from Kochi Port in Quantity Terms

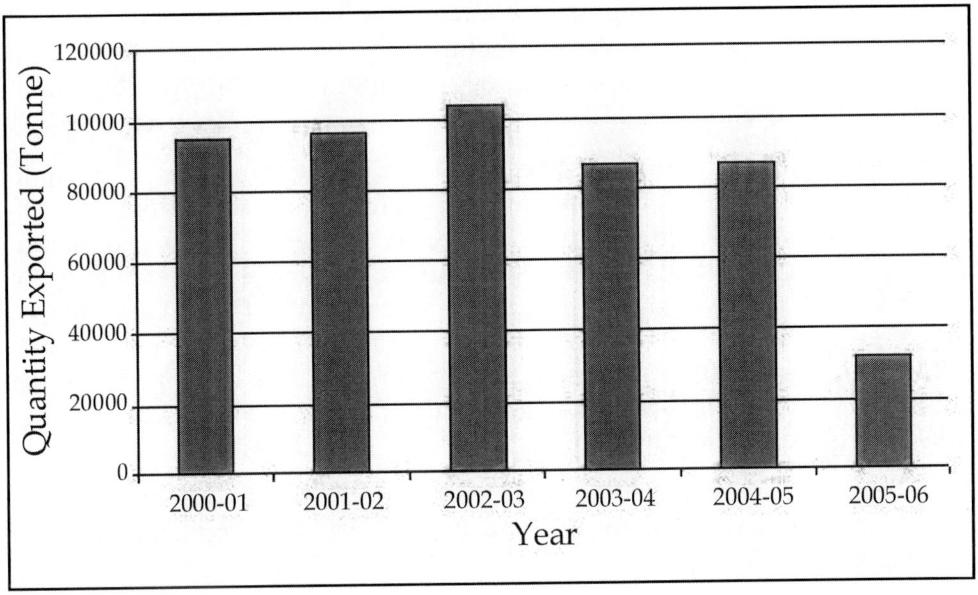

Source: Economic Review 2006

Siliguri, Amritsar, Coonoor, Coimbatore and Kochi. The Tea Board had taken steps for introduction of electronic trading of tea in major public tea auction centers. Electronic auction has been introduced partially in the tea auction centers at Guwahati, Siliguri, Coonoor, Coimbatore and Kochi. The Board has also launched an IT portal (www:indiateaportal.com) for effective dissemination of information relating to all aspects of tea industry.

Forecasts for the World Tea Economy to 2010/2011

Future Trends in Production Consumption and Trade

The production and trade projections presented in this section are derived from a dynamic times series model of the world tea market. This model quantifies key market relationships on the basis of observations on the past behaviour of volumes produced and traded, prices, and population and income growth.

Black Tea Production

World black tea production is projected to increase to 2.4 million tonnes in 2010, an annual average growth rate of 1.2% from 2.15 million tonnes in 2000. This growth would result largely from the improvement in yields.

Most African producers are likely to see significant production growth as tea bushes reach optimum production age, and production skills of small growers improve. For example, production in Kenya would grow by 2.3% a year from 236 300 tonnes in 2000 to 304 000 tonnes in 2010, while growth rates in Tanzania and Uganda are expected to be 1.7 % and 2.7 %, respectively.

Most producers in Asia would experience a steady growth in production. Indonesia is expected to achieve an annual growth of 1.1 %, from 130 600 tonnes in 2000 to 147 000 tonnes in 2010. Over the same period, production in India, the world's largest black tea producing country, is projected to grow by 2.5 % to 1.07 million tonnes, accounting for nearly 44% of global production, compared to 38 % in 2000. Tea production in Sri Lanka is projected to reach 329 000 tonnes by 2010, an annual average growth rate of 0.7%. Black tea production in China is expected to continue to decline to 54 000 tonnes as the balance of production shifts to other teas with stronger market prospects.

The three largest black tea producing countries, India, Kenya and Sri Lanka, are expected to account for 70% of the world tea production in 2010, compared to 63 % in 2000.

Black Tea Exports

World black tea exports in 2010 are projected at 1.14 million tonnes, reflecting an average annual increase of 1.1 % from 1 million tonnes in 2000.

Most of this increase would take place in Africa, where production is likely to continue to grow while domestic consumption remains small. Exports from Kenya would increase by 2.6% annually from 208 200 tonnes in 2000 to 275 000 tonnes in 2010, giving Kenya 32% of global exports. Over the same period, export availability in Malawi would remain unchanged at 38 000 tonnes. Major tea exporting countries in Asia are expected to experience slight declines in exports in line with expected growth in income and population that would foster domestic consumption. Exports from India and Indonesia would decrease by 2.4% and 1.1% respectively. Conversely, exports from Sri Lanka would increase with an annual average growth rate of 0.4%.

Black Tea Consumption

In 2010,world net imports of black tea, a proxy for consumption, would amount to 1.15 million tones, reflecting an average annual increase of 0.6% from 1.08 million tones in 2000. Net imports in the countries of the former Soviet Union would increase from 223 600 tones to 315 200 tones, an annual average growth rate of 3%. Pakistan would increase its net imports by 2.9 % /year from 109 400 tones to 150 000 tones. The United States is expected to increase net imports by 1.4 % a year to 94 300 tones, while Japan would increase its net imports from 18 000 to 22 000 tones, an annual average growth rate of 1.8%. On the contrary, net imports by the United Kingdom are expected to decrease by 0.6% annually to 125 500 tones. These major importers together would account for about 60% of global net imports.

The model does not take into account stock levels. Hence, the difference between production and exports is treated as a proxy for domestic consumption in producing countries. In 2010, the quantity of black tea production that would be consumed in these countries is expected to reach 1.27 million tonnes, or 52 % of global black tea production, compared to 1.14 million tonnes in 2000. Domestic consumption of black tea in India is expected to increase by an average annual rate of 3.7 % to 919 300 tonnes by 2010, or 86 % of the black tea produced in the country. During the same period, Indonesia is expected to increase its domestic consumption at an average annual rate of 4.0 % from 33 100 tonnes to 51 000 tonnes. Domestic consumption in Bangladesh and Sri Lanka would grow by 2.0% and 3.8 % to reach 45 000 tonnes and 36 000 tonnes, respectively.

Green Tea Market Trends

Projections for green tea are provided only for production and exports due to data limitations. World green tea production is forecast to increase from 680 700 tonnes in 2000 to 900 000 tonnes in 2010, reflecting an annual average growth rate of 2.6%. During this period, production in China would grow by 2.7 % per annum from 500 000 tonnes to 671 000 tonnes, accounting for 75 % of global green tea production in 2010, compared to 73.5% in 2000. Production in Japan would grow by an average rate of 0.1 % 90 800 tonnes, while production in Viet Nam is expected to increase by an average rate of 2.5% to 50 000 tonnes. Output in Indonesia would grow by 2.3 % annually to reach 49 000 tonnes.

Green tea exports are expected to exhibit a significant upward trend in keeping with production. Total exports would increase by 2.8% annually from 186 800 tonnes in 2000 to 254 000 tonnes in 2010. China would continue to be the world's dominant green tea exporter, with shipments reaching 210 000 tonnes by 2010, reflecting an annual average growth rate of 2.7 %. During the same period, exports from Indonesia are expected to increase by 3.8% per annum to 12 000 tonnes, while exports from Viet Nam would increase by 2.5 % a year to 25 000 tonnes. Japan would consume most of its domestic production.

Morocco, the world's leading green tea importer, is expected to increase imports from 35 200 tonnes in 2000 to 57 100 tonnes in 2010, an annual average growth rate of 4.5%.

Issues and Uncertainties

The projections indicate that over the next decade exports of black tea would increase at an annual growth rate of a little over 1%, mirroring a similar growth in production. However, the world market is expected to remain broadly in balance. In consequence, price levels should be maintained.

In contrast, with consumption outstripping the production of green tea, an upward trend would persist in the medium-term.

Several actions can be taken to enhance returns from black tea production. On the supply side, by reducing unit costs through productivity gains, building capacity of small growers, streamlining marketing channels and improving infrastructure, improved returns to growers may result.

On the demand side, consumption can be raised through effective marketing. Variations in demand among countries suggest that marketing activities need to be tailored to individual markets. Successful market-specific activities require in-depth knowledge and understanding of the target market, including consumers' preference and market structures. In addition, worldwide marketing efforts, such as the generic promotion of tea using the Tea Mark, could have a significant impact, if planned and implemented appropriately.

It is important that any action taken is done holistically to improve longer-term price prospects. Forming such a strategy requires better understanding of markets. Exchanging information and views between producers and consumers, as well as public and private sectors, could promote greater market transparency.

UPASI Tea Clones Approved by Tea Board and Released for Commercial Planting

S. No.	Clone	Nursery performance	Growth & field performance	Remarks
1	'ATHREY' (B/6/61) * Y.P.: 2.6 ** B	Excellent rooter with uniform growth	Vigorous and semi-orthotropic grower with an excellent spread; establishes readily in clearings; fairly tolerant to drought	Faring well at different altitudes and thrives even in soils with near neutral reaction (pH 6.8)

Contd...

Contd...

2	'BROOK LANDS' (B/6/24) Y.P. : 3.1 A	Early and easy rooter with uniform and vigorous growth	Vigorous orthotropic grower with a good spread and dense plucking table; establishes fast in the clearings; early training is helpful; fairly tolerant to drought	Faring well at mid and high elevations generally with humid climate and good rainfall pattern. Has tendency to flower during drought and when adequate fertilizer is not applied.
3	'EVER GREEN' (B/4/141) Y.P. : 3.1 A	Growth rather uneven but a good rooter	Very upright and vigorous growth, forming a compact bush; hardy, resistant to drought and fairly tolerant to wind	Requires an early and low centering; 120 x 60 cm or even closer spacing desirable; flourishing at mid-elevations
4	'GOL CONDA' (B/6/36) Y.P.: 3.4 A	Good rooter; though initially slow, vigorous grower	Vigorous grower with a very good spread and dense plucking table	Thriving well at all elevations, particularly in warm, humid zones
5	'JAYA RAM' (B/4/142) Y.P. : 3.5 A	Easy rooter with a good and fairly uniform growth	Very vigorous and orthotropic growth; establishes readily in the clearings; excellent spread with a dense plucking table; fairly hardy, tolerant to drought and wind	Thriving at all elevations in both warm and cold zones
6	'PANDIAN' (B/6/62) Y.P. : 2.8 A	Quick rooting with uniform growth	Vigorous and orthotropic with a bushy habit, branching from collar; forms a broad dense plucking table; hardy clone, resistant to drought and fairly so to wind	Flourishing at all elevations; thrives even in soils with pH up to 6.8. Has a tendency to flower under conditions of stress
7	'SINGARA' (S/6/99) Y.P. : 2.8 A	Very good and easy rooter	Though initially slow, is a vigorous orthotropic grower with excellent spread forming a broad, dense plucking table	Susceptible to blister blight at mid-elevation but flourishing at higher altitudes. Colour liquor and flavoury tea

Contd...

Contd...

8	'SPRING FIEL'D (SP/4/5) Y.P. : 2.5 A	Good rooter with average growth in the nursery	Fairly vigorous with semi-plagiotropic growth; hardy, resistant to drought and fairly tolerant to wind and mild frost	Flushes throughout the year even through the cold weather at high altitudes; desirable in certain situations, flavoury tea; infusion light, liquors bright even at mid-elevations; produces quality teas throughout the year
9	'SUN DARAM' (B/5/63) Y.P. : 3.3 A	Good rooter but sensitive to texture, pH and moisture of the rooting medium; sandy-loam with moisture of 20% and pH of about 4.8 is ideal for rooting	Orthotropic and vigorous grower with an excellent spread and dense plucking table	Response to manuring is remarkable and returns for a unit of input are higher than many other clones. Flourishing at all elevation. Bright infusion and liquor.
10	'SWARNA' (B/6/203) Y.P.:2.8 A	Easy rooter and grows well	Vigorous orthotropic grower with good spread and dense plucking table.	Flourishing well at mid and high elevations. Very bright liquor.
11	B/6/10 Y.P.:2.2 A	Good rooter	Good grower with semi-plagiotropic and profuse branching.	Faring well at mid and high elevations.
12	B/6/21 Y.P:2.8 A	Good rooter	with its acutely orthotropic growth forms a compact bush and a dense plucking table; vigorous grower.	A spacing of 120x60 cm or even closer is desirable because of the compact bush; flourishing at mid and high elevations.

Contd...

Contd...

13	B/6/29 Y.P.:1.2 A	Excellent rooter	Vigorous grower but with sparse branching	Requires early training: average yielder. Makes good, tippy, orthodox tea with good flavour and quality. Appearance of CTC teas is not good because of heavy pubescence of the flush shoot.
14	B/6/34 Y.P. : 3.4 B	Good rooter	Good grower with profuse branching and good spread with a dense plucking table; tolerant to drought and fairly so to wind.	Flourishing at mid and high elevations, both in warm and cold zones.
15	B/6/57 Y.P. :2.4 A	Good rooter	Semi-orthotropic and vigorous grower with good branching; fairly tolerant to drought.	Faring well at mid and high elevations.
16	B/6/27 Y.P. :3.0	Good rooter	Vigorous, orthotropic grower.	Good at mid-elevations, where it flushes through drought.
17	B/6/129 A	Good rooter	Semi-orthotropic, good grower.	Faring well at mid-elevations.
18	B/6/137 Y.P. 3.2 A	Good rooter	Good grower with profuse and orthotropic branching and good spread forming a dense plucking table.	Faring well at high-altitudes. Makes tippy teas with useful character.
19	B/6/182 Y.P. : 3.4 A	Good rooter	Good grower with profuse branching forming a dense plucking table with good spread; hardy, fairly tolerant to drought and mild frost.	Desirable at all elevation; flushes during winter at high elevations.
20	SP/4/6 Y.P. : 3.3 A	Rather slow in rooting	Though slow initially, a good grower with profuse, semiplagiotropic branching and a good spread; hardy resistant to drought and tolerant to mild frost.	Flushes throughout the year through both drought and winter; good for high elevations. Flavoury teas.

Contd...

Contd...

21	B/4/198 Y.P. : 2.46 A	Good rooter with vigorous and fairly uniform growth.	Vigorous grower with an excellent spread and dense plucking table.	Faring well at mid and high elevations. Tippy orthodox teas. Appearance of CTC not good owing to heavy pubescence of the flush, flavoury: bright liquor.
22	B/5/148 Y.P. : 2.83 A	Good rooter	Vigorous grower with excellent spread; establishes well and easily in the clearings; tolerates drought fairly well.	Desirable at mid and high elevations.
23	B/5/149 A Y.P. : 3.36	Good rooter	Good grower with good spread; fairly tolerant to drought.	Flushes throughout the year through both winter and summer, faring well at mid and high elevations. Scented flavour.
24	B/7/372 Y.P. : 2.98	Good rooter with uniform growth	Good semi-orthotropic grower; resistant to drought	Faring well at mid elevations. Bright liquor.
25	PICS-1 K/ 19/ 16 Y.P:2.5	Good rooter exhibit slow growth in the first two months, subsequently, growth is quick, highly vigorous, uniform growth	Semi-orthotropic with thick wood and good frame; recovery from pruning is quick with profuse budbreak; plucking table is fairly compact with a good spread and numerous plucking points; flush shoots large, light coloured and pubescent, prone to blister blight in one high elevation estate	Yield clone with overall quality, large and nearly horizontal leaves, fare well in favourable aspects and under regulated shade with high performance, Plucking cost should be low because of large crop shoots

Contd...

26	PICS-6 K/ 19/ 16 Y.P:3.8	Good rooter with uniform growth, very vigourous.	Semi-orthotropic with moderately thick wood and good frame; branches from the collar, recovery from formative prune is quick with an early and profuse bud-break; plucking table compact with an excellent spread and numerous plucking points; flush shoots small, dark green, glabrous; very highly tolerant to drought; flushes throughout the year	High yielding, drought tolerant clone with overall quality- and all purpose clone, consistent performance in different agro-climatic regions, for infilling and replanting, particularly in poor soils.
27	PICS-4 K/ 19/ 16 Y.P:2.4	Good rooter with uniform growth, very vigourous.	Acutely orthotropic with moderately thick wood and good frame, recovery from formative prune is quick with a profusion of buds; plucking table compact with a moderate spread and numerous plucking points; flush shoots are of medium size, dark and sparsely pubescent; tolerant to drought	Moderate to high yielding, drought tolerant clone with above average quality. Being acutely orthotropic, an early and low centering is essential to induce satisfactory spread. Higher population density of a little over 13,000 bushes/ha in a double hedge system should be useful to obtain an early and good ground cover as well as higher yields.

y.p = yield potential : higher the figure than 1, greater the yielding capacity.
Quality : A = above average
B = average for high elevation conditions.

Pest and Disease Management of Plantation Crops

Arecanut

Pests/diseases	Symptoms	Control Measures
Pests		
Mites (*Raoiella indica*) (*Oligonychus indicus*)	Adults and young ones suck the lower surfaces of leaves, causing them to turn yellow and bronzed in appearance	Spray the lower surface of leaves with dicofol (0.05%)
Spindle bug (*Carvalhoia arecae*)	Adults and young ones suck the sap from the tender spindle resulting in loss of vigour and consequent death	Place 2 g of phorate granules taken in perforated polybags in the inner most leaf axils.
Inflorescence caterpillar (*Tirathaba mundella*)	Caterpillars feed on the flowers and clamp the inflorescence into a wet mass of frass with silky threads	Infected spadices may be forced open and sprayed with malathion (0.05%)
Diseases		
Koleroga or mahali (*Phytophthora palmivora*)	Water-soaked lesions appearing on the nut surface near the perianth spread over the other parts giving the nut a dark green colour. Infected nuts shed without perianth.	SprayBordeaux mixture (1%) twice at 45 days interval during monsoon.
Bud rot (*Phytopththora palmivora*)	Affected spindles appear yellow, later changing to brown and finally the whole spindle rots.	Early removal of the infected tissue and treat the healthy tissue with Bordeaux paste. Drench the crown with Bordeaux mixture (1%) as a prophylactic measure.
Anabe roga (*Gonoderma lucidum*)	Small brown irregular patches appear on the stem and a brownish exudate oozes out from these patches	Provide better drainage, isolate the affected palms by trenches, drench with captan 0.3%.
Yellow leaf disease (*Mycoplasma like organisms*)	Leaves become yellow, smaller, stiff and pointed, crown gets reduced, palm remains stunted with a few or no nuts.	Regular manuring, ensure drainage, grow cover crops, remove the affected palms.

Cashew

Pests/diseases	Symptoms	Control Measures
Pests		
Tea mosquito bug (*Helopeltis antonii*)	Pest appears with the emergence of new flushes and panicle. Drying of inflorescence and dieback of shoots	Spray 0.1% carbaryl or 0.05% quinalphos or 0.03% phosphamidon. Spraying may be done once, twice or thrice depending upon necessity.First spraying is to be given synchronizing with the emergence of new vegetative flushes in October-November. The second spraying may be given synchronizing with the commencement of panicle emergence in December-January. The third spraying may be given at completion of flowering / initiation of fruit set in January-February
Stem and root borer (*Placaederus ferrugineus* L.)	Yellowing of leaves, drying of twigs, presence of holes at the base of stem with exuding sap and frass	Swab the trunk and exposed roots with carbaryl Remove the dead trees and trees in advanced stage of infestation at least once in 6 months. Swab the trunk region (up to 1 m height from the ground level) and exposed roots, with a suspension of mud slurry + carbaryl 0.2% or coal tar and kerosene (1:2) or 5% neem oil twice a year during March-April and November-December along with soil application of Sevidol 4 G@ 75 g/tree.
Diseases		
Dieback or pink disease	Appearance of white patches on branches followed by drying of twigs from the tip.	Chisel out the affected parts and apply Bordeaux paste. Give prophylactic sprays of 1% Bordeaux mixture during May-June and October

Coconut

Pests/diseases	Symptoms	Control Measures
Pests		
Rhinoceros beetle *(Oryctes rhinoceros)*	Adult beetle bores through the unopened fronds and spathes hampering the growth of the palm and predisposing the palm to infestation by red palm weevil and bud rot. The	Provide field sanitation to prevent breeding of beetles. Hook out beetles from attacked palms by using beetle hook. Topmost three leaf axils around spindle may be filled with any of the following

Contd...

	attacked frond when fully opened shows characteristic triangular cuts. The beetle attack is maximum during summer months from March to June and minimum in September-October.	mixtures as a prophylactic measure: a) Sevidol 8G 25 g + fine sand 200 g, which is to be done thrice in a year in April-May, September-October and December-January.(b) Naphthalene balls 10.5 g (approx. three balls) covered with fine sand, once in 45 days. Treat manure pits and other possible breeding sites with 0.01% carbaryl on w/w basis. Treatment will have to be repeated every three months. Release *Baculovirus oryctes* infected adults @ of 10-15 / ha to bring down the pest population. Inoculation of breeding sites with entomopathogenic fungus *Metarrhizium anisopliae* (@ 5 x 10 11 spores / ml) gives effective control.
Red palm weevil *(Rhyncophorus fereugineus)*	This weevil worst affects relatively young palms of age group between 5 to 10 years. The mode of entry, nature of infestation and colonization of this pest go unnoticed, defeating normal pest management strategies. The diagnostic symptoms are the presence of holes on the stem, oozing out of a viscous brown fluid and extrusion of chewed up fibrous matter through the hole, longitudinal splitting of leaf base and wilting of central shoot. Sometimes the gnawing sound produced by the feeding grubs inside is audible.	Field sanitation should be given prime importance. Avoid making steps or any other injury on the tree trunks to reduce the loci of infestation. Leaf axil filling as suggested in the case of rhinoceros beetle will be useful against the red palm weevil also. When green leaves are cut from the palms, stumps of not less than 120 cm may be left on the trees in order to prevent successful inward movement of the grubs through the cut end. In attacked palms, observe for bore-holes and seal them except the top most one. Through top most hole, pour 1% carbaryl or 0.2% trichlorphon suspension @ one litre per palm, using a funnel. Coconut log traps with fermenting toddy or pineapple or sugarcane activated with yeast or molasses can be set in coconut plantation to attract and trap the free floating population of red palm weevil. Incorporate any of the insecticide to each trap to kill the weevils trapped. Use of pheromone trap for attracting and killing adult weevils @ one trap per 2 ha.

Contd...

Leaf eating caterpillar *(Opisina arenosella)*	The caterpillar feeds on green matter from the lower leaf surface, remaining within galleries of silk and frass.	As a prophylactic measure, the first affected leaves may be cut and burnt during the beginning of the summer season. Arrange for release of larval / pupal parasitoids, *Goniozus nephantidis, Elasmus nephantidis* (brown species) and *Brachymeria nosatoi.* When infestation is very severe and if the biocontrol is not likely to be effective, spray the undersurface of the fronds with dichlorovos 0.02%, malathion 0.05%, quinalphos 0.05%, endosulfan 0.05% or phosalone 0.05%.
Coreid bug *(Paradasynus rostratus)*	The attacked buttons become deformed with characteristic crevices on the husk below the perianth with gum exudations and the tender nuts become barren. The attacked buttons do not develop resulting in immature nut fall.	Spraying has to be done three times a year. Apply 0.1% carbaryl suspension on the newly opened inflorescence after receptive phase of the female flowers and spray entire crown excluding the leaves and older bunches. Destruction of pollinating insects can be avoided if spraying is done in the afternoon hours.
Coconut Eriophyid mite*(Aceria [Eriophyes] guerreronis*	The earliest symptom on 2-3 month old buttons is pale yellow triangular patches seen below perianth. Later, these patches become brown. Severely affected buttons may fall. As the buttons grow, brown patches lead to black necrotic lesions with longitudinal fissures on the husk. Uneven growth results in distortion and stunting of nuts leading to reduction in copra yield. In severe cases, quality of fibre is reduced and distorted nuts increase labour requirement for dehusking.	Collect and destroy all fallen buttons of the affected palm. Apply 2% neem oil + garlic emulsion or commercial neem formulation of azadirachtin 0.004% (Neemazal T/S 1% @ 4 ml per litre of water) or micronized wettable sulphur 0.4% in the crown on young bunches.

Contd...

Diseases		
Root (wilt) disease	The characteristic symptom is flaccidity of leaflets. Yellowing of older leaves, necrosis of leaflets and deterioration and decay of root system are other salient features of the disease. The leaflets curve inwardly to produce ribbing so that the whole frond develops a cup like appearance. Abnormal shedding of buttons and immature nuts are also noticed.	Rogue out palms that are affected severely by root (wilt) and yield less than 10 nuts / palm / year Integrated nutrient management, pest and disease management.
Stem bleeding *Thielaviopsis Paradoxa*	Exudation of the reddish brown liquid through growth cracks mostly at the basal part of the trunk and bleeding patches higher up in the trunk are characteristic symptoms	Chisel out completely the affected tissues and paint the wound with tridemorph 5%. Apply coal tar after 1-2 days. Destroy the chiseled materials by burning. Avoid any mechanical injury to trunk. Apply neem cake @ 5 kg per palm in the basin along with other organics. Root feed with tridemorph 5%, thrice a year during April-May, September-October and January-February to prevent further spread of lesions. Apply tridemorph @ 25 ml in 25 litre of water as soil drenching once in four months.
Bud rot *Phytophthora Palmivora*	Palms of all age are liable to be attacked but normally young palms are more susceptible, particularly during monsoon when temperature is low and humidity is very high. In seedlings, the spear leaf turns pale and comes off with a gentle pull. In adult palms, first visible symptom is the colour change of the spear, which becomes pale and breaks at the base and hangs down.	In early stages of disease (when the heartleaf starts withering) cut and remove all affected tissues of the crown. Apply Bordeaux paste and protect it from rain till normal shoot emerges. Burn all disease-affected tissues removed from the palm. Spray 1% Bordeaux mixture on spindle leaves and crown of disease affected as well as neighbouring palms, as a prophylactic measure. Palms that are sensitive to copper containing fungicides can be protected by mancozeb. Small, perforated sachets containing 2 g of mancozeb

Contd...

		may be tied to the top of leaf axil. When it rains, a small quantity of the fungicide is released from the sachets to the leaf base, thus protecting the palm. Adopt control measures for rhinoceros beetle. Provide adequate drainage in gardens. Adopt proper spacing and avoid over crowding in bud rot prone gardens.

Cocoa

Pests/diseases	Symptoms	Control Measures
Pests		
Striped squirrel *Funambulus sp.*	Squirrels gnaw bronzing pods and extract beans along with mucilaginous pulp.	Mechanical protection of pods.
Red borer *Zeuzera coffeae*	Larvae bore into main stem of young plants and fan shoots of older trees causing drying up.	Prune off and burn affected shoots. Spray carbaryl 0.1% on main stem.
Rats *Rattus rattus*	Rats gnaw bronzing pods and extract beans along with mucilaginous pulp	Baiting with anticoagulant rodenticides.
Mealy bugs *Planococcus citri*	Seen in cherelles, developing pods and shoots and desap the tissues.	Spot application of quinalphos 0.025%
Diseases		
Phytophthora pod rot/black pod- *Phytophthora palmivora*	Fruits of all stage are affected. Circular brown spots appears which enlarges and finally becomes black.	Periodical remove and destroy all dried up and infected pods. Spray Bordeaux mixture (1%) with adhesive.
Colletotrichum pod rot *Colletotrichum gleosporioides*	Rotting of immature pods, turns to black in colour and remains on the tree.	Periodical remove and destroy all dried up and infected pods. Spray Bordeaux mixture (1%) with adhesive

Contd...

Pink disease *Corticium salmonicolor*	It appears as pinkish powdery coating on affected stem ,splitting of bark etc.	Remove all infected and dried branches. Spray Bordeaux mixture (1%) with adhesive before the onset of monsoon.
Vascular streak dieback(VSD)- *Oncobasidium theobromae*	Characteistic chlorosis of one or two leaves on second or third growth flush behind the tip, followed by leaves fall .Infected stem shows discoloured brown streak.	Remove all infected twigs. Grow VSD tolerant varieties.

Coffee

Pests/diseases	Symptoms	Control Measures
Pests		
Coffee berry borer(*Hypothenemus hampei)*	Female beetle bores into berries,make tunnels and lay eggs.Larvae feed on beans,making small tunnels.	Spray insecticide along with wetting agent 120-150 days after flowering.
White stem borer(*Xylotrechus quadripes)*	Adult beetle lay eggs in the crevices on main stem and larvae bore into stem, make tunnels causing die back.	Swabing main stem with 50WP@4kg/200l of water once or twice in April-May or Oct-Dec.
Shot hole borer(*Xylosandrus compactus)*	Caterpillars makes tunnels on secondary and tertiary branches leads to drying up of branches.	Prune affected twigs 5-8 cm beyond shot hole and burn,destroy infested suckers during summer.
Diseases		
Leaf rust(*Hemileia vastatrix)*	Yellow spots appear on lower surface during rainy season leads to premature defoliation	Spray with Bordeaux mixture (0.5%) 3-4 times a year.
Black rot (*Koleroga noxia)*	Affected bushes have blackening and rotting of leaves, twigs and developing berries leads to defoliation and berry drop.	Spray with Bordeaux mixture (1%) during break in monsoon.

Oil Palm

Pests/diseases	Symptoms	Control Measures
Pests		
Rhinoceros beetle	Damage emerging fronds and spindle.	Field sanitation and elimination of breeding sites. Pest can be suppressed by using virus *Baculovirus oryctes.*
Red palm weevil	Larvae tunnel into crown and feed on growing tissue, wilting ,chlorosis.	Treatment with 1% carbaryl.
Birds	Birds feed on mesocarp of oil palm	Scaring of birds.
Diseases		
Anthracnose	Disease in nursery.	Spraying mancozeb@200g/100l of water.
Spear rot	Affect oil palms of all ages. Yellowing starts from tip of innermost whorl of leaves. At advanced stages new leaves become rudimentary and show rotting.	Rouging of all affected palms.

Rubber

Pests/diseases	Symptoms	Control Measures
Pests		
Scale insect *(Saissetia nigra Nietn)*	Small insects with an outer black, dome-shaped covering, occur on leaflets, petioles and tender shoot portions and suck the sap. Severely affected portions dry up and die. Ants and sooty mould are associated with this.	Spray organophosphorus insecticides like malathion at 0.05% concentration
Mealy bug *(Ferrisiana virgata Ckli)*	These are soft-bodied small insects with white mealy outer covering. Damage is similar to scale insects	Spray organophosphorus insecticides or fish oil rosin soap.
Cockchafer grub *(Holotrichia Serrata F)*	The grub feeds on the roots of seedlings in the nursery and young plants. The affected plants droop and fall over.	
Mite *(Hemitarsonemus dorsalis)*	These are minute organisms with four pairs of legs. They suck sap from leaves resulting in crinkling and shedding.	Dust sulphur or spray suitable acaricides like Kelthane.

Contd...

Diseases		
Abnormal leaf fall *Phytophthora palmivora* Butl.	First the fruits rot, later infected leaves fall in large numbers prematurely, either green or after turning coppery red.	Prophylactic spraying of the foliage prior to the onset of south west monsoon with Bordeaux mixture (1%) using high volume sprayers.
Powdery mildew *Oidium heveae* Steinm.	Tender leaves with ashy coating curl, crinkle, and edges roll inwards and fall, leaving the petioles attached to the twigs like a broom-stick. After a few days, the petioles also fall. Die-back of twigs follows. On older leaves, white patches later causing necrotic spots reduce photosynthetic efficiency. Infected flowers and tender fruits are shed, affecting seed production.	Dusting during the refoliation period commencing from bud break in about 10% of the trees, giving 3 to 5 rounds at weekly to fortnightly interval using 11 to 14 kg 325-mesh fine sulphur dust per round per hectare.
Pink disease *Corticium salmonicolor* Berk	It is more damaging for plants in the age group of 2 to 12 years. The main site of infection is usually the fork region. White or pink coloured cobweb mycelial growth occurs on bark surface with streaks of latex oozing out from lesions. Rotting, drying up and cracking of affected bark will follow. Sprouts develop from below the affected portion. The distal portion of branches dry and dried leaves stick to the dead branches.	Spraying two rounds of 1% Bordeaux mixture (during May and August) on to the fork and branches reduces disease incidence.

Tea

Pests/diseases	Symptoms	Control Measures
Pests		
Nursery i.Nematodes	Infested roots develops galls, exhibit chlorosis and stunted growth	Heat treatment, Mix carbofuran 3G (700g/m3) with soil in nursery beds
ii.White/cockchafer grubs (*Holotrichia sp.*)	Grubs eats away roots causing ring barking of stem, chlorosis and defoliation	Heat treatment of soil and use of granular insecticides.
Red coffee borer *Zeuzera coffeae*	Young stems are bored and makes tunnels	Cut affected stem and pour quinalphos and holes are plugged with clay paste.

Contd...

Diseases		
Black root disease(*Rosellinia arcuata)*	Wilting,chlorosis.drying without defoliation and death of bush.	Drench soil with mancozeb 3g / 10l. Trichoderma(200 g/pit)
Blister blight(*Exobasidium vexans)*	Affects tender leaves and stems. Affected leaves distorted and irregularly rolled. Stem infection leads to goose neck shape, dieback and snapping at the point of infection	Copper oxychloride 350g+plantomycin 70g/ha,3-4 days interval.

REFERENCES

Abdul Salam, 1997. Strategies for Improved productivity through soil and Nutrient Management. In: *Indian Cashew Industry* (ed.) Singh, H.P., Balasubramanian, P.P., Venkatesh, N. Hubballi pp.92-100

Abdul Salam, M., Mohanakumaran, N. and Balasubramanian, P.P. 1998. *Cashew cultivation. Science and Techniques.* Directorate of Cashew and Cocoa Development, GOI, Kochi

Agricultural Statistics At A Glance (2003). Directorate of Economics and Statistics, Department of Agriculture and Cooperation, Ministry Of Agriculture, Government of India.

Ascherio, A. 2002. Epidemiologic studies on dietary fats and coronary heart disease. *American Journal of Medical Science* 113: 9-12

Balasimha, D. and Rajagopal, V 2004. *Arecanut.* ICAR. p.306

Bavappa, K.V.A. 2000. Plantation crops based cropping and farming system options for Asia and South East Asian Countries. Paper presented in XIV PLACROSYM from 12th to 15th Dec. 2000 at Hyderabad, India

Bavappa, K.V.A., Kailasam, C., Khader, K.B.A., Biddappa, C.C., Khan, H.H., Kasturi Bai, K.V., Ramadasan, A., Sundararaju, P., Bopaiah, B.M., George V. Thomas, Mishra, L.P., Balasimha, D., Bhat, N.T. and Sham Bhat, K. 1986. *Journal of Plantation Crops* 14: 41-47

Beena, B., Adbul Salam, M. and Wahid, P.A. 1995. Nutrient offtake in cashew (*Anacardium occidentale* L.) *The Cashew* 9: 9-16

Biddappa, C.C., Upadhyay, A.K., Hegde, M.R. and Palaniswami, C. 1996. *Journal of Plantation Crops* 24: 71-85

Bopaiah, B.M. 1988. *Microbiological studies in relation to high density multispecies cropping system in coconut.* Ph.D. thesis, University of Mysore, India, p.168

Clarke, R.J. 1985. *Coffee.* Elsivier Applied Science Publishers Ltd. New York.

FAO, 2000. FAO production database.http://apps.fao.org/page/collections.

Fernandopulle, M.N. 2000. Fertilizer use in cashew. *Cashew Bulletin* 3: 1-6.

Gajanana, T.M., 1988, Growth rates of Area, Productivity and Production of arecanut in Karnataka. *Mysore Journal of Agricultural sciences*, 22: 274-282.

George, P.J. and Kuruvilla Jacob, C (ed.) 2000. *Natural Rubber- Agromanagement and Crop processing.* Rubber Research Institute of India. Rubber board, Kottayam, India, p.648

Gonzalez O.N.1991 Coconut as food. *Philippine Journal of Coconut Studies* (Philippines) 16, 2.

Khan, H.H., Upadhyay, A.K. and Palaniswami, C. 2002. (*In*) *Plantation Crops Research and Development* (Eds. P. Rethinam) Coconut Development Board, Kochi pp.9-22

Liyanage, L.V.K., Jayasundhera, H.P.S., Mathes, D.T. and Fernando, D.N.S. 1989. Pasture cattle Coconut system CORD (Indonesia) 5: 53-66

Nagaraj Shetty. 2004, Cocoa and Arecanut offering rich biodiversity. The Hindu survey of Indian agriculture.pp: 86-87

Nagarajan, R., Manickam, T.S. and Kothandaraman, G.V. 1988. Integration of pasture fodder and cattle in coconut small holdings. *Journal of Plantation Crops* 16: 59-62

Nair, C.P.R., Mariamma Daneil and Ponnamma, K.N. 1997. Integrated Pest Management in palms (eds. Nambiar KKN and Nair, M.K.), Coconut Development Board, Kochi p.30

Nair, M.K., Bhaskara Rao, E.V.U, Nambiar, K.K.N. and Nambiar, M.C. (ed.) *Cashew* (*Anacardium occidentale*) L. ICAR 1979, p.169

Nair, M.K., Biddappa, C.C., Palaniswami, C. and Upadhyay, A.K. 1996. A critical review of plant nutrient supply needs, efficiency and policy issues for the plantation crops for the year 2000 and 2025 AD. In: *Proc. of symposium on plant nutrient needs, supply, efficiency and policy issues: 2000-2025 AD.* Kanwar, J.S. and Katyal, J.C. (Eds.). National Academy of Agricultural Sciences, New Delhi, India p.329

Nayar, K.G. 1995. Cashew: a crop with unlimited potential. *The Cashew*, Jan-March, 1995.

Peng-FamgRen, Jiamg-BaoLong, Tay-JuHana; Lim-Nam, Peng, F.R., Huang, B.L., Tay, J.H. and Lim, N. 1996. The structural characteristics and biomass productivity of coconut-pineapple inter plantation. Journal-of-Nanjing-Forestry-University 20(1): 68-72

Prabhu, S.R. and George, V. Thomas. 1998. Occurrence of *Azospirillum amazonense* in the root environment of coconut palm. 30th Annual Conf. Assn. Microbiologists, College of Fisheries, Mangalore

Prabhu, S.R., Subramanian, P., Biddappa, C.C. and Bopaiah, B.M. 1998. Prospects of improving coconut productivity through vermiculture technology. *Indian Coconut Journal* Aug. 1998. pp.79-94.

Prakash, T.N. 1997. *An evaluation of arecanut co-operative marketing system in Karnataka.* Ph.D.thesis submitted to University of Mysore.

Preeti, N.S. 1998. *Economic analysis of arecanut cultivation and its impact on farm Economy in Shimoga district, Karnataka.* M.Sc. thesis submitted to UAS, Bangalore.

Raghuramulu, Y. and Kamala Bai, S. 1998. Package of Practices of organic coffee. Coffee board India. p.72

Rethinam, P. and Sivaraman ,K. 2003. Economic significance of intercropping systems in arecanut plantations under two irrigation methods. *Journal of Plantation Crops* 31: 53-56

Rohini Iyer and Nambiar, K.K.N. 1998. Integrated disease management in coconut. *Indian Coconut Journal* 29: 90-94

Rothfos, B. 1980. Coffee production (*In*) *Coffee Technology* (2)(eds.) Clarke, R.J.and Macrae ,R. Elsivier Applied Science Publishers Ltd. New York

Sanchez and Rasco 1983. Coconut in white soft cheese. *Philippine Journal of Coconut Studies* (Philippines) 8: 1-2.

Satyavati Krishnankutty., Krishnaswamy., C., Jayalakshmi. A., Balachandran., C., Sreedhara, N., Thomas.P.P., Sreemula Nathen and Mathew. A.G. 1984 . Innovations in edible products from coconut. Proceedings of the seminar on coconut processing and utilization, TVM, 8th Feb.

Sivetz M.and Desrosier, N.W 1979.*Coffee Technology*, AVI Publishing Company Westport,Conncticut.

Thomas George, V. and Shantaram, M.V. 1984. *in Situ* cultivation and incorporation of green manure legumes in coconut basins: An approach to improve soil fertility and microbial activity. *Plant and Soil* 80: 373-380

Thomas George, V. and Shantaram, M.V. 1986. *Journal of Plantation Crops* 14: 42-48

Thomas George, V., Biddappa, C.C. and Prabhu, S.R. 2001. *In situ* cultivation and incorporation of green manure legumes in coconut basins: An approach to improve soil fertility and microbial activity. *Tropical Agriculture* (Trinidad) 78: 13-18

Thomas George, V., Prabhu, S.R., Reeny, M.Z. and Bopaiah, B.M. 1998. Evaluation of lignocellulosic biomass from coconut palm as substrate for cultivation of *Pleurotus safer caju* (Fr.) singer

Thomas George, V., Prabhu, S.R., Subramanian, P. and Rohini Iyer. 2001. Organic farming technologies in coconut. ATIC series publication No.4, CPCRI, Kasaragod p.14

Thomas George, V., Rohini Iyer and Bopaiah, B.M. 1991. Beneficial microbes in the nutrition of coconut. *Journal of Plantation Crops* 19: 127-138

Thomas, K.G., 2000, Crop diversification in arecanut gardens. *Indian Arecanut, Spices and Cocoa Journal*, 1: 97-99.

Varghese, P.T., Nelliat, E.V. and Balakrishnan, T.K. 1978. Beneficial interactions of coconut-cocoa, crop combination. (*In*) Nelliat, E.V., Ranganathan, V., Vishveswara, S., Potti, S.N. and Zachariah, P.K. (eds.). Proc. PLACROSYM I, Kasaragod, India, pp.383-392

Vikraman Nair, Mallika, V.K. and Prasannakumari Amma, S., Koshy Abraham and Balasubramanian. 2000. *Cocoa cultivation science and technique*. Published by Directorate of Cashewnut and Cocoa Development. GOI, Kochi.

Wable, K.W.J. and James, W.P.T. 1993. Isomeric fatty acids and human health. *European Journal of Clinical Nutrition* 47: 829-839

Yusof, B. and Chan, K.W. 2003. Going back to basics: Producing high oil palm yields sustainably. *Oil Palm Bulletin* 46: 1-14.

Chapter 5

SPICES

Introduction

Spices hold a special place in world trade and were a catalyst in many historical events, discoveries, wars and inventions. It has even been cited as an important element in the spread of knowledge and culture. Spices are non-leafy parts of plants used as a flavoring or seasoning. Asia is known as 'Land of Spices' as it is the place of origin, production, consumption and export of most spices. Out of the world's 70 plant species grown as spices, 13 are considered major spices produced in Asia. Spices had many uses, from adding flavour to food to medicinal properties, but only grew in a few places. In the last decade, the annual average growth in spice production was 4.3% and trade in spices grew at an average of 5.8% /year. More sophisticated consumers, health-conscious eating habits (but also the rise in the amount of processed foods that use herbs and spices as inputs), rising incomes, increased world trade and globalisation are driving this new era of profitability. Spices are ideal commodities for international trade. They are normally exported in raw or dried form and are usually easy to store and transport. Consequently they do not need expensive infrastructure such as cold chain management or advanced storage facilities and they are not particularly time sensitive. In addition, they are recognised as high value products that can be produced relatively easily on small farms and do not require investments in advanced technology and machinery.

Table 1. World Production of Spices by Variety, 1965-2005

World Production (mt)	Years				Average annual growth			Market share (%)
	1965	1995	2000	2005	00-05	95-05	65-05	
Capsicum and pimento	1,067,777	1,928,688	2,302,879	2,450,336	1.2	2.4	2.1	37.2
Other spices	356,759	900,016	1,863,160	1,899,000	0.4	7.8	4.3	28.9
Ginger	82,283	672,780	916,438	1,004,546	1.9	4.1	8.5	15.3
Anise, bedam, fennel	97,260	309,888	325,040	446,296	6.5	3.7	3.9	6.8
Pepper, white/long/ black	74,442	236,843	309,667	409,899	5.8	5.6	4.4	6.2
Cloves whole + stems	26,378	108,157	104,999	145,370	6.7	3.0	4.4	2.2
Cinnamon (canella)	19,692	81,263	108,247	134,410	4.4	5.2	4.9	2.0
Nutmeg, mace, cardamom	15,418	64,946	70,061	81,292	3.0	2.3	4.2	1.2
Vanilla	1,749	4,367	4,564	7,339	10.0	5.3	3.7	0.1
Total	**1,731,758**	**4,306,948**	**6,005,055**	**6,578,488**	**1.8**	**4.3**	**3.4**	**100.0**

Source: FAOSTAT

World production of spices

Global production of spices has rapidly increased since the 1960's, from 1.7-million metric tons in 1965 to 6.6-million metric tons in 2005 (Table 1). It shows that production techniques and therefore yields have increased and land cultivated has risen to meet growing global consumption of spices. India leads the world in spice production, with 3.1-million metric tones in 2004, or just under half of world production. China is the second largest producer, with 692,000 metric tones in 2004, or 10.5% of total world production, followed by Indonesia's 448,000 metric tones or 6.8%. South Asia (India, Pakistan and Nepal) is the leading region, producing 55% of world production. South East Asia (Indonesia, Vietnam and Thailand) follows with 13.2% and East Asia (China) produces 10.5%. The EU and the Middle East have been increasing their production of a variety of spices through new production techniques, including green houses, drip irrigation and farming technology improvements. Nonetheless, only some spices can be grown profitably in these environments (even with the help of technology), while others cannot be grown in developed countries at all. Developing countries have grown their production levels of spices from 1965 to 2005 by an annual average growth rate of 3.6%, while developed countries had an annual average growth of only 0.6%. However, developed nations tend to have colder climates that are less friendly to spice agriculture.

Ginger has been the best performing spice variety for the entire period 1965 to 2005, with an average annual growth rate of 6.5%, followed by cinnamon at 4.9% per annum and pepper and cloves at 4.4% per annum (Table 1). Although aniseed, cloves and especially vanilla have been good performers recently, vanilla is a rather 'light' commodity, with a small effect on overall spice production figures. When examining spices' production growth, one should also keep in mind the length to maturity of a particular spice.

Consumption

Almost half of the world's spices are produced in India, country also consumes most of the spices it produces. Indonesia, on the other hand, exports a third of its production. Europe, North America and Japan are large spice consumers and importers, and therefore the most promising markets for exporters. Countries like Malaysia, Netherlands and Germany are particularly important for raw spice producers, as they act as spice trading hubs. Germany is the world's leading importer of unground and unprocessed black pepper and one of the top exporters of processed pepper and mixed spices, mainly to its European neighbours. Even countries with massive domestic markets like India, China and Mexico are not necessarily able to produce the quantities of spices their markets demand, especially as tastes grow and diversify and more people move to the cities, creating opportunities for other exporting countries to access these markets.

Supply and Demand

Spices, are less susceptible to supply-side drivers like climate, quality of infrastructure, availability of human and financial capital and advanced technologies than agricultural commodities in general. Therefore spices be seen as almost 'perfect' export commodities for developing countries, they face other problems when trying to promote their produce for export. Since spices in general requires little capital outlay and technology, it is relatively easy for rural farmers to engage in their production. This results in oversupply of the commodity, declining prices and volatile markets. In addition, the tariff schedules of the developed world tend to encourage the export of raw, unprocessed commodities, with low or non-existent tariffs at the low-value end and higher tariffs for final products. This might initially promote greater production from developing countries, but ultimately it is likely to further depress the price of raw commodities while raising the prices of final commodities such as spice mixtures for the ordinary consumer. Spices are increasingly used in the food industry for flavouring and colouring, while growing trend for greater variety is fuelling the demand for spices in the production of 'exotic' meals. At the retail level, spice purchases are moving towards 'convenience' spice mixtures that can be used for instant flavouring, colouring and marinating.

The world market for imported spices and culinary herbs was valued at more than $2.8bn in 2004. The top varieties by value were capsicum and pimento (chilli peppers, cayenne, paprika and allspice) accounting for about 20% of total world trade in spices, followed by black/white pepper (17%), vanilla (14%), ginger (10.5%), 'other' spices (6.5%), mixed spices (5%), cinnamon (4.4%), cloves (4%), and cumin and nutmeg (2.8% each). The main importing regions are EU, North America and East Asia. Although East Asia experienced a decline in imports, mainly due to Japan's recession, demand for spices has recently lifted again. The South Asian region experienced rapid growth which tailed off slightly in 2004. The two biggest importing regions, EU and North America, following slow-downs in 2000 and 2001, have recovered strongly and posted record figures in 2003 and 2004. Growth for these two regions over the four-year period from 2000-2004 was 4.1% and 3.9% per annum, respectively, while for the eight-year period 1996-2004, imports into the EU grew at an average annual compounded growth rate of 6.6% while the US grew at 6.9% per annum. The import market expanded by 17.3% for the EU and16.5% for US.

The Middle East saw a significant decline in spice imports because of large investments in domestic spice production. The fastest growing importing region was South Asia, driven primarily by India, which is interesting given its status as the world's leading producer of spices. Here demand outstrips supply despite the vast amount of spices produced each year.

The EU imported $53.3m worth of spices from East Asia, $68m from South Asia, $91.1m from South America and $91.5m from Africa, of which $66.3m came from SADC. $394m was spent on intra-regional trade. In total the EU imported $994.5m worth of spices in 2004, or 35% of total world imports.

Table 2. World Spice Supply usage by Region and Selected Countries, 2003 (mt)

	Production	Imports	Exports	Domestic supply	% of world production	% of world exports	% of world consum-ption
World	6,440,704	1,554,977	1,543,067	6,461,522	100.0	100.0	100.0
Developed countries	227,220	813,140	248,137	793,233	3.5	16.1	12.3
Developing countries	6,213,484	741,837	1,294,930	5,668,289	96.5	83.9	87.7
South Asia	3,643,742	242,173	265,514	3,620,521	56.6	17.2	56.0
Africa	646,890	77,330	76,375	658,675	10.0	4.9	10.2
South East Asia	794,238	10,187	281,161	517,614	12.3	18.2	8.0
East Asia	663,769	200,527	402,911	461,585	10.3	26.1	7.1
NAFTA	102,909	302,278	53,123	352,064	1.6	3.4	5.4
Middle East	240,713	152,658	131,634	263,737	3.7	8.5	4.1
EU	13,020	299,577	140,296	174,916	0.2	9.1	2.7
Central America & Caribbean	136,569	45,616	59,736	122,234	2.1	3.9	1.9
South America	93,372	29,546	85,030	38,501	1.4	5.5	0.6
Oceania	4,191	13,251	5,027	10,600	0.1	0.3	0.2
SADC	111,343	26,536	45,684	101,905	1.7	3.0	1.6
Selected countries							
India	3,104,000	93,215	222,000	2,975,215	48.2	14.4	46.0
Indonesia	392,375	14,800	117,412	289,763	6.1	7.6	4.5
China	657,799	24,675	399,836	282,838	10.2	25.9	4.4
US	2,720	244,976	24,277	223,419	0.0	1.6	3.5
Japan	-	144,345	1,073	143,272	0.0	0.1	2.2
Malaysia	27,100	99,206	40,235	81,071	0.4	2.6	1.3
Germany	-	70,356	22,272	48,084	0.0	1.4	0.7
UK	-	46,599	7,447	39,452	0.0	0.5	0.6
The Netherlands	200	55,158	38,394	20,964	0.0	2.5	0.3

Contd...

Table 2. Contd...

Selected SADC countries							
Botswana	-	2,111	10	2,101	0.00	0.00	0.03
Congo, DRC	33,000	107	-	33,107	0.51	0.00	0.51
Lesotho	-	1,200	-	1,200	0.00	0.00	0.02
Madagascar	24,515	41	17,635	10,721	0.38	1.14	0.17
Mauritius	673	1,855	118	2,410	0.01	0.01	0.04
South Africa	10,000	17,395	9,636	17,962	0.16	0.62	0.28
Swaziland	-	1,250	299	956	0.00	0.02	0.01
Tanzania	19,700	158	5,957	17,001	0.31	0.39	0.26
Zimbabwe	15,535	185	9,408	8,367	0.24	0.61	0.13

Source: FAOSTAT

Leading Importers/exporters by Country

The world's leading importer of spices in 2005 was US, with more than \$500m worth of spice (Table 3). Japan (with \$216m worth of spice) and Germany (with \$200m) were the second and third largest importers. Together these three countries account for roughly a quarter of all spice imports in the world. But despite their large market size, the US, Japan and Germany have all been growing their imports slowly over the past five years, with annual average growth rates over the period 2000-2005 at -1.8%, 1.6% and 2.1%, respectively. The fourth largest importer in 2005 was India, who is also the world's largest producer and consumer of spices.

The majority of world's top importers are located in North America and Europe. Their imports for 2005 stood at \$129.5m, an amount which has shown rapid growth of over 23% per annum over the past five years. Among the leading 20 importing countries were a number of developing countries, including Singapore which imported \$110m worth of spices, Mexico (\$99m), Saudi Arabia (\$93.5m), Malaysia (\$90m), Pakistan (\$49.6m) and South Korea (\$36m). Most of these countries are also large spice producers and their considerable imports demonstrate a high degree of intra-industry trade. The leading African importer of spices was South Africa, which imported roughly \$20m worth of spices in 2005.

China was the leading exporter of spices, mainly due to its exports of ginger to Japan (Table 4). Ginger exports from China alone amounted to \$209m in 2005 and have been growing by over 22% per year from 2000, while Japan's imports from China amounted to \$105m in 2005. Total Chinese exports of

Table 3. Major Spice Importers by Country, 2005

Rank	Importer name	Imports 2005 (US $ '000)	Market share (%)	CGAR 2000-2005
	World	3,405,461	100	6.5
1	US	500,762	14.7	0.18
2	Japan	216,073	6.3	1.6
3	Germany	200,086	5.9	2.1
4	India	129,592	3.8	23.8
5	Netherlands	122,527	3.6	-3.3
6	Spain	111,569	3.3	6.7
7	UK	110,875	3.3	2.8
8	Singapore	110,311	3.2	-9.8
9	Mexico	99,335	2.9	11.3
10	France	97,133	2.9	-0.5
11	Saudi Arabia	93,588	2.7	-1.1
12	Malaysia	90,313	2.7	7.1
13	Canada	66,824	2.0	1.8
14	Belgium	64,797	1.9	10.0
15	Pakistan	49,645	1.5	-
16	Italy	37,434	1.1	-0.8
17	Austria	37,261	1.1	8.0
18	Korea Rep.	36,299	1.1	0.9
19	Poland	28,422	0.8	5.5
20	Denmark	27,766	0.8	9.2
Selected countries				
23	Thailand	24,750	0.7	19.9
24	Russia	24,464	0.7	7.1
25	Australia	24,025	0.7	0.2
27	Brazil	20,956	0.6	6.3
28	South Africa	20,245	0.6	3.4
37	China	11,410	0.3	-1.7
48	Argentina	7,156	0.2	-12.3
73	Mauritius	2,766	0.1	5.6
85	Zambia	982	0.0	13.3
90	Mozambique	737	0.0	67.3
101	Malawi	239	0.0	0.11

Source: UN Comtrade

Table 4. Leading Exporting Countries of Spices, 2005 (US$ - Million)

Rank	Exporter name	Exports 2005	Market share (%)	CGAR 2000-2005
	The World	3,291,531	100	4.4
1	China	431,422	13.1	14.4
2	India	316,677	9.6	2.3
3	Indonesia	282,864	8.6	-6.1
4	Vietnam	212,959	6.5	8.1
5	Brazil	170,742	5.2	10.4
6	Peru	145,468	4.4	80.3
7	Germany	123,624	3.8	10.8
8	Madagascar	119,210	3.6	-6.3
9	Sri Lanka	100,754	3.1	7.1
10	Spain	100,419	3.1	7.1
11	Netherlands	81,657	2.5	-0.7
12	France	71,275	2.2	10.8
13	Turkey	68,013	2.1	6.3
14	Guatemala	67,597	2.1	-7.7
15	Thailand	65,298	2.0	5.3
16	Malaysia	58,794	1.8	-16.1
17	Iran, Islamic Rep.	57,721	1.8	
18	Syrian Arab Rep.	56,839	1.1	
19	US	56,544	1.7	
20	Mexico	50,926	1.5	
Selected countries				
21	Israel	49,309	1.5	
29	South Africa	19,099	0.6	
32	Ittaly	15,580	0.5	
33	Morocco	15,430	0.5	
34	Nigeria	15,378	0.5	
37	Tanzania	14,193	0.4	
44	Zimbabve	10,035	0.3	
46	Uganda	9,267	0.3	
52	Australia	7,056	0.2	
68	Ethiopia	2,895	0.1	
69	Malawi	2,832	0.1	
71	Kenya	2,775	0.1	
78	Argentina	1,991	0.1	

Source: UN Comtrade

spices have been growing at 14.5% per annum from 2000 to 2005. Second on the top exporters list in 2005 was India, which exported $316m worth of spices. India's exports, unlike China's, spread over a range of different spices. In 2005, India exported over $90m worth of the capsicum and pimento families, an amount that has grown by nearly 20% per annum from 2000. India also exported pepper to the value of $57m, turmeric ($36m), cumin seeds ($19m), coriander seeds ($17m), curry ($14m), ginger ($14m), cardamom ($10m) and other spices ($52m). The majority of the country's exports are destined for the US ($55m), although Malaysia is also becoming an important destination. Indonesia is the third largest exporting country but its exports of $280m in 2005 are significantly lower than the $387m it recorded in 2000. This is mainly due to the loss of pepper exports, which have more than halved since 2000, although this has been mitigated somewhat by increases in the country's other major export commodities - nutmeg, cinnamon, mace and cloves improved by 7%, 6%, 13% and 20.5% respectively. Indonesia's main trading partners have historically been the US, the Netherlands, Germany and India. Indonesia is also a major exporter of cloves to Singapore, with the total value of its exports to that country topping $34m in 2005. Vietnam, the fourth largest exporter, trades mainly in pepper, where it is the world leader Roughly $44m of Vietnam's 2005 pepper exports went to the EU, $35m went to the US and $26m was destined for the Indian pepper markets.

Brazil is also a major exporter of pepper, exporting about $79m in 2005 which amounted to just under half of its total spice exports. Brazil is also emerging as a large exporter of other types of spices, including varieties from the capsicum family which currently account for $46m of its total spice exports. Brazil has enjoyed growth rates of over 38% per annum since 2000. While the majority of Brazilian spices were sold in the US markets in the past, the country has recently started to target the German market. Peru has phenomenal success with its spice exports, which have grown from $7m in 2000 to $145m in 2005. The country's success is mainly due to the growth of paprika and other capsicum spices, which for the most part go to Spain and the US. Spices from the capsicum family exported from Peru have grown by over 80% per annum for the last five years, catapulting Peru from an insignificant spice exporter to the number six position.

India's Spices Export - Volumes and Values

Spices exports have registered substantial growth during the last one decade. It has increased from 203398 tones valued MLN US $ 241million in 1995-96 to 350363 tones valued MLN US $ 593 million in 2005-06, registering an annual average growth rate of 9.4% in value terms. During the year 2006-

07, the spices export from India has registered an all time high both in terms of quantity and value. In 2006-07 the export of spices from India has been 373,750 tones valued MLN US $ 793 million registering an increase of 34% in value over 2005-06. India commands a formidable position in the World Spice Trade with 47% share in Volume and 40% in Value.

Trend in India's Spice Exports

Table 5. Export of Spice Crops from India

Item - Wise Export of Spices from India									
Item	2004-05			2005-06			2006-07 (E)		
	Qty	Value		Qty	Value		Qty	Value	
	(MT)	(Rs. Crs.)	(MLN. US$)	(MT)	(Rs. Crs.)	(MLN. US$)	(MT)	(Rs. Crs.)	(MLN. US$)
Pepper	14,148	121071	27.14	17,363	150.95	34.06	28,750	306.20	67.90
Cardamom (S)	642	23.62	5.27	863	26.82	6.05	650	22.36	4.96
Cardamon (L)	954	11.30	2.52	1,046	11.55	2.61	1,500	16.95	3.76
Chilli	138,073	499.03	111.28	113,174	403.01	90.93	148,500	807.75	179.13
Ginger	13,890	59.86	13.35	9,411	72.96	9.69	7,500	39.75	8.81
Turmeric	43,097	156.25	34.84	46,405	152.86	34.49	51,500	164.80	36.55
Corainder	33,582	82.09	18.31	23,756	67.71	15.28	20,500	74.62	16.55
Cumin	15,767	115.29	25.71	12,879	98.19	22.16	26,000	201.50	44.68
Celery	4,297	14.50	3.23	4,165	15.01	3.39	3,550	13.21	2.93
Fennel	7,590	27.50	6.13	5,725	27.82	6.28	3,575	23.80	5.28
Fenugreek	14,635	27.48	6.13	15,525	34.03	7.68	8,500	26.99	5.98
Other Seeds (1)	16,576	40.19	8.96	12,670	33.22	7.50	8,000	22.40	4.97
Garlic	2,929	7.36	1.64	34,688	47.98	10.83	11,500	21.28	4.72
Tamarind	5,944	18.34	4.09	14,101	30.78	6.95	10,200	30.00	6.65
Nutmeg & Mace	1,260	23.24	5.18	1,530	31.17	7.03	2,100	42.74	9.48
Vanilla	43	28.76	6.41	72	12.27	2.77	125	19.96	4.43
Other Spices	9,707	55.54	12.39	7,033	44.15	9.96	9,300	42.80	9.49
Curry Powder	8,415	66.97	14.93	9,340	78.38	17.69	9,500	86.93	19.28
Mint Products (3)	11,143	502.19	111.99	14,544	813.21	183.49	16,250	1100.95	244.15
Oils & Oleoresins	5,831	469.31	104.65	6,074	505.57	114.08	6,250	510.79	113.27
Total	348,524	2350.52	524.15	350,363	2627.62	592.90	373,750	3575.75	792.95

(1) Include Bishop Weed (Ajwanseed). Dill Seed. Poppy Seed. Aniseed, Mustard Etc.
(2) Include Asafoetida, Cinnamon, Cassia, Cambodge, Saffron, Spices (Nes) Etc.
(3) Include Menthol, Menthol Crystals and Mint Oils.

Source : DGCI & S., Calcutta / Shipping Bills / Exporters Returns.

India's Share in World Trade of Spices - 2006

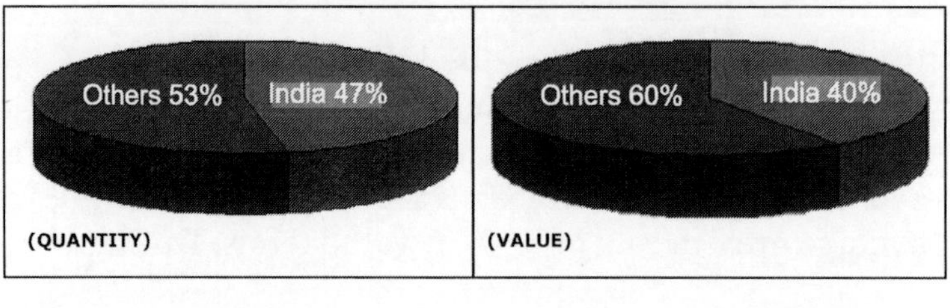

Constraints in Spices Production

The problem in successful production varies with location and crop. In general following are some of the constraints in production.

- In adequate supply of quality seed / planting materials of improved varieties
- Lack of appropriate inputs other than seed and their timely availability
- Climate and weather variations during production
- Pests and diseases
- Lesser fertilizer inputs, low fertilizer-use efficiency and micronutrient deficiency
- Market fluctuations
- Poor post harvest handling, processing and inadequate storage

Strategies to Improve Productivity of Spices

Though India has been acclaimed as the land of spices, Indian spices industry has been facing a stiff competition from other spices producing countries. There has been an alarming growth in the consumption pattern of spices in India itself. Indian spices are known for their intrinsic quality, flavour and aroma. The strong domestic market is a great booster for augmenting production of spices. As the international market becomes increasingly competitive, it is imperative to enhance the production and bring down cost of cultivation to make the Indian Spices globally Competitive.

The research and development bases of spices are very strong in India. A number of research organizations, viz. Indian Institute of Spices Research (IISR), Calicut, Kerala, National Research Centre for Seed Spices, Ajmer, Rajasthan, Spices Board, Indian Cardamom Research Institute, (ICRI), Myladumpara, Kerala, Central Food Technological Research Laboratory, Mysore, Karnataka, United Planters Association of South India (UPASI) and a number of State Agricultural Universities (SAUs) in India are playing an important role in evolving varieties, agrotechniques, plant protection measures, value additions and transfer of technologies on spices.

The Spices Board under Ministry of Commerce, Government of India with its head quarters at Cochin, Directorate of Spices and Arecanut, Calicut, Kerala and a number of Horticulture and Agriculture departments and Non Government Organisations (NGOs) have been playing pivotal role in development, promotion of spices and export.

Indian spices are popularly known for their flavour (Table 6) both in the domestic and international market. They are not only indispensable for the dietary pattern but are also known for their medicinal and curative effects since time immemorial (Table 7).

Table 6. Important Flavour Compounds in Spice Crops

Spice	Flavour compounds
All Spice	Eugenol, b-caryophyllene
Black Pepper	Piperine, S-3-Carene, b-caryophyllene
Cardamom	a-terpinyl acetate, 1-8-cinole, linalool
Cinnamon, Cassia	Cinnamaldehyde, Eugenol
Clove	Eugenol, Eugenol acetate
Ginger	Gingerol, Shagad, Nenal, geranial
Mace	a-pinene, Sabinene, 1-terpenin-4-ol
Nutmeg	Safrimine, a-pinene, myristiein
Turmeric	Turmerene, Zingeberene, 1,8-cineole
Vanilla	Vanilla, P-OH-benzyl methyl ether

Table 7. Medicinal Properties of Important Spices

S. No.	Spices	Part used	Medicinal properties	Uses
1	Black pepper (*Piper nigrum* Linn.)	Fruits	Carminative, antipyretic, diuretic, antihelminthic, anti-inflammatory, and antiepileptic	Cough, asthma, cholera, colic, diarrhea, fever, gas, gastric ailments, hemorrhoids, indigestion, chronic rheumatism, obesity, plegm, sinus congestion, skin diseases, sore throat
2	Cardamom (*Elettaria cardamomum* Maton)	Seeds	Antidepressive, carminative, appetizer, diuretic	Asthma, belching, biliousness, bronchitis, colds, cough, diarrhea, headache, hoarseness, indigestion, involuntary urination, nausea, respiratory disorders, stomach complaints to stimulate the mind vomiting
3	Ginger (*Zingiber officinale* Rosc.)	Rhizome	Carminative, anti nauseant, diurectic, antiflatulence, antihistaminic, aphrodisiac and cholesterol lowering	Atherosclerosis, heart disease, chemotherapy support, migraine headaches, morning sickness, motion sickness, nausea and vomiting following surgery, rheumatoid arthritis, bleaching,

Contd...

Table 7. Contd...

				Laryngitis, vomiting, constipation, incontinence, flatulence, colic, spasms, fever, eye diseases, asthma, colds, cough
4	Turmeric (*Syzygium aromaticum* Linn.)	Rhizome	Carminative, antibiotic, antiflatulence, antiseptic, and anti inflammatory	Anemia, atherosclerosis, bursitis, chron's disease, diabetes, edema, hemorrhoids, hepatitis, hysteria, indigestion, inflammation, inflammatory bowel syndrome, jaundice, psoriasis, rheumatoid arthritis, skin disorders, urinary diseases, wound and bruise healing
5	Clove (*Syzygium aromaticum* Linn.)	Flower buds		Muscular cramps, cholera, nausea, gum ailments, teeth ailments, headache, heaviness in head due to cough and cold, toothache, throat irritation due to coughing, bronchial afflictions
6	Nutmeg (*Myristica fragrans* Houtt)	Seed		Colicky pain and diarrhea, dehydration due to diarrhea particularly in cholera, diarrhea, dysentery, eczema, ringworm, pimples
7	Cinnamon (*Cinnomomum zeylanicum* Blume)	Bark		Diarrhoea, bad breath, loss of taste sensitivity in the tongue, hoarseness, pharyngitis and sore throat, headache, caused by exposure to cold air, cough, acne, blackheads and pimples, to improve the complexion, to improve memory, sleeplessness

Spices are grown in 3,220,000 ha with a production of 37,65,000 tones of which 2,64,106.5 lakh tonnes valued Rs. 20,8671.02 lakhs were exported during 2002-03. To meet the internal consumption and international demand, an annual growth rate of 8-10 per cent is envisaged. During the last one decade the international trade in spices has grown up substantially. An estimated 500,000 tonnes of spices and herbs valued at 15000 million now traded globally every year. An impressive 46 per cent of this supply comes from India. India's exports of spice extracts have been spectacular growth attaining over 50 per cent of the global market within a short span.

India is suitable for production of tropical black pepper to temperate saffron. International standard organisation (ISO) recognized 109 plants species as spices and they belong to 31 families. There is a need to increase the total production per unit area of all the spices and also to take up value addition on top most priority to augment income of farmers and to make Indian spices globally competitive. Value added items exported from India include spice oils and oleoresins and spice powders including curry powder. India enjoyed global monopoly in the export of spice oils and oleoresins.

The year 2004-05 for the first time export volumes grew by a very healthy 32 per cent in terms of volumes and crossed the magic figure of 3,00,000 tones and reached 3,35,488 tones. Exports values reached US $ 490.60 million, increasing by about 18 per cent compared to the previous year. Per unit export realizations continued to be under severe pressure throughout 2004-05 for many items. Cardamom (Small and Large), coriander, fennel, garlic and vanilla were the worst affected commodities. Only mint products and oils and oleoresins were generally free from the general depression in per unit values.

A significant contributor to the increased volumes of 04-05 was the spectacular performance of chilli exports at about 1,35,000 metric tones valued at Rs.499.00 crores against the export of 86,575 metric tones during the previous year. The decline of pepper as a major export item continued due to severe price competition from low cost origins with poor domestic demand and the continuance of huge carry over stocks with the producing countries. The export of value added pepper products continued to grow consistently. Value added pepper exports fetched Rs.133.52 crores during 2004-05 compared to Rs.129.00 crores in 2003-04.

Quality Certification

Spices Board has formulated and implemented a three tier quality certification programme conforming to HACCP. Award of Spice House Certificate for good manufacturing practices, award of Logo for quality of the product and accreditation under ISO 9000 for international acceptance are the three certification systems adopted by the Board. Yet another area of activity centered upon by the Board is Value Addition. India can now boast as the monopoly supplier of spice oils and oleoresins world over. In the case of curry powders, spice powders, spice mixtures and spices in consumer packs, India is in a formidable position. The consistent effort of the Board during the last one decade has improved the share of the value added products in the export basket to more than 60%.

With the support of the Spices Board, exporters have established adequate infrastructure for improving quality on a sustained basis. Quality improvement and technological upgradation are taken up by exporters as an on-going programme. These developments are in tune with the changing levels of market acceptance. Other areas focused upon by the Board are export promotion in identified markets, interaction with policy makers in the importing countries, development of new end uses, farm level training for farmers etc.

BLACK PEPPER

Among spices, black pepper is the king of spices. It has extensive culinary uses for flavouring and preserving processed foods and is important medicinally. Of the total spices traded internationally, pepper accounts for about 34% by value. South West India is the traditional home of this important spice, particularly the Western coastal regions of South Peninsular India (the Malabar Coast).

Black pepper was the first oriental spice to be introduced into the Western world, and was well known among the Romans and Greeks. In the middle ages, pepper assumed great importance in Europe. Its use resulted in revolutionary changes in Western cooking: together with other spices, pepper helped to improve flavour and preservation of food became easier. Pepper was also used in medicine, as a carminative and febrifuge, for aiding in digestion, and in curing the common cold. Black pepper, the most widely traded spice of the world is the dried mature berries of the species *Piper nigrum.*

Soil and Climate

Pepper is essentially a crop of humid tropics and requires a heavy, well distributed rainfall and high temperature. It is successfully cultivated in areas between 20° North and South latitude at elevations ranging from sea level to 1500 m amsl but lower altitudes are preferable. A well distributed rainfall of 125-150 cm is considered ideal for pepper. It can also be grown in areas receiving less rainfall provided distribution of rainfall is uniform. Prolonged drought conditions are detrimental to the crop. The crop tolerates temperature between 10°C and 40°C, the ideal being around 25°-35°C.

Soil the best suited for pepper is a friable loam with high humus content and good drainage. It is extensively grown in naturally well drained red laterite soils and alluvial soils rich in humus. It thrives in a wide range of soils with pH ranging from 4.5-6.6. In soils having pH less than 5, lime is to be applied to nuetralize aluminium toxicity Whatever may be the soil, it should have good drainage and water holding capacity, friable in structure with low acidity and sufficient nutrient reserves.

Nursery Management

Selection of Mother Plants

Cultivate only varieties which are proven to be highly productive. Select mother plant which gives regularly high yields and possesses other desirable attributes like vigorous growth, maximum number of spikes/ unit area, long spikes, close setting of berries and disease tolerance. Selected mother plants should be in the age group of five to twelve years. Mark and label selected mother plants in October-November.

Raising Rooted Cuttings

Pepper is propagated vegetatively from cuttings. Select runner shoots produced at the base of mother plants. Keep the vines coiled on to wooden pegs fixed near base of vine to prevent them from striking roots in the soil. Separate the runner shoots from vines in February-March. The middle one-third portion of runner shoots is preferred for planting. Very tender and too hard portions of the shoots are to be avoided.

The shoots are cut into pieces of two to three nodes in each. Leaves, if any, are to be clipped off leaving a small portion of the petioles on the stem. Dipping the lower cut end (up to 2 cm) of the cuttings in 3 - Indol Butyric Acid (1000 ppm) for 45 seconds will substantially increase root formation and development. Plant the treated cuttings in nursery beds or preferably in polythene bags or baskets filled with potting mixture. The potting mixture is prepared by mixing two parts of fertile top soil, one part of river sand and one part of well rotten cattle manure. In commercial nurseries, it is advisable to solarize the potting mixture before it is used for filling bags. For solarization, the potting mixture is spread as raised beds of 30 cm height, one meter width and convenient length in open area. Then the beds are thoroughly irrigated and covered by thick polythene sheets. The edges of polythene sheets should be sealed with soil. This has to be kept as such for about one month. During solarization, the temperature of potting mixture will go to 57°C and up, this helps to destroy all the microbes. When polythene bags are used sufficient number of holes (16-20) may be provided at the base to ensure good drainage. The cuttings should be planted at least one node deep in the soil. The cuttings after planting should be kept under good shade. In large nurseries, sheds are to be constructed for this purpose. The cuttings are to be well protected from direct sunlight and frequent watering is recommended in the nursery to maintain a humid and cool atmosphere around the cuttings. Two or three watering per day is sufficient. Heavy waterings which makes the soil slushy and causes water logging is to be avoided.

Rapid Multiplication Technique

An efficient technique (commonly called as bamboo method) was developed in Sri Lanka and is increasingly popular in India (Bavappa and Gurusinghe, 1978). In this method, a trench of 60 cm deep and 40 cm wide, having convenient length is made. The trench is filled with rooting medium (preferably forest soil, sand, farmyard manure mixture in the ratio 1:1:1). Split halves of bamboo with septa or split halves of PVC pipes having 1.25-1.5 m length and 8-10 cm diameter provided with plastic septa are fixed at 45° angle on a strong support. The bamboo/PVC pipes can be arranged touching one another. Rooted cuttings, one each/ bamboo, are planted. The lower portions of the bamboo or PVC pipe splits, are filled with a rooting medium (preferably weathered coir dust-farm yard manure mixture, 1:1) and the growing vine is tied to bamboo or PVC split pipes in such a way as to keep nodes pressed to rooting medium. The tying could be done with degradable jute or other vegetable fibres. The vines are irrigated daily. As the vines grow up, filling up the bamboo with rooting medium is to be continued regularly. For rapid growth, the following nutrient solution may be applied every fortnight, urea (1.0 kg), super phosphate (0.75 kg), muriate of potash (0.5 kg), magnesium sulphate (0.25 kg) in 250 litres of water. The solution is applied at the rate of 0.25 litre/vine two or three times, a month.

When the vines reach the top (the initially planted vine takes three to four months for this), the terminal bud is nipped off and the stem is crushed at about three nodes above the base, to activate the axillary buds. After about 10 days, each vine is cut at the crushed point and removed from rooting medium and each node is separated. The nodal cutting with bunch of roots intact is planted in polybags filled with potting mixture. Care should be taken to keep the axil above the soil. The polybags should be kept in a cool, humid place, or should be covered with thin polythene (200 guage) sheet to retain high humidity. The buds start developing in three weeks and then the polybags can be shifted to partial shade. The stumps regenerate giving out one or two shoots each and they also can be trained as in earlier case. Thus, continuous regeneration and harvesting of rooted cuttings are possible. About four harvests can be taken in an year, giving a multiple rate of about 1:40 in a year.

Pit Method of Black Pepper Propagation

A new technique for propagating black pepper from single nodes of field grown vines is developed at Indian Institute of Spices Research, Calicut. The technique is simple and suited to small and marginal farmers.

Cost estimates for a multiplication unit with an annual capacity of 20000 units. [Bamboo method].

S.No.	Items	Cost (Rs)
I. Non-recurring expenditure		
1.	Cleared, leveled nursery area with good drainage	
2.	Semi-permanent nursery structure of size 24 x 6m with GI pipe pillar and MS pipe trusses and purling	87000
3.	Masonry charges including cost of cement	12000
	Total	**99000**

II. Recurring Expenditure (once in 3 years)

1	Bamboo 200 numbers @ Rs.54/pc	10800
2	Arranging bamboo splits (10 mandays)@Rs.120/day	1200
3	Mother vines @Rs.6.00 for 600 nos	3600
4	FYM at Rs.20/cft (260 cft)	5200
5	Forest soil @6/cft (650 cft)	3900
6	Sand @Rs.9/cft) (260 cft)	2340
7	Fumigation of potting mixture	1300
8	Coir dust (60 cft)	120
9	Preparation of rooting medium 240cft (6man days)	720
10	Preparation of potting mixture 982 cft (15 man days)	1800
	Total	**30980**

III. Recurring Expenditure (every 3 year)

1.Fertilizers	
a. Urea 29kg@Rs.4.65/kg	134.85
b. Super phosphate 29kg @Rs.3.10/kg	89.9
c. Muriate of Potash 15kg @Rs.4.35/kg	65.25
d. Magnesium sulphate 7kg @Rs.3.50/kg	24.5
e. Application of cow dung slurry	100

2. Plant protection chemicals		
	a. Ridomil 1 kg @ Rs.1300/kg	1300
	b. Phorate 22 kg @ Rs.50/kg	1050
	c. Quinalphose 1lit @ Rs.320	320
	d. Copper Oxy chloride 10 kg @ Rs.14.61/kg	610
3	Charges for application of nutrient solution and pesticides/fungicides (15mandays) @ Rs.130/day	1950
4	Labour charges for tying vines, irrigation and maintenance	36000
5	Cost of temporary shed (1000 capacity 12m x 6m)	8500
6	Cost of shade net	8000
	Total	**58145**
	Annuity value @14%	32736
	Total cost of production	90881
	Cost of production/cutting (Rs)	5.00

Source: Spices Board of India.

A pit of 2m x 1m x 0.5m is to be prepared under a cool shaded area in the farm. Single nodes with approximately 8-10 cm length and their leaf intact, taken from runner shoots of field grown vines are to be planted in polythene bags (25x15cm, 200 gauge) filled with a mixture of sand, soil, coir dust and cow dung in equal proportions. The single nodes are to be planted in polythene bags in such way so as their leaf axil will rest above the potting mixture. The poly bags with the planted single nodes should be arranged in the pit. In a pit of above size, approximately 150 bags can be kept. After keeping the bags in the pit, the pit should be covered with a polythene sheet. This sheet may be secured in position by placing stones, weights etc on the corners. The cutting should be watered at least five times a day with a rose can. Every time immediately after watering, the pit should be covered with polythene sheet. It is advisable to drench cuttings two-three times with copper oxychloride (2g/litre).After 2-3 weeks of planting, the cuttings start producing roots. These roots will be visible through polythene bags. After the initiation of rooting, watering should be reduced to 3-4 times a day. After about one month, healthy shoots start emerging from leaf axil. At this stage, it is advisable to keep the cuttings inside the pit that they will not suffer from any 'shock' when they are taken out the pit.

The cuttings can be taken out of pit after two months of planting, kept in a shaded place and watered twice a day. These cuttings will be ready for field planting after about another 2 ½ months. By this method, 80-85% success can be obtained. Foliar application of nutrient solution enhances growth of cuttings.

Advantages of this method are that it is simple, cheap and quick as cuttings are ready to plant in about 4 months compared to six months in conventional method, and leads to more efficient utilization of runner shoots as single node cuttings are used instead four node cuttings in conventional method.

Bush Pepper

By planting lateral branches (fruiting branches), pepper could be grown as a bush. This growth pattern suits cultivation in pots. One year old lateral branches with three to five nodes are to be planted in nursery during March-April. The cut ends are treated with IBA (1000 ppm) solution for 45 seconds. Rooting percentage of laterals is less than 20. Well-rooted cuttings could be planted @ 3-5/ pit or pot. Potted plants are to be kept preferably under partial shade (Nybe *et al.*, 1997).

Bush pepper is an ideal under-storey crop in Coconut plantations and agro forestry systems and also suitable for terrace gardens in urban and rural areas since it can be grown in pots. Application of NPK @ 1.0, 0.5, 2.0 g/pot (10kg soil) at bi-monthly intervals, would increase the yield 239%. The chemical source of N can be substituted with neem cake (30g) or groundnut cake (14g).

Micropropagation

A viable protocol is developed for *in vitro* mass multiplication of black pepper (KAU, 2001). Techniques for direct and indirect organogenesis are standardized. Shoot tip and eye bud explants are the best explant for in vitro culture. Murashige Skoog (MS) medium containing NAA (0.5 ppm) and BA (3 ppm) resulted in speedier establishment of cultures.

Better rooting is observed in Knudson medium containing NAA(5 ppm) The plantlets establish well when they are kept in distilled water for eight hours followed by dipping in $1/10^{th}$ strong liquid MS medium for eight hours before they are planted outdoor. A mixture of 1:1 v/v vermiculite and sand was good for establishment of plantlets when sterilized with Emisan (0.1%). TC derived plants were also evaluated in farmers' fields. No significant difference was noted among the TC regenerants when compared with the conventional propagules. The bearing habit was stable and five-year-old plants now in the field have started giving steady yield.

Establishment of Pepper Plantation

Site Selection

While selecting the site for pepper plantation, avoid slopes facing south, so that the vines are not subjected to the scorching effect of southern sun during summer.

Water Table

Low lying areas with water stagnation are not suitable for pepper cultivation as water logging invites foot rot disease incidences. Land may be flat or sloppy, and in case of flat lands water table must remain 1m below the ground level. In top water tabled situations, raised beds are prepared for pepper planting. Adequate drainage to reduce water stagnation is to be always looked into. In sloppy lands, there must be provisions to check water and soil erosion, and water stagnation around base of plants should never be allowed.

Standards

Black pepper needs a reliable support tree. Black pepper vines are trailed on supports, which are called standards. Being a climber, it requires a support for its growth, development and yield. The most popular live standard is *Erythrina indica*. Other common standards are *Garuga pinnata* (Indian coral tree), *Gravillea robusta* (Silver Oak), *Glyricidia sepium, Leucaena leucocephala, Ailanthus sp.* or other similar trees (pure crop). Coconut, arecanut, jack, mango and subabul are the other supports used in homestead gardens.

Cut *Erythrina indica* and *Garuga pinnata* stems/stem cuttings in March-April and stack them in shade. The stacked stems start sprouting in May. After the first rain in May-June, the sprouted stems should be planted at the edge of pits [described below] dug for planting pepper vines. Whenever, *Erythrina indica* is used as standard, application of carbofuran (30g) may be done once in a year to control nematodes and root grubs.

Planting

Prepare a pit of size 50 x 50 x 50 cm before onset of monsoon and fill it up after onset of monsoon with top soil + FYM [@10 kg] + Neem cake [1 kg] + bone meal and Rock phosphate [70 g]. Wherever possible, a large pit filled with alternate layers of coconut husk and the above mixture is preferable. This helps in conserving soil moisture and help the young plants to survive hot summer. Planting can be done within 15 days of filling of pits.

Plant the cuttings in northern side of the standard. Two to three rooted cuttings of pepper are to be planted individually in pits on the Northern side of each standard. In case of cuttings, which have not developed roots, about 4-5 cuttings with 4-5 nodes are to be planted. In gardens with grownup shade trees, up to 10 cuttings may be planted all around the living support. To ensure proper rooting at least, two nodes of cuttings should be below the soil. The cuttings should not be positioned vertically, but semi-horizontally at an angle of 30 – 40° towards the standard. Artificial shade must be provided for newly planted cuttings.

Spacing

Optimum spacing is 3.0 x 3.0m in the plains. Such spacing accommodates 1100 standards/ha. The optimum spacing for planting pepper vines in sloppy land is 3 x 2m. A closer spacing of 2 x 2 can be adopted when non living standards are used. Water stagnation around the vines during rainy season should be avoided. The soil around the cutting should be formed into a small mound (5cm), slopping outward and away from vines.

Under Planting

After regular bearing for about 20 years, pepper vines of most varieties start declining in yield. The old and senile vines can be removed 3-5 years after under planting depending upon growth of young vine.

Mixed Cropping

For planting pepper on adult coconut and arecanut or on shade trees (silver oak) in coffee/tea plantations, the rooted pepper cuttings are planted one meter or half meter, respectively, away from trunk of palms/trees. Long vines at the rate of two-three/ standard are taken through a shallow channel of about 5 cm depth, filled with potting mixture and then taken to standard to a length of 20-25 cm from base and tied. This method helps in quicker growth and establishment of vine under mixed cropping system in coffee and tea garden.

Cultural Practices

Training the Vines

As the cuttings grow, the shoots are to be tied to the standards regularly.

Irrigation

With receipt of monsoon in May-June, primary stem cuttings/seedlings of support trees are planted. Rooted cuttings of pepper are also planted during the opportune time. The young vines are protected from hot sun during summer by providing artificial shade. Adequate mulch with green leaf or organic matter need to be applied around the base of vines, without disturbing the root system.

Black pepper trailed on standards should be irrigated with 7 litres of water/ day through drip between October to March. If drip system are not in use, application of 100 litres of water/ vine once in a week from October to March is necessary.

In India, black pepper is grown in lateritic soil, which is porous with poor water holding capacity. In view of the drought spell during summer months, protective irrigation is needed during initial two years. Studies carried out at Indian Institute of Spices Research Calicut, showed that burrying earthern pots of five liter capacity in the rooting region of pepper vines and refilling with water once in 10-15 days during the initial two years, about 90-95% of pepper vines survived. Water trickles and drips down slowly from the pot into root zone. The wetting zone was 30 cm. Evaporational loss of water was brought down. Further weed growth in the basin was minimized. Irrigation studies carried out at Pepper Research Station, Panniyur showed that irrigating pepper vines of Panniyur-1 variety at 1W/CPE ratio of 0.25 from November to end of March and withholding irrigation thereafter till monsoon breaks, increases pepper yield by about 90%. The depth of irrigation was 10 mm (100 litres of water/ irrigation at an interval of 8-10 days). The water is to be applied in basins taken around the vine at a radius of 75 cm. Mulching the basin of pepper vines with green leaves/coir dust @ 5 kg/ vine helps to tide over the summer drought. Keeping shade up to April end minimizes drought effect in black pepper and shade regulation help the pepper to receive solar radiation and initiation of spike.

Spraying of 2% lime solution on black pepper in summer months reduces leaf scorching in exposed areas.

In Karnataka, black pepper is trained on shade trees in coffee plantations. Due to delayed rains, spike shedding due to production of only female flower and anthracnose incidence are extensively noticed in coffee plantations. It can be managed by irrigating black pepper vines (basin) from April onwards @ 50 litres/ vine/ week upto May or good rains and shade regulation in April which results in good spike initiation and yield. It also reduces anthracnose incidence due to early maturity of leaf and spike.

To provide irrigation in summer, coffee / black pepper planters have to conserve and store rainwater in farm ponds and check dams so that they can be reutilized for sustainable yield.

Drainage and Sanitation

Pepper garden should be well drained. Water logging is harmful for pepper plant and high humidity promotes the incidence of foot rot disease.

Tillage

From fourth year onwards, usually two diggings are to be given, one during May-June and the other, towards end of southwest monsoon in October-November.

Provision of Shade to Young Vines

During the first one or two years, the young vines should be provided artificial shade by covering them with palm fronds or similar materials, loosely bound over them to protect them from the hot sun. These artificial shades should be removed at the onset of rainy season.

Shade regulation: Lopping of standards should be done carefully from fourth year onwards, not only to regulate height of the standards, but also to allow enough sunshine passes through to the plants underneath. Excessive shading during the flowering and fruiting encourages pest infestation.

Frequency of Shade Regulation

Lopping the branches of support tree/shade trees should be done at least twice a year, once before onset of monsoon (March-April) and the second one after the monsoon is over (July-August).

Mulching: Adequate mulch with green leaf @ 10kg or organic matter should be given towards end of North East monsoon by November-December and before onset of summer. Mulching conserves soil moisture in the basins and prevents sun scorching of young vines during summer. Any locally available material like dry leaves, dried grass, arecanut husk and saw dust could be used as mulch.

Weed Control

Two to three rounds of hand weeding in a year generally control weeds. Weed growth being intense during rainy months, a careful watch on flowering habit of weeds and pulling out of weeds before seed spread of weeds enable easy weed control. Weeded materials can be mulched around base of vines and excess weed material, if available can be spread all over the land in between the rows, of vines. During weeding, plant base should not be disturbed causing injury to the root system. During rainy season, weeds pulled out can be heaped and spread out once dried. Slash weeding is a cost-effective method and to keep a cover always over the soil. Chemical weed control is not advocated, which is harmful to micro flora and fauna although it is a labour saving method.

Soil and Water Conservation

Water, soil and gully erosion can occur in pepper fields depending upon slope of land and intensity of rains. Stone pitched 'kayyalas' built along the slope of land at regular spacing checks water and soil erosion. Terraces having 1m.width are also useful for checking erosion and to provide footpath through rows of pepper vines along the sloppy lands. Spread of mulch is a cost-effective method for controlling erosion as well. Digging the soil is to be avoided. Weeds pulled out or slash weeded can be spread and used as mulch to have erosion control. Growing cover crops like *Calapagonium* sp. in between rows of pepper vines will also check soil erosion effectively especially gullies not to be born with erosion. Cover crops also enrich soil fertility.

Nutrition Management

Manuring

At least 10 kg organic manure in the form of FYM/ green leaves/ compost/ well rotten cattle/ poultry manure should be given/ vine/ annum during May i.e. at the onset of South West monsoon. These may be applied in shallow basins taken around the plants, to be covered with a thin layer of soil after application of manure. Application of lime at the rate of 600g/ vine during April-May in alternate years should be done to increase soil pH, availability of nutrients and to reduce aluminium toxicity in soil.

Fertilizer Application

Adult Vines [3-4] Years

Fertilizer should be applied 10-15 days after pruning of living supports. The rate, composition and interval of fertilizer application should be suited to the vine requirement.

Recommended nutrient dosage for pepper (3 years and above) is:

N : P_2O_5 : K_2O g/vine/year

50:50:150 (general recommendation)

50:50:200 (for Panniyur and similar areas)

140:55:275 (for Kozhikode and similar areas)

Apply 1/3 dose for one year old plants and ½ dose for two year old plants.

The fertilizers may be applied in two split doses, the first in May-June with receipt of a few soaking rains and second in August-September.

Young Vines

1. During first year (on the year of planting) - 1/3rd of dosage recommended for adult vines should be applied during September.
2. During second year, two thirds of the dosage recommended for adult vines should be applied in two equal instalments, one during May-June, and the other during September-October.

Points to be Noted in Fertilizer Application

1. Apply fertilizers in two split doses
2. Fertilizer should be applied only when there is sufficient soil moisture
3. Fertilizers should be applied at a distance of about 30cm all around the vine and covered with a layer of soil
4. Care should be taken to avoid direct contact of fertilizers with the roots of pepper vine.
5. If soil is highly acidic, 500g lime/ vine is to be applied in alternate years.

Cost of Cultivation (Rs.)

Items	I year	II year	III year	IV year	Total
Land preparation	3000	0	0	0	3000
Digging of pits, Filling up of pits	4000	0	0	0	4000
Plant materials	4000	400	0	0	4400
Planting/staking	3000	0	0	0	3000
Pruning/Training	800	1200	1900	2000	5900
Cost of FYM and Fertilizers	1762	2960	3807	3907	12436
Irrigation	300	200	200	200	900
Plant protection measures	900	800	1900	2000	5600
Interculture	2000	2200	3500	3500	11200
Watch and ward	700	700	700	1400	3500
Harvesting	0	0	0	1000	1000
Cleaning, drying, packing	0	0	0	1000	1000
Others/Live hedge/Intercropping	2500	1500	1900	1500	7400
Total (Rounded off)	**23000**	**10000**	**14000**	**16600**	**63600**

Source: Spices Board of India

Harvest and Post-harvest Processing

The crop takes 6-8 months from flowering to harvest. The harvest season extends from November to January in plains and January to March in hills. When one or two berries in the spike turn bright orange or purple, it is time for harvest. However, pepper berries are harvested at different maturity levels depending on intended use or product preparation .Pepper powder is the best made from berries with maximum starch content and therefore has to be fully mature when harvested .The extraction industry prefers berries which contain the highest level of oleoresin and essential oils .High levels of both are usually found when the berries are picked, weeks before full maturity.

The following describes the intended use of pepper berries and the maturity level at which berries should be harvested for ensuring the best results.

Products	Maturity at harvest
Pepper in brine/ Canned pepper	Green and tender (4-5 months)
White pepper	Fully matured (ripened)
Black pepper	Fully mature and near ripe
Dehydrated Green pepper	10-15 days before full maturity
Oil and Oleoresin	15-20 days before maturity
Pepper powder	Fully matured with maximum starch

Harvesting, Processing and Storage

Quality of product depends on quality of raw materials, the methods used in processing and the packaging and marketing practices. Therefore, in order to ensure quality of the product, constant care should be maintained from time of harvesting till the time it reaches the consumer.

Despiking

If the harvested spikes are kept in a bag for 12 to 24 hours or heaped and covered overnight, despiking will be easy. Trampling with legs is a traditionally followed technique for despiking. However, use of mechanical pepper threshers is both hygienic and time saving.

1. Threshing should be done only on clean cement or concrete floors.
2. Never do threshing on floors smeared with cow dung.
3. Use mechanical threshers as far as possible.
4. Wash spikes which have accumulated dirt, dust and similar substances.

It is ideal to grade the green berries using a wire mesh to remove light berries and pin head.

It is the best to use plastic or aluminium vessels to handle green pepper. Bamboo trays smeared with cowdung, old gunny bags etc. are sure sources of contamination.

Blanching

Blanching is done by immersing threshed pepper berries collected in a basket in boiling water for one minute. Blanching gives a waxy coating to pepper and disinfection also takes place simultaneously.

Blanching provides following advantages:

1. The dried pepper gains a waxy coating and uniform black colour
2. Blanching removes dust particles, birds excreta and other extraneous matter from berries.
3. Pepper can be dried in 3-4 days as against 5-6 days required when following the traditional practice.
4. The pepper treated in this manner fetches better price in the market.

Drying

Green pepper contains 80-85% moisture. The sun-dried pepper contains only 10-12% moisture. Sun drying of pepper can be best done by spreading blanched berries, [after draining of water] on bamboo mats smeared with fenugreek, cemented drying yards or on polythene sheets. Pepper put for sun drying has to be turned over periodically to facilitate uniform drying. Generally black pepper recovery is about 1/3rd of green berries.

Environment and Personal Hygiene

Premises used to dry and store pepper should be clean and hygienic and free from dust, cobwebs, undesirable foreign matter bird and mammalian excreta. Drying yard should be preferably covered by nylon or other netting to keep off birds. Entry of rodents to storage area should be prevented.

White Pepper

White pepper is an important commodity with great demand in the world of spices. India, is placed far behind Indonesia, Malaysia and Brazil in white pepper production. Total annual white pepper production in the world is 30,000 to 40,000 tonnes. Indias annual production of white pepper is only lesser than 150 tonnes against the world demand of 155300 tonnes/ annum.

White pepper is preferred to black pepper in Europe, Malaysia and USA, because of the charming colour, and suitability to use in all food preparations, free from many contaminants and no substantial difference in pungent principles with that of black pepper (Gopalam *et al.*, 1991). The organoleptic properties of white pepper have been extensively studied and remain unchanged and chemically white pepper is more or less similar to black pepper. It has a milder, more delicate flavour than black pepper.

White pepper is prepared by removing pericarp of pepper berries, either before or after drying. The different techniques used for white pepper manufacture are

(1) Water Steeping Method

(a) From Ripened Fresh Berries

This is the old indigenous technique. Harvested berries are packed in gunny bags or steeped as such in water tanks or under running water for 7-10 days, stirred two or three times daily. At the end of the time they are heaped on a tarpaulin and rubbed by hand or trampled to remove the outer softened skin. The de skinned berries are then washed and put in a galvanized iron vessel containing a solution of bleaching powder for two days. They are removed, drained and dried in sun, cleaned and sold as white pepper.

(b) From Dried Berries

The dried black pepper berries are steeped in water for 10-15 days, then they are removed, rubbed, washed thoroughly, steeped in bleaching solution for two-three days, drained, dried in the sun and sold as white pepper. The white pepper is more easily prepared from fresh ripe pepper than from dried black pepper. The disadvantage of this method is that it takes more time and retting leads to development of off flavour.

This is an improved method developed by CFTRI, Mysore (Lewis *et al.*, 1969). This process in brief consists of steaming or boiling of mature, ripened green berries for 10-15 minutes. The boiled berries are then passed through a motorized fruit pulping machine to remove the skin. The deskinned berries are then washed and treated with sulfur dioxide or bleaching powder solution after which they are drained and dried in sun artificially. Skin of berries collected separately from the machine goes for steam distillation to collect essential oil. This method produces only inferior quality products and during boiling gelatinization of starch occurs and hence loss of volatile oil.

White pepper is also thoroughly steeped in bleaching solution for two-three days, drained, dried in the sun and sold as white pepper. However, white pepper is more easily prepared from fresh ripe pepper than from dried black pepper. The disadvantage of this method is that it takes more time and retting leads to development of off flavour. White pepper is also prepared by mechanical decortication of dried pepper in decorticating machines. The major problem is too much breakage of berries and lack of uniformity in shape and surface. Another method to produce white pepper powder by mechanical means was tried by Thomas *et al.* (1987). Selective grinding technique was tried. The disadvantage is that only pepper powder can be obtained by this method.

Matured green pepper is buried under damp ground for five days, the skin is degraded by microbes, washed, bleached and dried. Another bacterial method is by using fermentative bacteria to remove skin both from dried and fresh pepper for a period of four days and three days respectively, then passed through a pulping machine to remove skin, washed, bleached and dried (Gopinathan *et al.,* 1991). The disadvantage of this method is that costly fermentation media are required and hence economically not feasible.

A chemical process for white pepper was patented by Josi (1962), which consists of (1) steeping whole dried black pepper with about five times its weight of water for four days (2) treating with 4% sodium hydroxide solution and boiling the mixture (3) removing the skin by agitation (4) washing (5) bleaching with 2.5% hydrogen peroxide solution and (6) drying. The disadvantage is that the process involves usage of high concentration of alkali and has not been tried commercially.

Another chemical method patented by Omanakutty *et al.* (2006) comprises of soaking berries in a dilute solution of alkali, followed by blanching, decortication in a pulper, bleaching and drying. Advantages are that the decortication is made possible within 12-14 hours in black pepper and less than an hour in fresh green pepper. The skin portion can be utilized for oil recovery, thus giving this improved method an edge over other indigenous methods.

Among the various enzymes used for decortication purpose, pectinase was very effective for both dried and fresh pepper. After pretreatment of berries the enzyme will acts on it and decortication is made possible. Since the enzyme is costly, the process is not economically feasible.

Black Pepper

Black pepper is used in different culinary and pharmaceutical preparations for its aroma and pungency. Pepper is harvested at different maturity for following products.

Dark Shining Black Pepper

A good quality black pepper of uniform black colour is obtained by subjecting fresh green pepper berries to blanching. Harvest pepper at uniform maturity. Despike and clean the berries. Take green pepper in a gunny bag or in a perforated vessel and dip in boiling water for one minute. Drain the water and dry it in sunlight. This process called blanching gives the product a shining black colour. This also reduces the drying time and the product will be ready within 3-4 days. This fetches better price compared to normal sun dried pepper.

Sterilized Black Pepper

Spices are known to harbour various types of micro organisms which are of either spoilage nature or of public health significance. The CFTRI, Mysore, Karnataka has developed a method for preparation of sterilized black pepper, without use of chemicals.

Dehydrated Green Pepper

Harvest pepper 10-15 days before normal maturity, boil in water for 10 min. or steam to prevent blackening and dry it in an oven without exposure to sunlight. Exposure to sunlight will bleach the produce. Drying in a microwave oven or draft oven with fan is ideal. While drying, care should be given to avoid touching of berries, which may lead to browning. Dried green pepper will have wrinkled appearance and it will restore its size and taste, when it is put in water.

Pepper in Brine

Four to five months old pepper are preferred for this item. Harvest the immature spikes and clean it thoroughly. Light pepper, pinhead or broken berries are not considered as extraneous matter. Twelve to 14% common salt and citric acid not exceeding 0.6% by mass of the packing medium are used for the preparation. The drained mass of the content shall not be less than 50% (m/m) of the net mass.

Canned Green Pepper

Prepared using fresh green pepper berries of proper maturity. The material should be free from preservatives and artificial flavouring and colouring matter. About 1-2% common salt and 0.6% citric acid by mass of covering medium are used for the preparation. The berries should possess a tender nature, firm but not hard and shall be characteristic of green pepper of proper maturity. Materials shall be packed in cans made of tin or containers made of glass or food grade plastic. Cans shall be either plain or internally lacquered with food grade lacquer and hermetically sealed. Vaccum in the cans can be measured with electronic recording type machine without the can.

Bottled Green Pepper

Fresh pepper spikes (bunches) are preserved in 50 litre capacity HDP containers by using 10% brine containing 2% acetic acid as covering medium.

Dry Green Pepper

This product is made from fresh tender green pepper. The dry green pepper so obtained has fresh spicy flavour and reconstitutes easily in hot water.

Black Pepper and Green Pepper Powder

Dehydrated green pepper and black pepper can be powdered to uniform mesh.

Agmark Grade Specifications of Black Pepper

Grade designations	Trade name	Size, diameter of sieveholes on which retained(mm)	Extraneous matter not exceeding (% w/w)	Light berries not exceeding (% w/w)	Moisture content not exceeding (% w/w)
Tellicherry Garbled Black Pepper					
TGSEB	Special Extra Bold	4.75	0.5	3.0	11.00
TGEB	Extra Bold	4.25	0.5	3.0	11.00
TG	Tellicherry Garbled	4.25 (50% min) 4.00 (50% max)	0.5	3.0	11.00
Garbled Malabar Black Pepper					
MG Grade 1	-	-	0.5	2.0	11.00
MG Grade 2	-	-	0.5	5.0	11.00
UnGarbled Malabar Black Pepper					
MUG Grade 1	-	-	2	7.0	12.00
MUG Grade 2	-	-	2	10.0	12.00
MUG Grade 3L	-	-	3	15.0	12.00
MUG Grade 4L	-	-	4	20.0	12.00

Source: Agmark Grade specifications for Spices, Spices Board, 2001

Whole Black Pepper: Chemical and Physical Specification

Specifications	Suggested limits
ASTA cleanliness specifications	
Whole dead insects, by count	2
Mammalian excreta, by mg/lb	1
Other excreta, by mg/lb	5.0
Mold % by weight	
Insect defiled infested % by weight	
Extraneous % by weight	1.00
FDA DALs	
Insect infested and/or moldy pieces by weight	Ave. of 1%
Mammalian excreta	Ave. of 1 mg lb
Foreign matter pickings and siftings by weight	Ave. of 1%
Volatile oil	2.0% min
Moisture	12.0% max
Ash	5.0% max
Acid-insoluble ash	0.5% max
Average bulk index (mg/100 g)	165

Ground Black Pepper: Chemical and Physical Specifications

Specifications	Suggested limits
FDA DALs (6 sub samples)	
Insect fragments or	Ave. of 475 or more/50 g
Rodent hairs	Ave. of 2 or more/50 g
Volatile oil	1.5% min
Moisture	12.0% max
Total ash	5.0% max
Acid-insoluble ash	0.5% max
Military specifications (EE-S-631J, 1981)	
Volatile oil (ml/100 g)	2.0 min
Moisture	12.0% max
Total ash	7.0% max
Acid-insoluble ash	1.0% max
Crude fibre	12.5% max
Nonvolatile methylene chloride extract	7.5% min
Starch	30.0% max
Granulation	70% min through U.S.S. 30

Nutritional Composition of Black Pepper / 100g

Composition	USDA	ASTA
Water (g)	10.51	8.0
Food energy (kcal)	255	400
Protein (g)	10.95	10.0
Fat (g)	3.26	10.2
Carbohydrates (g)	64.81	66.5
Ash (g)	4.33	4.6
Calcium (g)	0.437	0.4
Phosphorus (mg)	173	160
Sodium (mg)	44	10
Potassium (mg)	1259	1200
Iron (mg)	28.86	17.0
Thiamine (mg)	0.109	0.070
Riboflavin (mg)	0.240	0.210
Niacin (mg)	1.142	0.8
Ascorbic acid (mg)		ND
Vitamin A activity (RE)	19	19

Composition of Foods, Spices and Herbs, USDA Agricultural Handbook, 8.2,1977.
The Nutritional Composition of Spices, ASTA Research Committee, 1977.
ND=Not detected.

Marketing

Demand and Supply Analysis

India continues to be the leading producer, consumer and exporter of black pepper. Apart from India, other pepper producing countries are Indonesia, Brazil, Malaysia, Cambodia, Sri Lanka, China and Vietnam. The global production of pepper fluctuates between 3-3.5 lakh tonnes. Area under pepper in India (1999-2000) was 131775 ha with production of about 58000 tonnes which shows a considerable change in area and production.

India with more than 40% share of the world area under pepper, contributes about 23% of the total production of pepper in the world. Given the high domestic demand due to internal consumption and high FOB prices, export fluctuated between 36,908 tonnes in 1988-89 to 16700 tonnes in 2005-06.There is considerable change in the area during 2004-05, (242441 ha) compared to 231886 ha in 2003-04.

State Wise Area and Production of Pepper

Area under pepper shows a slight change in 2000-01 (213865 ha) with production of 63667 t which slightly increased during 2004-05 (242441 ha) and production (76012.22 t).

State Wise Area and Production of Pepper
(Area in Hectares, Production in Tonnes)

State	2000-01		2001-02		2002-03		2003-04		2004-05	
	Area	Production	Area	Production	Area	Production	Area	Production	Area	Production
Andaman & Nicobar	430	83	450	86	550	116	573	120.23	655	120.22
Karnataka	7249	1760	9697	2219	9899	2259	10694	2356	0	0
Kerala	202133	60929	203956	58240	208607	67358	216440	69015	237669	74980
Tamil Nadu	4053	895	4114	907	4030	871	4179	933	4117	912
Total	213865	63667	218217	61452	223086	70604	231886	72424.23	242441	76012.22

Source: Spices Board of India

Supply of pepper has witnessed a dramatic increase over the last ten years. While prices have fallen over the last three years, the market has absorbed the supply of pepper.

Global Scenario

The global production of pepper fluctuates between 3-3.5 lakh tonnes, with a production of 2.95 lakh tonnes recorded in 2005. Vietnam (90000 tonnes), India (70000 tonnes), Indonesia (35000 tonnes), Brazil (35000 tonnes), Malaysia (19000 tonnes), Sri Lanka (14000 tonnes), Thailand and China are the major producers of pepper in the world. Vietnam's sudden increase in production resulted in the global production, increasing 3-3.5 lakh tonnes from 1.9-2 lakh tonnes in the nineties. Vietnam is the worlds largest producer and exporter of pepper now.

Global Export

The world export of pepper, neither crushed nor ground was 2.41 lakh tonnes in 2005. Vietnam accounted for 37.79%, Indonesia and Brazil contributed 14.15% and 9.65% respectively. India's share was only 5.27%. Due to emergence of Vietnam as a major producer and exporter, Indian exports declined significantly during the current decade. Another factor was the declining price competitiveness of Indian pepper in relation to Vietnam.

World Production of Black Pepper During 2005

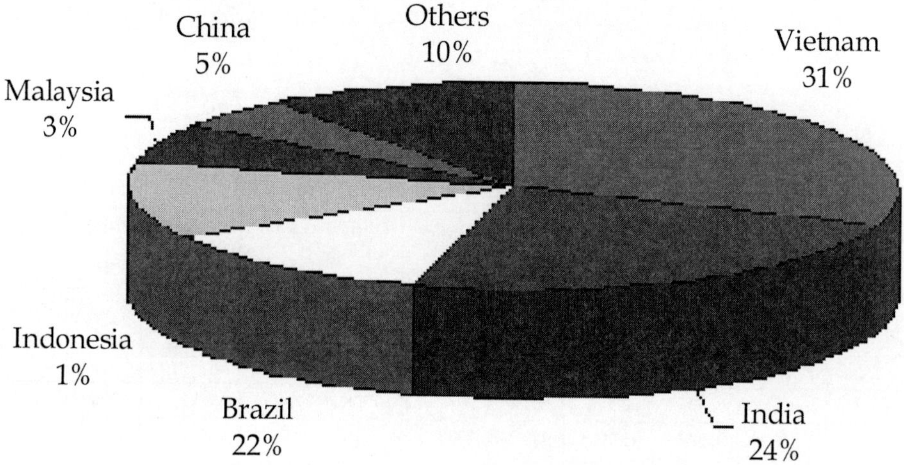

Source: *Pepper News & Market Review*, November 2005, p.8.

The world export of pepper, crushed nor ground was 37,895 MT in 2005. India accounted for 23.13%. Germany and Vietnam contributed 12.85% and 10.02% respectively. Other important exporters were USA (8.39%), Netherlands (7.79%), Brazil (6.20%), Thailand (3.17%), Malaysia (3.16%), Singapore (2.94%) and China (1.79%).

Import Pattern

The world import of pepper was 2.40 lakh tonnes in 2005, of which 24.73% was accounted by USA and 8.34% by Germany. India's share was 7.85% followed by Singapore (5.21%), Netherlands (4.66%), Russian Federation (3.59%), Spain (3.13%), France (3.02%), Japan (2.99%) and Poland (2.13%).

The world import of pepper, crushed or ground was 39.356 MT in 2005, of which 19.04% was accounted by USA and 9.06% by UK. Other important importers were Germany (8.02%), Canada (5.92%), France (4.92%), Japan (4.57%), Netherlands (4.06%), Saudi Arabia (4.01%), Belgium (3.14%) and Russian Federation (1.87%).

India's Export

India's pepper exports declined from 46,942 tonnes in 1996 to 35,114 tonnes in 1999, the year prior to the commencement of Indo-Sri Lankan FTA and emergence of Vietnam as a major producer and exporter. The highest export value was also realized in 1999 at Rs.703.25 crores which plummeted to Rs. 140.5 crores in 2004-05 along with steep decline in quantity from 35,114 tonnes to 15,800. However, there was slight improvement in exports in 2005-06. Export rose to 28,750 tonnes in 2006-07. Which brought an export revenue of Rs. 306.20 crores.

Pepper Exports from India

Year	Values (Rs.Crores)	Quantity (Tonnes)	Unit Price Rs/kg
1996-97	408.09	46,942	86.93
1997-98	481.94	34,877	138.18
1998-99	605.65	32,532	186.17
1999-00	703.25	35,114	200.28
2000-01	310.46	18,463	168.15
2001-02	160.2	16,319	98.17
2002-03	149.81	17,467	85.77
2003-04	142.76	16,635	85.82
2004-05	140.5	15,800	88.92
2005-06	150.95	17363	86.95
2006-07	306.20	28,750	106.50

Source: dgft and spices board

India's Import

India has been importing pepper in small quantities till 2000, ranging between 2000 - 3400 Mt. It almost doubled during 2001-02. Thereafter there was a surge in import from 15480 tonnes in 2002-03 to 18857 tonnes in 2005-06. Vietnam, Sri Lanka and Indonesia were the three major sources of India's import of pepper. The import pattern over the period 2003-04 to 2005-06 shows that approximately half the import is from Vietnam, one third from Sri Lanka and approximately one-eighth from Indonesia.

India's Import of Pepper During 1996-97 to 2005-06

Year	Import		
	Values (Rs.Crores)	Quantity (tonnes)	Unit Price Rs/kg
1996-97	18.42	2,292.35	80.34
1997-98	28.16	2,153.12	130.79
1998-99	59.51	3,365.39	176.82
1999-2000	50.23	3,124.29	160.79
2000-01	60.10	6,016.00	99.90
2001-02	56.39	6,319.16	89.23
2002-03	123.81	15,480.15	79.98
2003-04	100.08	14,583.98	68.63
2004-05	104.67	16,312.64	64.16
2005-06	115.25	18,856.61	61.12

Source: dgft.delhi.nic.in

Kerala's Exports and Imports

Kerala's share in total pepper export from India is around 94% both in terms of volume and value. Export of pepper from Kerala came down from 41332 tonnes in 1999-00 valued Rs. 856.99 crores, declined to 21448 M.T during 2001-02 and to 13358 tonnes in 2004-05. However it rose to 16120 tonnes in 2005-06. As a result, the export value fell from 187.21 crores in 2001-02 to Rs. 114.35 crores in 2004-05, which rose to Rs. 134.78 in 2005-06.

It is significant to note that major exporters were also major importers. Out of the 10 leading exporters in 2005, 9 were major importers who ranked from 1 to 9. While they exported 16120 MTs during 2005-06, they also imported 8297 tonnes through the Cochin port (Cochin port trust).

Export of Pepper from Kerala During 2001-2006

Year	Qty (MT)	Value Rs. Crores
2001-02	21448	187.21
2002-03	15094	139.90
2003-04	15710	134.47
2004-05	13358	114.35
2005-06	16120	134.78

Spices Board of India

Major World Markets

New York, Singapore and Rotterdam are major international trading centres for pepper. The primary international grade and their markers are Lampung at Panjang (Indonesia), Sarawak at Kuching (Malaysia), Vietnam at HCM city (Vietnam). Malabar grade of pepper from India traded at Kochi, Kerala is the premium grade of pepper, ruled above international grades.

Major Indian Markets

Kochi and Sulthan Bathery in Kerala are the major primary markets. Nagpur, Indore and Delhi have recently developed as the major up country markets for pepper.

Market Influencing Factors

- Indian pepper is at a premium against all the international grades. Production and export of pepper from other locations have a profound influence on Indian pepper price too.
- Weather and annual production in a year.
- Year ending stocks and stocks to consumption. Indian pepper arrives in the market in the beginning of the year. However, distress sale is not witnessed in pepper and producers hold back the stock in anticipation of better prices.
- Government policies with regard to imports and exports.

Traders allege large scale imports of pepper from Sri Lanka and re-export from India as a major price depressing factor and government has been requested to take measures to stop the practice.

CARDAMOM (Small)

Cardamom (*Elettaria cardamom* Maton) the "Queen of Spices", enjoys unique position in the international spices market, as one of the most sought after spices. From time immemorial, India is known as the home of cardamom. Cardamom is indigenous to the evergreen forests of Western Ghats in South India. It is the dried fruit of the tall perennial herbaceous plant, belonging to family Zingiberaceae.

E. cardamomum exhibits considerable variation under cultivation and the naming of commercial types after the places of production has led to confusion regarding identity of varieties. Two varieties based on size of fruits are recognized. They are: (1) *E.cardamomum* var. Major Thw. comprising the 'wild' indigenous cardamom of Ceylon (Sri Lanka) or Greater oblong cardamom or long cardamom and (2) *E.cardamomum* var. minor comprising all the cultivated races, particularly those included under the names Malabar and Mysore cardamoms. Var. major is the more primitive variety from which the cultivated var. minor is derived. All the varieties and races are inter fertile and the observed variations are probably due to natural crossing.

Climate and Soil

The natural habitat of cardamom is the evergreen forests of Western Ghats. It grows within an altitude between 600 and 1200 m above MSL. Considerable variations both in total rainfall pattern and its distribution are noticed in cardamom tracts. In most of the cardamom areas, the annual rain fall is between 1500 to 4000 mm. and the temperature ranges 10 to 35°C. Cardamom generally grows well in forest loamy soils. These soils are generally acidic in nature, with pH from 4.2 to 6.8 Analytical data of soil samples in cardamom growing tracts indicate that they are high in organic matter and nitrogen, low to medium in available phosphorous and medium to high in available potassium.

Nursery Management

To raise a cardamom plantation, seedlings or suckers of high yielding varieties are to be used. The different steps involved in raising nursery are given below:

1. Primary Nursery

a. *Site selection:* Select nursery sites on gentle sloppy area and preferably near to a perennial water source. Clean the area from all existing vegetation, stumps, roots and stones. In the cleared area, beds can be prepared with one meter width, 20 cm height and required length, generally six meters. Jungle top soil can be spread to a thickness of 2 to 3 cm on the beds. Fumigate the beds with 2% formaline (10 litres/ bed) which helps in eliminating pathogens, nematodes and other soil pests. Fumigation should be done in raised nursery beds before sowing. The beds should be made air tight by covering with polythene sheets and the fumigant is allowed to penetrate into the soil for two or three days. Later the treated beds should be kept open for another week before taking up sowing.

b. *Seed collection:* Fully ripened bold capsules from high yielding and disease-free mother clumps of known source can be collected from second and third harvests for seed extraction. Seeds after extraction should be washed using water to remove mucilage. It is then mixed with wood ash and dried in shade. Storage of seed is not advisable for longer period, because it is experimentally proved that 15 days of storage decreased germination by about 20% and 3 to 5 months storage decreased germination up to 94%. Therefore sowing of seeds after extraction should not be delayed much. Sowing in September is the best for good germination. Sowing in winter and during south west monsoon should be avoided.

c. *Seed treatment:* Cardamom has a hard seed coat which prevents good germination. Seed treatment with acid or similar chemicals improves the germination. Treatment with concentrated sulphuric or nitric acid for two minutes is sufficient to break seed coat. Planters can get their seeds treated in the Research stations of Spices Board or they can do it by themselves. The technique involved is very simple. The seed is to be cleaned with water if it is covered with ash and dry them in shade. Take the seed in a glass jar and place the jar in cold water. Then slowly pour acid in the jar so as to drench the seed, stir gently for about two minutes and drain the acid using a strainer. Transfer the seed immediately to a large volume of water. Wash the seed free of acid in running water and later soak in water and keep overnight. The seed is ready for sowing on next day.

d. *Sowing:* Sowing can be done in lines in rows at a distance of 10 cm. Seed rate is 30 to 50 g/ 6x1 m size bed. After sowing, cover the bed with thin layer of fine soil and then with mulch material, such as potha grass or paddy straw. Avoid contact of mulch materials with soil by supporting twigs laid across the bed. Water the beds to sufficient moisture conditions. Once sprouting is observed, remove the mulch and cover the bed with thinly sliced mulch material. To protect seedlings from direct sunlight, provide overhead pandal. Germination commences 20 to 25 days after sowing and continues for further 30 to 40 days.

2. Secondary Nursery

There are two methods of raising seedlings in secondary nursery. They are bed and poly bag nurseries.

a. *Bed Nursery :* Prepare beds as in primary nursery. A layer of cattle manure and wood ash may be spread on bed and mixed with soil. Seedlings of three to four leaf stage from primary nursery beds can be transplanted in the secondary nursery at a distance of 20 to 25 cm. Mulching and watering of beds should be done immediately after transplanting. Over head pandal can be erected to protect seedlings from direct sunlight.

b. *Poly Bag Nursery* : Black HM/HDP bags of 20 x 20 cm size and thickness of 100 guage with 3 to 4 holes at the bottom can be used for this purpose. Fill the bags with potting mixture in the ratio of 3:1:1 jungle top soil, cowdung and sand. The bags may be arranged in rows of convenient length and breadth for easy management. Seedlings at three to four leaf stage can be transplanted into each bag (One seedling/bag). Adequate space in between the bags may be provided for better tillering. The advantages of raising seedling in poly bags are:

1. Seedlings of uniform growth and tillering can be obtained.
2. Nursery period can be reduced to five to six months after transplanting the seedlings as against 10 to 12 months in secondary nursery.
3. Better establishment and growth of seedlings in the main field.

Cardamom plants from secondary nursery or poly bags can be transplanted to the main field during the last week of May after receipt of pre-monsoon showers or first week of June soon after commencement of South West monsoon.

3. Vegetative Propagation

Rapid Clonal Multiplication of Elite Cardamom Selections : Clonal multiplication which ensures genetically uniform planting material can be resorted to either by macro i.e. clonal multiplication through rhizome under intensive care in the clonal nursery, or through micro-propagation by tissue culture under aseptic conditions. The cost of tissue cultured plants is high and not within the reach of small and marginal farmers who constitute nearly over 70% of planters. Besides, the selections/ clones in cardamom are highly location specific. Under such limitations, clonal multiplication of elite clumps identified from farmer's field itself provides an easy way out. Moreover, plants raised from rhizomes bear earlier by one year compared to seedling population, thus recovering the initial investment of plantations. In view of these practical problems, a 'rapid clonal multiplication technique' was evolved at Cardamom Research Centre, Appangala (Korikanthimath, 1992).

This method is simple, reliable and facilitates easy multiplication of selected types. Plants raised from rhizome come to bear earlier than the seedlings raised from seeds by about a year. Vegetative propagation is advantageous in areas where viral diseases are not a problem.

Rhizome multiplication may be taken up from first week of March to the first fortnight of October. The site is selected in open gently slopping well drained areas near a source of water. Trenches of 45 cm width and 45 cm deep and convenient length are taken across the slope or along the contour at 1.8 m apart. They are filled with equal quantity of humus rich top soil, sand and cattle manure. Uproot a part of high yielding disease free mother clump identified in the plantation. Trim the roots and separate suckers so that the minimum planting unit consists of one grown up tiller and a growing young shoot. Plant them at a spacing of 1.80 x 0.60 m in filled up trenches. Provide sufficient mulch and stake each planting unit. Provide overhead pandal as in seedling nursery and remove shading material with onset of monsoon. Provide irrigation once in a fortnight and adopt necessary plant protection measures. Apply fertilizers 100:50:200 kg/ha NPK in six splits at an interval of two months. Application of DAP and MOP for first two rounds enables better establishment and growth. Apply neem cake @ 100-150g/ plant along with fertilisers. On an average, 20 to 30 suckers/initial planting unit can be produced within one year of planting. Care should be taken to identify and collect mother clumps only from areas totally free from viral diseases.

Micropropagation : Micropropagation techniques are reported by Rao *et al.* (1999). Shoot buds are harvested from young sprouting tillers. Shoot buds of cardamom are enclosed in many tightly overlapping leaf initials. By virtue of this genuine natural protection against surface contaminants, the buds are easily sterilized. The most appropriate explant size for culture initiation is two-three cm high and 1.3-1.6 diameter at base.

Primary Culture Establishment

This is the most time consuming and difficult stage. Stimulation of primary cultures can be achieved by culturing explants on medium supplemented with relatively high levels of cytokinins. This reduces dominance of apical meristem with the result that adventitious and/or axillary buds arise directly from explants.

Among various media formulation tried LS, MS, KC, SH etc., SH was more effective for cardamom shoot induction supplemented with BAP and

KN each at 0.5 mg/l as cytokinins and IAA at 2 mg/l as an auxin. Coconut water (CW) 5% (v/v) and the growth factors D-biotin and Calcium pantothenate were also added @ 0.1 mg/l. The buds elongated on MS medium within six-eight weeks, grew to a height of 30-40 mm within 12 weeks.

Multiplication of Propagules

Once the primary cultures are established, the cultures have to be transferred to shoot multiplication medium. At this stage, SH medium is substituted with MS basal medium, keeping all the other supplements the same. When the primary cultures are about 30-40 mm long, they are individually transferred to shoot multiplication medium.

In this medium, an average of three shoots/ explant developed within 100 days in about 60% of the tubes. Shoots are subcultured regularly at intervals of 27-35 days to obtain approximately three shoots/ explant.

Shoot Elongation and Rooting

Whites modified rooting medium 1963 without any hormones was more effective than any other media in promoting shoot elongation and subsequent rooting in cardamom.

Typically, white roots appear between four and 14 days after transferring to rooting medium, if the leafy shoots are one-three cm height (two-three leaf stage). Roots can take upto three weeks to appear. The shoots grow rapidly to a length of 50-60 mm during this period. The rooting (%) at this stage can be 86-93%. Once the shoots have rooted, two options are available.

1. The rooted shoots can undergo an *in vitro* hardening period prior to transplantation into soil.
2. The rooted shoots can be left for two-three weeks on rooting medium and transferred immediately (directly) to external environment.

Field Planting and Management

Before taking up the planting, field should be made ready. For planting in a new area, ground should be cleared and if it is replanting area, old plants should be removed. Shade regulation, terracing and preparation of pits should be done during summer.

a) Shade Regulation

Shade regulation is one of the important practices. It should be attended during summer (March-April) in the new planting areas and during May-

June after the receipt of summer showers in the existing plantation. If there is thick shade due to dense branches and bigger leaves, lopping off branches should be done to provide filtered light of 40 to 60% of the open area. Cut alternate side branches of tree in the lower one third to half portion of the total canopy height. Lopping should not be done on one side only. Cutting branches from all sides ensures a balanced canopy. South-Western slopes should be provided with more shade than North Eastern slopes. Shade trees should have small leaves, tap root system and in summer, it should not shed leaves. If area are open due to tree fall, planting of tree species viz: Karuna (*Vernonia arborea*) Corangati, Chandana Viambu and Njaval tree should be taken up immediately to protect plants from direct sun light. Too much shade or too much openness of area is not advisable for cardamom cultivation as it affects growth and yield.

b) Field Preparation

In areas having medium and steep slopes, soil preparation will be different from that of gentle slopes. In sloppy areas, soil should be protected from soil loss (erosion) due to rains for which planting should be taken up in terraces. Terraces should be made across the slope at required distances depending on spacing adopted. Almost 8 to 15 cm depth of top soil should be removed before making terraces and kept aside which can be used for pit filling. Width of terraces should be 1.5 to 1.8 m. Pits of 90 x 90 x 45 cm can be prepared before commencement of monsoon, about 1/3 of pit should be filled with top soil and 1/3 filled with 1:3 mixture of organic manure and top soil. In low rainfall areas, trenches of size 75 cm width and 30 cm depth may be taken and plants may be planted at a spacing of 1 to 1.5m.

c) Planting

Planting material of high yielding variety suitable for the areas may be selected for planting. They may be planted in the already prepared and filled pits and plants should be protected from wind by staking. For Mysore and Vazhukka cultivars, plant to plant distance can be 3x3 m or 2.4x 2.4 m when planted in high rainfall or irrigated areas. A spacing of 1.8 x 1.8 m or 1.2 x 1.8 m is suitable in Karnataka. Immediately after planting, the plant base should be mulched well with available dried leaves to protect soil from erosion and conservation of moisture. Planting should be done diagonally to the slope which will be helpful as a self protector of soil.

d) Weed Control

Weeds are potential competitors in the consumption of water and nutrients which will depress the cardamom growth. At the initial stage, if cardamom clump development is not enough, weed growth will be more. Two or three rounds of hand weeding at the plant base during May, September and December/January and slash weeding in other areas are advisable. Use of spade for weeding is to be avoided as it will loosen the soil and cause soil erosion. The weeded materials may be used for mulching.

Irrigation Management

Judicious irrigation during summer ensures increase in yield by at least 50%. Irrigation is required generally from February to April but at times from January to May depending upon availability of rainfall. But in Tamil Nadu, where the South-West monsoon is not very effective, irrigation during March-August is advisable. This is the period in which development of young tillers and panicles takes place. If plant suffer during this stage, yield will be reduced. Water may be stored during rainy season wherever possible by constructing check dams without causing much damage to the environment. Irrigation can be done through different methods like pot irrigation, hose irrigation, sprinkler irrigation and drip irrigation depending on facilities available in plantation. Pot irrigation or hose irrigation can be done at weekly intervals at the rate of 20-30litres/clump depending upon clump size. Sprinkler irrigation (25 mm) once in 12 days gives good yield. Mini sprinklers and drip irrigation daily (@ 8 liters) can be adopted where water and labour are limited.

Soil and Water Conservation

Conservation of natural resources like soil and water is very important for production of crop. Cultivation of agricultural crops on sloppy hills with intensive operations which loosen and expose the soil will increase soil erosion by water. Planting in trenches across the slope, mulching of soil, diagonal planting and opening of rectangular silt pit (1.8 x 0.5 x 0.6m) in between four plants will help in soil and water conservation in gentle slopes. If slope is steep, construction of stone pitching walls at 10-20 m intervals across the slope and also making water collecting trenches along the wall will be helpful.

Planting Method

In low to medium rainfall areas (1200-2000 mm), trench system of planting, i.e. planting cardamom in 1.5 x 1.5m dimension trenches by filling with farm yard manure, jungle soil and top soil to conserves more soil moisture and supplements the plants even during dry spells.

Contour Staggered Trenches

Contour staggered trenches of size (2 m x 45 cm x 30 cm) can be opened in alternate rows to conserve soil and water in cardamom plantations.

In cardamom plantations, average runoff and soil loss over the years ranged from 2219.74 kg/ ha to 495.56 kg/ ha. Minimum runoff (31.82 mm) and soil loss (495.56 kg/ha) was recorded in contour staggered trenches with pineapple.

Shelter Belts

It would be a good practice to grow trees as wind breaks along contours particularly in Western Ghat for better protection from desiccating wind in summer.

Vegetative Barriers

The vegetative barrier like pineapple can be planted in between rows to prevent soil erosion and restrict runoff in cardamom plantation. Addition of vegetative barriers like pine apple, french bean and ginger ensures additional income.

Earthing up and Mulching

Due to heavy rains in June, July and August, most of the mulch and soil on cardamom base get washed away leading to exposure of most of the active and fresh roots. Therefore, immediately after rainy season, the humus rich top soil beyond 75 cm radius from the clump base needs to be scrapped and applied as a thin layer to the base of clump just up to the collar region. It forms the soil mulch and covers the exposed roots and rhizomes from the rain splash and desiccating effect of sunlight and also conserves soil moisture besides providing essential micro-nutrients.

Forking the plant base to a distance up to 90 cm and to a depth of 9-12 cm enhances root proliferation and better growth of plants. As far as possible, the entire plantation and particularly the plant base are to be kept under mulch. It is very essential to keep the plant base mulched (5-10 cm thick) except during June to September to reduce ill effects of drought, for reducing evaporation loss and to maintain optimum temperature.

Trashing: Trashing consists of removing old tillers and dry leaves and leaf sheaths. This operation may be carried out once in a year at any time, one month after completion of final harvest. These materials can also be used as mulch.

Drainage

Drainage is very important in flat land, gentle slope and low-lying areas to avoid water stagnation and rhizome rot disease. Drainage may be provided based on the slope and run off can move towards farm ponds, check dams, stream and other storage structures.

Fertiliser Recommendation

Fertilizer use has become a regular operation in cardamom plantation However, this has to be carried out judiciously and with thorough knowledge. Indiscriminate application of chemical fertilizers will do more harm to the crop than applying no fertilizer at all. In addition, such an application becomes a wasteful and expensive exercise. Hence it is absolutely essential that the planters follow a judicious fertilizer schedule to achieve satisfactory return and also reduce cost of cultivation. Whether fertilizer is applied or not, organic manure is a must or cardamom crop.

Application of Organic Manures

Organic manures are essential in improving physical characteristics of soil, in addition to their nutrient value.

Application of organic manures such as neem cake (@ 1 kg/ plant) or poultry manure/farmyard manure/cowdung compost (@ 5 kg/plant) may be made once in an year, during May/June along with mussoriephos (180 g/ plant) and muriate of potash (90 g/plant). The manures should be thoroughly mixed with surface soil after application. For the subsequent application to be made in September, organic manures need not be applied.

In valley area and in high fertile soils, soil application of chemical fertilizers may be avoided. Instead, one round of organic manures may be applied (during May/June) and NPK fertilizers given as foliar sprays as per schedule given elsewhere.

Schedule for use of NPK fertilizers

Soil Application

After removing mulches around the plant base, fertilizers may be applied in a circular band of width 15 cm, leaving 30 cm from plant base and thoroughly mixed with the top 5 to 7 cm of the surface soil with hand fork. The fertilizer applied area may be covered again with mulches.

Fertilizer Recommended (NPK in kg/ha)

Situations	First year of planting			Second year of planting			Third year of planting		
	N	P	K	N	P	K	N	P	K
I Rainfed (2 split application)	25	25	50	40	40	80	75	75	150
II Irrigated (3 split application)	25	25	50	60	60	80	125	125	250

Source: (Krishnakumar and Potty, 2002)

Situations	Both rainfed and irrigated areas
1st year of planting	2% DAP (2 kg/100 litres of water - two applications during Sep./Oct. and Nov./Dec.
2nd year of planting	2% DAP + 1% MOP in two applications during Sep./Oct. and Nov./Dec.
3rd year of planting	2% DAP + 2% DAP in three applications during Aug./Sep., Oct./Nov. and Dec./Jan.

Source: Spice India, Vol.16(5);2003.

Foliar Application

The spray solution, containing fertilizers in the prescribed proportions may be applied to foliage of plant covering both sides of leaves, using hand sprayer or knapsack sprayer. The application should not be made in a rainy day. A clear sky or even a cloudy condition may be suitable for the application.

It may be noted that number of split applications may be increased to more than 2 or 3 because smaller doses give better results.

Soil Cum Foliar Application (for yielding plants only)

Rainfed areas	Irrigated areas
Soil application (kg/ha) Nitrogen - 40, Phosphorus - 40 and Potassium - 80 (Single application in May/June) Foliar application 2% DAP + 2% MOP (2 sprays in Sep./Oct. and Nov./Dec.)	Soil application (kg/ha) Nitrogen - 80, Phosphorus - 80 and Potassium - 160 (2 split doses in May/June and Sep./Oct.) Foliar application 2% DAP + 2% MOP (2 sprays in Sep./Oct. and Nov./Dec.)

Source: Spice India, Vol.16 (5); 2003.

Lime/Dolomite Application

Results of field experiments on effect of lime/dolomite showed that application of lime/dolomite did not have any significant effect on cardamom growth and yield. Moreover, correlation analysis of foliar nutrient content with yield of cardamom indicated that enhanced uptake of Calcium and Magnesium beyond certain limit may cause decline in cardamom production. On the basis of these findings, it is now considered that application of lime/ dolomite is not necessary for cardamom.

Use of Micronutrients

Micronutrient survey conducted recently showed that Zinc deficiency is widespread in cardamom soils and Boron deficiency is observed in certain areas. Application of Zinc to the foliage enhanced not only cardamom growth and yield but also quality of the produce. Hence it is recommended that Zinc,may be applied as a foliar spray as Zinc Sulphate @ 250 gm/ 100 litres of water during April/May and Sept/Oct. Approximately 1.5 kg Zinc Sulphate will be needed for one application/ha. Zinc should be applied alone and not mixed with any insecticide/fungicide/fertilizer.

Soil application of Boron in the form of commercial grade Borax at the rate of 7.5 kg/ha is recommended in Boron deficient areas. It may be applied in 2 doses along with NPK fertilizers.

The survey again showed that Iron and Copper are not deficient in soils. The results of a field experiment indicated that Boron may have antagonistic effect over Zinc if applied together on cardamom. Based on these information, it is considered not advisable to use multi-micronutrient formulations for cardamom.

A comparative report on the nature of the land labour and cardamom cultivation practices in Idukki (India) and Guatemala

Particulars	Idukki dt (India)	Guatemala
Area under Cardamom Cultivation	Kerala - 40,867 hectares Idukki - 32,139 hectares	40,000 hectares
Available Expansion Area of Virgin Soil (under Process)		30,000 hectares to 40,000 hectares
Soil	Forest loamy soil	Rich forest soil

Contd...

Rain fall	1500 to 4000 mm (approximately)	3225 mm approx. a) Even distribution b) January to March; once in 10 days
Altitude	600 to 1200 meters	900 to 1200 meters
Average temperature	10°C to 35°C	18°C to 23.5°C
Land scape	Mostly hilly areas	Many tiny hills of which leaving the 1/4th part on the top, the rest is cultivated
Land holders	Large/Medium/Small	85% small holders (1/2 to 10 acres)
Trees	Even distribution of tall and small trees	Tall trees, young trees are sparsely located
Honey bees		They activities are remarkable
Mixed crop		Pineapple, corn, banana, sugarcane and coffee

Source: Spice India, Vol.19 (10), 2006

Cultivation Practices (Replantation)

	India	Guatimala
A) Pits size	1.0x0.66m	Less than 15x15cm
B) Plant spacing	2.66x3.33m	1.66 to2 m It helps to decrease the humidity lose and inhibit growth of weeds
C) Number of plants/ha	1000 to 1250/ha	1500 to 2200/ha a) They keep one meter of the tiller above the Rhizome, chopping the rest of the plant.
D) Planting material	Single tiller as such	b) Seed nursery is also widely seen.
E) Life cycle		8 to 15 years
Picking	7 to 9 rounds	5 rounds
Weeding + pruning	3 to 4 rounds	One time in May, when the pruned matter and dry leaves are mulched around the plant.

Contd...

Yield	100 to 1000 kgs/acre	200 kgs average/acre
Soil nature around the plant		Soil not exposed; mostly covered with mulch
Pesticide spraying	8 to 12 rounds	Nil
Fungal control	Bordeaux, Coc & emission, around 4 times according to necessity	Nil
Fertilizer	3 to 7 rounds	Some 10 % of Planter once a year (NPK 150 g/plant)
Theft problem		5 to 15%
Labour a) Unions and Disputes b) All labour benefits	 Yes Yes	

Source: Spice India,Vol.19(10),2006

An approximate comparative analysis of expenditure for one ha of land between India and Guatemala with salary and cost of cardamom as constants

Particulars	India	Guatemala	Difference
REPLANT a) Pitting b) Closing of pits	(10 plant/labour) Rs.8,750 Rs.5,000	 Rs.1,250 ----	 Rs.7,500 Rs.5,000
PICKING (20 plants/day)	8 rounds x 50 labour = 400 x Rs.75 = Rs.30,000	5 round x 50 labour = 250 x Rs.75 = Rs.18,750	Rs.11,250
Weeding	3 rounds (25 plants/labour) = Rs,11,250	1 round (600 plants) = Rs.7,500 (15 plants/labour)	Rs.3,750
PESTICIDE (700 ml/round) (50 labour/10 rounds) = Rs.3750	10 rounds =17.5 Lts Rs.300 =Rs.5250+Rs.3750 =9000		Rs.9000
Fertilizer (125kg/round 15 labour/3rounds = Rs.1125	3 rounds =375 kgxRs.3.50 =Rs.1312.5+Rs.1125 =Rs.2438		Rs.2438

Contd...

Organic manure	(Rs.10/plantsx1250) =Rs.12500		Rs.12500
Fungicide (15 abour/3rounds) =Rs.1125	3 rounds =Rs.2500+Rs.1125 =Rs.3725		Rs.3725
Soil mulching (20 plants/labour)	Rs.4688		Rs.4688
Labour Bonus	Rs.12500		Rs.12500
Irrigation (25 labour/5 rounds)=Rs.1875	Rs.2500(Fuel) =Rs.2500+1875 =Rs.4375		Rs.4375
Total	Rs.1,04,226	Rs.27500	Rs.76726

Source: Spice India,Vol. 19(10),2006

Cost of cultivation for cardamom

Unit - 1 ha
No. of plants - 1100/ha
Wage rate - Rs.88/labour/day

I. Labour

S. No.	Particulars	I year		II year		III year	
		Labour	Expenses	Labour	Expenses	Labour	Expenses
1	Clearing site	40	3520	-	-	-	-
2	Marking lines	8	704	-	-	-	-
3	Making roads/ foot paths	10	880	4	352	-	-
4	Taking drain	7	616	-	-	-	-
5	Shade regulator	21	1848	11	968	14	1232
6	Taking pits	73	6424	-	-	-	-
7	Filling pits	36	3168	8	704	-	-
8	Planting/staking, mulching	29	2552	-	-	-	-
9	Manuring	26	2288	41	3608	46	4048
10	Weeding	61	5368	69	6072	58	5104
11	P.P. operation	32	2816	39	3432	66	5808
12	Digging	45	3960	54	4752	42	3696

Contd...

13	Irrigation	38	3344	58	5104	62	5456
14	Gap filling	-	-	6	528	4	352
15	Maintaining drainage and foot paths	-	-	4	352	5	440
16	Harvesting & processing	-	-	72	6336	204	17952
17	Trashing	-	-	-	-	39	3432
	Total	426	37488	366	32208	540	47520

II. Materials

S.No.	Name of components	I year	II year	III year
1	Cost of planting material @ Rs.13, 1100 Nos.	14300	858	338
2	Cattle manure @ 5 t during first year @ 5.5 t during 2nd and 3rd years	5700	6250	6250
3	Neem/uromel nut cakes, bone meal, vermicompost etc. @ 1 kg/ plant	7700	7700	7700
4	Fertilizers I year @ 25:25:50 NPK II year @ 40:40:80 and III year @ 75:75:150	1339	2123	4022
5	PP chemicals	550	1475	1625
	Total	29589	18406	19935

Source: Spices Board of India

Harvesting and Curing

Cardamom plants take about two years to bear capsules. Generally flowering starts in April-June with its peak during July-August. Cool humid weather is necessary during flowering season for fruit set. It takes about three months for fruit maturity. Harvesting of cardamom is the most important operation that requires special attention of growers. Proper harvesting will improve quality and quantity of the produce. Faulty harvesting of immature capsules reduces quality and realizes only lower price in the market. Judging the maturity of capsule is very important and some points to identify maturity stage are:

1. Distinctly developed shoulders of capsule.
2. Clear cut line in between two segments of the shell.

3. Triangular deep cut at point of attachment
4. Loose attachment of capsule to panicle.
5. Change of seed color to brownish black or black

Following points should also be noted while harvesting.

1. Harvesting on rainy days results in lower out turn
2. Pest affected capsules give low out turn
3. Capsules from younger plants will have higher recovery %
4. Plants growing in swamps give higher out turn
5. Out turn increases with progress of harvest up to December-January and there after declines.

Harvesting of capsules for the season should be started only at the right stage, otherwise it will have impact on entire crop harvest. There are two types of picking viz light picking and hard picking. When light picking is done, great care is to be exercised in harvesting only the green and mature capsules. This process will naturally give a lower green crop/pick. When hard picking is done, semi-mature capsules are also removed. This process reduces curing percentage. But it increases the picking average, reduces chances of fruit drop and gives green coloured capsules upon curing. When a light picking is adopted, gap between picking rounds can be maintained at 20 to 30 days, where as in hard picking, this can be maintained between 30 to 45 days. Choice of picking depends at times on availability of labour also. Usually when fourth round ends, dry weather conditions set in. If windy condition exist, closer round will have to be adopted. If showers are available the gap between rounds can be extended to 40-50 days to allow capsules to swell in size. Care may be taken to pick capsules only at a physiologically mature to fully ripened stage to allow proper development of seed and to obtain higher recovery. Over ripening of capsules should be avoided as it results in loss of capsules due to rodents and squirrels in the field and splitting at the time of curing, which results in lower market value. After harvest, washing the capsules with clean water and draining the water are essential to get good quality dried cardamom. Storing of harvested capsules is to be avoided.

Curing

Cardamom curing is a process in which the moisture of freshly harvested capsule is reduced from 80% to 10 to 12% at an optimum temperature of 50°C to retain green colour to maximum extent.

There are mainly two types of drying viz. Natural (sun drying) and artificial drying by using fire wood (traditional flue pipe system), electrical current or solar (sun light). Except the traditional flue pipe wood based drying system, all the other methods have major problems. Though fire wood based flue pipe system is a good system, small and marginal farmers who constitute nearly 70% of cardamom growers face difficulties in possessing their own curing house due to economic constraints. They solely depend on large growers to get their fresh capsules cured by paying high drying charge.

The traditional fire wood based curing house consists of wood burning furnace, flue pipes and drying racks in a chamber. The furnace is made up of fine bricks or ordinary bricks having about 2.5 x 0.6 m grate size to facilitate charging with big wood logs. The hot flue gases escape through the chimney after circulating through 25-30 cm diameter flue pipes, made of mild steel sheets placed in the centre of room. The racks containing harvested cardamom capsules are kept at 30-50 cm distance from flue pipe to avoid overheating. The roof of the curing chamber is insulated from inside with wooden planks. The capacity of the curing house varies from 200 to 500 kg fresh capsules. Providing proper air ventillation and opening of exhaust vent periodically avoid accumulation of moisture inside the room and facilitates better green colour during drying. However this type of curing has very low thermal efficiency of 6-8%. Hence Indian Cardamom Research Institute (Spices Board) is evaluating alternative methods of cardamom drying using Kerosene or Liquid Petroleum gas as source of fuel and has achieved considerable success. The cost of construction of drying house along with adoption of Kerosene or LP gas system works out to Rs. 30,000 to 35,000 for 100 kg drying and drying cost works out to Rs. 2 to 2.50 /kg fresh capsules.

After drying cardamom, polishing is done by rubbing against hard surface or using polishing machine. Later the dried cardamom is stored in gunny bags. For efficient retention of green colour of cardamom during storage, it is essential that cardamom should be dried down to a moisture content of 10 to 12%. Use of 300 guage black polythene lined gunny bags improves the storage efficiency. It is better to keep these bags in wooden boxes, which prevent damage of capsule by rodents.

To get prime price in market, grading of capsules by sieving under different diameter sieves is to be done. The cardamom with good green colour and above 7 mm fetches the highest price in the market.

A few tips to improve quality of cured cardamom are:

1. Dry cardamom immediately after harvest to retain the original colour at which they are harvested.

2. Maintain temperature inside the room between 40° and 50°C in the first 10 to 12 hours, then increase to 55°C for rest of the curing period.
3. Provide proper openings for expelling the moisture from the room while curing, which is essential to retain good green colour during drying.
4. Avoid rising of temperature above 65°C inside the room to reduce splitting of capsules and loss of vital volatile cardamom oil.
5. Polish the cured capsules when they are hot.
6. Use black polythene lined gunny bags for packing cured cardamom and store them in wooden boxes for better storage efficiency.

Grades of Cardamom (Small)

1. *Bold*: It is a popular export grade; 90% and above capsules will be having 6.5 mm and above diameter, matured and Greenish color. Lt. Wt. will be 415 gm.
2. *Super Bold*: It is a very special variety. All capsules will be matured greenish and having above 8 mm diameter size. Lt. Wt. will be more than 450 gm.
3. *Extra Bold*: Best in the Export market. All capsules will be matured, greenish and having 7 mm and above diameter. Lt. Wt. will be 435 gm.
4. *Bulk*: This is the grade of cardamom produced as it is. This grade will contain all size, mature and immature capsules, black, yellow and splited cardamom. This is to be graded.
5. *Small*: Small size cardamom having size between 5.5 mm and 6.5 mm. Cleaned and removed dust, husk and black capsules. Lt. Wt. will be around 385 gm.
6. *Open / Splits*: More than 60% capsules will be in open condition and the color may be partly greenish/pale yellow. All capsules will be matured and the sizes are 6.5 mm and above.
7. *Seeds*: Black / Brown colour seeds are the original content in every cardamom capsule. The husks were fully removed. Lt. Wt may be around 550 g to 600 g.
8. *Fruit*: Fruits are generally over matured capsules with slight yellowish in color and Lt. Wt. over 425 g.

Agmark Grade Specifications of Cardamom

Grade Designation	Trade name	Colour	Unclipped capsules (% by count), max	Empty and malformed capsules, (% by count), max	Immature and shriveled capsules, (% w/w), max	Blacks and splits, (% by count), max	Size, diameter of the sieve hole on which retained, mm	Weight in G/L by wt., max
Alleppey Green Cardamom								
AGEB	Cardamom extra bold	Deep green, green or light green	-	2.0	2.0	0.0	7.0	435
AGB	Cardamom bold	-do-	-	2.0	2.0	0.0	6.0	415
AGS	Cardamom Superior	-do-	-	3.0	5.0	0.0	5.0	385
AGS-1	Shipment green-1	-do-	-	5.0	7.0	10.0	4.0	350
AGS-2	Shipment green-2	-do-	-	7.0	9.0	12.0	4.0	320
AGL	Light	-	-	-	-	15.0	3.5	260
Coorg Green Cardamom								
CGEB	Extra bold	-	0.0	0.0	0.0	0.0	8.0	450
CGB	Bold	-	0.0	2.0	3.0	0.0	7.5	435
CG 1	Superior	-	0.0	3.0	5.0	0.0	6.5	415
CG 2	Coorg green or Motta green	-	3.0	5.0	10.0	0.0	6.0	385
CG 3	Shipment	-	5.0	10.0	15.0	10.0	5.0	350
CG 4	Light	-	-	-	-	15.0	3.5	280
Bleached or half bleached cardamom								
BL 1	-	-	-	0.0	0.0	-	8.5	340
BL 2	-	-	-	0.0	0.0	-	7.0	340
BL 3	-	-	-	0.0	0.0	-	5.0	300

Source: Agmark Grade specifications for Spices, Spices Board, 2001

Cardamom Products

Essential oil: produced commercially by steam distillation of crushed fruits. At least 4 hr distillation is needed to produce the full ester content of oil.

Oleoresin: made of two components viz., the volatile oil and resin. Solvent extraction yield oleoresin and 1.0 kg of oleoresin will replace 20 kg of ground spice.

Decorticated Seeds and Seed Powder

Commands a lower price than whole spice due to fairly rapid loss of volatile oil during storage and transportation. Seeds are also classified by weight into prime and light seeds, the former fetching a better price. Seed powder is also sold on a limited scale.

Other Products

1. Sugar cardamom mix - blend of sugar powder with encapsulated cardamom flavour with sunset yellow colourant and tricalcium phosphate as anticaking agent.
2. Cardamom flavoured cola beverage
3. Cardamom flavoured 'Flaur' milk + sugar + starch, added colour, flavour and gelling agents.
4. Cardamom chocolate - Cocoa mass, butter sugar, milk powder, encapsulated cardamom flavour and emulsifiers.
5. Cardamom plus: freshly ground cardamom with encapsulated cardamom flavour.
6. Cardamom tincture: This product is made by extraction of crushed cardamom seeds with other spices using alcohol as solvent (cinnamon, caraway, cochineal).
7. Cardamom coffee and tea.

Marketing

Demand and Supply Analysis

Cardamom is also referred as "grains of paradise". Guatemala, India, Tanzania, Srilanka, El Salvador, Thailand, Vietnam, Laos, Cambodia, Papua New Guinea and Nepal are the major producers. The world production of cardamom is around 35000 t/ annum. Guatemala produces nearly 2/3 of global cardamom production

In India, area under cardamom during 1997-98 was 72444 ha with a production about 7900 t, which remained steady and there was no considerable change in area and production due to stiff competition from Guatemala in the overseas market.

There is no considerable change in area during 2005-06, (73795 ha.) compared to 2004-05 (73725 ha). Due to lower cardamom prices in domestic and international markets since last three years, planters showed less interest in new plantings.

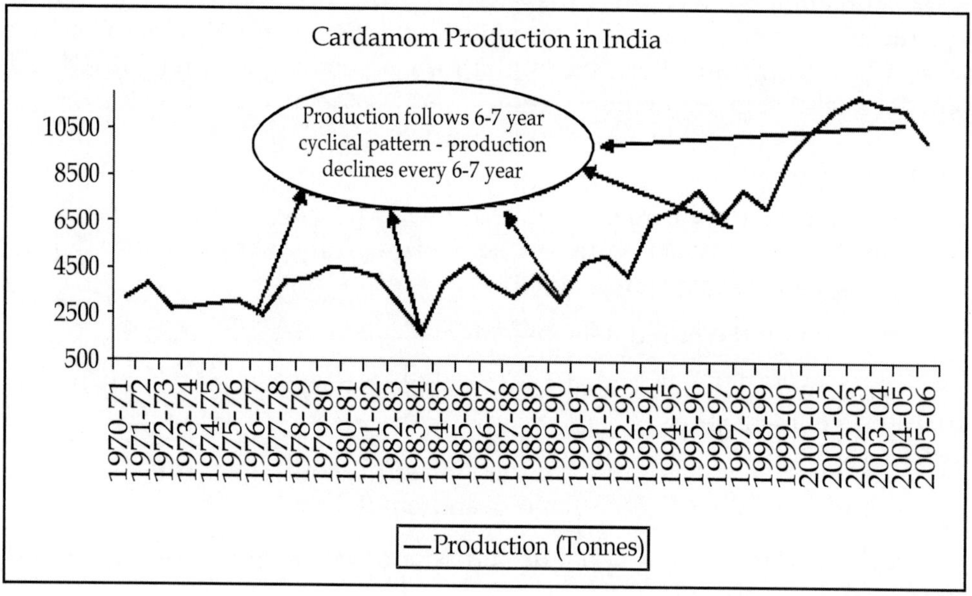

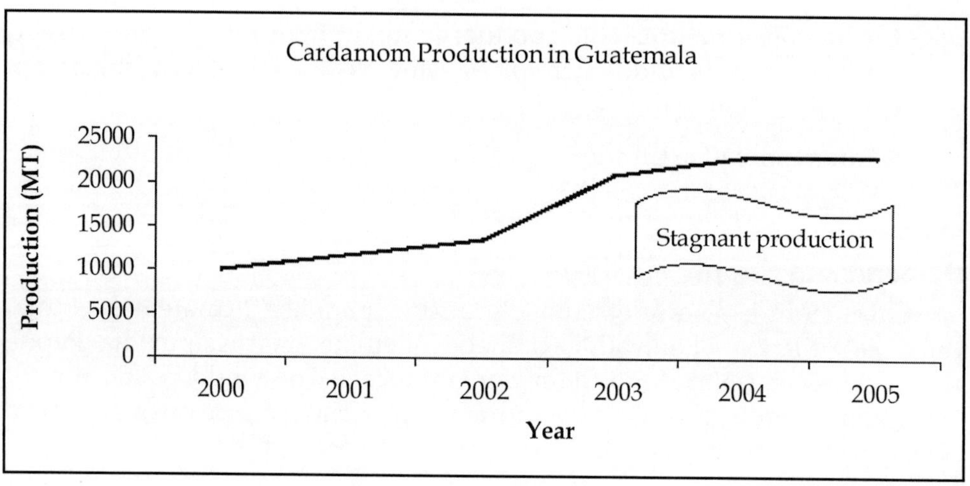

India stands second in world production of cardamom, production in 2005-06 estimated to tumble to 10,000 tonnes compared to 11415 tonnes in 2004-05. Uneven rain in Kerala and unfavorable weather conditions lead to further decline in production.

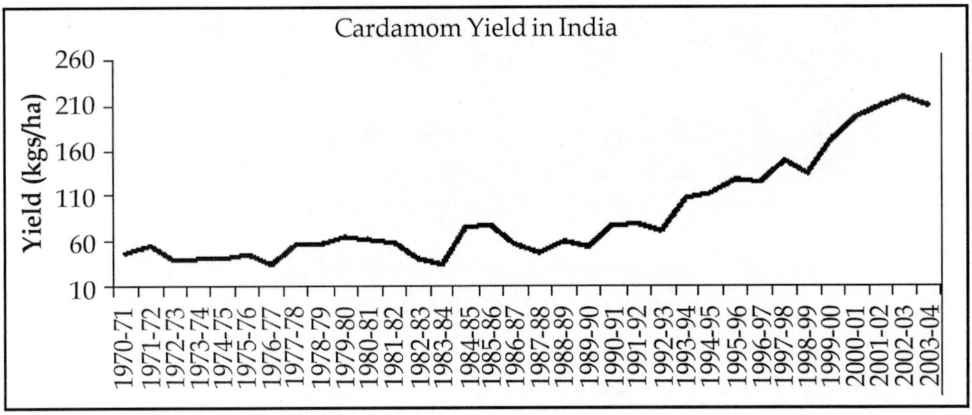

Cardamom Yield in India

Yield levels play an important role in total production of any commodity in spite of higher area. Over the period of time, yield levels in Cardamom have gone up sharply due to cultivation of high yielding and better varieties. Yield levels remained steady around 215 kg/ha during 2001-02 to 2003-04, due to uneven rain fall and exhausted old plantations.

State Wise Area and Production of Cardamom (Small) in India

Area under cardamom remained stagnant due to lower prices and lack of demand in the market since last four years. Area during 2001-02 was around 72663 ha with production of about 11365 tonnes which increased slightly during 2005-06 with area of 73795 ha and production of 12540 tonnes.

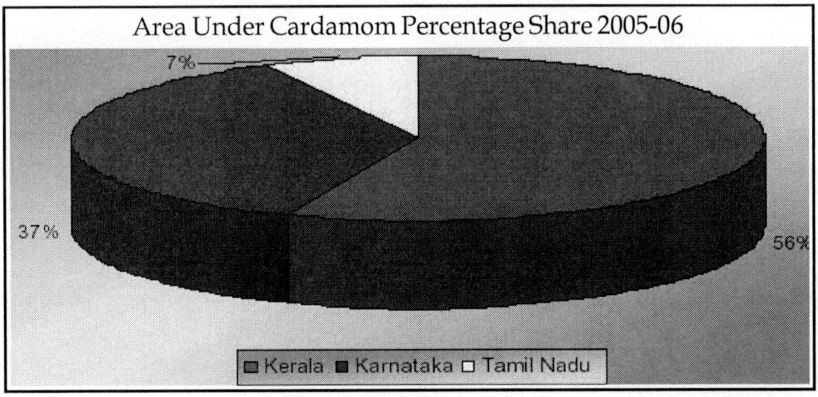

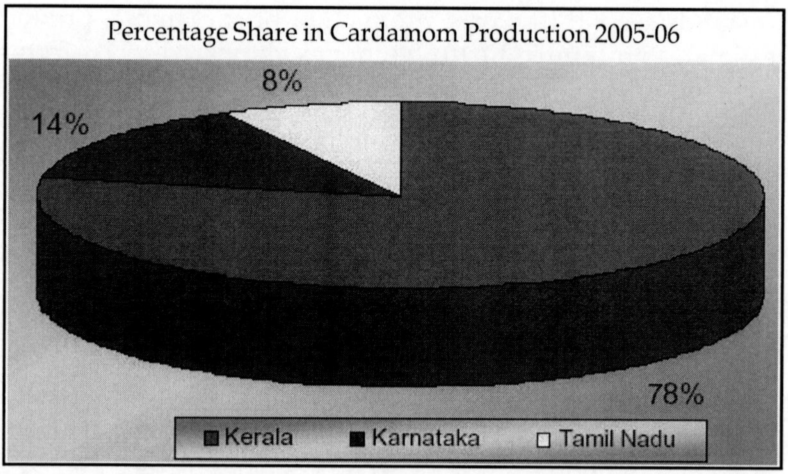

Kerala occupies first position in both area (56%) and production (78%) of cardamom followed by Karnataka with 37% share in area and 14% share in production. Although area under cardamom in Karnataka increased significantly, production is very less compared to Kerala.

Exports of Cardamom from India

Export of cardamom gradually declined after 2000–01, where the exports accounted around 1545 tonnes compared to 676 tonnes during 1999-00. During (2005-2006) export orders for Indian cardamom improved due to lower crop in Guatemala. Supreme quality of Indian Cardamom may attract overseas buyers.Emergence of Guatemala as a major competitor in export market hindered the Indian trade. Due to low cost of production, Guatemala offered cardamom at cheaper prices than Indian prices. Guatemala gradually grabbed the Indian export market.

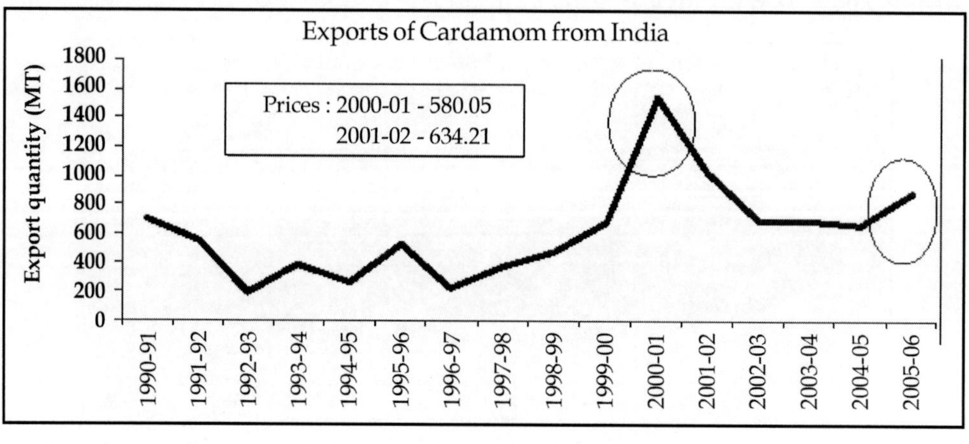

India roughly exports 5-8% of its total production. Saudi Arabia accounts for 42% and Japan 39% of India's exports (2004-2005).About 90% of the produce is consumed with in the nation. Exports of value added products from cardamom like cardamom oil and oleoresins are increasing to Germany, Netherlands and UK.

Export of cardamom from Kerala declined from 825 tonnes in 2000-01 to 454 tonnes during 2005-06. The corresponding fall in value was from Rs. 42.41 crores to Rs.14.46 crores. India's export of small cardamom also declined during this period from 1545 tonnes to 875 tonnes. The export value also declined from Rs.84.68 crores to Rs.27.01 crores. Kerala's share in India's cardamom export also declined from 53% in 2000-01 to 51.89% in 2005-06 although it had touched 70% in 2002-03. Kerala exported 4.87% of the total cardamom production (9329 tonnes) in 2005-06.

Export of Cardamom (Small) from Kerala and India

Year	Quantity (Tonnes)			Value (Rs.crores)		
	Kerala	India	Kerala's share (%)	Kerala	India	Kerala's share (%)
2000-01	825.00	1545.00	53.40	42.41	84.68	50.08
2001-02	630.00	1031.00	61.11	40.32	61.70	65.35
2002-03	475.00	682.00	69.65	32.78	47.74	68.66
2003-04	435.00	757.00	57.46	21.62	36.92	58.56
2004-05	387.00	650.00	59.54	13.99	23.90	58.55
2005-06	454.00	875.00	51.89	14.46	27.01	53.54

Source: Economic Review 2006

Global Export

World export of cardamom was 46,494 tonnes in 2005 of which Guatemala was the market leader with 68.08 % of volume and 64.81% of the value. Guatemala followed by Nepal with 11.89% of the quantity and 9.77 % of value. Other important exporters included Indonesia (6.23%), India (4.22%), Singapore (2.78%), UAE (1.77%), Saudi Arabia (1.58%), Colombia (0.51%), Netherlands (0.48%), and Germany (0.40%).

Import Pattern

World import of cardamom was 40,701 tonnes in 2005 of which Saudi Arabia was the market leader with 27.08 % of the volume and 37.75% of the value followed by India with 13.31% of the quantity and 8.64 % of value.

Exporting country	Qty in tonnes	% Share in world (Qty)	Value in Rs crores	% Share in world (Value)	Annual growth in value between 2001-2005, (%)	Annual growth in quantity between 2001-2005, (%)
Guatemala	31,654	68.08	316.1	64.81	-8	21
Nepal	5,527	11.89	47.6	9.77	1	5
Indonesia	2,898	6.23	16.7	3.42	3	5
India	**1,964**	**4.22**	**37.4**	**7.67**	**-10**	**-2**
Singapore	1,294	2.78	21.3	4.37	-29	-8
UAE	822	1.77	9.4	1.93	23	31
Saudi Arabia	736	1.58	10.1	2.09	-15	22
Colombia	239	0.51	4.3	0.90	-13	11
Netherlands	224	0.48	4.5	0.94	-9	10
Germany	184	0.40	3.3	0.70	-10	14
Others	952	2.05	16.6	3.40	-	-
World	**46,494**	**100.00**	**487.85**	**100.00**	**-10**	**12**

Source: ITC calculations based on COMTRADE statistics

Other important importers included United Arab Emirates (8.52%), Syrian Arab Republic (6.45%), Vietnam (5.36%), Pakistan (5.35%), Bangladesh (3.84%), Singapore (3.34%), Jordan (2.92%), and United Kingdom (1.95%).

India's Import

The import of cardamom into India has been increasing since 2003-04. It has increased from 4191 tonnes valued at Rs. 44.87 crores in 2003-04 to 5416 tonnes valued at Rs. 45.05 crores in 2005-06. Nepal and Gautimala are the leading exporters of cardamom into India. The imported cardamom is mainly used for the production of oleoresin. The share of large cardamom in imports has been decreasing from 94.03 % in 2003-04 to 58.20 % in 2006-07, while share of import of other cardamom, which includes cardamom small, cardamom powder, cardamom husk and cardamom oleoresins, has been increasing from 5.97 % in 2003-04 to 41.80 % in 2006-07.

Consumption of Cardamom

West Asia is the major consumer of Cardamom contributing 59% to total world consumption. It plays a significant role in trade of cardamom in

Import of Cardamom into India

Exporting country	Qty in tonnes	% Share in world (Qty)	Value in Rs crores	% Share in world (Value)	Annual growth in value between 2001-2005, %	Annual growth in quantity between 2001-2005, %
Saudi Arabia	11,022	27.08	200.45	37.75	-9	18
India	**5,416**	**13.31**	**45.90**	**8.64**	**-1**	**5**
United Arab Emirates	3,468	8.52	44.47	8.38	-12	19
Syrian Arab Republic	2,625	6.45	23.12	4.35	4	31
Viet Nam	2,183	5.36	12.48	2.35	95	43
Pakistan	2,178	5.35	27.78	5.23	-	-
Bangladesh	1,563	3.84	14.53	2.74	-9	-
Singapore	1,358	3.34	19.21	3.62	-30	-4
Jordan	1,190	2.92	9.86	1.86	7	42
United Kingdom	794	1.95	12.06	2.27	-14	15
Others	8,904	21.88	121.24	22.83	-	-
World	**40,701**	**100.00**	**530.98**	**100.00**	**-11**	**12**

Source: ITC calculations based on COMTRADE statistics

Import of Cardamom into India

Year	Quantity (tonnes)			Value (Rs. crores)		
	Large cardamom	Other cardamom	Total cardamom	Large cardamom	Other cardamom	Total cardamom
2003-04	3940.21	250.29	4190.5	41.6237	3.2416	44.8653
	(94.03)	(5.97)	(100.00)	(92.77)	(7.23)	(100.00)
2004-05	4386.40	373.3	4759.7	43.0206	3.8963	46.9169
	(92.16)	(7.84)	(100.00)	(91.70)	(8.30)	(100.00)
2005-06	4928.27	487.87	5416.14	40.2493	4.8026	45.0519
	(90.99)	(9.01)	(100.00)	(89.34)	(10.66)	(100.00)
2006-07 (Apr-Jun)	128.55	92.33	220.88	0.79	0.7844	1.5744
	(58.20)	(41.80)	(100.00)	(50.18)	(49.82)	(100.00)

Source: http://dgft.delhi.nic.in/

producing countries. Usually during Ramdhan festival, there will be huge import orders from these countries.

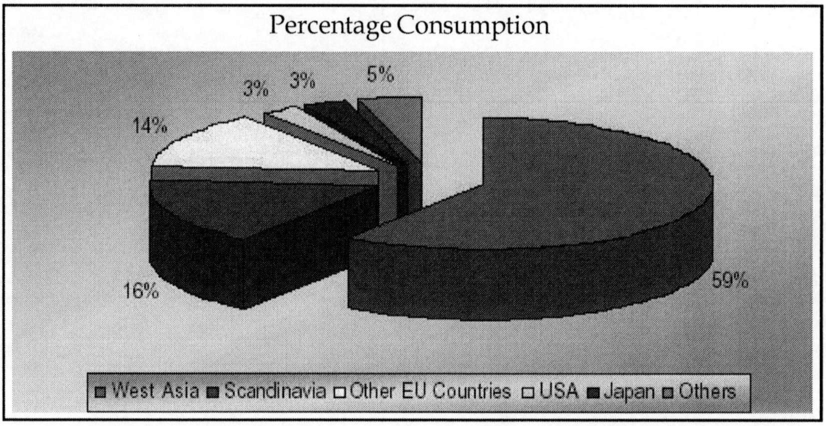

Major Indian Markets

Cardamom is sold at auction centres. Important markets for cardamom in the country are Vandanmedu, Bodinayakanar, Kumily, Thekkady, Kumbum and Pattiveeranpatti in Kerala.

Price Analysis

Cardamom prices are governed mainly by rain fall in producing centers. This crop requires continues rain and enough moisture in soil. Lack of rain in producing centers may lead to lower production.

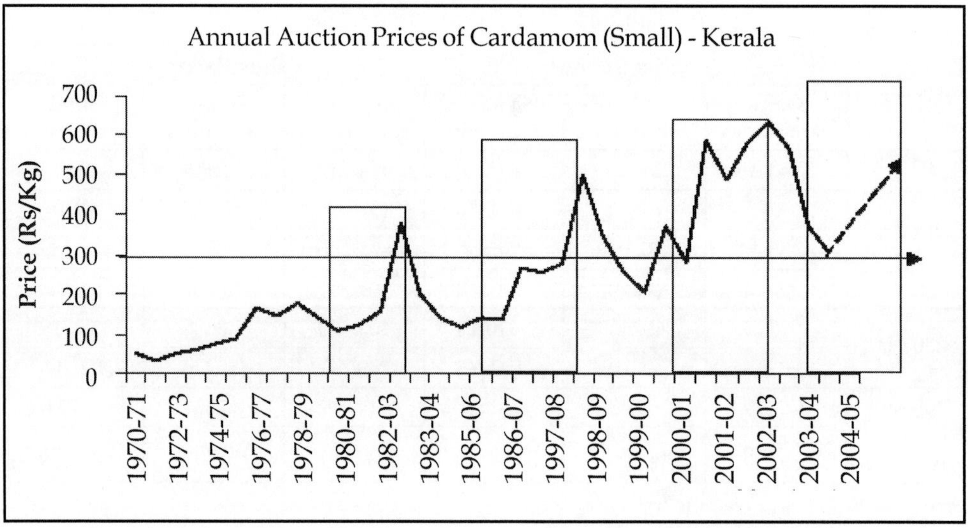

Cardamom prices follow usually 4 to 5 years cyclical pattern where prices move up continuously on every fourth to fifth year. Above graph depicts the historical price trend of cardamom signaling uptrend in market for rest of year. Lower production estimates (about 30% production is estimated to be less this season (2005-2006)

In Kerala, most of the planters cultivated the new variety (Njallani Green Gold) which yields or produces continuously nearly eight months depending on rain fall and water availability.

Harvesting of cardamom in Kerala starts from August and continues till January but percentage of picking declines from month to month. Although Augest-January is the harvesting season of cardamom, we can rally in the prices from above graph due to heavy demand from domestic market. Festival season between August and January usually leads to heavy buying from stockist and exporters.

Harvesting (Picking) Pattern in Cardamom

Arrivals mainly depend on percentage of picking in the growing regions

- August to September - 20 to 25%
- October to November - 15 to 20%
- December to January - 5%

Price Target

Cardamom prices depend on rain fall in Kerala and other producing centers. Prices to go up sharply on account of lower production and huge demand in the coming months.

Factors Influencing Cardamom Markets

- Fresh cardamom is green and has a characteristic aroma. Freshness, colour, aroma and size are the major factors that influence cardamom prices, in addition to the current supply-demand scenario.
- Indian cardamom specially Alleppey Green is a premium grade against all other international grades. Production and export from Guatemala have profound influence on Indian cardamom prices.
- Weather and annual production of a year.
- Production in other countries particularly Guatemala.
- Year ending stocks and stock to consumption ratio.
- Time of arrival of new crop in the market.

CARDAMOM (Large)

Large Cardamom (*Amomum subulatum Roxb.*), the greater Indian or Nepal cardamom, is a member of Zingiberaceae family under the order Scitaminae. It is a native of Eastern Himalayan region and is one of the main cash crops cultivated in sub-Himalayan state of Sikkim and Darjeeling District of West Bengal covering an area of 30,000 ha. The annual production varies from 5000-5500 tonnes. It is also cultivated in parts of Uttarakhand and in North Eastern States. Nepal and Bhutan are other countries where large cardamom is cultivated. It is used as a spice and in several ayurvedic preparations. It contains 2 to 3% of essential oil and possesses medicina1 properties like carminative, stomachic, diuretic, cardiac stimulant and antiemetic.

There is controversy over grouping of cardamom, the ISO (International Standards Organization) has officially recognized nine species under three main groups:

Group I: *Elettaria cardamomum*

Group II: 4 species of *Aframomum*

(a) *A. angustifolium* (Sonn) K.Schum - Madagascar cardamom

(b) *A. hanburyi* K.Schum - Cameroon cardamom

(c) *A. korarima* (pereira) Engler - Korarima cardamom

(d) *A. melegueta* (Roscol) K.Schum - Grains of paradise or Guinea grains.

Group III: 4 species of Amomum

(a) *A. aromaticum* Roxb- Bengal cardamom

(b) *A. kepulaga* Spraque et - Round cardamom of Java

(c) *A. krervanh* Pierre et Gagnipain - Cambodian cardamom

(d) *A. subulatum* Roxb - Greater Indian cardamom, Nepal cardamom or large cardamom.

Climate and Soil

The crop grows well under shade of forest trees in sub Himalayan mountains at altitudes from 1000 to 2000 m. amsl with rainfall of 3000 to 3500 mm distributed in about 200 days a year. The crop is at present grown in relatively drier shadow areas also where productivity is much low. Cultivars suited to higher altitudes tolerate lower temperatures and those suited to lower altitudes to marginally higher temperature regimes. Deep and well drained soils with loamy texture are the best suited. Large cardamom soil is generally rich in organic matter and nitrogen, medium in available phosphorous and medium to high in available potash. The pH ranges from 4.5 to 6.0. Even though the crop can be grown in undulating and steep terrains, land with moderate slope is preferred.

Cultivars

There are mainly five popular cultivars viz. Ramsey, Sawney, Golsey, Varlangey and Seremna. Bebo, Boklok Tali, Jaker and Belak are the other cultivars found in Arunachal Pradesh.

1. Ramsey

This cultivar is well suited to high altitudes above 1515 m amsl and can be cultivated even in steep slopes. The variety is identified by the maroonish colour of tiller and narrow leaves. Plants are 1.5 to 2.0 m tall, robust with large number of tillers. Flowering starts in May and crop is ready for harvest by October-November. Capsules are smaller in size with 25-40 seeds. It is susceptible to viral diseases like Chirke and Foorkey.

2. Sawney

It is a widely adapted cultivar, the most suited to medium and high altitude areas of above 1350 m amsl. Plants are 1.5 to 2.0 m tall, robust in nature, leaves ovate and broad and the colour of tiller is similar to Ramsey. Capsules are bigger and bold with 35-50 seeds. Flowering starts from March to May and harvest begins in September-October and extends upto November in high altitude areas. The cultivar is susceptible to both Chirke and Foorkey diseases.

3. Golsey (Dzongu)

It is suitable to areas below 1500 m amsl and is very specific in Dzongu area of North Sikkim and the plant height is 1.0 to 1.5m and not as robust like other cultivars. Unlike Ramsey and Sawney, the tillers are green in colour and leaves are narrow and erect. Capsules are big and bold and contain 50-70 seeds. Flowering starts in March and harvesting in September-October. It is relatively tolerant to Chirke but susceptible to Foorkey and leaf streak diseases.

4. Seremna

The cultivar is grown in a small pocket of the Hee-Gaon, West Sikkim, at low altitude and is known for its high yield potential. Plants are 1.5 to 2.0 m tall, tillers green and leaves mostly drooping, hence named as 'Seremna'. On an average, 2-3 spikes in each productive tillers with about 10 capsules in each spike and 65-70 seeds/ capsule are recorded.

5. Varlangey

This cultivar grows in mid and high altitude areas. Its yield is exceptionally high at high altitudes. Plant height is 1.5-2.5m, robust type and resembles Ramsey with narrow leaves having wavy margins. The spike and productive tiller ratio is relatively high in this cultivar. Size of capsule is bold with 50-70 bold seeds. Flowering starts in May at medium altitude and during June-July at high altitude. Consequently, harvesting is delayed upto end of November in high altitudes.

Improved Selection

The Regional Station of the Indian Cardamom Research Institute under the Spices Board of India has conducted exploratory survey of large cardamom plantations in Sikkim and Darjeeling district and identified a few high yielding selections. They are under evaluation in different locations. The following are the important ones.

Selection	Cultivars	Altitude'adaptability
*SBLC 5	Sawney	Medium to High
SBLC 47, SBLC 42, SBLC 44, SBLC	Varlangey, Golsey	Low
50,	(Dzongu) Seremna, Sawney	Low
SBLC 51	Green Golsey	Medium to High

Low: less than 975 m above MSL; Medium: 975-1515 m above MSL; High: More than 1515 m above MSL
* Spices Board Large Cardamom (SBLC)

Nursery Management

Propagation of large cardamom is done through seeds and suckers. Propagation through seeds enables production of a large number of seedlings. Virus diseases are not transmitted through seeds and therefore seedlings are free from viral diseases, if adequate care is taken to isolate and protect the nursery from fresh infection. Plants raised from seeds need not necessarily be high yielders even if they are collected from very productive plants due to cross-pollination. The major pollinator is wild Bees, even though honey bees also play a role in pollination. Planting suckers, on the other hand, ensures true to type and high productivity, if they are collected from high yielding plants.

Propagation Through Seeds

Seeds are collected from high yielding and well-maintained plantations, free from viral diseases. Well matured capsules from bottom and middle position of spikes are selected for extraction of seeds. After dehusking, the seeds are mixed with sand and rubbed with hand and then washed with water to remove mucilage completely. Once the water is completely drained, seeds are mixed with wood ash, dried in shade and sown in primary nursery. The seeds may be sown immediately after extraction for maximum germination. Normally, the germination % is only 20-30 and to be sown immediately after drying under shade. Treatment of seeds with acid gives early, uniform and better germination.

The seeds are treated with nitric acid (25%) after removing the mucilage (25 ml of concentrated nitric acid mixed with 75ml of water to make 25% nitric acid). The seeds may be completely immersed in acid and stirred well for 10 minutes. The acid is then drained out. The seeds are washed thoroughly in running water to make it free of acid.

i) Primary Nursery

Seeds are generally sown in September-October. Seed beds are prepared in well-drained soil dug to a depth of 30 cm and left for weathering. Raised beds with 15 to 25 cm height, 1 m width and convenient length preferably 6 m are prepared. Well decomposed cattle manure is mixed with soil and the surface of bed is made to fine tilth. 80-100 g of seeds/ bed are sown in lines across the bed at a distance of 10 cm. Seeds are covered with fine soil and mulched with paddy straw/dry grass (10-15 cm thick). Watering is done at regular intervals to keep surface of bed moist. Germination of acid treated seeds commences after 25 to 30 days of sowing. When average germination is noticed, mulch materials are removed. The inter space between rows is then remulched with chopped paddy straw. Shade pandals are immediately erected by using bamboo mats/reed mats or agro-shade nets. The beds are watered regularly and weeding is done as and when required. When seedlings attain 3 to 4 leaf stage, they are transplanted to secondary beds.

ii) Poly Bag Nurseries

Polythene bags of 15 x 15cm with perforations at base are used for planting seedlings. The bags are filled with potting mixture of soil, sand and cowdung in the ratio of 4: 1: 1. The bags filled with the mixture are arranged in rows of one metre width and convenient length under shade pandals. Seedlings with 3 to 4 leaves are planted in bags in April-May and watered regularly. They become ready for field-planting in 10-12 months.

iii) Secondary Nursery

Beds of size 15 cm in height and 100 cm width with convenient length are prepared and well-decomposed cattle manure is mixed with soil and an even surface is formed. Seedlings with 3-4 leaves are transplanted to beds in May-June at a spacing of 15 cm between them. The inter-space is mulched with chopped paddy straw or dried leaves. Overhead pandal is erected for providing shade; the soil is kept moist with irrigation. The seedlings once reach of 45 to 60 cm height with 2 to 3 tillers, are planted in main field during June-July of subsequent year.

Sucker Multiplication Nursery

As mentioned earlier, suckers should be generated only in sucker multiplication nursery where adequate precautions are taken to ensure that virus diseases are not transmitted through the suckers produced. The site for such a nursery should be located at least 500 m away from large cardamom plantations. They are established either under shade of forest trees or under shade pandals with 50% shade by using black agro shade nets. Trenches of 30 x 30 cm are prepared at convenient length with an inter space of 30 cm. The trenches are filled to the brim with well decomposed cattle manure or compost mixed with soil before planting suckers from high yielding, disease free plantations with one grownup shoot with an emerging bud planted at 30 cm apart in trenches. The time of planting is May-June. After planting, the plant base is mulched with dried forest leaves. The multiplication rate in this method is about 1:8 in one year time. The grown up tillers are split into units of one tiller with an emerging bud and planted in main field during June-July.

Micropropagation

For rapid multiplication of elite/high yielding clones, vegetative buds from disease free high yielding mother plants are collected and plantlets produced through Tissue Culture Technique. These plantlets are hardened either in poly bags or in secondary nurseries. Once sufficient growth is attained, they can be planted in main field during June-July.

Field Planting and Management

Large cardamom grows well in forest loamy soils with gentle to medium slopes. Water logged condition is detrimental to growth of plants. It performs well under shade. Utis (*Alnus nepalensis*) is the most common and preferred shade tree for large cardamom. The other species of shade trees are Panisaj (*Terminalia myriocarpa*), Pipli (*Exbuchlandia populnea*), Malato (*Macaranga denticulata*), Asarey (*Cole brookianum*), Bilaune (*Maesa chesia*), Karane (*Symplocos ramosissima*), Siris (*Albizzia lebbeck*), and Chilauni (*Schima wallichi*).

Land Preparation

Planting is done during June-July when there is enough moisture in the soil. The land selected for planting is cleared of all under growth and weeds for new planting or if it is replanting, old plants are removed. Pits of size 30 x 30 x 30 cm are prepared on contour at a spacing of 1.5 x 1.5 m after onset of monsoon showers. Wider spacing of 1.8 x 1.8 m is recommended for robust cultivars like Sawney, Varlangey and Ramsey. The pits are left for weathering for a fortnight and then filled with topsoil mixed with cowdung or compost @ 1-3 kg/ pit. Seedlings/Suckers are planted in the centre of pits. Care should be taken not to plant the seedling/rhizome very deep in the pit. After planting, the seedlings/ suckers may be staked and base of plant mulched with dried leaves.

Nutrient Management

For sustained production, the soil fertility should be maintained to its optimum. Well decomposed cattle manure/ compost or organic products like non-edible oil cakes may be applied @ 2 kg/plant atleast once in two years in April- May. If all the crop residues are recycled in the plantation, application of inorganic fertilizers may not be necessary. In plantations with very high productivity, fertilizers @ 20:30:40 kg NPK/ hectare may be applied in two split doses with full of P and half dose of N and K in April and half dose of N and K in September. Mulching base of plants after application of second dose, helps plants in intake of nutrients.

Weeding

Weed control in plantations is important for maximum utilization of available soil moisture and nutrients by the plant. Three rounds of weeding are required for effective control of weed growth in the initial two to three years. Weeding can be either hand or sickle depending upon intensity of weed growth. From around the plant base, the weeds can be pulled out by hand and weeds in the interspace need only be slashed with sickle. Clean weeding may be avoided. While weeding, dried shoots and other trashed materials can be used as mulch around the plant base to conserve soil moisture in the ensuing dry months, and to prevent weed growth around plant base.

Soil and Moisture Conservation

This crop is mainly grown in highly steep terrain. The topography and climatic condition permit soil erosion to a considerable extent. Intensive operations, which loosen and expose soil, increase soil erosion and therefore only minimum tillage operations should be followed. As far as possible, contour terraces may be made well before takingup planting operations. This helps in reducing soil erosion and soil moisture conservation. Though contour terrace making is expensive and requires high initial investment, the long term benefit will be more to compensate initial extra expenditure.

Irrigation

In some of the large cardamom plantations, water sources are available which can be exploited to irrigate the crop by gravity flow, either through pipes, sprinklers or flood irrigation through open channels. It is observed that productivity is higher in plantations where irrigation is provided. For sustainable and better yield, the plants may be watered during dry months. Depending on availability of water sources, hose or sprinkler or flood irrigation through channels can be adopted. Hose irrigation can be done @ 40-50 litres/ plant at fortnightly intervals. In case of sprinkler irrigation, equivalent to 35-45 mm of rain at fortnightly interval is recommended.

Cost Estimation for Large Cardamom in Sikkim under Irrigated Condition

S. No.	Particulars	Year				
		1	2	3	4	5
1	Planting materials	5880	888	444	-	-
2	Forest manure	750	750	750	750	750
3	Plant protection chemicals	200	400	400	400	400
4	Hand sprayer (Knap sack)	2500	-	-	-	-
5	Curing of capsule	-	-	240	900	1800
6	Irrigation equipment	7800	-	-	-	-
7	Labour	6480	3480	3480	4480	4480
	Total	26610	5518	5314	6530	7430
	Rounded	26610	5520	5310	6530	7430

Source: Spices Board of India

Harvesting and Curing

Indication of time of harvest is when seeds of topmost capsules turn brown. As soon as the said colour appears and to enhance maturity, bearing tillers are cut to a height of 30-45 cm and left for another 10-15 days for full maturity. The spikes are harvested by using special knives known as 'Elaichichhuri'. The harvested spikes are heaped and capsules separated and dried. The cured capsules are rubbed on wire mesh for clearing and removal of calyx (tail).

Traditionally cardamom is cured in Bhatti, where the capsules are dried by direct heating. Under this system, the cardamom comes in direct contact with smoke which turns the capsules to dark browner to black colour with a smoky smell. Improved curing techniques are presently available by which cardamom is processed to give better quality and appearance. One such method is (ICRI Spices Board) improved bhatti system of curing in which cardamom is dried by indirect heating at 45-50°C.

Packaging

The properly dried capsules should be allowed to cool and then packed in polythene lined jute bags. The bags may be stored on wooden platform to avoid absorption of moisture, which results in fungus growth damaging the stored produce.

Whole Cardamom: Chemical and Physical Specifications

Specifications	Suggested limits
ASTA cleanliness specifications	
Whole dead insects, by count	4
Mammalian excreta, by mg/lb	3
Other excreta, by mg/lb	1.0
Mold (%)by weight	1.00
Insect defiled infested (%) by weight	1.00
Extraneous (%) by weight	0.50
FDA DALs	None
Volatile oil	3.0% min
Moisture	12.0% max
Ash	10.0% max
Acid-insoluble ash	2.0% max
Average bulk index (mg/100 g)	
Bleached	320
Green	250

Source:www.indianspices.com

Ground Cardamom : Chemical and Physical Specifications

Specifications	Suggested limits
FDA DALs	None
Volatile oil	3.0% min
Moisture	12.0% max
Total ash	10.0% max
Acid-insoluble ash	2.0% max
Military specifications (EE-S-631J, 1981) (decorticated cardamom)	
Volatile oil (ml/100 g)	3.0 min
Moisture	12.0% max
Total ash	7.0% max
Acid-insoluble ash	3.0% max
Granulation	95% min through a U.S.S. 40
Bulk index (ml/100 g)	190

*Source:*www.indianspices.com

Average bulk index, Granulation will effect number

Nutritional Composition of Cardamom Seed/ 100 g

Composition	USDA (Ground)	ASTA
Water (g)	8.28	8.0
Food energy (kcal)	311	360
Protein (g)	10.76	10.0
Fat (g)	6.70	2.9
Carbohydrates (g)	68.47	74.2
Ash (g)	5.78	4.7
Calcium (g)	0.383	0.3
Phosphorus (mg)	178	210
Sodium (mg)	18	10
Potassium (mg)	1119	1200
Iron (mg)	13.97	11.6
Thiamine (mg)	0.198	0.18
Riboflavin (mg)	0.182	0.23
Niacin (mg)	1.102	2.3
Ascorbic acid (mg)		ND
Vitamin A activity (RE)	Trace	ND

Source: Composition of Foods, Spices and Herbs, USDA Agricultural Handbook ,8.2,1977.
The Nutritional Composition of Spices, ASTA Research Committee,1977.
ND=Not detected

Marketing

Demand and Supply Analysis

Amomum species are known in North East Indian and South East Asian countries, while the Aframomum species are known in African regions of Sierra Leone, Guinea Coast, Madagascar and Tanzania. Fruits of the Amomum and Aframomum are much larger in size in comparison to *Elettaria cardamomum* and it is easy to distinguish them, but the seed size and anatomy are similar in all the three genera. This species is cultivated in swampy places along mountain streams in Nepal, West Bengal, Sikkim and Assam (eastern Himalayas) and forms one of the cash crops of eastern India.

In India, area under large cardamom during 2000-01 was 26358 ha with production of 5200 tonnes, which remained steady and there was no considerable change in area and production during 2005-06 (30039 ha).

Exports of Large Cardamom from India

During 2005-06, production of large cardamom in India was 5185 tonnes against 5850 in 2001-02.About 80% of the produce is consumed within the country. India roughly exports 19-20% of its production to 17 countries. Pakistan, UK and UAE are the major importers.

Monthly Average Domestic Prices of Large Cardamom (Rs./kg)

Year	Siliguri Average	Gangtok Average
2000-01	209.31	186.84
2001-02	161.48	141.53
2002-03	161.19	141.36
2003-04	137.68	114.96
2004-05	110.65	91.88
2005-06	107.84	89.14
2006-07	135.53	116.73

Source: Spices Board of India

GINGER

Ginger (*Zingiber officinale* Rosc.) is a slender herbaceous perennial belonging to family Zingiberaceae, which are aromatic with unbranched aerial stems, distichous leaves, open sheaths and hypogeal germination, mainly confined to old world tropics with centre of distribution in Indo-Malaysia. It originated in Southeast Asia and was used in this continent for a very long time. The underground stem, or rhizome is used as a medicine in Asian, Indian, and Arabic herbal traditions since ancient times. It is an important cooking spice and is used as a carminative and aromatic stimulant to the gastrointestinal tract and externally as a rubefacient and counter-irritant. It is also famous as an aphrodisiac.

Climate and Soil

Ginger is cultivated in almost all states in India. Kerala is the major ginger growing state. Other major ginger growing states are Orissa, Meghalaya, Himachal Pradesh and Karnataka. Ginger grows in warm and humid climate. It is mainly cultivated in the tropics from sea level to an altitude of 1500 amsl and it can be grown both under rainfed and irrigated conditions. For successful cultivation of the crop, a moderate rainfall at sowing time till the rhizomes sprout, fairly heavy and well distributed showers during growing period and dry weather for about a month before harvesting are necessary. Ginger thrives the best in well drained soils like sandy or clay loam, red loam or lateritic loam. A friable loam rich in humus is ideal. Being an exhausting crop, it may not be desirable to grow ginger in the same site year after year.

High Yielding and Good Quality Ginger Cultivars/Varieties

Traits	Cultivars
High yield (fresh)	Rio-de-Janeiro, Suprabha, Suravi, Jugijan, Thing Pui, Wynad Local, Himachal, Karakkal, Varada, Mahima and Rejatha
Bold rhizome	China, Taffingiva, Sg-35, Mahima and Rejatha
High dry recovery	Zahirabad, Jorhat Local, Kuruppampady, Eranad, Chernad, Suruchi, Maran, Assam, China, Mowshom, Thing Pui, Varada, Mahima and Rejatha
Low fibre	Zahirabad, Kuruppampadi, PGS-16, China, Nadia, Poona, Jamaica, Varada and Mahima
High oleoresin	Wynad Kunnamangalam, Ambalavayalan, Eranad Chernad, Sawthing Pui, Rio-de-Janeiro
Nematode resistance	Mahima
High oil	Rejatha

Propagation

Ginger is propagated by portions of rhizomes known as seed rhizomes. Carefully preserved seed rhizomes are cut into small pieces of 2.5-5.0 cm length weighing 20-25 g each having one or two good buds. The seed rate varies from region to region and with the method of cultivation adopted. In Kerala, the seed rate varies from 1500 to 1800 kg/ha. The seed rhizomes are treated with mancozeb 0.3% (3 g in one litre of water) plus Monocil (2 ml/L) for 30 minutes, shade dried for one hour and planting at a spacing of 20-25 cm along the rows and 20-25 cm between the rows. The seed rhizome pits are prepared with a hand hoe and covered with well rotten farm yard manure and a thin layer of soil and levelled.

Preparation of Land and Planting

The land is to be ploughed four to five times or dug thoroughly with respect to early summer showers to bring the soil to fine tilth. Beds of about one meter width, 15 cm height and of convenient length are prepared with an inter space of 50 cm in between beds. In irrigated crop, ridges are formed 40 cm apart. Planting is done at a spacing of 20 x 20 cm or 25 x 25 cm and a depth of 4-5 cm with viable bud facing upwards. The crop prefers light shade for good growth, yield and dry ginger recovery (Jayachandran *et al.*, 1991).

Manuring

At the time of planting, well decomposed and dried cattle manure or compost @ 25-30 tonnes/hectare has to be applied either by broadcasting over the beds prior to planting or applied in pits at time of planting. Application of neem cake @ two tonnes/hectare at the time of planting helps in reducing incidence of rhizome rot disease and increasing yield.

The recommended dose of fertilizer for ginger grown under open condition (75 kg N, 50 kg P_2O_5 and 50 kg K_2O/ hectare). The recommendation varies with soil types. The fertilizers are to be applied in split doses. This is insufficient for ginger intercropped in coconut gardens. A higher fertilizer dose of NPK @150:100:100kg/ha) is necessary for increasing yield of ginger. The beds are to be earthed up, after each top dressing with fertilizer.

Fertilizer Schedule for Ginger (per ha)

Fertilizer	Basal application	After 40 days	After 90 days
N	-	37.5 kg	37.5 kg
P_2O_5	50 kg	-	-
K_2O	25 kg	-	25 kg
Compost/Cowdung	25-30 tonnes	-	-
Neem cake	2 tonnes	-	-

Water Management

It is grown successfully with an annual rainfall of 1500-3000mm.In the high rainfall tract, it is grown as a rainfed crop while in other areas, a few irrigations are given for commercial production during dry period. It thrives well under partial shade, hence it can be grown as an intercrop under coconut /arecanut and other plantations.

Water requirement of ginger is estimated between 1320-1520mm. Irrigation is given at fortnightly intervals, usually during middle of September to middle of November in India which would increase the yield by 56% and also improved the quality. Germination stage, rhizome initiation stage and rhizome development stage are the critical stages. Scheduling irrigation at 60 or 80mm cumulative pan evaporation enhanced green ginger production. Scheduling irrigation at IW/CPE ratio of 1.0 produced maximum rhizome yield in coastal areas and plains of Kerala.

Cultural Practices

Mulching the beds with green leaves is essential to prevent erosion of soil due to heavy rain. It also adds organic matter to the soil and conserves moisture during the latter part of cropping season. The first mulching is done at the time of planting with green leaves @ 10-12 tonnes/hectare at 40 and 90 days after planning, immediately after weeding and application of fertilizers.

Weeding is done just before fertilizer application and mulching; two-three weedings are required depending on intensity of weed growth. Proper drainage channels are to be provided when there is stagnation of water.

Crop Rotation and Mixed Cropping

Crop rotation is generally followed in ginger. The crops most commonly rotated with ginger are ragi, paddy, gingelly, groundnut, maize and vegetables. In Karnataka, ginger is also mix cropped with ragi, red gram and castor. Ginger is also grown as an intercrop in coconut, arecanut and coffee plantations in Kerala and Karnataka.

Cost of Cultivation of Ginger

S. No.	Components of different cost concepts	Cost/ha (Rs.)	(%) distribution of cost
1	Hired human labour	16367	36.13
2	Animal labour	6	0.01
3	Machine labour	429	0.95
4	Seed/ seedlings	9133	20.14
5	Farmyard manure and chemical fertilizers	9345	20.61
6	Plant protection	797	1.75
7	Land tax and irrigation cess	249	0.55
8	Repair and maintenance charges	4012	8.85
9	Interest on working capital	3736	8.24
10	Other expenses	1258	2.77
11	Total cost 'A' (1-10)	45352	100
12	Interest on fixed capital	816	
13	Cost 'B1' (11+12)	46170	
14	Interest on land value	194551	
15	Cost 'B' (13+14)	240721	
16	Imputed value of household labour	5621	
17	Cost 'C' (13+14)	246342	

Source: Dept. of Economics and Statistics

Input Wise Cost of Cultivation in Ginger (Rs./ha)

Items	Cost(Rs)
Human labour	18740
a) Family	13200
b) Hired	6540
Seeds	14091
Manure	4183.35
Fertilizer	2328.00
PP chemicals	1320.60
Interest on working capital	1631.70
Land revenue	25.00
Total cost Rs./ha	**62059.65**

Harvesting and Processing

For making vegetable ginger, harvesting is done from 6th month onwards. The rhizomes are thoroughly washed in water twice or thrice and sun dried for a day. For preparing dry ginger, the produce is kept soaked in water overnight. The rhizomes are then rubbed well to clean them. After cleaning, the rhizomes are removed from the water and the outer skin is removed with bamboo splinters having pointed ends. While scraping, care must be taken not to rupture oleoresin cells lying just below the outer skin. Iron knives should not be used for peeling as they leave black stains on the peeled surfaces affecting the appearance of rhizomes. The peeled rhizomes are washed and dried in sun uniformly for one week. Ginger should be dried on clean surfaces like clean bamboo mats, cemented / concrete drying yards to ensure that the product does not get contaminated by extraneous matter. Ginger should be dried to a safe moisture level of 8-10 %. The yield of dry ginger is 16-25 % of fresh ginger depending upon the variety.

Only clean and new gunny bags should be used for packing dried ginger. It is preferable to use polythene laminated gunny bags for packing dried ginger. Dried ginger should be stored ensuring protection from dampness. Tonnage of wooden crates should be used to stack the bags to prevent moisture ingress from the floor. Care should be taken to stack the bags 50–60 cm away from the walls. No insecticide should be directly used on dried ginger. Insects, rodents and other animals should be effectively prevented from getting access to the premises where ginger is stored. Stored ginger should be periodically exposed to sun. Prolonged storage of ginger results in deterioration of its aroma, flavour and pungency.

Two types of dried ginger are produced in India viz., bleached and unbleached. Fumigation is necessary before export, since the dried rhizomes are often subjected to infestation. Dry ginger, ginger powder, raw ginger, ginger in brine, salted ginger, crystallized ginger, syrupy ginger, ginger beer, ginger ale and ginger wine are a few of the products prepared from ginger. There is good export market potential for fresh ginger products.

Fresh ginger which is succulent in nature and having low fibre content is used for making products like ginger candy, preserved and salted ginger etc.

Ginger Powder

Ginger powder has a uniform mesh of 50-60. This is a major ingredient of curry powder. It is used in the preparation of ginger beer, ginger wine, soft drinks, cake, biscuit and ginger chicken.

Ginger Oil

It is prepared by distilling dry ginger powder or fresh ginger pieces. Ginger has 1-2% oil and the sesquiterpene 'Zingiberene' is the major ingredient.

Ginger Oleoresin

It is extracted from dry ginger powder using organic solvents like acetone, alcohol or ethylene dichloride. It is the total extract of the spice content.

Ginger Candy

This is prepared by crystallizing soaked fresh ginger pieces in sugar solution.

Ginger Beer

It is prepared by fermenting ginger extract. Six part dry ginger and 3 part hops and other spices are mixed and boiled for 20-30 minutes. Filter this syrup, mix with sugar and cool upto -10°C. Ferment with yeast and citric acid, if required.

Salted Ginger

Fresh ginger with relatively low fibre is harvested at immature stage, clean and soak in 30% salt solution containing 1% citric acid. After about 14 days, remove rhizomes from salt solution, clean and preserve in cold condition. This can be stored in 1-2% brine containing citric acid.

Agmark Grade Specifications of Ginger

Grade Designation	Trade name	Size of rhizome	Extraneous matter, (% w/w), max	Very light pieces, (% w/w), max	Lime as CaO, (% w/w), max
Calicut Ginger - Non Bleached					
NGK	Garbled, Non Bleached Calicut	Not less than 15 mm	2.0	-	-
NUCK	Ungarbled Non bleached Calicut	Not less than 15 mm	3.0	4.0	-
NUGK	Ungarbled Non-bleached Calicut Good	-do-	4.0	6.0	-
Calicut Ginger – Bleached					
BGK	Garbled Bleached Calicut	Not less than 15 mm in length	2.0	-	3.5

Contd...

BUGK	Ungarbled, Bleached Calicut special	-do-	3.0	4.0	4.0
BUGC	Ungarbled, Bleached Calicut Good	-do-	4.0	6.0	6.0
Cochin Ginger - Non Bleached					
NGC	Garbled Non Bleached Cochin	Not less than 15 mm	2.0	-	-
NUGC	Ungarbled Non-bleached Cochin special	Not less than 15 mm	3.0	4.0	-
NUGC	Ungarbled Non-bleached Cochin Good	-do-	4.0	6.0	-
Cochin Ginger – Bleached					
BGK	Garbled, Bleached Calicut	Not less than 15 mm in length	2.0	-	3.5
BUGK	Ungarbled, Bleached Calicut special	-do-	3.0	4.0	4.0
BUGC	Ungarbled, Bleached Calicut Good	-do-	4.0	6.0	6.0

Source: Agmark Grade specifications for Spices, Spices Board, 2001

Whole Ginger: Chemical and Physical Specifications

Specification	Suggested limits
ASTA cleanliness specifications	
Whole dead insects, by count	4
Mammalian excreta, by mg/lb	3
Other excreta, by mg/lb	3.0
Mold % by weight	
Insect defiled infested % by weight	
Extraneous % by weight	1.00
FDA DALs	
Moldy/or insect infested pieces	Ave. of 3% by weight

Contd...

Mammalian excreta	Ave. of 3 mg/lb
Volatile oil	2.0% min
Moisture	12.0% max
Ash	5.0% max
Acid-insoluble ash	1.0% max
Average bulk index (mg/100 g)	NA

Source: www.indianspices.com

Ground Ginger : Chemical and Physical Specifications

Specification	Suggested limits
FDA DALs	None
Volatile oil	1.5% min
Moisture	12.0% max
Total ash	5.0% max
Acid-insoluble ash	1.0% max
Military specifications (EE-S-631J, 1981)	
Volatile oil (ml/100 g)	1.5 min
Moisture	12.0% max
Total ash	7.0% max
Acid-insoluble ash	1.0% max
Crude fibre	8.0% max
Starch	42.0% max
Granulation	95% min through a U.S.S. 30
Average index (ml/100 g)	210

*Source:*www.indianspices.com

Marketing

Demand and Supply Analysis

Indian ginger especially Kochi ginger is one of the best in the world market. It has high esteem in global market because of its characteristic lemon like flavour and absence of fibre content. Hence, the demand is primarily for Kochi/ Kozhikode ginger. Kochi ginger usually trades at a premium.

The crop exhibited an annual growth rate of 4.6% in area, 7.4% in production and 2.7% in productivity. Though grown all over India, the finest quality ginger comes from Kerala endowed as it is with a congenial climate and a rich earthy soil. Indian dry ginger is known in the world market as 'Cochin Ginger' (NUGC) and Calicut Ginger (NUGK). However, four states viz. Kerala, Meghalaya, Arunachal Pradesh and Orissa together contribute over 60% to the country's total production.

State Wise Area and Production of Ginger in India

India is the largest producer of ginger in the world with 3,07,370 tonnes from an area of 85,930 ha in 2002-03 though in terms of area Nigeria and China are on top.

Ginger is cultivated in almost all the States in the country. However, the major producers in 2002-03 were Meghalaya 45,650 tonnes; Kerala (32,410 tonnes Arunachal Pradesh (32,330 tonnes); Mizoram (31,140 tonnes); Orissa (29,490 tonnes); Sikkim (23,610 tonnes); and West Bengal (20,210 tonnes).

World Scenario: According to the FAO, ginger production is looking bright, because harvesting areas and production have increased since 1999, and is estimated to enlarge more in 2002. The harvesting areas of ginger in the world were 310,100, 319,751 and 321,732 hectares in 1999, 2000 and 2001, respectively. Overall production of ginger around the world has also increased. Production was 835,173 tonnes in 2001, compared with 812,372 tonnes in 2000 and 775,717 tonnes in 1999. Nigeria is the world's biggest ginger producer. In 1999, Nigeria cultivated 166,800 hectares of ginger across the country and the area of cultivation extended to 174,000 hectares in 2000 and 2001. This is because of high demand from both domestic and foreign markets. India is another important producer. Harvesting areas across India remained stable at 80,000 hectares from 1999–2001. Producing areas in China have also enlarged each year, 20,700 hectares were used for ginger cultivation in 2001, compared with 17,750 and 19,170 hectares in 1999 and 2000, respectively.

World production of ginger during 2005 was at 1,004,546 tonnes from an area of 339,826 ha., mainly distributed over India, China, Nigeria, Indonesia, Bangladesh, Thailand, Philippines and Jamaica. The countries importing the amounts are United Kingdom, United States and Saudi Arabia. Nigeria ranks first with respect to area under ginger covering about 56.23% of world total area under ginger followed by India (23.60%), China (4.47%), Indonesia (3.37%) and Bangladesh (2.32%) which are the important countries producing ginger in the world.

State Wise Area and Production of Ginger in India
(Area in hectares, production in tonnes)

State	2000-01		2001-02		2002-03		2003-04		2004-05	
	Area	Production	Area	Production	Area	Production	Area	Production	Area	Production
Ginger										
Andaman & Nicobar	388	1220	390	1225	399	1278	508	1625	528	1828.1
Andhra Pradesh	2329	6516	2550	6263	2026	3313	1873	5344	1873	10906
Arunachal Pradesh	4774	35297	4616	38021	4448	32332	4680	37177	4451	32877
Bihar	711	1136	302	414	491	728	367	565	365	555
Chattisgarh	0	0	0	0	1382	1103	1291	1353	1330	1318
Gujarat	1142	18265	1051	16275	1152	19411	1911	22938	1831	26924
Himachal Pradesh	2299	0	0	0	0	0	2011	14658	2043	14567
Jammu & Kashmir	40	48	55	66	38	46	29	36	22	27.7
Karnataka	10758	14160	10760	14170	9760	12850	8281	10854		
Kerala	11612	42704	10706	14170	9760	12850	8516	32972	9991	45305
Maharashtra	1259	1187	1243	1182	1304	1249	813	573	1293	1239
Meghalaya	7871	44902	8397	46590	7941	45655	8882	49215	9234	47138
Mizoram	5130	25377	7287	46648	5097	31136	4481	29582	4532	38069
Manipur	1140	1880	2142	3530	1270	2100	1270	2100	2480	4090
Madhya Pradesh	4265	5353	4950	6801	5187	5325	0	0	0	0
Nagaland	1016	13208	0	0	2010	13330	2010	13330	0	0
Orissa	12060	23840	13040	25300	15340	29490	15480	30020	15224	29793
Rajasthan	361	980	260	1100	181	640	134	582	86	174
Sikkim	4700	25785	5102	24000	4950	23613	6470	33530	6515	34710
Tamil Nadu	522	9625	479	10265	416	6665	0	0	0	0
Tripura	0	0	1350	5400	1360	5450	1023	2700	1180	2774
Uttar Pradesh	625	1935	644	2170	665	1583	832	2367	0	0
Uttarakhand	1095	11084	1270	15815	1270	16015	1599	17079.07	0	0
West Bengal	9328	18908	9326	17832	9383	20206	9408	18269	9125	22074
Total	**83425**	**303410**	**85920**	**297237**	**85830**	**286368**	**81869**	**326869.07**	**59660**	**225366**

Area, Production and Yield of Ginger in 2005

Country	Area	Yield	Production
China	23,400	117,521	275,000
India	70,000	32,857	230,000
Nigeria	185,000	5,946	110,000
Bangladesh	8,094	59,303	48,000
Indonesia	17,000	88,824	151,000
Thailand	14,000	24,286	34,000
Philippines	3,610	62,812	22,675

Source: FAO Stat

Indian scenario: India ranks first with respect to ginger production contributing about 32.75%of total world's ginger followed by China (21.41%), Nigeria (12.54%) and Bangladesh (10.80%). World average ginger productivity is 2419 kg/ha. USA has the highest productivity/hectare (55636 kg) and India's is only 3367 kg.

Ginger Exports from India

Average annual export of ginger in India was about 8%of the domestic production from 1990-91 to 1998-99. Export of ginger showed a highly fluctuating trend during the last forty years. India is exporting ginger mainly to Pakistan, Bangladesh, Saudi Arabia, UAE, Morocco, and USA. Indian ginger is facing stiff competition from other producing countries like China, Nigeria and Thailand.

Exports of ginger from India crossed 13,000 tonnes in 2004-05 for the first time after 1997-98 following decline in supply from other major producing countries. The highest-ever export of ginger was in 1996-97 when the country exported 29,737 tonnes valued at Rs 59.24 crore. The unit value then was a meagre Rs 19.92/ kg.

Average annual export of ginger in India was about 8%of the domestic production during 1990-91 to 1998-99. Export of ginger showed a highly fluctuating trend during the last forty years. India is exporting ginger mainly to Pakistan, Bangladesh, Saudi Arabia, UAE, Morocco, and USA. Indian ginger is facing stiff competition from other producing countries like China, Nigeria and Thailand.

India emerged as a major producer and exporter of value-added products of ginger like ginger oil and oleoresin. The export of ginger oleoresin from India has increased substantially during the last five years.

Price Behaviour

During 2001-02, the price exhibited a downward trend due to excess supply and poor demand from exporters and local traders. The monthly average prices touched a low of Rs.2100/quintal in Kozhikode market of Kerala. Lack in demand amid bumper arrivals witnessed during December and January kept pressures on prices. Lower export demand has also weighed on the market

Shortage in the international market had also pushed up the prices. In fact, the prices were ruling high in 2003-04 due to short supply. The continuous low prices in the past years forced the major producer Nigeria to stop cultivation while the crop in China had been poor last fiscal and the year before.

In 2003-04, the unit (kg) value was Rs 46.81/kg, Rs 42.13 in 2004-05. and Rs.45 in 2005-06. In fact, during the period from 1991-92 to 1997-98, the exports remained at higher levels above 12,000 tonnes except in 1992-93 when it was 9,825 tonnes.

Major World Markets

The main overseas markets of the commodity are Australia, Pakistan, Bangladesh, Saudi Arabia, Yemen, UAE, Morocco, Canada, the Netherlands, Japan, the UK and the US.

Though there is tough competition from other producing countries like China, Nigeria and Thailand, the superior quality of Indian ginger attracted steady market, they said.

TURMERIC

Turmeric is one of the oldest spices and used in India since ages. This spice belongs to India and also called Indian saffron. It is used in many Hindu rituals and Indians tend to be attached to it emotionally. It is the dried rhizome of a herbaceous perennial tropical plant. Turmeric of commerce is the dried rhizome of the plant *Curcuma domestica* Val. syn. *C. longa* L. The genus Curcuma originated in the Indo-Malayan region. Considerable species diversity of Curcuma occurs in this region. About 40 species of the genus including *C. longa* are indigenous to India indicating Indian origin. Though turmeric is now grown in India, Pakistan, Malaysia, Myanmar, Vietnam, Thailand, Philippines, Japan, China, Korea, Sri Lanka, Nepal, South Pacific Islands, East and West Africa, Malagasi, Caribbean islands, and Central America, India is the major producer and exporter of turmeric at present.

Soil and Climate

Turmeric can be grown in diverse tropical conditions from sea level to 1500 m above MSL at temperature ranging from 20–30°C with a rainfall of 1500 mm or more/annum. It is grown in different types of soils from light black, ashy loam, clayey loam and red soils. However, it thrives the best in a well drained sandy or clay loam soils.

Planting

The land is prepared with receipt of early monsoon in May. The soil is brought to fine tilth by giving about four deep ploughings. Weeds, stubbles and roots are removed. Beds of size 1-1.5 m width, 15 cm height and of convenient length are prepared with spacing of 50 cm between beds.

Whole or split mother rhizomes or fingers are used for planting and whole mother rhizomes are superior contributing towards maximum yield. Well developed, healthy and disease free rhizomes are to be selected. Rhizomes are planted in small pits made with a hand hoe in the beds in rows with a spacing of 25 x 30 cm and covered with soil or dry powdered cattle manure. The optimum spacing in furrows and ridges is 45-60 cm between rows and 25 cm between plants. A seed rate of 2000-2500 kg of rhizomes is required for planting one hectare of turmeric.

Manuring

Farm yard manure (FYM) or compost @ 40 tonnes/hectare is applied by broadcasting at the time of preparation of land or as basal dressing by spreading over the beds to cover the seeds after planting. Fertilizers @ 60 kg, 50 kg K_2O_5 and 120 kg K_2O/ hectare are to be applied in split doses. Zinc @ 5 kg/ha may also be applied at the time of planting and organic manure like groundnut cake can also be applied. In such case, the quantity of FYM can be reduced.

Fertilizer Schedule for Turmeric (kg/ ha)

Schedule	N	P_2O_5	K_2O	Compost/ cowdung
Basal application	-	50	60	40 tonnes
After 40 days	30	-	-	-
After 90 days	30	-	60	

Mulching

The crop is to be mulched immediately after planting with green leaves @ 12-15 tonnes/hectare. Mulching may be repeated for a second time after 45 days with the same quantity of green leaves after weeding and application of fertilizers.

Weeding and Irrigation

Weeding has to be done thrice at 60, 120 and 150 days after planting depending upon weed intensity. In iirrigated crop, depending upon the weather and soil conditions, about 15 to 23 irrigations are to be given in clayey soils and 40 irrigations in sandy loams.

Mixed Cropping

Turmeric can be grown as an intercrop in coconut and arecanut plantations. It can also be raised as a mixed crop with chillies, colocasia, onion, brinjal and cereals like maize and ragi.

Cost of Cultivation of Turmeric

S. No.	Components of different cost concepts	Cost /ha (Rs.)	(%) distribution of cost
1	Hired human labour	12692	34.58
2	Animal labour	204	0.56
3	Machine labour	4556	12.41
4	Seed/ seedlings	4556	12.44
5	Farmyard manure and chemical fertilizers	7044	19.19
6	Plant protection	151	0.41
7	Land tax and irrigation cess	103	0.28
8	Repair and maintenance charges	2550	6.95
9	Interest on working capital	3096	8.43
10	Other expenses	1743	4.75
11	Total cost 'A' (1-10)	36705	100
12	Interest on fixed capital	1844	
13	Cost 'B1' (11+12)	38549	
14	Interest on land value	170421	
15	Cost 'B' (13+14)	208970	
16	Imputed value of household labour	7201	
17	Cost 'C' (13+14)	216171	

Source: Dept. of Economics and Statistics.

Harvesting and Processing

The processing of turmeric consists of four stages.

Curing: Fingers are separated from mother rhizomes. Mother rhizomes are usually kept as seed materials. The fresh turmeric is cured for obtaining dry turmeric. Curing involves boiling of fresh rhizomes in water and drying in the sun. The traditional method of curing is as follows:

The cleaned rhizomes are boiled in copper or galvanized iron or earthern vessels with water just enough to soak them. Boiling is stopped when froth comes out and white fumes appear jigging out a typical odor. The boiling lasts for 45-60 minutes when the rhizomes are soft. The stage at which boiling is stopped largely influences colour and aroma of final product. Over cooking spoils colour of final product while under-cooking renders the dried product brittle.

Improved Scientific Method of Curing Turmeric

The cleaned fingers / mother rhizomes are taken in perforated trough made of GI or MS sheet with extended parallel handle. The perforated trough contain the raw turmeric are then immersed in a pan of water which can hold 3-4 troughs at same time. Boil it till the fingers / mother rhizomes become soft. The cooked turmeric is taken out of the pan by lifting the trough and draining the water into pan itself. The same hot water in the pan can be used for boiling next set of raw turmeric. The cooking of turmeric is to be done with in 2-3 days of harvest.

The rhizomes may also be placed in baskets with perforated bottom and sides and dipped in coveted tanks when the quantity is large or may be put directly into the vessels when the quantity is small. The mother rhizomes and fingers are generally cured separately.

Drying: The cooked fingers are dried in the sun by spreading in 5-7 cm thick layers on bamboo mats or drying floor. A thinner layer is not desirable as colour of dried product may be adversely affected. During night time, the material should be heaped or covered. It may take 10-15 days for rhizomes to become completely dry. Artificial drying using cross flow hot air at a maximum temperature of 60°C also give a satisfactory product. The yield of dry turmeric varies from 20-30 % depending upon variety and the location where the crop is grown.

Polishing: Dried turmeric has a poor appearance and rough dull colour outside the surface with scales and root bits. The appearance is improved by smoothening and polishing the outer surface by manual or mechanical rubbing.

Manual polishing consists of rubbing the dried turmeric fingers on a hard surface or trampling them under feet wrapped with gunny bags. The improved method is by using hand operated barrel or drum mounted on a central axis, the sides of which are made of expanded metal mesh. When the drum filled with turmeric is rotated, polishing is effected by abrasion of surface against the mesh and by mutual rubbing against each other as they roll inside the drum. The turmeric is also polished in power operated drums. The yield of polished turmeric from raw material varies 12-25 %.

Colouring: The colour of turmeric always attracts the buyers. In order to impart attractive yellow colour, turmeric suspension in water is added to the polishing drum in the last 10 minutes. When the rhizomes are uniformly coated with suspension, they may be dried in the sun.

After polishing, it should be kept in clean sacks and stored over wooden pallets in stores. Stores should be clean, free from infestation of pests, spiders and harborage of rodents. Pesticides should not be applied on the dried / polished turmeric to prevent storage pest.

Agmark Grade Specifications of Turmeric

Grade Designation	Flexibility	Fingers broken or whole of 15 mm or less in length, (% w/w), max	Foreign matter (% w/w), max	Immature or defective bulb, (% w/w), max	Bulbs, (% w/w), max
Alleppey Finger turmeric					
Good	Should be hard to touch	5	1.0	3.0	4.0
Fair	-do-	7	1.5	5.0	5.0
Turmeric fingers (for varieties other than Alleppey)					
Special	Should be hard to touch and break with metallic twang	2	1.0	0.5	2.0
Good	-do-	3	1.5	1.0	3.0
Fair	Should be hard	5	2.0	1.5	5.0

Source: Agmark Grade specifications for Spices, Spices Board, 2001

Whole Turmeric: Chemical and Physical Specifications

Specification	Suggested limits
ASTA cleanliness specifications	
Whole dead insects, by count	3
Mammalian excreta, by mg/lb	5
Other excreta, by mg/lb	5.0
Mold % by weight	3.00
Insect defiled infested % by weight	2.50
Extraneous % by weight	0.50
FDA DALs	None
Curcumin content	5.0% min
Moisture	10.0% max
Ash	8.0% max
Acid-insoluble ash	1.0% max
Average bulk index (mg/100 g)	N/A

*Source:*www.indianspices.com

Ground Turmeric: Chemical and Physical Specifications

Specification	Suggested limits
FDA DALs	None
Curcumin content	5% min
Moisture	10.0% max
Total ash	8.0% max
Acid-insoluble ash	1.0% max
Military specifications (EE-S-631J, 1981)	
Volatile oil (ml/100 g)	3.5 min
Moisture	10.0% max
Total ash	8.0% max
Acid-insoluble ash	0.6% max
Crude fibre	9.5% max
Colour power expressed as curcumin	
Granulation	95% min through a U.S.S. 40
Average index (ml/100 g)	185

*Source:*www.indianspices.com

Nutritional Composition of Turmeric per 100 g

Composition	USDA (Ground)	ASTA
Water (g)	11.36	6.0
Food energy (kcal)	354	390
Protein (g)	7.83	8.5
Fat (g)	9.88	8.9
Carbohydrates (g)	64.93	69.9
Ash (g)	6.02	6.8
Calcium (g)	0.182	0.2
Phosphorus (mg)	268	260
Sodium (mg)	38	10
Potassium (mg)	2525	2500
Iron (mg)	41.42	47.5
Thiamine (mg)	0.152	0.090
Riboflavin (mg)	0.233	0.190
Niacin (mg)	5.140	4.8
Ascorbic acid (mg)	25.85	50
Vitamin A activity (RE)	Trace	ND

Source: Composition of Foods, Spices and Herbs, USDA Agricultural Handbook ,8.2,1977.
The Nutritional Composition of Spices, ASTA Research Committee,1977.
ND=Not detected

Marketing

Demand and Supply Analysis

India is the largest producer, consumer and exporter of turmeric in the world. Indian turmeric is the best in the world market because of its high curcumin content. World production of turmeric stands at 800000 tonnes in which India holds a share of approximately 75-80% of worlds turmeric production and 60% of world exports. Also India consumes around 80% of its own production. Indian turmeric industry size is roughly Rs2500-3000 Crores. Export of turmeric and its products is around Rs.120-170 Crores. Other major producers are Pakistan, China, Haiti, Jamaica, Peru, Taiwan and

Thailand. Asian countries consume much of their turmeric production. The total annual consumption of turmeric around the globe is approximately 38 lakh bags to 40 lakh bags (65 kg/bag).

Area and Production of Turmeric in India

The important turmeric growing states in India are Andhra Pradesh (34% of area), Tamil Nadu (12%), Orissa (13%), Maharashtra, Assam, Kerala, Karnataka and West Bengal. In case of production Andhra Pradesh accounted 60%of total turmeric production in India followed by Tamil Nadu (13%) and Orissa (12%). Production of turmeric in India is showing an increasing trend with 1,35,200 tonnes in 1975-76 to 6,60,000 tonnes in 2003-04 with a maximum of 7,07,400 tonnes during 1993-94. Decline in area since 2000-01 could be attributed to poor prices especially during 2000-01, shift towards other crops like sugarcane and drought during 2003-04.

State Wise Area and Production of Turmeric in India
(Area in Hectares, production in Tonnes)

State	2000-01		2001-02		2002-03		2003-04		2004-05	
	Area	Production	Area	Production	Area	Production	Area	Production	Area	Production
Turmeric										
Andhra Pradesh	73931	375760	61680	249487	56822	283541	58438	320683	60356	417818
Arunachal Pradesh	454	1872	573	2094	514	1950	447	1527	505	1794
Assam	11626	8032	11812	8164	12066	8315	12249	8420	11697	8401
Bihar	3173	3104	2918	3038	2895	2873	2978	2881	2823	2788
Chattisgarh	0	0	0	0	680	521	678	616	738	642
Gujarat	533	8707	689	12243	843	14688	949	1183	1017	14116
Himachal Pradesh	115	63	0	0	0	0	124	71	164	114
Jammu & Kashmir	40	48	55	66	38	46	1	1.00	0	0
Karnataka	9251	41040	6710	35600	6153	30147	5410	26381	0	0
Kerala	4127	9037	3558	7895	3140	6938	2774	5652	2881	6244
Maharashtra	6903	8671	6765	8507	6644	8220	6806	8527	6833	8539
Meghalaya	1523	8565	1523	8577	1535	8675	1561	8656	1673	9442
Mizoram	320	2915	280	2811	287	2785	275	2258	196	1648
Manipur	250	170	310	210	330	230	330	230	470	320
Madhya Pradesh	371	334	382	399	556	542	655	682	664	635

Contd...

Nagaland	600	3000	0	0	610	3050	610	3050	0	0
Orissa	25310	60630	27140	65830`	23640	55970	23780	56420	23554	55977
Rajasthan	157	536	134	481	129	213	96	366	91	196
Sikkim	400	1200	400	1200	428	1300	520	1700	520	1735
Tamil Nadu	33000	158643	23638	118257	17298	64536	16181	67250	21616	118446
Tripura	0	0	1512	6550	1512	6600	1470	4820	1102	3751
Uttar Pradesh	1054	1812	1159	1981	1111	2032	1130	2140	0	0
Uttaranchal	437	3062	467	3469	483	3469	654	7394.19	0	0
West Bengal	13885	22445	12794	22031	11731	21346	12619	24496	12400	24478
Total	**187460**	**719646**	**164499**	**153969**	**149445**	**527987**	**150735**	**555404.19**	**149300**	**677084**

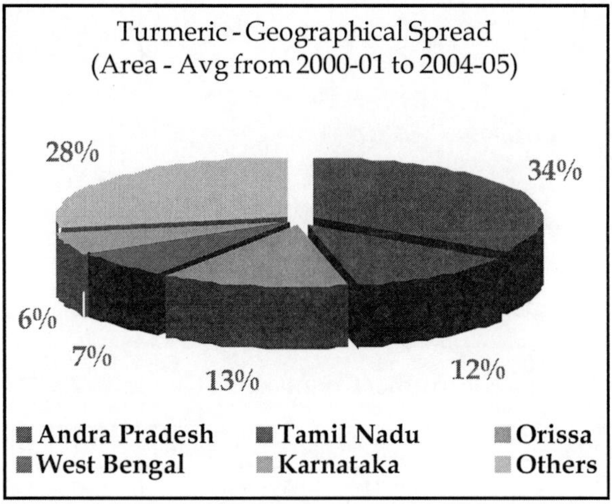

Source: Spices Board India

Export Trends

The Indian turmeric is the best in the world. The country contributes to the largest share in the world, total export around 40,000 tonnes. United Arab Emirates (UAE), is the major importer accounting for 15.17% of the total exports followed by United States of America (USA) with 10%. The other leading importers are Bagladesh, Japan, Srilanka, UK, Malaysia, South Africa, Netherland and Saudi Arabia. The processed forms of turmeric exported are dry turmeric, fresh turmeric, turmeric powder and oleoresin. Even though production in India is high, surplus is less because of higher domestic consumption. Although export demand would be around 40,000 tonnes annually, good export demand could drive the prices. Export demand is rising from last couple of years on account of increased non-food use and also food use in some countries.

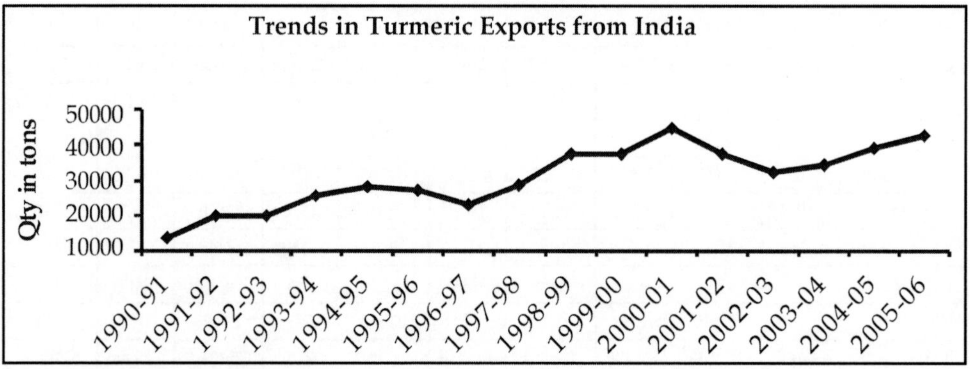

Key export markets

- UAE - 17%
- USA - 10%
- Bangladesh - 9%
- Sri Lanka - 7%
- Japan - 7%
- Malaysia - 6%
- UK - 6%

Note: Share based on exports for 2003-04 and 2004-05

Export of Turmeric to other Countries by Quantity (2003 to 2005)

S.No.	Country	Quantity in tonnes			
		2003-2004	**% to total**	**2004-2005**	**% to total**
1	U.A.E	7,209	19.46	5,663	15.17
2	U S A	3,880	10.47	3,731	10.00
3	Bangladesh	3,547	9.58	3,112	8.34
4	Japan	2,694	7.27	2,445	6.55
5	Sri Lanka	2,316	6.25	2,935	7.86
6	U K	2,060	5.56	2,289	6.13
7	Malaysia	2,051	5.54	2,538	6.80
8	Others	9,713	26.22	10,907	29.22
9	South Africa	1,353	3.65	1,298	3.48
10	Netherland	1,198	3.23	1,348	3.61
11	Saudi Arabia	1,020	2.75	1,057	2.83
	Total	**37,042**	**100.00**	**37,322**	**100.00**

Source: DGFT, Ministry of Commerce, GOI

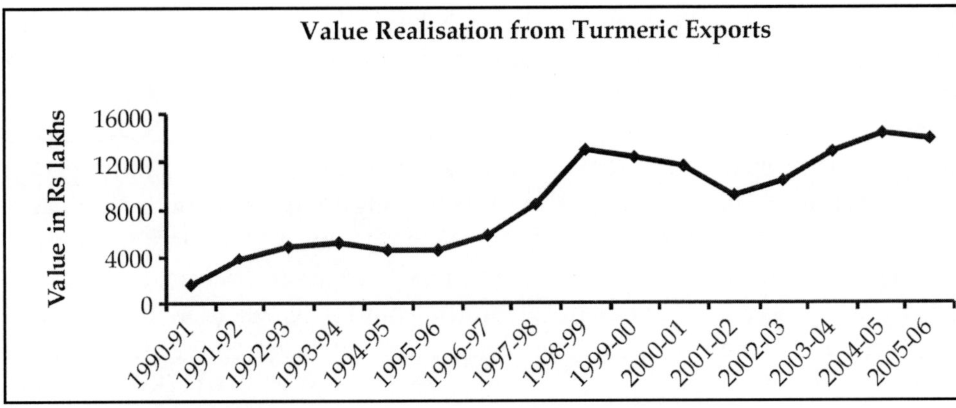

Source: Spices Board of India

Most of the produce shipped is dry turmeric after polishing. In terms of volume, turmeric oleoresin account for about 200 tonnes/ annum and turmeric powder constitutes a very small portion.

Major Trading Centers of Turmeric

Turmeric holds reputation of being auspicious in India. It is used in India since ages and also used in many Hindu rituals and ceremonies. Production exceeds the domestic demand and that gets India into an exporting country.

Turmeric is available in two seasons in India February to May and August-October. There are various varieties of turmeric produced in India namely Alleppey Finger (Kerala) Erode Turmeric (T.N) Salem Turmeric (T.N), Rajapore Turmeric (Maharashtra) and Sangli (Maharashtra).

In India, turmeric is traded at Nizamabad and Dugirala in A.P, Sangli in Maharashtra, Salem, Erode, Dharmapuri and Coimbatore in T.N. Also turmeric is traded in Indian commodity exchanges namely National Commodity and Derivatives Exchange Ltd and the Spices and oilseeds Commodity Exchanges.

TREE SPICES

Tree spices are spices originating from tree crops. Altogether 17 species are put under the group "tree spices" of which many of them are grown in India, Sri Lanka, Indonesia, Malaysia, China, Myanmar, Vietnam, Tanzania and Jamaica or occurring in the humid tropical forests of South East Asia, Australia, Pacific Islands and Tropical America (Krishnamoorthy *et al.*, 1997). Cinnamon (*Cinnamomum veerum* Bercht. & Presl.), nutmeg (*Myrstica fragrans*

Houtt.), clove (*Syzygium aromaticum* (L.). Merr. & Perry and allspice (*Pimenta dioica* L.) are the important tree spices grown in India. The commercial products of these crops are whole spice, ground spice, spice oil, oleoresin etc. In India, cinnamon is cultivated in Kerala, Karnataka and Tamil Nadu and is more prevalent in the hilly regions of Western Ghats. Clove is mainly grown in Kanyakumari, Tirunelveli, Salem and the Nilgiris districts of Tamil Nadu and a few pockets in Kerala, Karnataka, Maharashtra and Andaman Islands. Nutmeg cultivation is more prevalent in the districts of Trichur, Ernakulam. Kottayam and Trivandrum of Kerala and also a few pockets in Tamil Nadu and Karnataka. A few allspice trees are available at Nagarcoil, Kallar, Burliar (TamilNadu), Peruvannamuzhi, Ambanas Estates, Kollam (Kerala) and Ratnagiri (Maharashtra).

Clove (*Syzygium aromaticum* L.)

Clove of commerce is the aromatic, fully grown but unopened dried flower buds of *Syzygium aromaticum*. It is one of the ancient valuable spices of the orient, holding a unique position in international trade. It originated in Moluccas, the so called 'Spice Island' in Indonesia. The islands of Zanzibar and Pemba (now parts of Tanzania), Indonesia, Madagascar, Sri Lanka and Malayasia are the important producers of clove.

Clove was first introduced to India around 1800 AD by the East India company in its 'spice garden' in Courtallam, Tamil Nadu. Induced by the success of its introduction, cultivation of clove was extended during the period after 1850 AD to Nilgiris (Burliar), southern region of the erstwhile Travancore State and the slopes of Western Ghats. The important clove growing districts in India now are Nilgiris, Tirunelveli, Kanyakumari, Nagercoil and Ramanathapuram districts of Tamil Nadu, Kozhikode, Kottayam. Kollam and Thiruvananthapuram Districts of Kerala and South Kanara district of Karnataka.

Climate and Soil

Clove is a tropical plant and requires warm humid climate. Although there is a general belief that clove requires proximity to sea for proper growth and yield, experience in India showed that the trees did well in the hinterland conditions too. Clove thrives in all situations ranging from sea level up to an altitude of 1000 meters. Deep loam soil with high humus content found in the forest region is the best suited for its cultivation. It grows satisfactorily on laterite soil, loam and rich black soils having good drainage.

Varieties and Planting Materials

Clove plantations in India originated from a few seedlings obtained originally from Mauritius. The germplasm collections made from within the country have not therefore given appreciable variability in yield and growth factors.

Selection of Seed Garden

In clove plantations of about 40 years old with healthy and high yielding (above 3 kg dry buds/tree) trees are chosen as seed gardens.

Selection of Mother Trees

Mother trees are selected after surveying trees in the selected garden. The trees are evaluated for their yield and indexed for diseases and pests incidence. These selected trees would form the source of seeds for production of seedling planting materials. Mother trees should be high yielding producing about three kg dry buds/tree, regular bearer, bushy in nature and free from diseases. There are no released varieties in clove at present and hence identifying high yielding trees as parent trees is a pre requisite for production of quality planting materials.

Collection of Seeds and Sowing

Clove seeds are collected from fully ripe fruits popularly known as 'mother of clove'. The ripe fruits from the trees could be collected during July-August from the first a few harvests. As the viability is low, the seeds are to be sown immediately after collection. If they are to be transported, they have to be kept in a basket and sown immediately after reaching the site.

Nursery Techniques

The fruits are to be soaked in water and the pericarp removed before sowing. Only fully developed and uniform seeds with a pink radicle are selected for sowing. Raised nursery beds are prepared on fertile soil with high percentage of organic matter. The beds normally measure one metre width and two to three metre length. The seeds are sown at a spacing of two-three cm and depth of two cm in sand. The nursery has to be watered regularly and the nursery protected from sun.

Care and Management of the Nursery

The germination commences within two weeks of sowing and lasts for about six weeks. The sprouts are to be transplanted in polythene bags (size 30 x 15 cm), containing a mixture of soil, sand and well powdered cowdung (3:3:1 ratio). One year old seedlings are transplanted to bigger polythene bags (35 x 22 cm). The seedlings are ready for planting in the main field when they are about 18-24 months old; such seedlings should be free from diseases about 45 cm height, bushy and well branched.

Site Selection and Planting

The site for cultivation of clove should have good drainage since the crop cannot withstand water logging. It can be grown in coconut gardens of midland. At higher elevations, it can be mixcropped with pepper or coffee. Clove requires a location protected from wind. If the site is open, windbreaks must be provided. Eastern and North Eastern hill slopes, well-drained valleys and riverbanks are ideal for clove cultivation. The crop thrives well under open condition at high altitude where there is fair distribution of rainfall.

The area selected for raising clove plantation is cleared off wild growth before monsoon. Pits of size 75 cm cube are dug at a spacing of seven metre accommodating about 200 trees/ ha. If grown as an inter-crop, spacing is to be adjusted based on the main crop. Pits are filled with a mixture of compost or cattle manure and loose friable top soil. Seedlings are planted in the centre of pits in May-June with onset of monsoon and watered regularly. Banana may be planted to provide cool and humid atmosphere to tender plants. Watering may be done during summer.

Manuring

Clove trees are to be manured regularly for proper growth and flowering. About 15 kg of rotten cattle manure or compost is applied per plant in the initial years. The quantity is increased gradually so that a well grown tree of 15 years and more gets 40 to 50 kg of organic manure. Inorganic fertilisers are applied, starting with 20 g Nitrogen (N), 18 g Phosphorus (P_2O_5) and 50 g Potash (K_2O) per plant in the first year, 40 g Nitrogen (N), 36 g Phosphorus (P_2O_5) and 100 g Potash (K_2O) per plant in the second year and gradually increased to 300 g Nitrogen (N), 250 g Phosphorus (P_2O_5) and 750 g Potash (K_2O) per plant for trees of 15 years and more.

Organic manure is applied in May-June with commencement of monsoon. Fertilisers are given in two equal split doses, one in May-June along with the organic manure and the other in September-October. For manuring, shallow trench is dug around the tree about 50 to 160 cm away from base depending upon age.

No inter cultivation is usually done for clove. Weeds are removed at regular intervals. As the branches of full grown trees have tendency to overcrowd, thinning is done occasionally. Dead and diseased shoots should be removed once or twice a year.

Harvesting and Curing

Clove tree begins to yield from seventh year of planting and full bearing stage is attained after 15 to 20 years. The flowering season is September to October in plains and December to February at high altitudes.

Flower buds are formed on young flush. It takes about five to six months for buds to become ready for harvest. The optimum stage for picking clove buds is when the buds are fully developed and base of calyx has turned from green to pink. Such clove buds are carefully picked by hand. Care should be taken to collect buds at correct stage as otherwise, the quality of produce will be poor to a considerable extent. When trees are tall and branches are beyond reach, platform ladders are used for harvesting. Bending branches or knocking down the bud clusters with sticks is not desirable as these practices will affect future bearing.

Buds after separation from stalks are spread evenly to dry, in-the sun on mats or cement floors. During nights, buds should be stored undercover, otherwise they re-absorb moisture. Period of drying depends on prevailing climatic conditions. Normally, it is possible to dry cloves in four or five days under direct sun and in about four hours when they are heated on zinc trays over a regulated fire. Fully dried buds develop the characteristic dark brown colour and are crisp. Improperly dried and stored cloves have much darker colour and wheat wrinkled appearance. Such a produce is inferior in quality. About 8000 to 10,000 good quality clove buds would weigh one kilogram.

Yield

A well maintained fully grown tree under favourable conditions starts flowering from fourth year after planting. The average annual yield at the 15th year may be taken as two kg/tree or 400 kg/ hectare.

Processed Products

Clove bud oil (yield 16%), clove stem oil (yield 4-5%), clove leaf oil (yield 1-2%), oil of mother cloves (yield 6.5%) and clove root oil (yield 6%) are the main products of clove.

Whole cloves: Chemical and physical specifications

Specification	Suggested limits
ASTA cleanliness specifications	
Whole dead insects, by count	4
Mammalian excreta, by mg/lb	5
Other excreta, by mg/lb	8.0
Mold % by weight	1.00
Insect defiled infested % by weight	1.00
Extraneous % by weight	1.00
A 5% allowance for attached clove stems over and above the tolerance for other extraneous matter is permitted.	
FDA DALs	
Adulteration with stems by weight	Ave. of 5%
Volatile oil	16.0% min
Moisture	8.0% max
Ash	5.0% max
Acid-insoluble ash	0.5% max
Military specifications (EE-S-631J, 1981)	5% or more by weight of stems
Average bulk index (mg/100 g)	240

*Source:*www.indianspices.com

Ground cloves : Chemical and physical specifications

Specification	Suggested limits
FDA DALs	None
Volatile oil	14.0% min
Moisture	8.0% max
Total ash	5.0% max
Acid-insoluble ash	0.5% max
Military specifications (EE-S-631J, 1981)	
Volatile oil (ml/100 g)	15.0 min
Moisture	9.0% max
Total ash	6.0% max
Acid-insoluble ash	0.5% max
Granulation	95% min through a U.S.S. 30
Average index (ml/100 g)	180

*Source:*www.indianspices.com

Nutritional composition of cloves per 100 g

Composition	USDA (Ground)	ASTA
Water (g)	6.86	5.0
Food energy (kcal)	323	430
Protein (g)	5.98	6.0
Fat (g)	20.06	14.5
Carbohydrates (g)	61.22	68.8
Ash (g)	5.88	5.0
Calcium (g)	0.646	0.7
Phosphorus (mg)	105	110
Sodium (mg)	243	250
Potassium (mg)	1102	1200
Iron (mg)	8.68	9.5
Thiamine (mg)	0.115	0.11
Riboflavin (mg)	0.267	ND
Niacin (mg)	1.458	1.5
Ascorbic acid (mg)	80.81	81
Vitamin A activity (RE)	53	53

Source: Composition of Foods, Spices and Herbs, USDA Agricultural Handbook ,8.2,1977. The Nutritional Composition of Spices, ASTA Research Committee,1977.
ND=Not detected

Medicinal Use

According to traditional herbalist cloves are beneficial in the following problems:

- Treats indigestion
- Diarrhoea
- Hernia
- Ringworm
- Athlete's foot and other fungal infections.
- Respiratory.
- Used in anti-gout
- Relieves toothache.
- Helpful in insomnia and Curbs the desire for alcohol

Other Uses

- It is widely used for food flavoring and aroma in a variety of cuisines, used in confectionery and other products
- Personal Hygiene products: Widely used in toothpastes, mouthwash and tooth powders.
- Flavoring Industry
- Perfumery
- Cosmetics
- Toiletries

Use of Clove Oil

- *Medicine*: Clove has traditionally been used in various medical preparations because of its flavor, analgesic and other properties. It is also used as a topical germicide. Clove oil is extensively used in the medical industry because it is an effective antibiotic, antiseptic, antiviral, analgesic, antioxidant, carminative, expectorant, and stimulant
- *Aromatherapy*: Clove oil is widely used as a blend with other oils for its therapeutic value. The oil is stimulating, revitalizing, purifying, energizing, and thus used in aromatherapy
- Flavoring Industry
- Perfumery
- Personal Hygiene products

Trade in Cloves

Global Scenario

Indonesia is the largest producer of cloves, followed by Madagascar, Zanzibar, India and Sri. Lanka, Indonesia has further accelerated its production in the last 30 years. Madagascar, the world's second largest producer, has supplied around 10% of world consumption since 1975, but has not grown its market share significantly since 1980. Tanzania (Zanzibar), the world's third largest producer of cloves, previously the second biggest, has seen its share decline. Cloves are also being traded in the international circuit for its derivatives including essential oil and oleoresins. The worldwide clove market has been stagnant for the past few years according to statistics, global production has decreased by 12% and also productivity gone down by 4%.

In addition to Indonesia, cloves are grown in India, Malaysia, and Sri Lanka, but production in the Asian countries is quite small as compared to that in countries in other regions, namely Madagascar, Tanzania (especially in Zanzibar), and the West Indies. Statistics of production are available only from Sri Lanka that produced 1,900 t, 1,100 t, and 1,400 t in 1993 to 1995, respectively. During this period, the following amounts were exported: 1,650t (1993), 971 t (1993) and 1,116 t (1995). In addition to Indonesia that produces three quarters of the total world output and export, Grenada is second in rank, while other Southeast and South Asian countries produce small amount, mainly for domestic consumption. In 1990, Indonesia produced 15,000 t. The figures for 1992-94 are: 17,316t (1992), 20,911 t(1993), and 23,792 t (1994) Sri Lanka produced 850, 790 and 910 t during 1993 to 1995, respectively. Madagascar is also a leading clove producer, second only to Indonesia, accounting for roughly 11% of total world production by tonnage.

Source: Export Development Board, Colombo

World production of cloves, 1975-2005 (tonnes)

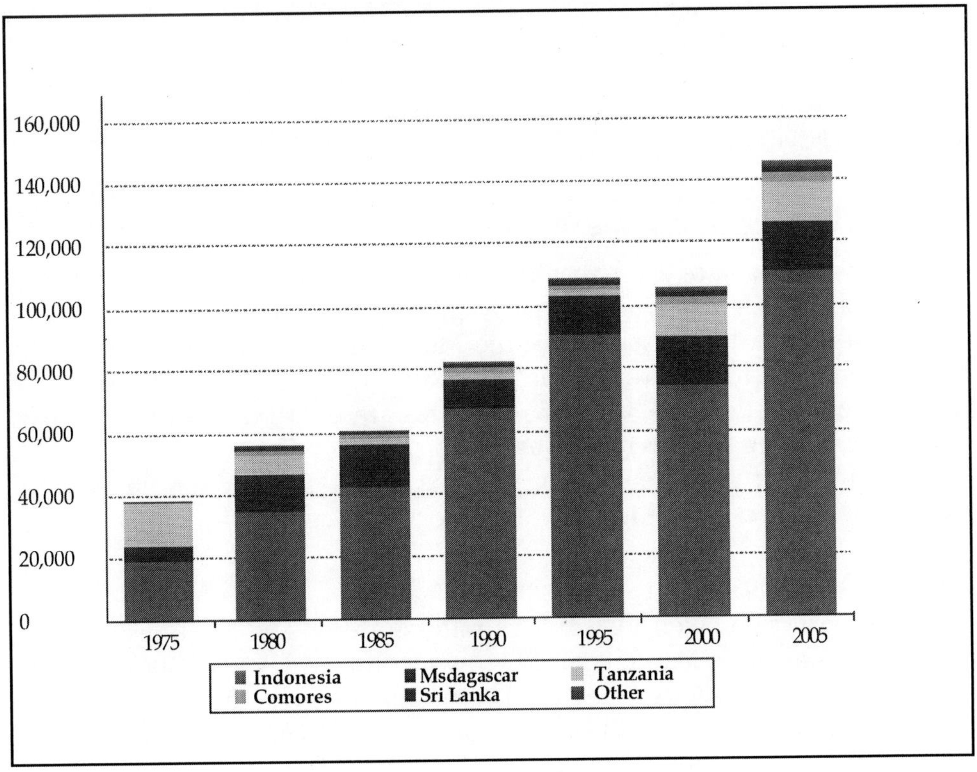

Source: FAOSTAT

Indian Scenario

Southern Region in India is known to produce large quantities of cloves, most of which is domestically consumed. Karnataka, Kerala and Tamil Nadu are the major cloves producing states. Farmers are using advanced technology for quality and consistent supply.

State wise Area and production of clove in India
(Area in Hectares, production in Tonnes)

State	2000-01		2001-02		2002-03		2003-04		2004-05	
	Area	Production	Area	Produ-ction	Area	Produ-ction	Area	Produ-ction	Area	Produ-ction
Clove										
Andaman & Nicobar	94	4	95	4	124	6	188	8.5	196	11.63
Karnataka	262	215	259	199	259	199	208	194	183	283
Kerala	802	55	754	52	737	53	9718	2427	10780	2700
Tamil Nadu	723	705	777	785	1007	1116	1304	1557	1011	1121

Source: Spices board of India

Nutmeg (*Myristica fragrans* Houtt.)

Nutmeg and mace are two different parts of the same fruit of nutmeg tree, (*Myristica fragrans* Houtt.) Family (Myristicaceae). The nutmeg tree is indigenous to the Banda islands in the Moluccas. The species of the genus *Myristica* are distributed from India and South- East Asia to North Australia and the Pacific Islands. A total of 72 species are distributed in these areas. The major nutmeg growing areas are Indonesia and Grenada (West Indies). It is also grown on a smaller scale in Sri Lanka, India, China, Malaysia, Western Sumatra, Zanzibar, Mauritius and the Solomon Islands.

Soil and Climate

Nutmeg thrives well in warm humid conditions in locations with an annual rainfall of 150cm-250cm. It grows from sea level to about 1000m altitude. Clay loam, sandy loam and laterite soil are ideal for its growth. Dry climate or water logged conditions are not suitable. River banks and sheltered valleys are the best suitable for this crop.

Nursery Technique

Seed Propagation

Common method of propagation is through seeds collected from regular bearing and high yielding trees, yielding more than 10,000fruits/tree/ year and having 30g weight/ fruit,1g wet mace/fruit and 10g wet weight of nuts/ fruit.

Nursery beds of appropriate size may be prepared in a cool and shady place. A mixture of garden soil and sand in 3:1 ratio may be used for preparing beds. The seeds are to be sown immediately after collection and sand may be spread to a thickness of 2-3cm .Seeds of nutmeg normally germinate slowly. Pre sowing treatment namely water soaking GA (1000 ppm) registered 85% germination. Watering is to be done judiciously. When seedlings attain 2-3 leaf and 10-15cm height they may be transplanted in pots or polythene bags containing a mixture of soil, sand and well decomposed cowdung at 3:3:1 ratio. Retain the seedlings for 18-24 months before planting in the main field.

Production of Quality Planting Materials

As the main problem in nutmeg cultivation is the segregation of seedlings into male and female plants resulting in about 50%of unproductive male trees, the high yielding trees have to be propagated through vegetative means.

Selection of Mother Plants

Twenty year old trees yielding above 2500 nuts/tree/year with an average nut weight of above 8 g (dry) and mace weight above 2 g (dry) could be identified as parent trees. The trees should be healthy and free from thread blight, horse hair blight, mealy bugs and scale insects.

Establishment of Scion Bank

An orchard of different elite cultivars/released varieties may be established for obtaining plant propagules for production of quality planting materials at the source of production and should be maintained free of pests and diseases.

Raising Rootstocks

Nutmeg is commercially propagated through grafts. For raising rootstocks, naturally split healthy fruits are harvested during June-July. The seeds are extracted from the pericarp and sown immediately in sand beds of convenient length, 1.0 to 1.5 m width and 15 cm height. Regular watering is necessary for good germination. Germination may commence from 30th day and last up to 90 days after sowing. About 20 days old sprouts are transplanted to polythene bags containing a mixture of good soil, sand and cow dung (3:3:1). The selected rootstock at the first leaf stage should have a thick stem (diameter of 0.5 cm or more) with sufficient length to give a cut of three cm long while grafting.

Bud Stick/Scion

Bud stick/scions are taken from high yielding female trees which are maintained at the research station or by private agencies which have been already identified as elite types and are healthy.

Grafting

Scions with two to three leaves, collected from high yielding parent trees can be used for grafting. The stock and scion should have approximately the same diameter. A 'V' shaped cut is made in the stock and a tapered scion is fitted carefully into the cut. Bandaging at the grafted region may be done with polythene strips. The completed grafts are to be placed in polythene strips. The scion is covered with a polythene bag and kept in a cool shaded place protected from direct sunlight. During grafting, precautions should be taken to prevent wilting of scions and to complete grafting as soon as possible. Maximum percentage of successful grafts (about 80%) could be obtained, if grafting is carried out during August-September.

Transplanting

After one month, the bags can be opened and those grafts showing sprouting of scions may be transplanted into polythene bags, containing a mixture of soil, sand and cow dung (3:3:1) and kept in shade for development. The potting mixture should be sterile and free from any contamination. The polythene bandage covering the grafted portion can be removed after three months.

The grafts can be distributed after they have attained 30cm height and have put forth a minimum of eight to ten leaves. The grafts can be planted in the field after 12-18 months of grafting.

Planting and Management

Nutmeg is a shade loving plant, young as well as grown up plants require certain amount of shade .Suitable fast growing shade trees like Albizia and Erythrina are planted in advance. Banana can also be grown as a shade crop in the early stages. Pits of 90 x 90 x 90 cm are dug at a spacing of 8x8m with onset of South West monsoon. The pits are filled with top soil and compost or well decomposed cattle manure and seedlings are planted.

As an intercrop in coconut plantations, nutmeg is planted in between two rows of coconut to accommodate 125-150 plants/ha. Regular weeding and mulching keep the field clean and conserve moisture. Cover crops like *Mimosa* sp. and *Stylosanthes sp.* may also be cultivated for suppressing weed growth. Periodical irrigation may be given during summer. For four year old plants, 20 litres of water/ plant thrice a week are given.

Manuring

Apply 10 kg cattle manure or compost/seedling during the first year. Increase the quantity gradually till a well grown tree of 15 years and above receives 50 kg of organic manures/ year. Apply $N:P_2O_5:K_2O$ @ 20:18:50 g/ plant during first year. This may be doubled in the next year and gradually increase the dose to 500:250:1000 g/plant/year to obtain full dose from 15th year onwards.

Harvesting and Processing

Fruiting commences from fifth or sixth year but may take even eight or nine years. Optimum productivity is attained in about 15 years. Fruit attains ripening with in 6-9 months after flowering. Fruits are available throughout the year, peak period of harvest is from Decmber-May. When fruits are fully ripe, the nuts split open. These are either plucked from the tree or allowed to drop. The fruits are composed of three parts: the pericarp or husk, the mace and the seed. After de-rinding the nutmeg fruit, red feathery aril (mace) is separated from pericarp. The mace is detached, flattened and dried in sun on mats for 3-5 days. Nutmeg lose 25% of their weight by drying.

Oleoresin

Nutmeg and mace oleoresins are prepared by extracting ground spice with organic solvents. Yield of oleoresin is 10-12% for nutmeg and 10-13% for mace.

Nutmeg Butter

Nutmeg contains 25-40% of fixed oil that can be obtained by pressing the crushed nuts between plates in presence of steam or by extracting with solvents. It is highly aromatic orange coloured fat with the consistency of butter at ambient temperature.

Nutmeg Oil

The unshelled nuts are coarsely crushed in a mechanical cracker and steam distilled to obtain a pale yellow to white volatile liquid possessing a fresh warm aromatic odour.

Mace Oil

The mace yields 4-17 % colourless to pale yellow liquid possessing organoleptic properties similar to nutmeg oil.

Weight percent (w/w%) composition of nutmeg oils by country

Compound	Grenada	Indonesia	Jamaica
α-pinene	13.2	26.5	19.9
β-pinene	8.0	15.0	18.8
myrcene	3.4	3.7	4.7
α-phellandrene	0.7	0.9	1.6
α-terpinene	4.2	2.0	2.1
limonene	4.4	3.6	4.8
p-cymene	0.8	0.6	<0.1
linalool	0.3	0.2	0.3
terpinen-4-ol	4.7	3.0	17.8
α-terpineol	0.3	0.6	0.4

Whole nutmeg: Chemical and physical specifications

Specification	Suggested limits	
ASTA cleanliness specifications	Broken	Whole
Whole dead insects, by count	4	4
Mammalian excreta, by mg/lb	5	0
Other excreta, by mg/lb	1.0	0.0
Mold % by weight		
Insect defiled infested % by weight		
Extraneous % by weight	0.50%	0.00
FDA DALs		
Insect infested and/or moldy pieces by count	Ave. of 10%	
Volatile oil	7.0% min	
Moisture	8.0% max	
Ash	3.0% max	
Acid-insoluble ash	0.5% max	
Average bulk index (mg/100 g)	NA	

*Source:*www.indianspices.com

Ground nutmeg: Chemical and physical specifications

Specification	Suggested limits
FDA DALs (6 subsamples)	
Insect fragments or	Ave. of 100 or more/10 g
Rodent hairs	Ave. of 1 or more/10 g
Volatile oil	6.0% min
Moisture	8.0% max
Total ash	3.0% max
Acid-insoluble ash	0.5% max
Military specifications (EE-S-631J, 1981)	
Volatile oil (ml/100 g)	7.5 min
Moisture	8.0% max
Total ash	3.0% max
Acid-insoluble ash	0.5% max
Nonvolatile ether extract	25.0% min
Granulation	95% min through a U.S.S. 20
Bulk index (ml/100 g)	180

*Source:*www.indianspices.com

Nutritional composition of mace per 100 g

Composition	USDA (Ground)	ASTA
Water (g)	8.17	4.5
Food energy (kcal)	475	565
Protein (g)	6.71	8.0
Fat (g)	32.38	38.8
Carbohydrates (g)	50.51	46.1
Ash (g)	2.23	2.3
Calcium (g)	0.252	0.2
Phosphorus (mg)	110	110
Sodium (mg)	80	70
Potassium (mg)	463	500
Iron (mg)	13.90	11.3
Thiamine (mg)	0.312	0.370
Riboflavin (mg)	0.448	0.560
Niacin (mg)	1.350	1.2
Ascorbic acid (mg)		ND
Vitamin A activity (RE)	80	80

Source: Composition of Foods, Spices and Herbs, USDA Agricultural Handbook ,8.2,1977. The Nutritional Composition of Spices, ASTA Research Committee, 1977. ND = Not detected

Nutmegs are used for preparing medical dosage that is useful in following health problems like

- Reduces flatulence
- Aids digestion
- Improves appetite
- Treats diarrhoea
- Useful in Vomiting and nausea
- Insomnia
- Common Cold
- Dehydration

Use of Nutmeg Oil

- *Aromatherapy*: The oil is used for relaxing body. It is used to cure overexertion and calm senses and muscles. The oil is also sometimes combined with thyme or rosemary essential oil.
- *Beverages*: The essential oil is used to flavor beverages and syrups (coca cola). It is also used to flavor food, sweets etc.
- *Personal Hygiene Products:* It is also used as a major ingredient in personal hygiene products like toothpaste, tooth powders and mouth wash.
- *Medicinal Use:* Oil is said to have therapeutic value and used for diseases related to nervous system and digestive system. It is also used in preparation of certain cough syrups.
- The oil finds application in cosmetic and pharmaceutical industry.

Trade in Nutmeg and Mace

Global Scenario

Indonesia and Grenada are the world's largest producers of high quality Nutmeg and Mace. Indonesia has close to 75% of the world market share with Granada having nearly 20% of the market share. The remaining 5% production is done by other countries including India, Malaysia, Sri Lanka etc. EU, USA, Japan are the major nutmeg and mace consuming countries. Singapore and Netherlands are largest exporting countries of the spice.

World Trade in Nutmeg Oil

Indonesia and Grenada are the two major producers of nutmeg and its derivatives. Both the countries contribute to around 75% and 20 % to the world production respectively. The extraction of nutmeg essential oil from the spice is increasing due to its potential use in a large number of industries. The main import markets of the essential oil are EU, USA, Japan and India. Singapore is one of the main re exporter of the nutmeg oil.

Source: Export Development Board, Colombo

Production of nutmeg, mace and cardamom, 1975-2005 (tonnes)

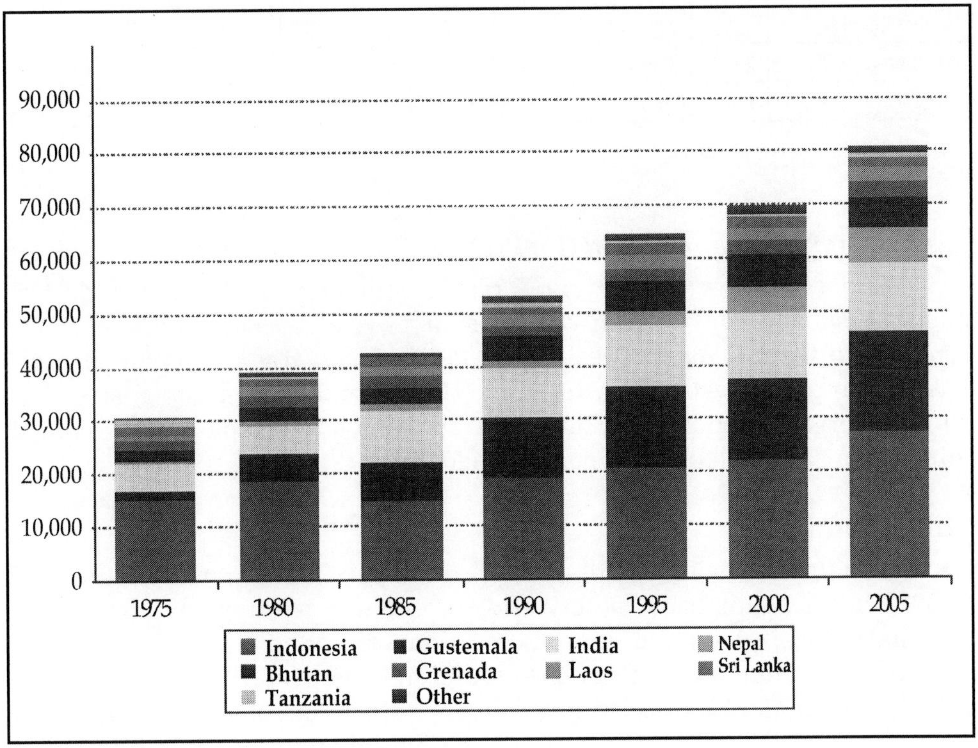

Source: FAOSTAT

Indian Scenario

Southern Region in India is known to produce large quantities of nutmeg and mace, most of which is domestically consumed. Kerala, Karnataka and Andaman and Nicobar islands are the major nutmeg and mace producing states. It is a profitable crop in mixed cropping system with coconut and arecanut.

State wise Area and production of nutmeg in India
(Area in Hectares, production in Tonnes)

State	2000-01		2001-02		2002-03		2003-04		2004-05	
	Area	Production	Area	Production	Area	Production	Area	Production	Area	Production
Nutmeg										
Andaman & Nicobar	-	-	21	2	30	2	44	3.37	60	4.55
Karnataka	227	95	228	96	228	96	221	119	-	-
Kerala	6950	1731	7601	1888	8418	2086	9718	2427	10780	2700

Source : Spices Board of India.

Cassia (*Cinnamomum cassia* (L.) Presl)

Cassia, also known as lignea or chinese cinnamon,is one of the oldest of spices.Cassia is obtained from various sources like *C.cassia* Bercht. & Presl., *C.burmannii* Bl., *C.loureirii* Nees and *C. tamala* Nees. It is an evergreen tree, aromatic in all its parts,which resembles C.verum, but is larger in size. The centre of origin is Southern China, Burma, Laos and Vietnam. Commercial cultivation is restricted to China and Vietnam.

Climate and Soil

Cassia flourishes in hot ,sunny and humid conditions, at temperatures of 15-30°C but is adaptable to a very wide seasonal range from the high 30°C-7°C, and also to very large differences in day-night temperatures.The main areas of cultivation in China have a summer maximum of 38°C and minimum of 28°C, winter 15°C and -0.5°C respectively. Cassia cannot withstand periods of very cold weather, but is resistant to sporadic light frost. A minimum annual rainfall of 1500 mm is necessary. Cassia is a forest tree, partially shade tolerant especially young trees, but mature trees grow well in full sun shine. The soils of cassia are usually well drained,severely leached, hill or mountain soils of low fertility and many strongly acidic pH 4.5-5.5.

Propagation

Cassia is propagated through seed, cuttings or layering, but seed propagation is the most common in China. Ripe fruits are obtained from selected mother trees, which have a thick bark of good aroma.The harvested fruits are usually fermented in heaps for 2-3 days, which assist in removal of pericarp by washing. The seeds,which have short viability,are dried in shade and sprinkled with ashes.Well ripened seed from the tree is also used for direct sowing which gives satisfactory germination in 5-15 days.

Nursery beds, slightly raised and about 1m wide are prepared in a fertile sheltered area with available water. Shades are provided using palm leaves. The seeds may be planted at a close spacing of 5 x 5 cm at a depth of about 1 cm and subsequently transplanted into baskets or polybags. Alternatively, the seeds may be sown at a wider spacing of 20 x 20 cm and plants left in position until being transplanted to the field, when they are 9-12 months old and 60-90 cm tall. The seedlings are hardened off by gradual removal of shade.

Air Layering

Air layering in two year old cassia plants gives a success of 87.5% and 50% rooting when carried out during July and November, respectively, with 100% field establishment.

Micropropagation Cassia

Successful micropropagation using nodal explants from seedlings on MS medium containing low concentrations of BA and NAA and transferring the explants to medium containing BA alone, for shoot elongation are observed. The elongated shoots were rooted in a third medium supplemented with only NAA.

Planting and Management

Preparation of planting holes is generally as for cinnamon, and seedlings are planted out at 15-18 cm high. A spacing of 50-60 cm in 100 cm rows is usual in China, but density is often regulated by shape and size of individual fields and terraces. Between 9500 and 10,000 trees/ha are considered the optimum for leaf oil production.

Trees are cut almost to ground level at 5-6 years, and subsequently to 15-25 cm every 3-4 years. The strongest three to four shoots are allowed to develop and cut when 2-4 cm in diameter. Some trees are left uncut to produce seed and cassia buds, usually 15 x 30 m apart, and these provide intermittent shade. Plantation life is 40-60 years before replanting, although dead or diseased trees are replaced. Weeding and plantation maintenance are generally as for cinnamon in Sri Lanka. Irrigation, especially of young trees, is common where water and topography allow. An improved system of management was developed in Hawaii to obtain maximum leaf and twig production, with a minimum population of 10,000 trees/ha. Trees are not cut back to near ground level, but are cut to provide a vertical frame work of stems and branches to maximize leaf and twig production. Since the climate in Hawaii is not strongly seasonal, trees are cut three or four times annually.

Harvesting

Bark collection, stripping and drying are basically as for cinnamon and is described in detail (Purseglove *et al.*, 1981). Cassia buds, *kui isz*, are the dried immature fruit including calyx and often pedicel. These are harvested from trees especially left in plantations, or collected from wild trees in October-November. Cassia buds are smooth, grayish or reddish brown, 6-10 x 5 mm in diameter. The attached calyx is hard and wrinkled. Bark, leaves, twigs, small stems and fruit buds can be used for oil production since characteristics are similar.

Leaves are collected when the bark is harvested, but prunings and excess foliage can be distilled at any time. The main harvest in China is June-July, a lesser in January-February; the highest oil yield is from green leaves harvested in May-July. Large mature leaves yield the best quality oil in China, and a high admixture of twigs and small branches reduces quality. Leaves are often wilted for 6-8 days before distilling, but the reason for this is unclear since fresh leaves generally give a higher oil yield.

Leaf Oil

The oil content of fresh leaves in China is 0.3-0.4%, seeds 1.5% and small twigs and stems 1-2% (Anon, 1976), the oil content of bark is 1.5%, buds 0.55%, budsticks 1.64% and a mixture of leaves, petioles and young twigs is 0.77%.

Chinese cassia leaf yields brownish-yellow to dark-brown viscous liquid which can deposit a resinous sediment with storage. The odour is similar to rectified oil, but more tenacious, and thus preferred by food processors and cola beverage makers to whom the dark colour is no disadvantage. Rectified oil is clear, pale to dark yellow, less viscous, with a persistent, sweet, spicy warm note. The flavour is warm, spicy, balsamic, and very sweet. Rectified oil is also used in the flavouring industry, but its clarity enables it to be used in perfumes, toiletries, soaps and similar products. The major constituent is cinnamaldehyde (70-95%), with unadulterated oils and Chinese official grade oils having a minimum content of 85 or 90%.

Main Constituents

Cassia contains max. 4% essential oil, 75 to 90% of which are composed of cinnamic aldehyde. There are only traces of eugenol , but significant amounts (7%) of coumarin. Therefore, analysis of eugenol and coumarin discriminates between Ceylon and Chinese cinnamon. Trace components of cassia oil are benzoic acid, cinnamic acid, salicylic acid and the corresponding esters and aldehydes. Cassia bark contains significantly more slime (11%) than Ceylon cinnamon bark. The leaf oil (0.3 to 0.8%) has a similar composition to the bark oil.

Whole cassia or cinnamon: Chemical and physical specifications

Specification	Suggested limits	
	Cassia	Cinnamon
ASTA cleanliness specifications		
Whole dead insects, by count	2	2
Mammalian excreta, by mg/lb	1	1
Other excreta, by mg/lb	1.0	2.0
Mold % by weight	5.00	1.00
Insect defiled/infested % by weight	2.50	1.00
Extraneous % by weight	0.50	0.50
FDA DALs (6 subsamples)		
Insect infested pieces by weight or	Ave. of 5%	
Moldy berries by weight	Ave. of 5%	
Mammalian excreta identified as to source when possible	1 mg/lb	
Volatile oil	Varies by quality Typical 1.5-3.0%	
Moisture	14.0% max	
Ash	5.0% max	
Acid-insoluble ash	1.0% max	
Average bulk index (mg/100 g)	N/A	

*Source:*www.indianspices.com

Trade in Cassia

Southern China is one of the major producers of Cassia Bark and its derivatives. It is cultivated in the country for large scale oil production. Indonesia is another significant producer of Cassia on a commercial basis.

Vietnam, India and Nepal are other Cassia Sticks producing countries. The leaves, twigs and fragments of bark are steam distilled to obtain cassia oil.

The annual world trade of cassia is between 20,000 to 25,000 t, of which Indonesia accounts for two-thirds and China for most of the remainder. About 2,000 to 3,000 t of cassia bark are exported from Vietnam annually (Sial 1995). It is quite difficult to obtain production statistics from these producing countries shown as blanks. Most of cassia entering international markets is used as spice. Only small quantities of cassia oil are exported. There is also some production of oleoresin for flavoring purposes. Barks also find local use as a herbal medicine in China and other Southeast Asian countries.

Source: Export Development Board, Colombo

Cinnamon (*Cinnamomum zeylanicum*)

Cinnamon, the earliest known spice is indigenous to Sri Lanka, introduced to India from Sri Lanka by European settlers. In Sri Lanka, distinct cultivars are prevalent such as sweet honey cinnamon, camphor cinnamon, snake cinnamon, astringent cinnamon, etc. Of these only sweet honey cinnamon is cultivated commercially. It is one of the economically important cultivated tree spices in South India especially in Kerala, Tamil Nadu, Karnataka and Union Territory of Andaman and Nicobar Islands. In ancient Egypt, cinnamon was used medicinally and as a flavouring for beverages. It was also used in embalming, where body cavities were filled with spiced preservatives. In the ancient world, cinnamon was more precious than gold.

Soil and Climate

Cinnamon flourishes well in humid regions, receiving an annual rainfall of 150-250 cm with an average temperature of 27°C. Prolonged dry spells without rains are vulnerable for their growth. They can come up in places from almost sea level to an elevation of 1000 m above sea level.

Cinnamon is a hardy plant which can be grown in almost all types of soils under a wide range of climates. The quality of bark is highly influenced by soil and ecological factors. In Sri Lanka, it is cultivated on the white sandy soil. Well drained deep sandy soils rich in humus are ideal for cinnamon cultivation. The best quality is obtained when grown in silicaceaous sandy soils. Waterlogged and marshy areas may be avoided.

Nursery Technique

Seed Propagation

Cinnamon can be propagated either through seed or clonally. Seed propagation is the easiest and the most widely adopted method. Seed is recalcitrant and loses viability quickly when stored. When sown immediately after harvest, seeds give 90-94% germination and the viability is completely lost after storage of five weeks.

Ripe fruits are collected from selected mother trees having the following desirable characters (i) erect stem with smooth bark; (ii) vigorous growth (iii) easiness of peeling stem bark (iv) free from pests and diseases; and (v)good quality characteristics viz. sweetness, pungency and flavour. Fully ripe berries are collected and left in heaps for two to four days in shade to soften and rot. Seeds have to be washed and cleaned and are sown without delay in shaded seed beds or in polythene bags. Germination starts in about 20 days. For field planting, one-year old seedlings are used.

Phytohormones were effective on germination. Soaking of seeds in GA_3 (150 ppm) or thiourea (1500 ppm) resulted in significantly high germination. GA_3 (50 ppm) reduced number of days taken for commencement of germination by ten days.

Vegetative Propagation

Since cinnamon is a cross pollinated species, vegetative propagation is necessary to produce uniform high yielding plantation. It can be propagated by cuttings. Cuttings with one or two leaves can be made to root within 40 days under humid conditions in a closed chamber with diffused light. Seradix could hasten their rooting. Phosphobacteria (soil application + dipping the cutting in phosphobacteria containing slurry) application gave a significantly higher rooting (%) percentage. Cuttings kept in moist coir dust, root within 60 days.

Air Layering

Air layering is also a successful method of vegetative propagation. Semi hardwood shoots are suitable for this purpose. Application of gallic acid during air layering resulted in 80% rooting in July and sphagnum moss was the best rooting media.

Micropropagation

Micropropagation is useful for clonal multiplication of selected elite lines of cinnamon to augment the conventional vegetative propagation methods.

Multiple shoots were induced from seeds on MS medium supplemented with Benzyladenine (BA) or kinetin. The shoots were multiplied by subculturing hypocotyl and internode segments on a second medium supplemented with BA or kinetin in combination with NAA. The *in vitro* generated shoots were rooted in Whites medium containing combination of IAA, IBA and IPA.

Planting and Management

Seedlings are transplanted in main field when they are about 10-12 months old at a spacing of 2 x 2 m. Planting is done in pits of 60 cubic cm which are dug earlier and filled with compost and top soil. The appropriate time for planting is June-July or October- November when North East monsoon prevails. In the early period, it is necessary to provide artificial shade. Weeding and mulching reduce weed growth, 3-4 weedings are required for first two years, and thereafter, 2 weedings in a year are sufficient.

Manuring and Intercultivation

Systematic manuring is rarely practiced in India for this crop. Since leaves and shoots are economic parts, nitrogenous fertilizers are advocated for quick and rapid growth. The common practice is to mulch the plant bases with the trash obtained during weeding. One earthing up is given at the time of shoot collection in June. For better growth, the plants should be manured systematically.

A fertilizer mixture of urea, rock phosphate and muriate of potash in 2:1:1 is recommended. The rate of application suggested is 200,400 and 600 kg/ha for first, second and third year. The Kerala Agricultural University recommended a fertilizer schedule of 20:20:25 g of NPK along with 20 kg of compost/ year during first year of planting, which gradually increased to 200:180:200 g NPK and 50 kg of compost/ year from 15th year onwards. Fertilizers may be applied in two splits during May-June and September-October.

The organic manure is given in one dose just at commencement of monsoon in May-June.

Pruning / Cropping

At the initial periods, the growth may be slow but as the plants advance in age, the growth is faster. When the plants are two years old, they are coppiced or cut back during June-July to a height of about 12 cm from the ground level. The stump, is then covered by earthing up. This is repeated for every side shoot, developing from main stem, during succeeding season so that the plant assumes shape of a low bush of about 2 m height, and a bunch of canes suitable for peeling would crop up in about 4 years time. Regular peeling operations can be commenced in the case of seedling bushes from fourth or fifth year, depending upon extent of development of peeler shoots.

Harvesting

The cinnamon tree attains a height of 10-15 metre in normal condition, but in cultivation, it is generally coppiced or cut back periodically for extracting the bark. When the plants are two years old, they are coppiced during June-July to a height of 12 cm from ground. The stump is then covered by earthing up. This operation encourages development of side shoots from stump. This is repeated for every side shoot, developing from main stem during succeeding season, so that the plant assumes shape of a low bush of about two metre height and a bunch of canes suitable for peeling would develop in a period of about four years. Coppicing for bark extraction can be commenced from fourth or fifth year of planting.

The shoots are harvested from September to November, under Kerala conditions. Coppicing is done in alternate years and shoots having 1.5-2.0 cm thickness and uniform brown colour are ideal for bark extraction. A 'test cut' can be made on stem with a sharp knife to judge suitability of time of peeling. If the bark separate readily, coppicing can be commenced immediately. The stems are cut close to the ground when they are about two years old, as straight as possible, 1.00 to 1.25 m length and a thickness of not less than 1.25 cm. Such shoots are bundled after removing leaves and terminal shoots and carried out in the peeling shed.

Regular peeling operations commence from fourth or fifth year. The best quality is obtained from thin bark of shoots in the centre from middle portion of shoots. Fully developed shoots of 1.5 - 3.0 cm diameter, harvested during rainy season give good quality bark and high yield. From the first cutting after 3-4 years, 56-67 kg of quills/ hectare can be obtained. In a mature plantation, the yield is around 180-200 kg of quills/ hectare.

Preparation and Curing of Bark

Preparation and curing of bark involve (i) cutting of branches of right size and shape in right season at right stage of maturity, (ii) scraping of outer rough cork layer (iii) peeling and removing the bark from inner wood, and (iv)piping and preparation of quills, quillings, feathering and chips.

Cutting and Peeling

Cutting is followed by scraping and peeling operations. Peeling is a specialized operation, requiring skill and experience. It is done by using a specially made knife, which has a small round end with a projection on one side to facilitate ripping of bark. The shoots cut in the morning are peeled on same day. The rough outer bark is first scrapped off. Then the scrapped portion is polished with a brass, wooden or an aluminium rod to facilitate easy peeling. A longitudinal slit is made from one end to other. The bark can be easily removed by working the knife between bark and wood.

The curled pieces of peeled inner bark are next assembled into compound quills or pipes by placing one quill inside the other. The best and longest quills are used outside and the inside is filled with smaller pieces of bark. The compound quills are then rolled by hand to press the outside edges together and the ends are neatly trimmed with a pair of scissors once the required length of 107 cm is attained.

Drying

The compound quills are placed on coir rope racks and dried in shade and sun alternately for three-four days to prevent warping. To obtain firm and compact quills, they are rolled by hand daily until judged sufficiently dry. A moisture content of 12% is recommended for dried quills. The quills are bleached, if necessary, by sulphur treatment for about eight hours and are sorted into grades according to thickness of bark.

Grading

The quills are graded from '00000', being the finest quality, to '0' the coarsest quality. Pieces of bark less than 107 cm long are categorized as quillings.

Quillings

The small pieces of bark (5-20 cm long), left after preparing the quills are graded as quillings.

Featherings

The very thin inner pieces of bark or twisted bark are graded as featherings.

Chips

Chips are trimmings of quills or bark of coarser canes that are scraped off, instead of peeling. The bark that is scraped off without removing outer bark is known as unscrapped chips and that scrapped after removing outer bark are scrapped chips.

Cinnamon Powder

The different grades of bark are powdered to get cinnamon powder.

Standards

Quality of product depends on thickness of bark, their appearance and aroma and flavour.

According to Sri Lankan system of grading, the cinnamon quills are graded into four groups:

Alba (< 6 mm in diameter); *Continental;* Mexican) and *Hamburg* grades. These groups are further divided into specific grades, as follows.

Fine/continental grades (6-19 mm diameter)

C-00000 special (Thinnest) (6 mm); C-00000 (10 mm); C-0000 (13 mm); C-000 (16 mm); C-00 (17 mm); C-0 (thickest) (19 mm).

Mexican grade (16 mm diameter)

M00000 special, M00000 and M0000 depending number of quills/kg. Mexican grade is intermediate in quality between Fine and Hamburg grades.

Hamburg grade (23-38 mm diameter)

H-1 grade (23 mm); H-2 (25 mm); H-3 (38 mm).

Marketing

The dried quills are tightened into small bundles, each bundle containing about 30-40 quills. The edges are then trimmed making them into ready for marketing. For whoselsale export, the quills are packed in compact cylindrical bales of 50 kg. Whole cinnamon should be packed in clean, sound and dry containers and stored in cool dry place to prevent loss of volatile oil and aroma.

Other Products

Bark Oil

Bark oil is extracted from broken pieces of bark by hydro or steam distillation. The bark oil contains cinnamaldehyde as major component.

Leaf Oil

Leaf oil is obtained by steam or hydro distillation of leaf that is removed from stem used for extraction of bark. Major component of leaf oil is eugenol.

Oleoresin

Oleoresin is extracted with organic solvents from bark of various grades. Generally chips are used for preparation of oleoresin.

Ground Spice

The different grades of bark are powdered to get ground spice which is used in spice mix and curry powders.

Ground cinnamon: Chemical and physical specifications

Specification	Suggested limits
FDA DALs (6 subsamples)	
Insect fragments or	Ave. of 400 or more/50 g
Rodent hairs	Ave. of 11 or more/50 g
Volatile oil	Varies by quality
	Typical 1.5 - 3.0%
Moisture	14.0% max
Total ash	5.0% max
Acid-insoluble ash	1.0% max
Military specifications (EE-S-631J, 1981)	
Korintje cinnamon	
Volatile oil (ml/100 g)	1.5 min
Moisture	11.0% max
Total ash	5.0% max
Acid-insoluble ash	2.0% max
Granulation	95% min through a U.S.S. 60
Average index (ml/100 g)	165

Source: www.indianspices.com

Nutritional composition of cinnamon/100 g

Composition	USDA (data for *C. cerum* and *C. cassia* combined)	ASTA
Water (g)	9.52	10.00
Food energy (kcal)	261	355
Protein (g)	3.89	4.5
Fat (g)	3.18	2.2
Carbohydrates (g)	79.85	79.8
Ash (g)	3.55	3.5
Calcium (g)	1.228	1.6
Phosphorus (mg)	61	50
Sodium (mg)	26	10
Potassium (mg)	500	400
Iron (mg)	38.07	4.1
Thiamine (mg)	0.077	0.14
Riboflavin (mg)	0.14	0.21
Niacin (mg)	1.30	1.9
Ascorbic acid (mg)	28.46	40
Vitamin A activity (RE)	26	26

Source: Composition of Foods, Spices and Herbs, USDA Agricultural Handbook ,8.2,1977. The Nutritional Composition of Spices, ASTA Research Committee, 1977. ND=Not detected

Trade in Cinnamon

Global Scenario

For cinnamon and cassia, total world production has increased by 65% over the last decade, driven by increases in Indonesian production, which has risen 60%, Chinese production which has doubled and Vietnamese production which has more than doubled. Madagascar, which produced 1,700 tons in 1995, now only produces 1,500 tonnes a decrease of 12%. (*Source:* Export Development Board, Colombo).

Global trade in cinnamon ranges between 8500-12000 tonnes annually. Sri Lanka contributes nearly 80% of the total cinnamon production followed by Seychelles, Madagascar and India. USA, Japan and EU are major import markets of the spice. The true cinnamon, *Cinnamomum verum*, which originated in Sri Lanka and south India, has been introduced to many other countries, notably Madagascar and Seychelles, but Sri Lanka still contributes 80 to 90% of the world trade (Sial 1995). Its annual production during 1993-95 were 9,600 t (1993), 10,500 t (1995) and 11,200 t (1995) while its exports were 8,728 t (1993), 9,016 t (1994) and 9,852 t (1995).

World Trade of Cinnamon Oil

Sri Lanka is the largest producer and exporter of quality cinnamon oil to other countries. It accounts for around 70% of cinnamon oil global production. Cinnamon bark oil is a highly demanded and a high value-essential oil, though the international trade volume is low. Initially Sri Lanka has a monopolistic control on cinnamon oil market and trade, but in the recent years other Asian countries have also started producing oil in response to the increase in demand.

The major market for cinnamon oil is France followed by USA. Hongkong is another significant exporter of cinnamon oil. India produces cinnamon leaf oil, mostly for domestic consumption. Madagascar and Seychelles are other important suppliers of leaf oil.

Source: Export Development Board, Colombo

Production of cinnamon and cassia, 1975-2005 (tonne)

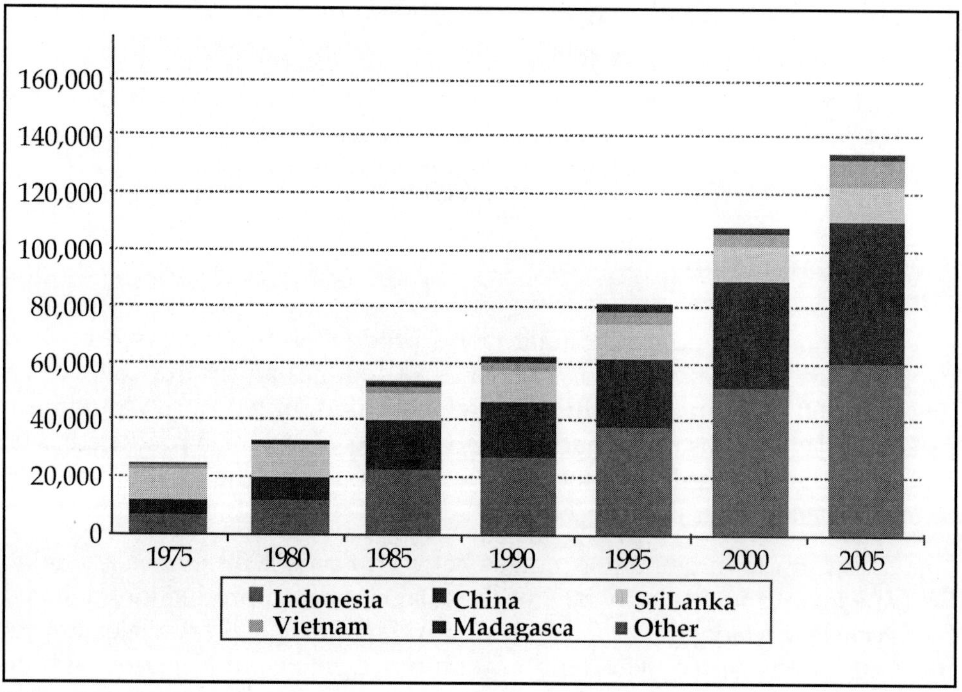

Source: FAOSTAT

Indian Scenario

Cinnamon production is a labour intensive process, requiring the bark of the plant to be cleaned thoroughly before being dried and finally used for consumption. Karnataka, Kerala and Andaman and Nicobar Islands are major production areas in India.

State wise Area and production of cinnamon in India (Area in Hectares, production in Tonnes)

State	2000-01		2001-02		2002-03		2003-04		2004-05	
	Area	Prod-uction	Area	Produ-ction	Area	Produ-ction	Area	Produ-ction	Area	Produ-ction
Cinnamon										
Andaman & Nicobar	50	7	53	7	57	8	80	11.27	103	14.43
Karnataka	166	69	-	-	-	-	48	69	-	-
Kerala	262	Neg.	285	Neg.	293	Neg	311	80.27	328	Neg.

Source: Spices Board of India

Allspice (*Pimenta dioica* (L.)

Allspice (*Pimenta dioica* (L.) Merr) (Myrtaceae) is a functionally dioecious evergreen tree, growing to a height of 15-20 meters and branching profusely. This tree with fragrant, white, small flowers does not set fruits in all parts of India. The climatic preferences of allspice are very specific to make it set fruits. The fully mature but unripe fruits are used as spice and are marketed as pimento of commerce.

Allspice is a native of Jamaica (West Indies) which is the major producer and exporter of the spice, accounting for 70% of the world trade. The remaining 30% is produced by Honduras, Guatemala, Mexico, Brazil, Costa Rica, Cuba and the Caribbean Islands.

Soil and Climate

The natural habitat of allspice in Jamaica is lime stone forest. Although allspice is planted on a wide range of soils, a well drained fertile loamy lime stone soil with a pH of 6-8 is ideal. Allspice grows well in semitropical low land forests with a mean temperature of 18-24°C, and a low of 15°C and high of 32°C. Allspice flourishes well up to 1000 meter above mean sea level. An annual rainfall of 150-170 centimeter evenly spread throughout the year is desirable but grows well with a rainfall of 120-250 centimeter.

Allspice grows abundantly where it occurs naturally, attempts to introduce it as a commercial crop into many countries outside this region have failed for unspecified reasons. In India, all the locations are not suited for the crop and can find varied performances in its fruiting behaviour. In India, a few trees are available in Maharashtra, Tamil Nadu, Karnataka and Kerala. The fruiting of allspice in different agro climatic regions of our country is not uniform.

Locations in India where all spice sets fruits

States	Locations	Flowering	Fruit set	Altitude (m)	Temperature (°C)	Rainfall (cm)
Kerala	Ponmudi, Trivandrum District	Good	Good	915	18-34	390
	Ariencavu, Kollam dist.	Good	Good	550-680	20-34	280
	Ambalavayal Wyanad dist.	Good	Good	974	18-30	230
	Peruvannamuzhi Calicut dist.	Very low	No fruit set	39	23-38	330
Tamil Nadu	Nagarkovil, Kanyakumari district	Very good	Very good	500-1000	15-30	250
	Yercaud, Salem District	Moderate	Moderate	1500	11-32	160
Karnataka	Coorg District (Sampaji, Somwarpet, Appangala)	Good	Good	800-1100	15-34	200
Maharastra	Ratnagiri	Good	Good	75	18-34	280

Source: Spice India, Vol.18 (2), 2005

Selection of Seed Garden

Since there are no released varieties of allspice, the seed must be produced from registered parent trees of seed garden. Allspice gardens of about 30 years old healthy trees may be selected as seed garden.

Collection and Sowing of Seeds

Fresh ripe fruits collected from high yielding and regular bearing trees producing well formed fruit bunches are used for seed extraction. Fruits will be ready for collection during August-September. Ripe fruits are light brown in colour.

Nursery Techniques

The seeds are extracted from ripe fruits after soaking them overnight and rubbing them in a sieve to remove pericarp. Freshly washed seeds should be sown as soon as possible or else germination will be greatly reduced. If sowing cannot be taken up, immediately after harvest, it is better to leave seeds in fruit itself till two-three days and extract at time of sowing. The freshly collected seeds may be sown in shaded nursery beds. If only very little seeds are available, they can be sown in small basins filled with sand.

Care and Management of Nursery

Watering is done using a fine spray. The beds must be mulched with dried leaves to hasten germination, which may begin nine to ten days after sowing and likely to continue over an extended period.

On germination, a short purplish cotyledon appears above the ground followed by a pair of small, opposite leaves. The seedlings are carefully lifted at four leaf stage from beds or basins and transplanted to polythene bags of 30 x 15 cm size, filled with usual potting mixture. The seedlings are ready for planting in the main field at nine to ten months old, when they are about 25-30 cm high.

Vegetative Propagation

January is the best season for air layering, which recorded 100 % success. It took about 18-28 months to separate rooted layers from mother plants.

Leafy soft wood cuttings treated with IBA (1000 to 1500 ppm) obtained up to 30% rooting.

Approach Grafting

Raising Rootstocks

Allspice seeds are collected from fully matured fruits during June-July. The seeds are sown immediately on sand beds at a spacing of 2-3 cm and 2 cm deep in raised beds of 15-20 cm high and 1 m wide and of convenient length. Watering is done regularly. Germination commences within two weeks and lasts up to 30-45 days. The seedlings are transplanted into polythene bags containing a mixture of soil, sand and well powdered cowdung (3:3:1). One year old seedlings are transplanted into bigger polythene bags (35 x 22 cm) and maintained in the nursery. Two year old seedlings are used as rootstocks for grafting scions.

Selection of Mother Trees

Healthy and vigorously growing high yielding and regularly bearing trees, free from diseases and pests attack are chosen as mother trees for grafting. Healthy shoots of mother trees with a terminal bud, 15-20 cm long and measuring about 5 mm diameter are selected for grafting. Benches of convenient height are erected around the tree for holding the bagged rootstocks for grafting.

Method of Grafting

Slice approach grafting was adopted for grafting allspice. Two years old rootstocks of about five-six cm in diameter are selected. A slice of bark and wood is cut from scion and rootstock and cut surfaces are secured together for close cambial contact with wax or polythene tape. Grafting can be carried out during August to December to get about 60% success. The rootstock and the mother tree are watered regularly. The graft union takes place in about six-eight months and the stock above the union and the scion below the union can be cut to separate the graft.

After care

The detached grafts are maintained in shade and watered regularly. New branches sprouting from the rootstock below the graft union are removed every 15-20 days. The successful grafts can be planted in the field in the next monsoon.

Planting and After Care

Seedlings of 6-9 months old having 25-30 cm are used for field planting. Seedlings are planted at a spacing of 6 x 6 m and in poor soils a closer spacing is preferred. Pits of size 60 x 60 x 60 cm are made, filled with top soil to which well rotten manure or compost can be incorporated. Transplanting should be done in wet weather. After planting, temporary shade, irrigation and weeding are provided for better establishment. Since the plant is dioecious, it is desirable to keep a male-female ratio of 1:10 in the plantation.

Harvest and Processing

Allspice is not a regular bearer. Paclobutrazol (1.25,2.5 and 3.75 g) treatment induced flowering in allspice plants, increased the number of panicles, flowers and setting of berries which in turn resulted in higher yield. Male trees flower early when compared to females and the usual flowering time in India is March to June. From flowering, 3-4 months are taken for fruit maturity. For bearing, 5-6 years are taken under good management and 12 years under poor management.

Economic bearing period of allspice is in between 15 to 25 years. For spice purpose, immature green berries are harvested during July-August. Then it is cured for three to twelve days, till they are completely dried. After the completion of curing, the berries become crisp and produce metallic sound. A tree yields about 20-25 kg of dry berries.

Main Constituents

The fruits contain 2 to 5% essential oil (the exact content depends much on the time of harvest). As main components, eugenol, eugenol methyl ether, and terpenes (myrcene, 1,8-cineol and á-phellandrene) are reported. In allspice, fruits from Jamaica, eugenol (65% to 90%) is the main constituent; methyl eugenol is found in minor (10%) and myrcene in trace amounts (1%). Allspice from México is dominated by methyl eugenol (50 to 60%) with smaller amounts of myrcene (15%) and eugenol (10%). The leaves contain less essential oil, but the content is high enough to make distillation profitable. In composition, it is similar to essential oil from the fruits.

Whole allspice: Chemical and physical specifications

Specifications	Suggested limits
ASTA cleanliness specifications	
Whole dead insects, by count	2
Mammalian excreta, by mg/lb	5
Other excreta, by mg/lb	5.0
Mold % by weight	2.00
Insect defiled infested % by weight	1.00
Extraneous % by weight	0.50
FDA DALs	
Moldy berries by weight	Ave. of 5%
Volatile oil	3.0% min
Moisture	12.0% max
Ash	5.0% max
Acid-insoluble ash	1.0% max
Average bulk index (mg/100 g)	265

Source: www.indianspices.com

Ground allspice: Chemical and physical specifications

Specification	Suggested limits
FDA DALs (6 subsamples)	
Insect fragments or	Ave. of 30 or more/10 g
Rodent hairs	Ave. of 1 or more/10 g
Volatile oil	2.0% min
Moisture	12.0% max
Total ash	5.0% max
Acid-insoluble ash	1.0% max
Military specifications (EE-S-631J, 1981)	
Volatile oil (ml/100 g)	3.0 min
Moisture	10.0% max
Total ash	5.5% max
Acid-insoluble ash	0.5% max
Granulation	95% min through a U.S.S. 25
Average index (ml/100 g)	190

Source : www.indianspices.com

Nutritional composition of allspice per 100 g

Composition	USDA (Ground)	ASTA
Water (g)	8.46	9.0
Food energy (kcal)	263	380
Protein (g)	6.09	6.0
Fat (g)	8.69	6.6
Carbohydrates (g)	72.12	74.4
Ash (g)	4.65	4.2
Calcium (g)	0.661	0.80
Phosphorus (mg)	113	110
Sodium (mg)	77	80
Potassium (mg)	1044	1100
Iron (mg)	7.06	7.5
Thiamine (mg)	0.101	0.100
Riboflavin (mg)	0.063	0.060
Niacin (mg)	2.86	2.9
Ascorbic acid (mg)	39.2	39
Vitamin A activity (RE)	54	54

Source: Composition of Foods, Spices and Herbs, USDA Agricultural Handbook ,8.2,1977. The Nutritional Composition of Spices, ASTA Research Committee,1977.
ND=Not detected

VANILLA

Vanilla planifolia Andrews (syn V. fragrans Salisb) is the second most expensive spice traded in the international market. It is native to the humid tropical rain forests of South-Eastern Mexico, Central America, the West Indies and Northern part of South America.

History of vanilla dates back to early sixteenth century. It.was Bernal Diaz, an Officer under Hernando Cortez, the Spanish conqueror of Mexico, who is believed to be the first European to recognize the flavour and immense value of vanilla, when the Aztec ruler, Montezuma offered him the vanilla flavoured cocoa beverage. Vanilla flavour was first introduced to France and England during the early part of seventeenth century. The cultivation of vanilla started in Mauritius in 1827 and in Madagascar in 1848. Even though, vanilla originated in Mexico, the major producing countries are Madagascar and Indonesia.

Vanillin is mainly responsible for fragrance, flavour and aroma of vanilla essence. It is used in the preparation of ice creams, chocolates, cakes, pastries, puddings, soft drinks, pharmaceuticals, liquors, perfumery and in nutraceuticals. At present, synthetic products such as ethyl vanillin and synthetic vanillin are used for above purposes. Such synthetic products are being rejected by discerning customers world over, and there is increasing demand for natural vanillin.

Soil and Climate

Vanilla needs warm and moist conditions and thrives well between 10° N and 20°S latitudes having a well distributed rainfall of 150 to 300cm with a temperature range of 25 to 32°C.It comes up well up to 1500m above MSL and prefers land with gentle slope and light porous soil with adequate drainage. Although vanilla thrives the best under moist conditions, excessive humidity and moisture lead to diseases. Ideal relative humidity is 80%. Forest soil rich in humus is ideal but plants survive well in sandy loam to even laterite soils. The crop cannot withstand extended droughts or strong winds but requires a dry spell for flowering.

Nursery Techniques

Vine Cuttings

Vanilla is propagated through stem cuttings. They can either be planted directly in the main field or can be initially grown in polythene bags and transplanted. Only vigorously growing, healthy, disease and pest free vines are to be selected from yielding plants for collection of cuttings for production of quality planting materials. Any part of the vine with dormant axillary vegetative buds can be selected for taking stem cuttings. Vines of current year's growth, which are in vegetative phase, having tender shoots at tip, are preferred for early sprouting. Already yielded or over matured vines from previous year's growth should not be used as vegetative buds which would put forth new shoots are nearly absent in such vines.

Rapid Multiplication Technique

For production of a large number of cuttings from a limited area, a simple rapid multiplication technique can be employed. For this purpose, a rapid multiplication nursery is to be established. The site for such a nursery should be ideally located with respect to accessibility, availability of water source, leveled land and deep and fertile soil of loamy nature with optimum natural shade. Exposed areas can also be selected which need to be provided with shade using agro-shade nets permitting 50% light intensity. After clearing the undergrowth, trenches of 60 cm wide and 60 cm deep are opened at convenient length, leaving 40 cm in between the trenches are filled with a mixture of top soil, well dried and powdered farmyard manure and sand in the ratio 3:1:1. Standards for trailing of vines are planted at a distance of one meter along the trenches at least two months before planting vanilla. For this purpose, species of quick growing trees possessing low branching habit, small leaves and rough bark are preferred. *Plumeria alba, Erythrina lithosperma* and *Glyricidia maculata* are commonly used. One meter cuttings with at least ten nodes are to be used for planting in the rapid multiplication nursery. Three or four basal leaves are cut retaining one third basal portion of leaf blade and dipped in Bordeaux mixture(1%) for a while and then should be kept in shade for about one week for partial wilting before planting.

The basal portion of vine is laid on the soil surface near the standards and covered with a thin layer of soil. The basal end of stem cuttings should be kept just above soil surface to prevent rotting. Growing tip of cuttings is then tied on to standards. The plant base should be mulched with partially decomposed wood debris or leaf litter. The vines are allowed to grow on the supports and later coiled around the branches of supports. The plants are to be irrigated as per need and do not overwater. Well decomposed farm yard manure or vermi-compost application as well as spraying of vermin-wash encourages good growth. Application of bio-agent *Trichoderma* along with FYM or vermi-compost and *Pseudomonas* or *Bacillus* helps to reduce occurrence of fungal infections in nursery and promotes vine growth. A well maintained plant would produce about three to five meters of growth/ year. On an average, one hectare of nursery would be sufficient to produce about 30-40 thousand meters of vine annually.

Poly-Bag Nursery

An alternative method to generate planting material is to raise cuttings in polythene bags and plant them in the field during favourable season. One and two node cuttings are being used. Bags of 18 x 12 cm size and 100-150 gauge thickness are usually used for planting stem cuttings. To facilitate proper drainage, five to six holes are given at lower half of each bag. The bags are filled with potting mixture prepared with jungle top soil, dry powdered farm yard manure and sand in the ratio of 3:1:1. Vermicompost can also be used in the rooting media. Cuttings with two or three nodes are planted in polybags. The exposed soil in polybag should be given a layer of mulch, preferably decomposed leaves to retain soil moisture and to serve as a source of plant nutrients. Partial shade should be provided either by using agro-shade nets or coir nets if natural shade is not available.

The plants are maintained with regular watering. Periodic application of vermicompost, vermi-wash and bio-agents promote better growth and reduce occurrence of diseases. The cuttings so raised in polybags will be ready for field planting in six to seven months by which time they should have produced about 50-70 cm of vine growth.

Impact of Trailing Material on Vine Growth

For production of planting materials, the growing vines may be vertically trailed as it favours rapid vine elongation to the tune of about 10 to 15 m/ year depending upon standards. Stem cuttings gave faster vine growth compared to tissue culture plantlets irrespective of standards used. Among the standards used, coir rope favoured maximum elongation (18 cm) followed by wooden sticks (11 cm) and Agro-shade net strips (9 cm)/ month.

Micropropagation

Micropropagation (Tissue culture) is an alternative method used for the large scale production of planting materials, which are 'true to type'. The protocol for micropropagation is standardized and large scale production of TC plantlets are being undertaken.

Protocol for Vanilla Micropropagation

1. Source Plants and Collection of Explants

Maintain actively growing shoots from healthy, disease free and high yielding plants in a protected area (net house/ glass house/ poly tunnels). Give regular prophylactic foliar spray once in two weeks with Bavistin or Indofil (0.2%).

Two-three cm long nodal segments (explants) should be collected from 2nd or 3rd node to the 5th or 6th nodes from shoot tip. November-May, August-September months are the most suitable for collection of explants from field grown plants.

2. Initiation of aseptic cultures/ Primary Culture Establishment

a) Surface sterilization and preparation of explants

Nodal segments/ explants should be thoroughly washed in a mild detergent solution before subjecting them to surface sterilization procedures. Surface sterilization is carried out by suspending the explants in 0.08% Mercuric Chloride solution for 12 minutes preferably under vacuum. Two to three drops of surfactants such as Tween-20 or Teepol should be added to the sterilant. If vacuum cannot be provided, swirl the contents well during the treatment. The sterilant is then drained off and the explants washed thoroughly with several changes of sterile distilled water. Subsequently the explants are transferred to sterilized Petri-plates containing sterile blotting papers, trimmed and blot dried prior to inoculation to nutrient medium.

b) Culture medium - Primary Culture Initiation and Establishment

Medium A: Murashige and Skoog (1962) / Knudson-C (1946) + BAP 1 mg/l, IAA 0.1 mg/l; D-Biotin 0.1 mg/l, Ca pantothenate 0.1 mg/l, Sucrose 3%, and agar 0.65 gm/l. pH of medium: 5.8.

The explants should be inoculated on to Medium A, followed by incubation under 18-hour photoperiod (1000 lux) at 25 ± 2°C. Axillary buds start sprouting by second week of culture initiation.

3. Shoot Multiplication

Composition of medium is the same as that for culture establishment i.e. Medium A. Sub cultures can be initiated after 30-35 days of initial inoculation/ primary culture initiation or when the new axillary sprout attains a length of more than one cm. For this, the new sprouts should be separated from original explant and transferred to fresh medium for further multiplication. Further subcultures can be carried out at 40-45 days intervals on fresh medium of the same composition.

4. Transfer to pots, hardening and acclimatization

The plantlets (8-10 cm long) are removed carefully from culture vessels and gently washed under tap water to remove agar. This is followed by suspension of plantlets in 0.1% bavistine solution for 15 min. The plantlets are then transferred to micropots containing soilrite (or) pure river sand and watered immediately. Subsequently they are covered with clear plastic and kept under green house conditions.

Regular sprays with water have to be provided in this case which is labour intensive. Alternatively, the plantlets are directly placed in a mist chamber where intermittent mist facility is available. If found necessary, the plantlets should be sprayed with Knop's or Hoagland's solution once in a week. Thereafter, they are progressively hardened by gradually reducing humidity and shade for four-five weeks.

After hardening, the plantlets should be transferred to potting mix and allowed to get acclimatized in the green house or nursery for 2 months. When the plantlets attain a length of 25-30 cm, they can be transferred to field conditions.

Planting and Management

Vanilla, being a climbing orchid requires standards for support and shade. A plant having low branches, rough bark and small leaves are ideal for support. Under Indian conditions, Glyricidia is a very good standard for vanilla. The support trees should be planted at least six months before planting of vanilla. Standards are planted at a spacing of 2.5-3.0m between rows and 2m within row with a planting density of 1600-2000 trees/ha.

The ideal month to plant vanilla in South Indian conditions is August-September when the intensity of the South West monsoon is low. By this time, the support trees should have grown well. Good quality vines from disease free plants, sufficiently grown rooted cuttings or secondary hardened tissue-cultured plantlets of at least 30cm height can be used for planting. If shorter cuttings/plantlets are used, the juvenile phase would be longer. Care should be taken to use cuttings collected either from nurseries or juvenile unflowered portion of vine from a garden.

Stem cuttings selected for planting should be kept in shade for about a week prior to planting. Generally, three to four basal leaves of the cutting are clipped away before they are put in shade. It is recommended to dip the basal tip in 1% Bordeaux mixture or Bordeaux paste or *Pseudomonas* paste before planting the cutting. While planting, the defoliated basal portion of cutting is to be placed in loose soil, near base of support, just below the surface, in a half loop in such a way that the basal tip is above the soil surface. The top end of cutting is to be tied to the support. Mulching base of support tree with partially decomposed organic matter is recommended. It takes about four to eight weeks for the cutting to take root and to show signs of initial growth.

Tissue cultured plantlets or rooted cuttings should be planted at base of support after removing the polythene bag carefully and without dislodging the soil around root zone. After planting, the soil at base of plant may be covered with a thick mulch of easily decomposable or partially decomposed material.

If the support tree do not provide adequate shade, the same may be provided using palm fronds or such other materials. Just as lack of shade is injurious to vanilla plant, too much shade and consequent lack of sunlight are also injurious to the plant. In mixed cropping with coconut and arecanut, shade regulation should take into account the shading provided by these trees.

After Care

a. Mulching

Mulching at base with easily degradable organic materials is good for conserving both moisture and soil. Mulch will add to fertility of soil when it disintegrates. Dried organic matter, leaves, weeds and coconut leaves/ husks can be used for mulching.

b. Manure Application

Vanilla is highly amenable to organic cultivation .Decomposed organic matter, bone meal, rotten cow dung, compost and fermented cakes can be used as manure for vanilla at least twice a year i.e. during June-July and September -October.

c. Watering

Adequate irrigation is to be provided for better vegetative growth of plant. During the first two - three years after planting, regular watering is required especially during summer once in two or three days. Mist and sprinkler irrigation are more effective and good for growth of vines. Since flowering requires stress for the vines, irrigation for mature vines of three years and more of growth should be stopped after cessation of North East monsoon. Irrigation should be resumed in such cases after initiation of flower bunches.

d. Shade Management

The thumb rule is that vanilla requires about 50% shade. Lopping of branches of living support is very important to regulate shade. The support trees, especially Glyricidia, should be lopped in May and November to initiate growth of new side branches for trailing of vines. Flowering and quality of beans are highly influenced by degrees of shade to which the vines are exposed.

Studies reveal that vanilla vines need a little more exposure to sunlight than shade during flowering and at the time of beans maturing. Over exposure to direct sunlight causes yellowing of vines and leaves. Similarly in heavily shaded plots, the stems are thin, leaves small and flowering delayed. So judicious shade management is very important for growth, timely flowering and fruit set.

e. Trailing of Vines

Trailing of vines is an important cultural operation. Growth of plants should be limited up to about 150 cm height. This is mainly to facilitate easy manual pollination. The vines are normally grown coiled around the lower branches of support trees and allowed to hang down or allowed to grow on the horizontal supports. The vines are trailed by coiling them around the branches of support trees or on horizontal supports without letting them to touch the ground.

The support trees may be pruned to form an umbrella shaped appearance about 1.5 -2.0 m above the ground. This is to give better shade and protection to the growing vines. The vines are allowed to grow up to a height of 1.2 -1.5 m. Then they are trained horizontally on branches and coiled round the standard. Bending of vines in this manner helps in accumulation of carbohydrate and other flower forming materials beyond the bend and to induce flower production at this portion of vine. If vegetative growth of vanilla vine are arrested by pruning, it turns to reproductive phase.

Vanilla plants are, in general, free from major pests and diseases. Among insect pests, a few small Lamellicorn beetles and ash gray weevil bite holes in flowers and often destroy the column. In addition, caterpillars, earwigs, snails and slugs live on tender parts of plant like shoot, flower buds and immature beans. Grasshoppers and crab are also found to cut growing tip of plants during establishment stage of plantation. Regular surveillance and removal of pests reduce their damage to a great extent.

Root rot, shoot tip rot, stem and bean rot and immature bean dropping are fungal diseases noticed. Phytosanitation measures are to be adopted in plantation for controlling the diseases. Application of organic manure in excess and heavy mulching in plant base are to be avoided. Fungicides such as Bordeaux Mixture (1%) and Copper Oxychloride (0.2%) may be applied as prophylactic measure. Application of pseudomonas (10-15 g/litre of water) will also control spread of disease.

Flowering and Pollination

Normally vanilla flowers during third year of planting. The flowering period varies from December to April depending on altitude.

Following practices may induce more flowering in vines that have reached sufficient growth.

- Stop watering the mature plants ready for flowering after North East monsoon.
- Clip off the shoot tip
- Lopping of shade to let in sunlight.

Flower bunches are produced in axils of leaves. Normally it takes 45 to 60 days from flower initiation to opening of flower. Only one or two flowers open in a day. Due to the peculiar structure of flower, natural pollination is not possible. Flowers are large, pale greenish yellow and bisexual. Sepals and petals look alike. The lower petal is short, broad and modified into a 'labellum'. The lower part of the labellum envelops a central structure called 'column' which is formed by fusion of pistil and stamen. The tip of column bears a single stamen with pollinia separated from stigma by a cap like structure called rostellum' which prevents natural pollination.

Pollination has to be done on the same day of opening of flower, preferably between 6-11 am. It is a very simple process once we understand the technique. Pollination is carried out by hand with help of a pointed bamboo splinter or a tooth pick. Flower is held in left hand facing us and thumb is let free. Rostellum of flower is lifted up using bamboo splinter held on right hand to open the stigma and with help of left hand thumb, the pollinia is pressed on to stigma so that pollen grains fall on it. Now the pollination process is over.

If the pollination process is not successful, the flower will fall off the next day. For commercial production pollinate only 10 to 12 flowers in a bunch to get quality beans. Be careful not to damage the parts which develop into the beans while carrying out pollination. It is found that one man can pollinate on an average 1500 flowers a day.

Unit Cost for Vanilla Cultivation

Area - 1 ha
Spacing - 3 x 2m
Plant population - 1600 nos.
Wage rate - Rs.85/manday

S. No.	Details	Ist year		IInd year		IIIrd year	
1	Preparation of land Layout, digging of pits, planting of support trees and vanilla cuttings	65 mandays	5525	5	425	-	-
2	Cost of planting materials for support trees	1600 nos.	3200	160	320	-	-
3	Cost of planting materials for vanilla	1600 nos.	19200	160	1900	-	-
4	Cost of manures, fertilizers, mulch materials etc.	-	1880	-	3760	-	5140
5	Cost of plant protection	-	320	-	420	-	460
6	Cultural operations including weeding manuring, trailing hand pollination etc.	50 mandays	4250	75	6375	100	8500
7	Harvesting, processing	-	-	-	-	40	3400
	Total		34375		13200		17500

Source: Spices Board of India

Vanilla

A. Cost of cultivation for 1st year - Rs. 34375.00

B. Cost of cultivation for 2nd year - Rs. 13200.00

C. Cost of cultivation plus processing of beans for the 3rd year - Rs. 17500.00

Total - Rs. 65075.00

Carbohydrates	-	7.1-9.1	Ash	-	4.5-4.7
Vanillin	-	1.48-2.90	Resines	-	1.5-2.6
Calcium	-	19.7	Potassium	-	16.2
Sodium	-	6.7	Phosphorus	-	9.5
Iron	-	0.3			

Quality Requirements

The primary quality requirement for cured vanilla beans is the aroma/ flavour character. Other factors of significance in quality assessment are the general appearance, flexibility, the length and vanillin content. The relative importance of these quality attributes depends upon the intended end use of cured beans.

Traditionally, appearance, flexibility and size characteristics are of importance since there is a fairly close relationship between these factors and aroma or flavour. Top quality beans are 17-25 cm long, fleshy, supple, very dark brown to black in colour, somewhat oily appearance, strongly aromatic and free from scars and blemishes. Low quality beans are usually hard, dry, thin, brown or reddish-brown in colour and possess poor aroma. The moisture content of top grade beans is high (30 to 40%), where it may be as little as 10% in lower grades.

Traditional Curing Methods

A number of procedures are evolved for curing of vanilla but they are all characterized by four phases (Theodose, 1972).

a) Killing or Wilting

This stops further vegetative development in fresh bean and initiates enzymatic reactions responsible for production of aroma and flavour. Killing is indicated by development of a brown colour of bean.

b) Sweating

This involves raising temperature of killed beans to promote desired enzymatic actions and to provoke at first, fairly rapid drying, to prevent harmful fermentation. During this operation, the beans acquire a deeper brown colouration and become quite supple, and development of an aroma becomes perceptible.

c) Slow drying

The third stage entails slow drying at ambient temperature, usually in shade, until beans have reached about one-third of their original weight. During this stage, production of different fragrances takes place.

d) Conditioning

In this final stage, the beans are stored in closed boxes for a period of three months or longer to permit full development of desired aroma and flavour.

The two most important methods are those of Mexico and the Indian Ocean Islands producing 'Bourbon' vanilla. Bourbon vanilla is considered the best quality vanilla in the world trade; hence this method of processing is suitably modified and followed in India.

Modified Bourbon Method

Modified Bourbon method of vanilla processing also involves four steps explained above, but the duration taken for different steps is different and the temperature and relative humidity required are quantified.

Killing

Various methods are employed for stopping the biological activity of beans by means of 'killing'. A modification of the Bourbon scalding technique is eventually advocated for convenience and to make products with consistent quality. The beans have to be sorted according to size before killing. The sorted beans are put in either a bamboo basket or gunny bag and immersed in hot water for scalding. Bigger beans are dipped for 5 minutes, medium sized beans for 4 minute and small beans for 3 minutes. The water is rapidly drained and beans are wrapped in woolen clothing and stored in woolen lined wooden box. Scalding is done at a temperature of 68°C, when beans are dipped temperature of water comes down by about 5°C. Temperature of water is a critical factor as higher temperature de-activates the enzymes responsible for conversion of vanillin and other flavour inducing components. Properly killed beans turn brown in colour with in 24 hours of scalding. The scalding also helps in disruption of cell structure and release of enzymes. Care should be taken not to attempt scalding at temperature more then 70°C.

Sweating

This is the most critical phase in vanilla processing. The enzymes responsible for conversion of vanilla and related compounds are the most active in this phase. The killed beans are spread on woolen cloth in sun and allowed to gather temperature. When temperature of beans attains 55°C (too hot to hold with bare hands), they are wrapped in woolen clothing and kept in sun for half an hour more and then stored in woolen lined wooden box for sweating for rest of the day. The sunning may take one and a half hour to three hours depending on location and also part of the day. Exposing beans to sun on a raised platform about 1 meter above the ground is better than exposing them directly on the ground. It is better to allow temperature to rise slowly than expose beans suddenly during the hottest part of day. This process of sunning and sweating has to be continued for about 8-10 days. At the end to sweating period, the beans attain chocolate brown colour, become very supple and loose about 40-50% of the moisture. Weight of beans is reduced by 50-60% by end of sweating. It helps to mildly massage the beans at this stage. The b-glycosidase located entirely on fleshy wall of beans comes in contact with most of the glucovanillin, present in the central segment or the placental mass.

Slow Drying

This step helps in removing the excess moisture and bringing it down to the desired level. At the end of sweating period, the moisture will be about 50-60% in the beans, this has to be brought down to 25-30%. This is achieved by spreading the beans after sweating, on the wooden racks in a well-ventilated room, at ambient temperature. The moisture is lost slowly; the beans become deep chocolate brown in colour. The beans have to be observed regularly during this stage, as the molds are likely to spoil the beans. Due to loss of moisture, weight of beans come down by 70-75% at the end of slow drying. Vanillin and related compound continue to be formed during this stage. The slow drying process may take 10-15 days.

Conditioning

Conditioning is done after moisture in the processed beans is brought to desired level by slow drying process. After slow drying, beans are bundled according to their size and into bundles of known weight. These are wrapped in butter paper and stored in airtight containers under ambient temperature for about two to three months. The beans under conditioning have to be examined at regular intervals, as they could get infested by molds or mites. When stored at low temperatures, vanillin crystals are formed on beans, hich is called frosting. Temperatures in the range of 35 to 45°C accelerated conditioning and the product had superior aroma compared to those conditioned at 13°C or 27°C.

Moisture content of beans during conditioning appears to have profound influence on aroma and flavour quality. Major moisture loss occurs during sweating/drying stages but some moisture reduction also occurs during first three months of conditioning, after which it becomes negligible. Beans with an average moisture content of 32% has a well developed, suave aroma and a high degree of flexibility.

Sorting and Grading

Vanilla is sold by growers either fresh or processed. The buyers of green vanilla, who are usually exporters, either cure it themselves or hire a curer. The exporters make a thorough examination of cured pods and resort them if necessary. The pods are usually kept under observation for a month before they are re-bundled and exported.

Grading

Harvested vanilla takes about four months for curing. Properly cured and dried beans are dark brown and supple enough to be twisted around the figure without rupturing. Processed beans are sorted and graded depending upon their length, colour, moisture content and flavour. Vanilla beans of commerce are divided for convenience into five geographic types.

Types	Places where grown
Mexican	Mexico
Bourbon	Island of Reunion, Madagascar, Comoros and Seychelles and India
South American	French West Indies
Tahiti	Tahiti
Java	Indonesia

Indian vanilla grades for international market (Bourbon grading procedure)

1st grade — whole beans > 14 cm length, moisture content < 25%, no scratches, blemishes, good flavour

2nd grade — whole beans > 14 cm length, moisture content 25 to 28%. Some scratches are permitted

3rd grade — whole beans > 14 cm length, moisture > 30%, more scratches/ blemishes

4th grade	whole beans with less flavour, length < 14 cm with more scratches, moisture > 30%
5th grade	broken, cut, short beans which do not have good flavour. Higher moisture level

Packing and Storage

Even after completion of curing, vanilla beans often continue to develop a desirable aroma and flavour, if stored under suitable conditions. After grading, beans are again sorted according to their length into bundles having length varying from 12 to 26 cm. Each bundle will have 70 to 100 beans and will weigh 150 to 500 g. These bundles are wrapped in butter paper and stored in air tight tin boxes.

Processed Products

1. *Vanilla extract or essence.* It is the extracted aroma and flavour characteristics of vanilla beans in an aqueous alcoholic solution.
2. *Vanilla flavouring.* It is similar to vanilla extract but with ethyl alcohol content always less than 35%.
3. *Concentrated vanilla extract/Flavouring.* These are products from which solvent is removed to concentrate the vanilla constituent keeping the level of alcohol as in the less concentrated form.
4. *Vanilla tincture.* It differs from the extract in alcohol content which according to use ,is not less than 35% and may be as high as 90%.
5. *Vanilla oleoresin.* Semi solid concentrate obtained by complete removal of solvent from vanilla extract, with inferior flavour and aroma.
6. *Vanilla-vanillin extract and flavouring:* It is prepared by fortifying diluted vanilla oleoresin with synthetic vanillin.
7. *Vanilla powder:* It comprises a mixture of vanilla powder or vanilla oleoresin with sugar, food starch or gum acasia.

Uses

Vanilla beans consist of a broad array of flavour profiles of about 300 flavour components, which are easily volatized by heat or chemical distillation processes.

Vanilla is by far the most popular food flavour at the present time in the ice cream, baking and chocolate industries in the United States. It is chiefly used in flavouring of chocolates, beverages, cokes, confections, cakes, custards, puddings and ice creams and in the manufacture of soaps, perfumes and sachet powders. Vanilla is generally used in the form of an extract from the beans. In the manufacture of chocolate, however, the beans are usually ground finely with sugar and included with chocolate. The bean is also sometimes used directly in the manufacture of ice creams and other confectionaries, flavouring for medicines and as a fragrance to conceal the strong smell of rubber tyres, paint, and cleaning products.

Marketing

Demand and Supply Analysis

The estimated area of vanilla cultivation in India is about 1600 ha and currently producing 100 tonnes of cured beans, which is likely to rise to 400 tonnes by 2006 as many growers in Karnataka, Kerala and Tamil Nadu have taken up vanilla cultivation as monocrop and as intercrop in 2001-02. In the absence of domestic demand for natural vanilla concentrates in India and lack of natural vanillin extraction facilities in India, almost entire production of cured beans is exported to USA, UK, France and Germany for further processing and extraction of natural vanillin. India's domestic market for vanillin is completely dominated by synthetic vanillin due to its low price and easy availability. India imports around 600-700 tonnes of synthetic vanillin, which is mostly consumed by food and beverages industry (200 tonnes), ice-cream industry alone consuming 130 tonnes.

The world export market for natural vanillin is valued at around $422 million with nearly 50% sourced from Madagascar. There is a strong global demand now between 16,000 and 18,000 tonnes / year, of which, 2000 tonnes are lignin vanillin. By comparison, total world demand for natural vanillin, at present, is below 2,000 tonnes

The US, France and Germany account for about 80% of world natural vanilla imports, the US absorbing 50-60%, and France and Germany between 10% and 15% each. These three countries are also major re-exporters of both vanilla beans and processed vanilla products.

Vanilla: International Market

The international vanilla market is characterized by its extreme volatility. Three features dictate the international vanilla market, viz., highly speculative cycles, raw material quality and competition from synthetic vanillin. Prices fluctuate enormously over a period of a few months. The world production was of the order of 5,400 tonnes in 2004. The production is increasing rather heavily in recent years with an average annual growth rate of 4 % over a 10-year period. Increase in production without corresponding growth in demand has pushed cured vanilla prices in the international market down to US$ 35/ kg from over US$ 500/ kg, two years ago. Shooting prices during 2003-04 was due to shortage of supply from Madagascar, owing to their political instability and crop failure as a result of repeated cyclones. High prices in the recent years enthused farmers in Uganda, Papua New Guinea, India, Costa Rica, China and Colombia to take up vanilla cultivation. The current fall of prices can be attributed to recovery of Madagascar crop, introduction of new entrants in the market and shift in demand to alternate products such as synthetic vanillin and natural identical vanillin as a result of skyrocketing prices of natural vanillin. Although vanilla extract from beans is still used by food industry, this accounts for less than 1 % of vanillin production. Remaining 99% is obtained synthetically. The current depression in world vanilla market is expected to prevail until companies who shifted to alternate products revert to using natural vanillin from beans. Traders expect vanilla price to settle down at US$ 60-70/ kg by 2007-08.

Indian Market

India entered the international vanilla market during a crisis period due to shortage of supply from Madagascar and resulting price rise. Indian vanilla, according to industrial sources is at par with Madagascar Bourbon vanilla in terms of vanillin content and quality, which made importers source vanilla beans from India in order to meet demand of natural vanillin. In 2003-04, India exported 26 tonnes of cured vanilla beans valued Rs.36.06 crores (US$ 8 million) registering a phenomenal annual average growth rate of 92% between 1999-2003. According to certain latest estimates by Spices Board of India, vanilla exports from India have posted a 90% increase in quantity and a higher average price than Madagascar at US$ 47.30/ kg of cured beans during last fiscal. India came into the market with a small production last year at about 50 tonnes only. Currently India has 100 tonnes of cured beans to offer to international vanilla market.

Since the last season, Indian growers were fetching Rs. 900 (US$ 20)/ kg of cured vanilla beans and on an average of Rs. 250 –275 (US$ 5.5-6) for one kg of green beans. As per industry sources, currently Indian vanilla is enjoying fairly good demand in the US, France, Germany, the Netherlands, UK and Japan. Considering the export potential, Spices Board is encouraging vanilla cultivation as an intercrop to coconut, arecanut, coffee, cardamom and other spices in Tamil Nadu, Karnataka and Kerala and also intends to provide support facilities for expansion of area under cultivation of vanilla, as an intercrop, to Andhra Pradesh, Maharashtra, Madhya Pradesh, Orissa, West Bengal, Andamans and North Eastern States. Vanilla cultivation has become a major activity for farmers in Karnataka particularly in Shimoga, Uttara Kannada, Dakshina Kannada, Udupi and Chikmagalur, which are declared as the Agri-Export Zone (AEZ) for vanilla by Government of India.

Vanilla Yield in India

Kerala exported 7 tonnes of vanila and earned Rs. 1175.61 lakhs during 2004-05 achieving 75% increase in quantity and 100% in value compared to 2003-04. But the price in domestic market decreased considerably during this period.

Vanilla Beans Production in Madagascar

Madagascar is the first vanilla beans exporter in the world. Its quality regarding perfume and taste cannot be compared. After the market realisation, the conquest of the international market was a success. Quality was much better than quantity, Vanipro's quality is known and well appreciated in Europe, USA and Japan, as a quality model, serious and realiable. Vanilla production in Madagascar is 1000 tonnes in (2002), 600 tonnes in (2003), 1700 tonnes in (2004), meaning 5 to 7 times more of green vanilla beans. Other producers are Indonesia; Mexico; Tahiti; Comoros; Reunion Island, Uganda, China, India and New Guinea.

ORGANIC SPICES

Global awareness of health and environmental issues is spreading fast in recent years, especially in developed countries. Sustainability in production has become the prime concern in agriculture development. The organic method of farming is the best option to ensure that the air, water and soil around us remain unpolluted, leaving the environment safe for present and future generations.

In many countries, exploitative agriculture using industrial inputs is the norm since 1960s, to cater to an increasing population and to combat occurrence of famine and natural calamities. Such a system of farming causes imbalances in the constituents of biosphere, bioforces, bioforms and biosources. As a result, health of 'Mother Earth' is deteriorating. Organic agriculture aims to tackle the above concern, and also aims at protecting the environment from continuous decline.

The concept of organic farming is based on holistic approach, where nature is perceived to be more than just an individual element. In this farming system, there is dynamic interaction among soil, humus, plant, animal, eco-system and environment. Hence organic farming differs from industrial agriculture as in the latter, biological systems are replaced by technical production systems with liberal use of chemicals.

Organic farming improves structure and fertility of soil through balanced choice of crops and implementation of diversified crop rotation systems. Biological processes are strengthened without recoursing to chemical remedies, like synthetic fertilizers and pesticides. In this farming system, control of pests, diseases and weeds is primarily preventative, and if required, adopting organic products, which will not adversely affect the environment. Genetically modified organisms are not normally acceptable because of manipulations made in their natural set up. Organic matter of various kinds, nitrogen fixing plants, pests and disease resistant varieties, soil improvement practices like mulching and fallowing, crop rotation, multiple cropping, mixed farming, etc., are freely adopted. In brief, organic farming merges traditional and respectable views on nature with modern insights.

Bio-dynamic agriculture is yet another approach to organic farming. It is based on anthropsophy on the ideas formulated by the Austrian expert, Rudolf Steiner, in 1924. In this system, maintenance and furtherance of life processes on Earth are achieved by harnessing cosmic energy and various influences of sun, stars, moon and other planets. Bio-dynamic agriculture, most often, combines animal husbandry and crop production and use of compost and bio-dynamic preparations to revitalize soil and plants and subsequently animals and human beings.

Organic Spices - Indian Capability

India grows over 50 different varieties of spices. Total production is around 2.7 million tonnes. Of this, about 0.25 million tonnes (8-10%) is exported to more than 150 countries. The Indian share of the world trade in spices is 45-50% by volume (25%by value).

Organic Cultivation

In recent years, organic agriculture is gaining considerable importance. Many farmers today show interest in organic cultivation. Several of them have begun switching to this traditional method of cultivation as a means to produce safe foodstuffs and preserve the environment. The concept of sustainable farming has caught on in India.

World Demand for Organic Foods

The world demand for organically produced foods is growing rapidly in developed countries like Europe, USA, Japan and Australia. The current estimated share of organic foods in these countries is approximately 1 to 1.5%. Worldwide, food trends are changing with a marked health orientation. Since organic foods are free from chemical contaminants, demand for these products should steadily increase in the new millennium.

According to the ITC, UNCTAD/GAT, more than 130 countries produce certified organic foods. 100 of them are from Asia and Africa.

Internationally, there is definite shift towards traditional / ethnic medicines. Since spices form part of many of these medicines, demand for organically produced spices should grow.

Organic cultivation is nothing new to India. The country has always been practising the traditional ways of using indigenous technologies and inputs mostly in line with modern organic farming principles. The per capita consumption of fertilizers and pesticides in India is far below that of developed countries, which means, it is very easy for Indian farmers to embrace organic spice farming in its true sense.

Spice Board of India has taken a major initiative in promoting the production and export of organic spices in a big way.

Spices Board supports production, processing, certification and marketing of organic spices. Assistance is provided for organic cultivation of paprika, chillies, ginger, turmeric, Kokkam, cumin, fennel, fenugreek, coriander and dill in Maharashtra, Orissa, Chattisgarh, Uttarakhand, Bihar, Gujarat and Rajasthan. The Board has also programmes to encourage production of organic pepper, herbal spices, Lakadong turmeric and ginger in the North Eastern States. Research is on to develop suitable package of practices for organic cultivation of chilli and cardamom. In certain spice crops, where organic production practices have not fully developed, Board encourages Integrated Pest Management (IPM) Programmes.

The National Programme for Organic Production (NPOP) generally conforms to guidelines of IFOAM, EU, NOP and JAS in major aspects.

Export of organic products is allowed only if they are produced, processed or packed under the approved production programme and certified by an accredited certifying agency. Spices Board is designated as one of the agencies empowered to accredit certification agencies.

The Indian Initiative

Export of organic spices from India started in right earnest. The country at present exports around 50 tonnes of different varieties of organic spices. Exports will get a significant boost in coming years as more farmers switch to organic methods.

Spices Board of India prepared a document on production of organic spices. It features organic concepts, principles, basic standards, production guidelines, documentation, inspection and certification. The document is published after approval by National Standards Committee constituted by the members of IFOAM in India.

Research programmes on organic cultivation of important spices have commenced. The work is carried out at the Spices Board's Indian Cardamon Research Institute at ldukki District in Kerala. Besides organising demonstrations to educate and motivate prospective organic spice growers, the Board is simultaneously involved in training programmes to existing spice growers on organic principles and practices.

Organic Farming - Guidelines

Organic farming is a crop production method respecting the rules of nature. Organic farming is targeted to produce nutritive, healthy and pollution free food. It maximises use of on farm resources and minimises the use of off-farm resources. It is a farming system that seeks to avoid use of chemical fertilisers and pesticides. It is not profit oriented but social profit oriented. Commitment to nature protection is a pre-requisite for practising organic farming. In organic farming, entire system ie. plant, animal, soil, water and micro-organism are to be protected.

Scope

Organic farming helps in rejuvenating degraded soil and ensure sustainability of crop production. Common man and farmers are aware of hazards from use of chemicals and pesticides. It is a common practice that farmers maintain part of their rice fields without pesticide application for their own consumption. When vegetables are grown in the Kitchen garden, no chemical fertilisers or pesticides are used since the house wife knows that the vegetables are meant for their own consumption. Now, the consumers prefer to consume natural/ethnic foods, particularly organic foods across the world. Moreover, they are ready to pay a premium price for such foods. The demand for organic agricultural products is increasing day by day.

Minimum Requirements

In organic farming system, certain minimum requirements are to be met to fulfill its objectives. Steps should be taken to maintain bio-diversity, viz. swamps, gross lands, forests, etc.

i) *Mixed farming:* Animal husbandry, poultry, fisheries, etc. should be practised in addition to agricultural farming. Shifting cultivation is not allowed. Then only the farm is certified as organic.

ii) *Conversion:* When a farmer switches over to the system of organic farming from the conventional system of farming, it is known as conversion. The time between the start of organic management and certification is called conversion period. The farmers should have a conversion plan prepared, if the entire field is not converted into organic at a time. In that case, it is necessary to maintain organic and non-organic fields separately. In the long run, entire farm including livestock should be converted into organic. The conversion period is decided based on past use of land and ecological situation. Generally, conversion period is two years for annual crops and three years for perennial crops. However, the conversion period can be relaxed based on verification by certification agency if the requirements are fully met.

iii) *Cropping Pattern:* Crop rotation should be followed if annual crops are grown. Intercropping should be practised when perennial crops are grown. Crop rotation should cover green manure and fodder crops. In case of perennial crops, cover crops like Kolinji should be grown to protect the soil. Monocropping should be avoided.

iv) *Planting*: Species and varieties cultivated should be adapted to soil and climatic condition and resistant to pests and diseases. Seeds/planting materials should be procured from organic sources. If not available, chemically untreated seeds/planting materials can be used .Even one time use of genetically engineered seeds or planting materials such as tissue culture, pollen culture, transgenic plants is not allowed.

v) *Manurial Policy*: Soil fertility should be maintained/ enhanced through raising green manure crops and leguminous crops. The residues of plants after harvest should be incorporated into soil as far as possible. Bio-degradable materials of microbial, plant or animal origin shall be applied as manures. (eg. compost, vermin compost, farm yard manure, sheep penning etc.) Use of synthetic/chemical fertilisers is not permitted. The mineral based materials like rock phosphate, gypsum and lime, can be applied in limited quantities when there is absolute necessity.

Following products are permitted for use in manuring/soil conditioning in organic fields:-

- Farm yard manure, slurry, green manures, crop residues, straw and other mulches from own farm.
- Saw dust, wood shaving from untreated wood, calcium chloride, lime stone, gypsum and chalk, rock magnesium chloride, sodium chloride, Bacterial preparations (Bio-fertilisers), eg. azospirillum, rhizobium, Bio-dynamic preparations, preparation and extracts, eg. Plant ,neem cake, Vermicompost.
- Following products shall be used when they are absolutely needed and taking into consideration of factors like contaminations, depletion of natural resources, and nutritional imbalances. If proposing for certification, the certification agency may be consulted before using these inputs such as farm yard manure, slurry, urine and straw from other farms, blood meal, bone meal, fish meal without preservatives, minerals like basic slag and sulphate of potash, wood ash from untreated wood and vermi compost from other farms.

vi) *Pest, Disease and Weed management:* Use of synthetic/chemical pesticides, fungicides and weedicides is prohibited. Natural enemies shall be encouraged and protected. (eg. raising trees in the farm attracts birds which kill pests of crops and nest construction). Products collected from the local farm, animals, plants and micro-organisms and prepared at the farm are allowed for control of pests and diseases. (eg. Neem Seed Kernel Extract, cow urine spray). Use of genetically engineered organisms and products are prohibited for controlling pests and diseases. Similarly, use of synthetic growth regulators is not permitted.

Slash weeding is to be done between plants. Weeds under base of the plants shall be cleaned and put as mulch around the plant base. The weeded materials should be applied as mulch in the ground itself.

The products that are permitted for control of pest and diseases are:

Neem oil and other neem preparations like Neem Seed Kernel Extract, Chromatic traps, Mechanical traps, Pheromone traps, Plant based repellants, Soft soap Clay.

The following products shall be used when they are absolutely necessary and taking environmental impact into consideration. The certification agency shall be consulted before using these inputs such as copper salts e.g. Bordeaux mixture, Plant and animal preparations Cow urine spray,

Garlic extract, Chilli extract, Light mineral oils like Kerosene, release of parasite predators of insect pests e.g. Trichogramma.

vii) *Soil and water conservation*: Measures like stone pitching/contour wall construction are to be taken up to prevent soil erosion. In saline soils, saline resistant varieties may be grown. Judicious irrigation is to be practised. Mulching is required. Pollution of surface and ground water shall be prevented. Clearing of primary forest is prohibited. Cleaning of land through straw burning should be restricted to minimum.

viii) *Contamination control:* It is necessary to take up following measures to minimise contamination from outside and within farm.

a) If neighbouring fields are non-organic, a buffer zone should be maintained. The height of buffer crop shall be twice the height of organic crop and width of buffer shall be 8-16m. (When chilli is grown as the main organic crop, castor or Agathi (*Sesbania)* can be grown as buffer crop. The crops from buffer zone should be sold as non-organic).

b) If the farm is under conversion, equipments used in conventional areas shall be well cleaned before using for organic areas.

c) Products based on polythene, polypropeline and other polycarbarnates are allowed to cover protected structure, insect netting, nursery and drying subject to the condition that these materials shall be removed from field after use and they shall not burnt or put in the soil. Use of polychloricle based products like PVC pipe is prohibited

ix) *Processing* Processing technologies like solar drying, freeze drying and hot air chambers are permitted. Irradiation of agricultural produce is not permitted. No synthetic additives/dyes are to be added during processing. Lot no. should include the crop, country, field no, date of harvest (in Julian Calendar) and production year.

x) *Labelling*: The label should convey clear and acurate information on organic status of product. (i.e. conversion in progress or organic). The labels for organic and conversion in product, quantity of product, name and address of producer, name of certification agency, certification and lot number are to be given in the label.

Lot number is helpful in tracing back the product particularly the field no. in which it is grown in case of contamination. Calendar is ranging from 1 to 365 or 366, starting 1st January, as I and December 31st as 365/366.

Crop	OC (organic chillies)
Country	I (India)
Field No.	05
Date of harvest	32 (1 st Feb)
Year	1999
Lot number OC 1 0532 1999.	

xi) *Packaging*: For packing, recycling and reusable materials like clean jute bags, shall be used. Use of bio-degradable materials shall also encouraged. Unnecessary packaging material should be avoided. Organic and non-organic products shall not be stored and transported together except when labelled.

xii) *Social Justice*: Social right and justice are integral part of organic agriculture. The laws relating to labour welfare and rights of children should be honoured. All employees and their families should have access to potable water, food, housing, education, transportation and health services. All employees should have equal wages when doing same job. They must have equal opportunities irrespective of colour, creed and gender. Social security needs (include maternity, sickness and retirement benefits) should be met. Labour conditions regarding noise, dust, light and exposure to chemicals should be within acceptable limits; and they should have adequate protection. The rights of indigenous people should be respected.

xiii) *Documentation*: Documentation of farm activities is a must for acquiring certification especially when both conventional and organic crops are raised. The following documents/records are to be maintained.

a) Field map
b) Field history sheet
c)Activity register
d) Input record
e) Output record
f) Harvest record
g) Storage record
h) Sales record
i) Pest control records
j) Movement record
k) Equipments cleaning records
l) Labelling records.

Certification Process

Certification of organic farms is required to satisfy the consumers that the produce is totally organic. Certification agency conducts inspection that minimum requirements prescribed for organic agriculture are fully met and certificates issued

The producer makes contact with certifying agency. Certification agency provides information on standards, fees, application, inspection, certification and appeal procedures. The producer then submits application along with field history, farm map and record keeping system. Then the contract, indicating scope, obligation, inspection and certification, sanction and appeals, duration and fee structure, is executed.

Then the Inspector of agency comes and carries out inspection. The Inspector gives inspection report with his recommendation to the agency, the agency issues approval or denial of certificate. Certificate is given for current year's harvest only and hence annual certification is required.

Production of Organic Pepper

General Guidelines

When pepper is grown in mixed cultivation system, it is essential that all the crops in the field are maintained following organic methods of production. It is also advisable that the entire farm is converted to organic. Livestock should be reared according to organic principles. All crop residues and farm wastes should be recycled through composting so that soil fertility is restored and maintained at a very high level. Soil and nutrient loss through soil wash, run off and percolating water should be minimized through proper agronomic practices. Weeding is to be limited to slashing. Slashed materials should be used for mulching the plant base. The plantation should have a green cover with leguminous crops. They could be cut and mulched during summer to prevent moisture loss. Shade and support trees, leguminous and green manuring shrubs to provide bio-mass and neem and other sources for plant protection agents should find a place with in or on the borders of such organic farm.

Guidelines for Organic Pepper Production

An isolation belt of at least 25 m wide should be maintained around the organic plantation. The produce from this isolation belt shall not be treated as organic. Precaution should be taken to avoid the entry of run off water and drift from the neighbouring farms. While converting an existing plantation to organic cultivation, a minimum three years conversion period is a must. The conversion period can be relaxed for the organic farm raised on a land where chemicals were not used previously. In all these cases, proper records must be maintained for submission to the certifying agency.

Sources of Planting Material for Organic Production of Pepper

Initially, cuttings from conventional plantations can be used in the absence of purely organic sources. The production of rooted cutting should be done according to organic principles. In subsequent years, the runner shoots or aerial shoots collected from elite mother vines grown organically only shall be used for generation of planting material.

Adoption of following practices in the nursery is recommended. The soil used for potting mixture should be solarised prior to use. Solarised soil should be fortified with cultures of VAM and Trichoderma (250 g mass multiplied in 25 kg. compost).

The vines in the rapid multiplication units may be sprayed with vermi wash (50 ml/unit) for enhancing growth.

Disease Management in Nursery

The two important nursery diseases viz., leaf rot caused by *Rhizoctonia solani* and basal wilt caused by *Sclerotium rolfsi* can be minimum, if solarised soil inoculated with VAM and Trichoderma are used. If isolated incidence of these diseases is noticed, timely adoption of phytosanitary measures and spot application of Bordeaux mixture (1%) may be done. In areas, where nematode problems occur, addition of crushed neem seed or neem cake is recommended.

Prevention of Run off and Drifts from other Farms

In sloppy lands, adequate soil and water conservation measures are necessary at the time of preparing the land for planting to avoid entry of run off water and drift from neighbouring farms.

Use Variety of Standards

As many varieties of standards as possible should be used. Use of *Erythrina sp.* should be minimized as the plant harbours root knot nematode.

Nutrients

Application of 2 kg compost of rotten cowdung mixed with 125 g rock phosphate at the time of planting rooted cuttings should be done as a basal dose.

Cultural Operations

The same set of cultural operations recommended for conventional production can be adapted for organic pepper production also.

Manuring

Compost or farmyard manure may be applied from second year onwards@ 4kg/vine which can be gradually increased to10 kg/vine as it grows during May-June. This can be partially or completely substituted by vermicompost in which case, the quantity needed will be half. If found necessary, based on soil test, application of rock phosphate, bone meal, lime and dolomite may be carried out. Wood ash may be used in potash deficient areas. Compost made from green loppings, crop residues, grasses, cow dung and poultry droppings, fortified wood ash and/ or rock phosphate should be used regularly instead of farmyard manure alone. Such compost can be further enriched with non-edible oil cakes and right microbial cultures prior to withdrawal from compost pit and before applying to field. Crushed neem seed @ 2 kg/vine/year may be applied in areas infested with nematodes. Use of bio-fertilizers can also be resorted to in a restricted manner.

Ginger

Ginger (*Zingiber officinale)* is one of the important spices grown in India. Ginger of commerce is the dried rhizome. It is marketed in different forms raw ginger, dry ginger, bleached dry ginger, ginger powder, ginger oil, ginger oleoresin, gingerale, ginger candy, ginger beer, brined ginger, ginger wine, ginger squash and ginger flakes.

Climate and Soil

Ginger is cultivated in almost all states in India. Kerala is the major ginger growing state. Other major ginger growing states are Orissa, Meghalaya, Himachal Pradesh and Karnataka. Ginger grows in warm and humid climate. It is mainly cultivated in the tropics from sea level to an altitude of above 1500 amsl and it can be grown both under rainfed and irrigated conditions. For successful cultivation of the crop, a moderate rainfall at sowing time till the rhizomes sprout, fairly heavy and well distributed showers during growing period and dry weather for about a month before harvesting are necessary. Ginger thrives the best in well drained soils like sandy or clay loam, red loam or lateritic loam. A friable loam rich in humus is ideal. However, being an exhusting crop, it may not be desirable to grow ginger in the same site year after year.

Maintenance of Buffer Zone

To cultivate ginger organically, a buffer zone of 8 to 16m feet is to be left all around from conventional farm, depending upon location of farm. The produce from this buffer zone belt shall not be treated as organic. Being an annual crop, the conversion period required will be two years. Ginger can be cultivated organically as an inter or mixed crop provided all other crops are grown following organic methods. It is desirable to include a leguminous crop in rotation with ginger. Ginger-banana- legume or ginger-vegetable-legume can be adopted.

Sources of Planting Material

Carefully preserved seed rhizomes free from pests and diseases which are collected from organically cultivated farms can be used for planting. To begin with, seed material from high yielding local varieties may be used in the absence of organically produced seed materials. Seed rhizomes should not be treated with any chemicals. The seed rate varies from region to region and with method of cultivation adopted. The seed rate varies from 1500- 2500 kg/ hectare.

Preparation of Land and Planting

While preparing the land, minimum tillage operations may be adopted. Beds of 15 cm height, 1m width and of convenient length may be prepared giving at least 50 cm spacing between beds. Solarisation of the beds is beneficial in checking the multiplication of pests and disease causing organisms. Solarisation is a technique by which polythene sheets are covered over moist beds of field, reaching all the sides and exposing to sun for a period of 20-30 days. The polythene sheets used for soil solarisation should be kept away safely after the work is completed.

At the time of planting, apply 25g powdered neem cake and mix well with the soil in each pit taken at a spacing of 20-25 cm within and between rows. Seed rhizomes may be put in shallow pits and mixed with well rotten cattle manure or compost mixed with *Trichoderma*, an antagonistic (parasitic) fungi (10g compost inoculated with *Trichoderma*). The best time for planting in West Coast is first fortnight of May with receipt of monsoon. Under irrigated conditions, it can be planted well in advance during middle of February or early March.

Cultural Practices

Mulching ginger beds with green leaves is an essential operation to enhance germination of seed rhizomes and to prevent washing off of soil due to heavy rain. This also helps to add organic matter to soil and conserve moisture during the later part of cropping seasons. The first mulching is to be done with green leaves @ 10-12 t/ha at the time of planting. It is to be repeated @ 5 t/ha at 40th and 90th day after planting. Use of *Lantana camara* and *Vitex negundo* as mulch may reduce infestation of shoot borer. Cow dung slurry or liquid manure may be poured on bed after each mulching to enhance microbial activity and nutrient availability. Weeding may be carried out depending on intensity of weed growth. Such materials may be used for mulching. Proper drainage channels are to be provided in the inter rows to drain off stagnant water.

Manuring

Application of well rotten cow dung or compost @ 5-6 t/ha may be made as basal dose while planting the rhizomes in pits. In addition, application of neem cake @ 2 t/ha is also desirable

Plant Protection

Pests

Shoot borer is the major pest infesting ginger. Regular field surveillance and adoption of phytosanitary measures are necessary for pest management. It appears during July -October. Spot out shoots infested by borer and cut open the shoot and pick out caterpillar and destroy. Spray neem oil (0.5%) at fortnightly intervals, if necessary. Light traps will be useful in attracting and collecting the adult moths.

Diseases

Soft rot or rhizome rot is a major disease of ginger. While selecting the area for ginger cultivation, care should be taken to see that the area is well drained as water stagnation pre- disposes plants to infection. Hence provide adequate drainage. Select seed rhizomes from disease free areas since this disease is seed borne. Solarisation of soil done at time of bed preparation reduces the fungus inoculum. However, if the disease is noticed, the affected clumps are to be removed carefully along with the soil surrounding the rhizome to reduce the spread. *Trichoderma* may be applied at the time of planting and subsequently, if necessary.

Restricted use of Bordeaux mixture (1%) in disease prone areas as spot application may be made to control it.

Harvesting and Post Harvest Operations

The crop is ready to harvest in about eight to ten months depending upon maturity of variety. When fully mature, leaves turn yellow and start drying up gradually. Clumps are lifted carefully with a spade or digging fork and rhizomes are separated from dried leaves, roots and adhering soil. The average yield of fresh ginger/ hectare varies with varieties ranging from 15 to 25 tonnes. For making vegetable ginger, harvesting is done from 6th month onwards. The rhizomes are thoroughly washed in water twice or thrice after harvest and sun-dried for a day. For preparing dry ginger, the produce is kept soaked in water overnight. Rhizomes are then rubbed well to clean them. After cleaning, rhizomes are removed from water and outer skin removed with a bamboo splinter or wooden knife having pointed ends. Iron knife is not recommended, as colour will be faded. In order to get rid of last bit of the skin or dirt, dry rhizomes are rubbed together. Peeled rhizomes are washed and dried in sun uniformly for one week. Rhizomes are to be dried to a moisture level of 11% and they are stored properly to avoid infestation by storage pests. Storage of dry ginger for longer periods is not desirable. The yield of dry ginger is 16-25% of the fresh ginger depending upon the variety and location where the crop is grown.

Preservation of Seed Rhizomes

The rhizomes to be used as seed material should be preserved carefully. The indigenous practices like spreading layers of leaves of *Glycosmis pentaphylla* being followed by farmers can very well be adopted for this purpose. To get good germination, seed rhizomes are to be stored properly in pits under shade. For seed materials, big and healthy rhizomes from disease-free plants are selected immediately after harvest. For this purpose, healthy and disease-free clumps are marked in the field when the crop is 6-8 months old and still green. Seed rhizomes are stored in pits of convenient size, made inside the shed to protect from sun and rain. Walls of pits may be coated with cow dung paste. Seed rhizomes are stored in these pits in layers along with well-dried sand/saw dust (i.e. put one layer of seed rhizomes, then put 2 cm thick layer of sand/saw dust). Sufficient gap is to be left at top of pits for adequate aeration. Seed rhizomes in pits need inspection once in twenty days to remove shrivelled and disease affected rhizomes. Seed rhizomes can also be stored in pits dug in the ground under shade of a tree provided there is no chance for water to enter pits. In some areas, rhizomes are loosely heaped over a layer of sand or paddy husk and covered with dry leaves in a thatched shed.

Turmeric

Turmeric (*Curcuma longa*), the ancient and sacred spice of India is a major spice produced and exported from India. Turmeric is used as condiment, dye and cosmetic in addition to its use in religious ceremonies. India is the leading producer, consumer and exporter of turmeric in the world.

Climate and Soil

Turmeric can be grown in diverse tropical conditions from sea level to 1500mm amsl at temperature of 20-30°C with a rainfall of 1500 mm or more/ annum or under irrigated conditions. It is grown on different types of soils from light black, loam and red soils to clay loams. It thrives the best in a well-drained sandy or clayey loam.

Maintenance of Buffer Zone

To cultivate turmeric organically, a buffer zone of 8 to 16m shall be maintained if the neighbouring farms are non-organic. The produce from this zone shall not be treated as organic. Turmeric being an annual crop, the conversion period required will be two years, Turmeric can be cultivated organically as an intercrop with other crops provided organic methods of cultivation are followed for all the companion crops.

Sources of Planting Material

Carefully preserved seed rhizomes free from pests and diseases collected from organically cultivated farms, should be used for planting. To begin with, seed material from high yielding local varieties may be used in the absence of organically produced seeds. A seed rate of 2500 kg rhizomes is required for planting one hectare.

Preparation of Land and Planting

While preparing land, minimum tillage operations may be adopted. Beds of 15 cm height, 1 m width and of convenient length may be prepared giving at least 50 cm spacing between beds. Solarisation of such beds is beneficial in checking the multiplication of pests and disease causing organisms. The polythene sheets used for soil solarisation should be kept away safely after the work is completed.

At the time of planting, apply 25 g powdered neem cake and mix well with the soil in each pit taken at a spacing of 20-25 cm within and between rows. Seed rhizomes may be put in shallow pits and covered with well rotten cattle manure or compost mixed with *Trichoderma* (10 g compost inoculated with *Trichoderma).* Turmeric can be planted during April-May with receipt of pre monsoon showers.

Cultural Practices

Mulching turmeric beds with green leaves is an essential operation to enhance germination of seed rhizomes and to prevent washing off of soil due to heavy rain. This also helps to add organic matter to soil and conserve moisture during the later part of cropping season. Judicious mix of leguminous leaves with high nitrogen content and leaves rich in phosphorous like *Acalypha* weed and leaves rich in potassium like *Calatropis* can be used according to availability. The first mulching is done at planting with green leaves @ 10-12 tonnes/ha. It is to be repeated again @ 5 tonnes/ha at 50th day after planting. Cow dung slurry may be poured on bed after each mulching to enhance microbial activity and nutrient availability. Weeding may be carried out depending on intensity of weed growth. Such materials may be used for mulching. Proper drainage channels are to be provided in inter rows to drain off stagnant water.

Manuring

Application of well rotten cow dung or own compost from own form @5-6 tonnes/ha may be made as basal dose while planting rhizomes in pits. In addition, application of neem cake@2 tonnes/ ha is also desirable.

Plant Protection

Pests

Regular field surveillance and adoption of phytosanitary measures are required for pest management. If shoot borer incidence is noticed, such shoots may be cut open and larvae picked out and destroyed. Spray neem oil (0.5%) at fortnightly intervals, if necessary.

Diseases

No major disease is noticed in the crop. Leaf spot and leaf blotch can be controlled by restricted use of Bordeaux mixture (1%). Application of *Trichoderma* at time of planting checks incidence of rhizome rot.

Harvesting and Post Harvest Operations

Turmeric is to be harvested at correct maturity. Depending upon variety, crop becomes ready for harvest in 7-9 months, medium varieties in 8-9 months and late varieties after 9 months.

Usually the land is ploughed and rhizomes are gathered by hand picking or clumps are carefully lifted with a spade. Harvested rhizomes are cleaned of mud and other extraneous matter adhering to them. Average yield/hectare comes to 20-25 tonnes of green turmeric.

Fingers are separated from mother rhizomes. Mother rhizomes are usually kept as seed material. The fresh turmeric is cured for obtaining dry turmeric. Curing involves boiling of rhizomes in fresh water and drying in sun.

No chemical should be used for processing. The cleaned rhizomes are boiled in copper or galvanized iron or earthen vessels, with water just enough to soak them. Boil till the fingers/mother rhizomes become soft. The cooked turmeric is taken out of pan by lifting troughs and draining the water into pan itself. The same hot water in the pan can be used for boiling next set of raw turmeric which is already filled in troughs. The cooking of turmeric is to be done within 2-3 days after harvest.

Rhizomes may also be cooked using baskets with perforated bottom and sides. The mother rhizomes and fingers are cured separately.

The cooked fingers/mother rhizomes are dried in sun by spreading in 5-7 cm thick layers on bamboo mats or cement floor. A thinner layer is not desirable as colour of dried product may be adversely affected. During night time, the material should be heaped or covered. It takes 10-15 days for rhizomes to become completely dry. Artificial drying using cross-flow hot air at a maximum temperature of 60°C gives satisfactory product. In sliced turmeric, artificial drying has clear advantages in giving brighter coloured product than sun drying which tends to suffer due to surface bleaching. Recovery of dry product varies from 20-30% depending upon variety and location where the crop is grown.

Dried turmeric has poor appearance and rough dull colour outside the surface with scales and root bits. Smoothening and polishing outer surface by manual or mechanical rubbing improve appearance.

Manual polishing consists of rubbing dried turmeric fingers on a hard surface. The improved method is by using hand-operated barrel or drum mounted on a central axis, sides of which are made of expanded metal mesh. When the drum filled with turmeric is rotated, polishing is effected by abrasion of surface against mesh by mutual rubbing against each other as they roll inside the drum. The turmeric is also polished in power-operated drums. The yield of polished turmeric from raw material varies 15-25%.

Colour of turmeric always attracts buyers. To impart attractive yellow colour, turmeric suspension in water is added to polishing drum in the last 10 minutes. When rhizomes are uniformly coated with suspension, they may be dried in sun.

Rhizomes for seed purpose are generally stored after heaping under shade of tree or in well ventilated shade and covered with turmeric leaves. Sometimes, the heap is plastered with earth mixed with cow dung. The seed rhizomes can also be stored in pits with sawdust. The pits can be covered with wooden planks with one or two holes for aeration.

Small Cardamom

Cultural Practices

Clean weeding is to be limited to plant bases (50 cm) and inter rows are maintained by slash weeding. The weeded materials should be used for mulching. Trashed materials and fallen leaves may also be used for mulching. Trashing the dry leaves and leaf sheaths removal of yielded old suckers may be carried out once in a year after completion of final harvest, which can be used for composting. The inter rows should not be dug at any cost. Clean water free from insecticides, fungicides, other chemicals and fertilizer leaches should only be used for irrigation under organic system of cultivation. This implies that watersheds for irrigation sources should also be maintained following organic methods of production. In areas, where adequate soil conservation measures and mulching are practiced, there will not be any necessity for earthing up.

To facilitate penetration of sufficient light, restricted lopping of shade tree branches may be made. However, even under such situations, no tree top shall be cut. In areas, which are over exposed, planting of shade trees is an essential operation and while doing so, maximum bio-diversity suited to local situation may be considered. Trees having desirable characters like defoliation during rainy season, self-pruning habit, flowering during summer and medicinal values may be considered. If such trees belong to leguminous species, they are preferred. Restricted loppings and leaf litter may be used for green leaf manuring or composting. Preservation of bee fauna is an integral part of organic cultivation. Integration of apiculture will not only ensure bio-diversity, but also help in increasing production through assured pollination. Application of organic manures like neem cake@ one kg or poultry manure/ farm yard manure/compost/ vermi or compost@ 2 kg/ plant may be done May-June. Well rotted cattle manure or FYM, sheep manure, bone meal, fish meal and caster cake enhances yield. Applications of cow dung slurry and slurry of ground nut oil cake and neem cake are widely practiced for better growth. Applications of bio-agents like *Trichoderma* or *Pseudomonas* help to check disease.

Application of neem cake @ 1kg/plant or bio-agents like *Metarrhizium* is effective for control of nematodes and root grubs.

Large Cardamom

Large cardamom of Sikkim and Darjeeling region largely are eco friendly and are grown without use of chemical fertilizers. Controlling weed growth in plantations is important for maximum utilization of available soil moisture and nutrients by the crop. Weeding may be done as and when necessary either by hand weeding or sickle weeding depending upon intensity of weed growth. Around the plant base, weeds can be pulled out and weeds in inter space need only slashed with sickle. Along with weeding, dry shoots and other trash materials are also cut and removed preferably in September-October and the trashed materials can be used as mulch around the plant base to a thickness of about five cm. Intensive operations which loosen and expose soil increase soil erosion and therefore, only minimum tillage operations should be practiced. As far as possible, contour terraces should be made only at planting points well before taking up planting operations. For sustainable and better yield, plants may be watered during dry months.

For sustained production, soil fertility should be maintained at a fairly high level. Well rotten cattle manure, compost or organic products or non edible oil cakes should be applied @ 2 kg/plant. If all the crop residues are recycled in the plantation itself, it is possible to maintain fertility of soil. Mulching base of plants after application of second dose of manuring helps better intake of nutrients by plants.

Vanilla

Vanilla prefers loamy laterite soil with plenty of organic matter with proper perfect drainage. Organic matter and decomposed mulch, are main sources of nutrients for plants. A thick layer of organic debris also helps to retain enough moisture and gives a loose soil structure for roots to spread. Hence, it is very important that easily decomposable organic matter is applied around plant base at least three to four times a year. Timely irrigation should be provided in initial two to three years of growth. Weeds growing at base of support tree around the plant within a radius of 50 cm should be pulled out and left there itself to decompose. Any operation done in plantation should not disturb roots, which are mainly confined to mulch and surface layer of soil. Any new shoot of vanilla should be allowed to attach to support tree with their own roots. Vines should grow upwards on support tree. By bending/nipping off tip of vines, one can produce more number of shoots or branches in a particular plant.

Support tree such as Glyricidia, Jatropha, Plumeria, Casuarina should be allowed to form branches to different directions to have an umbrella shaped appearance about 150 cm above ground. The support trees should be pruned to allow 50% shade. If the trees do not shed leaves, they may be pruned before onset of heavy rains to allow more sunlight. Another light pruning may be done just before beans are harvested to facilitate their maturity and initiate flower bud formation for next crop.

Vines are allowed to grow up to a height of 1.20 to 1.50 meters and allowed to hang down the branches. Such vines should be brought back to ground and a portion of it, is placed under mulch. Later it is allowed to grow up on same support again. Thus the vines should be lopped up and down adjusting height in such a way that it is not more than the shoulder height of workers. This makes operations such as pollination and harvest easy and will make the crop more productive. Trailing is an important cultural operation as it helps accumulation of carbohydrate and other flower forming compounds at leaf axils, which facilitate early induction of flowering.

Vanilla, being an orchid, prefers organic manures. Decomposed organic materials may be applied two to three times a year. Well rotten cowdung, compost or vermicompost @4 to 5 kg/plant/year may be applied. Application of *Azospirillum, Phosphobacteria* @ 25-50 g/vine is useful. Use of *Trichoderma* and *Pseudomonas* helps to prevent/suppress fungal diseases.

Cloves and Nutmeg

Cloves and nutmeg are tree spices grown under appropriate climatic situations for maximum yields. Clove is planted in medium elevated places without shade and near river basin with abundant water facilities for irrigation to yield well. Slash weeding may be resorted with intensified mulching practices to arrest weed growth and favour retention of moisture. Apply well rotten cattle manure or compost @ 15 kg/plant/year during May/June in initial years. The quantity may be increased gradually so that a well grown tree of 15 years and above gets 40 to 50 kg of organic manure. Inter cropping is also advisable in cloves and nutmeg plantations. Clove and nutmeg are crops very much easier for organic management.

Clove

Seedlings are planted during monsoon in pits of 1 m^3 size filled with soil + FYM. Mulching during initial years is done and adopting cover crops Calopogonium and Peuraria are beneficial. Banana, albizzia or jack may be planted along boarders. Compost of FYM at four kg/ plant may be applied during initial years at onset of monsoon. For bearing trees, 50 kg compost, half dose at on set of monsoon and the leaf as top dressing in September may be beneficial. Basins are mulched with dry leaves, 50 g Azospirillum and Phosphobacteria.

Nutmeg

Well drained, deep rich alluvial loamy and laterite soil is the best suited. Semi shaded conditions in coconut and arecanut gardens are desirable. If shade trees are not available, fast growing Erythrina, Glyricidia or banana are planted prior to planting seedlings. For permanent shading, jack and mango are suitable.

Manuring

Every year, 15 kg compost (Cattle manure) in split doses of two at outset of monsoon and during weeding are applied. For bearing trees 50 kg compost is beneficial. Prawn dust and ground cake are also suitable. Bio agents Azospirillium and Phosphobacteria at 50 g each are also beneificial. Mulch with dry leaves during summer to conserve moisture. No water logged condition is allowed.

Good Agricultural Practices

This Code of Good Agricultural Practices (CGAP) is a general guide to production and handling of spices. The objectives are to ensure that the spices are (a) of the highest quality, (b) cultivated or processed hygienically minimizing microbial contamination and (c) produced with care minimizing physical and chemical contamination. There are no effective and legally permitted methods for reducing microbial load of dried spices. Therefore, these guidelines are intended to help minimize such contamination during growing, harvesting, drying, packing and storage stages. Farmers, processors and distributors are encouraged to devise practical measures for their workers to implement the code. It is intended that buying company should circulate these GAP guidelines to producers and processors. Application of the GAP guidelines can be advised when preparing agreements with growers/ processors of organic spices.

Spices are exposed to microbial and other contamination from a wide variety of sources in the field. Such contaminations cannot be removed effectively by washing and peeling techniques which are applicable to many other crops, nor can they be reduced significantly by low temperature, drying, necessary for conservation of colour and flavour characteristics of spices. This code of practices set out here are guidelines to minimize such contaminations at primary producer/ processor's level.

Cultivation

Spices should not be cultivated in soils contaminated with sewage sludge, heavy metals, pesticides and other industrial chemicals. The soil should be well drained and irrigation (if necessary) should be regular and uniform to avoid water logging and high humidity microclimates which promote mould growth. The source of irrigation water should be free from human as well as animal faecal contamination. Organic manures (no human faecal material) should be well composted before use and generally applied after the final harvest and before next planting for seasonal crops. No cattle should be allowed to move freely in the cultivated area. Plants should be spaced to minimize weed growth. Slash weeding should be regular and dead weeds with other plant debris should be removed from the cultivated area and destroyed, if they promote fungal infection and pest damage.

Organic pesticides and herbicides should be approved by the customer whenever possible and applied by the trained personnel. Application should be carried out at pre-harvest intervals specified by customer.

Harvesting

Harvesting should not be carried out in wet (dew or rain) or high humidity conditions, i.e., wherever possible harvesting should be carried out in dry, low humidity conditions. Harvesting equipment should be clean and well maintained.

All containers used for primary collection of produce must be kept free from previously accumulated plant materials, and when not in use, they must be kept in a dry place, free from vermin and inaccessible to farm and domestic animals.

Damaged and spoiled produce should be sorted and discarded. Harvested produce should be collected in dry sacks, baskets, trailers or hoppers. It must not be collected on ground. Mechanical damage, high compaction and storage which promote composting should be avoided for which plastic sacks should not be used during harvesting. Sacks must not be overfilled and stacking should be done avoiding compaction.The time between harvest and transport of the produce to the drying site should be kept as short as reasonably practicable.The harvested produce should be protected from pests and domestic as well as farm animals.

Drying

The produce should be unloaded and unpacked as soon as possible on arrival at drying facilities. It must not be allowed to stand for extended periods in direct sunlight and must be protected from rain. Buildings used for drying the produce should be well ventilated and never used for keeping livestock. The building should be constructed to protect the produce from birds, insects, domestic and farm animals. Drying racks, if any, should be kept clean and regularly maintained. Drying on the floor and in direct sunlight are not adopted.

Dried produce should be inspected and sieved or winnowed to remove discoloured, mouldy and damaged material, soil, stones and other foreign bodies. Sieves should be kept clean and maintained well.

Clearly marked waste bins should be provided, emptied daily and cleaned. Dried as well as drying produces should be protected from pests and domestic and farm animals. Dried produce should be packed as soon as possible for protection and also to lessen opportunity of pest infestation.

Packing

After removal of damaged materials and foreign bodies, the sound dried produce should be packed in clean dry sacks, bags or boxes, preferably new. Packaging materials should be stored in a clean dry place free from pests and inaccessible to animals. Re-usable packaging materials like jute sacks and plastic bags should be well cleaned and dried before re-use. The packed produce should be stored in a dry place away from wall and off the ground and be protected from pests, domestic and farm animals. Whenever possible, the packaging materials used should be agreed between supplier and buyer.

Storage and Transport

Packed dried produce should be stored in a dry, well ventilated building with minimal variation in diurnal temperature avoiding high heat and with good air ventilation. Shutter and door openings should be protected by wire screens to keep out pests and domestic animals.

Packed and dried produces should be stored in buildings with concrete floors, on pallets, away from wall and well separated from all other produces. For bulk deliveries, use of vented containers for transport and storage in temporary warehousing is highly recommended to minimize contamination risks. Alternatively, suitably vented transport vehicles and temporary storage facilities are recommended. Whenever possible, conditions for transport and temporary storage should be agreed between supplier and buyer.

Equipments

Equipments used for cultivation and handling of produces should be well cleaned to minimize contamination. Dry cleaning is recommended. Where the use of water is unavoidable, equipments should be dried as quickly as possible after washing.

All equipments should be installed to allow easy access and should be well maintained and cleaned regularly.

Personnel and Facilities

Working practices should comply with General Principles of Food Hygiene of Codex Alimentarius Commission.Personnel handling of food material should maintain a high degree of personal hygiene and be provided with suitable changing facilities and toilets with hand washing facilities. Personnel should not be permitted to work in the material handling area if they are suffering from, or carriers of diseases likely to be transmitted through food, including diarrhoea. Personnel with open wounds, sores and skin infections should be transferred away from handling areas until completely recovered.

Training

Training of personnel, whether engaged in crop production, handling or processing is highly recommended. This can be achieved by using experts from local agricultural institutes or certifying agencies.

Market Potential for Organic Spices

With the trend towards pre-processed foods (convenience foods), the demand for organic spices increases likewise. Processed food can only be labelled as organic, if 95% of all ingredients originate from organic agriculture. The remaining 5% can be products, which are listed in a special annexure of EU-regulation. Spices and culinary herbs are not listed and must, therefore, always be of organic origin.

Assuming that within the next 10 years, organics will have a market share of 10% in Europe, US and Japan. The current production is still small and demand much higher.

Pest and Disease Management of Spices

Black pepper

Pests/diseases	Symptoms	Control Measures
Pests		
Leaf gall (*Liothrips karny)*	Formation of marginal galls on leaves. Infested leaves become thick, malformed and crinkled	Spray dimethoate 0.05% on tender flushes.
Top shoot borer(*Cydia hemidoxa)*	Shoots turn black and dry up	Spray quinalphos 0.05%; repeat spray to protect emerging new shoots.
Scale insects *Lepidosaphes piperis* *Aspidlotus destructor*	Infested leaves and stems turn yellow, wilt and dry up.	Spray dimethoate 0.1%. Repeat spray after 21 days. Prune-off severely affected portions.

Contd...

Pollu beetle *Longitarsus nigripennis*	The grub bores into the berries, feeding on the internal contents and making them hollow. The infested berries turn yellow initially and then black and crumble when pressed.	Spray quinalphos (0.05%) twice a year during June-July and September – October.
Slow decline (slow wilt) Nematodes: *Radopholus similes Meloidogyne incognita.* Fungus: *p.capsici*	1.Leaf yellowing and or shedding from October onwards 2.Root degeneration 3.Blackening of branches and eventual death of the vine.	a) Use nematode free rooted cuttings raised in soil mixture sterilized by methyl bromide fumigation. b) Remove the severely affected vines. c) Treat the pits with phorate @15g or carbofuran @ 50g at the time of planting. d)Soil application of phorate @30g or carbofuran @ 100/vine twice in a year. First application during May/June with the onset of South West monsoon and second application during Sept/Oct) Drench basin with 1% Copper Oxychloride @ 5-10 lit/vine, 2-3 times during the monsoon season.
Diseases		
Foot rot [Quick wilt] *Phytopthora capsici [Phytopthora palmivora*	Dropping of infected leaves. Tender leaves and fresh runner shoots turn black. Gradual infection of entire vine by rain splash. Collar infection Rotting or wilting of entire vine Sometimes the leaves and spikes develop black spots. Branches break up at nodes. Vine collapses within a short time of infection normally four weeks. Root damage, yellowing , defoliation, and drying up of the wine wholly or in parts.	Use disease free planting material. Drench planting pits with 0.2% copper oxychloride 15 days prior to planting. Fortify nursery mixture with bio agents(1 gm of *T.harzianum* or *G. virens* and 100 cc of VAM inoculums per kg of nursery mixture. Conduct regular pruning to regulate shade. Apply neem cake @ 1 Kg per vine. Apply bio control agents along with FYM and neem cake 50 gm of *T.harzianum* or G.virens mixed with 500 gm compost or 1 kg neem cake into the pit before planting,1MAP and 2MAP. Practice minimum tillage. Uproot and burn infected vines. Drench the base of the vines with 0.2% copper oxychloride or 1% Bordeaux mixture@ 5-10 liters per vine and spray the foliage with 1% Bordeaux mixture during may-June. Repeat chemical control measures during August September and

Contd...

		October. Repeat chemical control measures during October if the north east monsoon prolongs. Uproot and burn infected vines before the onset of the monsoon and adopt chemical control measures described earlier.
Little Leaf disease Cucumber mosaic virus (CMV)	1. The vines show shortening of internodes. 2. The leaves become small and abnormally narrow and give a sickle shaped appearance. 3. Leaves appear leathery, puckered and crinkled. 4, The affected branches give a typical 'Witches broom' appearance in advanced stages.	Eradicate infected plants.
Phyllody disease Phytoplasma Anthracnose	Malformation of spikes and flowers, leaves turn small, chlorotic, and the internodes become small give a 'witches broom appearance' from a distance.	Eradicate infected vines
Fungal Pollu (Anthracnose) *Colletotrichum gloeospoioides*	Affected berries show brown sunken patches during the early stages. Further development of berries is affected. In later stages, the discoloration gradually increases and the berries show the characteristic cross splitting. Finally, the berries turn to black in colour and dry. Leaf show angular to irregular brownish spots with chlorotic halo or circular spots	Spray 1% Bordeaux mixture

Large Cardamom

Pests/diseases	Symptoms	Control Measures
Pests		
Leaf Eating Caterpillar, *Artona chorista*	Feeds on the leaf lamina from under surface of the leaf and finally defoliate the leaves completely leaving only the midribs.	Quinalphos 25EC @0.05% Use of insecticide is not advisable
Diseases		
Chirke and Foorkey	Symptoms are characterized by mosaic appearance on the tender leaves with pale streaks, which slowly turns into' brown, resulting in withering and drying of the plants.	Uproot and destroy affected plants Repeat detection and roughing at regular intervals. Use suckers from certified nurseries

Small Cardamom

Pests/diseases	Symptoms	Control Measures
Pests		
Cardamom thrips *Sciothrips cardamomi* (Ramk.)	Thrips colonise and breed in different parts of the plant such as unopened leaves, spindles, leaf sheaths, flower bracts, perianths and flower tubes. Adults and numphs of the insect cause damage to panicles, flowers and capsules .Tender capsules develop as scabby growth on capsules as they mature. Affected capsules appear malformed, shrivelled and sometime with gaping slits	Removal of collateral host plants of thrips. Remove dry drooping leaves, dry leaf sheaths, old panicles and other dry plant parts immediately before the commencement of first application of insecticide. Apply insecticide Quinalphos 0.025% ; Fenthion 0.03%; Phosalone 0.05%. Dimethoate 0.05%
Shoot/Panicle/Capsule borer -*Conogethes. Punctiferalis Guen.*	Larvae bore into tender panicles or unopened leaf spindle or immature capsule, decaying of the central spindle and the characteristic dead heart symptom develops.	Monocrotophos or Fenthion 0.05%
Early capsule borer *Jamides alecto*	Caterpillars bore and feed flower buds, flowers and capsules. Attakced capsules are completely emptied leaving a circular hole on the capsule, which eventually turn yellowish brown, decay and drop off	Methylparathion or Monocrotophos 0.05%
Cardamom whitefly *Dialeurodes cardamomi*	Larvae crawls on the leaf and finds out a feeding site. nymphs secrete a sticky honey dew which drops on to lower leaves. On these, black sooty mould develops, which interrupts photosynthesis of the leaves.	Spraying on lower surface of leaves a mixture of neem oil (500 ml) and Triton (500 ml) in 100 lit. of water. Acephate 0.075% and Triazophos 0.04% are equally effective, 2 or 3 times at 15 days interval
Lace wing bug *Stephanitis typicus*	They suck the cell sap from leaves, and later greyish yellow spots develop on leaves, in severe infestation plant growth is retarded and yield is adversely affected.	Spraying of insecticides recommended for thrips control may be adopted for its management.

Contd...

Diseases		
Katte or mosaic(Virus disease)	Affected leaf shows discontinous stripes of light green colour running papallel to each other from mid rib to margin of leaves form a mosaic pattern.	Use only disease free planting material. Regular application of insecticide for control of vector.
Azhukal (*Phytophthora sp.)*	It affects leaves, tender shoots, panicles and capsules leads to rotting and shedding of leaves and decay of capsules.	Adopt phytosanitary measures. Spray shoots with Bordeaux mixture (1%)
Clump rot or rhizome rot	yellowing of leaves and decay of tillers starting from the collar region. Rotten rhizomes become soft, dark brown coloured and ultimately result in the total death of the plant	Plant sanitation. Drench with(0.2%) copper oxy chloride(2-3 l/ plant)

Ginger

Pests/diseases	Symptoms	Control Measures
Pests		
Shoot borer *Conogethes punctiferalis*	Larvae bore into pseudo stems and feed on internal tissues resulting in yellowing and drying of leaves of infested pseudo stems. The presence of a borehole on the pseudo stem through which frass is extruded and withered and yellow central shoot is a characteristic symptom of pest infestation	Spraying dimethoate or quinalphos 0.05 %.
Rhizome scale *Aspidiella hartii*	Feed on sap and when rhizomes are severely infested, they become shrivelled and desiccated affecting its germination	Treat seed material with quinalphos 0.075% (for 20-30 minutes) before storage and also before sowing in case infestation persists.
Rhizome fly	Flies penetrate into rhizomes and feed the internal content completely	Soil application of carbofuran 3G granules @ 20-25 kg/ha. at 10-15 cm distance from plants and 5-7 cm deep in the soil and covered with soil. They can be applied at 105, 135 days after sowing.
Diseases		
Soft rot or rhizome rot	Infection starts at collar region of pseudo stem and progresses upwards as well as	Treatment of seed rhizomes with mancozeb 0.3% for 30 minutes

Contd...

P. vexans and P. myriotylum	stem and progresses upwards as well as downwards. The collar region of affected pseudo stem becomes water soaked and rotting spreads to rhizome resulting in soft rot. At later stage, root infection is also noticed. Foliar symptoms appear as light yellowing of the tips of lower leaves, which gradually spreads to the leaf blades.	before storage and once again before planting reduces the incidence of the disease. Application of *Trichoderma harzianum* along with neem cake @ 1 kg/ bed helps in preventing the disease.
Bacterial wilt *Ralstonia solanacearum*	Water soaked spots appear at the collar region of the pseudo stem and progresses upwards and downwards. The first conspicuous symptom is mild drooping and curling of leaf margins of the lower leaves which spread upwards. Yellowing starts from lowermost leaves and gradually progresses to upper leaves. In the advanced stage, the plants exhibit severe yellowing and wilting symptoms. The vascular tissues of the affected pseudo stems show dark streaks. The affected pseudo stem and rhizome when pressed gently extrudes a milky ooze from the vascular strands. Ultimately rhizomes rot.	The seed rhizomes may be treated with streptocycline 200 ppm for 30 minutes and shade dried before planting. Once the disease is noticed in the field, all beds should be drenched with Bordeaux mixture 1 % or copper oxychloride 0.2%.
Leaf spot *Phyllosticta zingiberi.*	The disease starts as a water soaked spot and later turns as a white spot surrounded by dark brown margins and yellow halo. The lesions enlarge and adjacent lesions coalesce to form necrotic areas. The disease spreads through rain splashes during intermittent showers. The incidence of the disease is severe in ginger grown under exposed conditions.	Spraying Bordeaux mixture 1 % or mancozeb 0.2%.

Turmeric

Pests/diseases	Symptoms	Control Measures
Pests		
Shoot borer *Conogethes punctiferalis*	The larvae bore into pseudostem and feed on internal tissues. The presence of a borehole on the pseudostem through which frass is extruded and the withered central shoot is a characteristic symptom of pest infestation	Spraying malathion 0.1% or monocrotophos 0.075% at 21 days intervals during July to October
Rhizome scale	Feed on sap and when the rhizomes are severely infested, they become shrivelled	Treat seed material with quinalphos 0.075% (for 20-30

Contd...

Aspidiella hartii	desiccated affecting its germination	minutes) before storage and also before sowing in case the infestation persists.
Rhizome fly	Flies penetrate into rhizomes and feed the internal content completely	Soil application of carbofuran 3G granules @ 20-25 kg per hectare at 10-15 cm distance from plants and 5-7 cm deep in the soil and covered with soil. They can be applied at 105, 135 days after sowing.
Diseases		
Leaf blotch *Taphrina maculans*	The leaves also turn yellow. In severe cases the plants present a scorched appearance and the rhizome yield is reduced	Spraying mancozeb 0.2%.
Leaf spot *Colletotrichum capsici*	Spots appears as brown spots of various sizes on the upper surface of the young leaves. Fungus Produce toxic metabolites which cause necrosis on leaves. The spots are irregular in shape and white or grey in the centre. Later, two or more spots coalesce and form an irregular patch covering almost the whole leaf. The affected leaves eventually dry up. The rhizomes do not develop well.	Spraying zineb 0.3% or Bordeaux mixture 1 %
Rhizome rot *Pythium graminicolum.*	Collar region of pseudo stem becomes soft and water soaked, resulting in collapse of plant and decay of rhizomes.	Treating the seed rhizomes with mancozeb 0.3% for 30 minutes prior to storage and at the time of sowing prevents the disease. When the disease is noticed in the field, the beds should be drenched with mancozeb 0.3%.

Vanilla

Pests/diseases	Symptoms	Control Measures
Pests		
Vanilla bug *Hayomorpha, Picus sp.*	Suck sap from buds and inflorescence leading to mottling	Collect egg mass at first nymph stage in Nov-Feb Spray 0.1% monocrotophos

Contd...

Vanilla vine weevil *Sipalus sp*	Black coloured weevil bore into and eat tender leaves and stem.	Collect and destroy adult weevil during Nov-Feb.
Beetle -*Saula ferruglnea*	Adults cut through the leaves from lower surface eat the entire tissue except for the thick transluscent cuticle of epidermis.	Collect and destroy or spray 0.1% malathion
White grubs	Feed on roots and destroy them	Drench soil at the base with 0.05% chlorpyriphos
Catterpillar-*Plusia aurifera*	Light greenish caterpillar feed on vegetative shoots and first leaf forming web in between	Spray 0.1% monocrotophos
Achatina - *Achatina sp.*	Feeds on young shoots and leaves which turn whitish.	Collect and destroy, avoid mulch material that do not decompose easily, spray sevin dust.
Diseases		
Vanilla mosaic Vanilla mosaic virus	Mosaic pattern or mottling on foliage, stunting, sterility and leaf distortion	Uproot and destroy
Vanilla necrosis Vanilla necrosis virus	Brown necrotic lesion on leaves leading to drying and crinkling	Uproot and destroy, avoid use of affected planting material
Cymbidium mosaic- Cymbidium mosaic virus	Mild mottled or mild chlorotic streaks on leaves	Uproot and destroy, avoid use of affected planting material
Odonto glossam ring spot Odonto glossam ring spot virus	Mostly symptomless mild mottling of leaves	Uproot and destroy, avoid use of affected planting material

IMPROVED VARIETIES OF SPICE CROPS

Improved varieties of black pepper and their salient features

Name	Pedigree	Released from	Average yield. Dry (kg/ha)	Oleo-resin (%)	Piperine (%)	Essen-tial oil (%)	Salient features	Recommended state/region
Panniyur-1	F_1 of Uthirankotta x Cheriyakaniya kadan	Pepper Research Station, Panniyur (KAU)	1242.0	11.8	5.3	3.5	Vigorous growing vine, do not tolerate shade, medium maturity group, long spikes, close setting of berries, bold berries	Kerala, Karnataka and Tamil Nadu
Panniyur-2	Open pollinated progeny of Balankotta	Pepper Research Station, Panniyur (KAU)	2570.0	10.9	6.6	3.4	Shade tolerant, suitable for intercropping, medium maturity group, medium quality	Kerala and Karnataka
Panniyur-3	F_1 of Uthirankotta x Cheriyakaniya kadan	Pepper Research Station, Panniyur (KAU)	1953.0	12.7	5.2	3.1	Performs well under open situation, late maturity group, long spike and bold berries	Kerala, Karnataka and Tamil Nadu
Panniyur-4	Clonal selection from 'Kuthiravally'	Pepper Research Station, Panniyur (KAU)	1277.0	9.2	4.4	2.1	Stable yielder, performs well under adverse condition also, tolerant toshade, late maturity	Kerala, Karnataka and Tamil Nadu
Panniyur-5	Open pollinated progeny of 'Perumkodi'	Pepper Research Station, Panniyur (KAU)	1098.0	12.3	5.5	3.8	Suitable for both mono and mixed cropping in coconut/ arecanut gardens, shade tolerant, medium maturity, tolerant to nursery disease, long spike	Kerala

Contd..

Panniyur-6	Clonal selection from Karimunda	Pepper Research Station, Panniyur (KAU)	2127.0	8.27	4.94	1.3	Vigorous vine, tolerant to drought and adverse climatic conditions, stable and regular bearer, medium maturity group, suitable for open as well as partial shade condition, spike 6 - 8 cm, more number of spikes per unit area, close setting and attractive bold berries	Kerala and Karnataka
Panniyur-7	Open pollinated progeny of Kalluvally	Pepper Research Station, Panniyur (KAU)	1410.0	10.6	5.57	1.5	Vigorous vine, regular bearer, long spike, hardy type vine, tolerate adverse climatic condition, suitable for open and shaded conditions, very long spike (16-24 cm).	Kerala, Karnataka and Tamil Nadu
Subhakara	Clonal selection from 'Karimunda'	IISR, Kozhikode	2352.0	12.4	3.4	6.0	Wide adaptability to all pepper growing tracts, suitable for intercropping as well as for high elevations, high quality medium maturity group	Kerala, South Karnataka, Tamil Nadu
Sreekara	Clonal selection from 'Karimunda'	IISR, Kozhikode	2677.0	13.0	5.1	7.0	Suitable for all pepper growing regions including high elevations as well as for intercropping with high quality medium maturity group	Kerala, South Karnataka, Tamil Nadu
Panchami	Clonal selection of 'Aimpiriyan	IISR, Kozhikode	2828.0	12.5	4.7	3.4	High yielding, spikes twisted, late maturing, suitable for high elevations, excellent fruit set	Kerala and Karnataka
Pournami	Clonal selection of 'Ottaplackal'	IISR, Kozhikode	2333.0	13.8	4.1	3.4	Tolerant to root knot nematode and shade, medium maturity, suitable for intercropping with arecanut and banana	Kerala and Karnataka
PLD-2	Clonal selection from 'Kottanadan'	CPCRI, RS, Palode	2475.0	15.5	3.3	3.5	Late maturity, high quality cultivar suitable for plains and higher elevations	Kerala

Contd...

IISR Thevam	Selection of 'Thevanmundi	IISR, Kozhikode	1787.0	8.15	1.6	3.1	Stable yielding, vigorously growing, field tolerant to *Phytophthora*, medium maturity, suitable to high elevations in coffee and tea estates under rainfed situations	Kerala and high altitude areas of South India upto 3000 ft. MSL
IISR Malabar Excel	F_1 of 'Cholamundi' x Panniyur-1	IISR, Kozhikode	1065.0	13.5	2.96	3.2	Recommended for rainfed conditions including coffee and tea plantations suitable for high elevations and plains	Kerala and High altitude areas of South India.
IISR Girimunda	F_1 of 'Narayakodi' x 'Neelamundi'	IISR, Kozhikode	2112.0	9.65	2.2	3.4	Recommended for rainfed conditions including tea and coffee estates, medium maturity group	Kerala and High altitude areas of South India.
IISR Shakthi	OP of Perambra-mundi	IISR, Kozhikode	5755.0	10.2	3.3	3.7	Tolerant to *P. capsici*.	Kerala, Karnataka, plains and high ranges

Source: Spice India,Vol.18(4);2005.

Improved ginger cultivars and their salient features

Variety	Pedigree	Released from	Average yield (Fresh t/ha)	Maturity (days)	Dry recovery (%)	Crude Fibre (%)	Oleoresin (%)	E. oil (%)
Suprabha	Clonal selection from 'Kunduli local'	High altitude Research Station (Orissa University of Agriculture & Tech. Pottangi, Orissa)	16.6	229	20.5	4.4	8.9	1.9
Suruchi	Clonal selection from 'Kunduli local'	"	11.6	218	23.5	3.8	10.0	2.0
Suravi	Induced mutant of 'Rudrapur local'	"	17.5	225	23.0	4.0	10.2	2.1
IISR Varada	Germplasm selection	Indian Institute of Spices Research, Kozhikode, Kerala	22.6	200	19.5	3.29	6.73	1.75
Himagiri	Clonal selection	University of Horticulture & Forestry, Solan, H.P.	13.14	230	20.2	6.05	4.29	1.63
IISR Mahima	Germplasm selection	Indian Institute of Spices Research, Kozhikode, Kerala	23.2	200	21.12	3.26	4.48	1.72
IISR Rejatha	Germplasm selection	Indian Institute of Spices Research, Kozhikode, Kerala	22.4	200	20.8	4.0	6.3	2.4

Source: Spice India,vol.18(2);2005

Improved varieties of turmeric and their salient feature

Name	Pedigree	Released from	Average yield (Fresh t/ha)	Duration (days)	Dry recovery (%)	Curcu-min (%)	Oleoresin (%)	E. oil (%)	Suitable for
Co.1	Mutant of Erode local	TNAU, Coimbatore	30.5	285	19.5	3.2	6.7	3.7	TN. Drought tolerant
BSR-1	Mutant of Erode Local	TNAU, Coimbatore	30.7	285	20.5	4.2	4.0	3.7	TN. (drought prone area & problem soil)
BSR-2	-do-	-do-							
Krishna	Clonal sel. from Tekur	Maharashtra AU, Kasha	9.2	240	16.4	2.8	3.8	2.0	Maharashtra. Tolerant to major pests
Sugandham	Germplasm selection	Gujarat AU, Jagudan	15.0	210	23.3	3.1	11.0	2.7	Gujarat Tolerant to pests
Suvarna (PCT-8)	Germplasm selection	IISR, Kozhikode	17.4	200	20.0	4.3	13.5	7.0	Kerala, Karnataka & AP
Suguna (PCT-14)	-do-	-do-	29.3	190	12.0	7.3*	13.5	6.0	Kerala & AP Short duration
Sudarsana (PCT-14)	-do-	-do-	28.8	190	12.0	5.3*	15.0	7.0	-do-
IISR Prabha	Open pollinated progeny	-do-	37.5	205	19.5	6.5	15.0	6.5	Kerala & TN, High curcumin type
IISR Prathiba	-do-	-do-	39.1	225	18.5	6.21	16.2	6.2	-do-
Roma (PTS-10)	Clonal selection	HARS (OUAT) Pottangi	20.7	250	31.0	9.3*	13.2	4.2	Orissa, TN, HP, AP & Kerala
Suroma (PTS-24)	-do-	-do-	20.0	253	26.0	9.3	13.1	4.4	Orissa, TN, HP
Rasmi (PTS-9)	Germplasm selection	-do-	31.3	240	23.0	6.4	13.4	4.4	Orissa, TN, AP

Contd...

Ranga	Clonal Sel. (Rajapuri)	-do-	29.0	250	24.8	6.3	13.5		-do-
Rajendra Sonia (RH-10)	Germplasm selection	Rajendra AU, Dholi, Bihar	4.8	225	18.0	8.4		5.0	North Bihar
Kanti	Clonal sel. (Mydukur)	KAU, Vellanikkara	7.9(**)	240 to 270	20.2	7.2	8.3	5.2	Kerala
Sobha	Sel. From 'Methala Local'	-do-	5.74(**)	240 to 270	19.4	7.4	9.7	4.3	Kerala
Megha Turmeric-1	Sel. From 'Lakadong' type	ICAR Res. Complex for NEH Region, Shillong	20.0	300 to 315	16.37	6.8	---	---	Meghalaya
IISR Alleppey Supreme***	Selection of Alleppey Finger Turmeric	IISR, Kozhikode	5.58**	220	18.9	6.0	16.0	-	Suitable for Kerala, Karnataka, Maharashtra & North Bengal
IISR	Germplasm selection	IISR, Kozhikode	5.28**	220	19.3	5.5	13.7	-	Suitable for Kerala, Karnataka, Maharashtra & North Bengal
Sona	Germplasm selection	KAU Vellaikkara	4.02**	240 to 270	18.9	7.1	10.3	4.4	Kerala
Varna	Germplasm selection	- do -	4.16**	240 to 270	19.1	7.9	10.8	4.6	Kerala

* Provisional *Source*: Spice India ,Vol.18(2);2005.
** Dry yield
*** Proposed for release

Improved varieties of cardamom

Variety	Source	Yield (kg/ha)	E. oil (%)	1,8-cineole (%)	α Terpenyl acetate (%)	Capsule shape	Area (s) recommended	Special distinguishing characteristics
PV-1	Kerala Agricultural University	260	6.8	33	46	Long	All cardamom tracts of Kerala & Karnataka	An early maturing variety, long panicle, oblong, bold and parrot green capsules
Mudigere-1	University of Agricultural Sciences (B), Mudigere	275	8.0	36	42	Oval	Malanad region of Karnataka	Compact plant, suitable for high density planting, short panicle, oval, bold, pale green capsules
Mudigere-2	University of Agricultural Sciences (B), Mudigere	476	8.0	45	38	Round	Traditional cardamom, growing tracts of hill zones of Kerala	Ealy maturing type, round/ oval and bold capsules.
ICRI-1	Indian Cardamom Research Institute, Myladumpara	375	8.3	29	38	Round	South Idukki zone of Kerala	Early maturing profusely flowering variety, medium sized panicle with globose extra bold dark green coloured capsules
ICRI-2	Indian Cardamom Research Institute, Myladumpara	375	9.0	29	36	Oblong	Vandanmetu & Nelliampathi zone of Kerala & Annamalai & Meghamalai hills of Tamil Nadu	Performs well under irrigated conditions, suitable for higher altitude, medium long panicles, oblong bold and parrot green capsules (tolerant to azhukal disease)
ICRI-3	Indian Cardamom Research Institute, Myladumpara	439	6.6	54	24	Oblong	Hill zones of Karnataka	Early maturing type, non-pubescent leaves, oblong bold, parrot green capsules, tolerant to rhizome rot disease

Contd...

IISR Suvasini	Indian Institute Spices Research, Cardamom Research Centre, Appangala	745	8.7	42	37	Oblong	Kodagu & Hassan districts of Karnataka	Early maturity, long panicle, tolerant to rhizome rot, thrips shoot / panicle / capsule borer, suitable for high density planting
ICRI-4	Indian Cardamom Research Insitute, Thadiyankudisai	455	6.4	---	---	Globose	Lower pulney hills of Tamil Nadu & Wynad of Kerala	Early maturing medium sized paincle globos bold capsule relatively tolerant to rhizome rot and capsule borer
*IISR Avinash (RRI)	Indian Institute of Spices Research, Cardamom Research Centre, Appangala	847	6.7	30.4	34.6	-	Rhizome rot infested areas	Selection from hot spot areas. Resistance to rhizome rot
*IISR Vijetha (NKE-12)	Indian Institute of Spices Research, Cardamom Research Centre, Appangala	979	7.9	44.9	23.4	-	Suited to mosaic infested areas	For highly shaded and mosaic disease areas. Capsule oblong virus resistant
PV 2	KAU, Pampadumpara	982	10.45	-	-	Long		Capsules are ovoid and parrot green, tolerant to biotic stresses like thrips, clump rot and capsule rot.
Farmer's selection "Njallani Green Gold"	Farmer's Selection from Jdukki district of Kerala	2475 kg/ha					Kerala	Clonal selection by a cardamom grower in Idukki district. High yielding bold capsule over 70% of the cured cardamom about 7 mm.

* Recommended for release; *Source*: Spice India, Vol.16(5);2003.

Cinnamon varieties released in India

Variety	Organization responsible for release	Bark yield	Bark oil (%)	Leaf oil (%)	Bark oleoresin (%)	Bark (%)	Cinnamaldehyde in		Eugenol in		Special characters
							Bark oil	Leaf oil	Bark oil	Leaf oil	
IISR Navashree	IISR, Kozhikode	200 kg dry/ha	2.7	2.8	8.0	40.6	73.00	15	6.0	62	A selection with high shoot regeneration capacity. Higher cinnamaldehyde and oleoresin in bark
IISR Nithyashree	IISR, Kozhikode	250 kg dry/ha	2.7	3.0	10.0	30.7	58.00	14	5.0	78	A selection with high shoot regeneration capacity. Gives quality quills. Bark oil, leaf oil and oleoresin. Contents high, giving good aroma and taste.
Konkan Tej	Konkan Krishi Vidya Peeth, Dapoli, Maharashtra	85 g dry/plant	3.2	2.3	-	29.16	70.23	-	6.93	75.5	Resistant to major pests & diseases
YCD-1	Horticultural Station, Yercaud (TNAU)	360 kg dry/ha	2.8	3.1	-	35.30	-	-	-	-	Drought tolerant, high yielding. Resistant to major pests & diseases
PPI(C)-1	Horticultural Research Station, Pechiparai (TNAU)	980 kg fresh/ha	2.9	3.3	-	34.22	-	-	-	-	No major pests & diseases. Released mainly for leaf oil purpose.

Contd...

Sugandhini	Aromatic & Medicinal Plants Research Station, Odakkali, KAU	-	-	3.8% on dry weight	-	51.00	45	-		93.7	
RRL(B)C-6	Regional Research Laboratory, Bhubaneswar, Orissa	-	-	0.9							

Improved varieties of Nutmeg and their salient features

S. No.	Variety/ year of release	Pedigree/ parentage	Institution/ University	Average yield	Salient features	Recommended State/region
1	Konkan Sugandha 1998	Single plant selection from local seedling population	Regional fruit Research Station (BSKKU) Vengurla, Maharashtra	200-526 fruits/ tree	Adaptable in konkan region. Tree canopy is conical and compact	Maharashtra
2	Konkan swad 2003	Selection from nutmeg seedlings from Ratnagiri dist.	Regional Coconut Research Station Bhatye (Dr. BSKKU), Ratnagiri district	761.38 fruits/ tree	Adapted to konkan region with warm humid conditions as well as shade provision. Canopy erect, conical shape, contain 39.8% essential oil in seed and 10.96% in mace. No incidence of pest and disease are noticed	Konkan region of Maharashtra
3	Viswashree 2001 (identified)	Clonal selection from elite germplasm	IISR, ICAR, Calicut, 673012, Kerala	1000 fruits (1.33 kg mace, 9 kg dry 3122 kg/ha)	Bushy and compact canopy, low incidence of fruit rot. Nut recovery 70.0%, mace recovery 35.0%, oil 7.14%, mace oil 7.13%, oleoresin 2.48% and mace 13.8% respectively buffer 30.9%, myristicin 12.48% and mace 20.03% respectively	Kerala & other areas of similar climatic features

REFERENCES

Bavappa, K.U.A. and Gurusinghe. 1978. Rapid multiplication of black pepper for commercial planting. *Journal of Plantation Crops* 6: 92-95

Biswas, A.K., Deka, T.N. and Thomas, J. 2003. Large cardamom cultural operations. 16: 26-29

Gopalam, A., Zachariah, J.T., Baker, N.K., Sudanandan, A.K. and Ramadasan, A. 1991. Chemical quality of black and white pepper *Spice India* 4: 8-10

Gopinathan, K.M. and Manilal, V.B. 2005. White pepper preparation through Bacterial Fermentation. *Spice India.*

Jayachandran, B.K., Meera Bai, M., Abdul Salam, M., Mammen, M.K. and Mathew, K.P. 1991. Performance of ginger under shade and open conditions. *Indian Cocoa Arecanut and Spices Journal* 15: 40-41

KAU, 2003. *Package of Practices Recommendations Crops*-2003, Kerala Agricultural University, Kerala, India.

Korikanthimath, V.S. 1992. Large scale multiplication of cardamom, the NRCS experiment. (*In*) Sarma, Y.R., Devasahayam, S. and Anandaraj, M. (eds.). *Proceedings of the National seminar on black pepper and cardamom problems and prospects*. 17-18 May 1992, Indian society for Spices, IISR, Calicut, pp.73-76

Krishnakumar, V.K. and Potty, S.N. 2002. Nutrition of cardamom. In: Ravindran, P.N. and Madhusoodanan, K.J. (eds.) *Cardamom - the Genus Eletteria* Taylor and Francis, London, pp.129-142

Krishnamoorthy, B., Rema, J. and Mathew, P.A. 2004. For the best in cinnamon. *Spice India* 17: 19-21

Krishnamoorthy, B., Sasikumar, B., Rema, G., George, G.K. and Peter, K.V. 1997. Genetic resources of tree spices and their conservation in India. *Plant Genetic Resources News Letter* 3:53-58

Lewis, Y.S., Nambudiri, E.S., Krishnamoorthy, N., Natarajan, C.P. 1969. White pepper. *Perfume and Essential Oil Research.* 60: 53-56

Murashige, T. and Skoog, F. 1962. A revised medium for rapid growth and bioassay with tobacco tissue culture *Physiologia Plantarum* 15: 473-497

Narong Chomchatav. 1996. Spices production in Asia - An overview. *Proceedings of the Conference IBC's Asia Spice Markets 96*, 27-28 May, 1996, Singapore.

Nybe, E.V., Mammooty, K.P. and Sujatha, V.S. 1997. *Nursery Manual for pepper.* Directorate of Extension, Kerala Agricultural University, Mannuthy, Kerala, India

Nybe, E.V., Mini Raj, N. and Peter, K.V. 2007. *Spices*. Peter, K.V. (ed.) New India publishing agency, New Delhi, p.316

Omanakutty, M. 2006. White pepper. *Spice India* 19: 34-36

Purseglove, J.W., Brown, E.G., Green, C.L. and Robbins, S.R.J. 1981. *Spices* (2). pp.644-813

Rao, V.S., Mathew, M.K., Kumar, P.K., Madhusoodanan, K.J., Potty, S.N. and George, C.K. 1999. *A Manual for Micropropagation of Selected Spices.* Indian Cardamom Research Institute, Spices Board Cochin, Kerala, India

Ravindran, P.N. and Madhusoodanan, K.J. 2002. *Cardamom. The Genus Elettaria.* Taylor and Francis. London, p.368

Ravindran, P.N. and Nirmal Babu, K. 2005. *Ginger-the Genus Zingiber..* CRC Press,Boca Ratou, USA, p558.

Ravindran, P.N., Nirmal Babu, K., Shiva, K.N. and Johny A. Kallupurackal .2006. *Advances in Spices Research. History and Achievements of Spices Research in India since Independence.* Agrobios India, Jodhpur, p.994

Rema, J., Krishnamoorthy, B. and Mathew, P.A. 2004. Making cloves, Allspice, Nutmeg productive. *Spice India.* 17: 22-25

Rema, J., Krishnamoorthy, B. and Mathew, P.A. 2005. Flowering in all spice. *Spice India* 18: 13-14

Spices Board. 1996-2007. Various miscellaneous publications in spice India. *Indian Spices and Spice Bulletin*

Spices Board. 2001. *Agmark Grades Specifications for Spices,* Spices Board, Cochin, India. pp.27-41

Spices Board. 2001. Guidelines for production of organic spices in India. Cochin. p.125

Spices Board. 2005. *Spices Statistics.* Spices Board, Kochi, Kerala, p.282

Theodose, R. 1972. Traditional methods of vanilla preparation and their improvement. *Tropical Science* 15: 45-57

Thomas, P.P., Nirmala Menon, A., Bhat, A.V. and Mathew, A.G. 1987. Selective grinding as a basis for separating white pepper. *Journal of Food Science and Technology* 24:126-27.

Important websites in Horticulture

Australian Society for Horticultural Science	http://www.aushs.org.au/
Agricultural & Processed Food Products Export Development Authority (APEDA)	http://www.apeda.com/
American Society for Horticultural Science	http://www.ashs.org/
Asian Vegetable Research and Development Center (AVRDC)	http://www.avrdc.org.tw/
Australian Society for Horticultural Science	http://www.aushs.org.au/
Central Food Technological Research Institute (CFTRI)	http://www.cftri.com/
Central Institute of Medicinal & Aromatic Plants (CIMAP)	http://www.cimap.org/
Central Institute of Post harvest Engineering and Technology	http://www.icar.org.in/ciphet.html
Central Plantation Crops Research Institute (CPCRI), Kasaragod, Kerala	http://cpcri.nic.in/
Central Tuber Crops Research Institute (CTCRI), Thiruvananthapuram, Kerala	http://www.ctcri.org/
Consultative Group on International Agricultural Research, CGIAR	http://www.cgiar.org/
Coffee Board, India	http://indiacoffee.org/
Department of Agriculture and Co-operation, India	http://agricoop.nic.in/
Department of Bio-technology, India	http://dbtindia.nic.in
Department of Scientific and Industrial Research, India	http://dsir.nic.in/
FAO	http://www.fao.org/
Global Agribusiness Information Network:	http://www.fintrac.com/gain/:

Greenhouse Vegetable Information:	http://www.ghvi.co.nz/
Indian Agricultural Research Institute (IARI)	http://www.iari.res.in/
Indian Council of Agricultural Research (ICAR)	http://www.icar.org.in
Indian Institute of Horticultural Research (IIHR)	www.iihr.res.in
Indian Institute of Spices Research (IISR), Calicut, Kerala	http://www.iisr.org/
Indo-American Hybrid Seeds	www.indamseeds.com
Institute of Vegetable and Ornamental Crops	http://www.igzev.de/
Institute for Horticultural Development, Victoria, Australia	http://www.nre.vic.gov.au/agvic/ihd/
Kerala Agricultural University	www.kau.edu
Iowa State University Department of Horticulture	http://www.hort.iastate.edu/
National Bureau of Plant Genetic Resources (NBPGR), India	http://nbpgr.delhi.nic.in/
National Horticulture Board (NHB), India	http://hortibizindia.nic.in/
National Institute of Agricultural Extension Management (MANAGE), India	http://www.manage.gov.in/
National Research Centre for Cashew (NRCC), India	http://kar.nic.in/cashew/
National Research Centre for Mushroom (NRCM), India	http://www.nrcmushroom.com/
National Research Centre for Oil Palm (NRCOP), India	http://www.ap.nic.in/nrcop
North Carolina State University, Dept. of Horticulture	http://www2.ncsu.edu/cals/hort_sci/
Oregon State University, Dept. of Horticulture	http://osu.orst.edu/dept/hort
Pineapple News	http://agrss.sherman.hawaii.edu/pineapple/pineappl.htm
Pomology Resources Center	http://www.bsi.fr/pomologie/english/pomology:
Rubber board, India	http://rubberboard.org.in/
Spice Paprika web site	http://www.paprika.deltav.hu/:
Spices Board, India	http://www.indianspices.com/
Sri Lanka Agribisness on-line	http://www.agro-lanka.org/
Sustainable Apple Production:	http://orchard.uvm.edu/

Tea Board, India	http://tea.nic.in/
The Horticultural Taxonomy Group	http://www.hortax.org.uk/
The International Society of Citriculture	http://www.lal.ufl.edu/isc_citrus_homepage.htm
The Internet Garden	http://www.internetgarden.co.uk/
The Rose Resource	http://rose.org/
The USDA Agricultural Research Service	http://www.ars.usda.gov/
University of Florida, Dept. of Environmental Horticulture	http://hort.ifas.ufl.edu/
University of California, Fruit&Nut Research	http://fruitsandnuts.ucdavis.edu/
US Environmental Protection Agency	http://www.epa.gov/:
USDA	http://www.usda.gov/

Important websites in Horticulture

Australian Society for Horticultural Science	http://www.aushs.org.au/
Agricultural & Processed Food Products Export Development Authority (APEDA)	http://www.apeda.com/
American Society for Horticultural Science	http://www.ashs.org/
Asian Vegetable Research and Development Center (AVRDC)	http://www.avrdc.org.tw/
Australian Society for Horticultural Science	http://www.aushs.org.au/
Central Food Technological Research Institute (CFTRI)	http://www.cftri.com/
Central Institute of Medicinal & Aromatic Plants (CIMAP)	http://www.cimap.org/
Central Institute of Post harvest Engineering and Technology	http://www.icar.org.in/ciphet.html
Central Plantation Crops Research Institute (CPCRI), Kasaragod, Kerala	http://cpcri.nic.in/
Central Tuber Crops Research Institute (CTCRI), Thiruvananthapuram, Kerala	http://www.ctcri.org/
Consultative Group on International Agricultural Research, CGIAR	http://www.cgiar.org/
Coffee Board, India	http://indiacoffee.org/
Department of Agriculture and Co-operation, India	http://agricoop.nic.in/
Department of Bio-technology, India	http://dbtindia.nic.in
Department of Scientific and Industrial Research, India	http://dsir.nic.in/
FAO	http://www.fao.org/
Global Agribusiness Information Network:	http://www.fintrac.com/gain/:

Greenhouse Vegetable Information:	http://www.ghvi.co.nz/
Indian Agricultural Research Institute (IARI)	http://www.iari.res.in/
Indian Council of Agricultural Research (ICAR)	http://www.icar.org.in
Indian Institute of Horticultural Research (IIHR)	www.iihr.res.in
Indian Institute of Spices Research (IISR), Calicut, Kerala	http://www.iisr.org/
Indo-American Hybrid Seeds	www.indamseeds.com
Institute of Vegetable and Ornamental Crops	http://www.igzev.de/
Institute for Horticultural Development, Victoria, Australia	http://www.nre.vic.gov.au/agvic/ihd/
Kerala Agricultural University	www.kau.edu
Iowa State University Department of Horticulture	http://www.hort.iastate.edu/
National Bureau of Plant Genetic Resources (NBPGR), India	http://nbpgr.delhi.nic.in/
National Horticulture Board (NHB), India	http://hortibizindia.nic.in/
National Institute of Agricultural Extension Management (MANAGE), India	http://www.manage.gov.in/
National Research Centre for Cashew (NRCC), India	http://kar.nic.in/cashew/
National Research Centre for Mushroom (NRCM), India	http://www.nrcmushroom.com/
National Research Centre for Oil Palm (NRCOP), India	http://www.ap.nic.in/nrcop
North Carolina State University, Dept. of Horticulture	http://www2.ncsu.edu/cals/hort_sci/
Oregon State University, Dept. of Horticulture	http://osu.orst.edu/dept/hort
Pineapple News	http://agrss.sherman.hawaii.edu/pineapple/pineappl.htm
Pomology Resources Center	http://www.bsi.fr/pomologie/english/pomology:
Rubber board, India	http://rubberboard.org.in/
Spice Paprika web site	http://www.paprika.deltav.hu/:
Spices Board, India	http://www.indianspices.com/
Sri Lanka Agribisness on-line	http://www.agro-lanka.org/
Sustainable Apple Production:	http://orchard.uvm.edu/

Tea Board, India	http://tea.nic.in/
The Horticultural Taxonomy Group	http://www.hortax.org.uk/
The International Society of Citriculture	http://www.lal.ufl.edu/isc_citrus_homepage.htm
The Internet Garden	http://www.internetgarden.co.uk/
The Rose Resource	http://rose.org/
The USDA Agricultural Research Service	http://www.ars.usda.gov/
University of Florida, Dept. of Environmental Horticulture	http://hort.ifas.ufl.edu/
University of California, Fruit&Nut Research	http://fruitsandnuts.ucdavis.edu/
US Environmental Protection Agency	http://www.epa.gov/:
USDA	http://www.usda.gov/

INDEX

B

E

F

M

N

T

U

V

W

Y

Z